HFI/NQI 2004

HFI/NQI 2004

Proceedings of the 13th International Conference on Hyperfine Interactions and 17th International Symposium on Nuclear Quadrupole Interactions (HFI/NQI 2004) Bonn, Germany, 22–27 August 2004

Edited by

K. MAIER

University of Bonn, Bonn, Germany

and

R. VIANDEN

University of Bonn, Bonn, Germany

Reprinted from *Hyperfine Interactions*
Volume 158, Nos. 1–4 (2005)
Volume 159, Nos. 1–4 (2005)

Springer

A C.I.P. Catalogue record for this book is available from the Library of Congress.

ISBN 3-540-30923-3

Published by Springer
P.O. Box 990, 3300 AZ Dordrecht, The Netherlands

Sold and distributed in North, Central and South America
by Springer
101 Philip Drive, Norwell, MA 02061, U.S.A.

In all other countries, sold and distributed
by Springer
P.O. Box 990, 3300 AZ Dordrecht, The Netherlands

Printed on acid-free paper

Printed in The Netherlands

Table of Contents Volume I

Magnetism and Magnetic Materials – Bulk and Thin Layers

HFI Probes in Semiconductors, Metals and Insulators

Hyperfine Interactions (2004) 158:1–3
DOI 10.1007/s10751-005-9133-0

Preface

In August 2004 more than 200 scientists from 27 nations attended the joint meeting of the XIIIth International Conference on Hyperfine Interactions and the XVIIth International Symposium on Nuclear Quadrupole Interactions. The meeting was hosted by the Helmholtz-Institut für Strahlen- und Kernphysik of the Rheinische Friedrich-Wilhelms-Universität Bonn.

The programme covered the theory of hyperfine fields in solids, as well as the traditional scientific subjects connected to hyperfine interaction detected by nuclear radiation. A major section was dedicated to nuclear quadrupole resonance measurements. Two hundred twenty abstracts were submitted to the conference from which the joint programme committee selected 27 for oral presentation. Additionally 17 scientists were invited to present their latest results or give overviews of various new developments in the field.

One hundred thirty-one refereed papers were finally accepted for publication and are presented in this volume of Hyperfine Interactions. It is dedicated to the late Prof Dr. Erwin Bodenstedt, who started this line of research in Bonn and made many important contributions to its development over more than 25 years.

The Editorial office is grateful to the authors for their cooperation by submitting the electronic versions of their contributions largely in time, and to the referees for their effort to ensure the scientific quality of these proceedings.

Further, we would like to thank the staff and students of the Helmholtz-Institut für Strahlen- und Kernphysik for their dedication in the preparation and organisation of the conference.

Finally, we acknowledge the support of the University of Bonn, the Stadt Bonn, the International Union of Pure and Applied Physics, Springer-Verlag, and especially thank the DFG – Deutsche Forschungsgemeinschaft for their generous financial contribution.

Bonn, January 2005

K. Maier and R. Vianden
Guest editors

Chair:

R. Vianden
K. Maier

Programme committee:

K. Bharuth-Ram
D. Brinkmann
W. Evenson
G. Langouche
K. Maier
H. Petrilli
G. Schatz
G. Sprouse
R. Vianden

Local organising committee:

D. Eversheim
M. Forker
P. Herzog
K. Maier
I. Müller
R. Nédélec
C. Schwenk
K.-H. Speidel
T. Staab
R. Vianden

International advisory committee:

N. Achtziger
H. Akai
E. Alp
E. Baggio-Saitovitch
Z. Berant
J. Billowes
P. Blaha
P. Boolchand
D. Brinkmann
J. Budnick
T. Butz

S. Campbell
G. Catchen
D. Chaplin
L. Chow
G.S. Collins
R. Coussement
S. Cox
P. Dalmas de Réotier
T.P. Das
S.K. Date
P. Dederichs
J. Gardner
H. Haas
H. Jaeger
R. Kiefl
K.-P. Lieb
M. Mekata
T. Minamisono
A. Pasquevich
H. Petrilli
J. Ramakrishna
G.N. Rao
P. Riedi
R. Rüffer
G. Savard
J.C. Soares
K.H. Speidel
N. Stone
R. Wäppling
T. Wichert
P. Wodniecki
M.O. Zacate
S. Zhu

Hyperfine Interactions (2004) 158:5–7
DOI 10.1007/s10751-005-9136-x

Erwin Bodenstedt – In Memoriam

This contribution is dedicated to the late experimental physicist, teacher, and colleague, Professor Erwin Bodenstedt, who passed away suddenly in March 2002, about 10 years after his retirement. Its relevance to this International Conference is that Erwin Bodenstedt was one of the founding fathers of the 'Hyperfine Interactions' field, on which he left a decisive imprint with his own research accomplishments, spanning an extended period from 1960 to 1992.

E. Bodenstedt was born on January 25, 1926 in the city of Cologne. He received his scientific education at the University of Bonn, where he graduated in 1951. He embarked on a PhD Thesis at the 'Physikalische Institut' under the leadership of Wolfgang Riezler, and was awarded his Degree in 1952. In that year, when Wolfgang Paul became director of the Institute, Bodenstedt was involved in the design of a 500 MeV electron synchrotron, which subsequently became a key facility for research and training of students.

In 1954 Bodenstedt went as a postdoctoral fellow for one year to the U.S., where he worked in the group of R. R. Wilson at Cornell University on setting up a 1.5 GeV electron accelerator. Upon his return to Germany, he worked for a short period in the research laboratories of Siemens at Erlangen before joining, in 1956, Walter Jentschke, who became director at the Physikalische Staatsinstitut in Hamburg. It was there that Bodenstedt started his research in experimental nuclear physics, profoundly inspired by the outstanding personality of Jentschke and the excellent atmosphere prevailing in that Institute. In several pioneering experiments, Bodenstedt performed a systematic investigation of magnetic moments of excited nuclei, which earned him widespread recognition, both in Germany and in the international community. With a summary of this work he received his Habilitation in 1960.

Two years later, in 1962, Bodenstedt became Professor at the University of Bonn in the 'Institut für Strahlen- und Kernphysik', where he taught and conducted research for almost 30 years. His work there commenced with a continuation of his Hamburg activities, which were substantially intensified due to increased financial support and the engagement of numerous, highly motivated students. In addition, the Institute housed a cyclotron, of which he made extensive use in producing a broad spectrum of radionuclides, thereby broadening the scientific activities considerably.

Bodenstedt's research was primarily devoted to nuclear spectroscopy in all its multi-faceted aspects, employing sophisticated and original techniques, as well as state-of-the-art detectors. The focal point of the methods he applied devolve on angular correlations between electromagnetic radiations and/or conversion electrons emitted from radioactive sources. Very early on, he had recognized that only with top-class instrumentation could such measurements be performed with high precision and reliability. He therefore never hesitated to acquire the very best equipment available. As one of his students, I was responsible for installing in 1965 the first germanium detector, as a replacement for the widely-used scintillators. Due to the superior energy resolution of this novel detector, a new fascinating world became apparent in the nuclear laboratory: "We will have to question and eventually revise many of our former observations," was his immediate response upon seeing such a gamma-ray spectrum for the first time.

Bodenstedt's many-fold contributions to nuclear structure were the result of numerous, highly-original investigations. One of the highlights was his early, deep involvement in the measurement of circularly-polarized gamma-rays, as a means for the detection of parity mixing in nuclear states. Also noteworthy are his extensive activities in applying a variety of nuclear spectroscopy techniques to the determination of solid-state properties on an atomic scale. The systematic study of electric field gradients in non-cubic metals and their model interpretation was a long-standing subject in which Bodenstedt invested all his ingenuity and profound knowledge of physical phenomena in general. The experimental aspects of this activity were reinforced by setting up an isotope separator which, in combination with the existing cyclotron, became a unique facility that permitted the production and subsequent implantation of radioactive nuclei in essentially any material required. Bodenstedt considered this branch of solid-state physics with nuclear probes as a very promising field with a bright future, based on its close relation to interdisciplinary applications. However, in spite of the progressive and fascinating activity pursued by many of his students and collaborators in his laboratory, he remained a nuclear physicist at heart up until his retirement. After all, nuclear structure and nuclear moments were the roots of his scientific upbringing.

As a teacher, Bodenstedt was a consummate master, due to his admirable capability for explaining difficult and complex relationships in simple language, and with infectious enthusiasm. His 3-volume pocket-book on nuclear physics entitled 'Experimente in der Kernphysik und ihre Deutung' (published in 1979) exemplifies his uncommon scientific insight and appreciation of significant and unconventional measurements. This great talent and flair of his naturally attracted many students to work towards MSc and PhD theses in well-organized working groups. Bodenstedt's creativity and unwavering optimism made a profound and lasting impression, not only on the young students, but also on his collaborators in and beyond the Institute.

Bodenstedt was also an excellent administrator, taking extremely seriously his personal responsibility for all employees, as well as for all operational aspects of the complex technical installations in the Institute, a laboratory well-known for its diversity in structure and organization. All who knew him remain full of admiration for Bodenstedt's great achievements, none-the-least with respect to his unique style and effectiveness in negotiating with the administration and organs of the University. He was the spiritual father of many scientists who currently occupy leading positions in education, society and politics. Erwin Bodenstedt remains an unforgettable personality.

K.-H. Speidel
Helmholtz-Institut für Strahlen- und Kernphysik
University of Bonn

Hyperfine Interactions (2004) 158:9–18
DOI 10.1007/s10751-005-9018-2

Ab Initio Calculation of Hyperfine Interaction Parameters: Recent Evolutions, Recent Examples

STEFAAN COTTENIER[1,*], VEERLE VANHOOF[1], DORU TORUMBA[1], VALERIO BELLINI[2], MEHMET ÇAKMAK[1,3] and MICHEL ROTS[1]
[1]*Instituut voor Kern-en Stralingsfysica, K.U.Leuven, Celestijnenlaan 200 D, B-3001, Leuven, Belgium; e-mail: stefaan.cottenier@fys.kuleuven.ac.be*
[2]*INFM National Research Center on nanoStructures and bioSystems at Surfaces (S3) and Dipartimento di Fisica, Università di Modena e Reggio Emilia, Via Campi 213/A, I-41100, Modena, Italy*
[3]*Department of Physics, Gazi University, Teknikokullar, TR-06500, Ankara, Turkey*

Abstract. For some years already, *ab initio* calculations based on Density Functional Theory (DFT) belong to the toolbox of the field of hyperfine interaction studies. In this paper, the standard *ab initio* approach is schematically sketched. New features, methods and possibilities that broke through during the past few years are listed, and their relation to the standard approach is explained. All this is illustrated by some highlights of recent *ab initio* work done by the *Nuclear Condensed Matter Group* at the K.U.Leuven.

Key Words: *ab initio* calculations, Density Functional Theory, hyperfine interactions.

1. Introduction

Prior to 1985, accurate *ab initio* calculations with predictive power for hyperfine interaction properties in solids were hardly available. In the decade 1985–1995, a few theory groups applied their newly developed *ab initio* methods with increasing success to relatively simple systems (J. Kanamori in Osaka, K. Schwarz in Vienna, P.H. Dederichs in Jülich, H. Petrilli in São Paulo, etc.). Around 1995, user-friendly general-purpose *ab initio* codes started to become publicly available. For the hyperfine interaction community, the WIEN code [1] developed by P. Blaha and K. Schwarz (Vienna) is the most famous example. Helped by the ever increasing performance-to-cost ratio of computational resources, *ab initio* calculations got to be used at many places outside the traditional theory groups, and ever more complex systems were tackled. This is called 'computational condensed matter physics', an approach to condensed matter problems that is situated between theory and experiment.

---

* Author for correspondence.

In this short review, Section 2 will give a schematic overview of the standard *ab initio* approach to (nuclear) condensed matter physics. Recent evolutions in the available methods allow new kinds of physics to be studied, and a few of these recent evolutions are summarized in Section 3. All of this will be illustrated by some highlights from recent *ab initio* work done in Leuven in Section 4. Some thoughts on the future of the field of hyperfine interaction studies and the role of *ab initio* methods therein are given as a conclusion (Section 5).

2. The standard *ab initio* approach

The fields of chemistry, atomic physics, molecular physics, condensed matter physics and molecular biology can all be summarized under one header: the study of different manifestations of a quantum gas of electromagnetically interacting positive nuclei and negative electrons. Figure 1 sketches one (not the only) way how this problem can be tackled *from first principles* or *ab initio*, i.e. starting from basic quantum mechanics without using experimental information or tuneable parameters.[1] Writing down the Hamiltonian that describes a given system (say an O_2 molecule, or bcc-Fe) is not difficult: both the positive nuclei and the negative electrons move (kinetic terms), and they Coulomb-interact with each other (electron-nucleus, electron–electron and nucleus–nucleus interactions). The real problem is *solving* this Hamiltonian, which without approximations takes an almost infinite amount of time. A first simplification is the Born–Oppenheimer approximation: the heavy nuclei move much slower than the light electrons, and can be considered to be instantaneously at rest. The nuclear kinetic energy vanishes, and the nucleus–nucleus interaction becomes a constant. The remaining electron Hamiltonian (moving electrons, interacting with the fixed nuclei and with other electrons) is still way too difficult to solve. The culprit is the electron–electron interaction. Here Density Functional Theory (DFT) enters the stage. DFT transforms the problem of interacting electrons that interact with nuclei into a rigorously equivalent problem of imaginary particles that do *not* interact among each other. They interact only with the nuclei and with an imaginary external potential, called the exchange-correlation potential V_{xc}. If you would know V_{xc} e.g. for bcc-Fe, then you could in a rather straightforward numerical way solve the Hamiltonian for the imaginary problem, which would give you the same result as for the original bcc-Fe Hamiltonian. Such numerical solution methods bear names as LAPW, APW + lo, LMTO, PAW, etc., and they are implemented in computer codes such as WIEN2k, VASP, FPLO-3, etc. Unfortunately, these exchange-correlation potentials (or more precisely: the unique functional from which they are derived) are not known. Instead of the unknown, true V_{xc}, a parameter-free guess V_{xc}^{guess} is used (with names as LDA or GGA) that is believed/hoped to be not too

[1] See [2] for a basic introduction, written for a non-theoretical audience.

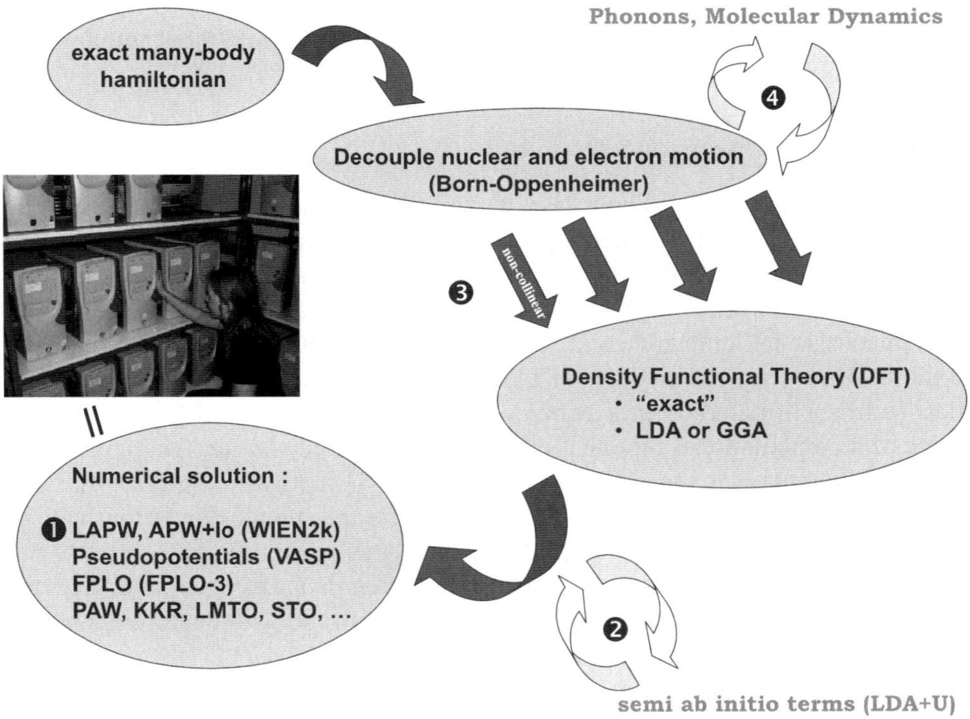

Figure 1. Schematic overview of the DFT-based *ab initio* approach, starting from the exact many body hamiltonian and ending with numerical calculations on a computer (Section 2). New evolutions – which are extensions of the standard scheme – are indicated by *white numbers in black circles*: (1) the transition from LAPW to APW + lo (Section 3.1), (2) the semi *ab initio* LDA + U method (Section 3.2), (3) a non-collinear formulation (Section 3.3), and (4) introducing time-dependence by phonons and/or Molecular Dynamics (Section 3.4). The photograph shows part of the Leuven home-built PC-cluster, dedicated to computational HFI work (also on the picture: Veerle Vanhoof).

different from the exact one. DFT can be formulated for magnetic and non-magnetic systems, non-relativistic, partly relativistic or fully relativistic, etc. which is why there are several arrows towards the DFT stage in Figure 1.

3. New evolutions

3.1. FROM LAPW TO APW + LO

The full-potential LAPW method that allows to solve the DFT equations in such a way that hyperfine information can be retrieved from them, has recently been improved into the APW + lo method [2–4]. The accuracy of the latter is the same as the accuracy of LAPW, but it is numerically much more efficient. As a result, the calculation time needed for a typical problem has dropped by a factor 5–10. As the time for a typical calculation is of the order of days to weeks, APW + lo allows to treat more complex and/or more realistic systems without running into unreasonably lengthy calculation times.

3.2. THE LDA + U METHOD

Applying the scheme of Figure 1 with the LDA and GGA approximations to many different solids, has demonstrated that the agreement between calculated and experimental properties becomes especially bad for solids with particularly strong electron–electron interactions (strongly correlated materials). Examples of such materials are lanthanide and actinide based compounds, and transition metal oxides. More advanced methods that go beyond the sketch of Figure 1 are able to deal with such materials in a better way (e.g. the GW method [5]), but at the price of a computational time that is orders of magnitude larger. An increasingly popular alternative is the semi *ab initio* LDA + U method, where the missing correlation effects are expressed by a single tuneable parameter U that can be estimated from experiment. LDA + U has the same execution time as LDA and GGA. Although not fully *ab initio* any more – there is a tuneable parameter that is not fixed by quantum mechanics – LDA + U allows to gain better quantitative insight in strongly correlated materials.

3.3. NON-COLLINEAR MAGNETISM

Implementing the scheme of Figure 1 in a computer code is considerably simplified if all magnetic moments are fixed to be parallel with each other (then there is a unique quantization axis at every point in the crystal). This restricts the application of such a code to non-magnetic, ferromagnetic and antiferromagnetic materials only. More and more general-purpose codes are extended nowadays to include also non-collinear magnetism, using an approach developed around 1990 by L.M. Sandratskii and others (reviewed in [6]). For the WIEN-code, this implementation was done recently by R. Laskowski [7]. It is now possible to treat non-collinear magnets and spin spirals.

3.4. PHONONS

In the usual Born–Oppenheimer approximation, the nuclei are at rest and phonons are effectively neglected. The effect of phonons can be reintroduced, however, by calculating the total energy for various positions of the nuclei (for every set of positions one static *ab initio* calculation needs to be done) and then deriving from this information the full phonon dispersion spectrum. An efficient way to do this has been developed by K. Parlinski (Cracow) [8, 9], and has been seamlessly integrated with *ab initio* codes as WIEN and VASP. Knowledge of the phonon dispersion relations is interesting in its own right, but can also serve as a starting point to derive information about properties of a solid at elevated temperatures (DFT itself is a ground state method, i.e. all results are for 0 K). For some applications this might be a more efficient approach than *ab initio* Molecular Dynamics, another rapidly developing field [10].

4. Highlights from work in Leuven

In this section, the impact of the above-mentioned new evolutions on computational hyperfine interaction research will be illustrated by some recent research topics of the *Nuclear Condensed Matter Group* in Leuven. Some of these results are published in 2004, other topics are in various stages of progress. Where appropriate, references will be given to more extensive discussions.

4.1. COORDINATION DEPENDENCE OF THE Cd HYPERFINE FIELD ON Ni SURFACES

The sensitivity of hyperfine interaction techniques to specific atomic sites make them a unique tool for the study of the diffusion behaviour of small amounts of radioactive tracers on metallic surfaces. This line of research has been followed since about 1985 mainly by groups in Groningen (L. Niesen), Konstanz (G. Schatz) and Berlin (H. Haas, H. Bertschat). In particular, the electric-field gradient (EFG) and magnetich hyperfine field (HFF) of Cd at various sites of low-index fcc surfaces (Ni, Cu, Pd, Ag) have been very well documented. Recently, H. Bertschat suggested that the Cd HFF on a Ni surface depends only on the number of Ni neighbours [11, 12], and smoothly increases from negative values at the highest coordination (bulk Ni, 12 neighbours) to positive values at the lowest coordination (adatom site on Ni(111), three neighbours). We examined this conjecture by *ab initio* calculations [13, 14]. Impurities at surfaces require long calculations times, and therefore the efficient APW + lo method discussed in Section 3.1 was of vital importance. We conclude [14] that *as a main trend* the conjecture is true, but also that there is no fundamental reason why there could not have been more scatter around this trend. Rather accidentally, Nature seems to behave smoothly here. The Cd HFF is not only sensitive to the *number* of Ni neighbours, but also to their exact position, and to the number and position of more distant neighbours. We do understand the mechanism behind this increasing trend (caption of Figure 2). For the Cd EFG on Ni, Cu, Pd and Ag surfaces, we can explain [13] why adatom sites have a lower EFG than sites in the surface layer – an experimental observation that is used as a practical site identification rule but which was not yet properly understood. Both the coordination-dependence rules of HFF and EFG for Cd are generalized for all other impurities in the range Cd–Ba [13, 14].

4.2. HYPERFINE INTERACTIONS OF LANTHANIDE IMPURITIES IN Fe

Understanding the HFF of all elements as substitutional impurities in Fe is a prototype problem in the field of hyperfine interactions, and decades of experimental [15] and theoretical [16–24] efforts have been devoted to it.

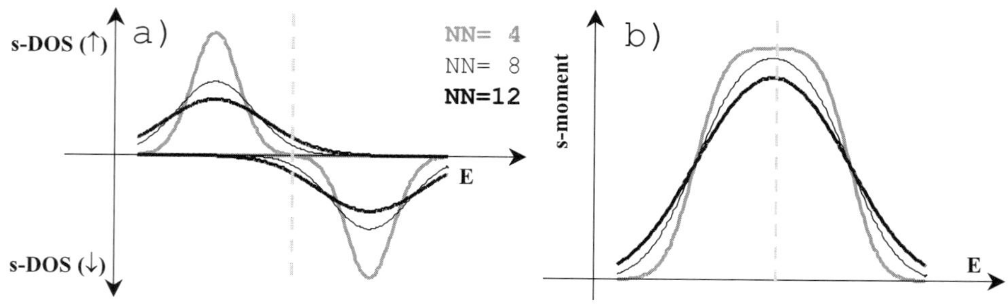

Figure 2. Mechanism behind a positive contribution to the Cd HFF, growing with reducing nearest neighbour coordination number NN. (a) Cartoon of a detail of the Cd s-DOS near the Fermi energy (*vertical dashed line*) at fcc Ni surfaces. Reducing the coordination number results in band-narrowing. (b) Spin *s*-moment as a function of energy, derived from (a) by integrating $s_\uparrow$ and $s_\downarrow$, and subtracting them. The value of this integral at the Fermi energy is the actual spin moment. As picture (b) shows, this region of the DOS is responsible for a contribution to the *s*-moment (and hence to the contact HFF) that becomes more positive if the coordination number is reduced.

Nowadays, the mechanism behind the HFF's of all sp- and d-elements in Fe is well-understood. For f-elements, this was not yet the case, mainly because LDA and GGA (Section 2) fail to describe the strong electron correlations in the 4f and 5f shells. With the LDA + U method (Section 3.2), this problem can be largely overcome. We have calculated [24] the HFF of all 15 elements from La to Lu as substitutional impurity in Fe, and find a reasonable agreement with the experimental values (Figure 3a). The 4f spin moment couples antiferromagnetically to the Fe 3d moment, and the change of sign between the light and the heavy lanthanides reflects Hund's third rule. The large HFF's for lanthanides are almost entirely of orbital nature, with small contributions from the dipolar field and the Fermi contact field. From arguments not discussed here (see [24]), it is concluded that La, Ce *and* Pr have delocalized 4f electrons, while the currently available experiments do not exclude that this is the case also for lanthanides up to Sm. All other lanthanides with localized 4f electrons are trivalent, except for Yb which is divalent. Due to the strong spin–orbit coupling, some lanthanides feel a very strong EFG (Figure 3b), which could be calculated correctly as well. Similar work on actinides in Fe is in progress, with the main aim to study the influence on the HFF of the delocalization → localization transition around Am.

4.3. SPIN FLUCTUATIONS AND IMPURITY COMPLEXES

As soon as impurity atoms in a host material start to aggregate into small clusters, the number of possible sites for a probe atom quickly grows. This makes the interpretation of experimental measurements much more difficult – if not impossible. For a Ag host where single Fe-impurities, Fe–Fe dimers and larger Fe clusters were supposed to be present, ^{57}Fe Mössbauer parameters have been reported [25] (Table I). We have calculated these Mössbauer parameters for

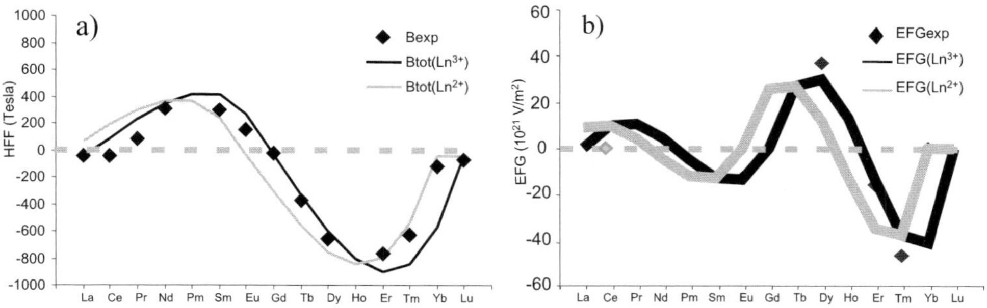

Figure 3. a) HFF of lanthanides in Fe: experiment (*black diamonds*), and as calculated by LDA + U in a divalent (*gray*) and trivalent (*black*) configuration. b) Idem, for the EFG of lanthanides in Fe (the *gray point* for Ce is an approximate value).

Table I. Experimental [25] and calculated Mössbauer parameters for a single Fe and a Fe–Fe dimer in Ag

	IS^{exp}	$\lvert QS \rvert^{exp}$	IS^{calc}_{NM}	QS^{calc}_{NM}	IS^{calc}_{FM}	QS^{calc}_{FM}	HFF^{calc}_{FM}
Fe	0.52	–	0.78	–	0.60	–	−9
Fe–Fe	0.45	0.85	0.40	1.55	0.47	−0.76	−14
Fe–Fe distance	?		2.16 Å		2.44 Å		

Isomer Shift (IS) and Quadrupole Splitting (QS) are in mm/s, the HFF is in Tesla. Non-magnetic (NM) and ferromagnetic (FM) calculations are listed. The last line gives Fe–Fe distances in Å.

single Fe and for Fe dimers in large 32-atom supercells, where all atoms were allowed to relax (again the efficient APW + lo method (Section 3.1) was crucial). As is shown in Table I, this results in a rather bad agreement with experiment, both for the single Fe and for the dimer. The agreement is very much improved if the two Fe atoms are allowed to be magnetic (with a ferromagnetic coupling between both). The associated HFF is not seen in the experiment, however. This can be explained by fast spin fluctuations: they average out the HFF over the life time of the Mössbauer level, but, as the dimer remains always ferromagnetically coupled, the Fe–Fe separation that corresponds to the magnetic case is conserved, and hence also the corresponding quadrupole splitting. We suggest that the EFG can be used as a sensitive indicator for the presence of spin fluctuations. With hindsight, indications for spin fluctuations are seen also in the Mössbauer data (line broadening, effect of external field). Similar effects are studied for Fe–Co dimers and Co–Co dimers [26].

4.4. TEMPERATURE DEPENDENCE OF THE Cd-EFG IN HCP Cd

The temperature dependence of the EFG in metals [27] is another long-standing issue in the field of hyperfine interactions. By elimination of all other properties, the major contribution to this temperature dependence is most likely due to phonons. A quantitative proof of this is lacking, however. Also, it is not clear

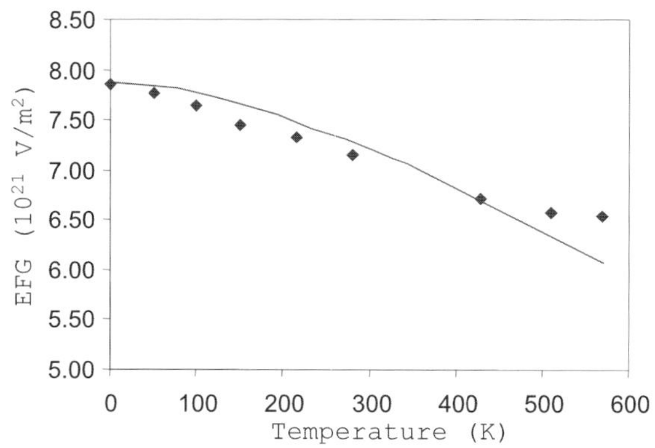

Figure 4. Line: experimental EFG of Cd in hcp-Cd (shifted upward by 0.55 10^{21} V/m^2 to account for a systematic deviation in the calculations which is due to limitations of the GGA functional). Dots: EFG calculated by means of the *ab initio* phonon dispersion curve.

why in some metals the EFG follows a $T^{3/2}$ dependence, while in other materials it is linear. Together with K. Parlinski (Cracow), we have embarked on an *ab initio* study of the EFG of Cd in hcp-Cd, with the goal to answer the above-mentioned questions. The effect of temperature is taken into account by means of the *ab initio* calculated phonon dispersion curve of hcp Cd. With this approach, we obtain the correct order of magnitude of the temperature dependence (Figure 4), showing for the first time in a *quantitative* way that phonons indeed determine the temperature dependence of the EFG in metals. Instead of the $T^{3/2}$ dependence, however, we find something that is rather linear. We suspect this is due to anharmonic effects, which have not yet been taken into account. Work is in progress to consider anharmonicity in a proper way.

4.5. NON-COLLINEAR MAGNETISM AT THE Fe/Ag INTERFACE

The Fe/Ag interface in Fe/Ag(001) multilayers offers the interesting possibility to study the magnetism on a local scale at both sides of the interface by hyperfine interaction methods (^{57}Fe Mössbauer and ^{110}Ag Low Temperature Nuclear Orientation). It has been shown experimentally [28] that the small magnetic moment of Ag in the interface layer (a moment induced by the neighbouring Fe layers) is nearly perpendicular to Fe moments, the latter lying in the interface plane. This is a somewhat surprizing observation. Preliminary non-collinear *ab initio* calculations (Section 3.3) show that the Ag-moment prefers to align parallel with the Fe-moment. This result suggests that the experimentally observed behaviour is not a property of a sharp and flat interface.

5. The future of hyperfine interactions

Hyperfine interaction methods can give detailed information on nanometer and nanosecond length- and time-scales, that often cannot be obtained by any other means. This outstanding advantage is counter-balanced by some disadvantages, that limit the applicability of HFI methods, and might eventually threaten the survival of the field: HFI methods are inherently slow (one data point can take days or weeks), and it is hard – if not impossible – to interpret spectra for more complicated samples in an unambiguous way. One cannot do much about the slowness; but when considering the unique information that hyperfine methods can deliver, this is not too severe a disadvantage. The problem of ambiguity has been avoided for a long time by studying very simple systems: Fe compounds, 1 isolated substitutional impurity in a perfect host lattice, etc. Problems of this type have been almost exhausted now, and the field is confronted with the question of how to provide useful information about the imperfect, complex materials that are in the focus of the broader field of condensed matter physics. In our opinion, part of the solution lies in a continued intimate integration of *ab initio* and experimental work (other parts of the solution are e.g. supplementing – some say: diluting – HFI experiments with more conventional experimental methods, and exploiting huge statistics and new possibilities of synchrotron experiments). Although far from being perfectly reliable yet, *ab initio* calculations can provide crucial pieces of information that help to obtain useful information from otherwise ambiguous experimental data. Furthermore, by calculations one can obtain *more* information from an experiment than is possible with the experiment only (e.g. the connection between spin fluctuations and the EFG in Section 4.3). With increasing computer speed, the cases that can be calculated will better and better approach the experimental complexity. Continued efforts using this type of combined experimental/computational approach might help to make HFI results a more appreciated source of information for the mainstream condensed matter community.

Acknowledgements

The computational work described in this paper was made possible by the projects G.0239.03 and G.0447.05 of the *Fonds voor Wetenschappelijk Onderzoek – Vlaanderen* (FWO), the Concerted Action of the KULeuven (GOA/2004/02) and the Inter-University Attraction Pole (IUAP P5/1). The authors are indebted to L. Verwilst and J. Knuts for their invaluable technical assistance concerning the computer resources (Figure 1).

References

1. Blaha P., Schwarz K., Madsen G., Kvasnicka D. and Luitz J., In: *WIEN2k, An Augmented Plane Wave + Local Orbitals Program for Calculating Crystal Properties*, ISBN 3-9501031-1-2 (1999).
2. Cottenier S., In: *Density Functional Theory and the Family of (L)APW-methods: a step-by-step introduction*, Instituut voor Kern – en Stralingsfysica, K.U.Leuven, Belgium, ISBN 90-807215-1-4 (freely available at http://www.wien2k.at/reg_user/textbooks).

3. Sjöstedt E., Nordström L. and Singh D. J., *Solid State Commun.* **114** (2000), 15.
4. Madsen G. K. H., Blaha P., Schwarz K., Sjöstedt E. and Nordström L., *Phys. Rev., B* **64** (2001), 195134.
5. Aryasetiawan F. and Gunnarsson O., *Rep. Prog. Phys.* **61** (1998), 237.
6. Sandratskii L. M., *Adv. Phys.* **47** (1998), 91.
7. Laskowski R., Madsen G. K. H., Blaha P. and Schwarz K., *Phys. Rev., B* **69** (2004), 140408(R).
8. Parlinski K., *Computer code Phonon* (http://wolf.ifj.edu.pl/phonon/).
9. Shukla A., Calandra M. and d'Astuto M. *et al., Phys. Rev. Lett.* **90** (2003), 095506.
10. Tuckerman M. E., *J. Phys. Condens. Matter* **14** (2002), R1297.
11. Potzger K., Weber A., Bertschat H., Zeitz W.-D. and Dietrich M., *Phys. Rev. Lett.* **88** (2002), 247201.
12. Prandolini M. J., Manzhur Y., Weber A., Potzger K., Bertschat H. H. and Dietrich M., *Appl. Phys. Lett.* **85** (2004), 76.
13. Cottenier S., Bellini V., Çakmak M., Manghi F. and Rots M., *Phys. Rev. B* **70** (2004), 155418.
14. Bellini V., Cottenier S., Çakmak M., Manghi F. and Rots M., *Phys. Rev. B* **70** (2004), 155419.
15. Rao G. N., *Hyperfine Interact.* **24–26** (1985), 1119.
16. Akai H., Akai M., Blügel S., Zeller R. and Dederichs P. H., *J. Magn. Magn. Mater.* **45** (1984), 291.
17. Akai M., Akai H. and Kanamori J., *J. Phys. Soc. Jpn.* **54** (1985), 4246.
18. Akai H., Akai M. and Kanamori J., *J. Phys. Soc. Jpn.* **54** (1985), 4257.
19. Akai H., Akai M., Blügel S., Drittler B., Ebert H., Terakura K., Zeller R. and Dederichs P. H., *Prog. Theor. Phys., Suppl.* **101** (1990), 11.
20. Ebert H., Zeller R., Drittler B. and Dederichs P. H., *J. Appl. Phys.* **67** (1990), 4576.
21. Korhonen T., Settels A., Papanikolaou N., Zeller R. and Dederichs P. H., *Phys. Rev., B* **62** (2000), 452.
22. Cottenier S. and Haas H., *Phys. Rev., B* **62** (2000), 461.
23. Haas H., *Hyperfine Interact.* **151–152** (2003), 173.
24. Torumba D., Cottenier S., Vanhoof V. and Rots M., submitted to *Phys. Rev., B* (2004) [preprint on request].
25. Morales M. A, Passamani E. C. and Baggio-Saitovitch E., *Phys. Rev., B* **66** (2002), 144422.
26. Vanhoof V., Cottenier S., l'Abbé C. and Rots M., manuscript in preparation.
27. Christiansen J., *et al., Z. Phys., B* **24** (1976), 177.
28. Phalet T., Prandolini M. J., Brewer W. D., De Moor P., Schuurmans P., Severijns N., Turrell B. G., Van Geert A., Vereecke B. and Versyck S., *Phys. Rev. Lett.* **86** (2001), 902.

Hyperfine Interactions (2004) 158:19–23
DOI 10.1007/s10751-005-9002-x

Magnetic Properties and the Electric Field Gradients of Fe_4N and Fe_4C

M. OGURA[1,2] and H. AKAI[1]
[1]*Department of Physics, Graduate School of Science, Osaka University, 1-1 Machikaneyama Toyonaka, Osaka, 560-0043, Japan*
[2]*Japan Science and Technology Agency, 4-1-8 Honcho, Kawaguchi, Saitama, 332-0012, Japan*

Abstract. The magnetic properties and the hyperfine interactions of Fe_4N and Fe_4C under high pressures are discussed on the basis of first-principles electronic structure calculations performed by the KKR Green's function method and its extention into a full potential scheme (FP-KKR). The results show that both Fe_4N and Fe_4C undergo a weak first order transition from high to low spin states under pressure. This transition is sensitively reflected not only by H_{hf} but also by EFG. The calculated Curie temperature, magnetic moments and H_{hf} well reproduce the experimental observations.

Key Words: binding surface, electric field gradient, hyperfine field, full potential KKR, magnetic state.

1. Introduction

It is well known that nitrogen sometimes plays an important role in the magnetism of iron-based ferromagnets such as $Sm_2Fe_{17}N_3$. For this reason the magnetism and the magnets related somehow to nitrogen are sometimes called nitromagnets. In *nitromagnets*, N affects the magnetic moments of adjacent Fe atoms in a complex way. One can see such an example in Fe_4N with anti-perovskite structure. The theoretical calculations [1] showed that in Fe_4Z, where Z is a hypothetical atoms whose atomic number Z is a continuous variable, the increase of the atomic number Z changes the nature of bonding between Z and adjacent Fe atoms from more covalent to more ionic. The transition occurs rather suddenly near $Z = 7$, i.e. N. At the same time, the local magnetic moment of Fe also changes its magnitude from around 2 μ_B typical for metallic systems to around 3 μ_B typical for insulating systems. This indicates that the magnetism of Fe_4N is not very stable: changing the crystal structure, lattice constant, chemical composition, etc., may easily cause the transition between high and low spin states.

Ishimatsu and his collaborators performed the Mössbauser and K-edge X-ray magnetic circular dichroism (XMCD) measurements of Fe_4N under high pressure

* Author for correspondence.

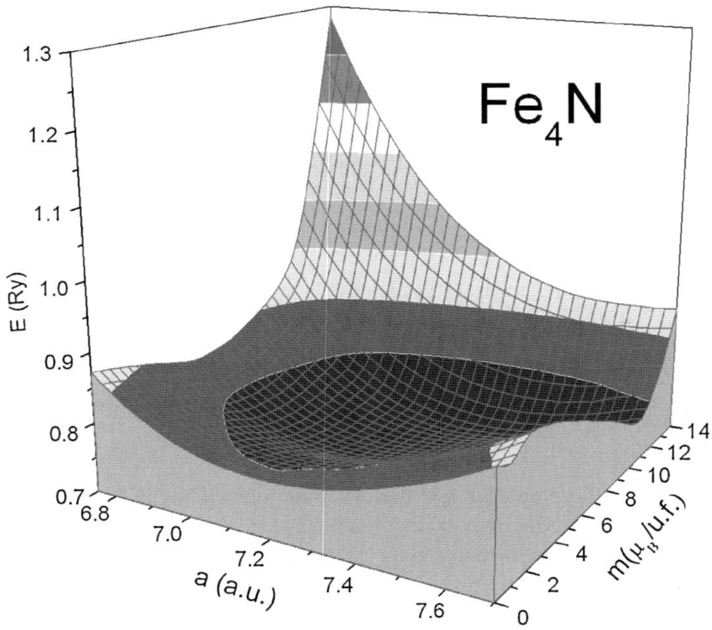

Figure 1. Binding surface of Fe_4N.

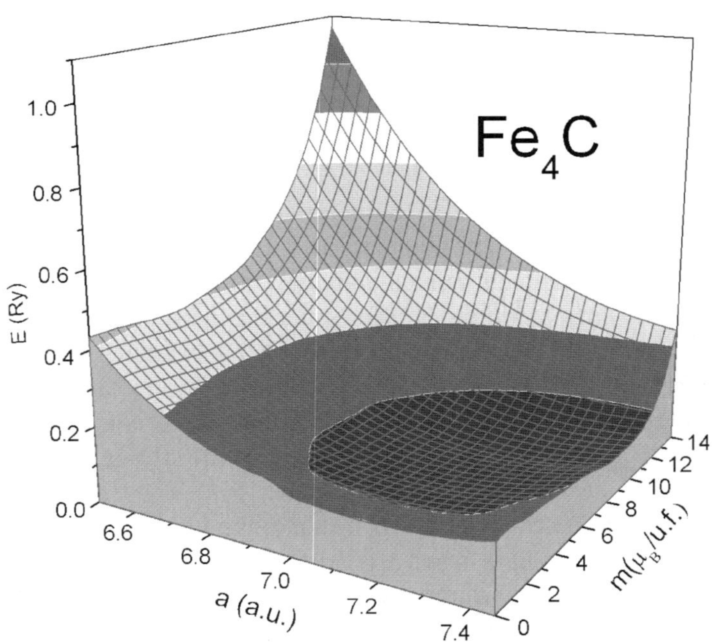

Figure 2. Binding surface of Fe_4C.

up to 24 GPa [2, 3]. They observed that the system undergoes a rather moderate transition from the ferromagnetic to paramagnetic state [3]. It is not quite clear, however, that whether this gradual transformation might be related to the above high/low spin states transition or not.

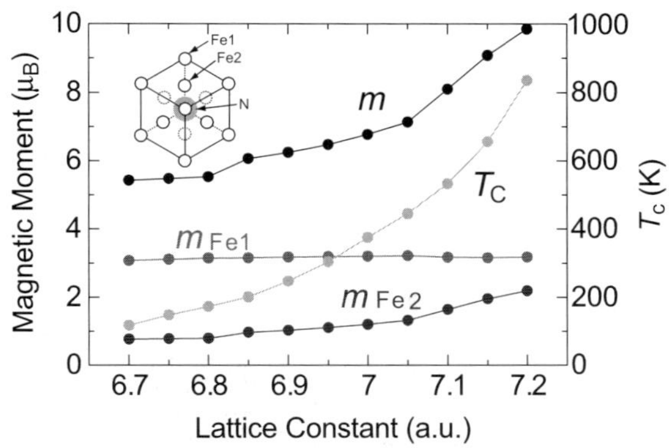

Figure 3. Calculated magnetization (*left scale*), the local magnetic moments of Fe1 (*corner site*) and Fe2 (*face centered site*) (*left scale*), and the Curie temperature T_C (*right scale*) of Fe_4N.

In this paper, we compare our calculated magnetic and electronic properties under pressure with experiments for Fe_4N and Fe_4C, the latter being off the boundary between the covalent and ionic region, and discuss the origin of the general trends.

2. Theoretical framework

We use the KKR Green's function method for the whole calculation of the electronic structure of Fe_4N and Fe_4C. The main part is performed by use of the full potential version of the KKR method newly developed by one (M.O.) of the present authors [4–6]. The magnetic transition temperature T_C is calculated in the framework of the mean field theory. We regard the well developed Fe local magnetic moments as classical spins, treating the paramagnetic state above T_C, a random arrangements of spins, by use of the KKR-CPA (coherent potential approximation). The binding surface, corresponding to the free energy as a function of the lattice constant a and magnetization m at 0 K, is calculated by the fixed spin moment method.

3. Results and discussion

Figures 1 and 2 shows the binding surfaces of Fe_4N and Fe_4C. The experimental lattice constant of Fe_4N and Fe_4C at the zero pressure is 7.16 (a.u.). Though both the surfaces seem rather structureless, they actually have a double minimum structure if one looks at the curves along a fixed lattice constant. This means when we change the lattice constant by applying pressure, two stable electronic structures are obtained for different values of the lattice constant. We may call one of them, with larger magnetic moment, the high spin state, and the other the low spin state. The larger the lattice constant the more the high spin state becomes stable. The double minimum structure, however, is rather subtle for Fe_4N and hence the transition from the high spin to low spin state is nearly

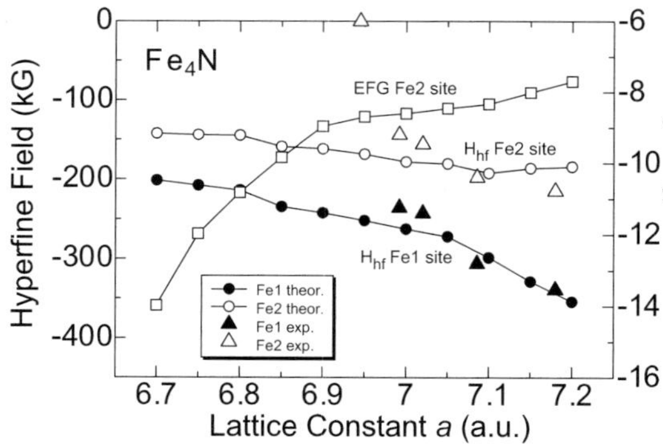

Figure 4. Calculated hyperfine field H_{hf} (*left scale*) and the electric field gradient V_{zz} (*right scale*) of Fe_4N as a function of the lattice constant.

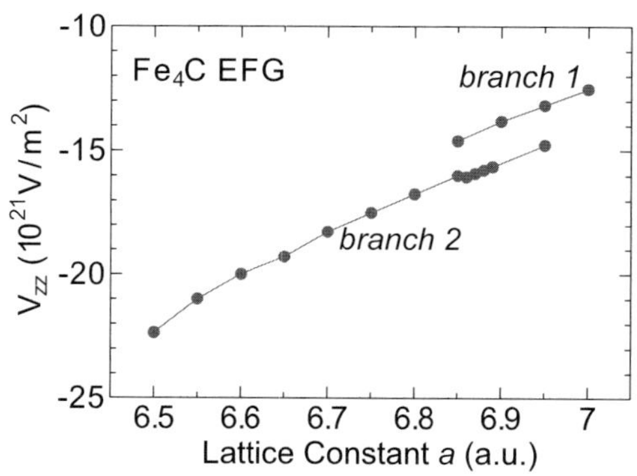

Figure 5. Calculated electric field gradient V_{zz} of Fe_4C as a function of the lattice constant.

continuous. The hysteresis associated with the first order transition is hardly seen with increasing/decreasing the lattice constant in the present calculation.

Figure 3 shows the calculated magnetization, the local magnetic moments of Fe_1 (corner site) and Fe_2 (face centered site), and the Curie temperature T_C. While the local magnetic moment of Fe_1 is quite insensitive to the lattice constant, it is not the case for Fe_2. This indicates that the behavior of Fe_2 is responsible to the major part of the feature seen in the binding surface. The Curie temperature decreases with decreasing lattice constant and at the lattice constant corresponding to the pressure of around 20 GPa, the Curie temperature becomes lower than the room temperature.

The calculated hyperfine field H_{hf} and the electric field gradient V_{zz} as a function of the lattice constant is given in Figure 4. The agreement between the theoretical and experimental H_{hf} is satisfactory for the low pressure region. However, at the high pressure (> 20 GPa) where the Curie temperature becomes lower than the room temperature, the hyperfine fields are not observed any more. EFG behavior as a function of the lattice constant shows a rather slow change at the low pressure region and a rapid change at the high pressure region. This reflects the fact that at high pressure, due to the shorter distance between Fe_2 and N atoms, N atoms start to affect the adjacent Fe_2 atom considerably at around a = 6.9 (a.u.). This change is related to the high/low spin transition mentioned above.

Finally, in Figure 5, EFG of Fe_2 in Fe_4C is shown. Here it turns out that EFG has two branches. They correspond to the high/low spin states. Thus together with the results for Fe_4N, we can conclude that the EFG of Fe_2 reflect the magnetic and electric state sensitively, giving rise to an important clue to understanding the role of typical elements in Fe based ferromagnets.

Acknowledgements

The present study was partly supported by the 21st century COE program "Towards a New Basic Science" and by Special Coordination Funds for the Promotion of Science and Technology, Leading Research "Nanospintronics Design and Realization". This work is also supported by NEDO.

References

1. Akai H., Takeda M., Takahashi M. and Kanamori J., *Solid State Commun.* **94** (1995), 509.
2. Ishimatsu N., Ohishi Y., Suzuki M., Kawamura N., Ito M., Maruyama H., Nasu S., Kawwakami T. and Shimomura O., *Nucl. Inst. Meth. Phys. Res. A* **467** (2001), 1061.
3. Ishimatsu N., Maruyama H., Kawamura N., Suzuki M., Ohishi Y., Ito M., Nasu S., Kawakami T. and Shimomura O., *J. Phy. Soc. Jpn.* **72** (2003), 2372.
4. Ogura M., *Dissertation in Physics*, Osaka University, 2004.
5. Ogura M., Akai H. and Minamisono T., see the article in this volume (2005).
6. Ogura M., *J. Phys.: Condens. Matter Phys.* (2005) to be published.

Hyperfine Interactions (2004) 158:25–28
DOI 10.1007/s10751-005-9003-9

Spin-orbit Induced Electric Field Gradients in Magnetic Solids

H. EBERT* and M. BATTOCLETTI
Department of Chemistry, University of Munich, Butenandtstr. 5-13, D-81377, München, Germany; e-mail: hubert.ebert@cup.uni-muenchen.de

Abstract. The spin–orbit induced electric field gradient in cubic ferromagnets has been observed experimentally in the past for many systems. Even its dependence on the orientation of the magnetisation with respect to the crystallographic axes could be convincingly demonstrated. A fully relativistic description is presented that is based on the Korringa–Kohn–Rostoker (KKR) method of band structure calculation. Application of this approach to substitutional 5d-transition metals in Fe led to a satisfying agreement with available experimental data. To allow for a more detailed discussion of the results an analytical model has been developed, that treats spin-orbit coupling as a perturbation.

Key Words: electric field gradient, ferromagnets, spin-orbit coupling.

1. Introduction

In magnetic solids spin-orbit coupling (SOC) gives rise to a number of interesting phenomena that are caused by a reduction in symmetry. On a microscopic scale this is manifested in a very convincing way by the occurrence of an electric field gradient (EFG) on nuclear sites even for systems having a cubic lattice. This paradox finding has been observed for the first time around 1970 by several groups (see for example [1]) who investigated Ir dissolved substitutionally in the cubic hosts bcc-Fe and fcc-Ni. Later corresponding results have been found for many other systems [9]. In addition, the anisotropy of the effect; i.e. the variation of the EFG with the relative orientation of the magnetisation and the crystal axis, that had to be expected on the basis of symmetry considerations [3], could be demonstrated.

All previous theoretical investigations (for an overview see e.g. [10]) concentrated on the SOC as the dominating source for the observed EFG. In general a tight binding Hamiltonian has been used for that purpose with the SOC being treated as a perturbation. Here a parameter-free alternative scheme based on the spin polarised relativistic Korringa–Kohn–Rostoker (SPR-KKR) band

* Author for correspondence.

structure method is presented. Results of an application to 5d-transition metal impurities in Fe are presented and compared with experiment.

2. Theoretical framework

We restrict ourselves to cubic systems with the magnetisation along the z-axis. In this case the symmetry is effectively tetragonal [3]. Therefore it is sufficient to consider only the zz-component Φ_{zz} of the EFG tensor. Expanding the electronic charge density $\rho(\vec{r})$ within a sphere of radius S around the central atom in terms of the spherical harmonics one expresses the on-site contribution to Φ_{zz} by the non-spherical charge density term $\rho_{20}(r)$ ($l = 2$, $m_l = 0$) [7]. Within the SPR-KKR formalism $\rho_{20}(r)$ can be expressed with the help of the corresponding Green's function [4] leading to

$$\Phi_{zz} = \frac{8\pi}{5} e \frac{-1}{\pi} \Im \int^{E_F} dE \sum_{\Lambda\Lambda'} \tau_{\Lambda\Lambda'} \sum_{\Lambda''\Lambda'''} \left[B_{\Lambda''\Lambda'''} \int_0^S g_{\Lambda''\Lambda} g_{\Lambda'''\Lambda'} \frac{1}{r} dr + B_{-\Lambda''-\Lambda'''} \int_0^S f_{\Lambda''\Lambda} f_{\Lambda'''\Lambda'} \frac{1}{r} dr \right]. \tag{1}$$

Here e is the electron charge and Λ and $-\Lambda$ stand for the set of quantum numbers (κ, μ) and ($-\kappa$, μ), resp., with the spin-orbit and magnetic quantum numbers κ and μ. Furthermore, $\tau_{\Lambda\Lambda'}$ is the so-called scattering path operator and the functions $g_{\Lambda\Lambda'}$ and $f_{\Lambda\Lambda'}$ are the major and minor components of the regular solutions Z_Λ to the Dirac equation for a spin-dependent potential [4], (the energy argument E of τ, g, f, and Z has been omitted). Finally, the coefficients $B_{\Lambda\Lambda'}$ are angular matrix elements. If one neglects the very small cross terms connecting angular momentum channels with $l - l' = \pm 2$ [2] the relativistic angular matrix elements $B_{\Lambda\Lambda'}$ adopt a shape familiar from non-relativistic theory [6]:

$$B_{\Lambda\Lambda'} = \frac{5\delta_{\mu,\mu'}\delta_{l,l'}}{4\pi(2l-1)(2l+3)} \sum_{m_s} (-1)^{\frac{1}{2}+m_s} C_\Lambda^{m_s} C_{\Lambda'}^{m_s} \left[3(\mu - m_s)^2 - l(l+1) \right]. \tag{2}$$

with the magnetic spin quantum number m_s and $C_\Lambda^{m_s}$ standing for the Clebsch–Gordan coefficients $C\left(l \tfrac{1}{2} j;\ (\mu - m_s) m_s\right)$ [8].

3. EFG of 4d- and 5d-transition metal impurities in Fe

The KKR-GF scheme briefly sketched above has the very appealing feature that it can be applied for numerical investigations as well as for analytical considerations. For the later purpose one sets up the Green's function on a

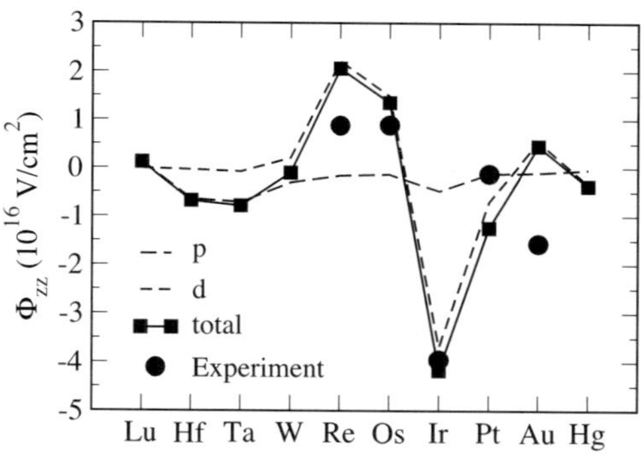

Figure 1. EFG parameter Φ_{zz} of 5d-transition metals as a substitutional impurity in bcc-Fe with the magnetisation along the [001]-direction. The experimental data have been taken from [9].

non-relativistic level for a spin-polarised solid. SOC is then treated as a perturbation by means of the Dyson equation [5, 10], that can be expanded in a power series with respect to the SOC. Evaluating the expression for the density term $\rho_{20}(r)$ one finds this way for cubic solids – as shown before by other authors [10] – that non-vanishing contributions to $\rho_{20}(r)$ occur starting with second order with respect to the SOC and only if the solid is spin-polarised, i.e. magnetic. For non-magnetic cubic system, on the other hand, the contributions due to the spin-diagonal and spin-off-diagonal parts of the SOC exactly cancel each other. These qualitative results can be confirmed numerically by model calculations that use the SPR-KKR-GF scheme with the SOC manipulated accordingly [4].

The numerical SPR-KKR-GF scheme has been applied to 4d- and 5d-transition metal impurities dissolved substitutionally in bcc-Fe. For this purpose their electronic structure has been calculated self-consistently on the basis of local spin density functional theory making use of the atomic sphere approximation (ASA) and the single-site approximation; i.e. the distortion of Fe-atoms in the vicinity of the impurity has been ignored. The corresponding EFG parameter Φ_{zz} for the impurity site has been determined on the basis of Equation (1). This implies that the wave functions g and f in this equation are those for the impurity potential and that τ is the scattering path operator projected onto the impurity site.

Figure 1 shows the calculated EFG parameter Φ_{zz} for all 5d-transition metals as an impurity in bcc-Fe. The decomposition of Φ_{zz} into contributions of the p- and d-electrons shows that the p-part varies rather smoothly with the atomic number Z and dominates Φ_{zz} at the beginning of the row. The d-part, on the other hand, varies very strongly with Z, shows changes in sign and dominates in the middle and end of the row. Corresponding results for 4d-impurities (not shown here) are – as expected because of the weaker SOC – an order of magnitude smaller. Interestingly, the p-contributions show a change in sign and are of the same order of magnitude as the d-contributions.

In Figure 1 experimental data for Φ_{zz} have been added that have been deduced from measurements on single crystal measurements with the magnetisation oriented along the [001]-axis [9]. Earlier experimental work was done in general on polycrystalline samples leading in general to smaller values for Φ_{zz}. As one can see, a rather satisfying agreement between theory and experiment have been obtained. Obviously, the present implementation of our formalism, that among others uses the ASA and ignores lattice relaxations, already allows to explain the observed spin-orbit induced EFG in a convincing way.

Acknowledgements

This work was supported by the DFG (Deutsche Forschungsgemeinschaft) within the programme *Theorie relativistischer Effekte in der Chemie und Physik schwerer Elemente.*

References

1. Aiga M. and Itoh J., Nuclear electric quadrupole interaction of iridium nuclei in dilute alloys of iron and nickel, *J. Phys. Soc. Japan* **31** (1971), 1844.
2. Blaha P., Schwarz K. and Dederichs P. H., First-principles calculation of the electric-field gradient in HCP metals, *Phys. Rev. B* **37** (1988), 2792.
3. Cracknell A. P., Time-reversal degeneracy in the electronic band structure of a magnetic metal, *J. Phys. C: Solid State Phys.* **2** (1969), 1425.
4. Ebert H., Fully relativistic band structure calculations for magnetic solids – Formalism and Application, In: H. Dreyssé (ed.), *Electronic Structure and Physical Properties of Solids, Vol. 535 of Lecture Notes in Physics*, Springer, Berlin Heidelberg New York, 2000, p. 191.
5. Ebert H., Minár J. and Popescu V., *Magnetic dichroism in electron spectroscopy, Vol. 580 of Lecture Notes in Physics*, Springer, Berlin Heidelberg New York, 2001, p. 371.
6. Gehring G. A. and Williams H. C. W. L., On the effects of spin-orbit coupling on impurity atoms in cubic ferromagnetic alloys, *J. Phys. F: Met. Phys.* **4** (1974), 291.
7. Herzig P., Electrostatic potentials, fields and field gradients from a general crystalline charge density, *Theoret. Chim. Acta. (Berlin)* **67** (1985), 323.
8. Rose M. E., *Relativistic Electron Theory*, Wiley, New York, 1961.
9. Seewald G., Hagn E., Zech E., Kleyna R., Voß M. and Burchard A., Spin-orbit induced noncubic charge distribution in cubic ferromagnets. I. Electric field gradient measurements on 5d impurities in Fe and Ni, *Phys. Rev. B* **66** (2002a), 174401.
10. Seewald G., Zech E. and Haas H., Spin-orbit induced noncubic charge distribution in cubic ferromagnets. II. Tight-binding analysis, *Phys. Rev. B* **66** (2002b), 174402.

Hyperfine Interactions (2004) 158:29–35
DOI 10.1007/s10751-005-9004-8

Ab initio Study of the Temperature Dependence of the EFG at Cd Impurities in Rutile TiO_2

L. A. ERRICO
Instituto de Física La Plata (IFLP-CONICET), Departamento de Física, Facultad de Ciencias Exactas, Universidad Nacional de La Plata, C.C. 67, 1900 La Plata, Argentina; e-mail: errico@fisica.unlp.edu.ar

Abstract. We report a FLAPW study of the temperature dependence of the electric-field gradient (EFG) tensor at Cd impurities replacing cations in TiO_2. In order to simulate the temperature effect, we considered experimental temperature-dependent lattice parameters, and we obtain the structural relaxations introduced by the Cd impurities and the EFG tensor for each lattice parameters set. We found that these relaxations depends on the temperature and are at the origin of the strong temperature dependence of the EFG. We also showed that a correct description of the electronic structure of the impurity-host system is essential for the description of the EFG and its temperature dependence.

1. Introduction

The electronic charge density $\rho(r)$ in a solid and its temperature dependence can be studied by measuring the electric-field gradient tensor (EFG), which is very sensitive to small changes in $\rho(r)$. In this sense, the Perturbed-Angular Correlation technique (PAC) [1] is specially suited to study the temperature dependence of the EFG because its sensitivity is temperature independent. The temperature dependence of the EFG in semiconducting and insulating oxides were study using this technique. Several behaviours have been observed, which have been explained (with more or less success) in terms either of host properties or processes induced by the presence of the hyperfine probes (generally, impurities in the systems under study) [see, e.g., [2] and references therein]. In the particular case of the wide band-gap semiconductor TiO_2 (rutile structure), the strong temperature dependence of the EFG tensor at substitutional Cd impurities was attributed to the nonisotropic thermal expansion of the structure, but no means to make a quantitative connection between the two effects was found at that time [3].

In this work we show that an *ab initio* calculation can explain (at least at first order) the temperature dependence of the EFG at Cd impurities in TiO_2. Calculations were performed with the Full-Potential Linearized-Augmented Plane Waves (FLAPW) method, that allows us to treat the electronic structure and the processes induced by the impurity in the host-lattice (atomic relaxations,

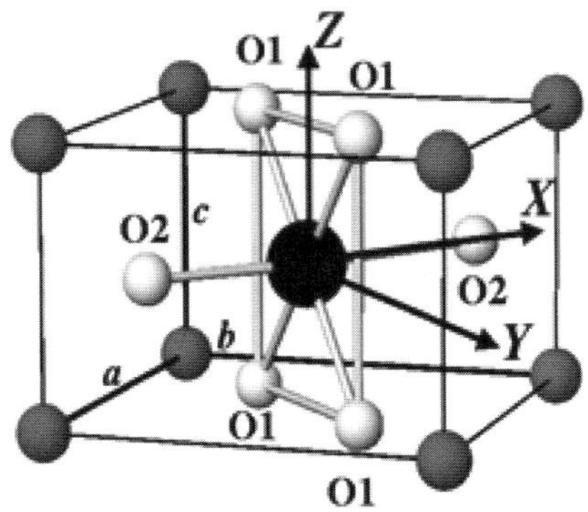

Figure 1. Unit cell of rutile TiO_2 (Ti *gray balls*, O *white balls*). The results discussed in this work are referred to the indicated axes system, assuming that Cd replaces the central Ti black atom.

impurity levels) in a fully self-consistent way, without the use of external parameters and arbitrary suppositions. Previous results have shown that, at 0 K, the Cd impurities introduce strong lattice distortions in the TiO_2 lattice and impurity levels in the band-gap of this semiconductor [4, 5]. To study the temperature dependence of the EFG tensor we use in this work temperature-dependent lattice parameters. For a given temperature we calculate (using the thermal expansions coefficients) the corresponding lattice parameters. These lattice parameters were used in the FLAPW calculations in order to obtain the structural relaxations introduced by the Cd impurities and the EFG tensor at each temperature. We also demonstrate that a correct description of the electronic structure of the impurity-host system is essential for a correct description of the EFG and its temperature dependence.

2. Method of calculation and results for the EFG tensor at 0 K

Rutile TiO_2 is tetragonal ($a = b = 4.5845_1$ Å, $c = 2.9533_1$ Å [6]). The unit cell (Figure 1) contains 2 Ti at (0,0,0; 1/2,1/2,1/2) and four O at $\pm(u,u,0;\ 1/2+u,\ 1/2-u,1/2)$ with $u = 0.30493$ [6]. In order to simulate the isolated impurities we used a super-cell (SC) of 12 unit-cells with one Ti replaced by the Cd atom. The resulting 72-atoms SC has dimensions $a' = b' = 2a$, $c' = 3c$ and is also tetragonal with $a'/c' = 0.97$. Concerning the thermal expansion of the structure, the expansion along the c axis is higher than in the a axis. The corresponding thermal expansion coefficients are respectively 9.6×10^{-6} K^{-1} and 8.0×10^{-6} K^{-1} (270 K < T < 470 K), and 9.2×10^{-6} K^{-1} and 8.4×10^{-6} K^{-1} (270 K < T < 1270 K) [7].

Previous to the discussion of the temperature dependence of the EFG tensor, it is important to describe the calculation of the EFG at 0 K. To deal with this problem we considered the mentioned SC, repeated periodically, and we performed FLAPW [8] calculations in order to determine the self-consistent po-

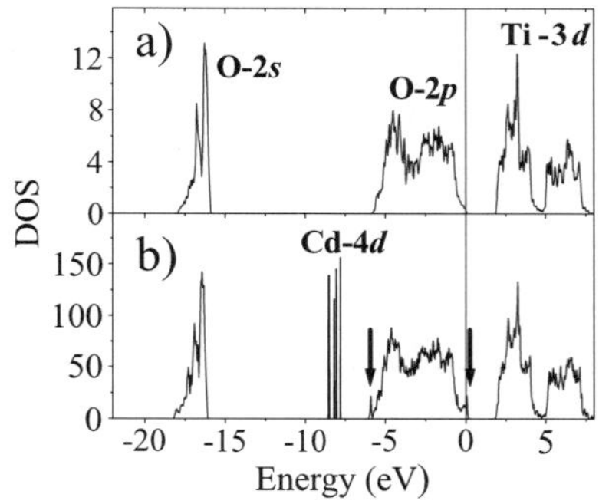

Figure 2. DOS for a) pure TiO_2; b) Cd^0 in TiO_2 (72-atom SC). The arrows indicate impurity states in the valence band. The vertical lines indicate the Fermi level.

tential and the charge density inside the SC. We studied the relaxation introduced by the impurity computing the forces on the Cd neighbors and moving them until the forces vanished. Calculations were performed with the WIEN97.10 code [9] within the LDA [10] approximation. In this method the unit cell is divided into non-overlapping spheres and an interstitial region. The spheres radii used for Ti, Cd, and O were 1.01, 1.11, and 0.85 Å, respectively. The parameter R_{KMAX}, which controls the size of the basis-set, was chosen as 6 (4500 LAPW functions) and we introduced local orbitals to include Ti-3*s* and 3*p*, *O*-2*s* and Cd–4*p* orbitals. Integration in reciprocal space was performed using the tetrahedron method taking an adequate number of *k*-points in the irreducible Brillouin zone (IBZ). The choice of parameters was checked by performing calculations increasing the number of *k*-points and the R_{KMAX} value. Details of the method of calculation and the way to deal with the impurity are extensively explained in [5].

There is a last point to be considered in the calculation of the electronic structure of the system: the charge state of the impurity-host system, the double acceptor Cd^{+2} in $Ti^{+4}O_2^{-2}$ (see [5]). TiO_2 is a semiconductor with the oxygen *p*-band filled. When a Cd atom replaces a Ti in the SC, the resulting system is metallic due to the lack of two electrons necessary to fill the oxygen *p*-band. Comparison of Figure 2a and b shows that the presence of Cd in the SC produces the appearance of Cd-*d* levels and impurity states at the top and the bottom of the valence band. The wave function of the impurity state at the Fermi level (E_f) has character Cd–d_{yz}, O1–p_y, O1–p_z. Then, to provide two electrons to the system implies a drastic change in the symmetry of the electronic charge distribution in the neighborhood of the impurity. We performed calculations for two physical situations: a) assuming that extra electrons are not available (neutral impurity, Cd^0); b) the system provides the lacking two electrons (charged impurity, Cd^{-2}). In situation (a) we used the described SC. To describe situation (b) we added two electrons to the SC that we compensated with an homogeneous positive background to obtain a neutral SC to compute energy and forces. The cal-

Table I. EFG largest principal component at Cd site, V_{33}, and asymmetry parameter η for the relaxed structures of the two charge states considered compared with experimental results at 300 K in single crystals [4] and polycrystalline samples [3, 11] d(Cd–O1), d(Cd–O2) are the distances in Å (after relaxation) from Cd to O1 and O2 atoms, respectively. The unrelaxed Cd–O_{NN} distances are d(Cd–O1) = 1.94 Å, d(Cd–O2) = 1.98 Å [6].

System		d(Cd–O1) (Å)	d(Cd–O2) (Å)	V_{33} (10^{21} V/m^2)	V_{33} direction	η
TiO_2:Cd	Cd^0 [4, 5]	2.15	2.11	+7.16	X	0.91
	Cd^{2-} [4, 5]	2.18	2.11	+4.55	Y	0.26
	Exp. [4]			5.34(1)	X or Y	0.18(1)
	Exp. [3,11]			5.23(5)/5.34(1)	–	0.18(1)/0.18(1)

culations were performed taking 8 k-points in the IBZ for the metallic system (situation (a)). In the case of situation (b) the system has an insulating character, and only two k-points in the IBZ are necessary to obtain well converged results [5]. In Table I we show the results for the relaxation of the six nearest oxygen neighbors (O_{NN}) to the Cd atom. We found that for both charge-states the relaxations are quite anisotropic, with the Cd–O1 distance larger than the Cd–O2, opposite to the un-relaxed structures [4, 5]. The results for the EFG tensor for Cd^{-2} agree very well with the experimental results at 300 K (see Table I) and are very different from the ones obtained for Cd^0. The high η value obtained for Cd^0 is too far from experimental data, showing that even at 300 K Cd in TiO_2 is in a charged state. To conclude, we want to mention that the point-charge model (PCM) predicts for TiO_2(Cd) $V_{33} = -2.27\times10^{21}$ V/m^2 (pointing along the X direction) and $\eta = 0.40$ when a value of -29.27 is used for the Sternheimer antishielding factor γ_∞, in clear contradiction with the experimental and FLAPW results. Even if the relaxed coordinates from our FLAPW calculations are introduced in the PCM, this model still fails in the description of the EFG [5].

3. Temperature dependence of the EFG. Results and conclusions

We can discuss now the temperature dependence of the EFG tensor. In Figure 3 we show the experimental results for V_{33} and η. In order to explain these results we considered, in a first approximation, a simple PCM considering lattice parameters calculated using the thermal expansion coefficients. This model present several assumptions, being the first one the validity of the ionic model. In addition, the electronic structure of Cd is computed using the γ_∞ factor, and the impurity character of Cd and the electronic effects that it induces in the host are not taking into account. The only effect of temperature is to enlarge the structure, without changes in the relative positions of the atoms in the host lattice (structural relaxations are not considered). This model predicts nearly constant

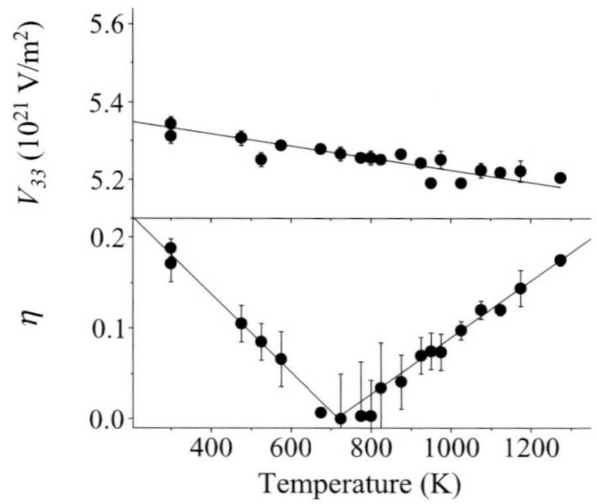

Figure 3. PAC results for V_{33} and η at Cd impurities located at cation sites in TiO_2 as a function of the measuring temperature (recreated from [3]). To obtain V_{33} from the experimental quadrupole frequencies a value of the quadrupole moment $Q(Cd) = +0.83_{13}$ b was used [12].

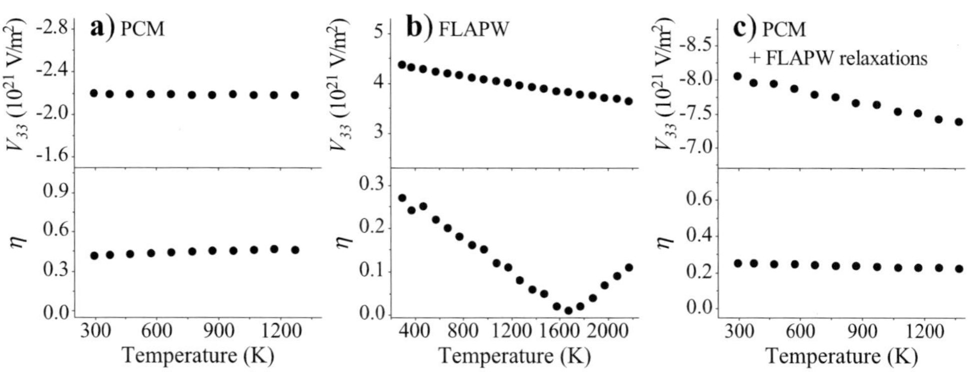

Figure 4. Predictions of the three different models used for the calculation of the temperature dependence of the EFG tensor at Cd sites in rutile TiO_2. a) PCM considering the thermal expansion of the structure. b) FLAPW calculations (see text). c) PCM considering the relaxations predicted by FLAPW.

values of V_{33} and η (Figure 4a), in bad agreement with the experimental results, shown in Figure 3.

In order to explain the temperature dependence of the EFG tensor we performed the FLAPW calculations. The temperature dependence was introduced in this calculations through the lattice parameters, which were calculated (for each temperature) using the thermal expansion coefficients. For each set of lattice parameters we obtain the equilibrium position (by a force minimization) of the Cd–O_{NN}. In these FLAPW calculations we considered that the impurity level at E_f is in a charged state (Cd^{-2}, see previous section). In this model we are taking into account the electronic structure of the Cd impurity and the structural and electronic process that Cd induces in the host lattice in a self-consistent way, without the use of external parameters or arbitrary suppositions. We found that the structural relaxations introduced by the Cd-impurities change with the lattice parameters or, in other words, with temperature. In the temperature range 300–2173 K we found that Cd–O1(Cd–O2) distances changes from 2.18 Å to 2.23 Å (2.11 Å to 2.17 Å). The results for V_{33} and η are shown in Figure 4b. As can be seen, our calculations reproduce, at least qualitatively, the experimental trends. The differences found between FLAPW predictions and the experimental results can be associated to different processes induced by the temperature (for example, lattice vibrations) that cannot be taken into account in our model.

Finally, we performed PCM calculations taking into account the thermal expansion of the lattice parameters and the structural relaxations predicted by FLAPW. These results for V_{33} and η are shown in Figure 4c. Again, PCM fails in the description of the thermal dependence of the EFG tensor, showing that, in addition to the structural effects, a correct description of the electronic structure of the host perturbed by the presence of the impurity (without the use of external parameters and arbitrary suppositions) is also essential for the calculation of the EFG tensor at Cd impurities in rutile TiO_2 and its temperature dependence.

In conclusion, in the present work we can explain, at least at first order, the temperature dependence of the EFG at Cd sites in rutile TiO_2. In our model, based in 0 K *ab initio* FLAPW calculations, the temperature effects are introduce in the calculations through the nonisotropic thermal expansion of the rutile structure and considering the charge state of the impurity levels introduced by the Cd impurity. Our calculations show that the presence of the Cd impurity induces structural relaxations in the TiO_2 host. These relaxations depends on the expansion of the structure and are at the origin of the temperature dependence of the EFG tensor.

Acknowledgements

This work was partially supported by CONICET (PIP 6032), Fundación Antorchas, UNLP, and ANPCyT (PICT98 03-03727), Argentina, and TWAS, Italy. L. A. E. is member of CONICET. The author want to thanks to Dr M. Rentería and G. Fabricius for his comments and fruitful discussions.

References

1. Frauenfelder H. and Steffen R., In: Siegbahn K. (ed.), *α–, β–, and γ–Ray Spectroscopy*, Vol. 2, North-Holland, Amsterdam, 1968, p. 917.
2. Shitu J., Pasquevich A. F., Bibiloni A. G., Rentería M. and Requejo F., *Mod. Phys. Lett. B* **12** (1998), 281.
3. Adams J. M. and Catchen G. L., *Phys. Rev.* **B50** (1994), 1264.
4. Errico L. A., Fabricius G., Rentería M., de la Presa P. and Forker M., *Phys. Rev. Lett.* **89** (2002), 55503.
5. Errico L. A., Fabricius G. and Rentería M., *Phys. Rev.* **B67** (2003), 144104.
6. Hill R. J. and Howard C. J., *J. Appl. Cryst.* **20** (1987), 467.

7. Grant F. A., *Rev. Mod. Phys.* **31** (1959), 646.
8. Wei S. H. and Krakauer H., *Phys. Rev. Lett.* **55** (1985), 1200.
9. Blaha P., Schwarz K. and Luitz J., WIEN97 (K. Schwarz, T. U. Wien, Austria, 1999), ISBN 3-9501031-0-4.
10. Perdew J. P. and Wang Y., *Phys. Rev.* **B45** (1992), 13244.
11. Wenzel T., Bartos A., Lieb K. P., Uhrmacher M. and Wiarda D., *Ann. Phys.* **1** (1992), 155.
12. Herzog P., Freitag K., Reuschenbach M. and Walitzki H., *Z. Phys. A* **294** (1980), 13.

Hyperfine Interactions (2004) 158:37–40
DOI 10.1007/s10751-005-9005-7

Impurity-Vacancy Complexes in Si and Ge

H. HÖHLER, N. ATODIRESEI, K. SCHROEDER*,
R. ZELLER and P. H. DEDERICHS
Institut für Festkörperforschung, Forschungszentrum Jülich, 52425 Jülich, Germany;
e-mail: k.schroeder@fz-juelide.de, p.h.dederichs@fz-juelide.de

Abstract. We examine the electronic and geometrical structure of impurity-vacancy complexes for 11 sp-impurities in Si and Ge, using the pseudopotential plane wave (PPW) and the all-electron Kohn–Korringa–Rostoker (KKR) methods. We find that all impurities of the 5sp and 6sp series prefer the split-vacancy configuration. For Cd and Sn we obtain good agreement of the calculated hyperfine parameters with experimental PAC and EPR data. Impurities of the 3sp and 4sp series form distorted substitutional complexes (except Al, which forms a split complex in Si). This trend strongly correlates with the lattice relaxations of nearest neighbors around the isolated (without vacancy) substitutional impurities.

Key Words: germanium, hyperfine fields, impurities, silicon, vacancy complexes.

Intrinsic defects and their complexes with impurities play an important role in semiconductor physics. Here we present ab initio calculations for vacancy-impurity complexes in Si and Ge. From experiments on Sn in Si (EPR [1]) and on Cd in Ge (PAC [2]) it was suggested that in these systems an exotic configuration is preferred: a split-vacancy complex with the Sn (Cd) atom on the bond-center position and the vacancy split into two half-vacancies on the nearest neighbor (NN) sites. Previous ab initio calculations [3] for Sn in Si found a very small energy preference (0.045 eV) for the split-vacancy complex compared to the substitutional-Sn-vacancy pair on NN sites.

We have used two different density functional methods (for details see [4]). The PPW method has been used to investigate the configuration and stability of the impurity-vacancy complexes. The stable configurations for the Cd and Sn impurity-vacancy complexes and the relaxations of the neighboring atoms have been recalculated by the KKR-Green-function method, which as an all-electron method allows to calculate the electric field gradients, isomer shifts and hyperfine fields.

We find that both in Si and Ge the substitutional Cd (and Sn) vacancy-complex is instable and relaxes into the highly symmetrical split-vacancy complex, being about 1 eV lower in energy than the substitutional configuration (see Figure 1 and Table II). The local density of states (LDOS) at the impurity site of the split-vacancy complex in Ge is shown in Figure 2. The different peaks are

* Author for correspondence.

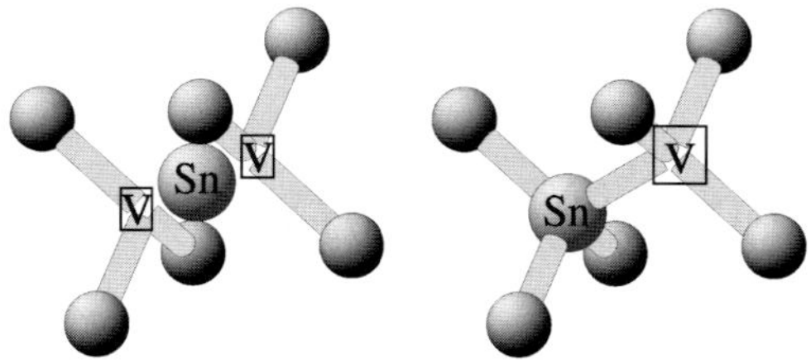

Figure 1. The Sn-split–vacancy configuration (*left*) and the substitutional Sn-vacancy configuration (*right*).

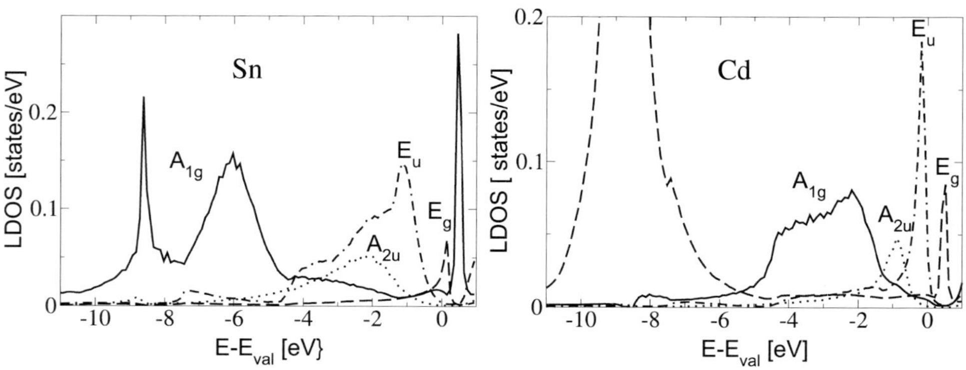

Figure 2. LDOS projected on the irreducible subspaces of the D_{3d} group at the Sn site (*left*) and Cd site (*right*) for the impurity split–vacancy complexes in Ge.

labeled by the irreducible subspaces A_{1g}, A_{2u}, E_u and E_g of the D_{3d} point group symmetry of the complex. Compared to the Cd complex (*d*-level at about -9 eV), the *d*-level of Sn is fully localized and at much lower energies, otherwise the level scheme is very similar: the A_{1g} state at relative low energies, the A_{2u} and the doubly degenerate E_u state slightly below the gap, the doubly degenerate E_g state in the gap and, for Sn, a second A_{1g} state at higher energies. Since the occupied A_{1g}, A_{2u} and E_u states accommodate eight electrons, of which the six neighboring host atoms provide six (one dangling bond each), the vacancy complex with Cd (with two valence electrons) is neutral ($[CdV]^0$), and the complex with Sn (with four valence electrons) is doubly positively charged ($[SnV]^{2+}$) provided the Eg level is not occupied. As discussed in [4], the level sequence is basically the same as for the divacancy. The Cd and Sn-atoms can be considered as Cd^{2+} and Sn^{4+} ions inserted in the center of the divacancy and only weakly hybridize with the six nearest neighbors. The attractive ionic potentials shift the divacancy states to lower energies, in particular the fully symmetrical A_{1g} state. This effect is naturally stronger for the Sn^{4+} ion than for the Cd^{2+} ion. The same level sequences is expected for other 5sp impurities.

By occupying the E_g state in the gap, different charge states of the complexes are realized, i.e., $[SnV]^+$, $[SnV]^0, \ldots$, and analogously $[CdV]^-$, $[CdV]^{2-}$. Both

Table I. Hyperfine fields HF and electric field gradients EFG at the Cd site for the Cd-split-vacancy complexes in Si and Ge

System	Si host		Ge host	
	HF(in kG)	EFG (in MHz)	HF(in kG)	EFG (in MHz)
$[V\|Cd\|V]^0$		−6.97		−32.69
$[V\|Cd\|V]^{1-}$	21.8	−27.99	8.5	−55.69
$[V\|Cd\|V]^{2-}$		−49.47		−79.64
EXP [2, 6]		±28.00		±54.00

Table II. Relaxations of neutral impurity-vacancy complexes and substitutional impurities in Si and Ge

	Si host			Δr	Ge host			Δr
	force	ΔE	pos.		force	ΔE	pos.	
Al	−22.4	−0.15	0.50	+2.03	−0.96	+0.03	0.82	−0.51
Si					−4.50	+0.30	0.88	−1.51
P	−11.5	+1.27	0.95	+0.04	−11.7	+0.81	0.93	−0.73
Ga	+5.37	+0.22	1.04	+0.32	+14.6	+1.11	1.04	−1.65
Ge	−21.7	+0.30	0.87	+1.13				
As	−41.7	+0.82	0.90	+2.52	−44.1	+0.41	0.87	+1.80
Se	−44.9	+0.55	0.92	+7.09	−37.6	+0.35	0.91	+5.65
Cd[a]	−85.2[a]	−1.04[a]	0.50[a]		−69.6[a]	−1.01[a]	0.50[a]	
In	−77.2	−0.69	0.50	+6.02	−43.2	−0.47	0.50	+3.17
Sn	−105	−0.83	0.50	+6.24	−79.0	−0.60	0.50	+4.18
Sb	−129	−0.68	0.50	+7.91	−109	−0.64	0.50	+6.32
Bi	−182	−0.92	0.50	+10.37	−152	−0.86	0.50	+8.68

The forces (in mRy/a.u.) on the impurity atom are given for the undistorted substitutional-vacancy complex, a negative sign means forces directed toward the bond center. The energy differences ΔE (in eV) are between the fully relaxed split-vacancy and distorted substitutional-vacancy complex. The stable positions of the impurities in (111) direction are also given; 0.5 is the bond-center and 1.0 the substitutional position. The NN relaxation Δr (in % of the NN-distance) refers to the isolated impurities.

[a] Cd results were obtained by the PAW method.

the $[SnV]^+$ and the $[SnV]^0$ complexes are magnetic with total moments of 1 and 2 μB, and with small local moments of 0.045 μB (0.008 μB) and 0.091 μB (0.016 μB) on the Sn site in Si (Ge). The calculations yield sizable relaxations of the neighboring host atoms toward the Sn (Cd) atoms, which increase with the electronic charge.

The hyperfine parameters for the Cd-split-vacancy complexes are listed in Table I. In the split configuration the weak hybridization of the impurity atoms with the six NN atoms results in a nearly isotropic charge density and a small EFG. We find good agreement with PAC measurements [2] and can uniquely assign the measured Cd EFG values to the Cd-split-vacancy complex in the charge state $[CdV]^-$.

For the neutral Sn-split-vacancy complex we obtain good agreement with available EPR data [1, 5]. The measured hyperfine field (−91.02 kG) [1] at the neighboring Si atoms adjacent to a SnV complex agrees well with our calculated value −82.96 kG for the neutral complex. The measured hyperfine field at the Sn atom [5] is close to our calculated value for the neutral Sn-split-vacancy complex [4]). The calculated isomer shifts agree well with unpublished measurements of R. Sielemann.

We have calculated the structure of impurity-vacancy complexes for other impurities, in particular for heavy impurities with larger sizes than Si or Ge atoms. For the elements Cd, In, Sn and Sb of the 5sp series and the even heavier element Bi we find that the split-vacancy impurity complex is preferred over the substitutional one by energies between 0.5 to 1 eV (see Table II). This suggests that this complex is the stable one for all oversized impurities in Si and Ge.

On the other hand, we find that impurities of the 3sp and 4sp series, with the exception of Al in Si, prefer the substitutional complex, in most cases rather distorted by a sizable relaxation of the impurity toward the bond center (see Table II). In order to discuss the importance of impurity size for the relative stability of the substitutional and split configurations we also considered isolated impurities, i.e., without vacancy, from the 3sp, 4sp, and 5sp series. In general, we find a good qualitative correlation of sign and size of the NN relaxations with the stability of the two configurations, although a strict one-to-one correspondence is not valid. As a rule of thumb we can define an impurity as "oversized," yielding a stable split-vacancy complex, if the NN relaxation exceeds 3%. There are two exceptions: substitutional Al shows only 2% NN relaxation in Si, yet we find the Al-split-vacancy complex stable, and Se shows a NN relaxation larger than 5% in both Si and Ge, yet the distorted substitutional complex is stable.

Acknowledgements

We thank R. Sielemann for helpful and motivating discussions. We gratefully acknowledge financial support by BMBF-Verbundforschung, project 05KK1CJA/2.

References

1. Watkins G. D., *Phys. Rev. B* **12** (1975), 4383–4390.
2. Haesslein H., Sielemann R. and Zistl C., *Phys. Rev. Lett.* **80** (1998), 2626.
3. Larsen A. N. *et al., Phys. Rev. B* **62** (2000), 4535; Kaukonen M. *et al., Phys. Rev. B* **64** (2001), 245213.
4. Höhler H., Atodiresei N., Schroeder K., Zeller R. and Dederichs P. H., *Phys. Rev. B* **70** (2004), 155343; and *Phys. Rev. B* **71** (2005), 035212.
5. Fanciulli M. and Byberg J. R., *Phys. Rev. B* **61** (2000), 2657.
6. Forkel D., Meyer F., Witthuhn W., Wolf H., Deicher M. and Uhrmacher M., *Hyperfine Interact.* **35** (1987), 715.

Hyperfine Interactions (2004) 158:41–46
DOI 10.1007/s10751-005-9006-6

The Electric Field Gradient Produced by a Gaussian Charge Density Distribution

T. BUTZ
Fakultät für Physik und Geowissenschaften, Universität Leipzig, Linnéstr.5, 04103 Leipzig, Germany; e-mail: butz@physik.uni-leipzig.de

Abstract. An analytical formula for the distance dependence of the electric field gradient produced by a Gaussian charge density distribution $n(r)$ is derived. This charge density is displaced by z_0 along the z-axis. The system has cylindrical symmetry; hence it suffices to calculate $V_{zz}(0)$. It turns out that $V_{zz}(0)$ is always smaller than the value with the total charge shrunk into a point. For distances larger than about four times the Gaussian width σ the expression approaches the point charge value. For $z_0 \rightarrow 0$, i.e., a spherically symmetric charge distribution around the origin, $V_{zz}(0)$ vanishes quadratically, as required by symmetry. A slab-wise calculation in cylindrical coordinates is presented which shows the contribution to $V_{zz}(0)$ for infinitesimally thin slabs as a function of distance from the origin. This analytical formula allows for a fast computation of electric field gradients from a given charge density distribution for Gaussian expansions of Slater-type orbitals. An example for a hydrogen atom will be given.

1. Introduction

Gaussian-type expansions of Slater-type orbitals have proven very useful in ab-initio calculations of electron densities in molecules containing 1s-, 2s-, and 2p-orbitals only [1]. In other words, the electronic charge density is expressed in terms of a series of products of Gaussian-type orbitals centerd around the nuclei of the atoms constituting the molecule. It should be pointed out that even for 1s-orbitals only the charge density around the nuclei is no longer spherically symmetric because of the overlap term, i.e., the cross-term between the orbitals at different centers, which arises when the absolute value squared of the wave function is calculated. To illustrate the principle, we restrict ourselves to s-orbitals and a single atom. Hence, for the calculation of nuclear electric quadrupole interactions it is desirable to have a compact analytical formula which expresses the electric field gradient tensor at the origin resulting from a Gaussian charge density distribution centerd at a given distance.

2. Results

Let the Gaussian charge density $n(r, \theta) = \exp(-(r^2 + z_0^2 - 2rz_0\cos(\theta))$ in spherical coordinates be located along the z-axis at z_0. For the sake of simplicity, we will measure distances in units of $\sqrt{2}\sigma$ and account for normalization at the end. We should not forget a minus sign when dealing with electron densities. The problem has axial symmetry, hence it suffices to calculate $V_{zz}(0)$ and the integration over the angle φ merely yields 2π. Mathematica [2] was used to carry out the angular integration over θ first which gives:

$$V_{zz}(0) = \exp(-r^2)\exp(-z_0^2)\pi(-6rz_0\cosh(2rz_0) + (3+4r^2z_0^2)\sinh(2\,r\,z_0))/(r^3z_0^3) \quad (1)$$

Next, Mathematica [2] was used to carry out the integration over r which yielded:

$$V_{zz}(0) = 2\pi\left[-2\exp(-z_0^2)z_0(3+2z_0^2) + 3\sqrt{\pi}\,\mathrm{erf}(z_0)\right]/(3\,z_0^3) \quad (2)$$

Normalized to unit charge and introducing the Gaussian width σ we finally get:

$$V_{zz}(0) = 2\,\mathrm{erf}\left(z_0/\sqrt{2}\sigma\right)/z_0^3 - 2\sqrt{2/\pi}\exp\left(-0.5(z_0/\sigma)^2\right)\left(1/3 + (\sigma/z_0)^2\right)/\sigma^3 \quad (3)$$

The result resembles that obtained for a periodic distribution of Gaussian charge density distributions in two dimensions [3]. For large z_0 it is immediately evident that the second term rapidly fades away whereas the first term approaches $2/z_0^3$, the value we would get for a unit charge shrunk into a point. It is also clear that $V_{zz}(0)$ is always smaller than the point charge value $2/z_0^3$ because erf(x) is approaching unity only for $x \to \infty$ and the second term is always negative. For the further discussion it is convenient to derive the series expansion of Equation (2):

$$V_{zz}(0) = 2\sqrt{2/\pi}\exp\left(-0.5(z_0/\sigma)^2\right)/\sigma^3\sum (z_0/\sigma)^{2n+2}/(2n+5)!! \quad (4)$$

Here, the sum runs from $n = 0$ to ∞ and the double exclamation mark denotes the factorial of odd integers. Form this expression it is clear that $V_{zz}(0)$ vanishes quadratically for $z_0 \to 0$, as required by symmetry. It is also clear that $V_{zz}(0)$ never changes sign.

Figure 1 shows a typical example for three different values of σ with z_0 = 1.425 a.u., a value which is close to the interatomic distance in the H_2-molecule. This has to be considered a rather short interatomic distance. It is interesting to see that the faster the decrease of the Gaussian, i.e., the smaller σ, the larger its contribution to $V_{zz}(0)$. On the other hand, the long tails of a Gaussian with large

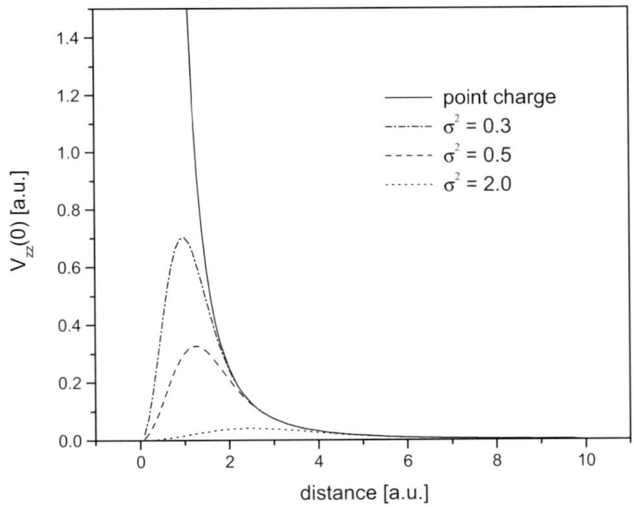

Figure 1. $V_{zz}(0)$ of a Gaussian charge density distribution with unit charge vs. distance of the centre for various values of the width σ^2; the 'point charge' line represents the result for a unit charge shrunk to a point.

σ contribute very little. Of course, it has to be kept in mind that the positively charged nuclei contribute $+2/z_0{}^3$ and the contribution from the electrons is the negative of Equations (3) or (4).

In order to obtain distances in metres multiply a.u. by Bohr's radius $a_0 = 0.52917 \times 10^{-10}$ m. For the conversion of a.u. to V/m^2 for $V_{zz}(0)$ multiply a.u. by $e/(4\ \pi\ \varepsilon_0\ a_0^3) = 9.7166 \times 10^{21}$ V/m^2. However, we will stick to a.u. in the following for convenience.

It is interesting to calculate the contribution of an infinitesimally thin slab a distance d away from the x–y-plane. For this purpose we return to measure distances in units of $\sqrt{2}\sigma$ and account for normalization at the end. We also skip the minus sign because we actually have electron densities. We switch to cylindrical coordinates ρ and z. The integration over ρ for fixed d is carried out using Mathematica [2] which gives:

$$-2\mathrm{abs}(d) + \sqrt{\pi}\left(1 + 2d^2\right) \exp\left(d^2\right)[1 - \mathrm{erf}(\mathrm{abs}(d))] \tag{5}$$

This function is plotted in Figure 2 for $z_0 = 1$. The cusp at the origin results from the absolute signs. The function is clearly asymmetric because the centre of the Gaussian distribution was located at $z_0 = 1$. Interestingly enough, the slabs closest to the origin on the side of the Gaussian distribution centre contribute most whereas the function of Equation (5) passes smoothly over the position of the centre at $z_0 = 1$. The asymptotic behaviour of the error-function ensures that the expression rapidly approaches zero for large d of either sign.

The integration over d using Mathematica [2] yields:

$$2\ \pi\left[-2\ z_0 \exp\left(-z_0^2\right) + \sqrt{\pi}\ \mathrm{erf}(z_0)/z_0^3\right] \tag{6}$$

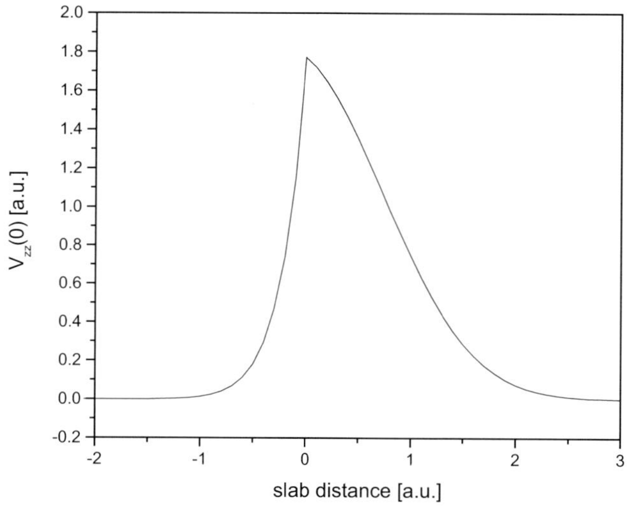

Figure 2. Contribution to $V_{zz}(0)$ of an infinitesimally thin slab of a Gaussian charge density distribution with unit charge centred at a distance of 1 a.u. versus slab distance to the x–y plane.

The central slab containing the origin requires special attention: for infinitesimal thickness we can use the usual 'cheese-hole' procedure of de Wette [4], i.e., we cut a spherical hole around the origin and let the radius shrink towards zero. For a homogeneous density of $\exp(-z_0^2)$ and infinite lateral extension of the slab we get a contribution to $V_{zz}(0)$ of $8\pi/3 \cdot \exp(-z_0^2)$. In essence, we subtract a homogeneous density of $\exp(-z_0^2)$, the density at the origin, from the true density, with infinite dimension in all directions, but slab-shaped. Thus $8\pi/3 \cdot \exp(-z_0^2)$ has to be subtracted from the result of Equation (6). We finally end up again with Equation (2).

3. An example

As a simple example – perhaps the simplest one – we calculate $V_{zz}(0)$ for a hydrogen atom located at $z_0 = 1.425$ a.u. The numerical integration, again with Mathematica [2], is compared with the Gaussian expansion of ref. [1] for $K = 2$ up to $K = 6$ Gaussians using Equation (3) which we rewrite in terms of expansion coefficients d_{1s} and exponents α_{1s} for the wave functions as follows (dropping the subscript "1s"):

$$V_{zz}(0) = 4\sqrt{2}\Big/\pi^2 \sum\sum \quad (\alpha_i\alpha_j)^{3/4} d_i d_j \times \left[\sqrt{\pi}\,\mathrm{erf}\left(X_{ij}\right) - \exp\left(-X_{ij}^2\right) X_{ij}\left(4/3X_{ij}^2 + 2\right)\right] \tag{7}$$

Here, $X_{ij} = z_0\,(\alpha_i + \alpha_j)^{1/2}$. The double sum runs over i and j from 1 up to K. An inspection of the individual contributions to the double sum reveals that while some contributions are negligibly small, many terms contribute, especially the cross-terms for $i \neq j$. The reason for that is simple: those narrow Gaussians which are required to model the cusp could in principle contribute strongly but have small expansion coefficients whereas those broad Gaussians which are required to model the tails contribute little even with large expansion coefficients. The results are given here:

$V_{zz}(0)$ (a.u.) exact	$K = 2$	$K = 3$	$K = 4$	$K = 5$	$K = 6$
−0.220625	−0.060423	−0.105348	−0.126965	−0.144209	−0.154193

Obviously, two Gaussians only are by no means sufficient to model the charge density cusp of an 1s-wave function (any s-wave function) whereas six are already doing reasonably well. The discrepancy is about 30%, not bad for the notoriously 'difficult' hydrogen atom at a rather close distance. The contribution from the nucleus is +0.69117 a.u.

4. Outlook

In principle, it is straightforward to transfer the above formalism to molecules. Let us consider H_2, the simplest molecule. We can use Gaussian expansions for the hydrogen 1s-wave function centred at each atom and calculate the resulting charge density. Now we get cross-terms between the individual Gaussians centred at each atom but also cross-terms between Gaussians centred at atom 1 with those centred at atom 2 which actually constitute the covalent bond. Thus there are K^2 intra-atomic terms from the neighbouring atom and K^2 inter-atomic terms which can all be calculated by Equation (7) with suitable definitions of X_{ij} and prefactors. In essence, the effective distances z_0 for the Gaussians which model the chemical bond will vary as a function of the exponents α_i and α_j. This will be discussed in a separate paper.

A further extension will include angular triatomic molecules. For H_3 it seems that it is adequate to calculate $V_{zz}(0)$ at the central atom using atom pairs only even for angles as small as 90°. This means that there is no problem in adding the fragment $V_{zz}(0)$ tensorially. There is some hope that metal centres in metalloproteins are also good candidates for this procedure, as suggested by the atomic overlap model used by the late Rogert Bauer [5].

Acknowledgements

It is a pleasure to thank L. Paditz and F. Heinrich for expert help and fruitful discussions. During the conference H. Wolf presented a very elegant derivation of Equation (3) without the help of Mathematica.

References

1. Hehre W. J., Stewart R. F. and Pople J. A., *J. Chem. Phys.* **51**(6) (1969), 2657 and a series of subsequent related papers.
2. Wolfram Research, Inc. 2004.
3. Butz T., *Z. Naturforsch.* **57a** (2002), 518.
4. de Wette F. W., *Phys. Rev.* **123** (1961), 103.
5. Bauer R., *Hyperfine Interact.* **39** (1988), 203.

Hyperfine Interactions (2004) 158:47–51
DOI 10.1007/s10751-005-9010-x

Electric Field Gradients at Hf and Fe Sites in Hf_2Fe Recalculated

J. BELOŠEVIĆ-ČAVOR, B. CEKIĆ*, N. NOVAKOVIĆ,
V. KOTESKI and Z. MILOŠEVIĆ
Institute of Nuclear Sciences Vinca, Belgrade, Serbia and Montenegro;
e-mail: cjeca@vin.bg.ac.yu

Abstract. The electric field gradients (EFG) of the Hf_2Fe intermetallic compound were calculated using the full-potential linearized augmented plain-wave (FP-LAPW) method as embodied in the WIEN 97 code. The obtained values are compared with other ab-initio calculations and on a qualitative basis with the previously reported experimental data obtained from TDPAC. The calculated results, $-23.1 \cdot 10^{21}$ V/m^2 and $2.7 \cdot 10^{21}$ V/m^2 for Hf 48f and Fe 32e position, respectively, are in excellent agreement with experimental data ($23.4 \cdot 10^{21}$ V/m^2 and $2.7 \cdot 10^{21}$ V/m^2), better than those reported in earlier calculations. The calculated EFG for Hf 16c position ($4.2 \cdot 10^{21}$ V/m^2) is stronger than the experimental one ($1.1 \cdot 10^{21}$ V/m^2).

Key Words: Hf_2Fe, electric field gradient, FLAPW, WIEN 97.

1. Introduction

Electric field gradient (EFG) can be used as a powerful tool to provide structural information of solids on the atomic scale. It arises from the interaction of a nucleus and the electric field created by its chemical environment. Since the EFG varies from site to site, it can be regarded as a way to label non-equivalent sites in an investigated solid [1].

The Hf_2Fe has attracted much attention in the past, mostly because of its ability to absorb a large amount of hydrogen [2]. It crystallizes in the cubic Ti_2Ni structure, with 96 atoms per unit cell. The 32 Fe atoms are in positions 32e, the 64 Hf atoms are distributed on positions, 16c (Hf1) and 48f (Hf2) [3]. The striking feature of the EFG observed at the 48f site is its high value, probably the largest ever observed for the ^{181}Ta probe ion in any metallic medium. The value of the EFG at the 16c site is about one order of magnitude smaller thus approaching the usual values in most metallic systems [4]. These interesting properties, along with the existence of high quality experimental data concerning the EFG problem in this compound, stimulated us to treat Hf_2Fe theoretically.

* Author for correspondence.

The precise determination of the EFG parameters is important from theoretical point of view, since they are extremely sensitive to the charge distribution around the probe atom and the charge transfer in the binary intermetallics can play an essential role in the determination of the character of the chemical bonds via hybridisation effects [5]. In addition, by using the FP-LAPW method we were able to check the accuracy of the earlier calculations performed in the atomic sphere approximation (ASA) [6].

2. Details of calculation

Electronic structure and EFG calculations of the Hf_2Fe were performed using WIEN 97-FLAPW code [7]. The calculations were carried out with 47 k points in the irreducible wedge of Brilloun zone and plain wave cut-off parameter $R_{MT}K_{max} = 8.0$. Muffin-tin (MT) radii for Hf and Fe were 2.3 and 2.15 a. u., respectively. Exchange and correlation effects were included within generalized gradient approximation (GGA) with the parameterisation given by [8]. Core-valence states separation was settled at -7.0 Ry. Core states were treated fully relativistically, while valence states were treated within the scalar relativistic approximation.

3. Results and discussion

The total density of states (DOS) for the Hf_2Fe structure, along with the site projected and l-decomposed DOS is shown in Figure 1. Hf 4f bands situated about 12 eV below the Fermi level (E_f) are not presented in Figure 1. Their energy dispersion is small, indicating low hybridisation with the other valence states. They also retain almost all of their electrons. The DOS in the energy region from -8 to -4 eV is predominantly of s character and from about -4 eV up to the E_f is of pronounced d character. There is a strong mixing of Fe 3d and Hf 4d orbitals, as well as the p orbitals of all constituent atoms throughout this region. Figure 1. reveals that Fe 3d electrons give the major contribution to the DOS of the central valence region. As we approach the E_f, the contribution from Hf2 d and p electrons becomes more prominent and of the same order as that of the Fe. Since the number of Hf1 atoms in unit cell equals one third of the number of Hf2 atoms, their contribution to the DOS is correspondingly smaller.

Table I reveals the site projected partial charge inside the MT spheres around Hf1, Hf2 and Fe. One can see that the s charge of Hf1, Hf2 and Fe, as well as the d charge of Hf1 and Hf2 is redistributed, filling mostly the Fe p states and the interstitial region. This charge transfer is an indication that the bonds in this metallic compound could also have partly ionic character.

After determing the self-consistent charge density we obtain the EFG tensor V_{ij} using the method developed in [9]. The usual convention is to designate the largest component of the EFG tensor as V_{zz}.

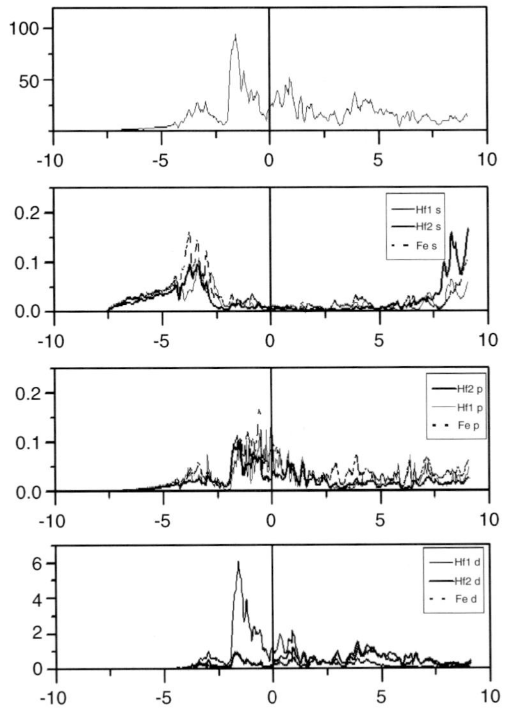

Figure 1. Total (*top*) and site-projected (*bottom*) density of states for the Hf_2Fe.

In Table II the decomposition of the EFGs at Hf and Fe lattice sites in the Hf_2Fe is presented. It is performed according to the relation:

$$\rho_{LM}(r) = \sum_{E<Ef} \sum_{lm} \sum_{l'm'} \boldsymbol{R}_{lm}(r)\boldsymbol{R}_{l'm'}(r)\boldsymbol{G}_{Ll l'}^{Mmm'}$$

where $\boldsymbol{R}_{lm}$ (r) are the LAPW radial wave functions and $\boldsymbol{G}_{Lll'}^{Mmm'}$ are Gaunt numbers. The density coefficients enter the relations for calculating the valence EFG. Two radial fuctions with angular momenta l and l' can contribute to the total EFG, but the number of non-zero combinations is limited, mostly to s–d, d–d and p–p combinations.

The EFG at Hf sites in Hf_2Fe was investigated by three experimental groups [4, 10, 11], which have been using the TDPAC method with ^{181}Ta probe ions. They observed a large difference between the magnitudes of the EFGs at Hf1 ($1.1 \cdot 10^{21}$ V/m^2) and Hf2 ($23.4 \cdot 10^{21}$ V/m^2) sites, at $T = 78$ K [4]. The signs for the EFGs have not been determined.

The experimental EFG measurements for ^{181}Ta impurities in Hf_2Fe should reflect the situation occurring at Hf atoms in the same compound, at least in a qualitative manner. Our theoretical calculations, performed for Hf (not Ta) atoms, confirmed that assumption. Although a direct quantitative comparasion of our results with the experimental ones is not possible, we have found that the basic conclusions extracted from the experiment are valid in the case of the pure Hf_2Fe compound also. The justification for this approach comes from the fact

Table I. *l*-decomposed site-projected charge inside the muffin-tin spheres around Hf1, Hf2 and Fe in Hf_2Fe

Atom	s	p	d	f
Hf1	2.214	5.900	1.236	13.948
Hf2	2.190	5.894	1.075	13.943
Fe	2.318	6.236	5.949	0.007

Table II. Decomposition of the calculated V_{zz} values in units of 10^{21} V/m^2

	s–d	p–p	d–d	Others	Total
16c	−0.084	5.367	−1.001	−0.084	4.197
48f	−0.120	−19.890	−2.942	−0.116	−23.069
32e	0.007	−0.118	2.805	0.007	2.702

that the major EFG contribution in most systems with Hf comes from its p electrons. Since Ta has only one d electron more than Hf, their radial p functions should be similar [12]. In ref. [1] was shown that the inclusion of the Ta impurity doesn't change much the EFG at the Hf sites.

Our calculations predicted a large difference between the EFGs at inequivalent Hf sites. We also predicted the signs of the EFGs, which have not been determined from the experiment. The EFG at Fe site in Hf_2Fe was investigated by Mossbauer study [13] and our calculated value is in excellent agreement with experimental one ($2.7\cdot10^{21}$ V/m^2).

Considering comparation with the earlier reported calculations using FP-LMTO ASA method [6], there are some differences. Firstly, for the Hf1 atom, the absolute value for V_{zz} is similar but has opposite sign. For the Hf2 and Fe atom, the obtained EFGs have the same sign as in the [6], but their absolute values are much bigger (closer to the measured values). Besides that, according to our calculations not only at Hf1 site but also at Hf2 site, the major contribution to the EFG comes from the Hf p electrons.

4. Conclusion

We have presented ab-initio calculations of the Hf_2Fe structure using FP-LAPW method, up to recently little investigated, probably because inherent complexity of the interactions in this system. It was shown that the main contribution to the EFG at Hf sites comes from the p electrons. In order to make an accurate comparison between the theoretical and experimental results, EFG calculations at the Ta impurity are necessary. The excellent agreement of the calculated and measured EFG, at Hf2 and Fe positions, represents an indication of the quality of our calculations. For the 16c (Hf1) position, the calculated value substantially differs from the corresponding experimental one, denoting the difference between the real and ideal structure.

Acknowledgement

We acknowledge the financial support from the Serbian Ministry of Science and Ecology (Grant No. 2021).

References

1. Terrazos L. A., Petrilli H. M., Marszalek M., Saitovich H., Silva P. R. J., Blaha P. and Schwarz K., *Solid State Commun.* **121** (2002), 525.
2. Mukai D., Miyata H. and Aoki K., *J. Alloys Compd.* **293–295** (1999), 417.
3. Cekić B., Prelesnik B., Koički S., Rodić D., Manasijević M. and Ivanović N., *J. Less-Common Met.* **171** (1991), 9.
4. Koički S., Cekić B., Ivanović N. and Manasijević M., *Phys. Rev., B* **48** (1993), 9291.
5. Pettifor D. G., *Solid State Phys.* **40** (1987), 43.
6. Lalić M. V. Popović Z. S. and Vukajlović F. R., *J. Phys. Condens. Matter* **11** (1999), 251.
7. Blaha P., Schwarz K. and Luitz J., In: WIEN 97 (Vienna University of Technology, Vienna 1977). Improved and updated UNIX version of the original copyrighted WIEN code, which was published by Blaha P., Schwarz K., Sorantin P. and Trickey S. B., *Comput. Phys. Commun.* **59** (1990), 399.
8. Perdew J. P., Burke S. and Ernzerhof M., *Phys. Rev. Lett.* **77** (1996), 3865.
9. Blaha P. and Schwarz K., *Phys. Rev. Lett.* **54** (1985), 1192.
10. Akselrod Z. Z., Komissarova B. A., Kryukova L. N., Ryasnyi G. K., Shpinkova L. G. and Sorokin A. A., *Phys. Status Solidi, B* **160** (1990), 255.
11. Van Eek S. M. and Pasquevich A. F., *Hyperfine Interact.* **122** (1999), 317.
12. Terrazos L. A., Petrilli H. M., Marszalek M., Saitovich H., Silva P. R. J., Blaha P. and Schwarz K., *Solid State Commun.* **122** (2002), 317.
13. Aubertin F., Schneider B., Gonser U. and Campbell S. J., *Hyperfine Interact.* **41** (1988), 547.

Hyperfine Interactions (2004) 158:53–57
DOI 10.1007/s10751-005-9007-5

Comparative Theoretical Study of Hyperfine Interactions of Muonium in A- and B-Form DNA

R. H. SCHEICHER[1,*,†], E. TORIKAI[2], F. L. PRATT[3], K. NAGAMINE[4] and T. P. DAS[1,5]
[1]*Department of Physics, State University of New York at Albany, Albany, NY 12222, USA; e-mail: ralph@fysik.uu.se*
[2]*Faculty of Engineering, Yamanashi University, 4-3-11 Takeda, Kofu, Japan*
[3]*ISIS Facility, Rutherford Appleton Laboratory, Chilton, Didcot, OX11 0QX, UK*
[4]*KEK-MSL, 1-1 Oho, Tsukuba, Ibaraki 305-0801, Japan*
[5]*Department of Physics, University of Central Florida, Orlando, FL 32816, USA*

Abstract. Muon Spin Relaxation (μSR) experiments in A- and B-form DNA have shown evidence for an enhanced electron mobility in the more closely-packed A-form. Besides dynamic effects (electronic diffusion) that could cause the observed difference in muon spin relaxation, one should also carefully examine the difference in the strengths of the hyperfine interactions of the muon (μ^+) with the moving electron in the two forms of DNA, since this could contribute to the observed difference in the muon spin relaxation rates as well. We have therefore investigated the (static) trapping properties of muon and muonium (μ^+e^-) in A-form and B-form DNA from first-principles with the aim to understand how the different structural geometries of A- and B-form DNA can influence the hyperfine interaction of trapped muonium.

Key Words: DNA, first-principles calculations, muon spin relaxation.

1. Introduction

A series of comparative muon spin relaxation (μSR) experiments in A-form and B-form DNA has recently been carried out [1] to investigate the dependence of electron mobility upon base pair separation. The results of this study suggest that electron mobility is enhanced in the drier A-form DNA where base pairs are more densely packed, and reduced in B-form DNA where adjacent base pairs are separated further due to a higher level of hydration.

This phenomenon could be described as a dynamic effect, caused by the movement of the electron which was brought in by the muon (μ^+), attached to the latter forming a muonium (Mu = μ^+e^-), which loses its electron after being

* Author for correspondence.
† Present Address: Dept. of Physics, Uppsala University, Box 530, S-751 21, Uppsala, Sweden.

trapped in the DNA system.★ The conclusion regarding the relative electron mobilities in A- and B-form DNA is implicitly based on the assumption that the fluctuating magnitude of the hyperfine fields at the trapped μ^+ site, that are produced by the moving electron and are affecting the muon spin relaxation, do not differ significantly between A-form and B-form DNA. One would however expect that these fluctuating hyperfine fields should in fact be different for A- and B-form DNA due to the different geometrical structure of the two forms. In the present paper we give an account of our first-principles study of the magnetic hyperfine interactions for Mu trapped in Adenine in A-form and B-form DNA in order to assess how significant the difference in the hyperfine interactions of Mu between these two forms is. This will provide insights into the influence of the difference in the fluctuating magnitudes of the hyperfine fields at the μ^+ due to the moving electron and the contribution of these hyperfine fields to the differences in the observed muon spin relaxation rate between A- and B-form DNA.

2. Procedure

We studied the magnetic hyperfine interaction of Mu and general trapping properties of μ^+ and Mu in Adenine in three configurations, namely Adenine from A-form DNA, B-form DNA, and from the isolated base. The structural data of Adenine as it occurs in A-form and B-form DNA has been taken from experimental X-ray diffraction analysis [3]. Only those atoms were included in the cluster that belong to the base and sugar ring of Adenine. The link to the phosphate has been terminated with a methyl group. To evaluate the importance of the sugar ring for hyperfine interactions, we included the isolated Adenine molecule (i.e., just the base) in our investigation, with the only modification from its original stoichiometry being the replacement of the H atom on N9 by a methyl group (Figure 1), so that better comparison could be made with Adenine in DNA which possesses a link to the sugar–phosphate backbone at N9.

The three systems with μ^+/Mu added to the sites under investigation were then each studied for their energetic properties and magnetic hyperfine interaction of Mu with its local environment.★★ In particular, following the motivation of the study, we were interested to see how large the difference in the hyperfine interaction would be between Adenine from A-form DNA and B-form DNA. In addition, we tested for distortions in the geometry of the isolated

★ For a review of the muon spin relaxation method, see Ref. [2].

★★ At all trapping sites considered, Mu is located at the end of a bond, thus placing it in a rather steep potential. Zero-point motion and the resulting vibrational averaging of the hyperfine interaction should therefore be negligible, and was consequently not considered in the present study.

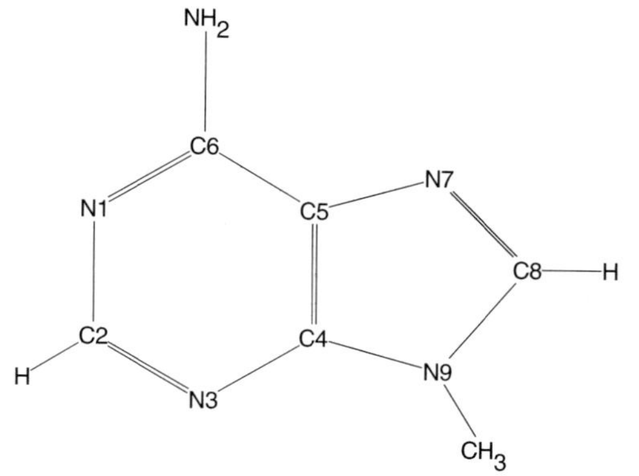

Figure 1. Chemical structure of Adenine with atomic labels for reference. The H atom on N9, that is present in the natural form of Adenine, was replaced by a methyl group.

Adenine molecule when μ^+/Mu was added, especially whether the presence of μ^+/Mu causes any deviation from the planar shape of isolated Adenine.

For our theoretical investigations we employed the Unrestricted Hartree–Fock Cluster procedure which allows for spin polarization. To incorporate correlation effects, second-order Møller–Plesset perturbation theory (MP2) has been used, and the convergence with respect to basis set size was studied using the standard split–valence basis sets 6–31G(d,p), 6–31++G(d,p), 6–311++G(d,p) and cc-pVDZ.

3. Results and discussion

We first report our results on the effect that μ^+ and Mu have on the geometry of isolated Adenine when trapped at different sites. The results are summarized in Table I. When Mu or μ^+ is attached to C4 or C5 (Figure 1), the geometry of Adenine is severely altered from its planar shape. The two rings of Adenine appear tilted, with an angle of about 25°. For Mu attached to N3 or N7, the same qualitative behavior is observed; however, quantitatively the change is less drastic; the tilt angle between the rings is about 10° for N3 and about 7° for N7. This kind of change in geometry does not occur when μ^+ is attached to N3 or N7. For those sites, the planar geometry of Adenine is preserved. For the trapping sites C2, C6 or C8, Adenine keeps its planar geometry when either μ^+ or Mu are attached. This behavior can be explained as follows. The hydrogen atoms at C2 and C8 (Figure 1) were observed to move out of the plane of the molecule, thus compensating the distorting effect that Mu or μ^+ have on the geometry of Adenine. Similarly, at C6, the NH_2 group is bending out of the plane, thereby compensating for the perturbing effect of Mu or μ^+. At the current stage of our investigation, it is not clear whether these predicted geometrical distortions would have a significant effect on the electron mobility, or whether μ^+ can effectively be treated as a passive probe in this system.

Table I. The effect of Mu/μ^+ on the geometry of isolated Adenine when trapped at different sites (see Figure 1 for labels)

Trapping site	C2	C4	C5	C6	C8	N3	N7
Mu: planar geometry?	yes	no	no	yes	yes	no	no
μ^+: planar geometry?	yes	no	no	yes	yes	yes	yes

The two last rows indicate whether or not Adenine kept its original planar shape after Mu/μ^+ was added.

Table II. Isotropic hyperfine coupling constant for Mu trapped at various sites in isolated Adenine, Adenine from A-form DNA and Adenine from B-form DNA

Trapping site	*A* [MHz] for isolated Adenine	*A* [MHz] for A-form Adenine	*A* [MHz] for B-form Adenine
C2	309.8	545.8	445.3
C4	313.6	966.0	894.1
C5	380.8	786.7	742.2
C6	303.8	652.0	665.5
C8	280.2	354.2	342.5
N3	−2.3	256.1	432.1
N7	17.7	363.6	435.8

Following the main motivation of our theoretical investigation, the isotropic hyperfine coupling constant *A* was calculated for Mu trapped at the various sites in the isolated molecule, and in Adenine from A-form and B-form DNA. The results are summarized in Table II.

The first thing to notice is that the hyperfine coupling constant is in general significantly lower for Mu trapped in isolated Adenine as compared to the case when it is trapped in either A-form or B-form Adenine. Since there is no sugar ring present in the isolated molecule, this implies that the sugar ring has a significant effect on the hyperfine interaction of Mu trapped in the base. Secondly, comparing the results for A-form DNA and B-form DNA, we see that the numbers are still quite different for most trapping sites, while not as different as in the previous comparison to the isolated Adenine system. This leads to the conclusion that there is indeed a significant difference between Mu hyperfine interaction in A-form and B-form Adenine. Lastly, we note that the results for both A- and B-form show substantial variation, so that in principle, a distinction of the Mu trapping site based on a measurement of the hyperfine coupling constant could be made.

4. Conclusions and future investigations

We have shown from first-principles that there exist significant differences between the magnetic hyperfine interaction of Mu trapped in Adenine from A-form DNA and B-form DNA. This result should have important implications for the interpretation of μSR experiments on the dependence of electron mobility on the separation of base pairs in DNA. In addition to a genuine effect on the muon spin relaxation rate due to an actual difference in electron mobility between A-form and B-form DNA, one should therefore also consider the relaxation-rate altering influence of the discovered difference in hyperfine interaction of Mu between the two forms. It remains to be seen what the relative importance of these two effects is, and whether they fortify or impair each other.

The observed substantial variation in the isotropic hyperfine coupling constant at different sites could furthermore open up the possibility to experimentally distinguish between Mu trapping sites in Adenine in either A- or B-form DNA. Lastly, we found that μ^+ and Mu can have profound effects on the geometry of Adenine when trapped at various sites in the base, leading to a tilt between the two rings. For some sites, however, the distorting effect of μ^+ and Mu is compensated by a reorientation of an H atom or an NH_2 group at that site.

Besides repeating this study for the other three bases of DNA, we are also planning to expand the clusters in size, so that they would contain several connected base pairs and thus could help to evaluate how important the direct effects of neighboring bases are. Furthermore, the influence of the differing amount of water that A-form and B-form DNA contain was already taken indirectly into account in the present investigation by using the experimentally determined geometries. It would however be beneficial to also study the direct effect that the different hydration environments have on the hyperfine interaction of Mu and on the general trapping properties of μ^+ and Mu, by incorporating an appropriate distribution of water molecules in the surrounding of the cluster.

References

1. Torikai E., Nagamine K., Pratt F. L., Watanabe I., Ikedo Y., Urabe H. and Grimm H., *Hyperfine Interact.* **138** (2001), 509.
2. Nagamine K., *Introductory Muon Science*, Cambridge University Press, 2003.
3. Berman H. M., Olson W. K., Beveridge D. L., Westbrook J., Gelbin A., Demeny T., Hsieh S.-H., Srinivasan A. R. and Schneider B., *Biophys. J.* **63** (1992), 751.

Hyperfine Interactions (2004) 158:59–62
DOI 10.1007/s10751-005-9008-4

Hyperfine Fields of Light Interstitial Impurities in Ni

C. ZECHA[1], H. EBERT[1,*], H. AKAI[2], P. H. DEDERICHS[3] and R. ZELLER[3]
[1]*Department of Chemistry, University of Munich, Butenandtstr. 5-13, D-81377 München, Germany*
[2]*Osaka University, Osaka, Japan; e-mail: hubert.ebert@cup.uni-muenchen.de*
[3]*Institute für Festkörperforschung, Forschungszentrum Jülich, Jülich, Germany*

Abstract. The magnetic hyperfine interaction of light interstitial impurities in Ni have been studied by means of the Korringa–Kohn–Rostoker (KKR) band structure method. This method allows to deal with the impurity problem by solving the corresponding Dyson equation for the Green's function. It also allows to account for lattice relaxations. For this purpose a new technique was developed that allows to handle in principle arbitrary lattice distortions. Corresponding calculations have been performed for the magnetic hyperfine fields of the light interstitial impurities H to Ne in Ni. By minimising the force on the nearest neighbour host atoms their equilibrium position was determined. The resulting hyperfine fields for the equilibrium configuration are found to be in rather good agreement with available experimental data.

Key Words: ferromagnets, hyperfine field, interstitial impurities.

1. Introduction

Magnetic hyperfine interaction has been exploited extensively in the past to study the local electronic and magnetic properties of impurities in magnetic transition metals. Concerning the negative hyperfine field (HFF) of early sp-impurities – in particular of the alkali metals – a qualitative explanation was given already in the 1960s on the basis of scattering theory [4]. A more general description was achieved later by pointing to the importance of the hybridisation of impurity s-states and host d-states [6]. Corresponding numerical investigations led to a very deep understanding of the various mechanisms responsible for the HFF of impurities in ferromagnets. While for substitutional impurities several extensive theoretical investigations can now be found in the literature [1, 5] only few work has been done so far on interstitial impurities [2, 7]. One of the main reasons for this situation seems to be the importance of lattice relaxations that are quite difficult and demanding to be accounted for.

* Author for correspondence.

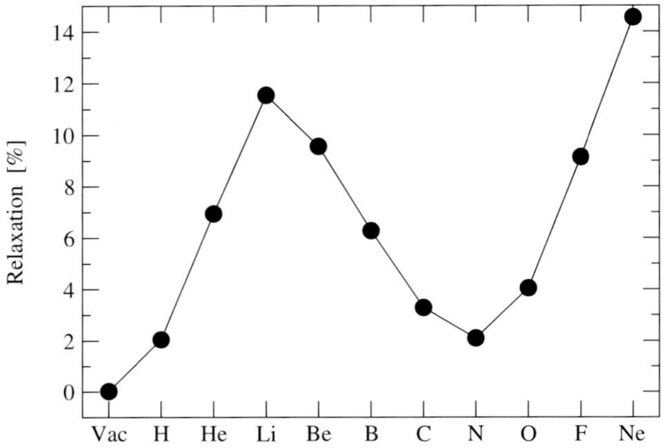

Figure 1. Equilibrium relaxation of the nearest neighbours for interstitial impurities in Ni expressed as a fraction of the nearest neighbour distance.

In the following, a new computational scheme to deal with lattice distortions is briefly introduced. The approach has been applied to calculate the HFF of light interstitial impurities in fcc-Ni.

2. Theoretical framework

Theoretical investigations on impurity systems are in general done by making use of the super-cell technique in connection with an adequate band structure method [9] or using the KKR (Korringa–Kohn–Rostoker) Green's function method to solve the corresponding Dyson equation. The later approach avoids the problem of interacting impurities in a periodic system and proved to be very powerful and flexible in the past. In particular lattice distortions can be accounted for by the so-called U-transformation technique [8]. In practice, however, this approach is limited to relatively small lattice distortions of a few percent with respect to the nearest neighbour distance. This problem can be overcome by means of the so-called Augmented Basis method. In case of an interstitial impurity the impurity site and the shifted atomic site positions of the host are repeated periodically. This allows to calculate the electronic Green's Function of the host system using the proper underlying lattice plus the auxiliary lattice essentially in the standard way by a corresponding Brillouin zone integration. In a next step the Dyson equation for an embedded impurity and its distorted host surrounding is solved. Calculation of the forces on atoms on the basis of the Hellmann–Feynman-theorem allows in principle to find the equilibrium geometry for any distorted system.

All calculations to be presented below have been done scalar-relativistically in a full-potential way on the basis of local spin density functional theory [10]. The HFF has been determined using the proper scalar-relativistic expressions [3].

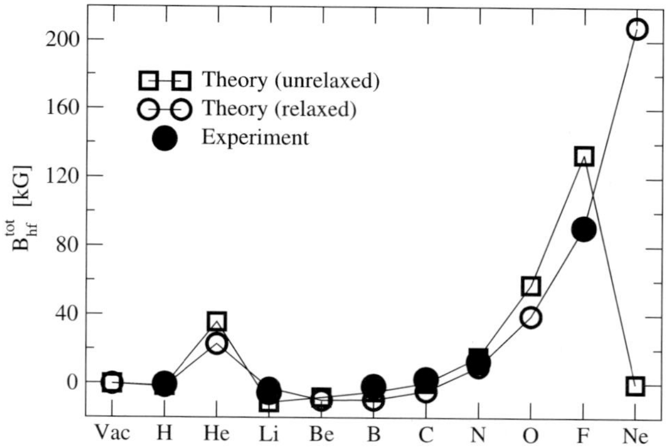

Figure 2. HFFs of the interstitial impurities in Ni calculated for unrelaxed and relaxed nearest neighbour positions in comparison with experiment.

3. Results and discussions

The approach sketched above has been used to investigate the magnetic properties of light interstitial impurities in Ni that occupy a octahedral position. To keep the numerical effort on a reasonable level only the positions of the nearest neighbour Ni-atoms around an impurity atom have been relaxed. As it turned out that the force on a nearest neighbour atom varies for moderate shifts rather well linearly with the shift from its original (unrelaxed) position. The equilibrium relaxations shown in Figure 1 have been determined by interpolation. For this purpose the force has been calculated for 0 and 8% relaxation and from this the equilibrium position for which the force vanishes has been found. The relaxation found this way is quite large (see Figure 1) and shows a minimum as a function of the atomic number (Z_{imp}).

As the HFF of the impurity atoms was also found to vary linearly with the lattice relaxation (for the range considered here) the HFF for the relaxed nearest neighbour positions were also determined by interpolation. Figure 2 shows the corresponding results in comparison with the HFFs obtained for the unrelaxed geometry (see also [7]). As one can see there are quite pronounced changes for the HFF due to the relaxation. In general, these changes lead to a much better agreement with experiment and demonstrate the need to account for the lattice relaxations when dealing with the hyperfine interaction of interstitial impurities.

The variation of the HFF with Z_{imp} shown in Figure 2, which is in particular characterised by an increase at the end of a period, is very similar to that observed for an Fe host. In both cases the HFF is dominated by its valence band contribution that is primarily determined by the population of the spin-resolved impurity s-like states. Accordingly, the variation of the HFF with Z_{imp} can be qualitatively explained by analysing the impurity s-like density of states (DOS) or – more appropriately – the local s-like DOS at the site of the impurity nucleus.

For noble gas impurities one observes peaks in the majority and minority DOS at the Fermi energy E_F. As the majority peak lies above E_F this results in a strong difference in the population of spin-up and -down impurity states leading to a maximum for the HFF for the noble gases. Proceeding to the next period the spin-up and -down peaks are both shifted below E_F leading to an abrupt drop in the HFF for the alkali atoms. Towards the end of the period an exchange split peak in the s-DOS builds up again above E_F. With increasing Z_{imp} this passes E_F leading to a continuous increase of the HFF until the end of the period.

In addition to the valence band contribution to the HFF there is an appreciable core contribution present for the second period which is largest for F (about -20 kG). As found in general for core polarisation fields this contribution is proportional to the local spin magnetic moment of the impurity. For the systems studied here the proportionality constant was found to by -200 kG/μ_B.

References

1. Akai H., Akai M. and Kanamori J., Electronic structure of impurities in ferromagnetic iron. II. 3d and 4d impurities, *J. Phys. Soc. Jpn.* **54** (1985), 4257.
2. Akai M., Akai H. and Kanamori J., Electronic structure of impurities in ferromagnetic iron. III. Light interstitials, *J. Phys. Soc. Jpn.* **56** (1987), 1064.
3. Blügel S., Akai H., Zeller R. and Dederichs P. H., Hyperfine fields of 3d and 4d impurities in nickel, *Phys. Rev. B* **35** (1987), 3271.
4. Daniel E. and Friedel J., Sur la polarisation de spin des electrons de conductibilite dans les metaux ferromagnetiques, *J. Phys. Chem. Solids* **24** (1963), 1601.
5. Dederichs P. H., Zeller R., Akai H., Blügel S. and Oswald A., *Ab initio* calculations for impurities in Cu and Ni, *Phil. Mag. B* **51** (1985), 137.
6. Katayama-Yoshida H., Terakura K. and Kanamori J., A calculation of the electronic structure of an interstitial and a substitutional impurity atom of boron in ferromagnetic nickel, *J. Phys. Soc. Jpn.* **46** (1979), 822.
7. Katayama-Yoshida H., Terakura K. and Kanamori J., Systematic variation of hyperfine field and relaxation time of impurity nuclei in ferromagnetic nickel. II. Case of light nuclei, *J. Phys. Soc. Jpn.* **49** (1980), 972.
8. Korhonen T., Settels A., Papanikolaou N., Zeller R. and Dederichs P. H., Lattice relaxations and hyperfine fields of heavy impurities in Fe, *Phys. Rev. B* **62** (2000), 452.
9. Seewald G., Hagn E., Zech E., Kleyna R., Voß M. and Burchard A., Spin–orbit induced noncubic charge distribution in cubic ferromagnets. I. Electric field gradient measurements on 5d impurities in Fe and Ni, *Phys. Rev. B* **66** (2002), 174401.
10. Vosko S. H., Wilk L. and Nusair M., Accurate spin-dependent electron liquid correlation energies for local spin density calculations: A critical analysis, *Can. J. Phys.* **58** (1980), 1200.

Hyperfine Interactions (2004) 158:63–69
DOI 10.1007/s10751-005-9009-3

FLAPW Study of the EFG Tensor at Cd Impurities in In_2O_3

L. A. ERRICO*, M. RENTERÍA, G. FABRICIUS and G. N. DARRIBA
Departamento de Física, Facultad de Ciencias Exactas, Universidad Nacional de La Plata, C.C. 67, 1900 La Plata, Argentina; e-mail: errico@fisica.unlp.edu.ar

Abstract. We report an *ab initio* study of the electric-field gradient tensor (EFG) at Cd impurities located at both nonequivalent cationic sites in the semiconductor In_2O_3. Calculations were performed with the FLAPW method that allows us to treat the electronic structure of the doped system and the atomic relaxations introduced by the impurities in the host in a fully self-consistent way. From our results for the EFG (in excellent agreement with the experiments), it is clear that the problem of the EFG at Cd impurities in In_2O_3 cannot be described by the point-charge model and antishielding factors.

1. Introduction

Perturbed-Angular Correlations (PAC) and other hyperfine interaction measurements are widely used experimental techniques that provide local information on the interaction of a probe-nucleus with the surrounding electronic charge distribution [1]. An interpretation of such measurements can lead to a detailed knowledge of structural, electronic and magnetic properties of a solid (see, e.g., Ref. 2). One of the measured quantities, the quadrupole coupling constant ν_Q, is proportional to the major principal component (V_{33}) of the electric-field gradient tensor (EFG). Due to the r^{-3} dependence of the EFG from the charge sources, the EFG "felt" by a nucleus reflects sensitively the non-spherical electronic charge distribution around the nucleus. Therefore, the EFG is one of the most important clues for the understanding of the electronic structure in solids and the nature of the chemical bonding.

Usually, the probe-atom is an impurity dopant, which is present in either trace or small amounts in the host crystal. For this reason, interpreting experimental results involves the understanding of chemical differences between the probe atom and the indigenous ion replaced by the impurity. For a long time, approaches for a quantitative calculation of the EFG were often based on the point-charge model (PCM). Since such calculations did not account for any onsite

* Author for correspondence.

polarization, Sternheimer antishielding factors were introduced, whereas chemical bonding was usually neglected. The accuracy of such calculations was often quite limited. For an accurate calculation of the EFG, the electronic configuration of the host, perturbed by the presence of the impurity, has to be determined. This can be done in the frame of the Density-Functional Theory. In this sense, in 1999 we began a systematic study of the EFG in doped oxides [3–5] using the Full-Potential Linearized-Augmented Plane Waves (FLAPW) method.

Among the binary oxides, those that crystallize in the bixbyite structure were the subjects of several PAC investigations using ^{111}Cd as probe (see, e.g., [6] and references therein). Due to the high degree of ionicity of these oxides, it was assumed in the past that PCM was accurate enough to exactly predict the EFG in these compounds [7, 8]. We report in this work a FLAPW study of the bixbyite In_2O_3 doped with Cd. The excellent agreement between the FLAPW predictions for V_{33} and the asymmetry parameter η and the experiments demonstrate that this approach is well suited to calculate accurately structural and electronic properties of the system and to provide a deeper understanding of them.

2. Method of calculation

In_2O_3 crystallize in the cubic bixbyite structure (a = 10.117$_1$ Å [9]). In this structure, the cations form a nearly cubic face-center lattice in which six out of the eight tetrahedral sites are occupied by oxygen atoms. The unit cell consists of eight such cubes, containing 32 In and 48 O atoms. Two nonequivalent cation sites, called *C* and *D*, both with 6-fold O coordination, characterize the structure. Site *D* is axially symmetric and can be described as an In surrounded by six O atoms (O_{NN}) at the corners of a distorted cube, leaving two corners of the same diagonal free. In site *C* the cube is more distorted and the O atoms leave free two corners on a face diagonal. The positions of all atoms in the cell are determined by four parameters: the coordinate u fixes the position of the *C*-type In atoms [In(*C*)], while the O positions are given by the parameters x, y, and z [9].

In order to calculate from first-principles the EFG tensor at a Cd diluted impurity in In_2O_3 taking into account the structural and electronic effects introduced by the impurity in the host lattice, we considered the unit cell of In_2O_3 periodically repeated containing a single Cd-atom. Calculations were performed with the WIEN97.10 implementation of the FLAPW [10] method as embodied in the WIEN97 code [11]. Exchange and correlation effects were treated using the generalized-gradient approximation (GGA) [12]. In this method, the unit-cell is divided into non-overlapping spheres with radius R_i and an interstitial region. The atomic spheres radii used for Cd, In and O were 1.11, 1.05 and 0.85 Å, respectively. We took for the parameter RK_{MAX}, which controls the size of the basis-set in these calculations, the value of 7. The correctness of the choice of these parameters was checked by performing

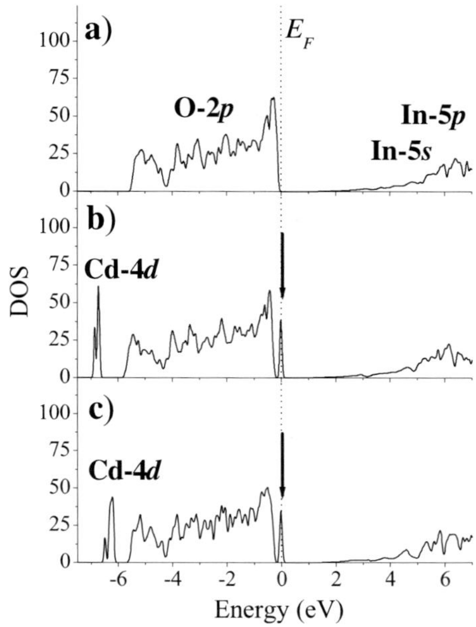

Figure 1. Density of states (DOS) for (a) pure In_2O_3; (b) one Cd impurity replacing a *D*–type In atom in In_2O_3. (c) One Cd impurity replacing a *C*–type In atom. In (b,c) the DOS correspond to the unrelaxed structures. Energies refer to the Fermi level (E_f). The arrows indicate impurity states in the valence band.

calculations for $RK_{MAX} = 6$ and 8. We also introduced local orbitals to include In-4*p* and 4*d*, O-2*s* and Cd-4*p* and 4*d* orbitals. Integration in reciprocal space was performed using the tetrahedron method taking 26 *k*-points in the first Brillouin zone. Once self-consistency of the potential was achieved, the forces on the atoms were obtained [13], the ions were displaced [14] and the new positions for Cd neighbors were obtained. The procedure was repeated until the forces on the ions were below a tolerance value of 0.025 eV/Å. At each step, the V_{ii} elements of the EFG tensor were obtained from the V_{2M} components of the lattice harmonic-expansion of the self-consistent potential [15].

There is an important point to be taken into account: the charge state of the impurity system, the single acceptor Cd^{2+} in ${In_2}^{3+}{O_3}^{2-}$. In_2O_3 is a semi-conductor with the oxygen *p*-band filled. When a Cd-atom replaces an In-atom, the resulting system is metallic because of the lack of one electron necessary to fill the oxygen *p*-band. Comparison of Figure 1(a)–(c) shows that the presence of Cd produces the appearance of Cd-*d* levels and impurity states at the Fermi level (E_f). The question is if the real system we want to describe provides the lacking electron or not, and if this point is relevant for the calculation. Here we will present results assuming that the system provides the lacking electron (the impurity level at E_f is filled). To describe this situation we added one electron to the cell that we compensated with a homogeneous positive background to obtain a neutral cell. A detailed study of the EFG as function of the charge-state of the impurity will be presented in a forthcoming longer paper.

3. Results and discussion

First, we performed calculations in undoped In_2O_3. Since u, x, y, and z are free internal parameters, which turned out to strongly affect the EFG at both cationic sites, we refined the In(C) and O positions by force minimization. The results obtained ($u = -0.0336$, $x = 0.3910$, $y = 0.1545$, $z = 0.3813$) are in very good agreement with the two X-ray diffraction results reported in the literature [9]. We also calculated the EFG at both cationic sites in pure In_2O_3. As can be seen in Table I, the obtained results for the EFGs at sites C and D are in relatively good agreement with those experimentally determined by PAC at Cd impurities [16]. This similarity can be understood from the similar electronic structure of Cd and In atoms.

After the study of undoped In_2O_3, we replace one D-type In atom [In(D)] by a Cd atom. As can be seen in Table I, the substitution of a indigenous In(D) atom by a Cd impurity does not produce significant changes in the EFG. Again, this result can be understand from the fact that In and Cd present similar electronic structures. But, beside this similarity in the EFGs, the substitution of a In(D) atom by a Cd impurity produces not negligible forces on the Cd–O_{NN}. In order to study the relaxation introduced by the impurity we considered the O_{NN} displacements (assuming that structural relaxations preserve the point group symmetry of the cell in its initial configuration) until forces vanished. Displacements of atoms beyond the O_{NN} are very small and none of the conclusions of this work are affected by them, thus we will not mention them in what follows. As can be seen in Table I, the Cd–O_{NN} relaxed outwards along the Cd–O directions. The displacement of each of these O atoms is of about 0.1 Å (5% of the unrelaxed Cd–O_{NN} distance). The magnitude of this relaxation was unexpected, and contradicts the assumption that Cd does not cause local lattice distortions in bixbyites [7, 8]. Concerning the EFG at Cd impurities located at site D, the only effect of the relaxation is a small reduction of V_{33}. The results for the EFG are in excellent agreement with the PAC results [16]. We want to mention that the EFG orientation in Table I corresponds to the case of Cd in the bixbyites Er_2O_3 and Ho_2O_3 [8]. Due to the similarities between the structures, we expect that the EFG orientation in In_2O_3 should be nearby the same as in Er_2O_3 and Ho_2O_3.

We performed a similar study for the case of a Cd impurity located at cationic site C. The substitution of an In(C) by a Cd atom produce a change of 10% in V_{33} (see Table I), but does not produce a change in the symmetry of the EFG. In addition, if we let the atoms moves (according to the forces acting on them), a strong relaxation of the six Cd–O_{NN} appears. The magnitude of this relaxation is different for each of the three pairs of distances Cd–O_{NN} but in average is similar to those produced at the site D (5%). The structural relaxation changes the magnitude of V_{33}, the symmetry of the EFG tensor, and produces a small reorientation of V_{33} (see Table I). The change in the symmetry of the EFG tensor reflects that the relaxation is non-isotropic. Finally, when the relaxed positions of the Cd–O_{NN} are taken into account, the agreement between our FLAPW results and the experimental ones [16] is excellent.

Table I. Major components, V_{33}, of the EFG tensor at Cd site (in units of 10^{21} V/m^2), asymmetry parameter η and V_{33} orientation obtained in the FLAPW calculations (this work) compared with experimental PAC results at 300 K

Site *D*

	d(Cd–O_{NN})	V_{33}	η	V_{33} orientation
In_2O_3	2.18	+8.2	0.00	[1;1;1]
In_2O_3:Cd unrelaxed structure	2.18	+7.8	0.00	[1;1;1]
In_2O_3:Cd relaxed structure	2.27	+7.6	0.00	[1;1;1]
Experimental [16]	–	7.7_1	0.00_5	[1;1;1]

Site *C*

	d(Cd–O_{NN1})	d(Cd–O_{NN2})	d(Cd–O_{NN3})	V_{33}	η	V_{33} orientation
In_2O_3	2.13	2.19	2.23	+5.4	0.98	[0;−0.70;1]
In_2O_3:Cd unrelaxed structure	2.13	2.19	2.23	+4.8	0.96	[0;−0.70;1]
In_2O_3:Cd relaxed structure	2.20	2.29	2.33	+5.6	0.68	[0;−0.75;1]
Experimental [16]	–	–	–	5.9_1	0.69_1	[0;−1;1]

d(Cd–O_{NN}) are the distances in Å from Cd to its O_{NN}. In the case of In_2O_3, d(Cd–O_{NN}) refers to the In–O_{NN} distances. The experimental EFG tensor orientations correspond to Er_2O_3 and Ho_2O_3 (see text). V_{33} orientation is referred to an axes system parallel to those defined by the crystalline axes.

It is important to compare our results with those predicted by the PCM. The results of these calculations are presented in Table II. As can be seen, the agreement between PCM and FLAPW predictions is very good for the case of site *D*. From these results, it could be stated that the ionic model can be used for the description of the EFG at cationic sites in In_2O_3. But the PCM predicts a negative sign of the EFG for site *C*, in contradiction with the FLAPW results. If the relaxed positions obtained by FLAPW are introduced in the PCM, the discrepancy in the sign of the EFG at site *C* is removed, but the agreement between the PCM, FLAPW predictions and experimental results for V_{33} at both sites and the symmetry at site *C* is not good. From these results, it is clear that the discrepancies between PCM and FLAPW (a method that takes into account the electronic structure of the Cd impurity and the structural and electronic changes that its presence induces in the In_2O_3 host in a self-consistent way without the use of external parameters) cannot be associated only to the structural relaxations. We have to conclude that the purely ionic model with the use of the Sternheimer factor fails in the description of the EFG tensor at cationic sites in In_2O_3.

Table II. V_{33}, η, and V_{33} orientation obtained in our FLAPW calculations (relaxed structures), PCM, and PCM considering the structural relaxations predicted by FLAPW

	Site *D*			Site *C*		
	V_{33}	η	V_{33} orientation	V_{33}	η	V_{33} orientation
FLAPW	+7.6	0.00	[1;1;1]	+5.6	0.68	[0;−0.75;1]
PCM	+7.5	0.00	[1;1;1]	−4.9	0.83	[0;−0.75;1]
PCM + FLAPW relaxation	+6.7	0.00	[1;1;1]	+4.6	0.55	[0;−0.75;1]

In conclusion, *ab initio* calculations using the FLAPW method successfully predict the experimental EFGs at Cd impurities in In_2O_3 and yield quantitative information on the lattice relaxation around the impurity. From our results it is clear that the problem of the EFG at impurities in this semiconductor cannot be described by a point-charge model and antishielding factors. A proper theoretical description of electronic properties of Cd in bixbyites should consider self-consistently the electronic structure of Cd and the impurity-induced distortions in the hosts without external parameters or arbitrary suppositions.

Acknowledgements

This work was partially supported by CONICET, Fundación Antorchas, ANPCyT (PICT98 03-03727), Argentina and TWAS, Italy. G.D. is a fellow of CONICET. L.E., G.F. and M.R. are members of CONICET. G.F. is associate researcher of the ICTP.

References

1. Schatz G. and Weidinger A., In: *Nuclear Condensed Matter Physics – Nuclear Methods and Applications*, John Wiley & Sons, Chichester England, 1996, p. 63.
2. Proc. of the 12th International Conference on Hyperfine Interactions, Park City, Utah 2001, by Evenson W. E., Jaeger H. and Zacate M. O. (eds.), *Hyperfine Interact.* **136/137** 2001.
3. Errico L. A., Fabricius G. and Rentería M., *Hyperfine Interact.* **136/137** (2001), 749.
4. Errico L. A., Fabricius G., Rentería M., de la Presa P. and Forker M., *Phys. Rev. Lett.* **89** (2002), 55503.
5. Errico L. A., Fabricius G. and Rentería M., *Phys. Rev. B* **67** (2003), 144104.
6. Errico L. A., Rentería M., Pasquevich A. F., Bibiloni A. G. and Freitag K., *Eur. Phys. J. B* **22** (2001), 149.
7. Bartos A., Lieb K. P., Uhrmacher M. and Wiarda D., *Acta Crystallogr. B* **49** (1993), 165.
8. Lupascu D., Bartos A., Lieb K. P. and Uhrmacher M., *Z. Phys. B* **93** (1994), 441.
9. Marezio M., *Acta Crystallogr. B* **20** (1966), 723.
10. Wei S. H. and Krakauer H., *Phys. Rev. Lett.* **55** (1985), 1200.
11. Blaha P., Schwarz K. and Luitz J., WIEN97 (K. Schwarz, T.U. Wien, Austria, 1999), ISBN 3-9501031-0-4.

12. Perdew J. P., Burke K. and Ernzerhof M., *Phys. Rev. Lett.* **77** (1996), 3865.
13. Yu R., Singh D. and Krakauer H., *Phys. Rev. B* **43** (1991), 6411.
14. Kohler, B., Wilker, S., Scheffler, M., Kouba, R. and Ambrosch-Draxl, C., *Comp. Phys. Commun.* **94** (1996), 31.
15. Schwarz, K., Ambrosch-Draxl, C. and Blaha, P., *Phys. Rev. B* **42** (1990), 2051.
16. Habenicht, S., Lupascu, D., Uhrmacher, M., Ziegeler, L., Lieb, K. P. and ISOLDE Collaboration., *Z. Phys. B* **101** (1996), 196.

Hyperfine Interactions (2004) 158:71–78
DOI 10.1007/s10751-005-9011-6

Are LCAO-MO Models Useful Estimators for Electric Field Gradients in Simple Molecules?

F. HEINRICH* and T. BUTZ
Universität Leipzig, Fakultät für Physik und Geowissenschaften, Linnéstr.5, 04103 Leipzig, Germany; e-mail: mail@frank-heinrich.de

Abstract. In this tutorial paper we compare the *ab-initio* calculations of electronic charge densities and related electric field gradients in simple molecules like H_2 as well as the triangular H_3 with variable bond distance and bond angle using the Amsterdam Density Functional (ADF) code with calculations based on a simple linear combination of hydrogen 1s-orbitals. In order to gain more insight into ADF – or other *ab-initio* – calculations it is rather useful to vary structural parameters. In addition to geometry optimisations we propose to vary bond distances and bond angles over extended ranges in order to arrive at a better interpretation of the results.

Key Words: *ab-initio* calculations, Amsterdam density functional, electric field gradients.

1. Introduction

Modern *ab-initio* calculations of electronic charge densities and related electric field gradients (EFGs) in molecules using density functional methods as implemented, e.g., in the Amsterdam Density Functional (ADF) code yield reliable results even for relatively large molecules, but usually a simple interpretation of results is lacking. Geometry optimisations in order to minimize the total energy or the forces normally explore a small area of the parameter landscape around the equilibrium position only. They are indispensable for a meaningful comparison between experiment and theory, but otherwise of little help for the interpretation of results.

In this paper we compare the ADF-calculations of simple molecules like H_2 (or HD or D_2) – for which extensive calculations have been performed for the derivation of the nuclear quadrupole moment of the deuteron – and the triangular molecule H_3 as a function of bond distance and bond angle with estimates based on the point charge model which is adequate for the nuclear contribution to the EFG. In this way we derive the electronic contribution and discuss it.

* Author for correspondence.

2. Computational details

Ab-initio calculations of the EFG have been performed with the Amsterdam Density Functional (ADF) code [1–3]. Basis functions in ADF are Slater type orbitals, which are well suited for a quantum mechanical treatment of molecules. Slater type orbitals are capable to describe the cusp behaviour of atomic wave functions as well as their long range decay. This correct approximation of atomic wave functions is essential for a correct calculation of the EFG.

All geometry optimisations and calculations of EFGs have been performed using all electron Slater basis sets of different qualities from single-zeta quality to triple-zeta quality. All calculations have been performed with the generalized gradient approximation (GGA) using the PBE or revPBE functional which have been shown to be well suited for EFG calculations with ADF of a large number of model compounds [4]. Relativistic corrections for light elements like H can be neglected.

3. Results

3.1. THE H_2 MOLECULE

For H_2, a molecule with a closed shell configuration, there have been extensive calculations of the electric field gradient (EFG) in an attempt to obtain a reliable value for the quadrupole moment of the deuteron. From a molecular beam experiment [5] including Sternheimer corrections an EFG calculations a value of Q = +0.002860(15)b was obtained. This compares well with the result from Coulomb interaction of aligned nuclei [6] Q = 0.00282(19)b. There is axial symmetry, hence $V_{zz}(0)$ suffices to describe the EFG-tensor. Table I lists all previous calculations of the electronic contribution to the EFG in atomic units (multiply by 9.7166×10^{21} to obtain V/m^2) and compares them with ADF calculations using various basis sets. The contribution from the nucleus with a charge of $Z = +1$ is $V_{zz}(0) = 0.69068$ a.u. for a distance of 1.425 a.u. (multiply by 0.52917×10^{-10} to obtain m), a distance which comes out of the ADF structural optimisation with the TZP basis set.

The addition of atomic hydrogen 1s-wave functions yields already a good estimate, provided a multiplicative factor $\zeta = 1.24$ is used in the exponent of the 1s-wave function [7]. This is the SZ basis set. The separation of the electronic contribution into that coming from the neighbouring hydrogen atom (−0.19707 a.u.) and the overlap term which constitutes the bond (−0.17312 a.u.) was carried out with 1s-wave functions by numerical integration. The DZ, DZP, and TZP basis sets give slightly different values for the total electronic contribution to the EFG (see Table I) but do not allow for an easy separation into neighbour and overlap contributions. It is interesting to see that both terms are of equal magnitude, the sum of both (−0.34963 a.u.) is roughly half the nuclear contribution and of opposite sign. It compares well with the latest reported

Table I. Calculations of V_{zz} due to electrons in H_2

Reference	$V_{zz}(0)$ due to electrons in a.u.
This work ADF SZ[a]	−0.36900
This work ADF DZ[a]	−0.34576
This work ADF DZP[a]	−0.34887
This work ADF TZP[a]	−0.34963
Nordsiek (R = 1.425) [9]	−0.36458
Bishop (R = 1.4) [7]	−0.33792
Nordsiek [1940][b]	−0.298
Ishiguro (1948)[b]	−0.314
Newell (1950)[b]	−0.336
Auffray (1961)[b]	−0.34752
Narumi (1966) + Code (1971)[b]	−0.34602
Kolos (1968)[b]	−0.34895
Reid (1972) [8]	−0.34894

[a] ADF = Amsterdam density functional code: SZ = single zeta; DZ = double zeta; DZP = double zeta plus polarization function; TZP = triple zeta plus polarization function.
[b] These values are obtained by averaging quoted values as a function of distance R over the zero point vibrations as reported by Reid [8].

value of −0.34894 a.u. of Reid [8]. The total EFG is +0.34105 a.u. for the TZP basis set.

A few side remarks:

(1) point charge estimates with an adapted effective charge Z_{eff} = +0.4936 in order to obtain the total EFG are not very meaningful because there is no charge on the hydrogen atoms; one could, of course, formally take half an electron from each atom and assign it to the bond, but this would mean that this electron in the bond does not contribute to the total EFG, in contradiction to the abovementioned contribution from the overlap term; a better approach would be to use Z_{eff} = +0.7142 in order to account for the nuclear contribution plus the neighbouring electronic contribution which can be interpreted as a partially screened nuclear charge. This leaves a total of 0.57 electrons for the overlap term. However, this is no more than playing with numbers.

(2) ADF calculations show that there is very little p_z-admixture (2.55% for the TZP basis set) which would contribute very little to the total EFG anyway because $\langle r^{-3}\rangle$ for the hydrogen 2p-wave function is 1/24 a.u. only. This shows that the p_z-character of the overlap term which results from combining two 1s-wave functions at different centers plays an important role, as does the partially screened nuclear charge.

3.2. THE ANGULAR H_3 MOLECULE

The H_3-molecule is the simplest tri-atomic molecule. There are three electrons which occupy the lowest energy level (2 electrons) and the next higher energy

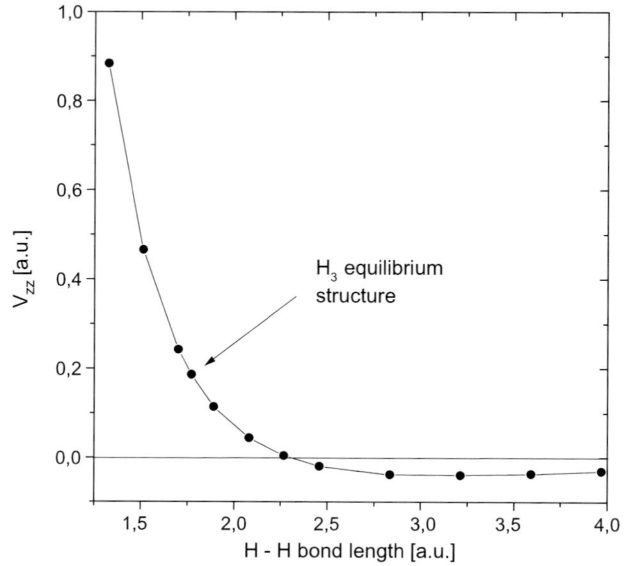

Figure 1. $V_{zz}(0)$ vs. bond distance in the linear H_3-molecule.

level (1 electron). For the linear form we varied the bond distance between 1.33 a.u. and 4 a.u. with the TZ2P basis set. The result is shown in Figure 1. Interestingly enough, $V_{zz}(0)$ changes sign at a distance of 2.3 a.u., a result which is completely unexpected in a point charge model. There is no change in sign of the charge at the central atom vs. bond distance. ADF structural optimisations using a TZ2P basis yield a linear molecule with an interatomic distance of 1.7708 a.u. and a bond energy of −7.11854363 eV. We obtained for the total $V_{zz}(0)$ a value of 0.18665 a.u. which results from the nuclear contribution of +0.72036 a.u. and the electronic contribution of −0.53371 a.u. These values can be compared with those for the H_2-molecule. Based on a r^{-3}-dependence, we would expect for the nuclear contribution a value a factor of 1.919 smaller than that for H_2, but a factor of two larger because of two hydrogen atoms. Both factors roughly compensate each other and, hence, the nuclear contributions are about the same. The situation for the electronic contribution is quite different: their ratio for H_3/H_2 is about 1.53 and certainly does not follow a simple r^{-3}-dependence. For linear H_3 we have roughly half the EFG of H_2 because the electronic contribution compensates the nuclear one to a much larger extent. Thus r^{-3}-arguments are not helpful. It is interesting to note that in this case a SZ basis leads to convergence problems and vastly different results, even with opposite sign.

In the following, we kept the bond distance fixed at 1.7708 a.u. Next, we varied the bond angle θ between 180°, i.e., from the linear form with the central atom and two equivalent terminal atoms, and 60°, i.e., towards the cyclic molecule with three equivalent atoms. The Walsh-diagram for the energy levels and the bond energy vs. bond angle are shown in Figure 2. For the linear molecule, the

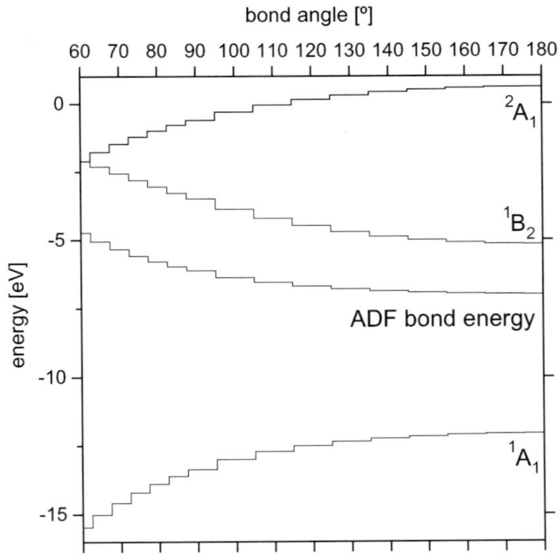

Figure 2. Walsh-diagram of energy levels and ADF-bond energy vs. bond angle.

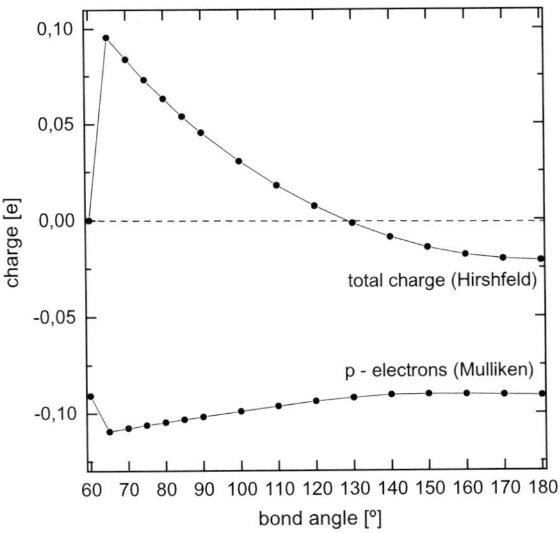

Figure 3. Total charge at the central atom (Hirshfeld approach) and charge at central atom due to *p*-wave functions (Mullikan approach) vs. bond angle.

lowest energy levels are labelled $^1A_1$, $^1B_2$, and $^2A_1$ (point group symmetry) in ascending order.

For the angular molecule, the symmetry is no longer axial. There is a smooth variation of the energy levels with bond angle with a degeneracy for the cyclic molecule. The charge at the central atom vs. bond angle is calculated in the Hirshfeld [10] approach using the TZ2P basis set. The Mulliken approach using a TZ2P basis set was used to calculate the charge with p-character at the central atom vs. bond angle. Both charges (1e corresponds to 1 a.u.) are displayed in

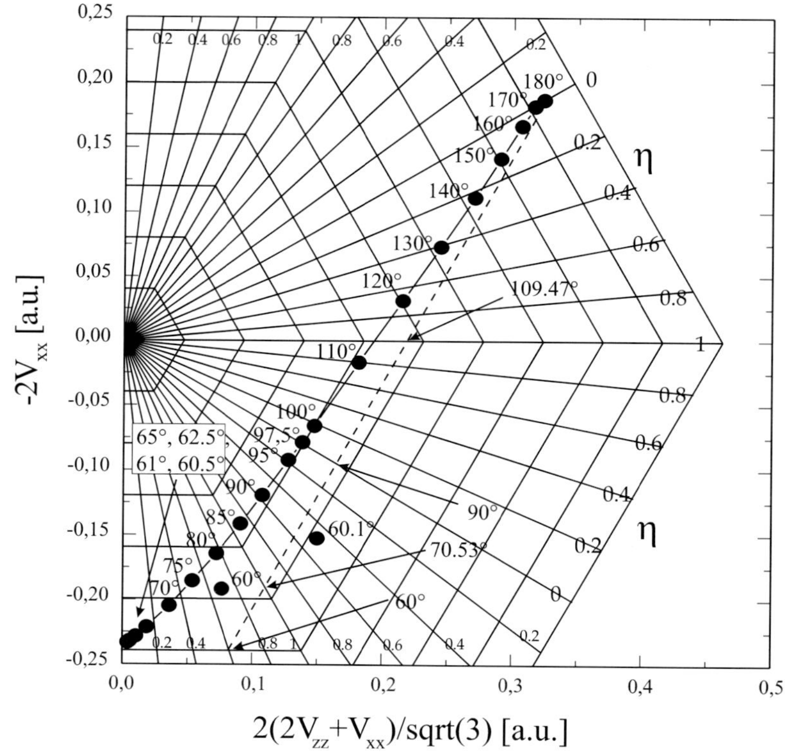

Figure 4. Extended Czjzek-plot of ADF results vs. bond angle. The point charge prediction with constant charges is shown as *dashed line* for comparison (scaled to coincide with the ADF point at 180°).

Figure 3. The charge at the terminal atoms is one half of that and has opposite sign. Generally, the charge separation is rather small. The Mulliken approach yields a negative charge with p-character of about 0.1 electron at the central atom, almost independent of bond angle. The Hirshfeld approach yields a negative charge at the central atom for the linear molecule which turns positive for bond angles below about 120°. It rises sharply towards smaller bond angles with a $\cos^2 \theta/2$-dependence and than abruptly drops to zero for the cyclic molecule, as required by symmetry. This is somewhat surprising because the bond between the terminal atoms obviously does not build up gradually, as might be expected.

The calculated EFG-tensor components $V_{xx}(0)$ and $V_{zz}(0)$ (or rather a linear combination thereof) are displayed in Figure 4 in the so-called extended Czjzek-plot. For comparison, the point charge prediction with constant charges is also displayed as a dashed line (normalized to coincide with the ADFpoint for the linear molecule). This trajectory is a straight line which forms an angle of 60° with the horizontal and is isometric in $\cos\theta$ [11]. The $\eta = 1$ line is crossed at 109.47°, the lower $\eta = 0$ line in the Czjzekplot is reached at 90°. The trajectory then continues and reaches the next $\eta = 1$ line at a bond angle of 70.53°. The nuclear contribution to the EFG tensor follows such a trajectory exactly. The

ADF-trajectory with the bond angle being the implicit parameter (the calculated points are labelled with the bond angle) starts at a positive $V_{zz}(0)$ and axial symmetry for the linear molecule and proceeds with decreasing bond angle nearly linearly as would be expected within the point charge model. The small irregularity at the beginning of the trajectory reflects the symmetry breaking and associated computational effects. At a bond angle of about 115° the $\eta = 1$ line is crossed where the largest component of the EFG-tensor changes orientation from being within the molecular plane to perpendicular to the molecular plane. The lower $\eta = 0$ line of the plot is reached for a bond angle of about 97.5°. After that the trajectory deviates more strongly from linearity. The next $\eta = 1$ line is crossed for a bond angle of about 84°. Here, the orientation of the largest component of the EFG-tensor flips back to in-plane. The trajectory reaches the lower $\eta = 0$ line of the Czjzek-plot for a bond angle of 60.5° and than abruptly jumps erratically near ring closure. The data points for bond angles of 60.3° and 60.1° are due to the erroneous population of the $^2A_1$ level and reflect a computational problem. Interestingly enough, a large value for η is obtained for the regular triangle which is considered topologically unstable for the H_3^+-molecule [12].

It is interesting to mention that the electronic contribution to the EFG-tensor components can be well fitted with a $\cos^2 \theta/2$-dependence as is the case for the charge at the central atom. The fact that the ADF trajectory lies left of the constant point charge trajectory means that the electronic contribution to the EFG decreases less rapidly with bond angle than the nuclear contribution. This is well interpreted by the charge redistribution within the molecule: since the population of orbitals with p-character is never large and since $\langle r^{-3} \rangle$ is 1/24 a.u. only, it essentially means that the s-electron density of the central atom is reduced with decreasing bond angle and transferred to the overlap term and the 1s-wave function of the terminal atoms.

It would be interesting to investigate the stability of the cyclic structure against distortions. The conjecture would be that a subtle distortion suffices to break up the ring and thus the molecule adopts its linear conformation. This is supported by the results for the bond energy vs. bond angle in Figure 2.

Summing up, the calculation of the EFG for a large range of bond distances revealed a zero EFG point at about 2.3 a.u., a quite unexpected result. The calculation of the trajectory over a very large angular range has revealed that (1) for moderate deviations from linearity point charge predictions with adjusted effective charges are still reasonable, and (2) the small charge separation leads to the small curvature of the trajectory.

4. Conclusion

The ADF code, based on a LCAO-MO scheme and using a SZ basis only, works reasonably well for H_2 but already fails for H_3. Here, an extended basis, i.e., the

TZ2P basis is required for convergence although the *p*-wave function admixture is never large. Hence, LCAO-MO models can be useful estimators for the EFG provided the basis is adequate.

The above example has shown that it is highly elucidating to explore a large parameter landscape around the equilibrium values both as far as bond distances as well as bond angles are concerned in order to learn more about the nature of the chemical bond. EFGs are much better tools than bond distances in this respect. A typical example is bonding to metals in metalloproteins: based on ligand distances often there is ambiguity whereas EFGs allow for an unambiguous differentiation between bonding and non-bonding [13]. The interpretation of EFGs in terms of coordination geometries requires both ab-initio density calculations as well as their separation into partial or fragment contributions, very much in the spirit of Bader [12]. Once this is accomplished, a detailed interpretation of the results of *ab-initio* calculations for more complicated molecules is possible.

Acknowledgement

It is a pleasure to thank J. Reinhold for fruitful discussions.

References

1. ADF2004.01, SCM, Theoretical Chemistry, Vrije Unversiteit, Amsterdam, The Netherlands, http://www.scm.com.
2. te Velde G., Bickelhaupt F. M., van Gisbergen S. J. A., Fonseca Guerra C., Baerends E. J., Snijders J. G. and Ziegler T., Chemistry with ADF, *J. Comput. Chem.* **22** (2001), 931.
3. Fonseca Guerra C., Snijders J. G., te Velde G. and Baerends E. J., *Theor. Chem. Acc.* **99** (1998), 391.
4. van Lenthe E. and Baerends E. J., *J. Chem. Phys.* **112** (2000), 8279.
5. Bishop D. M. and Cheung L. M., *Phys. Rev.* **A20** (1979), 381.
6. Kammeraad J. E. and Knutson L. D., *Nucl. Phys.* **A435** (1985), 502.
7. van Lenthe E. and Baerends E. J., *J. Chem. Phys.* **112** (2000), 8279.
8. Reid R. V. Jr. and Vaida M. L., *Phys. Rev. Lett.* **29** (1972), 494.
9. Nordsiek A., *Phys. Rev.* **58** (1940), 310.
10. Hirshfeld F. L., *Theor. Chim. Acta* **44** (1977), 129.
11. Butz T., *Hyperfine Interact.* **151/152** (2003), 49.
12. Bader R. F. W., Tung Ngyen-Dang T. and Tal Y., *J. Chem. Phys.* **70**(9) (1979), 4316.
13. Butz T. and Tröger W., In: Trautwein A. X. (ed.), *Bioinorganic Chemistry: Transition Metals in Biology and their Coordination Chemistry*, Priority Programme Report, VCH, Weinheim, Germany, 1997.

Hyperfine Interactions (2004) 158:79–88
DOI 10.1007/s10751-005-9012-8

The Electric Field Gradient of ^{111}Ag in Macrocyclic Crown Thioethers

FRANK HEINRICH[1], BERND CTORTECKA[2] and WOLFGANG TRÖGER[1,*]
[1]*Fakultät für Physik und Geowissenschaften, Universität Leipzig, Linnéstr. 5, 04103 Leipzig, Germany; e-mail: troeger@physik.uni-leipzig.de*
[2]*Garching Innovation GmbH, Marstallstr. 8, 80539 München, Germany*

Abstract. Time differential perturbed angular correlation experiments and *ab-initio* density functional theory calculation were used to determine the electric field gradients of the metal centres of the macrocyclic crown thioethers Ag(15S5)[BF_4], Ag(18S6-CH_2OH)[CF_3SO_3], Ag(18S6)$^+$, Ag(19S6-OH)[Tosylat] and Ag(20S6-OH)[CF_3SO_3]. The density functional theory calculations have been performed with the Amsterdam Density Functional code ADF. A "fingerprint system" is introduced, which allows to assign electric field gradients to certain Ag coordinations in these crown thioether complexes.

Key Words: Amsterdam density functional (ADF), Ag crown thioether, density functional theory (DFT), electric field gradient (EFG), PAC, time differential perturbed angular correlation (TDPAC).

Abbreviations: TDPAC – time differential perturbed angular correlation; EFG – electric field gradient; NQI – nuclear quadrupole interaction; ACCN – acetonitrile; THF – tetrahydrofuran; DMSO – dimethyl sulfoxide.

1. Introduction

The electric field gradient tensor (EFG) at a metal centre in a biological molecule can be determined by various hyperfine interaction spectroscopy methods, e.g., time differential perturbed angular correlation (TDPAC) [1]. The EFG, which arises from all extra-nuclear charges, reveals information about the coordination sphere of the metal probe, e.g., binding angles, bond distances, valence state and dynamics [1]. With the Amsterdam Density Functional code (ADF) [2–4] a density functional program package is available which allows to calculate reliable electric field gradients in molecules with a reasonable computational effort [5]. Therefore, *ab-initio* EFG calculations serve to validate proposed structures of metal centres or to establish "fingerprint systems" which allow to assign an experimentally determined EFG to a certain metal coordination.

* Author for correspondence.

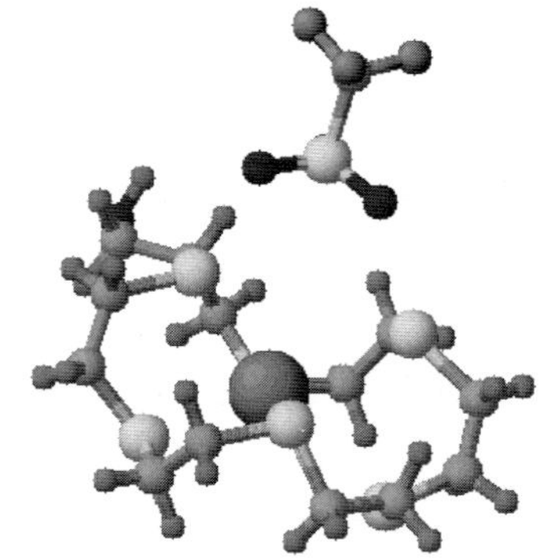

Figure 1. The structure of $Ag(18S6\text{-}CH_2OH)[CF_3SO_3]$.

Here, we present TDPAC measurements and ADF calculations on macrocyclic crown thioethers. Crown thioethers consist of a cyclic ligand of carbon and sulphur atoms saturated by hydrogen which coordinates an Ag^+ ion in its centre. A negatively charged counter ion assures the neutral charge of the whole complex. Figure 1 shows the $Ag(18S6\text{-}CH_2OH)[CF_3SO_3]$ crown thioether which consists of a ring with 12 carbon and six sulphur atoms. A CH_2OH group is connected to the ring and as the negatively charged counter ion serves CF_3SO_3. These crown thioethers are used for the formation of complexes with radioactive ^{111}Ag, a β^--emitter with some potential for the application in nuclear medicine [6]. For this purpose the *in-vivo* stability of the ^{111}Ag complex is essential. In order to achieve this in vivo stability several macrocyclic crown thioethers with different ringsizes have been prepared and investigated. Since ^{111}Ag can also be used for TDPAC studies, this family of molecules is ideally suited to test the capabilities and limits of the ADF calculations of the electric field gradient by TDPAC experiments.

2. Experimental

The theory of TDPAC is well described in standard textbooks [7], the data analysis in [8]. All TDPAC measurements were performed with a 6-detector TDPAC-spectrometer equipped with BaF_2 detectors [9]. In the present study, we make use of the radioactive isotope $^{111}Ag(\beta^-)^{111}Cd$ (half-life $t_{1/2}$ = 7d) which decays by the successive emission of two γ-quanta via an intermediate state of the nuclear spin $I = 5/2$ (half-life $\tau_{1/2}$ = 84 ns, quadrupole moment Q = 0,83(13) barn). Prior to the γ–γ-cascade used for the TDPAC measurement a β^--decay occurs. The $^{111}Ag(\beta^-)^{111}Cd$ was extracted from neutron irradiated Pd [10]. The nuclear quadrupole frequency ω_Q and the asymmetry parameter η, which are obtained by the TDPAC experiment are related to the main tensor components of the EFG V_{xx}, V_{yy} and V_{zz} as follows: $\omega_Q = 2\pi eQV_{zz}/(4I(2I\text{-}1)h)$, $\eta = (V_{xx} - V_{yy})/V_{zz}$. The Lorentz-type line broadening δ of the TDPAC signal takes into account

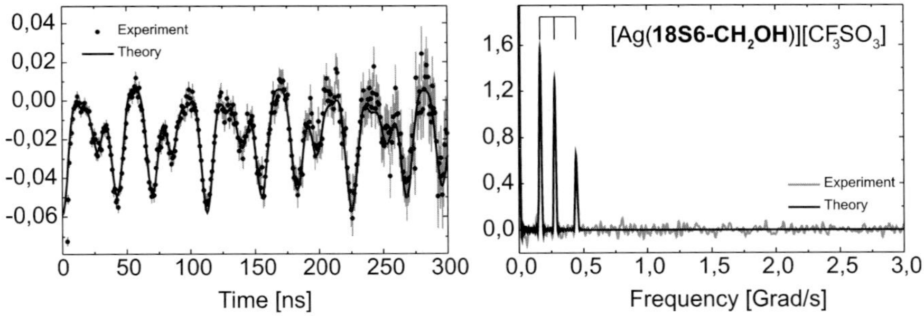

Figure 2. The TDPAC spectrum and its cosine transform of Ag(18S6-CH_2OH)[CF_3SO_3] obtained at room temperature. The frequency bar in the cosine transforms indicate the frequencies of the fitted NQIs (see Table II).

sample inhomogeneties, i.e., slight deviations of the coordination sphere of the TDPAC probes in the same type of molecules. Generally, it is quite instructive to present the NQI data in a Czjzek-plot (see Figure 3) as described in [11].

X-ray diffraction data of Ag crown thioethers can be found in literature [12–15]. Table I lists the Ag–S bond distances and the Ag coordinations of Ag crown thioethers which are structurally related to the Ag crown thioethers of this work. These data also serve as start-up geometries for ADF geometry optimisations. The two Ag(18S6) crowns of Table I exhibit a distorted octahedral coordination: all six sulphur atoms coordinate the Ag ion. The Ag(19S6OH) and Ag(20S6OH) crowns exhibit a distorted tetrahedral coordination: four sulphur atoms bind to the Ag ion. The Ag(15S5) crown shows a "4 + 1" one geometry: four sulphur atoms bind the Ag ion in square planar coordination, the fifth sulphur atom is used to build up large polymer structures by binding the Ag ions of the neighbour complexes.

Typically, small Ag crown thioether crystals have been prepared for TDPAC measurements grown from 1 to 2 ml solution containing the positively charged Ag complex (^{111}Ag and stable Ag as carrier) as well as the negatively charged counter ion ($\approx 1*10^{-3}$ M, solvents: ethanol, ACCN, or THF). The crystals have been grown by a continuous addition of diethyl ether for two to three days; hence, it is likely that the crystals contain diethyl ether. TDPAC measurements of frozen solutions of Ag crown thioethers have been performed by re-dissolving the crown thioether crystals in aqueous solutions of ACCN, DMSO or THF.

All experimentally determined NQIs either with frozen solutions of organic solvents or small crystals at different temperatures are presented in Table II. Figure 2 shows a typical TDPAC spectrum. Figure 3 displays all NQIs in a Czjzek-plot.

Almost all TDPAC spectra of the Ag(18S6-CH_2OH)[CF_3SO_3] crown under different conditions exhibit a very small line broadening of about 0.1%. Only the two measurements of this complex dissolved in DMSO at −261°C and −96°C show a larger line broadening of about 6%. There is also no significant difference

Table I. Ag–S bond lengths of Ag crown thioethers and the Ag coordinations determined by X-ray diffraction

Crown	Ag–S bond lengths [Å]	Ag coordinations by S	Ref.
Ag(15S5) [PF_6]	2.717; 2.561; 2.561; 2.717; (2.492)	4 + 1	[15]
Ag(18S6) [I_3]	2.726; 2.726; 2.726; 2.726; 2.801; 2.801	Distorted octahedral	[12]
Ag(18S6) [PF_6]	2.683, 2.683, 2.774, 2.774, 2.774, 2.774	Distorted octahedral	[13]
Ag(19S6-OH) [CF_3SO_3]	2.644; 2.573; 2.655; 2.551; (3.553; 3.269)	Distorted tetrahedral	[14]
Ag(20S6-OH) [BF_4]	2.685; 2.725; 2.582; 2.604; (4.502; 3.719)	Distorted tetrahedral	[14]

between the NQIs of dissolved and crystalline samples. These experimental findings indicate an excellent match of the metal ion with the ligand complex and a large rigidity of the metal centre in this Ag crown.

The two TDPAC spectra of the $Ag(18S6)^+$ crown dissolved in THF at different temperatures exhibit a line broadening of 4 and 8%. The experimentally determined NQIs are comparable to those of the Ag(18S6-CH_2OH)[CF_3SO_3] crown. The TDPAC spectra of Ag(19S6-OH)[Tosylat] show different NQIs for the dissolved molecules ($|V_{zz}| \approx 0.75$ a.u., $\eta = 0.35$) and the crystalline samples ($|V_{zz}| \approx 0.9$ a.u., $\eta = 0.55$). Consequently, the structures of the metal centres in the crystal and in solution have to be regarded as different. The TDPAC signals of crystalline Ag(20S6-OH)[CF_3SO_3] samples show a large line broadening of 9–24% and a significant temperature dependence of $|V_{zz}|$. Obviously, the largest ligand of this series of molecules is not capable to bind rigidly the Ag^+ ion in a well defined coordination geometry. The TDPAC spectra of crystalline and dissolved Ag(15S5)[BF_4] show the same V_{zz} but differ in η by a factor of 2. The TDPAC spectrum of the dissolved molecule exhibits a large line broadening of 22(3)% compared to a very small line broadening of the crystalline sample of only 0.1(2)%.

The experimental determined NQI can be classified into three groups of NQI with $|V_{zz}| \approx 0.55$ a.u., $|V_{zz}| \approx 0.75$ a.u. and $|V_{zz}| \approx 0.9$ a.u. which can be easily seen in the Czjzek-plot (see Figure 3). Since the structure of the crystalline complexes are known by the X-ray diffraction (see Table I) $|V_{zz}| \approx 0.55$ a.u. can be assigned to a "4 + 1" coordination of the Ag^+ ion and $|V_{zz}| \approx 0.9$ a.u. to a distorted tetrahedral coordination. For $|V_{zz}| \approx 0.75$ no definite assignment to a certain Ag coordination is possible: Distorted tetrahedral and distorted octahedral coordinations generate the same $|V_{zz}|$. This "fingerprint system" allows the identification of unknown Ag site structures in crown thioethers, albeit with an ambiguity for $|V_{zz}| \approx 0.75$.

Table II. The results of the TDPAC experiments

Ag crown	TDPAC measurement	ω_Q [Mrad/s]	δ [%]	η	$\lvert V_{xx}\rvert$ [a.u.]	$\lvert V_{yy}\rvert$ [a.u.]	$\lvert V_{zz}\rvert$ [a.u.]
Ag(18S6)$^+$	THF −29°C	27.7(2)	8(1)	0.35(1)	0.259	0.543	0.802
	THF −70°C	29.0(5)	4(2)	**0.50(3)**	**0.176**	**0.587**	**0.763**
Ag(18S6-CH_2OH) [CF_3SO_3]	DMSO −261°C	28.3(3)	6 fix	0.36(3)	0.263	0.554	0.817
	DMSO −96°C	28.5(3)	6 fix	**0.42(2)**	**0.229**	**0.561**	**0.789**
	DMSO −51°C	25.0(2)	0(1)	0.19(1)	0.321	0.468	0.789
	DMSO −27°C	25.2(2)	0.1(1)	0.22(1)	0.305	0.475	0.780
	Cryst. −262°C	26.5(2)	0.3(2)	0.39(1)	0.231	0.522	0.753
	Cryst. −51°C	27.7(2)	0.1(1)	0.40(1)	0.233	0.554	0.777
	Cryst. 21°C	27.7(3)	0.1(1)	0.40(1)	0.235	0.547	0.782
Ag(19S6-OH) [Tosylat]	DMSO −60°C	26(1)	6(6)	**0.34(9)**	**0.249**	**0.504**	**0.753**
	Cryst. −266°C	36.5(2)	9(1)	0.52(1)	0.229	0.718	0.947
	Cryst. −193°C	38.5(7)	8(2)	0.57(2)	0.208	0.757	0.964
	Cryst. −60°C	35.7(2)	7.0(6)	0.53(1)	0.216	0.701	0.917
	Cryst. 22°C	35.7(2)	5.8(4)	0.54(1)	0.211	0.696	0.907
Ag(20S6-OH) [CF_3SO_3]	Cryst. −261°C	41(11)	24(7)	0.52(7)	0.227	0.718	0.945
	Cryst. 24°C NQI 1 (50%)	29.7(7)	12(3)	**0.60(3)**	**0.147**	**0.582**	**0.729**
	Cryst. 24°C NQI 2 (50%)	38.0(7)	9(2)	0.66(2)	0.153	0.740	0.894
	Cryst. 50°C	37.2(7)	10 fix	0.75(2)	0.104	0.716	0.820
Ag(15S5) [BF_4]	Cryst. 24°C	19.08(5)	0.1(2)	0.32(1)	0.190	0.371	0.561
	ACCN −180°C	23.3(7)	22(3)	0.59(5)	0.118	0.457	0.575

Those NQIs which could be reproduced at best by *ab-initio* EFG calculations are set in bold (see also Table III).

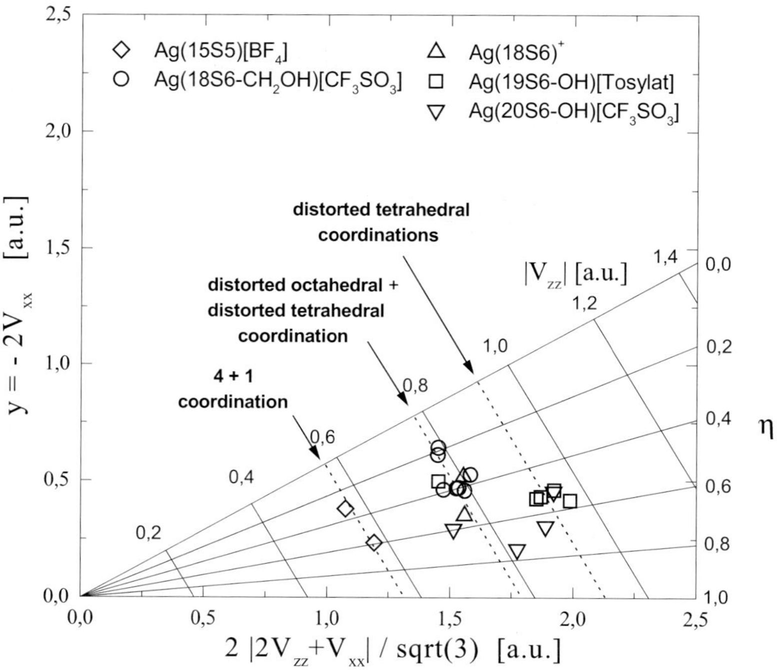

Figure 3. Czjzek plot of all experimentally determined NQIs. The axes are linear combinations of the main tensor components of the EFG. *Lines* of constant $|V_{zz}|$ and constant η are labelled ("hering bone grid"). The NQIs form groups according to their Ag coordination derived from X-ray diffraction data.

3. DFT calculations

Ab-initio DFT calculations of the EFG were performed with the Amsterdam Density Functional code (ADF) [2–5]. In ADF molecular orbitals are build up by Slater type functions. Slater type functions are capable to describe the cusp behaviour of atomic wave functions as well as their long range decay. This correct approximation of atomic wave functions is essential for a correct calculation of the EFG.

For small molecules it was shown by [5, 16] that most reliable EFGs with ADF can be obtained using large basis sets of at least triple-zeta quality plus two polarisation functions (TZ2P) and including relativistic effects via the ZORA algorithm [17–19]. Relativistic effects were taken into account on the scalar relativistic level for geometry optimisations and for the EFG calculations spin-orbit coupling was additionally included. Geometry optimisations of larger molecules (ca 50 atoms) were performed with frozen core basis sets to reduce the computational effort. We exclusively used the LDA density functional with GGA corrections of Perdew [20] or its revised version [21] for the correlation part and the corrections of Becke [22] for the exchange part. It was shown by geometry optimisations of a set of small molecules and transition metal complexes that these functionals obtain most accurate geometries [16]. Due to the molecular size of the studied crown thioethers a systematic investigation into the dependence of the calculated EFG from the basis set and different GGA functionals was not yet performed. An alteration of the GGA exchange and correlation functional results in a variation of the calculated EFG of typically 5% and, thus, this variation is much smaller than the uncertainty of the quadrupole moment $Q = 0.83(13)$ barn of the intermediate state of the TDPAC-probe $^{111}Ag(\beta^-)^{111}Cd$ and, therefore, much smaller than the uncertainty of the experimental EFGs. A systematic investigation into solvent effects was of higher importance for us. DFT calculations were performed with and without the simulation of the solvent using the COSMO solvation model as implemented in ADF [23]. The EFG was always calculated for the Ag complexes. The ADF code used here is not able to treat crystal structures. For the Ag(20S6-OH)[CF_3SO_3] crown all experimental NQI data come from crystalline samples. In this case, calculations of Ag(20S6-OH)[CF_3SO_3] were performed on molecules assuming diethyl ether as a solvent, since the crystalline samples were prepared with diethyl ether as described earlier.

Geometry optimisations of Ag crown thioethers including the counterion were performed before each EFG calculation. As start-up geometries, the structures derived by X-ray diffraction (see Table I) were used; in the case of a different counterion, it was replaced by the appropriate one.

The performed EFG calculations with and without solvent are shown in Table III. A calculation of Ag(15S5)[BF_4] was not performed due to the high computational effort of the treatment of the polymer chains.

Calculations not including solvent effects were not able to reproduce any experimental EFG (the bond lengths for these calculations are omitted in Table III). Therefore, in the following only the calculations including solvent effects are discussed.

Calculations using the COSMO model show an excellent agreement with the experimentally determined NQI with less than 10% mean deviation Δ of the main components of the EFG. There is only one exception: the EFG calculation of Ag(19S6-OH)[Tosylat] shows only a fair agreement with the experimentally determined EFG of the dissolved crown (Δ = 19,6%). For $Ag(18S6)^+$, Ag(18S6-CH_2OH)[CF_3SO_3] and Ag(19S6-OH) [Tosylat] the experimental NQIs of the dissolved crown ethers have been reproduced at best. It has to be mentioned that the NQIs of the crystalline sample of Ag(19S6-OH) [Tosylat] which are very different to those of the dissolved molecule could not be reproduced by the ADF calculations. For the crystalline sample of Ag(20S6-OH) [CF_3SO_3] two different NQIs have been detected at room temperature and one of these NQIs could be reproduced by ADF calculations.

The performed geometry optimisations conserved the coordination of the metal ion of the start-up geometry with the exception of the $Ag(18S6)^+$ crown

Table III. The results of the EFG calculations with and without solvent effects

Crown ether	Solvent	η	$\lvert V_{xx}\rvert$ [a.u.]	$\lvert V_{yy}\rvert$ [a.u.]	$\lvert V_{zz}\rvert$ [a.u.]	**Δ** [%]	Ag–S bond lengths [Å]
$Ag(18S6)^+$	None	0.174	0.190	0.254	−0.444	41.5	2.5508; 2.5795; 2.6327; 2.6795; (3.2394; 4.0967)
	THF	**0.450**	**0.207**	**0.546**	**−0.753**	**8.6**	
Ag(18S6-CH_2OH) [CF_3SO_3]	None	0.508	−0.175	−0.377	0.552	39.4	0.2670; 0.2755; 0.2883; 0.2906; 0.3018; 0.3224 (obt. from DMSO calc.)
	DMSO	**0.514**	**−0.180**	**−0.563**	**0.743**	**9.2**	
	Diethyl ether	0.587	−0.153	−0,585	0.738	15.0	
Ag(19S6OH) [Tosylat]	None	0.594	0.123	0.483	−0.606	24.8	2.6936; 2.6995; 2.9429; 2.9558; (3.2960; 3.4986)
	DMSO	**0.096**	0.311	0.377	**−0.688**	**19.6**	
Ag(20S6OH) [CF_3SO_3]	None	0.357	0.224	0.462	−0.686	26.3	2.6392; 2.6463; 2.6489; 2.6503; (4.1842; 4.7863)
	Diethyl ether	**0.675**	**0.116**	**0.595**	**−0.710**	**8.6**	

Those calculations which reproduced TDPAC experiments are set in bold (see also Table II). Δ gives the mean deviation of the calculated EFG tensor components to the experimental determined EFG tensor components (best case). The Ag–S bond lengths are derived from geometry optimisations including solvent effects. Distances for which no binding is assumed are set in brackets.

(see Tables II and III). The $Ag(18S6)^+$ coordination changed from a distorted octahedral to a distorted tetrahedral Ag coordination. Thus, Ag(18S6-CH_2OH)[CF_3SO_3] remains as the only crown with a distorted octahedral Ag coordination. All performed EFG calculations of distorted tetrahedral Ag coordinations result in a negative V_{zz}, whereas the EFG calculation of the distorted octahedral Ag(18S6-CH_2OH) [CF_3SO_3] crown yields positive V_{zz} (see Table III). Furthermore, the absolute value of all calculated $|V_{zz}|$ is in the range of 0.69 a.u. to 0.77 a.u.

4. Discussion

All ADF calculations which include solvent effects resulted in EFGs of appr. $|V_{zz}| \approx 0.75$ a.u. and reproduce those NQIs which were determined in the experiments with the dissolved Ag crown thioethers. Calculations not including solvent effects were not able to reproduce any experimental data.

Since the optimised structures of the crown thioehters showed distorted tetrahedral and octahedral metal coordinations, the calculations support the postulated "fingerprint system" as described above. A closer inspection of the calculated EFGs shows that octahedral coordinations exhibit positive V_{zz} values, whereas tetrahedral coordinations have negative V_{zz} values. This important extension of the postulated "fingerprint system" is unfortunately irrelevant for γ–γ-TDPAC studies, since there the sign of V_{zz} cannot be determined for principal reasons, i.e., the conservation of parity of the electromagnetic decay. Here, γ–β^- perturbed angular correlation spectroscopy is a possibility to validate this theoretical approach.

The experimentally determined NQI of the crystalline samples of Ag(19S6-OH)[Tosylat] and Ag(20S6-OH)[CF_3SO_3] with $|V_{zz}| \approx 0.9$ a.u. could not be reproduced by *ab-initio* ADF calculations. With the ADF package used here no calculations of larger periodic structures are possible. Therefore, it remains unclear whether a more suited theoretical approach or a modified structure is required for a better agreement of experiment and theory. The postulated fluxionality [14] of the Ag–S bond length in Ag(19S6) and Ag(20S6) crowns might influence the electric field gradient, too.

References

1. Tröger W., Nuclear probes in life sciences, *Hyperfine Interact.* **120/121** (1999), 117.
2. ADF2004.01, SCM, Theoretical Chemistry, Vrije Universiteit, Amsterdam, The Netherlands, http://wwww.scm.com.
3. te Velde G., Bickelhaupt F. M., van Gisbergen S. J. A., Fonseca Guerra C., Baerends E. J., Snijders J. G. and Ziegler T., Chemistry with ADF, *J. Comput. Chem.* **22** (2001), 931.
4. Fonseca Guerra C., Snijders J. G., te Velde G. and Baerends E. J., *Theor. Chem. Acc.* **99** (1998), 391.

5. van Lenthe E. and Baerends E. J., *J. Chem. Phys.* **112** (2000), 8279.
6. Schubiger P. A., Alberto R. and Smith A., *Bioconjug. Chem.* **7** (1996), 165.
7. Frauenfelder H. and Steffen R. M., In: Siegbahn K. (ed.), *Alpha-, Beta-, Gamma-Ray Spectroscopy*, Vol. 2, North-Holland, Amsterdam, 1965.
8. Butz T., Saibene S., Fraenzke T. and Weber M., *Nucl. Instrum. Methods, A* **284** (1989), 417.
9. Butz T., *Z. Naturforsch.* **51a** (1996), 396.
10. Alberto R., Bläuenstein P., Novak-Hofer I., Smith A. and Schubiger P. A., *Appl. Radiat. Isotopes* **43** (1992), 869.
11. Butz T., Ceolín M., Ganal P., Schmidt P., Taylor M. A. and Tröger W., *Phys. Scr.* **54** (1996), 234.
12. Blake A. J., Gould R., Parson S., Radek C. and Schröder M., *Angew. Chem., Int. Ed.* **34** (1995), 2374.
13. Ctortecka B., PhD Thesis, Faculty of Physics and Earth Science, Universität Leipzig, 1999.
14. Alberto R., Nef W., Smith A., Kaden T. A., Neuburger M., Zehnder M., Frey A., Abram U. and Schubinger P. A., *Inorg. Chem.* **35** (1996), 3420.
15. Blake A. J., Collison D., Gould R. O., Reid G., Schröder M., *J. Chem. Soc. Dalton Trans.* **4** (1993), 521.
16. Swart M. and Snijders J. G., *Theor. Chem. Acc.* **110** (2003), 34.
17. van Lenthe E., Ehlers A. E., Baerends E. J., *J. Chem. Phys.* **110** (1999), 8943.
18. van Lenthe E., Baerends E. J. and Snijders J. G., *J. Chem. Phys.* **99** (1993), 4597.
19. van Lenthe E., Baerends E. J. and Snijders J. G., *J. Chem. Phys.* **110** (1994), 9783.
20. Perdew J. P., Burke K. and Ernzerhof M., *Phys. Rev. Lett.* **77** (1996), 3865.
21. Zhang Y. and Yang W., *Phys. Rev. Lett.* **80** (1998), 890.
22. Becke A. D., *Phys. Rev., A* **38** (1988), 3098.
23. Pye C. C. and Ziegler T., *Theor. Chem. Acc.* **101** (1999), 307.

Hyperfine Interactions (2004) 158:89–93
DOI 10.1007/s10751-005-9013-7

Correlation between the EFG Values Measured at the Cd Impurity in a Group of Cu-based Delafossites and the Semiconducting Properties of the Latter

M. V. LALIC[1] and J. MESTNIK-FILHO[2]
[1]*Universidade Federal de Sergipe, Departamento de Física, P.O. Box 353, 49100-000, São Cristóvão, SE, Brazil; e-mail: mlalic@fisica.ufs.br*
[2]*Instituto de Pesquisas Energeticas e Nucleares, P.O. Box 11049, 05422-970, São Paulo, SP, Brazil*

Abstract. In this paper we analyze trend of EFG values measured at Cd impurity in a group of semiconducting delafossites with chemical formula $CuBO_2$ (B = Al, Fe, Cr, Nd). We conclude that this trend reveals one of the most subtle details in electronic spectrum of the compounds: if impurity states are formed within or out of the band gap. In $CuAlO_2$ and $CuFeO_2$ the Cd EFG exhibits larger value than in $CuCrO_2$ and $CuNdO_2$, when Cd substitutes the Cu atom. This occurs because in the first two compounds the Cd forms shallow band within the gap, and in the second two compounds does not. When Cd occupies the B position it exhibits almost the same EFG in all delafossites. In this case, Cd does not form its states within the gap in none of the compounds. To arrive to these conclusions we analyzed and calculated various systems (Cd-doped $CuAlO_2$ and $CuCrO_2$ compounds, fictitious molecules), using the FP-LAPW method.

Key Words: delafossites, density functional theory, electric field gradient.

1. Introduction

Ternary compounds ABO_2 (where A = Cu, Ag and B is trivalent metal ion) with delafossite structure (space group *R-3m*) attracted much attention recently owing to discovery that might be utilized as transparent semiconductors [1]. Knowledge of how the electronic spectrum of these materials changes under the doping is therefore of great importance, being a subject of many experiments so far [2]. In one of them [3], a group of $CuBO_2$ (B = Al, Cr, Fe, Nd) delafossites was studied by perturbed angular correlation (PAC) technique. Electric field gradient (EFG) was measured at Cd probes which substituted either Cu or B site, and results revealed interesting differences in EFG trends depending on position of Cd. When Cd occupies the B site its EFG exhibits almost the same value (6–7) in all compounds, while the EFGs of Cd residing at the Cu site oscillate, being 27–28 when B = Al, Fe and 21 when B = Cr, Nd, in units 10^{21} V/m^2 (see Table II in ref. [3]).

In this paper our aim is to interpret this EFG trend and to connect it with the semi-conducting properties of compounds if possible. The task was too complicated to be done directly, by performing the calculations for all compounds, since all of them except the $CuAlO_2$ are anti-ferromagnets with complicated alignments of magnetic moments (a fact which is difficult to treat accurately in the theory). Instead, we achieved the goal indirectly, in three consecutive steps:

1) by analyzing results of our previous theoretical study of doped $CuAlO_2$ compound [4] which exhibits an elevated EFG when Cd substitutes the Cu site, and performing some additional EFG decompositions;
2) by calculating EFGs in group of molecules constructed in a way which simulate the Cd neighborhood in delafossites;
3) by calculating electronic structure and the EFG in Cd-doped $CuCrO_2$ compound, which exhibits the smaller EFG at Cd occupying the Cu site.

All calculations were performed using the full potential linear augmented plane wave (FP-LAPW) method [5], as embodied in WIEN2k computer code [6].

2. Calculations and interpretation of the results

In ref. [4] we presented band-structure calculations for $CuAlO_2$ compound with an isolated Cd impurity substituting either Cu (Cd → Cu) or Al (Cd → Al) atom. In both cases Cd changes interatomic distances in its neighborhood. In Cd → Cu case the calculations performed with non-optimized interatomic distances (like in the pure compound) showed that Cd states are formed within the conduction band. By letting the atoms in the vicinity of Cd to relax (i.e., to reach optimized distances), the shallow band within the gap appears, near the conduction band bottom, converting the system into the *n*-type semiconductor. When Cd substitutes the Al atom, no states within the gap are observed, even with the optimized interatomic distances (figure 3 in ref. [4]).

In present paper we further analyzed the Cd EFG for both non-optimized and optimized geometries in Cd → Cu case, separating contribution from shallow band, and did the same for optimized geometry in Cd → Al case. We also calculated decomposition of these EFGs, tracing the information from which Cd shell, s- (s–d term), p- (p–p term) or d- (d–d term) the main contribution to the EFG arises. The results are summarized in the Table I.

Analysis of data in the Table I provides a clue to interpret experimental trend of Cd EFG in delafossites. When Cd occupies the Cu site in $CuAlO_2$, its EFG is approx. -26 (in 10^{21} V/m^2). From this number, approximately -5 originate from donor band within the band gap. If this band was formed as part of the conduction band, the resulting EFG would be approx. -21, like in $CuCrO_2$ and $CuNdO_2$. Thus, introduction of Cd impurity into the Cu site produces different effects in $CuAlO_2$ and $CuFeO_2$ on one hand, and in $CuCrO_2$ and $CuNdO_2$ on the other. It makes sense to assume that the first two compounds change their electric properties due to presence of Cd states within their fundamental gaps, and the last two compounds do not, since Cd forms its states within their conduction

Table I. Calculated principal values V_{zz} of the EFG tensor at the Cd nuclei in $CuAlO_2$ compound

	Non-optimized structure		Optimized structure	
	V_{zz} (at Cd)	V_{zz} decomposition	V_{zz} (at Cd)	V_{zz} decompositoion
Crystal $CuAlO_2$ (Cd → Cu case)	−33.90	s–d: 0.78 p–p: −34.29 d–d: −0.39	−25.73	s–d: 0.26 p–p: −20.12 d–d: −5.87
Shallow band →			−4.92	s–d: −0.42 p–p: 0.64 d–d: −5.14
The rest →			−20.81	s–d: 0.68 p–p: −20.76 d–d: −0.73
Crystal $CuAlO_2$ (Cd → Al case)			5.63	s–d: −0.12 p–p: 6.12 d–d: −0.39

V_{zz} units: 10^{21} V/m^2.

bands. We postulated that the observed EFG trend just reflects this fact. When Cd substitutes the B atom, its EFG is stable. In this case formation of the shallow band within the gap is not expected, a fact which is confirmed by our calculations.

In order to support above interpretation we performed some additional calculations. Our aim was to investigate formation of Cd EFG in delafossites by studying the electrostatic influence of its closest neighborhood. Thus we constructed molecules which simulate the closest surroundings of Cd occupying either Cu or B site in delafossites, and performed series of molecular calculations. The molecules were designed using both types of $CuAlO_2$ supercells (Cd → Cu and Cd → Al) in which we removed all the atoms except the central Cd and its nearest neighbors (NN). Constructed this way, the molecular supercells preserved delafossite's symmetry, still containing enough empty space to substantially reduce interaction with molecules in neighboring supercells. Other calculational details were the same as in ref. [4]. The calculated EFGs at Cd we confronted with the Cd EFGs calculated in $CuAlO_2$ (Table I), and withdrawn conclusions.

To analyze situation in which Cd substitutes the B atom in $CuBO_2$ (Cd → B case) we constructed a molecule containing just Cd and six NN oxygens around B site, with the same Cd–O distances as in relaxed Cd-containing $CuAlO_2$. In order to simulate Cd → Cu case we considered two molecules. First one consisted of Cd and two NN oxygens around Cu site, and second one included six second NN coppers also. In both molecules interatomic distances were set to be the same as in Cd-doped $CuAlO_2$ with both optimized and non-optimized interatomic distances. Calculated EFGs at the Cd nuclei are presented in Table II.

In Cd → B case the molecular calculations succeeded to reproduce almost completely the EFG value calculated at Cd occupying Al site in $CuAlO_2$

Table II. Calculated EFGs at Cd nuclei in molecules simulating the Cd environment in Cd-doped delafossites. First line corresponds to Cd → B case, second and third to Cd → Cu case

Molecules	Non-optimized structure		Optimized structure	
	V_{zz} (at Cd)	V_{zz} decomposition	V_{zz} (at Cd)	V_{zz} decomposition
$Cd + 6O^{(1)}$			5.65	s–d: −0.17 p–p: 6.74 d–d: −0.94
$Cd + 2O^{(1)}$	−41.02	s–d: 1.09 p–p: −53.38 d–d: 11.28	−28.74	s–d: 0.87 p–p: −34.46 d–d: 4.85
$Cd + 2O^{(1)} + 6Cu^{(2)}$	−33.35	s–d: 0.41 p–p: −42.66 d–d: 8.90	−18.97	s–d: 0.42 p–p: −21.25 d–d: 1.86

Interatomic distances are the same as in non-optimized and optimized geometries in the Cd-containing $CuAlO_2$. V_{zz} units: 10^{21} V/m^2.

(Table I). Both EFG values have practically the same decomposition. It implies that Cd situated at the B site in any delafossite mainly senses electrostatic influence of just six NN oxygens around it, and negligible impact of more distant neighbors. This fact explains why Cd residing at the B position exhibits very similar EFG in all delafossites.

In Cd → Cu case molecular calculations reproduced the EFG of Cd residing at the Cu site in $CuAlO_2$ taking into accounts just Cd first and second neighbors, for non-optimized geometry. For optimized geometry, molecular EFG corresponds closely to the Cd EFG value in $CuAlO_2$ when the latter does not contain the contribution from the shallow band. This result is not surprising since creation of the shallow band is consequence of collective effects occurring in the crystal. We could not therefore expect that molecular calculations account for these effects exactly. A fact that just first and second Cd neighbors were considered in the molecule (and this Cd neighborhood is the same in any delafossite) leads to following interpretation of the experimental EFG trend. When Cd occupies the Cu site in Cu-based delafossites, it exhibits "regular" EFG of approx. −21 (10^{21} V/m^2), as measured in $CuCrO_2$ and $CuNdO_2$ compounds. This EFG corresponds to situation when impurity states are positioned within the conduction band. The difference between "regular" EFG values and those measured in $CuAlO_2$ and $CuFeO_2$ arises from contribution of donor bands, formed within the fundamental gaps of these compounds.

Finally, in order to confirm our conclusions, we performed calculations for $CuCrO_2$ compound, with Cd substituting the Cu atom. In this case we should not expect formation of Cd states within the gap since measured EFG exhibits a "regular" value (approx. −21). In calculations we assumed ferromagnetic order of Cr moments although in the nature they are aligned antiferromagnetically. This approximation, however, was sufficient to open a gap in the pure compound and to provide a good starting point for a treatment of the defect problem. Other computational details were the same as in ref. [4]. After the self-consistency has been reached, we analyzed the resulting band structure details and verified the fact that no states were present within the band gap [7]. The calculated EFG at Cd (approx. −23) is found in good agreement with the experimental value (approx. −21). This result additionally enforced the evidences in favor of our explanation of the EFG trend in Cu-based delafossites.

Acknowledgements

M. V. Lalic kindly acknowledges financial support provided by FAP-SE, CAPES and CNPq (Brazilian Foundation Agencies).

References

1. Kawazoe H., Yasukawa M., Hyodo H., Kurita M., Yanagi H. and Hosono, H., *Nature* **389** (1997), 939.
2. Yanagi H., Hase T., Ibuki S., Ueda K. and Hosono H., *Appl. Phys. Lett.* **78** (2001), 1583.
3. Attili R. N., Saxena R. N., Carbonari A. W., Mestnik-Filho J., Uhrmacher M. and Lieb K. P. *Phys. Rev. B* **58** (1998), 2563.
4. Lalic M. V., Mestnik-Filho J., Carbonari A. W., Saxena R. N. and Moralles M. J., *Phys.: Condens. Matter* **14** (2002), 5517.
5. Andersen O. K., *Phys. Rev. B* **12** (1975), 3060.
6. Blaha P., Schwarz K., Madsen G. K. H., Kvasnicka D. and Luitz J. *WIEN2k, An Augmented Plane Wave + Local Orbitals Program for Calculating Crystal Properties*, Techn. Universität Wien, Austria, Karlheinz Schwarz, 2001.
7. Lalic M. V. and Mestnik-Filho J. to be published.

Hyperfine Interactions (2004) 158:95–98
DOI 10.1007/s10751-005-9014-6

Electric Field Gradients of Fluorides Calculated by the Full Potential KKR Green's Function Method

M. OGURA[1,2,*], H. AKAI[1] and T. MINAMISONO[1]
[1]*Department of Physics, Graduate School of Science, Osaka University, 1-1 Machikaneyama, Toyonaka, 560-0043, Osaka, Japan; e-mail: ogura@phys.sci.osaka-u.ac.jp*
[2]*Japan Science and Technology Agency, 4-1-8 Honcho, Kawaguchi, 332-0012, Saitama, Japan*

Abstract. We report on the calculation of electric field gradients (EFGs) of MgF_2, MnF_2, CoF_2, NiF_2 and ZnF_2. EFGs were calculated by the full potential Korringa–Kohn–Rostoker (KKR) Green's function method in the framework of the density functional theory. While EFG calculation of these fluorides are rather difficult in the muffin-tin-potential model, due to its sensitivity to the muffin-tin radius, no difficulties arise in the present full potential treatments. EFGs calculated by this method well reproduce experimental data.

Key Words: electric field gradient, full potential KKR method, nuclear quadrupole moment.

1. Introduction

Due to the development of various ab initio methods of electronic structure calculation, discussions of the electric field gradients (EFGs) from the view points of microscopic electronic structures are now feasible. Actually, many attempts to clarify the origin of the EFGs as well as their relation to the electronic properties of the system have been made in these decades. Among them are the calculations using the Korringa–Kohn–Rostoker (KKR) method in the framework of the local density approximation (LDA) of the spin-density functional theory. Though those calculations have succeeded in calculating EFGs for various systems [1, 2], they are yet limited to certain classes of materials due to the restriction of the muffin-tin potential model used in the normal KKR method. In this potential model, the atomic potentials are assumed to be spherically symmetric in inscribed spheres, called muffin-tin sphere, and zero in the interstitial region. This approximation, when the electronic structure depends much on the shape of the potential, makes the calculation of the EFGs rather unreliable. The problem is in particular serious in the case of F in MgF_2 since the determination of EFGs of this system has been one of the important issues in the nuclear physics.

* Author for correspondence.

Dritteler *et al.* developed the full potential KKR method and applied it to the EFG calculation for dilute Cu alloys, showing that the dominant contribution was the Cu d electrons [4]. In that method, the potential in each cell is treated as non-spherical and hence the method is considered of full potential in the exact meaning. However, the method is not yet completely practical for the calculation of electronic structure of arbitrary solids mainly due to its complicated analytical properties.

In the present study, we have independently developed the full potential KKR method based on the formalism of Dritteler's and calculated EFGs of MgF_2, MnF_2, CoF_2, NiF_2 and ZnF_2.

2. EFG Calculation

We calculated EFGs in MgF_2, MnF_2, CoF_2, NiF_2 and ZnF_2 using the full potential KKR method. The basic idea of the full potential KKR method is similar to Drittler's [4]. All those fluorides have the rutile structure. The lattice constants are referred to ref. [5]. In the calculation of MnF_2, CoF_2 and NiF_2, we took account of the antiferromagnetic spin alignment of metal atoms [3]. In the full potential method, wave function $R_L(r)$, potential $V(r)$, and charge density $\rho(r)$ are expanded into the real harmonics as

$$R_L(r) = \sum_{L'} R_{L'L}(r) Y_{L'}(\hat{r}), \tag{1}$$

$$V(r) = \sum_{L} V_L(r) Y_L(\hat{r}), \tag{2}$$

$$\rho(r) = \sum_{L} \rho_L(r) Y_L(\hat{r}), \tag{3}$$

where L is the set of angular momentum l and magnetic quantum number m. Here, the angular momentum l up to $l = 2$ were taken for the wave functions and up to $l = 4$ for the potentials and the charge densities. The states in the energy region down to -3.5 Ry, -4.0 Ry, -4.5 Ry, -5.0 Ry and -2.5 Ry from the Fermi level were treated as valence states for MgF_2, MnF_2, CoF_2, NiF_2 and ZnF_2, respectively, corresponding to the $2p3s$ states for Mg, the $3p3d4s$ states for Mn, Co and Ni and the $3d4s$ states for Zn. The other states were assumed to form core states and were assumed not to directly contribute to its own EFGs.

The experimental and theoretical EFGs are summarized in Table I. Here, z-axis is in the direction of the c-axis, y-axis along the <110> (towards the F atom) and x-axis is perpendicular to them. Theoretical EFGs in these fluorides well reproduce the experimental ones. In the table, the theoretical EFGs in MnF_2

Table I. EFGs in MgF_2, MnF_2, CoF_2, NiF_2 and ZnF_2 in 10^{19} V/m^2

			V_{xx}	V_{yy}	V_{zz}	η	Reference
MgF_2	Mg	exp.	71.3(7)	22.8(14)	48.5(15)	0.36(4)	[7]
		theor.	+71	−36	−35	0.14	present
	F	exp.	436(14)	296(12)	140(8)	0.36(3)	[8]
		theor.	−515	+345	+170	0.34	present
MnF_2	Mn	exp.	147(6)				[9]
		theor.	+194	−76	−118	0.21	[3]
			+235	−86	−149	0.27	present
	F	exp.	483(14)	338(12)	145(8)	0.40(3)	[8]
		theor	−581	+404	+177	0.39	[3]
			−676	+481	+195	0.42	present
CoF_2	Co	theor.	+2	−199	+197	0.98	present
	F	exp.	665(15)	492(15)	173(11)	0.48(3)	[8]
		theor.	−715	+419	+224	0.37	present
NiF_2	Ni	theor.	−84	−117	+201	0.17	[3]
			−41	−126	+167	0.51	present
	F	exp.	827(16)	596(17)	232(13)	0.44(3)	[8]
		theor.	−1181	+881	+300	0.49	[3]
			−1232	+923	+308	0.50	present
ZnF_2	Zn	exp.	+218(22)	−141(14)	−78(8)	0.29(3)	[10]
		theor.	+204	−136	−68	0.33	present
	F	exp.	768	549	219	0.43	[11]
		theor.	−966	+686	+280	0.42	present

Experimental EFGs are recalculated with the more recent reference values for the quadrupole moments of the nuclei [6].

and NiF_2 calculated by Dufek *et al.* using the full-potential linear augmented-plane-wave (FLAPW) method [3] are also listed. This method has widely been used and has been shown to reproduce the experimental EFGs of many crystals. Despite that, it is not quite clear whether FLAPW using spherical potentials inside the muffin-tin spheres in generating the basis functions is fully reliable for EFG calculations or not. However, the EFGs calculated with the FLAPW method turns out to agree well with the ones obtained by the full potential KKR method. Therefore, the treatment using the spherical potential in the muffin-tin sphere for the basis function seems acceptable in this situation.

The agreement of the theoretical and experimental EFGs at F sites is not satisfactory both in the cases of the full potential KKR and FLAPW methods. We think that the assumed nuclear quadrupole moment of ^{19}F*, $Q(^{19}\mathrm{F}^*) = -94.2(9)$ mb [12] might contain some errors. If we adopted the theoretical EFGs obtained by the present calculation, we would conclude $Q(^{19}\mathrm{F}^*) = -73(9)$ mb with the reliability of 10%. In order to make more convincing arguments about the quadrupole moments of F isotopes, we should further calculate the EFGs of much wider range of systems and check the reliability. Reliable predictions of quadrupole moments from nuclear structure theories are also desired.

Acknowledgements

We acknowledge Dr. K. Sato of Osaka University, Prof. Dr. P.H. Dederichs, Prof. Dr. S. Blügel and Dr. Zeller of Forschungszentrum Jülich for fruitful discussions. The present study was partly supported by the 21st century COE program "Towards a New Basic Science" and by Special Coordination Funds for the Promotion of Science and Technology, Leading Research "Nanospintronics Design and Realization".

References

1. Akai H., Akai M., Blügel S., Drittler B., Ebert H., Terakura K., Zeller R. and Dederichs P. H., *Prog. Theor. Phys., Suppl.* **101** (1990), 11.
2. Sato K., Akai H., Maruyama Y., Minamisono T., Matsuta K., Fukuda M. and Mihara M., *Hyp. Interact.* **120/121** (1999), 145.
3. Dufek P., Schwarz K. and Blaha P., *Phys. Rev. B* **42** (1993), 12672.
4. Drittler B., Weinert M., Zeller R. and Dederichs P. H., *Phys. Rev. B* **42** (1990), 9336.
5. Wyckoff R. W. G., *Crystal Structure* 2nd Edition. Krieger, 1986.
6. Pyykkö P., *Mol. Phys.* **99** (2001), 1617.
7. Hiyama Y., Woyciesjes P. M., Brown T. L. and Torchia D. A., *J. Magn. Res.* **72** (1987), 1.
8. Richter F. W., Wiegandt D., *Z. Phys.* **217** (1968), 225.
9. Yasuoka H., Ngwe T., Jaccarino V. and Guggenheim H. J., *Phys. Rev.* **177** (1969), 667.
10. Steiner M., Potzel W., Köfferlein M. *et al. Phys. Rev. B* **50** (1994), 13355.
11. Barfuss H., Böhmlein G., Hohenstein H., Kreische W., Meinhold M., Niedrig H. and Reuter K., *J. Mol. Struct.* **58** (1980) 203.
12. Halkier A., Christiansen O., Sundholm D. and Pyykkö P., *Chem. Phys. Lett.* **271** (1997), 273.

Hyperfine Interactions (2004) 158:99–103
DOI 10.1007/s10751-005-9015-5

Electric Field Gradients of Light Impurities in TiO_2 Calculated by the Full Potential KKR Green's Function Method

M. OGURA[1,2,*] and H. AKAI[1]
[1]*Department of Physics, Graduate School of Science, Osaka University, 1-1 Machikaneyama, Toyonaka, Osaka, 560-0043, Japan; e-mail: ogura@phys.sci.osaka-u.ac.jp*
[2]*Japan Science and Technology Agency, 4-1-8 Honcho, Kawaguchi, Saitama 332-0012, Japan*

Abstract. The electric field gradients (EFGs) of B, N, O and Na in TiO_2 were calculated by the full potential Korringa–Kohn–Rostoker (KKR) Greens function method in the framework of the density functional theory. The agreement with the experiments was much improved from the previous calculations that were based on the muffin-tin potential model.

Key Words: electric field gradient, full potential KKR, impurity, TiO_2.

1. Introduction

Nuclear methods using unstable nuclei, combined with various new experimental techniques, enable us to measure hyperfine interactions of impurities in a broad range of host crystals. The data obtained in such ways provide us with microscopic information of electronic structures of condensed matter systems. The information is unique because virtually unlimited combinations of probes and hosts are possible.

Recently, the electric quadrupole coupling constants of impurities in TiO_2 have been systematically measured by use of the β-NMR technique, which utilizes the asymmetric β-ray distributions from the polarized nuclei, and some interesting behaviors have been reported [1–3]. In order to interpret the experimental EFGs and to extract information about the electronic structure of impurities, first-principles calculations using the Korringa–Kohn–Rostoker (KKR) method [4] were performed [1, 5]. Though these calculations well reproduced the experimental EFGs in general, they failed in some cases to reproduce experiments. The reason was thought of to lie in the muffin-tin potential model, in which the atomic potential is assumed to be spherical in the inscribed sphere, called muffin-tin sphere, and is assumed to be zero in the

* Author for correspondence.

interstitial region. This approximation makes calculated EFGs unreliable when the electronic structure depends too much on the assumption on the shape of the potential.

Dritteler *et al.* developed the full potential KKR method and applied it to the EFG calculation [6]. In this method, a potential in a muffin-tin sphere is treated as non-spherical. However the method is not yet completely practical for calculation of the electronic structure of arbitrary solids due to its complicated analytical properties.

In the present study, we independently have developed the full potential KKR method based on the formalism of Dritteler's and calculated EFGs at the impurity sites in TiO_2.

2. EFG calculation

We calculated EFGs of B, N, O and Na in TiO_2 using the full potential KKR method based on the local density approximation (LDA) of the spin-density functional theory. The basic idea of the full potential KKR method is similar to that of Drittler's [6]. The exchange-correlation potential was calculated with the parameters after Moruzzi *et al.* [7]. The charge density distributions are obtained self-consistently for all electrons including the core states of all atoms.

TiO_2 has a rutile structure. The lattice constants are referred to ref. [8]. In the full potential method, wave functions, potentials, and charge densities are expanded into the spherical harmonics. Here, the wave functions were expanded into the real harmonics up to $l = 2$ and the potential and the charge density up to $l = 4$. The states in the energy region down to -3.0 Ry from the Fermi level were treated as the valence states, which correspond to the $3p3d4s$ states for Ti and the $2s2p$ states for O. The other states were assumed to form core states and were assumed not to directly contribute to its own EFGs.

From the experiments and the theoretical predictions of the previous calculations, B atoms are expected to be at the Ti substitutional and octahedral interstitial site [1], N atoms at the O substitutional and octahedral interstitial site [2], and O atoms at the O site and an unknown interstitial site [3]. In the present study, the calculation was performed for these impurity sites. For Na, the interstitial site is assumed for the implantation site. We simulated the impurities using a super cell which consists of two TiO_2 unit cells stacked along the c-axis. In the super cell, the impurity concentration is much higher than experimental situations. However, it is confirmed that the EFGs do not change for larger super cells. For the system with impurity at the substitutional and interstitial site, 21 and 30 k-points, respectively, in the irreducible wedge of the 1st Brillouin zone were used.

In order to estimate the lattice relaxation around the impurity, we calculated the electric field, namely the force at the neighboring atoms. The neighboring six

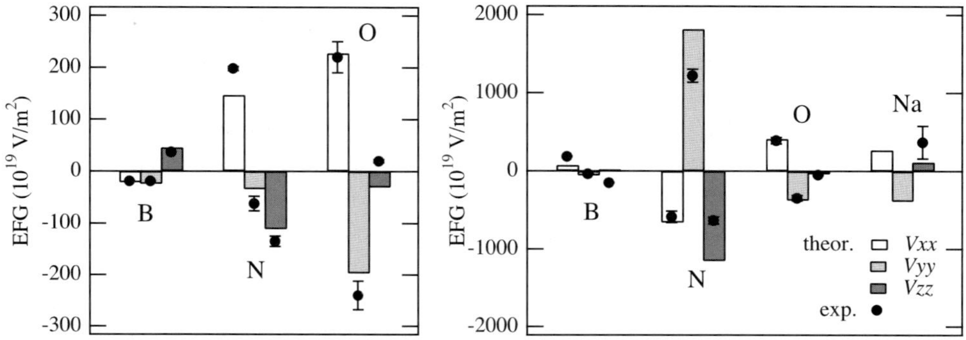

Figure 1. EFGs of impurities in TiO_2 calculated by the full potential KKR method. (*left*): substitutional site, (*right*): interstitial site.

Table I. EFGs of impurities in TiO_2

			V_{xx}	V_{yy}	V_{zz}	Ref.
B						
Ti sub.	Exp.		−18.8(4)	−18.8(4)	+37.1(5)	[1]
	Theor.	KKR	−54	−9	+63	[1]
		FP-KKR	−22	−25	+46	Present
Oct.	Exp.		+185(5)	−35(5)	−150(4)	[1]
	Theor.	KKR	+143	+36	−179	[1]
		FP-KKR	+77	−58	+19	Present
N						
O sub.	Exp.		±198(4)	±62(14)	±135(10)	[2]
	Theor.	KKR	+185	−178	−6	[5]
		FP-KKR	+147	−35	−112	Present
Oct.	Exp.		±586(73)	±1220(85)	±633(44)	[2]
	Theor.	KKR	−449	+1304	−855	[5]
		FP-KKR	−662	+1816	−1155	Present
O						
O sub.	Exp.		+220(30)	−240(28)	+20(3)	[3]
	Theor.	KKR	+209	−215	+6	[5]
		FP-KKR	+228	−197	−31	Present
Oct.	Exp.		390(40)	343(35)	47(5)	[3]
	Theor.	KKR	+4358	−2284	−2074	[5]
		FP-KKR	+414	−375	−40	Present
Na						
Oct.	Exp.				370(210)	[9]
	Theor.	KKR	+157	−187	+30	Present
		FP-KKR	+270	−389	+119	Present

O, three Ti and two O atoms were relaxed for the Ti substitutional, O substitutional and octahedral sites, respectively. In addition to the lattice relaxation, the charge state of the impurity atom might influence the result of the

EFG calculation. The calculated Fermi level was adjusted after the self-consistent calculation in such a way that it reproduced the charge states. Then, the charge density distribution was calculated for EFGs [5]. Here, B^{3+}, O^{2-} and Na^{+} were considered for the B, O and Na impurities, respectively. For the N impurity, N^{3-}, N^{-} were considered for the O substitutional site and the octahedral interstitial site.

The experimental and theoretical EFGs are summarized in Figure 1 and Table I. Here, z-axis is in the direction of the c-axis, y-axis along the $< 110 >$ (towards the O atom) and x-axis is perpendicular to them. The present calculation well reproduces most of the EFGs at these impurity sites. For B at the Ti substitutional site and N at the O substitutional site, the full potential calculation improves the agreement between the experimental and theoretical EFGs. For O at the octahedral interstitial site, the theoretical EFG agrees well with the experimental one whose site has not been identified. Based on our calculation, we can safely assign that this EFG is associated to the octahedral interstitial site. However, for B, the present calculation assuming the octahedral interstitial site failed to reproduce EFGs that in our previous calculation were safely assigned to be at the octahedral interstitial site. As for Na in TiO_2, it is reported that the direction of the largest component of the EFG tensor is parallel to the c-axis of the crystal [9]. This is not consistent to the result of the present calculation. Further improvements are desired for the calculation of impurities at interstitial sites.

Acknowledgements

We acknowledge Dr. K. Sato in Osaka University, Prof. Dr. P.H. Dederichs, Prof. Dr. S. Blügel and Dr. Zeller in Forschungszentrum Jülich for fruitful discussions. The present study was partly supported by the 21st century COE program "Towards a new basic science" and by Special Coordination Funds for the Promotion of Science and Technology, Leading Research "Nanospintronics Design and Realization".

References

1. Ogura M., Minamisono K., Sumikama T. and Nagatomo T. *et al., Hyperfine Interact.* **136/137** (2001), 195.
2. Minamisono T., Sato K., Akai H., Takeda S. and Maruyama Y. *et al., Z. Naturforsch.* **53a** (1998), 293.
3. Minamisono T., Nojiri Y., Matsuta K., Fukuda M. and Sato K. *et al., Phys. Lett. B* **457** (1999), 9.
4. Akai H., Akai M., Blügel S., Drittler B., Ebert H., Terakura K., Zeller R., and Dederichs P. H., *Progr. Theoret. Phys. Suppl.* **101** (1990), 11.
5. Sato K., Akai H., Maruyama Y., Minamisono T., Matsuta K., Fukuda M. and Mihara M., *Hyperfine Interact.* **120/121** (1999), 145.

6. Drittler B., Weinert M., Zeller R. and Dederichs P. H., *Phys. Rev. B* **42** (1990), 9336.
7. Moruzzi V. L., Janak J. F. and Williams A. R., *Calculated Electronic Properties of Metals*, Pergamon, U.S.A, 1978.
8. Wyckoff R. W. G., *Crystal Structure*, 2nd ed., Krieger (1986).
9. Minamisono K. TRIUMF Annual Report (2003), 48.

Hyperfine Interactions (2004) 158:105–110
DOI 10.1007/s10751-005-9016-4

Peculiarities of Quadrupolar Relaxation in Electrolyte Solutions

MARIA PAVLOVA* and VLADIMIR CHIZHIK
Saint-Petersburg State University, St. Petersburg, Russia; e-mail: mariapavlova@mail.ru

Abstract. Peculiarities of quadrupolar relaxation in electrolyte solutions were established via comparison of the data obtained from proton and deuteron resonances. It has been shown that quadrupole coupling constants (QCC) of deuterons depend not only on internal electron structure of molecule or ion, but on solution structure as well. To interpret the experimental results quantum-chemical calculations of QCC of deuterons in different molecular complexes simulating different solution substructures were carried out. Density functional theory (DFT) method with hybrid B3LYP functional was used for all calculations.

Key Words: electrolyte solutions, quadrupolar relaxation, quadrupole coupling constant, water clusters.

1. Introduction

Electrolyte solutions have been investigated for many years but their internal molecular structure is still out of the complete understanding. The investigation of the electrolyte solutions becomes rather complicated because of the presence of different substructures with fast molecular exchange among them. Nuclear magnetic relaxation method is one of the most powerful tool for the investigation of such systems. It provides an information about the microstructure in the close vicinity near ions and about the behaviour of fluctuating electromagnetic fields in solutions. The observation of quadrupolar relaxation process of deuterons belonging to ions and solvent molecules is especially fruitful. The relaxation method for the investigation of the microstructure of aqueous salt solutions was described in [1]. The good agreement between experimental results obtained from proton and deuteron resonances was found, but application of the method to studying deuterated solutions of some mineral acids and solutions containing 'strong' ions (like Ca^{2+} or SO_4^{2-}) revealed new peculiarities of quadrupolar relaxation in such molecular systems. This article reflects the first attempt to interpret the concentration dependence of the spin-lattice relaxation rates of the deuterons in acid solutions.

* Author for correspondence.

2. Qudrupolar relaxation in electrolyte solutions

If the condition of fast exchange of solvent molecules among all possible substructures in the electrolyte solution is realized, one can write for the spin-lattice relaxation rate of the solvent nuclei [2, 3]:

$$\frac{1}{T_1} = \sum_i \frac{p_i}{T_{1i}}; \qquad \sum_i p_i = 1, \tag{1}$$

where p_i is the relative concentration of the substructure i and T_{1i} is the spin-lattice relaxation time of the nuclei disposed in the substructure i. Usually, only one relaxation mechanism is predominant (the quadrupole interaction for deuterons) Equation (1) can be rewritten in the more detailed form:

$$\frac{1}{T_1} = \frac{1}{T_{1N}} + \sum_i^{N-1} \frac{mn_i L_\pm}{55.5}\left[\frac{1}{T_{1i}} - \frac{1}{T_{1N}}\right], \tag{2}$$

where m is the aquamolality of the electrolyte solution; n_i is the number of solvent molecules in the substructure i per one ion; $L_\pm$ is the number of ions in the solute molecule; N is the number of possible substructures in the solution (N is not known beforehand and can be determined in the experiments).

Equation (2) is written for the case of complete dissociation of the electrolyte and predicts the linear concentration dependence of $1/T_1$, but experimentally distinct deviations from linearity is observed. This effect allows us to get the information on the details of the microstructure of the hydration shells. It was shown in the work [1] that the bends of the concentration dependences were connected with disappearance of certain substructures in the solution and it allowed the estimation of coordination numbers of ions from this data.

The relaxation mechanism via electric interaction is a predominant mechanism for nuclei with spin $I \geq 1$ ($I = 1$ for the deuteron) and as usual it is much stronger than the mechanisms based on magnetic interactions. The orientation of quadrupole nuclei depends on the direction of inhomogeneous electric field. The quadrupolar relaxation process are governed by the fluctuations of electric field gradients in the site of the nucleus, which occur due to intermolecular and intramolecular electric interactions. Assuming the fast thermal motion, the relaxation rate for the quadrupolar relaxation mechanism is given by the expression [4, 5]:

$$\frac{1}{T_1} = const\left(\frac{eqQ}{\hbar}\right)^2 \tau_c, \tag{3}$$

where eQ is the quadrupole moment of nuclei; q is the electric field gradient at the site of the nucleus; $eqQh$ is the quadrupole coupling constant and its value is connected with the structure of the electric fields; τ_c is the correlation time which characterizes the time scale of the reorientations of the solvent molecules.

3. Experimental and calculations details

The resonance of the ^2H nuclei was observed at 14 MHz. Concentration dependences of spin-lattice relaxation rates in aqueous salt and mineral acid solutions at different temperatures were measured using home-modernized relaxometer BRUKER SXP 4-100. The measurement accuracy within 2% was achieved. The precision of sample temperature was within 1 K. All concentrations are given in units of aquamolality (mol of solute per 55.5 mol of solvent). The samples were prepared gravimetrically using acids of 98% deuteration and heavy water of 99.8% deuteration.

Quantum-chemical calculations of QCC of deuterons in different molecular complexes consisting of water molecules and ions were performed using GAMESS 2003 program [6]. DFT method with hybrid functional B3LYP was chosen to take into consideration non-local electronic correlation. In all cases molecular geometry was previously optimized at the same level of theory. It is known that results of such calculations are very sensitive to a choice of a basis set (see for example [7]). In order to get reliable information about the influence of the basis set on the QCC value the systematic investigation of QCC of deuterons in a separate water molecule and water pentamer was made using different flexible basis sets 6-31G**, 6-311G** and cc-pVXZ, X = D,T,Q and 5. The similar calculations were carried out for all basis sets augmented by diffuse functions that was supposed to be important for molecular systems with hydrogen bonds. The B3LYP/6-31G** combination was found to be a compromise between high-quality and time-consuming calculations. As a result the calculations of the deuteron QCC for $(D_2O)_5$, $(D_3O^+)(D_2O)_3$ and $Ca^{2+}(D_2O)n$ (n = 6, 8) clusters were carried out at B3LYP/6-31G** level of theory. The effective core potential basis set LACVP** was used for the Ca^{2+} ion.

4. Results and discussion

In Figure 1 the experimental concentration dependences of the relative relaxation rates of the deuterons in the solutions of nitric and perchloric acids are presented. To explain a great difference in the behavior of the concentration dependences of the proton and deuteron relaxation rates it is necessary to take into consideration the change of the deuteron QCC caused by the deuteron transfer from the water molecule to the D_3O^+ ion.

According to Equation (3) the relaxation rates are proportional to the reorientation times of solvent molecules and it seems being convenient to introduce the relative reorientation times:

$$\lambda_i = \frac{\tau_i}{\tau_0} = \frac{T_{10}}{T_{1i}}, \tag{4}$$

where τ_0 and T_{10} are the reorientation time of molecules and relaxation rate in the structure of pure solvent, respectively.

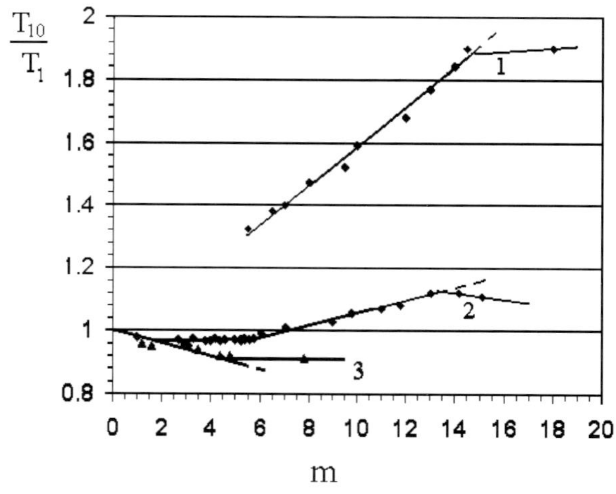

Figure 1. Experimental concentration dependence of the spin-lattice relaxation rate in aqueous acid solutions: the ^{1}H resonance in the $HClO_4$ solution [8] (1); the ^{2}H resonance in the HNO_3 solution (2); the ^{2}H resonance in the $HClO_4$ solution (3). T_{10} is the spin-lattice relaxation time in pure water. All dependences are given for 22°C.

Using Equations (2) and (3), one can derive the expression for relative spin-lattice relaxation rate (in comparison with pure water) at the concentration points corresponding to disappearance of certain substructures:

$$\frac{T_{10}}{T_1} = \lambda_\Sigma = \frac{1}{n_\Sigma} \sum_{i=1}^{N} n_i a_i^2 \lambda_i, \tag{5}$$

where n_Σ is the total number of solvent molecules coordinated by the ions; a_i describes the change of the deuteron QCC for solvent molecules in the substructure i in comparison with one for the pure solvent structure:

$$a_i = \frac{eq_iQ}{eq_0Q}. \tag{6}$$

In order to conciliate the data obtained from proton and deuteron resonances the electric field gradients and the QCC of the deuterons from different molecular complexes were estimated from quantum-chemical calculations. The results of quantum-chemical calculations of the deuteron QCC are summarized in Table I. According to the calculations for the isolated water molecule and isolated hydroxonium ion $a_i = 0.8$. But if one takes into account the change of bond lengths and angles in the water molecule and hydroxonium ion surrounding by their first hydration shells the value a_i reduces to 0.65. This value was used for the calculations of the spin-lattice relaxation rate in the nitric acid aqueous solution according to Equation (5). The comparison of the results of the calculations with the experimental data of the work [1] allows the revision of the reorientation time of water molecules near the D_3O^+ ion: $\lambda_1^+ = 1.3$

Table I. Results of the calculations of the deuteron QCC for different structures

	Cluster	QCC (KHz)	a_i[a]
D_2O	Isolated molecule	310[b]	
	+4 D_2O	226	1.00
D_3O^+	Isolated ion	247	
	+3 D_2O	146	0.65
Ca^{2+}	+6 D_2O	291	1.28
	+8 D_2O	218	0.96

[a] Referred to the 'hydrated' water molecule.
[b] The experimental value $eqQ = 308$ KHz [9].

(instead of $\lambda_1^+ = 1.9$ for the H_3O^+ ion in [1]). The difference can be explained by the slight inequality in the molecular mobilities in the protonated and deuterated aqueous solutions (about 20%) and besides in the work [1] the data for acid solutions were taken from the relaxation measurements for undergassed samples.

The change of the deuteron QCC for water molecules in the hydration shells of 'strong' ions Ca^{2+} and SO_4^{2-} were also detected on the basis of the relaxation rate measurements: $a_i = 0.6$ for the hydration shells of both ions. The calculations for the Ca^{2+} ion was fulfilled taking into account (1) one symmetrical hydration shell consisting from six water molecules and (2) the same structure with two additional water molecules in the second layer. It would be good if the tendency of the QCC decrease keeps with the increase of the amount of water molecules in the second layer. These calculations and the calculations for the SO_4^{2-} hydration shell are now in progress.

5. Conclusion

In earlier works (see, for example, [1, 10]) the equivalence of the proton and deuteron resonances for the investigation of electrolyte solutions was supposed. In contrast with earlier results in this work it is shown that one must take into account the possible change of the quadrupole coupling constants of deuterons that belong to hydrogen-containing ions or even water molecules bound up with some ions (SO_4^{2-}, Ca^{2+},..). The calculations of the deuteron QCC in $(D_2O)_5$ and $(D_3O^+)(D_2O)_3$ clusters allowed the re-evaluation of the reorientation time of water molecule near the D_3O^+ ion.

Acknowledgements

The financial support of the Russian Foundation for Basis research (Grant 04-03-32639) and the Ministry of Education of Russian Federation (Grants E02-5.0-83 and UR-01.01.046).

References

1. Chizhik V. I., *Mol. Phys.* **90** (1997), 653.
2. Zimmerman J. R. and Brittin W. E., *J. Phys. Chem.* **61** (1957), 1328.
3. Woessner D. E., *J. Chem. Phys.* **35** (1961), 41.
4. Abragam A., In: *The Principles of Nuclear Magnetism*, Oxford University Press, London, 1961.
5. Slichter C. P., In: *Principles of Magnetic Resounance*, Springer, Berlin Heidelberg New York, 1980.
6. Schmidt M. W., Baldridge K. K., Boatz J. A., Elbert S. T., Gordon M. S., Jensen J. H., Koseki S., Matsunaga N., Nguyen K. A., Su S. J., Windus T. L., Dupuis M. and Montgomery J. A., *J. Comput. Chem.* **14** (1993), 1347.
7. Olsen L., Christiansen O., Hemmingsen L., Sauer S. P. A. and Mikkelsen K. V., *J. Chem. Phys.* **116** (2002), 1424.
8. Chizhik V. I., Egorov A. V., Komolkin A. V. and Vorontsova A. A., *J. Mol. Liq.* **98–99** (2002), 173.
9. Garvey R. M. and DeLucia F. C., *Can. J. Phys.* **55** (1988), 1115.
10. Hertz H. G., Stalidus G. and Versmold H., *J. Chim. Phys. Phys-Chim.* **66** (1966), 177.

Hyperfine Interactions (2004) 158:111–115
DOI 10.1007/s10751-005-9017-3

Ordering Effects and Hyperfine Interactions in Fe–N Austenites

A. N. TIMOSHEVSKII* and B. Z. YANCHITSKY
Institute of Magnetism, National Academy of Science Ukraine, 36-b Verdansky St., 03142 Kiev, Ukraine; e-mail: tim@ukron.kiev.ua

Abstract. Using the high accurate *ab-initio* FLAPW method and the cluster expansion technique, interatomic N–N interactions for Fe–N austenite were calculated. The interactions were used for calculation of temperature dependence of the short range order for $Fe_{10}N$ austenite. For two model structures with different nitrogen distributions, the hyperfine interactions were calculated. It was revealed, that EFG might be nonzero on nuclei even for Fe atoms that do not have nitrogen atoms at the first coordination shell. This finding has to be considered for interpretation of Mössbauer spectra of austenite Fe–N.

1. Introduction

Nitrogen as a doping element is widely used in creation of various steels, for example, in the high nitrogen steels. That is why a large number of investigations have been performed to clarify an influence of the nitrogen on the fcc matrix of iron. Many of the changes, caused by the nitrogen, originate from nitrogen distribution over iron matrix. A basic method for studying distribution of nitrogen is Mössbauer spectroscopy. There are a large number of investigations of the problem [1–3]. But there is no single point of view in what concerns interpretation of the experimental data. Mainly this is because that there is no unique technique to decompose the experimental spectra into components, which are determined by the short range order in the austenite. A majority of investigators interpret the spectra within framework of a model which includes only three components. These components comes from three kinds of the iron atoms, the first kind corresponds to iron atom that does not have a nitrogen atom at the first coordination shell, the second and third kinds correspond to the case when one and two nitrogen atoms are located at the first shell. For complete understanding of how the quadrupole splittings on iron nuclei are formed, investigations of the short range order and the electronic structure of Fe–N system are needed. Now, many of similar problems are studying using high accurate *ab initio* methods. For Fe–N systems only a minor number of such

* Author for correspondence.

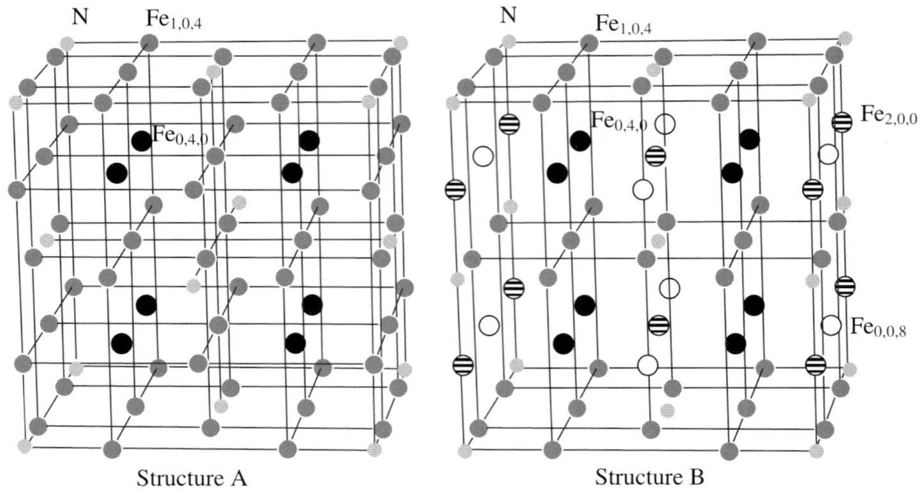

Figure 1. The model structures of the stoichiometry Fe_8N.

investigations have been performed [4, 5]. In the present paper, an influence of the nitrogen distribution in austenite Fe–N on formation of quadrupole splittings on the iron nuclei is investigated.

2. Results and discussion

For detailed investigation of dependence of different nitrogen distributions on quadrupole splittings on the iron nuclei in austenite Fe–N we used a high accurate FLAPW method [6]. For two model structures with different nitrogen distributions we calculated electronic structure and hyperfine interactions. Since most of investigators believe that in Fe–N system exists energetically favourable called as 'bell-like', configuration N–Fe–N [7] (iron atoms in this configuration will be denoted as Fe_2), we chose two model structures in that way so the second structure contains Fe_2 atom, but the first one does not. These structures can be seen in the Figure 1. Technical details of this *ab initio* calculation are shown in [8]. It should be stressed, that for both structures the optimization of geometry was performed; such optimization includes variations of the parameters of the unit cell and atomic positions within unit cell to achieve minimum of the total energy. On the base of the performed calculations, we investigated an influence of nitrogen atoms on two structural phases in fcc iron (low spin LS and high spin HS). Details of this investigation are discussed in [8]. Our calculations revealed that the total energy of the structure (B) is less than the total energy of the structure (A), thus confirming that formation of bell-like configurations is energetically favourable.

EFG on the iron nuclei was calculated using *ab initio* basis and electron density by means of the method developed in [9]. The results are presented in Table I. The first structure (A) contains two kinds of iron atoms: $Fe_{0,4,0}$ (four

Table I. Quadrupole splittings (mm/s) in Fe–N.

	Structure	$Fe_{1,0,4}$	$Fe_{2,0,0}$	$Fe_{0,4,0}$	$Fe_{0,0,8}$
Theory	Fe_8N (A) *No* 225 $Fm\bar{3}m$	0.18	–	0.00[a]	–
	Fe_8N (B) *No* 123 P4/mmm	0.28	0.17	0.27	0.48
	Fe_4N (model) [4]	0.03	0.32	–	–
	Fe_4N (nitride) [4]	–	0.50	–	–
Exper	Fe_4N [10]	–	0.50	–	–
	$Fe_{10.2}N$ [3]	0.25	0.40	–	–
	$Fe_{10.2}N$ [1]	0.39	0.72	–	–
	$Fe_{11}N$ [2]	0.39	0.75	–	–

[a] Quadrupole splitting is zero because the point group of this atom is $Td(\bar{4}3m)$.

Table II. The interatomic interactions (eV) by the least square fit method.

N	Interatomic interactions								
	w_0	w_1^1	w_1^2	w_2^2	w_3^2	w_4^2	w_5^2	w_6^2	w_1^3
5	−34634.7004	−1490.8849	0.127	−0.041	0.027	0.018	0.032	–	−0.014
6	−34634.6996	−1490.9087	0.129	−0.034	0.028	0.022	0.031	−0.007	−0.014

N is a number of pair potentials used in the fit.

nitrogen atoms at the second coordination shell) and $Fe_{1,0,4}$, these atoms at the first coordination shell have zero and one nitrogen atoms respectively. The second structure (B) has in addition $Fe_{2,0,0}$ atoms (two nitrogen atoms at the first shell) and atoms $Fe_{0,0,8}$ (eight atoms at the third shell). Modern interpretations believe that Fe_1 atoms in structures (A) and (B) must have equal quadrupole splitting. As it follows from Table I this assumption is in contradiction with our results. In our opinion there are as least two kinds of atoms Fe_1, one kind of them exists due to presence of Fe_2 and the second kind has nothing common with Fe_2, these kinds cannot be symmetry equivalent. As it is seen from our calculations, for structure (A) on nucleus of $Fe_{1,0,4}$ the quadrupole splitting is 0.18 mm/s, whereas for structure (B) the splitting is 0.28 mm/s. A similar picture may be observed for atoms $Fe_{0,m,n}$. For these atoms the quadrupole splittings are of the same order of magnitude as for atoms $Fe_{1,0,4}$ and $Fe_{2,0,0}$ (Table I). An important question is how many atoms of each kind does real austenite contain? To investigate the problem it is necessary to calculate temperature dependence of the short range order in Fe–N system. To do this, a calculation of interatomic potentials was performed.

We constructed 10 ordered structures with various distribution of nitrogen in fcc iron matrix [4]. All calculations were performed with spin polarization. Clusters corresponding to interatomic potentials are given in [4]. The values of calculated interatomic potentials are given in Table II. The symbols used have the following meaning: $w^{(0)}$ – is energy of the iron atom, $w^{(1)}$ – is energy of the nitrogen atom plus the energy needed to introduce the atom into iron matrix,

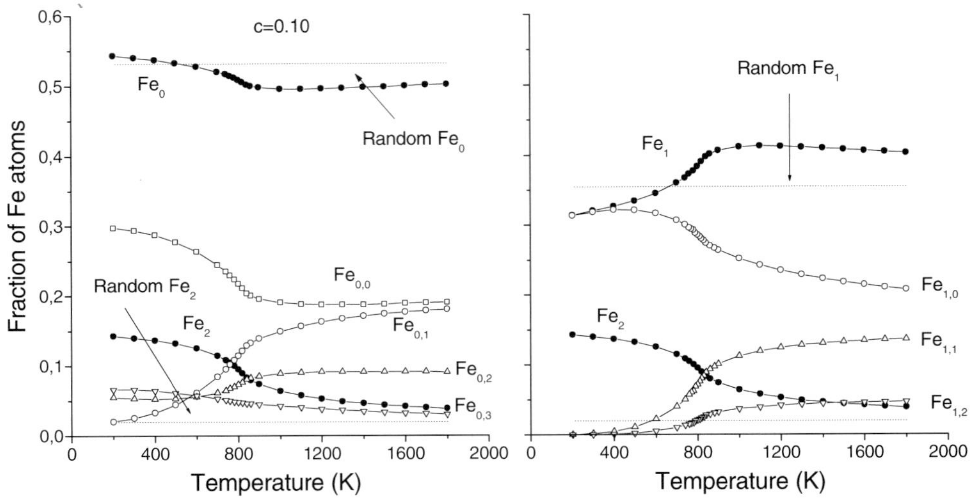

Figure 2. Temperature dependence of fraction of iron atoms of different kinds for $Fe_{10}N$ alloy. Label 'random' means that nitrogen atoms dot not interact.

$w_1^{(2)}$, $w_2^{(2)}$, $w_3^{(2)}$, $w_4^{(2)}$, $w_5^{(2)}$, $w_6^{(2)}$ – are energies of pair N–N interactions for six coordination shells, $w_1^{(3)}$ – is energy of three-body N–N–N interaction. Fit by the least square method gives the maximal residual for the total energy 0.0036 eV/ atom Fe.

The short range order was calculated by the Monte Carlo method for canonical ensemble. Temperature dependences of the fraction for various kinds of iron atoms for experimentally studied concentration of nitrogen 0.1 are given in Figure 2. It can be seen from the Figure 2 that the fraction of $Fe_{0,0}$ atoms is nearly equal to that for $Fe_{0,1}$. Atoms $Fe_{0,0}$ can contribute into singlet only, but for the nuclei of $Fe_{0,1}$ it is possible a formation of significant value EFG, thus adding into Mössbauer spectra a doublet. A similar situation exists for atoms $Fe_{1,m}$. It is seen in Figure 2, that the numbers of atoms $Fe_{1,0}$ and $Fe_{1,1}$ are close, and the nuclei of this atoms might contribute significantly into the spectrum of austenite.

3. Conclusion

As it follows from the presented investigation, in contradiction with existing interpretations of Mössbauer spectra, a more number of iron atoms of different kinds must be included. Even iron atoms, which do not have the nitrogen atoms at the first coordination shell, might contribute into spectra.

References

1. Oda K., Umezu K. and Ino H., *J. Phys. Condens., Matter* **2** (1990), 10147.
2. Gavriljuk V. G., Nadutov V. M. and Gladun O., *Phys. Met. Metallogr.* **3** (1990), 128.
3. Foct J., Rochegude P. and Hendry, *Acta Mater.* **36** (1988), 501.
4. Timoshevskii A. N., Timoshevskii V. A. and Yanchitsky B. Z., *J. Phys. Condens., Matter* **13** (2001), 1051.

5. Timoshevskii A. N., Yanchitsky B. Z. and Bakai A. S., *Fiz. Nizk. Temp.* **30** (2004), 626.
6. Blaha P., Schwarz K. and Luitz J., (1999), ISBN 3-9501031-0-4.
7. Gavriljuk V. G. and Berns H., *High Nitrogen Steels*, Springer, Berlin, 1999, p. 378.
8. Timoshevskii A. N., Timoshevskii V. A., Yanchitsky B. Z. and Yavna V. A., *Comput. Mater. Sci.* **22**(1–2) (2001), 99.
9. Blaha P., Schwarz K. and Herzig P., *Phys. Rev. Lett.* **54** (1985), 1192.
10. Rochegude P. and Foct J., *Phys. Status Solidi,* A **98** (1986), 51.

Hyperfine Interactions (2004) 158:117–124
DOI: 10.1007/s10751-005-9019-1

Mössbauer *In Situ* Studies of the Surface of Mars

G. KLINGELHÖFER
Inst. Anorganische & Analytische Chemie, Johannes Gutenberg Universität, 55099 Mainz, Germany; e-mail: klingel@mail.uni-mainz.de

Abstract. For the first time in history, a Mössbauer spectrometer was placed on the surface of another planet. Our miniaturized Mössbauer spectrometer MIMOS II [1–4] (Figure 2) is part of the instrument payload of NASA's twin Mars Exploration Rovers (MER) "Spirit" and "Opportunity" (see Figure 1), which in January 2004 successfully landed at the Gusev crater and the Meridiani Planum landing sites, respectively. MIMOS II determines the Fe-bearing mineralogy of Martian soils and rocks at the Rovers' respective landing sites [5]. The main goals of this planetary twin mission are to: (1) identify hydrologic, hydrothermal, and other processes that have operated and affected materials at the landing sites; (2) identify and investigate the rocks and soils at both landing sites, as there is a possible chance that they may preserve evidence of ancient environmental conditions and possible prebiotic or biotic activities. With MIMOS II, besides other minerals the Fe silicate olivine has been identified in both soil and rocks at both landing sites. At the Meridiani site the Fe sulfate jarosite has been identified by MIMOS II which is definitive mineralogical proof of the presence of water at this site in the past.

Key Words: Hematite, Jarosite, Mars, miniaturised Mössbauer spectrometer, Mössbauer effect, instrumentation.

1. Introduction

In January 2004 the US–American space agency NASA landed successfully two rovers on the surface of Mars. *Spirit* landed in Gusev Crater on January 4 (UTC), 2004. It was followed 21 days later by *Opportunity*, which landed on Meridiani Planum. The primary scientific objective of the MER mission is to explore two sites on the Martian surface where water may once have been present, and to assess past environmental conditions at those sites and their suitability for life. The Mars Exploration Rovers (MER) (Figure 1) Spirit and Opportunity are both carrying our Mössbauer spectrometer MIMOS II, part of the MER instrument suite consisting of remote sensing instruments [4], and the In Situ instruments mounted on an robotic arm (IDD) (Figure 3): (1) Rock Abrasion Tool (RAT), (2) Mössbauer (MB) spectrometer Mimos II [1], (3) Microscopic Imager [4], and (4) Alpha Particle X-ray Spectrometer (APXS) [4]. The IDD instruments are used to determine the chemistry and mineralogy of rocks and soils.

MIMOS II operates in backscatter geometry, detecting the reemitted 14.4 keV Mössbauer and 6.4 keV X-ray radiation. Because of the complexity of sample

Figure 1. NASA Mars Exploration Rover (MER), artist view (courtesy of NASA, JPL, Cornell). On the front side of the Rover, the robotic arm carrying the Mössbauer spectrometer and other instruments can be seen.

preparation, this is the choice for an *in situ* planetary Mössbauer instrument [1–3]. No sample preparation is required, the instrument is simply presented to the sample for analysis. Because of mission constrains for minimum mass, volume, and power consumption, the MIMOS II is extremely miniaturized (without loss in capability) compared to standard laboratory Mössbauer spectrometers and is optimized for low power consumption and high detection efficiency. Because of restrictions in data transfer rates, most instrument functions and data processing capabilities, including acquisition and separate storage of spectra as a function of temperature, are performed by an internal dedicated microprocessor and memory. High detection efficiency is extremely important in order to minimize experiment time. Experiment time is also minimized by using as strong a main 57Co/Rh source as possible. At landing the source intensities were about 150 mCi each. Physically, the MIMOS II Mössbauer spectrometer has two components, the sensor head (Figure 2) located at the end of the IDD, and the electronics printed-circuit board located in an electronics box inside the rover body. The sensor head contains the electromechanical transducer (mounted in the centre), the main and reference 57Co/Rh sources, multilayer radiation shields, detectors and their preamplifiers and main (linear) amplifiers, and a contact plate and contact sensor. The contact plate and sensor (Figure 3) are used in conjunction with the IDD to apply a small preload when it places the sensor

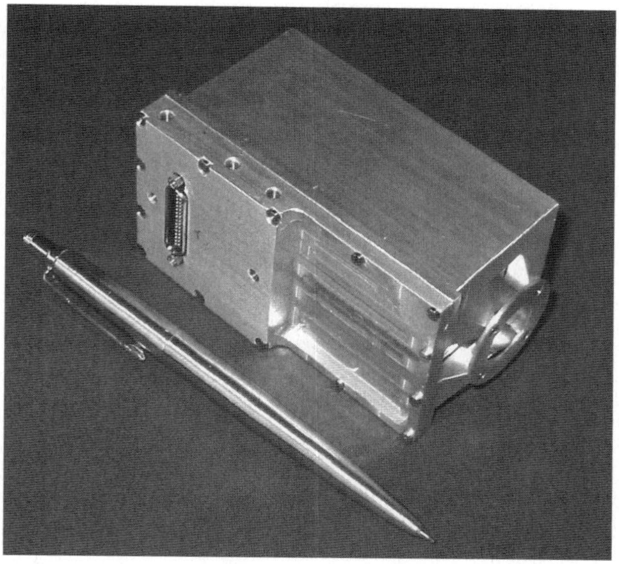

Figure 2. Mössbauer spectrometer MIMOS II sensor head (flight unit); dimensions about 45 × 50 × 95 mm; at the front side (left) the circular ring with an opening of 16 mm diameter is seen, determining the distance between sample surface and radiation detectors and carrying a temperature sensor.

head, holding it firmly against the target. The contact plate also carries a temperature sensor measuring the sample temperature allowing to perform temperature-dependent measurements.

Both spectrometers performed successfully and without any problems during the first 8 months of operation. Until August 2004, corresponding to about 240 Martian days of operations (the design goal of the mission was 90 days !), about 85 Mössbauer data sets at Gusev crater and about 100 data sets at Meridiani Planum have been acquired, respectively. Some selected data will be discussed in the following sections.

2. MER-A Spirit at Gusev crater

The Mars Exploration Rover Spirit landed at Gusev Crater, a large flat-floored crater with a diameter of about 160 km. The southern rim of Gusev is breached by Ma'adim Vallis, one of the largest branching valley networks on the planet. Sediments carried by the water that cut Ma'adim before it overflowed and exited through a gap in the northern rim of the crater, are expected to have settled on the crater floor. Exploration of this floor therefore was anticipated to offer opportunities to study sediments that were derived from the southern highlands of Mars.

Figure 3. Mössbauer spectrometer MIMOS II sensor head mounted at the robotic arm of the Mars Exploration Rovers. The circular contact plate (to the right of the center) can be seen, with an opening of 16 mm diameter. Picture has been taken on Mars, at the Spirit landing site right before taking the first Mössbauer spectrum on the Martian surface.

During the first 8 months of operation at the Gusev crater landing site the *Spirit* Mössbauer instrument analysed rocks and soils at Bonneville crater, a small crater close to the landing site, and the ejecta plains between Bonneville and the Columbia Hills. As determined by MB already in the very first spectrum (Figure 4) and confirmed by many others [7], the soils around Bonneville and in the cratered plains have a basaltic MB signature, and their Mössbauer spectra are dominated by an olivine doublet [7]. The MB results on rocks at the Gusev crater landing site show a primarily olivine–basalt composition. The presence of abundant olivine in rocks and in surrounding soil as determined by MB suggests that physical rather than chemical weathering processes currently dominate the plains at Gusev crater.

For some of the rocks a weathering rind has been detected using the RAT (brushing and abrading tool) and subsequently APXS and MIMOS II. In the outer layers of those rocks a significantly more intense Fe^{3+} doublet relative to the interior was found. Magnetite has been identified in both soils and rocks at Gusev. Comparing the soil spectrum (Figure 4) with the spectra of rocks (Figure 5; Adirondack, as a typical example) reveals a similar mineralogical composition, comprising of the dominant olivine Fe^{2+} doublet, a second silicate Fe^{2+} doublet, and less intense Fe^{3+} doublet. The Mössbauer parameters for the second Fe^{2+} doublet are consistent with pyroxene. The rock spectra show two small additional sextets, from octahedral and tetrahedral sites of the mineral

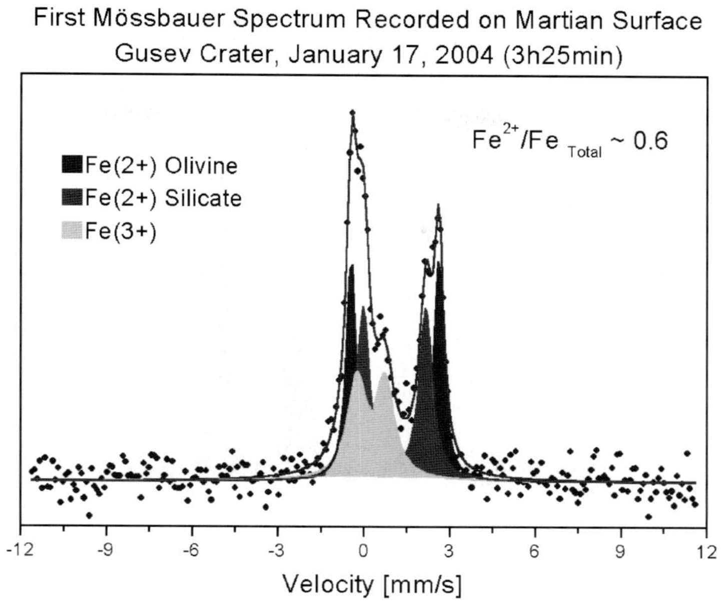

Figure 4. Very first Mössbauer spectrum ever taken on an extraterrestrial surface. The data from the Martian soil at the Spirit landing site were taken on sol (= Martian day) 13 of operations.

magnetite or low Ti magnetite. Magnetite is also detected in soil, but the pattern is not well defined because of inadequate counting statistics in the soils analyzed so far. We plan to have a longer integration on soil to improve the counting statistics and to determine the nature of the magnetite in soil. The average $Fe^{2+}/Fe_{(total)}$ ratio for the Gusev rocks is ~0.8.

3. MER-B opportunity at Meridiani Planum

The Meridiani Planum landing site, which is opposite to Gusev crater on the other side of the planet, shows from Orbit a smooth and flat topography which would support a save landing. Also the Mars Global Surveyor Orbiters TES data indicated the presence of ~15–20% by fractional area of the mineral hematite. The hematite-bearing unit is the top stratum of a layered sequence approximately 600 m thick that overlies Noachian cratered terrain. Hematite very often (but not always) forms in the presence of water which directed the MER-B rover to this place.

The Meridiani Planum landing site looks very different from Gusev crater. Opportunity landed inside a shallow crater named Eagle crater, with an outcrop covering part of the crater interior close to the rim. In contrast to Gusev crater the surface within Eagle and the surrounding plains are not covered with any rock debris. The surface is very flat and smooth with some wind ripples on the top.

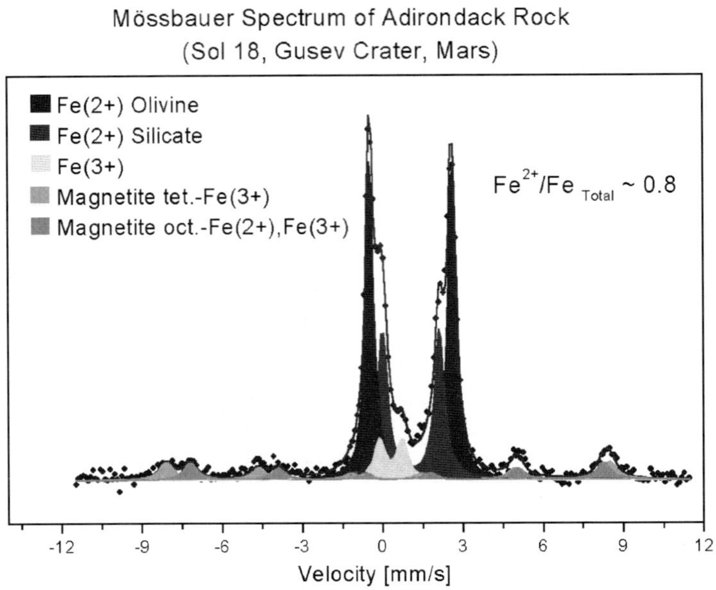

Figure 5. Mössbauer spectrum of the rock Adirondack at Spirit landing site. The data taken before and after brushing the surface are identical with the data taken after subsequent abrading of 3–4 mm of the surface material. It shows an olivine–basalt composition consisting of the minerals olivine, pyroxene, an Fe^{3+} doublet, and nonstoichiometric magnetite.

The composition of the soil at Meridiani is found by Mössbauer spectroscopy to be basaltic, dominated by olivine similar to the Gusev site.

Mössbauer measurements (Figure 6) show that the outcrop material consists predominantly of the Fe sulfate jarosite, hematite, and a basaltic component (olivine, pyroxene) (Figure 7). The same material was found again a couple of hundred meters away at craters Fram and Endurance suggesting that the whole area is covered with this jarositic material. The mineral jarosite ($(K,Na)Fe_3(SO_4)_2(OH)_6$) contains hydroxyl and is thus direct mineralogical evidence for aqueous, acid–sulfate alteration process on early Mars. This strongly supports the presence of large amounts of water at this site in the past.

The plains and a large portion of Eagle crater are covered by spherules, nicknamed Blueberries, with a diameter of several mm. Mössbauer data clearly show that the composition of these spherules is dominated by the Fe oxide hematite [6]. The composition of the soil at Meridiani is found to be basaltic, dominated by olivine similar to the Gusev site, but in some areas the soil is enriched in the mineral hematite. The reason for this is not clear yet. The $Fe^{2+}/Fe_{(total)}$ ratio at Meridiani Planum, as determined so far, varies from $\sim$0.8 for the basaltic soil to $\sim$0.1 for the outcrop.

The Morin transition of hematite is observed in the Mössbauer spectra of the hematite-rich samples, as shown by the outcrop spectra in Figure 7, by the

Figure 6. Mössbauer spectrometer MIMOS II sensor head placed on the outcrop at Eagle crater, landing place at Meridiani Planum, Mars. The robotic arm can be seen extending out to the outcrop, as well as the two front wheels of the rover Opportunity.

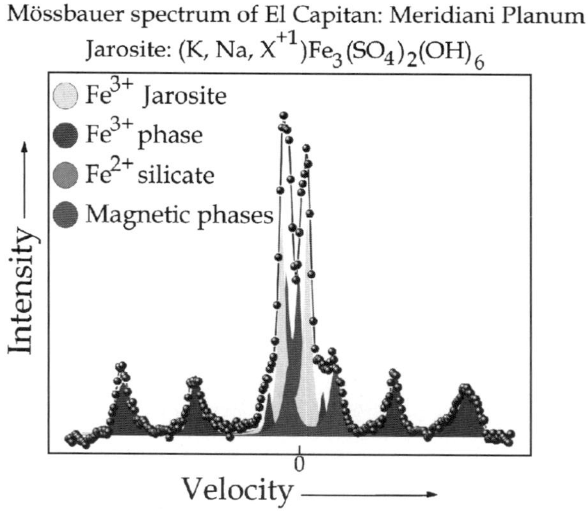

Figure 7. Mössbauer spectrum of the outcrop target 'El Capitan' at Eagle crater landing site. The velocity range is ±11 mm/s. The minerals Jarosite, hematite (magnetic phase), an Fe^{2+} doublet (probably pyroxene), and a Fe^{3+} doublet have been identified.

difference in peak positions of the sextets at temperatures above and below 250 K. These spectra were obtained utilizing the diurnal temperature variation on Mars and the temperature binning capability of the Mössbauer spectrometer [1]. This indicates that hematite is well crystalline.

4. Outlook

The Mars Exploration Rovers and their instrumentations are operating now for more than 10 months on the surface of Mars. They still behave very well. It is expected that the MERs and their instruments will operate successfully on Mars for a couple of more months. They will continue driving towards areas of geological and mineralogical interest and will continue to collect more Mössbauer and other data.

Acknowledgement

This project was funded by the German Space Agency DLR under contract 50QM9902.

References

1. Klingelhöfer G., Morris, R. V., Bernhardt B., Rodionov D., de Souza P. A. Jr., Squyres S. W., Foh J., Kankeleit E., Bonnes U., Gellert R., Schröder Ch., Linkin S., Evlanov E., Zubkov B. and Prilutski O., *J. Geophys. Res.* **108**(E12) (2003), 8067.
2. Klingelhöfer G., In: Miglierini M. and Petridis D. (eds.), *Mössbauer Spectroscopy in Materials Science*, Kluwer Academic Publishers, Dordrecht, 1999.
3. Klingelhöfer G., *Hyperfine Interact.* **113** (1998), 369–374.
4. Squyres S. W. *et al.*, *J. Geophys. Res.* **108**(E12) (2003), 8062.
5. Several papers in special volume of: Science, vol. 305 (6 August 2004).
6. Klingelhöfer G., Morris R. V., Bernhardt B., Schröder C., Rodionov D. S., de Souza P. A. Jr., Yen A., Gellert R., Evlanov E. N., Zubkov B., Foh J., Bonnes U., Kankleit E., Gütlich P., Ming D. W., Renz F., Wdowiak T., Squyres S. W. and Arvidson R. E., *Science* **306** (3 December 2004), 1740–1745.
7. Morris R. V., Klingelhöfer G., Bernhardt B., Schröder C., Rodionov D. S., de Souza P. A. Jr., Yen A., Gellert R., Evlanov E. N., Foh J., Kankleit E., Gütlich P., Ming D. W., Renz F., Wdowiak T., Squyres S. W. and Arvidson R. E., *Science* **305** (6 August 2004), 833–836.

Hyperfine Interactions (2004) 158:125–129
DOI 10.1007/s10751-005-9020-8

Temperature Dependence of the Magnetic Hyperfine Field at ^{140}Ce on Gd Sites in GdAg Compound

F. H. M. CAVALCANTE, A. W. CARBONARI*, R. N. SAXENA and J. MESTNIK-FILHO
Instituto de Pesquisas energéticas e Nucleares - IPEN-CNEN/SP; e-mail: carbonar@ipen.br

Abstract. Perturbed gamma–gamma angular correlation technique was used to measure the magnetic hyperfine field at Gd sites in the intermetallic compound GdAg using the ^{140}La→^{140}Ce nuclear probe. A major and well-defined magnetic interaction is observed at ^{140}Ce substituting Gd sites in GdAg below 130 K, corresponding to a ferromagnetic ordering of Gd moments. The temperature dependence of magnetic hyperfine field, however, shows a sharp deviation from an expected Brillouin-like behavior for temperatures below 75 K. This additional magnetic interaction is believed to result from the polarization of Ce spin moments induced by the magnetic field from Gd atoms.

Key Words: ^{140}Ce, magnetic hyperfine field, PAC spectroscopy, rare earth magnetism.

1. Introduction

The magnetic hyperfine field (mhf) at rare earth atom sites in intermetallic compounds is still a very interesting subject specially if the probe nuclei are rare earth elements where the orbital contribution to the mhf can be important. For closed shell probe nuclei as impurity in rare earth intermetallic compounds, the magnetic hyperfine field is proportional to the conduction electron spin polarization (CEP) at the probe site, as has been shown in literature for several compounds using, for instance, ^{111}Cd. In the case of rare earth probe nuclei, the 4f electrons are expected to contribute significantly to the magnetic hyperfine field. However, there are only few experimental results available for rare earth probe nuclei in rare earth magnetic compounds to permit a systematic investigation of the mhf contribution due to 4f electrons. Moreover, it is not well understood yet whether the influence of 4f electrons of the probe on the magnetic hyperfine field in rare earth compounds is always present. In this work we have investigated the mhf at ^{140}Ce probe nuclei substituting Gd sites of the cubic intermetallic compound GdAg in order to quantify the 4f contribution and

* Author for correspondence.

compare the results with previous measurements of mhf on ^{140}Ce in other rare earth compounds.

The cubic rare-earth intermetallic compounds of the type RM where M is a noble metal have shown interesting magnetic properties. These compounds usually show antiferromagnetic behavior when M is a monovalent metal Cu or Ag, the exception being PrAg. The antiferromagnetic (π,π,0) structure in these compounds is built up by ferromagnetic (1,1,0) planes coupled antiferromagnetically. The presence of phenomena like quadrupole interactions and crystal field splitting, spin fluctuations and incommensurate magnetism can turn the magnetic behavior in these compounds very complex. The magnetism in GdAg is less complex, however. Quadrupolar and crystal field effects are absent in this compound which also has the highest Néel temperature in the family of compounds with T_N ~132.8 K [1].

2. Experimental

Samples of GdAg were prepared by repeatedly melting the constituent elements (Gd 99.99% and Ag 99.9985%) under argon atmosphere purified with a hot titanium getterer in an arc furnace. The samples used for TDPAC measurements were prepared in a similar way but with radioactive ^{140}La (obtained by neutron irradiation of lanthanum metal) substituting about 0.1% of Gd atoms. Samples were annealed under an atmosphere of ultra pure Ar for 48 h at 700 °C. The structure of the samples was checked by X-ray diffraction, which indicated a single phase and the cubic CsCl-type structure with the $Pm\bar{3}m$ space group. Magnetization measurements were carried out in the temperature range of 4.2–200 K using a superconductor quantum interference device (SQUID).

The TDPAC measurements were carried out with a conventional fast-slow coincidence set-up with four conical BaF_2 detectors. The gamma cascade of 329–487 keV populated from the decay of ^{140}La with an intermediate level with spin I = 4^+ at 2083 keV ($T_{1/2}$ = 3.45 ns) in ^{140}Ce was used to measure the magnetic hyperfine field at Gd sites. The samples were measured in the temperature range of 10–295 K by using a closed-cycle helium cryogenic device. The time resolution of the system was about 0.6 ns for the ^{140}Ce gamma cascade. A detailed description of the method can be found elsewhere [2, 3]. The experimental data for temperatures below T_N were analyzed for a pure magnetic dipole interaction.

3. Results and discussion

Some of the TDPAC spectra measured with ^{140}Ce probe nuclei are shown in Figure 1. The solid curves are the least squares fit of the experimental data to the appropriate function in each case. The quadrupole moment of the 2083 keV 4^+ state of ^{140}Ce is known to be very small [4], consequently one expects to observe

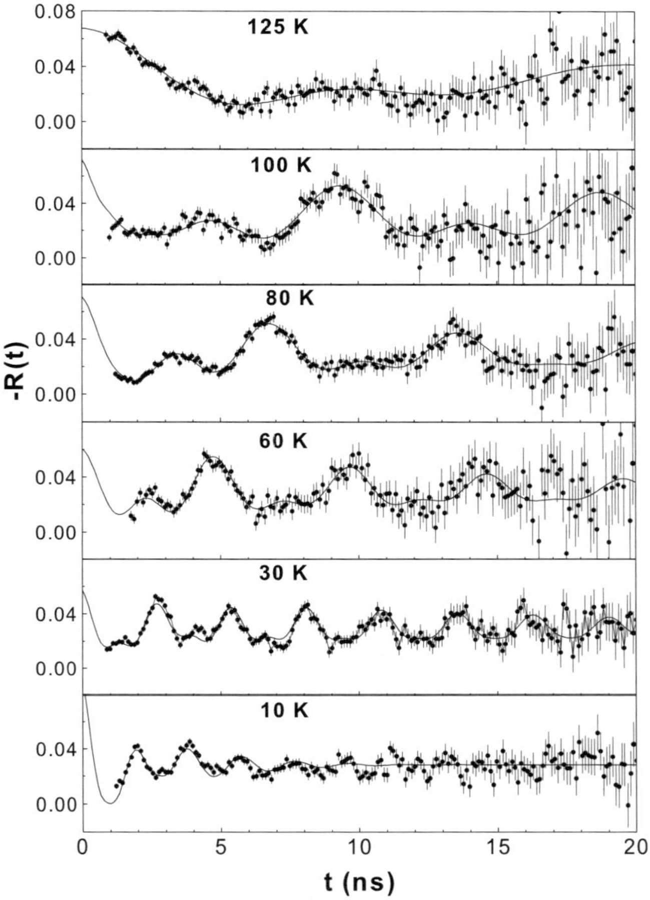

Figure 1. TDPAC spectra for ^{140}Ce at Gd sites in GdAg.

an almost pure magnetic dipole interaction in the antiferromagnetic phase of the sample. Below 130 K, a unique and well-resolved magnetic interaction is observed at ^{140}Ce in GdAg. The temperature dependence of B_{hf} is plotted in Figure 2. The observed magnetic interaction corresponds to the antiferromagnetic ordering of the Gd moments. The measurements below ~75 K, however, show a sharp deviation from the expected standard behavior for a simple antiferromagnetic ordering. The Ce hyperfine field, instead of approaching a saturation value, increases sharply at lower temperatures. The measured hyperfine field at 80 K is 20 T while at 10 K it is 73 T. The five points immediately below T_N in Figure 2 were least-square fitted to the modified Curie–Weiss law $B_{hf}(T) = B_{hf}(0)(1 - T/T_N)^{\beta}$. The results of this fitting yielded β = 0.41(2), B_{hf} (0) = 25(1)T, T_N = 131.5(3) K. The value of T_N is somewhat smaller while the value of β is higher than the values of 132.8(1) K and 0.33, respectively obtained from neutron diffraction measurements [1]. It is well known that neutron diffraction data for magnetic materials are affected by

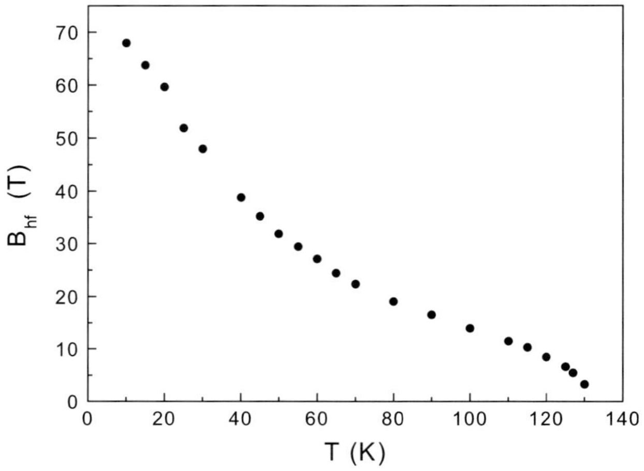

Figure 2. Temperature dependence of the magnetic hyperfine field at Gd sites.

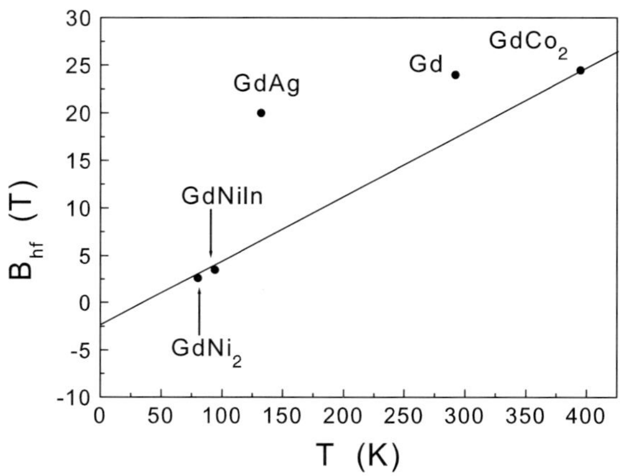

Figure 3. The extrapolated magnetic hyperfine field $B_{hf}(0)$ at ^{140}Ce in some Gd compounds as a function of the respective magnetic transition temperatures.

the temperature dependent extinction effects, which cause a reduction in the critical exponent value. The present value of β is higher than the theoretical value expected for a three-dimensional isotropic Heisenberg critical exponent ($\beta \approx 0.38$).

The explanation for the additional magnetic interaction is similar to that given for the temperature dependence of the B_{hf} at ^{140}Ce in $CeMn_2Ge_2$ compound [5]. The additional field is believed to result from the polarization of Ce spin moments induced by the magnetic field from Gd moments. Previous measurements of the mhf at ^{140}Ce in Gd [6] also shows a similar behavior for the temperature dependence of B_{hf}, with a sharp deviation from the standard magnetization curve below a certain temperature.

In [7] the saturation values of B_{hf} measured with ^{140}Ce for some Gd compounds are compared to the respective magnetic transition temperature. The linear dependence obtained shows that the main contribution to the B_{hf} comes from the conduction electron spin polarization via RKKY interaction. In order to insert the points for GdAg and Gd in such comparison, we have estimated the saturation values by visually extrapolating the B_{hf} curve from the points between the temperature where the deviation starts and the ordering temperature, e.g., we have disregarded the probe effect from the magnetization curve. The results are shown in Figure 3 and indicate that both GdAg and Gd do not follow the same proportionality. Therefore, we can assume that the B_{hf} for these compounds have additional contributions besides those from the conduction electrons spin polarization, even for the temperature region free of the probe effect.

Acknowledgements

Partial support for this research was provided by the Fundação de Amparo á Pesquisa do Estado de São Paulo (FAPESP). Authors FHMC and AWC thankfully acknowledge the support provided by CNPq in the form of research fellowships.

References

1. Chattopadhyay T., McIntyre G. J. and Kiiblerb U., *Solid State Comm.* **100** (1996), 117.
2. Pendl W. Jr., Saxena R. N., Carbonari A. W., Mestnik-Filho J. and Schaft J., *J. Phys. Condens. Matter.* **8** (1996), 11317.
3. Attili R. N., Saxena R. N., Carbonari A. W., Mestnik-Filho J., Uhrmacher M. and Lieb K. P., *Phys. Rev. B* **58** (1998), 2563.
4. Królas K., Wodniecka B. and Niewodniczanski H., Institute of Nuclear Physics, Kraków, Poland, Report No. 1644/OS-1993 (unpublished).
5. Carbonari A. W., Mestnik-Filho J., Saxena R. N. and Lalic M. V., *Phys. Rev. B* **69** (2004), 1444251.
6. Thiel T. A., Gerdau E. and Böttcher M., *Hyperfine Interact.* **9** (1981), 459.
7. Lapolli A., Carbonari A. W., Saxena R. N. and Mestnik-Filho J., this conference.

Hyperfine Interactions (2004) 158:131–136
DOI 10.1007/s10751-005-9021-7

Thermal Behaviour of the μSR Relaxation Rate at High Temperature in Insulators

P. DALMAS DE RÉOTIER[1,*], A. YAOUANC[1], P. C. M. GUBBENS[2], S. SAKARYA[2], E. JIMENEZ[3], P. BONVILLE[4] and J. A. HODGES[4]
[1]*Commissariat à l'Energie Atomique, DRFMC, Grenoble, France; e-mail: pdalmas@cea.fr*
[2]*Technische Universiteit Delft, IRI, Delft, The Netherlands*
[3]*Universidad Complutense de Madrid, Fac. C. Quimicas, Madrid, Spain*
[4]*Commissariat à l'Energie Atomique, DRECAM, Saclay, France*

Abstract. We show that, in rare earth based insulators, measurement of the thermal dependence of the muon spin-lattice relaxation rate at high temperature provides information on the nature of the magnetic correlations and on the crystal-field energy splitting, if any.

Key Words: magnetic correlations, μSR spin-lattice relaxation, Orbach process.

1. Introduction

The spin dynamics in rare earth based insulators at high temperature is expected to be controlled by two physical mechanisms. One is associated with the exchange interaction between the rare earth spins (mutual spin-flips) and the second is due to the magneto-elastic coupling between these spins and the phonons (single spin-flip with absorption/emission of phonons). The muon spin relaxation (μSR) technique is a well adapted tool to probe the spin dynamics in a broad temperature range.

We provide here the framework for interpreting μSR data recorded at high temperature, where the thermal energy is bigger than the characteristic exchange energy in the system. As an illustration, we present experimental data that we have recently obtained in geometrically frustrated magnetic systems. Our interpretative framework is not intended to specifically address the issues of geometrical magnetic frustration, which are actually most easily observed at low temperature [8]. It holds for any magnetic insulator well above the magnetic transition temperature, if any.

* Author for correspondence.

2. Theoretical framework for the muon relaxation rate

We consider the case where the zero-field μSR spectra are well accounted for by an exponential relaxation function characterised by a decay (or relaxation) rate λ_Z. When a magnetic compound is cooled down from a temperature high enough so that the thermal energy is large compared to the exchange energy, short-range pair-correlations are expected to progressively build up. They were first detected by nuclear magnetic resonance as an increase (decrease) of the spin-lattice relaxation rate for a ferromagnetic (antiferromagnetic) compound as it is cooled down and the pair correlations become increasingly important [9]. This is qualitatively understood as ferromagnetic (antiferromagnetic) correlations correspond to the enhancement (reduction) of the local field at the nuclear probe.

In the limit $T \to \infty$, there is no correlation between different spins. At finite temperature, one can expand the static correlation function of the total momentum J of the rare earth $\left\langle \delta J_i^\alpha \delta J_j^\beta \right\rangle$ to first order in $\mathcal{J}/(k_B T)$, where $\mathcal{J}$ is the exchange constant [3, 9]:

$$\left\langle \delta J_i^\alpha \delta J_j^\beta \right\rangle = \delta_{\alpha,\beta} \frac{2\mathcal{J}}{k_B T} \left[\frac{1}{3} J(J+1) \right]^2. \tag{1}$$

A cubic compound is assumed for simplicity. A key feature is evident from this formula: the pair-correlation depends on the sign of $\mathcal{J}$, being positive (negative) for ferromagnetic (antiferromagnetic) exchange. Taking into account the short-range correlations to first order in $\mathcal{J}/(k_B T)$, λ_Z can be written [3]:

$$\lambda_Z(T) = \lambda_Z^{(ss)} \left(1 + \frac{T_0}{T} \right), \tag{2}$$

where $\lambda_z^{(ss)}$ is the value of the muon spin-lattice relaxation rate in the high temperature limit, where cross-correlations vanish, and T_0 is a characteristic temperature, the sign and order of magnitude of which are those of the paramagnetic Curie–Weiss temperature θ_{CW}. The absolute value of θ_{CW} is proportional to the exchange constant $\mathcal{J}$, its sign is positive for ferromagnets and negative for antiferromagnets. It must be emphasized that $|T_0|$ is not only a function of the pair–correlation, but that it depends drastically on the rare earth spin configuration around the muon, and that it may even vanish. To illustrate this point, let us assume that the muon interacts only with near neighbour spins. If there is only one such spin nearby the muon, pair correlations cannot influence the muon spin and $T_0 = 0$. On the other hand, if two magnetic ions interact with the muon spin, $T_0 \neq 0$.

In addition to exchange driven spin relaxation, an additional mechanism can be operative in rare earth systems with well-defined crystal-field energy levels: the Orbach process, a two-phonon real process *via* an excited crystal-field level [2, 7]. The associated inverse muon relaxation rate is:

$$\lambda_Z^{-1} = B_m \exp\left(\frac{-\Delta_e}{k_B T}\right), \tag{3}$$

where B_m is a magnetoelastic coupling constant of the ionic spin with the phonon bath.

Combining the exchange and phonon driven relaxation mechanisms, we expect the inverse muon relaxation rate λ_Z^{-1} for an insulator to be given by the expression:

$$\lambda_Z^{-1} = \left[\lambda_Z^{(ss)}\left(1 + \frac{T_0}{T}\right)\right]^{-1} + B_m \exp\left(\frac{-\Delta_e}{k_B T}\right). \tag{4}$$

3. The spin-lattice relaxation rate data

We now present experimental results obtained in four geometrically frustrated systems: $Gd_2Ti_2O_7$, $Gd_2Sn_2O_7$, $Yb_2Ti_2O_7$ and $Yb_3Ga_5O_{12}$. The data have been recorded at the Swiss Muon Source and at the ISIS muon facility in UK.

θ_{CW} for $Gd_2Ti_2O_7$, $Gd_2Sn_2O_7$ and $Yb_2Ti_2O_7$ are, respectively, -9.9 (1) K, -8.6 (1) K [1] and 0.75 K [5]. That for $Yb_3Ga_5O_{12}$ is very small and consequently its sign is difficult to determine. However, at low temperature, the susceptibility deviates from the Curie–Weiss behaviour extrapolated from high temperature in such a way that the presence of antiferromagnetic interactions is suggested [2]. Both $Gd_2Ti_2O_7$ and $Gd_2Sn_2O_7$ display an antiferromagnetic phase transition at about 1 K [1]. The Yb-based compounds are characterised by a specific heat anomaly at low temperature (0.25 and 0.054 K for $Yb_2Ti_2O_7$ and $Yb_3Ga_5O_{12}$, respectively) but do not show long range magnetic order [2, 5].

Figures 1 and 2 show the thermal dependence of λ_Z together with fits to Equation (4). Our model, which describes the onset of pair correlations, is derived from an expansion of the correlation function with respect to the ratio of the exchange and thermal energies. It is therefore expected to break down when the temperature is low enough ($k_B T \sim \mathcal{J}$), where critical dynamics prevails. The fits to Equation (4) were therefore performed in the temperature range above ~10 K, except for $Yb_3Ga_5O_{12}$ (above 0.3 K). Table I gathers the values obtained for the various parameters.

For $Gd_2Ti_2O_7$ and $Gd_2Sn_2O_7$, no Orbach process was included in the fit ($B_m = 0$) because the Gd^{3+} ion has a vanishingly small crystal field splitting. The antiferromagnetic nature of the exchange interaction in these compounds is seen in the slight decrease of $\lambda_Z(T)$ on cooling. The corresponding T_0 values are therefore negative. Below 10 K, λ_Z increases as the critical regime is approached.

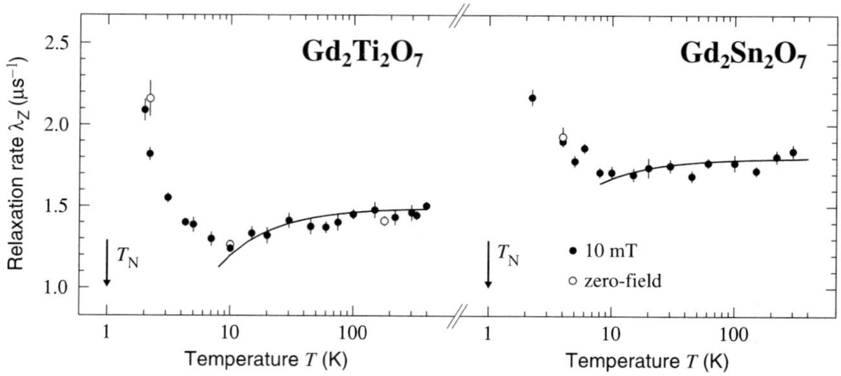

Figure 1. Relaxation rate λ_Z measured in $Gd_2Ti_2O_7$ and $Gd_2Sn_2O_7$ using zero- or longitudinal-field μSR. To emphasize the slight temperature dependence observed deep in the paramagnetic phase we show only data for $T \geq 2$ K. Below 2 K, λ_Z continues to rise to reach a maximum at T_N [10] as expected when approaching a phase transition. The lines are fits to Equation (4). They point to the presence of antiferromagnetic pair correlations.

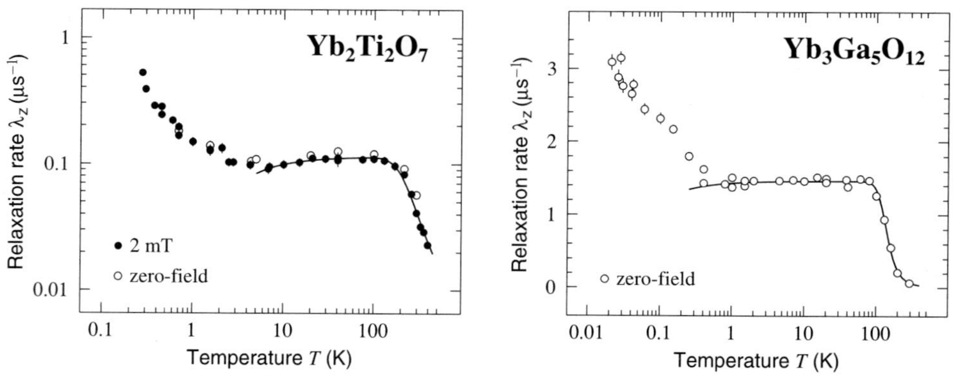

Figure 2. Relaxation rate λ_Z measured in $Yb_2Ti_2O_7$ and $Yb_3Ga_5O_{12}$ using zero- or longitudinal-field μSR. The lines are fits to Equation (4). Above ~100 K, they reflect the presence of crystal-field levels and at lower temperature the onset of antiferromagnetic correlations. In the case of $Yb_2Ti_2O_7$, only the 2 mT data have been used for the fit.

Table I. Parameters obtained from the fits shown in Figures 1 and 2.

Compound	T_0 (K)	$\lambda_Z^{(ss)}$ (μs^{-1})	Δ_e/k_B (K)	B_m (μs)
$Gd_2Ti_2O_7$	−2.0 (2)	1.49 (3)	–	–
$Gd_2Sn_2O_7$	−0.7 (2)	1.80 (5)	–	–
$Yb_2Ti_2O_7$	−1.4 (2)	0.115 (2)	900 (50)	310 (80)
$Yb_3Ga_5O_{12}$	−0.02 (1)	1.46 (5)	825 (15)	215 (20)

The absence of crystal field splitting in the Gd compounds implies that relaxation by the Orbach process is ineffective ($B_m = 0$ in Equation (4)).

In $Yb_2Ti_2O_7$ and $Yb_3Ga_5O_{12}$, $\lambda_Z(T)$ also points to antiferromagnetic interactions. As expected from macroscopic measurements, T_0 is found to be close to 0 in the latter compound. As to $Yb_2Ti_2O_7$, the results are surprising. The sign found for T_0 is opposite to that of θ_{CW}. An explanation for this discrepancy could arise from the (unknown) muon localisation site. This site could be such that only specific pair correlations are probed by μSR, whereas θ_{CW}, derived from the macroscopic susceptibility, is sensitive to the sum of the exchange interactions. We recall at this point that frustrated systems are essentially characterised by competing exchange interactions.

Turning now to phonon induced relaxation, the fits give for the relevant crystal field energies: Δ_e/k_B = 900 (50) and 825 (15) K for $Yb_2Ti_2O_7$ and $Yb_3Ga_5O_{12}$ respectively. The combination of results obtained from ^{170}Yb Mössbauer spectroscopy, ^{172}Yb perturbed angular correlation, magnetisation and susceptibility measurement has allowed to establish the crystal-field level scheme of Yb^{3+} in $Yb_2Ti_2O_7$. It consists of three doublets located at 620, 740 and 950 K above the ground state doublet [4]. Using a similar procedure, it was determined that the three excited doublets of Yb^{3+} in $Yb_3Ga_5O_{12}$ are closely spaced and lie at an average energy corresponding to 850 K above the ground state [6]. The values found for Δ_e/k_B in the two Yb compounds compare therefore very well with the known crystal field levels. Consequently we conclude that the marked drop observed in λ_Z above ~100 K is the signature of a muon spin relaxation associated with a two-phonon real process through the excited crystal field states (Orbach process).

In conclusion, we have shown that, for the magnetic insulators of interest here, two physical mechanisms allow to interpret the thermal dependence of the μSR relaxation rate in a high temperature regime compared to the exchange energy. First, the building up of spin pair correlations is characterised by an increase (decrease) of λ_Z when lowering the temperature in the case of ferromagnetic (antiferromagnetic) interactions. The nature of the pair correlations which are probed by the muon, e.g. between nearest neighbours or next nearest neighbours, depends on the muon localisation site. Second, the presence of crystal field levels induces a marked temperature dependence of λ_Z at high temperature.

References

1. Bonville P., Hodges J. A., Sanchez J. P., Vulliet P., Sosin S. and Braithwaite D., *J. Phys. Condens. Matter.* **15** (2003), 7777.
2. Dalmas de Réotier P., Yaouanc A., Gubbens P. C. M., Kaiser C. T., Baines C. and King P. J. C., *Phys. Rev. Lett.* **91** (2003), 167201.
3. Hartmann O., Karlsson E., Wappling R., Chappert J., Asch A. Y. L. and Kalvius G. M., *J. Phys. F. Metal Phys.* **16** (1986), 1593.
4. Hodges J. A., Bonville P., Forget A., Rams M., Królas K. and Dhalenne G., *J. Phys. Condens. Matter.* **13** (2001), 9301.

5. Hodges J. A., Bonville P., Forget A., Yaouanc A., Dalmas de Réotier P., André A., Rams M., Królas K., Ritter C., Gubbens P. C. M., Kaiser C. T., King P. J. C. and Baines C., *Phys. Rev. Lett.* **88** (2002), 077204.
6. Hodges J. A, Bonville P., Rams M. and Królas K., *J. Phys. Condens. Matter.* **15** (2001), 4631.
7. Orbach R., *Proc. Phys. Soc. Lond. Sect. A* **264** (1961), 458.
8. Ramirez A. P., In: Buschow K. H. J. (ed.), *Handbook of Magnetic Materials*, Vol 13, Elsevier, 2001.
9. Silbernagel B. G., Jaccarino V., Pincus P. and Wernick J. H., *Phys. Rev. Lett.* **20** (1968), 1091.
10. Yaouanc A., Dalmas de Réotier P., Bonville P., Hodges J. A., Gubbens P. C. M., Kaiser C. and Sakarya S., *Physica. B* **326** (2003), 456.

Hyperfine Interactions (2004) 158:137–143
DOI 10.1007/s10751-005-9022-6

Magnetic Texturing of Xenon-Irradiated Iron Films Studied by Magnetic Orientation Mössbauer Spectroscopy

G. A. MÜLLER[1,*], K. P. LIEB[1], E. CARPENE[1], K. ZHANG[1], P. SCHAAF[1], J. FAUPEL[2] and H. U. KREBS[2]
[1]*II. Physikalisches Institut and SFB 602, Universität Göttingen, D-37077 Göttingen, Germany; e-mail: g.a.mueller@gmx.de*
[2]*Institut für Materialphysik, Universität Göttingen, D-37077 Göttingen, Germany*

Abstract. Modifications of magnetic properties upon heavy-ion irradiation have been recently investigated for films of ferromagnetic 3d-elements (Fe, Ni, Co) and alloys (permendur, permalloy), in relation to changes of their microstructure. Here we report on Xe-ion irradiation of a highly textured iron film prepared via pulsed-laser deposition on a MgO(100) single crystal and containing a thin ^{57}Fe marker layer for magnetic orientation Mössbauer spectroscopy (MOMS). We compare the results with those obtained for a polycrystalline Fe/Si(100) sample produced by electron evaporation and premagnetized before Xe-irradiation in a 300 Oe external field. Characterization of the samples also included magneto-optical Kerr effect (MOKE), Rutherford backscattering spectroscopy (RBS) and X-ray diffraction (XRD).

Key Words: ion implantation, magnetic anisotropy, magneto-optical Kerr effect, Mössbauer spectroscopy, RBS, XRD.

1. Introduction

The high energy density of tens of keV per nm, which an energetic heavy ion deposits near the end of his range, produces strong radiation damage and thermal spikes and that can significantly modify the structural and magnetic properties of thin ferromagnetic films. In a comprehensive study, we recently reported on ion-induced magnetic texturing effects in Ni [1–4], Co [5], $Ni_{19}Fe_{81}$ (permalloy [6]), FeCo (per-mendur [7]) films and Co/Fe bilayers [8] irradiated with ions of Ne, Fe, Kr, Xe or Au. Inverse magnetostriction (Ni, permendur [2, 3, 6]) and ion-induced phase transfor mations (Co-hcp → Co-fcc in Co [5] and Co/Fe [8]) were observed.

The present experiments were carried out in order to investigate what is the influence of the defect structure and the pre-magnetization on the magnetic anisotropy of iron films *before* the irradiation. To this end, we used an epitaxial

*Author for correspondence.

Fe film prepared by pulsed-laser deposition on a MgO single crystal, and a polycrystalline Fe-film prepared by electron-beam evaporation and pre-magnetized before irradiating it with Xe-ions. Comparison with other ions will also be made [9, 10].

2. Experiments

A 49 nm thick Fe-film, 10×10 mm^2 in size, was prepared by pulsed laser deposition on a MgO(100) substrate, at a growth rate of 3.4 pm per pulse, using a KrF laser with a wavelength of 248 nm and a pulse duration of 30 ns. The base pressure in the chamber was 10^{-9} mbar. The film of natural isotope composition was interlayered with a 15 nm thin ^{57}Fe film (isotope enrichment 95%) in order to have access to depth-sensitive Mössbauer spectroscopy, either in the usual conversion electron Mössbauer spectroscopy (CEMS) geometry with the γ-beam hitting the sample at normal incidence, or in the MOMS geometry [11, 12] with the γ-beam hitting the sample at $\alpha = 45°$. MOMS has the advantage that it provides information on the magnitude and orientation of the hyperfine field(s) as function of the depth and without applying an external magnetic field (like for MOKE), which may disturb the domain structure.

The as-deposited film was checked by XRD and RBS (using the 900-keV beam of α-particles provided by the Göttingen implanter IONAS [13]) and was magnetically analysed by MOMS [12] and MOKE [9]. The sample was then irradiated homogeneously at room temperature with 1×10^{16} Xe-ions/cm^2 of 150 keV, in an external magnetic field of 104 Oe, which was oriented along one of the sample side ($\varphi = 0°$). At this ion energy, the Xe-ions are deposited in the middle of the Fe-film thus avoiding mixing at the Fe/substrate interface [14, 15] as confirmed by RBS. After ion irradiation, the films were again examined by RBS, MOMS, MOKE and XRD. For details see [9].

A similar procedure was applied to a 60 nm thick 10×7 mm^2 *polycrystalline* Fe film deposited via electron evaporation on Si(100) and containing a 13 nm ^{57}Fe marker layer close to the surface. After deposition, the film was checked by RBS, XRD, MOMS and MOKE and was then magnetized along the $\varphi \approx 100°$ axis in an external field of 300 Oe, which was sufficient to saturate the film. The film was re-analysed with MOMS (but not with MOKE) and then irradiated with 1×10^{16} Xe-ions/cm^2 of 200 keV. Finally, MOMS and MOKE analyses were carried out again in this order to check the effect of ion irradiation.

3. Results for Fe/MgO

Figures 1(a,b) show the XRD rocking curves of the Fe(200) reflex before (a) and after (b) irradiation, while Figure 1(c) illustrates the XRD φ-scan of the Fe(211) and MgO(311) reflexes after irradiation. One notes that the already very sharp Fe(211) reflex after deposition (FWHM = 1.03°) narrowed to FWHM = 0.61° after the Xe-irradiation. Furthermore, a slight decrease of the lattice constant by

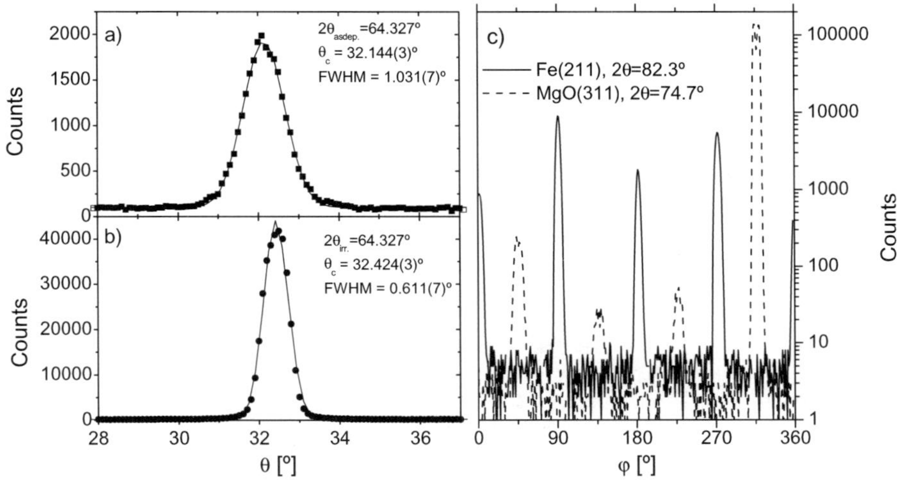

Figure 1. θ–2θ-rocking curves of the Fe(200) reflex for the sample Fe/MgO, measured before (a) and after (b) Xe-ion irradiation; φ-scans of the Fe(211) and MgO(311) reflexes after irradiation (c).

0.01 Å was visible indicating a relaxation of the compressive stress in the as-deposited film by the Xe-ions [16]. As expected the (100) direction of the Fe-film was oriented along the (110) direction of the MgO substrate, in agreement with a strong in-plane texture in the Fe-layer, producing an easy axis of magnetization along the angles $\varphi_0 = 45°$ and 135°.

The results obtained from the MOMS and MOKE analyses are illustrated in Figure 2, as-deposited (left) and irradiated (right). The angle φ here denotes the orientation of the sample in the MOMS apparatus [11]. Note that the MOMS and MOKE analyses were carried out in this order, i.e., first the as-deposited state was characterized by MOMS, in the absence of any external magnetic field. As to the interpretation of these data, the parameter I_2/I_3 denotes the intensity ratio of the second and third Mössbauer line corresponding to ^{57}Fe nuclei on substitutional, defect-free lattice sites. In the case of a fraction c_a of spins aligned in the direction ψ_a, a fraction c_b aligned in the direction ψ_b and a third out-of-plane fraction c_{op}, the angle dependence of the parameter I_2/I_3 for $\alpha = 45°$ follows the equation [11, 12]

$$I_2/I_3 = 4c_a\left[1 - 0.5\sin^2(\varphi - \psi_a)\right]/\left[1 + 0.5\sin^2(\varphi - \psi_a)\right] + 4c_b\left[1 - 0.5\sin^2(\varphi - \psi_b)\right]/\left[1 + 0.5\sin^2(\varphi - \psi_b)\right] + 4/3c_{op} \tag{1}$$

Figure 2 (left) illustrates the effects of *magnetization* in the as-deposited film. The ratio I_2/I_3 was nearly isotropic indicating that almost all the ^{57}Fe atoms had their spins oriented in-plane, but otherwise distributed isotropically ($c_a \approx c_b \approx 0.5$). The MOMS data taken after the Xe *irradiation* showed that 83% of the

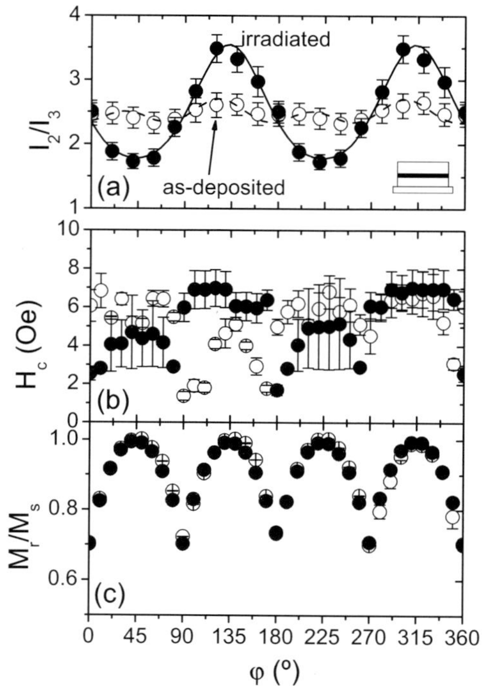

Figure 2. MOMS (a) and MOKE (b,c) analyses of the sample Fe/MgO in the deposited (*open circles*) and Xe-irradiated state (*dots*). The layer structure $^{nat}Fe(17\ nm)/^{57}Fe(15\ nm)/^{nat}Fe(17\ nm)/$ MgO is indicated in (a). For details see text.

spins were now aligned along $\psi_a = 133(3)°$, i.e., in-plane along the easy axis of magnetization.

Figure 3 illustrates MOKE magnetization curves taken along the hard axis ($\varphi = 0°$) and near the easy axis ($\varphi = 40°$) of the specimen. From such curves the relative remanence M_r/M_s (M_s being the saturation value) and the coercivity H_C shown in Figure 2 were deduced as function of φ. Under the influence of the MOKE field, a fourfold pattern of the remanence M_r/M_s and the coercivity H_C developed, and the magnetization turned into the easy axis already for the as-deposited sample. According to Brockmann *et al.* [17], the magnetic energy density normalized to the saturation magnetization M_s, can be expressed as the sum of an isotropic part (K_0/M_s), a uniaxially anistropic part (K_u/M_s), and a fourfold contribution, (K_1/M_s):

$$E_m/M_s = (K_0/M_s) + (K_u/M_s)\sin^2(\varphi - \varphi_0) + (K_1/4M_s)\sin^2(2\varphi - 2\varphi_1). \tag{2}$$

A fit to the MOKE data (left) gave $K_0/M_s \approx K_u/M_s < 3$ Oe, $K_1/M_s = 180(10)$ Oe and φ_1=133°. After the Xe-irradiation, the MOKE analysis essentially reproduced the pattern found in the as-deposited state.

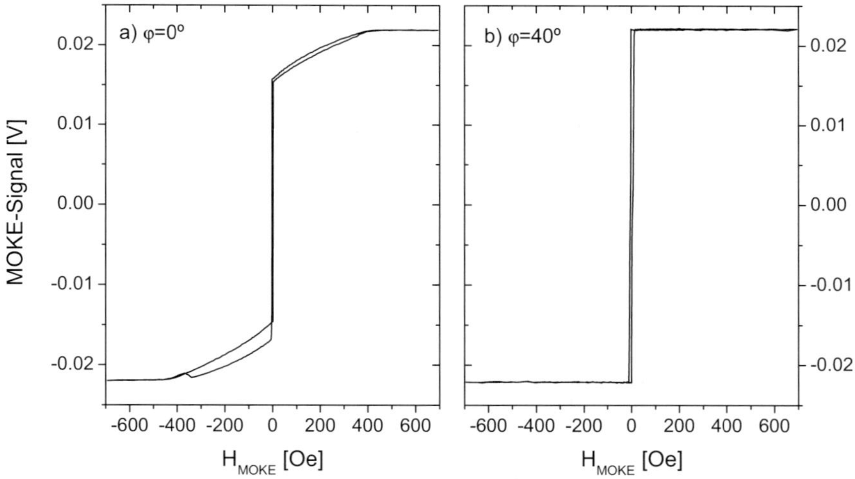

Figure 3. MOKE hysteris curves of the Xe-irradiated Fe/MgO sample taken along the hard axis ($\varphi = 0°$) and near the easy axis ($\varphi = 40°$).

In conclusion, for this epitaxial Fe-film on MgO, the originally isotropic spin alignment after deposition was turned into the easy axis by the MOKE field and did not change by the subsequent Xe-ion irradiation.

4. Results for polycrystalline Fe/Si and discussion

Let us now compare the above results with those gained for the electron-evaporated, polycrystalline Fe/Si(100) film. The results of the various MOMS and MOKE scans are summarized in Figure 4. The as-deposited sample exhibits isotropic MOKE signals M_r/M_s and H_c (see Figure 4(b,c)), but in MOMS a modulated I_2/I_3 ratio (see Figure 4(a)). The pre-magnetization in the external field turned the main hyperfine field direction from $\psi_a = 11(5)°$ to $\psi_a = 90(4)°$ and the subsequent irradiation kept this orientation ($\psi_a = 88(4)°$). Hence, this hyperfine field component $c_a = 0.66(2)$ did not change during this treatment, but the second component decreased from $c_b = 0.26(3)$ as-deposited to $c_b = 0.12(2)$ after pre-magnetization and ion irradiation. Consequently, again pre-magnetization had a much larger effect on the magnetic properties than ion-irradiation.

These results for both types of iron films are very different from those obtained in similar investigations for 75 nm polycrystalline nickel films deposited by electron evaporation on Si(100) and irradiated with 200-keV Xe-ions [1–4, 18]. Here, the ion irradiations were found to (i) turn the originally isotropic magnetization into the film plane and (ii) to orient it along an axis determined by either the stress field (inverse magnetostriction) or the external magnetic field present during the implantation. In the absence of any external magnetic or stress field, the ion implantation itself produces a sufficient number of extended defects and internal stress to induce a pronounced uniaxial magnetic anisotropy.

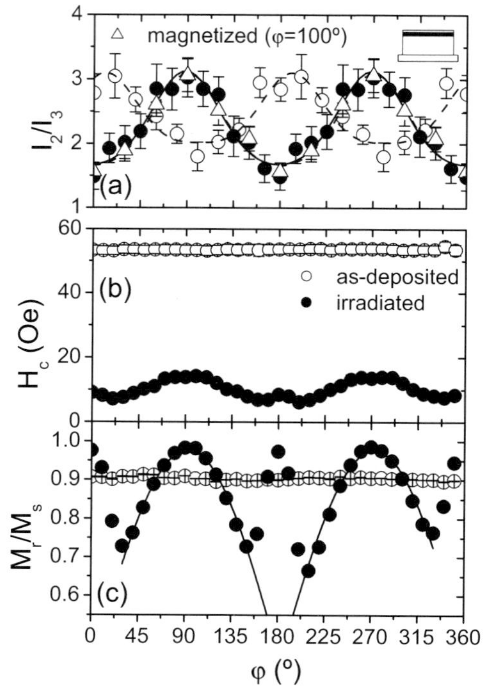

Figure 4. MOMS (a) and MOKE (b,c) analyses of a polycrystalline Fe-sample on Si, after electron-evaporation deposition (*open symbols*) and Xe-ion irradiation (*dots*). The layer structure was ^{nat}Fe(4 nm)/^{57}Fe(13 nm)/^{nat}Fe(41 nm)/Si(100). For further details see text.

Due to the much smaller magnetostriction constant in Fe compared to in Ni, magnetostriction is rather inefficient in Fe-films. However, the "magnetic history," i.e., pre-magnetization before the ion irradiation, in combination with the magneto-crystalline anisotropy (in the highly textured Fe/MgO film) and oriented defect structures [19], play an important role in iron and explain the data presented here. However, the effect of ion irradiation on the magnetic anisotropy is much weaker. A comprehensive paper of this work is in preparation [10].

Acknowledgements

The authors acknowledge fruitful discussions with Drs. R. Gupta and M. Uhr-macher and the help of D. Purschke during the implantations and RBS analyses. This work was funded by Deutsche Forschungsgemeinschaft.

References

1. Zhang K. *et al., Nucl. Instrum. Methods* **B161–163** (2000), 1016.
2. Zhang K., Doctoral thesis, University of Göttingen (2001).

3. Müller G. A. *et al., Hyperfine Interact.* **151/152** (2003), 223;
Kulinska A. *et al., J. Magn. Magn. Mater.* **272–276** (2004), 1149.
4. Lieb K. P. *et al., Acta Phys. Pol.* **100 A** (2001), 751.
5. Zhang K. *et al., Europhys. Lett.* **64** (2003), 668.
6. Gupta R., Lieb K. P., Luo Y., Müller G. A., Schaaf P. and Zhang K., submitted.
7. Gupta R. *et al., Nucl. Instrum. Methods* **B216** (2004), 350.
8. Zhang K., Gupta R., Müller G. A., Schaaf P. and Lieb K. P., *Appl. Phys. Lett.* **84** (2004), 3915.
9. Müller G. A., Doctoral thesis, Universität Göttingen (2003).
10. Müller G. A., Lieb K. P., Carpene E., Faupel J., Gupta R., Zhang K. and Schaaf P., in preparation.
11. Müller G. A., Gupta R., Lieb K. P. and Schaaf P., *Appl. Phys. Lett.* **82** (2003), 73.
12. Schaaf P., Müller G. A. and Carpene E. In: Miglierini M., Mashlan M. and Schaaf P. (eds.), *Mössbauer Spectroscopy in Materials Science II*, NATO Science Series Vol. 94, Kluwer Academic, Dordrecht, 2003, p. 127.
13. Uhrmacher M. *et al., Nucl. Instrum. Methods* **B9** (1985), 234.
14. Milosavljevic M. *et al., J. Appl. Phys.* **90** (2001), 4474.
15. Dhar S. *et al., Appl. Phys.* **A76** (2003), 773.
16. Krebs H.-U., *Int. J. Non-Equilib. Process.* **10** (1997), 3.
17. Brockmann M. *et al., J. Appl. Phys.* **81** (1997), 5047.
18. Lieb K. P. *et al.,* In: Jokic S., Milosevic I., Balaz A. and Nikolic Z. (eds.), *Proceedings of the 5th General Conference of the Balkan Physical Union, Vrnjacka Banja,* Serbian Physical Society, 2003, pp. 495–512.
19. Antonov L. I. *et al., Fiz. Met. Metalloved.* **48** (1977), 518.

Hyperfine Interactions (2004) 158:145–149
DOI 10.1007/s10751-005-9023-5

Mössbauer and TDPAC Studies on Fe/Mo Multilayers

Y. MURAKAMI[1,*], Y. OHKUBO[2], D. FUSE[3], Y. SAKAMOTO[3], T. ONO[3], S. KITAO[2], M. SETO[2], M. TANIGAKI[2], T. SAITO[4], S. NASU[3] and Y. KAWASE[2]
[1]*Graduate School of Science, Kyoto University, Kyoto 606-8502, Japan; e-mail: yukihiro@post3.rri.kyoto-u.ac.jp*
[2]*Research Reactor Institute, Kyoto University, Kumatori, Osaka 590-0494, Japan*
[3]*Graduate School of Engineering Science, Osaka University, Toyonaka, Osaka 560-0043, Japan*
[4]*Radioisotope Research Center, Osaka University, Toyonaka, Osaka 560-0043, Japan*

Abstract. Hyperfine fields at Fe and Mo layers in polyimide/Fe(10 nm)/[Mo(1.1 nm)/Fe(2.0 nm)]$_{120}$ and [Mo(1.3 nm) /Fe(2.0 nm)]$_{120}$ multilayers prepared by the electron-beam evaporation technique were measured at room-temperature by Mössbauer spectroscopy and perturbed-angular-correlation spectroscopy. The hyperfine fields in the Fe layers do not show a clear dependence on the Mo layer thickness. On the other hand, the hyperfine fields in the Mo layers show different magnetic structures in these samples. The difference suggests a variation of electron spin polarization in the Mo layers.

1. Introduction

Magnetic properties of metallic multilayers consisting of ferromagnetic layers separated by a nonmagnetic spacer have attracted considerable interest for both basic and technological researches. Oscillatory interlayer exchange coupling (IEC) is known as one of the interesting phenomena. The IEC oscillates alternately between ferromagnetic and antiferromagnetic as a function of the spacer thickness [1]. The mechanism of the oscillatory IEC across the spacer is basically explained by the spin polarization of conduction electrons in the interlayer. Since the existence of the electronic spin polarization causes a magnetic field which varies spatially in spacer layer, elucidation of the magnetic structure in the spacer of multilayers with different spacer thicknesses is important to understand the origin of IEC.

There have been few microscopic studies on oscillatory IEC. Mössbauer spectroscopy and time-differential perturbed-angular-correlation (TDPAC) spectroscopy are powerful techniques for investigating the local electronic states of materials through the hyperfine interaction parameters for probe nuclei, thus

* Author for correspondence.

applicable to such microscopic studies on oscillatory IEC [2, 3]. In this paper, we report on the local magnetic fields in Mo and Fe layers of Fe/Mo multilayers measured with TDPAC and Mössbauer techniques.

We prepared two Mo/Fe multilayer samples with different thicknesses of Mo layers: Fe(10 nm)/[Mo(1.1 nm)/Fe(2.0 nm)]$_{120}$ (for short Mo(1.1)/Fe(2.0)) and Fe(10 nm)/[Mo(1.3 nm)/Fe(2.0 nm)]$_{120}$ (for short Mo(1.3)/Fe(2.0)), the 10 nm Fe layer being used as a buffer layer. The subscripts are the number of the bilayers. In Fe/Mo multilayers, the IEC is at the maximum when the thickness of the Mo layers is 1.1 nm and oscillates with a period of 1.1 nm Mo layer thickness [4]. The Fe layers are coupled with each other ferromagnetically for Fe/Mo with Fe(2.0 nm) [5].

2. Experimental procedures

The multilayer samples were grown on polyimide films by the electron-beam evaporation technique in a chamber under a residual pressure of 10^{-7}-10^{-8} Pa at room temperature. The thickness of each layer was controlled with a quartz oscillator thickness monitor during the deposition. From the high-angle X-ray diffraction patterns, these Fe/Mo were confirmed to have bcc(110) texture. Magnetization measurements were performed at 300 K using a superconducting quantum interface device (SQUID) with a magnetic field applied parallel to the multilayer surfaces. The obtained magnetization curves for Mo(1.1)/Fe(2.0) and Mo(1.3)/Fe(2.0) are almost the same, and suggest that the adjacent Fe layers are coupled across Mo spacer ferromagnetically.

^{57}Fe-Mössbauer measurements were performed with a source of ^{57}Co in Rh. The sample was placed in a configuration such that the multilayer plane was perpendicular to the γ ray direction. The calibration spectra were obtained with an iron foil as a reference.

TDPAC experiments were performed with the 181 keV level of ^{99}Tc ($T_{1/2} = 3.6$ ns, $I = 5/2$, $\mu = 3.3\ \mu_N$). ^{99}Tc was obtained through the beta decay of ^{99}Mo, produced by irradiating ^{98}Mo in the Mo layers with thermal neutrons at the research reactor of Kyoto University. The TDPAC data were recorded using a fast-slow electronic circuit system with four BaF_2 detectors placed in a plane at 90° angular separations. The detectors plane was parallel to the multilayer plane. If the probe nuclei experience only one kind of magnetic hyperfine field, the constructed time dependent anisotropy function $R(t)$ is expressed as

$$\begin{aligned} R(t) &= (W_{180} - W_{90})/2^{-t/T_{1/2}} \\ &= 0.75 A_{22}[a_0 + a_1 \cos(\omega t) + a_2 \cos(2\omega t)]. \end{aligned} \tag{1}$$

Here $T_{1/2}$ is the half life of the intermediate state of the TDPAC probe, A_{22} is the angular correlation coefficient, $\omega = g\mu_N H/\hbar$ is the Larmor precession frequency for a magnetic field H, and the values of a_0, a_1, and a_2 depend on the

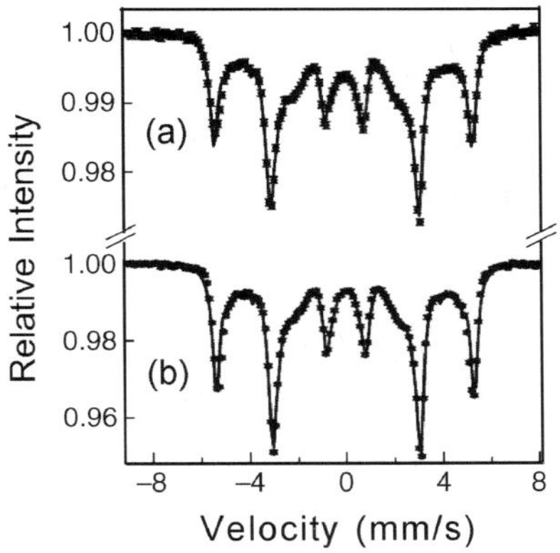

Figure 1. Room-temperature ^{57}Fe-Mössbauer spectra for (a) $[Mo(1.1\ nm)/Fe(2.0\ nm)]_{120}$ and (b) $[Mo(1.3\ nm)/Fe(2.0\ nm)]_{120}$.

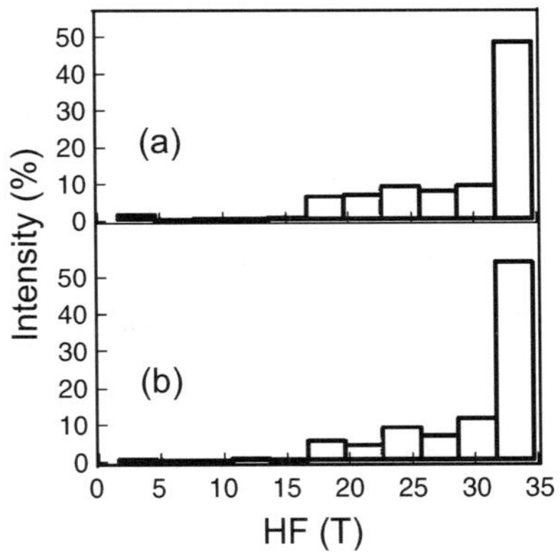

Figure 2. Distribution of the magnetic hyperfine fields in the Fe layers for (a) $[Mo(1.1\ nm)/Fe(2.0\ nm)]_{120}$ and (b) $[Mo(1.3\ nm)/Fe(2.0\ nm)]_{120}$.

orientation of the hyperfine field (HF) with respect to the detectors, with $a_0 + a_1 + a_2 = 1$.

Sample activation and TDPAC measurements were repeated several times for both samples. Migration of Mo atoms into near Fe layers by recoil during sample activation is considered to be negligible since the TDPAC spectra do not show any noticeable difference for the samples repeatedly irradiated with thermal neutrons.

3. Results and discussion

The ^{57}Fe-Mössbauer spectra for the Mo(1.1)/Fe(2.0) and Mo(1.3)/Fe(2.0) at room temperature are shown in Figure 1. For our spectrum analysis, we used essentially the same conditions as Kalska et al. used [2], i.e., the magnetic spectra were analyzed assuming a superposition of 11 sextets with an equidistant magnetic field step of 3 T. The Lorentzian linewidth for all individual lines and the relative intensity for all individual sextet were kept the same.

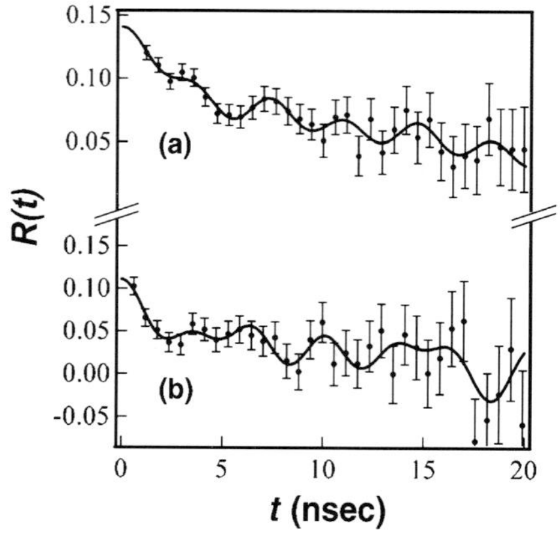

Figure 3. Room-temperature ^{99}Tc-TDPAC spectra for (a) [Mo(1.1 nm)/Fe(2.0 nm)]$_{120}$ and (b) [Mo(1.3 nm)/Fe(2.0 nm)]$_{120}$.

Table I. Results of the room-temperature ^{99}Tc-TDPAC measurements of the Fe/Mo multilayers

		H_1	H_2	H_3	H_4
[Fe(2.0 nm)/Mo(1.1 nm)]$_{120}$	*HF* (T)	13.6 (3)	7.6 (10)	5.5 (6)	1.3 (1)
	Fraction	1	1	1	2
[Fe(2.0 nm)/Mo(1.3 nm)]$_{120}$	*HF* (T)	14.9 (4)	9.6 (4)	7.0 (8)	1.4 (3)
	Fraction	1	1	2	1

The corresponding magnetic HF distributions are presented in Figure 2. The magnetic HF at Fe decreases linearly with the increasing number of Mo atoms at the first and second nearest neighbors of the Fe [6]. The intensity of each HF component roughly corresponds to the fraction of Fe atoms with a different number of surrounding Mo atoms. If there are differences in the interface structures (e.g. interface roughness), they should appear as different HF distributions through different fractions of Fe atoms with various Mo configurations. However, the HF distributions obtained for the two samples are almost the same. Therefore, the interface structures of Mo(1.1)/Fe(2.0) and Mo(1.3)/Fe(2.0) are the same.

Figure 3 shows the ^{99}Tc-TDPAC spectra at room temperature for the Mo(1.1)/Fe(2.0) and Mo(1.3)/Fe(2.0) samples. There is a difference in the pattern between the two TDPAC spectra. $R(t)$ is expressed as a superposition of Equation (1) by assuming several magnetic interactions:

$$R(t) = A_{22} \sum_i P_i [a_{i0} + a_{i1} \cos(\omega_i t) + a_{i2} \cos(2\omega_i t)] \tag{2}$$

Here P_i is the population of the probe nuclei perturbed by H_i. The solid curves in Figure 3 are the results of the fits with Equation (2). The obtained results are summarized in Table I. Under the assumption that the strongly perturbed region in the Mo layer is near the Mo/Fe interface, the values of H_1 and H_2 are considered to represent the interface Mo layers. Although it is confirmed from the Mössbauer spectra that the interface structures in Mo(1.3)/Fe(2.0) and Mo(1.1)/Fe(2.0) are the same, there are differences between the two samples in the values of H_1 and H_2.

We expect that the magnetization in the Mo layer is a superposition of the magnetic profiles induced by its two interfaces. For a large spacer thickness, the magnetization from one side of the interface can be regarded to oscillate with $\sin(2\pi x/\Lambda + \Phi)$ and is exponentially attenuated as the distance x from the interface increases [7]. The Λ and Φ are, respectively, the oscillation period of IEC and a constant given by the electronic structure of interlayer material. Hence, different spacer thicknesses cause different coherence in the superposition of the spatially oscillating magnetization from the two interfaces. Assuming that this model can be applied to the present Fe/Mo multilayers with the small spacer layer thicknesses (few monolayers), different Mo layer thicknesses should lead to different magnetizations (and HF). For the Fe(110)/Mo(110) multilayers, Λ is 1.1 nm and Φ is 0 or π [8]. From the observation that the Fe layers couple with each other ferromagnetically in both samples, it is expected that the magnetic field near the interface in the Mo layer of Mo(1.3)/Fe(2.0) is larger than that of Mo(1.1)/Fe(2.0). This expectation is consistent with the present TDPAC results. Therefore, we consider that the difference in the hyperfine fields observed in the Mo layers between Mo(1.1)/Fe(2.0) and Mo(1.3)/Fe(2.0) reflects a change of spatially oscillating electron spin polarization depending on the Mo spacer thickness.

Acknowledgement

This work was supported by the Ministry of Education, Culture, Sports, Science, and Technology of Japan [Grant-Aid for Scientific Research on Priority Areas (B) No. 751].

References

1. Brubaker M. E. *et al., Appl. Phys. Lett.* **58** (1991), 2360.
2. Kalska B. *et al., J. Magn. Magn. Mater.* **226–230** (2001), 1782.
3. Meersschaut J. *et al., Phys. Rev. Lett.* **75** (1995), 1638.
4. Bruno P. and Chappert C., *Phys. Rev. Lett.* **67** (1991), 1602.
5. Yan M. L., Sellmyer D. J. and Lai W. Y., *J. Phys.: Condens. Matter.* **9** (1997), L145–L149.
6. Vincze I. and Champbell A., *J. Phys. F. Met. Phys.* **3** (1973), 647.
7. Luetkens H. *et al., Phys. Rev. Lett.* **91** (2003), 017204.
8. Koelling D. D., *Phys. Rev., B* **50** (1994), 273.

Hyperfine Interactions (2004) 158:151–156
DOI 10.1007/s10751-005-9024-4

Mössbauer Studies on $(Zn, Cd, Cu)_{0.5}Ni_{0.5}Fe_2O_4$ Oxides

J. Z. MSOMI, K. BHARUTH-RAM, V. V. NAICKER and T. MOYO
School of Physics, Howard College Campus, University of KwaZulu-Natal, Durban 4041, South Africa; e-mail: moyo@ukzn.ac.za

Abstract. The temperature dependence of the magnetic hyperfine fields on ^{57}Fe nuclei at the tetrahedral (A) and octahedral (B) interstitial sites in $(Zn, Cd, Cu)_{0.5}Ni_{0.5}Fe_2O_4$ oxides have been investigated in Mössbauer measurements at sample temperatures ranging from 79 K to 850 K. The samples were prepared by solid-state reaction, and their phase determined from XRD measurements, which also provided information on their X-ray densities, grain sizes and lattice constants. The magnetic transition temperatures between ordered and paramagnetic states were obtained from zero-velocity measurements. The $Cu_{0.5}Ni_{0.5}Fe_2O_4$ sample is found to be antiferromagnetic with Néel temperature of 823 K, and its super-exchange interactions, J_{AB}, J_{AA} and J_{BB} were determined to be -24.14 k_B, -13.10 k_B and $+11.68$ k_B, respectively. The $(Zn, Cd)_{0.5}Ni_{0.5}Fe_2O_4$ are ferrimagnetic with Curie points of 543 and 548 K, respectively, and corresponding average exchange integrals $J_o = 186$ k_B and 188 k_B.

Key Words: Mössbauer spectroscopy, spinel oxides, super-exchange interaction, transition temperatures.

1. Introduction

The nickel substituted zinc ferrites are soft ferrimagnetic materials with low coercivity, high resistivity values and little eddy current losses [1, 2]. These properties make these ferrites excellent core materials for power transformers in electrical and telecommunication applications. Under controlled conditions, the ferrites crystallize in the spinel structure AB_2O_4 (B = Fe), with the A^{2+} and B^{3+} cations contained in tetrahedral and octahedral interstices, respectively. The magnetic order in these ferrites, thus, depends on competition between different super-exchange interactions among the tetrahedral (A) and octahedral (B) site cations. These interactions, between Fe ions, are mediated by an intervening oxygen ion, and are characterized by the exchange integrals J_{AB} (A-O-B), J_{AA} (A-O-A) and J_{BB} (B-O-B). Ferrimagnetic ordering takes place when the inter-sub-lattice interaction (A-O-B) is stronger than the intra-sub-lattice interactions (A-O-A and B-O-B). Two sets of data, for example hyperfine fields and magnetic moments are required to determine the super-exchange integrals [3, 4]. Provided the hyperfine fields at ^{57}Fe nuclei at the A and B sites are clearly resolved, Mössbauer measurements allow accurate determination of these fields, and hence provide two series of data which, without the use of external fields, can be used to infer reliable information on super-exchange interactions. Of interest are the changes in the magnetic interactions when Cd or Zn ions replace Cu ions. Cd ion has larger size (0.97 Å) than Zn (0.74 Å) ion while Cu (0.72 Å) is slightly smaller. However, while both Zn and Cd have full outer *s* orbitals, Cu has an unpaired electron in its outermost orbital. In this work we investigate the effects of size and electronic configuration differences between Cu, Zn and Cd ions on the magnetic interactions.

2. Experimental

Samples of $(Zn, Cd, Cu)_{0.5}Ni_{0.5}Fe_2O_4$ were prepared by solid-state reaction technique from high purity starting oxides, following preparation procedures reported previously [5, 6]. X-ray diffraction measurements were made on the samples using monochromatic beam of Co (K_α) radiation (wavelength λ = 1.78897 Å), to ensure that the samples crystallized in the cubic spinel phase, and to obtain information on the lattice constant, mean grain size, density and porosity of the samples. ^{57}Fe-Mössbauer measurements were made in constant acceleration, transmission mode, using a ^{57}Co source in a rhodium matrix. Curie/Néel points of the samples were determined from temperature-dependent, zero-velocity measurements. The bulk densities (ρ_b) of the sintered samples were measured using Archimedes' principle.

3. Results and discussion

The X-ray diffraction patterns (XRD) confirmed that the samples crystallized with single-phase cubic spinel structure. The X-ray densities, grain sizes and lattice parameters of the samples, determined from the XRD patterns, as well as the bulk densities and porosity, are listed in Table I.

The 'X-ray' densities (ρ_x) were calculated from the lattice constants derived from the XRD patterns associated with the unit cell of the spinel lattice. The percentage porosity (P) of the samples were deduced from the bulk and X-ray densities [5, 6], while the grain sizes were calculated from the broadening of the line width of (311) X-ray diffraction line using the Scherrer formula. The values of lattice parameters obtained here compare very well with those reported for similar compounds [2, 4, 5]. Replacement of Cu with the larger sized Zn (0.74 Å) or Cd (0.97 Å) ions result in increase in lattice parameter and densities, which is consistent with the systematic trends observed previously [6].

Mössbauer spectra of the Zn–Ni and Cu–Ni oxides at selected absorber temperatures are shown in Figure 1, and the variation of the hyperfine field with temperature in Figure 2. The spectra for $(Zn, Cd)_{0.5}Ni_{0.5}Fe_2O_4$ oxides were very similar, and are closely related due to the isoelectronic configurations of Zn and Cd ions. The spectra can be fitted with two sextets in the range 79–180 K; above

Table I. Bulk (ρ_b) and X-ray (ρ_x) densities, porosity (P), lattice parameter (a), grain size (d) and for (Zn, Cd, Cu)$_{0.5}$Ni$_{0.5}$Fe$_2$O$_4$ oxides

Sample	ρ_b (g cm^{-3})	ρ_x (g cm^{-3})	P (%)	a (Å)	d (Å)
Cu–Ni	4.43(2)	5.36(3)	17.4(3)	8.37(2)	444.5(2)
Zn–Ni	5.11(2)	5.35(3)	4.7(3)	8.39(2)	479.6(2)
Cd–Ni	5.17(2)	5.63(3)	8.2(3)	8.51(2)	450.6(2)

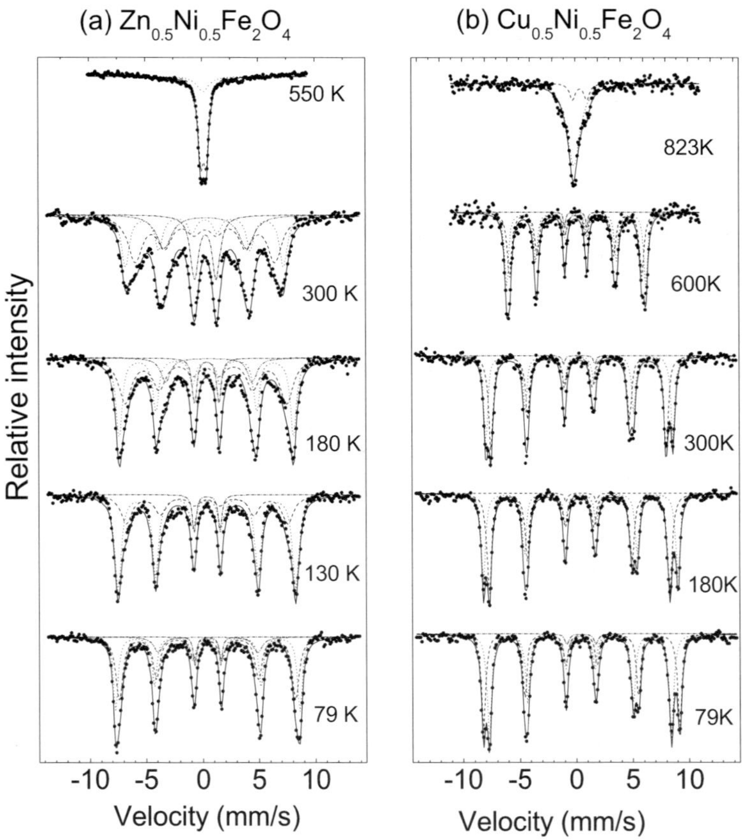

Figure 1. Mössbauer spectra of the (Cd, Zn, Cu)$_{0.5}$Ni$_{0.5}$Fe$_2$O$_4$ ferrites.

180 K two additional doublets are required. For Cu$_{0.5}$Ni$_{0.5}$Fe$_2$O$_4$ the two sextets were well resolved up to 550 K. The sextet components represent Fe ions in ordered spin state on tetrahedral (A) and octahedral (B) sites. Doublets reflect the Fe ions in paramagnetic states. The criterion used to assign sextets or doublets on A or B sites, is based on isomer shifts and quadrupole splitting deduced from the Mössbauer fits. The tetrahedral (A) site is expected to have lower isomer shifts and quadrupole splitting due to its higher symmetry compared to the roomier octahedral (B) sites [5].

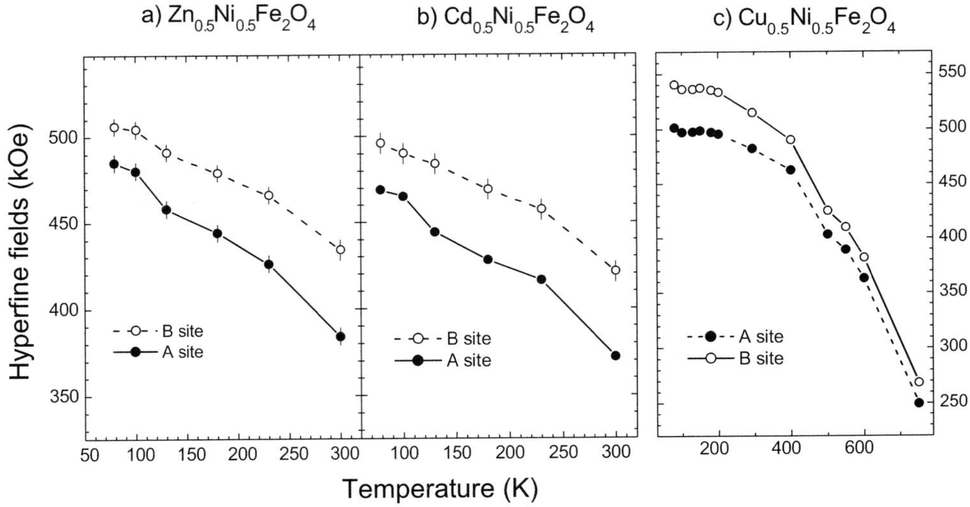

Figure 2. Variation of hyperfine field of Fe at the A and B sites with temperature in the (Cd, Zn, Cu)$_{0.5}$Ni$_{0.5}$Fe$_2$O$_4$ ferrites.

Table II. Isomer shifts (δ), hyperfine fields B_{hf} and site fractions (f) of Fe at tetrahedral (a) and octahedral (B) sites in the (Cu, Zn, Cd)$_{0.5}$Ni$_{0.5}$Fe$_2$O$_4$ oxides at 79 K

Sample	Site	δ (mm/s)	B_{hf} (kOe)	f (%)
Cu–Ni	A	0.35(2)	501(1)	52(2)
	B	0.46(2)	540(1)	46(2)
Zn–Ni	A	0.44(2)	484(2)	48(4)
	B	0.48(6)	506(2)	39(4)
Cd–Ni	A	0.39(2)	483(2)	61(5)
	B	0.44(2)	501(3)	33(5)

The Mössbauer parameters of the components required to fit the data of the different samples at 79 K are listed in Table II. The high values of the hyperfine fields are consistent with the ferric character of the Fe ions at both sites. The isomer shifts values also reflect the 3^{+}-oxidation state of the Fe ion. The quadrupole splittings were found to be zero within experimental error below the critical temperatures, which is consistent with the cubic crystal structure confirmed by the XRD data.

The magnetic splitting in the spectra for Cu$_{0.5}$Ni$_{0.5}$Fe$_2$O$_4$ is well resolved and is attributed to antiferromagnetic character of this ferrite. In the (Cu, Zn, Cd)$_{0.5}$Ni$_{0.5}$Fe$_2$O$_4$ oxides, the magnetic components are not as well resolved and the presence of partially paramagnetic component in the Mössbauer spectra indicates ferromagnetic behaviour. Our results, thus, show a change from ferrimagnetism to antiferromagnetism when Cu ions replace Zn or Cd ions, an effect that may be due to the effect of the unpaired electron in the valence shell of Cu.

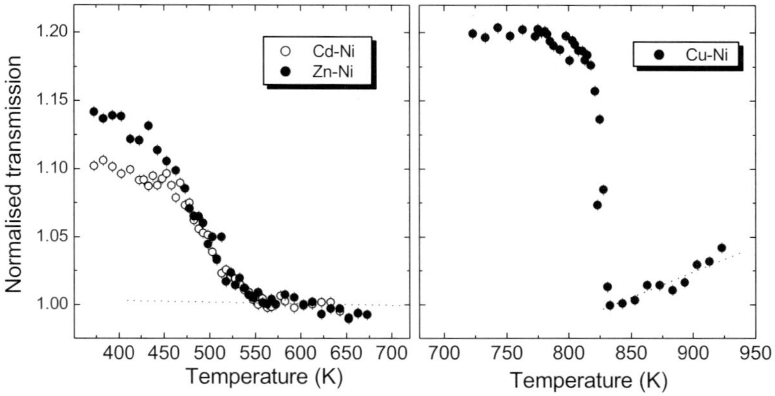

Figure 3. Zero-velocity transmission intensities as a function of temperature for the (Cd, Zn, Cu)$_{0.5}$Ni$_{0.5}$Fe$_2$O$_4$ ferrites.

The magnetic transition temperatures (Curie, T_C or Néel, T_N points) from ordered to paramagnetic states were measured using zero velocity Mössbauer technique in which the variation of transmitted γ-rays through an absorber sample with temperature is monitored. At the ordered to disordered magnetic transition point, the transmitted intensity of γ-rays drops significantly. T_C and T_N are taken at the discontinuity of rapid drop in transmitted intensity (low transmission rate edge) as shown in Figure 3. The Zn–Ni and Cd–Ni samples are ferrimagnetic, with Curie temperatures of 548(3) and 553(3) K, respectively, while the Cu–Ni displays clear antiferromagnetic behaviour, with a Néel temperature of 823(3) K. This value is in good agreement with the value of 820(2) K obtained by Kim et al. [4]. The low Curie points for (Zn, Cd)$_{0.5}$Ni$_{0.5}$Fe$_2$O$_4$ compared to the high transition temperature for Cu$_{0.5}$Ni$_{0.5}$Fe$_2$O$_4$ indicate a change from soft ferrimagnetic ordering in Zn–Ni or Cd–Ni to antiferromagnetic ordering in Cu–Ni oxide.

The zero velocity transmission intensity for the Zn or Cd based oxides remains constant, within experimental uncertainty, beyond the transition temperature. In the Cu–Ni oxide, on the other hand, the count rate increases linearly with temperature beyond T_N. Similar effects have been observed in a sputtered Fe$_{0.91}$Zr$_{0.09}$ amorphous alloy [7], and in measurements on NiFe$_2$O$_4$. This behaviour could be characteristic of antiferromagnetic interactions in material, and is being investigated further.

In the case of Cu$_{0.5}$Ni$_{0.5}$Fe$_2$O$_4$, the two sextet components in the spectra are well-resolved, and allow unambiguous determination of the hyperfine fields at the Fe ions in the tetrahedral and octahedral interstices. This, coupled with the good determination of the Néel temperature of the sample, yield super-exchange interaction integrals J_{AB}, J_{AA} and J_{BB} of -24.14 k$_B$, -13.10 k$_B$ and $+11.68$ k$_B$, respectively. These values are in good agreement with those determined by Kim et al. [4]. The Zn- and Cd-substituted samples are ferrimagnetic with Curie points of 543(3) and 548(3) K, respectively. From mean field theory the average exchange integral can be estimated to be $J_o \approx 3\ k_B T_C/S(S+1)$, which with $S = 5/2$ for Fe, gives $J_o = 186\ k_B$ for the Zn-sample, and $188\ k_B$ for the Cd-substituted sample.

4. Conclusions

The Cd- and Zn-substituted Ni ferrites are ferrimagnetic, while the Cu based sample is antiferromagnetic. This result implies that antiferromagnetism may be due to the unpaired electron in the outer shell. Work is in progress to investigate this further. The larger size of the unit cell of Cd based oxide compared to Zn and Cu mixed ferrites relates well with the larger size of the Cd atom. There is no significant change in the magnetic properties that can be attributed to the size difference between Cd, Zn and Cu ions. These changes can be attributed to their electronic configuration.

Acknowledgement

This work was supported by the National Research Foundation (South Africa) under grant GUN203350.

References

1. Singh K. A., Verma A., Thakur O. P., Prakash C., Goel T. C. and Mendiratta R. G., *Mater. Lett.* **57** (2003), 1040.
2. Costa A. C. F. M., Tortella E., Morelli M. R., Kiminami R. H. G. A., *J. Magn. Magn. Mater.* **256** (2003), 174.
3. Oak H. N., Baek K. S. and Kim S. J., *Phys. Status Solidi, B* **208** (1998), 249.
4. Kim W. C., Kim S. J. and Kim C. S., *J. Magn. Magn. Mater.* **239** (2002), 82.
5. Msomi J. Z., Moyo T. and Bharuth-Ram K., *Hyperfine Interact.* **C5** (2002), 181.
6. Moyo T., Msomi J. Z. and Bharuth-Ram K., *Hyperfine Interact.* **136**(3) (2002), 579.
7. Read D. A., Moyo T., Jassim S., Dunlap R. A. and Hallam G. C., *J. Magn. Magn. Mater.* **82** (1989), 87.

Hyperfine Interactions (2004) 158:157–161
DOI 10.1007/s10751-005-9029-z

Investigation of Hyperfine Interactions in GdNiIn Compound

A. L. LAPOLLI, A. W. CARBONARI*, R. N. SAXENA and J. MESTNIK-FILHO
Instituto de Pesquisas energéticas e Nucleares – IPEN-CNEN/SP; e-mail: carbonar@ipen.br

Abstract. Perturbed gamma–gamma angular correlation technique was used to measure the hyperfine interactions in the compound GdNiIn using the $^{111}In \rightarrow ^{111}Cd$ and $^{140}La \rightarrow ^{140}Ce$ probe nuclei at the In and Gd sites, respectively. A unique quadrupole frequency with asymmetry parameter $\eta = 0.78$ was observed for ^{111}Cd probe at In sites for the measurements above Curie temperature. Below T_C, the spectra for ^{111}Cd show combined magnetic dipole and electric quadrupole interaction. Below 85 K, a unique magnetic interaction is observed at ^{140}Ce. A linear relationship between the saturated magnetic hyperfine field and the magnetic transition temperature was observed for both probes, indicating that the main contribution to the mhf comes from the conduction electron polarization.

Key Words: magnetic hyperfine field, PAC spectroscopy, quadrupole interaction, rare earth magnetism.

1. Introduction

An important group within the series RTX (R = rare earth metal, T = transition metal, X = sp-element) is formed by compounds, which crystallize in the ZrNiAl-prototype structure (hexagonal structure for space group $P\bar{6}2m$) and show interesting magnetic properties and a variety of magnetic structures [1, 2]. This structure is formed by magnetic RT layers alternated with non-magnetic TX layers. The magnetic R atoms occupy the positions: $x, 0, 1/2; 0, x, 1/2; \bar{x}, \bar{x}, 1/2$ and form a triangular structure, which is a deformed Kagomé lattice. One of the characteristics of such a triangular arrangement of magnetic atoms is the frustration of the magnetic interactions when an antiferromagnetic order is present. The RNiIn family of compounds, which also crystallize in the ZrNiAl-type structure, has been very little studied so far. In this family, GdNiIn orders ferromagnetically below around 94 K as reported by Merlo *et al.* [3] using magnetization measurements. Canepa *et al.* [4] in the measurements of the Magnetocaloric properties observed a ferromagnetic transition at 96 K. Tyvanchuk *et al.* in AC

* Author for correspondence.

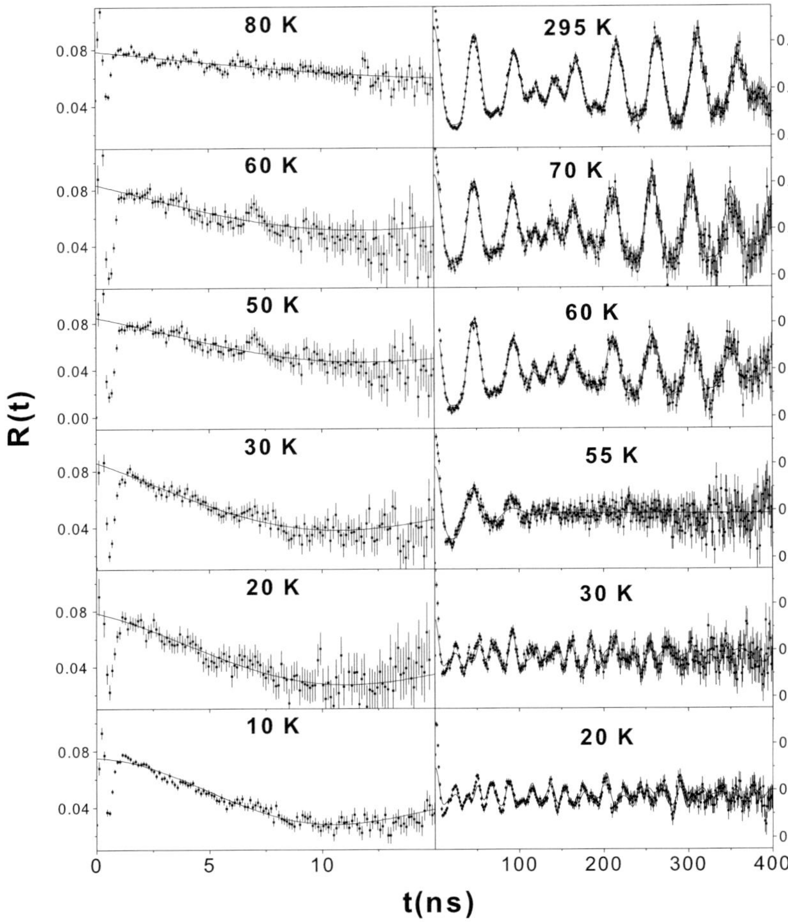

Figure 1. TDPAC spectra of the hyperfine interactions at Gd and In sites with ^{140}Ce (left) and ^{111}Cd (right) probes, respectively.

and DC bulk magnetic measurements in the compound GdNiIn reported an anomalous behavior at 80 K [5]. In this reference it was reported that because RNiIn compounds are good electrical conductor and there is a large interatomic distance between *R* atoms, the observed magnetic ordering would be caused by interactions via conduction electrons described by the RKKY model. In the present work, we have investigated the temperature dependence of the magnetic hyperfine field (mhf) on both Gd and In sites using ^{140}Ce and ^{111}Cd probes, respectively, as well as the behavior of the electric field gradient efg at In sites.

2. Experimental

The polycrystalline GdNiIn samples were prepared by repeatedly melting the constituent elements (Gd 99.99%, Ni 99.998%, In 99.9999%) in an arc furnace

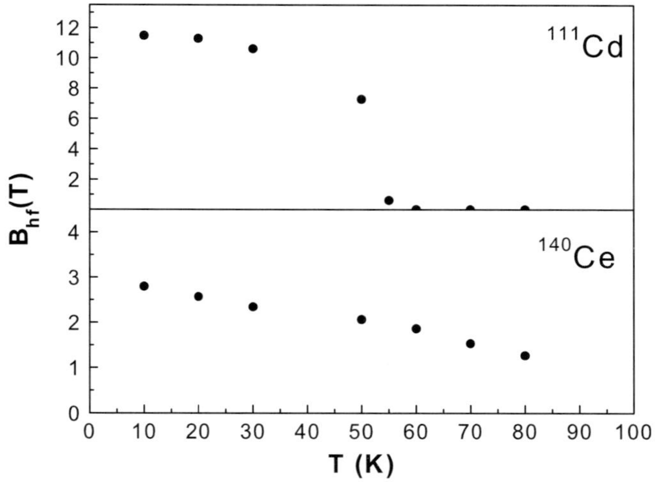

Figure 2. Temperature dependence of the magnetic hyperfine field at In (top) and Gd (bottom) sites in GdNiIn with ^{111}Cd and ^{140}Ce probes, respectively.

under argon atmosphere purified with a hot titanium getterer. Carrier free ^{111}In nuclei were introduced into the sample by diffusion. Another sample was prepared in a similar way but with radioactive ^{140}La nuclei obtained by neutron irradiation of lanthanum metal substituting about 0.1% of Gd atoms melted along with the constituent elements. Samples were annealed in vacuum for 72 h at 800°C. The structure of the samples were checked by X-ray diffraction measurement, which indicated a single phase and ZrNiAl-type structure with the $P\bar{6}2m$ space group for the compound.

The TDPAC measurements were carried out with a conventional fast–slow coincidence set-up with four conical BaF2 detectors. The well known gamma cascade of 172−245 keV, populated from the decay of ^{111}In with an intermediate level with spin $I = 5/2^+$ at 245 keV ($T_{1/2}$ = 84.5 ns) in ^{111}Cd, was used to investigate the hyperfine interactions in GdNiIn samples. The gamma cascade of 329−487 keV populated from the decay of ^{140}La with an intermediate level with spin $I = 4^+$ at 2083 keV ($T_{1/2}$ = 3.45 ns) in ^{140}Ce was used to measure the magnetic hyperfine field (B_{hf}) at Ce. The samples were measured in the temperature range of 10–420 K by using a closed-cycle helium cryogenic device. The time resolution of the system was about 0.6 ns for both gamma cascades. A detailed description of the method can be found elsewhere [6, 7].

3. Results and discussion

Some of the TDPAC spectra measured with ^{140}Ce and ^{111}Cd probe nuclei are shown in Figure 1. The solid curves are the least squares fit of the experimental data to the appropriate function in each case. The quadrupole moment of the

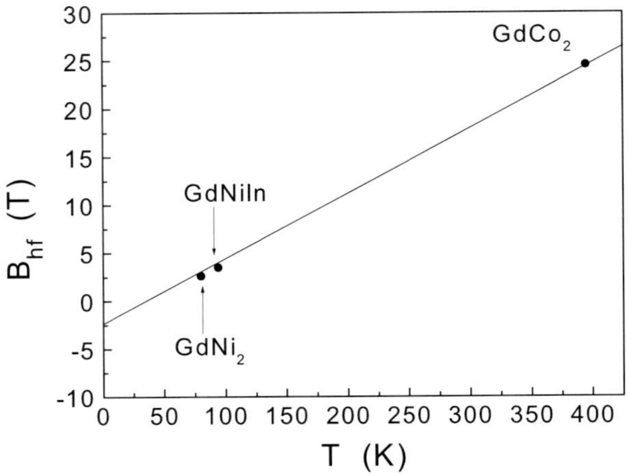

Figure 3. The extrapolated magnetic hyperfine field B_{hf} (0) at ^{140}Ce in some Gd compounds as a function of the respective magnetic transition temperatures.

2083 keV 4^+ state of ^{140}Ce is known to be very small [8], consequently one expects to observe an almost pure magnetic dipole interaction in the antiferromagnetic phase of the sample. Below approximately 85 K, a unique magnetic interaction is observed at ^{140}Ce at Gd sites of GdNiIn. The temperature dependence of B_{hf} for ^{140}Ce is shown in the lower part of Figure 2. One can observe that there is a slight increase in the B_{hf} values at low temperature where the hyperfine field should be saturated. It is not clear if this increase is real because the B_{hf} values are quite low and, consequently, less precise.

The TDPAC spectra for ^{111}Cd at room temperature shows a unique quadrupole interaction with a sharp frequency ν_Q = 82.9(3) MHz, δ = 1% and η = 0.78(5). Below 60 K, the spectra for ^{111}Cd show combined magnetic dipole and electric quadrupole interaction. These spectra are characterized each by a single quadrupole frequency (ν_Q ~84 MHz) with δ = 1% and η = 0.78, and a temperature dependent magnetic dipole interaction shown in Figure 2. The angle between the efg and mhf changes from 60° at 50 K to 30° at 20 K. A comparison of the temperature dependence of the B_{hf} for both probes shows a significant difference in the transition temperature, around 55 K and ~85 K for ^{111}Cd and ^{140}Ce, respectively. We have no explanation for this difference yet. The values of the mhf at low temperature for ^{111}Cd are four times greater than the values for ^{140}Ce. This fact is mainly due to a difference in the distance of the probe to the magnetic Gd ion (3.2 Å for ^{111}Cd-Gd and 3.9 Å for ^{140}Ce-Gd).

In [9], the saturation values of B_{hf} at ^{111}Cd on Gd sites in $GdNi_2$, $GdAl_2$ and Gd are compared to the respective Curie temperature and the result showed a linear dependence where B_{hf}/T_C = 0.116 T/K. Using 11.5 T as the saturation value for B_{hf} in GdNiIn, the ratio B_{hf}/TC ~0.12 T/K and follows the same linear behavior reported in [9]. The saturation values of B_{hf} on Gd sites in GdNiIn, $GdNi_2$ (Carbonari A. W., private communication), and $GdCo_2$ [10] are compared to the respective transition temperature of each host in Figure 3. One can observe that to a good approximation B_{hf} is a linear function of the transition temperature. According to the RKKY theory of indirect coupling the ratio between the conduction electron spin polarization (CEP) and the order temperature is expected to be proportional to $[J_{sf}(g-1)(J+1)]^{-1}$, where J_{sf} is the s-f coupling constant, g the Landé factor and J the total angular momentum. The linear relation between B_{hf} at ^{140}Ce and the magnetic transition temperature in Figure 3 thus may imply that the main contribution to the B_{hf} comes from the CEP at the probe site and the coupling constant J_{sf} has the same value in GdNiIn, $GdNi_2$ and $GdCo_2$ compounds. Therefore, the ^{140}Ce probes in this case behaves as closed shell nuclei like ^{111}Cd. Preliminary *ab initio* calculations using the WIEN2K code have shown that the main contribution to the mhf in GdNiIn using ^{111}Cd as impurity at In sites comes from valence electrons.

Acknowledgements

Partial support for this research was provided by the Fundação de Amparo á Pesquisa do Estado de São Paulo (FAPESP). AWC thankfully acknowledges the support provided by CNPq in the form of a research fellowship.

References

1. Szytula A. and Leciejewicz J., *Handbook of Crystal Structures and Magnetic Properties of Rare Earth Intermetallics*, CRC, Boca Raton, FL, 1994.
2. Ehlers G. and Maletta H., *Physica, B* **234–236** (1997), 667.
3. Merlo F., Fornasini M. L., Cirafici S. and Canepa F., *J. Alloys* **267** (1998), L12–L13.
4. Canepa F., Napoletano M., Palenzna A., Merlo F. and Cirafici S., *J Phys, D, Appl Phys* **32** (1999), 2721–2725.
5. Tyvanchuk Y. B. *et al.*, *J Magn Magn Mater* **277** (2004), 368.
6. Carbonari A. W., Saxena R. N., Pendl W. Jr., Mestnik Filho J., Attili R., Olzon-Dionysio M. and de Souza S. D., *J. Magn. Magn. Mater.* **163** (1996), 313.
7. Dogra R., Junqueira A. C., Saxena R. N., Carbonari A. W., Mestnik-Filho J. and Morales M., *Phys. Rev., B* **63** (2001), 224104.
8. Królas K. and Wodniecka B., *H. Niewodniczanski Institute of Nuclear Physics*, Kraków, Poland, Report No. 1644/OS-1993.
9. Müller S., de La Presa P. and Forker M., *Hyperfine Interact.* **133** (2001), 59.
10. Mestnik-Filho J., Carbonari A. W., Saitovitch H., Silva P. R. J., Investigation of the Magnetic Hyperfine Field at ^{140}Ce on Gd sites in $GdCO_2$ Compound, in Proceedings for the HFI/NQI 2004, Bonn, Germany, R. Vianden, ed., unpublished.

Hyperfine Interactions (2004) 158:163–167
DOI 10.1007/s10751-005-9031-5

The Magnetic Hyperfine Field of ^{111}Cd in the Rare Earth–Nickel Laves Phases RNi_2

S. MÜLLER, P. DE LA PRESA and M. FORKER*
Helmholtz Institut für Strahlen- und Kernphysik der Universität Bonn, Nussallee 14-16, D-53115 Bonn, Germany; e-mail: forker@iskp.uni-bonn.de

Abstract. The magnetic hyperfine field of ^{111}Cd in the C15 Laves phases RNi_2 has been investigated by perturbed angular correlation (PAC) spectroscopy as a function of temperature for the rare earth constituents R = Nd, Sm, Gd, Tb, Dy, Ho, Er, and Tm.

1. Introduction

The rare earth (R) – nickel intermetallic compounds RNi_2 present a C15 superstructure (space group F-43m) in which the 4a sites of the R sublattice are only partially occupied. The concentration of R vacancies decreases from a few percent for R = Pr to zero for R = Lu [1, 2] We have recently shown by a ^{111}Cd perturbed angular correlation (PAC) study [3] that these vacancies are highly mobile at room temperature and that for R = Pr, Nd, Sm and Gd they can be trapped by the PAC probe ^{111}Cd. Although ^{111}Cd resides on the cubic R site of the C15 structure, in some of the paramagnetic RNi_2 ^{111}Cd is therefore subject to an axially symmetric quadrupole interaction (QI) produced by the trapped vacancies. In the present contribution we report an extension of these ^{111}Cd PAC studies to the magnetically ordered phases of RNi_2.

2. Measurements

The PAC measurements were carried out with the 171–245 keV cascade of ^{111}Cd between 4.2 K and 290 K. The compounds RNi_2 with R = Nd, Sm, Gd, Tb, Dy, Ho, Er and Tm were produced by arc melting of the metallic components with the stoichiometry ratio 1 : 2 in an argon atmosphere. The samples, characterized by X-ray diffraction, were doped with ^{111}Cd by diffusion (800°C, 12 h) of carrier-free radioactive ^{111}In into the host lattice.

Figure 1 shows typical spectra of ^{111}Cd in RNi_2 observed at 290 K and 4.2 K. In the case of the heavy R constituents R = Tb, Dy, Ho, Er and Tm, the angular

* Author for correspondence.

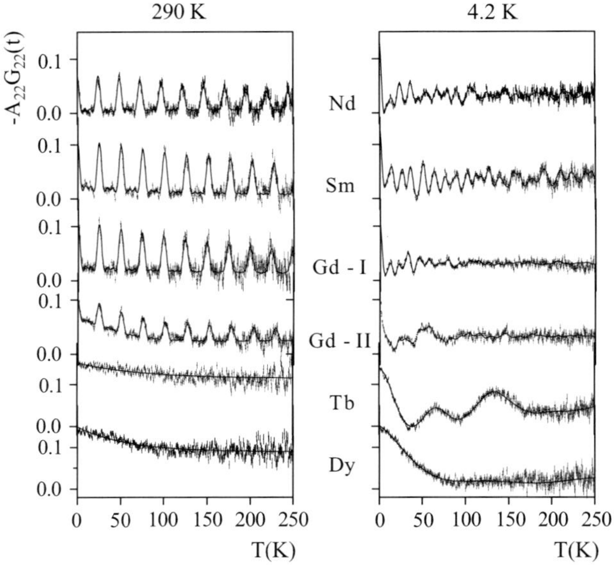

Figure 1. PAC spectra of ^{111}Cd in RNi_2 at 290 K and 4.2 K

correlation in the paramagnetic phase was unperturbed, as expected for ^{111}Cd on the cubic R site (the room temperature spectra of R = Ho, Er, Tm are comparable to those of R = Tb and therefore not included in Figure 1). The room temperature spectra of R = Nd, Sm and Gd, however, are characterized by an axially symmetric QI caused by trapped vacancies. In the case of R = Nd and Sm, the fraction of ^{111}Cd decorated with a vacancy always reached 100 percent at low temperatures. In the case of R = Gd, in one sample (Gd I) all ^{111}Cd had trapped a vacancy, in another sample (Gd II) the majority of probes (~60%) was unperturbed i.e., residing in a cubic environment without a nearest neighbour vacancy. The spectra at 4.2 K can be divided into 2 groups. For R = Nd, Sm and Gd I, the trapped vacancies lead to a perturbation by a combined magnetic and electric interaction. The magnetic hyperfine field B_{hf} was extracted by fitting the perturbation function of a combined interaction to the experimental data.

In the case of R = Tb, Dy, Ho and Er, one has a pure magnetic interaction. Only in $TbNi_2$ the magnetic hyperfine field is sufficiently strong to produce a clearly visible magnetic precession in a 300 ns time window. For R = Dy, Ho, Er the hyperfine field is obviously much weaker, only the initial decrease of the anisotropy could be observed. In these cases, we abstained from measuring the temperature dependence of B_{hf}. In the case of Gd II, the dominant cubic fraction is perturbed by a pure magnetic interaction leading to a clearly visible magnetic precession; the non-cubic fraction is subject to a combined interaction. In $TmNi_2$ at 4.2 K, no magnetic interaction was observed.

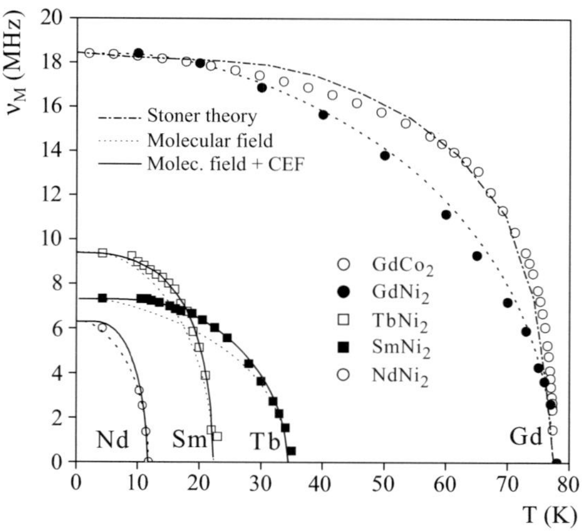

Figure 2. Temperature dependence of the magnetic hyperfine frequency of ^{111}Cd in *R*Ni2; *R* = Gd, Tb, Sm and Nd. The dotted lines show the temperature dependence of the magnetisation, calculated by the molecular-field theory of localized spins. The solid lines represent the molecular-field theory modified by crystal-field interactions. For a comparison with $GdNi_2$, the frequency of ^{111}Cd in $GdCo_2$ (normalized to $GdNi_2$) and the prediction of the free-electron Stoner theory have been included.

3. Results

Figure 2 shows the temperature dependence of the magnetic frequency $\nu_M(T)$ for R = Gd, Tb, Sm and Nd with Curie temperatures of T_C = 78 K, 34 K, 22 K and 11.5 K, respectively. For a comparison, the data for ^{111}Cd in RCo_2 (ref. [4]) have been included. While the experimental values of $GdNi_2$ follow the prediction of the molecular field model for localized spins, the temperature dependence $\nu_M(T)$ of ^{111}Cd in $GdCo_2$ is well described by the free-electron Stoner theory. This is consistent with the fact that in $GdCo_2$ the main part of the hyperfine field B_{hf} comes from the spin polarization of the itinerant Co 3d-electrons. This is also seen in the relatively weak 4f-spin dependence of B_{hf} of ^{111}Cd:RCo_2 (see Figure 3). In RNi_2, however, the 3d-constituent Ni of carries no 3d magnetic moment and B_{hf} is caused by the localized 4f spins alone. The deviations of $\nu_M(T)$ from the molecular field curve observed in RNi_2, R = Tb, Nd, Sm can be attributed to the crystal field interaction of the 4f shell. This is shown by the solid lines in Figure 2 which are obtained when the CEF interaction with the parameters of refs. [5, 6] is included in the molecular field calculation.

Figure 3 compares the spin dependences of the magnetic hyperfine field of ^{111}Cd at 4.2 K in RNi_2, in RCo_2 and in the rare earth metals. The magnetic order of RNi_2 is sustained by indirect 4f–4f exchange, as in the case of the R metals. According to the RKKY theory, the indirect interaction is mediated by a spin-

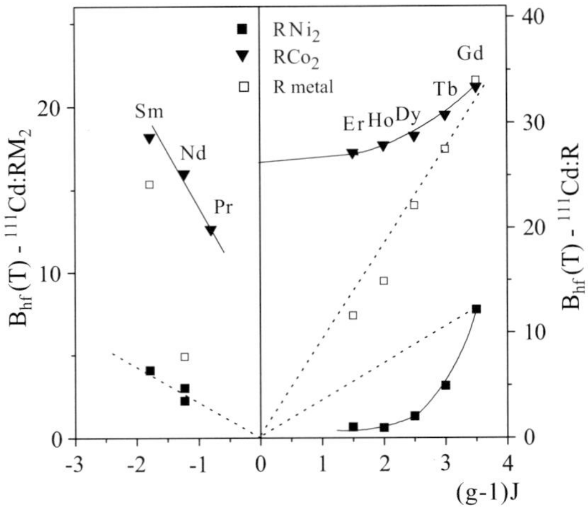

Figure 3. The spin dependence of the saturation value of the magnetic hyperfine field B_{hf} of ^{111}Cd in RNi$_2$ and RCo$_2$ (left-hand scale) and in the R metals (right-hand scale). The dotted lines correspond to the linear relation $B_{hf} \propto \Gamma\,(g-1)\,J$ predicted by the RKKY theory for the constant s-f exchange parameter Γ. The solid lines are to guide the eyes.

polarization of the s-conduction electrons induced by the localized 4f-spins. In another concept, the coupling is provided by intra-atomic 4f–5d exchange and 5d–5d-interaction between the spin polarized 5d-electrons of neighbouring R atoms [7]. In both cases, a spin polarization arises which leads – via Fermi contact interaction – to a magnetic hyperfine field B_{hf} at the nucleus of a probe atom. As long as spin exchange is the dominant mechanism, one expects the hyperfine field to be proportional to the spin projection (g-1)J (dotted lines in Figure 3) and in most 4f-ferromagnets one finds that the relation B_{hf} vs. (g-1)J is in fact approximately linear. As an example of a nearly linear spin dependence of the hyperfine field, the B_{hf} values of ^{111}Cd in the rare earth metals are included in Figure 3. The hyperfine field of ^{111}Cd in RNi$_2$, however, shows a completely different behaviour. In RNi$_2$ with heavy R constituents, B_{hf} vanishes almost completely between Gd and Dy. On the other hand, the B_{hf} values of the light R = Nd and Sm are roughly proportional to (g-1)J. The anomaly of the heavy RNi$_2$ cannot be attributed to the presence of vacancies in RNi$_2$. The analysis of the spectra of Gd I (combined interaction; 2 sites with B_{hf} = 7.95 and 7.54 T, ref. [8]) and Gd II (dominant pure magnetic interaction; B_{hf} = 7.8 T) shows that the presence of a vacancy at the probe changes the hyperfine field by less than 10%. Part of this change is probably a consequence of the fact that a nearest-neighbour vacancy gives rise to a finite dipolar field because the symmetry of the R site is no longer cubic.

The Curie temperatures of the heavy RNi$_2$ show the same anomalous spin dependence as B_{hf}. With T_C = 78 K for GdNi$_2$, one expects T_C ~52 K for TbNi$_2$ (exp. value 34 K) and T_C = 35 K for DyNi$_2$ (exp. value 26 K), if bilinear

Heisenberg exchange is the dominant interaction in the ferromagnetic state. CEF interactions, however, play an important role in RNi_2 with non-zero R orbital angular momentum. Inelastic neutron scattering has shown [5] that due to the CEF the ground state of paramagnetic $TbNi_2$ is non-magnetic. Magnetic order then requires a mixing of the ground and excited CEF levels. This interplay between exchange and CEF interaction reduces the saturation magnetic moment by more than 15 % and is probably the main factor responsible for the sharp decrease of B_{hf} from Gd to Tb.

Acknowledgements

This work has been supported by Deutsche Akademischer Austauschdienst (DAAD) and Deutsche Forschungsgemeinschaft (DFG).

References

1. Latroche M., Paul-Boncour V., Percheron A. and Achard J. C., *J. Less-Common Met.* **161** (1990), L227.
2. Gratz E., Kottar A., Lindbaum A., Mantler M., Latroche M., Paul-Boncour V., Acet M., Barner C., Holzapfel W. B., Pacheco V. and Yvon K., *J. Phys., Condens. Matter* **8** (1996), 835.
3. Forker M., de la Presa P., Müller S., Lindbaum A. and Gratz E., *Phys. Rev., B* **70** (2004), 014302.
4. Forker M., Müller S., de la Presa P. and Pasquevich A. F., *Phys. Rev., B* **68** (2003), 014409.
5. Gratz E., Goremychkin E., Latroche M., Hilscher G., Rotter M., Müller H., Lindbaum A., Michor H., Paul-Boncour V. and Fernadez-Dias T., *J. Phys., Condens. Matter* **11** (1999), 7893.
6. Goremychkin E. A., Natkaniec I., Mühle E. and Chistyakov O. D., *J. Magn. Magn. Mater.* **81** (1989), 63.
7. Campbell I. A., *J. Phys. F* **2** (1972), L47.
8. Müller S., de la Presa P. and Forker M., *Hyperfine Interact.* **133** (2001), 59.

Hyperfine Interactions (2004) 158:169–173
DOI 10.1007/s10751-005-9025-3

Magnetic Order in HoF_3 Studied via Ho Nuclear Spin Probes

W. D. HUTCHISON[1,*], D. H. CHAPLIN[1] and G. J. BOWDEN[2]
[1]*School of Physical, Environmental and Mathematical Sciences, Australian Defence Force Academy, The University of New South Wales, Canberra, Australia;*
e-mail: w.hutchison@adfa.edu.au
[2]*School of Physics and Astronomy, University of Southampton, S017 1BJ Southampton, UK*

Abstract. It is shown that *in-situ* ^{166m}Ho ($I = 7$) in a spherical single crystal of HoF_3 can be used as sensitive internal thermometer to thermally detect NMR (NMR-TDNO) from the 100% abundant stable ^{165}Ho ($I = 7/2$) nuclei. In addition, new ^{166m}Ho NMRON results are reported. Both the ^{166m}Ho NMRON and ^{165}Ho NMR-TDNO spectra show three distinct quadrupolar split sub-resonances, in zero applied field. The data is used to make estimates of the Ho magnetic moments and quadrupole parameters for the ^{166m}Ho and ^{166m}Ho sites.

Key Words: dipolar interaction, enhanced nuclear magnetism, NMR, nuclear orientation.

1. Introduction

In orthorhombic holmium tri-fluoride (HoF_3), the ground and first excited electronic states of the Ho^{3+} ions are both singlets [1]. As a result the Ho nucleus in HoF_3 exhibits pronounced enhanced nuclear magnetism [2, 3]. The compound orders ferromagnetically at $T_C = 0.53$ K primarily due to the dipolar interaction with the hyperfine interaction between electronic and nuclear moments playing a significant role [4]. In HoF_3 there are four Ho^{3+} ions per unit cell and neutron diffraction has been used to show that the four enhanced moments are symmetrically canted at $\sim\pm 25°$ in the *a*–*c* plane, such that the compound exhibits ferromagnetism along the a-axis but is antiferromagnetic along the *c*-axis [3].

In an earlier paper, NMRON studies using *in-situ* ^{166m}Ho ($I = 7$) in a spherical single crystal of HoF_3 were presented and discussed [5]. This new study extends the previous work to show that the ^{166m}Ho ($I = 7/2$) nuclei can be used as an ultra sensitive internal thermometer to thermally detect NMR (NMR-TDNO) of the 100% abundant stable ^{165}Ho nuclei in HoF_3. In addition, new ^{166m}Ho NMRON

* Author for correspondence.

results are presented. In both this NMRON and the ^{165}Ho NMR-TDNO spectra it has been possible to resolve three sub resonances, in zero applied magnetic field. These results are used to determine the magnetic and hyperfine field parameters for both the ^{166m}Ho and ^{165}Ho sites.

2. Experimental details

In this work we used the same 6 mm diameter single crystal HoF_3 sphere, and specially machined cold copper finger, as described in [5]. The sphere was mounted so that a DC magnetic field could be applied along the *a*-axis with a radio frequency field close to the *b*-axis. Low temperature nuclear orientation (LTNO) was monitored via the anisotropies of the daughter ^{166}Er 712 keV and 810 keV γ-rays, using a HP Ge detector aligned along the *a*-axis. The temperature of the cold copper finger was also monitored with a ^{60}CoCo LTNO thermometer.

3. Results

In [5] it was shown that the magnitude of the Ho electronic moments and hence the hyperfine field strength and the LTNO γ-ray anisotropy, depend strongly on the applied magnetic field. It was also observed that in applied fields close to the Lorentz field of the sphere (~0.4 T), and at approximately 70 mK, the rank 4 810 keV γ-ray emission showed superior NMRON signal compared with the rank 2 712 keV γ-ray. The NMRON result of the current study is shown in Figure 1. In this case, the data was collected in zero applied field and at a higher temperature of ~85 mK. Three of the quadrupolar split subresonances are visible via both gamma-rays, although once again the 810 keV has superior signal to noise. The data of Figure 1 was collected with a frequency modulation (FM) amplitude of ±4.0 MHz, but a similar result was obtained with zero FM. In practice, it is quite unusual for a dilute radioactive species to show homogeneous NMR line widths arising from spin–spin coupling. In this case the enhanced nuclear magnetism leads to strong electron mediated spin-spin coupling. Indeed the temperature increase apparent after traversal of the most populated sub resonance is likely to be a result of this coupling allowing heating of the crystal while the ^{60}CoCo thermometer showed no corresponding movement in the cold finger temperature and hence no variation in non-resonant heating via the cold finger. The resonant frequencies for the three lowest subresonances are 1534(1), 1464(2) and 1392(4) MHz, respectively. Thus the average quadrupolar splitting for the ^{166m}Ho is $P/h = -35.4(15)$ MHz with the magnetic centre frequency at 1074(20) MHz. The later leads to an estimate for the Ho^{3+} moment, coupled to a ^{166m}Ho nucleus, of 3.78(8) μ_B under the conditions of Figure 1, where use has

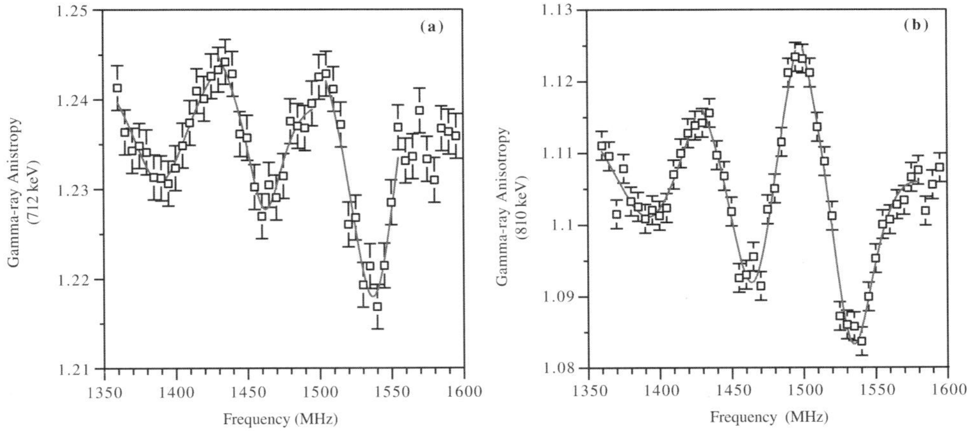

Figure 1. NMRON spectra for ^{166m}Ho in HoF_3 collected in zero applied magnetic field with 5 MHz steps and 600 s counting periods: (a) 712 keV γ-ray and (b) 810 keV γ-ray.

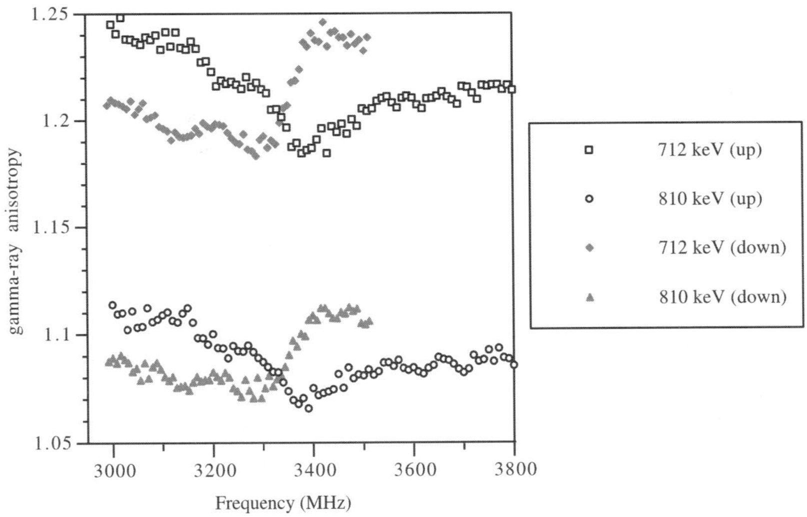

Figure 2. NMR-TDNO on HoF_3 in zero applied magnetic field. Counting with 800 s per point, 10 MHz step up in frequency, and 8 MHz step down.

been made of, (i) the universal applicability of the free ion ratio of hyperfine field to electronic moment [5], and (ii) a Ho^{3+} free ion value of 724.1 T ≡ 10 μ_B [7].

Some ^{165}Ho NMR-TDNO results can be seen in Figure 2. Here the integrated NMR lines distorted somewhat by thermal lag. The stable ^{165}Ho nuclei have a large heat capacity and the ability to remove heat from the insulating HoF_3 lattice is the primary bottle neck. The ^{166m}Ho relaxation rate is estimated to be a few hundred seconds which translates to ~30 s for ^{165}Ho (using g^2 ratio). With thermal lag the step direction of the NMR-TDNO experiment is important. In the

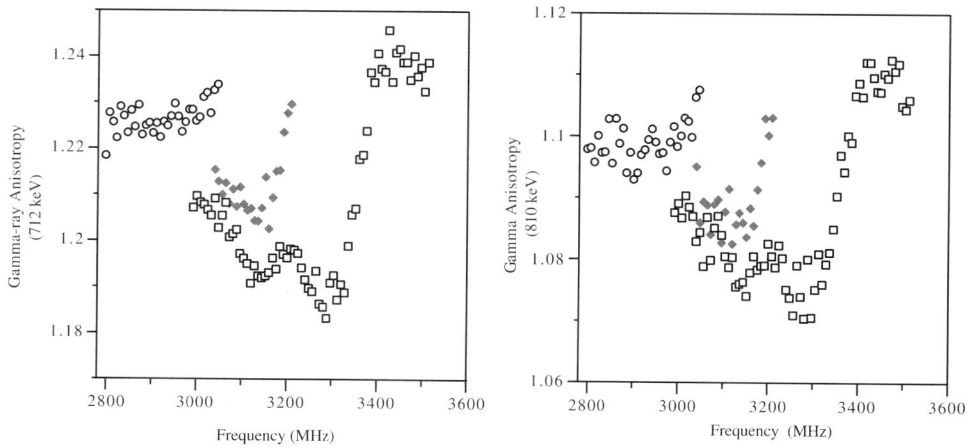

Figure 3. NMR-TDNO measurements on HoF_3 in zero applied magnetic field. Counting with 800 s per point and in −8 MHz steps.

experiment shown in Figure 2 using a step up sequence the second lowest ^{165}Ho sub-resonance is entered first, leaving residual heating prior to entering the lowest sub resonance. While in the step down case it is the second sub-resonance which is obscured by the heating of the dominant lowest sub-resonance. As for the ^{166m}Ho NMRON experiments, the application of FM was found to be inconsequential.

In order to better resolve three of the ^{165}Ho NMR sub resonances, three separate step down sequences were used, allowing cooling between each and with a judicious choice of starting frequencies. Results are shown in Figure 3, from which the three most populated sub resonance frequencies are estimated to be 3360(5), 3183(5) and 3017(15) MHz, respectively. The derived average value of the quadrupolar splitting is $P/h = -87(4)$ MHz and the magnetic centre frequency is 2838(22) MHz. The later leading to a Ho^{3+} moment of 4.31(6) μ_B. for the ^{165}Ho sites.

4. Discussion

Comparable NMRON and NMR-TDNO spectra of the ^{166m}Ho and ^{165}Ho nuclei with frequencies of 1534 MHz and 3360 MHz for the most populated sub-resonances, respectively have been presented. The Ho^{3+} moment at a ^{166m}Ho site is 0.53(10) μ_B less than that measured at a ^{165}Ho nucleus. The difference is due to the different contribution to the total moment from the nuclear enhanced magnetism in the two cases and compares to an estimated maximum difference of 0.8 μ_B at 0 K.

Finally, on the assumption that the total quadrupolar splitting is composed of a pseudo-quadrupolar P_{PQ} and a lattice (including 4*f*) component P_{EQI}, then a comparison of the results for the two isotopes using, g^2 scaling for P_{PQ} and, the published Q moments [8] gives $P_{PQ}(165) = -132$ MHz and $P_{EQI}(165) = +47.4$ MHz. This outcome is in broad agreement with calculations that suggest the P_{PQ} term is the dominant quadrupolar term with a value of ~ -140 MHz. We note that the P_{EQI},(165) value is somewhat larger than earlier HoF_3 estimates [1] but is consistent with scaled Mössbauer effect measurements of EQI's for TmF_3 [9].

Acknowledgements

Assistance with neutron irradiations was provided by the Australian Institute of Nuclear Science and Engineering. The authors acknowledge Drs S.J. Harker and M.J. Prandolini for discussions and assistance with earlier experiments.

References

1. Ram K. and Sharma K. K., *J. Phys. C. Solid State Phys.* **18** (1985), 619.
2. Bleaney B., Gregg J. F., Hill R. W., Lazzouni M., Leask M. J. M. and Wells M. R., *J. Phys. C. Solid State Phys.* **21** (1988), 2721.
3. Brown P. J., Forsyth J. B., Hansen P. C., Leask M. J. M., Ward R. C. C. and Wells M. R., *J. Phys. Condens. Matter* **2** (1990), 4471.
4. Leask M. J. M., Wells M. R., Ward R. C. C., Hayden S. M. and Jensen J., *J. Phys. Condens. Matter* **6** (1994), 505.
5. Hutchison W. D., Chaplin D. H., Harker S. J. and Bowden G. J., *Hyperfine Interact.* **136/137** (2001), 307.
6. Bleaney B. In: Elliott R. J. (ed.), *Magnetic Properties of Rare Earth Metals*, Chapter 8, Plenum Press, London and New York, 1972.
7. Stewart G. A., *Materials Forum* **18** (1994), 177.
8. Ravaghan P., *At. Data Nucl. Data Tables* **42** (1989), 189.
9. Stewart G. A., *Hyperfine Interact.* **13** (1990), 413–418.

Hyperfine Interactions (2004) 158:175–179
DOI 10.1007/s10751-005-9026-2

Spin Flop Studies in the AF Mixed Halide $(^{54}Mn)Mn(Br_xCl_{1-x})_2{\cdot}4H_2O$ via Low Temperature Nuclear Orientation

S. J. HARKER[1,a], W. D. HUTCHISON[1,*], D. H. CHAPLIN[1] and G. J. BOWDEN[2]
[1]*School of Physical, Environmental and Mathematical Sciences, The University of New South Wales at the Australian Defence Force Academy, Canberra, Australia; e-mail: w.hutchison@adfa.edu.au*
[2]*School of Physics and Astronomy, University of Southampton, SO17 1BJ, Southampton, UK*

Abstract. The spin flop behaviour of low anisotropy Mn ions within the mixed halide $(^{54}Mn)Mn(Br_xCl_{1-x})_2{\cdot}4H_2O$ is shown to be interpolative with that of the two terminal compounds, in striking contrast with the dynamics of nuclear spin lattice relaxation.

Key Words: antiferromagnets, mixed halides, nuclear orientation, spin flop.

1. Introduction

Westphal and Becerra [1] were the first to point out that the $Mn(Br_xCl_{1-x})_2{\cdot}4H_2O$ system provides a controllable, electronic magnetic anisotropy, antiferromagnet (AF), exhibiting increasing anisotropy with increasing x. The terminal compound $MnCl_2{\cdot}4H_2O$ ($T_N = 1.62$ K) has been widely studied at and around ^{3}He temperatures and its magnetic unit cell is well determined. Subsequently, it was also the first AF insulator to be resonated in a nuclear magnetic resonance on oriented nuclei (NMRON) environment using in situ gamma anisotropy from ^{54}Mn probes [2]. A comprehensive low temperature nuclear orientation (LTNO) and NMRON study was undertaken by [3] who showed that a magnon energy gap bottleneck was the primary cause for the severely limited LTNO, which occurred even for very weakly doped samples, but could be defeated by applying magnetic fields at or just below the spin flop field of $B_{SF} \sim 0.7$ T. By exploiting very large γ-anisotropies obtained via this novel magnon heat switch they were able to show in detail [3] that the spin flop transition was single stage, notwithstanding four Mn ions per unit cell with the Cl's in low symmetry cis

* Author for correspondence.

[a] Present address: School of Physics and Materials Science, Monash University, Melbourne, Australia.

geometry, and continuous, with no observable hysteresis for field sweeps in opposite directions; that is, second order in character. This unequivocal result was in contrast to the earlier conclusion of Rives and Benedict [4] who obtained substantial evidence for first order, discontinuous behaviour of the spin flop transition at their lowest temperature of 0.297 K, using differential magnetic susceptibility. On the other hand, the ~ ×4 higher anisotropy terminal compound (^{54}Mn)$MnBr_2 \cdot 4H_2O$ ($T_N = 2.12$ K) exhibits a clear two stage, smaller angle, spin flop, both transitions being continuous and independent of field sweep direction [5]. It is the purpose of this note to demonstrate that the spin flop behaviour, depending on internal static exchange and anisotropy fields, is interpolative, for varying x. The dynamics of nuclear spin lattice relaxation, however, are far from interpolative as has been demonstrated in [6] via dramatically enhanced nuclear spin cooling in the mixed halide system. Additionally, we demonstrate the single stage character of the spin flop is robust in that it does not split due to misalignment between field and easy axis as one proceeds towards the Br rich end. The pseudoquadrupolar split NMRON lines from one of the mixed halide crystals is explored in a companion paper [7]. In anticipation of those results, the robustness of the spin flop field in the presence of strong magnon stirring via non-resonant radiofrequency fields is also demonstrated.

2. Experimental details

The very dilute, isoelectronic radioactive ^{54}Mn probes were incorporated substitutionally into the single crystal host by growing the crystal from doped saturated solutions using a much smaller nonactive seed of good morphology. (^{54}Mn)$Mn(Br_xCl_{1-x})_2 \cdot 4H_2O$ single crystals with $x = 0.46(0.37)$ and $0.80(0.74)$ by mass (density) were prepared by slow evaporation from shallow, almost completely covered thin slots in Perspex blocks [8]. Many crystals were grown in total over the duration of the project. Typical radioactive strengths ranged from 3.4 to 19.4 μCi with masses from 16 mg up to 67 mg. The post growth density analyses of the mixed halides indicated that the final crystals could deviate significantly from starting halide compositions by up to 50%, with the more common, but not exclusive, tendency to end up chlorine rich. A second single crystal with x ~0.46 was fortuitously grown at a much later date and was employed to check on reproducibility of the spin flop transition. The single crystals were mounted with Balzers silver paint on the Cu cold finger of a ^{3}He–^{4}He dilution refrigerator with 8 mK base temperature in the absence of radioactive heating. Samples were usually, but not always, bottom loaded to be assured of alignment between detectors, magnetic field and easy axis of magnetisation. All the γ-ray anisotropies shown refer specifically to the 835 keV ^{54}Cr daughter transition. In addition, the temperature of the Cu cold-finger was monitored using a ^{60}CoCo thermometer. Finally, every care was taken to

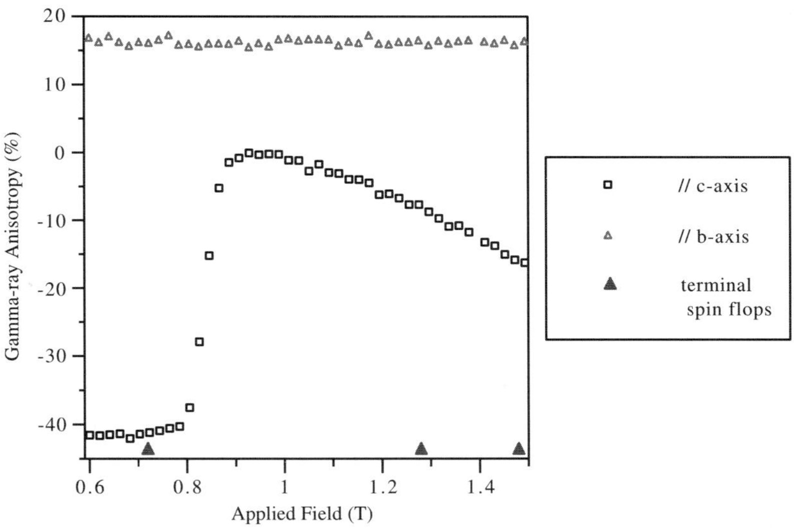

Figure 1. A slow sweep upwards through the spin flop transition for the first $(^{54}Mn)Mn(Br_{0.46}Cl_{0.54})_2 \cdot 4H_2O$ single crystal. Arrows indicate spin flop fields for the terminal compounds.

prevent dehydration of the hydrated salts during initial cool-down of the cryostat, and by generously covering them in Apiezon N grease.

3. Results and discussion

Figure 1 shows the γ-anisotropies from two orthogonal HPGe detectors monitoring a $(^{54}Mn)Mn(Br_{0.46}Cl_{0.54})_2 \cdot 4H_2O$ single crystal subject to a very slowly rising applied magnetic field through the spin flop field. The surprisingly sharp transition is monitored from a detector placed coaxial with the c axis and applied magnetic field, and hence close to the magnetic easy axis. The featureless upper γ-anisotropy is for a orthogonal detector (along the *b*-axis), and as such, is always orthogonal to the geometrically constrained co-planar spin flop. The three arrows on the abscissa illustrate where the two terminal compounds spin flop; 0.72 T for $(^{54}Mn)MnCl_2 \cdot 4H_2O$ [3] and 1.28 T and 1.48 T for the lower symmetry, higher magnetic anisotropy $(^{54}Mn)MnBr_2 \cdot 4H_2O$ [5], all uncorrected for demagnetising fields. The corresponding sweep down (not shown) provided no evidence for hysteresis.

Figure 2 shows the results obtained for the Br-rich $(^{54}Mn)Mn(Br_{0.80}Cl_{0.20})_2 \cdot 4H_2O$ single crystal. The sweep up (Figure 2(a)) and sweep down (Figure 2(b)) anisotropies were monitored using two HPGe detectors. Here, the lowermost γ-anisotropy $W(\theta = 0)$ was obtained with a detector coaxial with the magnetic field and the *c*-axis of the crystal, and therefore misaligned, by a few degrees, with the easy magnetic axis. The second detector was deliberately placed at $\theta = 114°$, in order to monitor the spin-flop transition with both detectors. The data convey two important effects. Firstly, the broadened transition, with a reduced spin flop

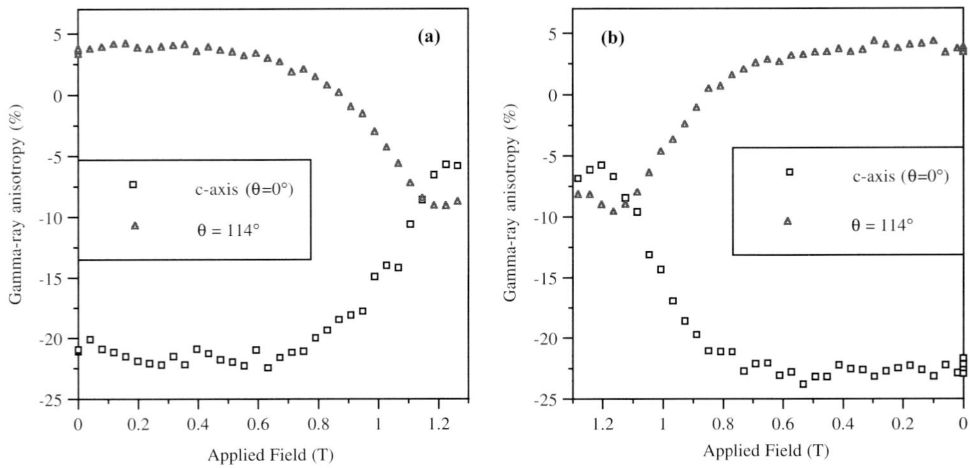

Figure 2. (a) A slow sweep upwards through the spin flop transition for the $(^{54}Mn)Mn(Br_{0.80}Cl_{0.20})_2 \cdot 4H_2O$ single crystal. (b) A slow sweep downwards for the same crystal. Note that the second detector is at $\theta = 114°$ in this case.

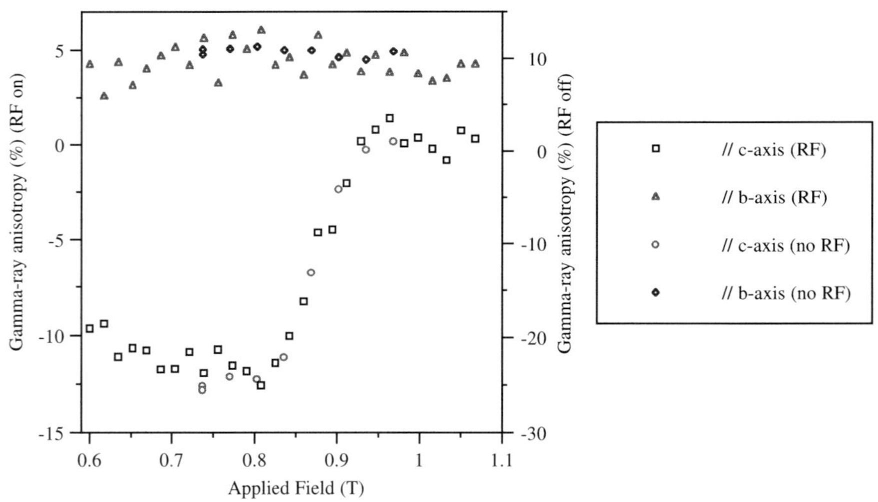

Figure 3. A slow sweep upwards through the spin flop field for a $(^{54}Mn)Mn(Br_{0.46}Cl_{0.54})_2 \cdot 4H_2O$ single crystal in the presence of strong non-resonant RF fields. A traversal of the spin flop region in the absence of RF is shown for comparison.

angle, is moved upwards to a spin flop field of ~1.0(1) T, as expected. Secondly, there is no detectable splitting into two distinct steps, as evident in the terminal compound $(^{54}Mn)MnBr_2 \cdot 4H_2O$ [5]. But it must be acknowledged that the transition is rather broad.

Figure 3 shows the spin flop field for a $(^{54}Mn)Mn(Br_{0.46}Cl_{0.54})_2 \cdot 4H_2O$ single crystal when the crystal was subject to intense, non resonant radiofrequency (RF) fields, +3 dBm, 2.5 GHz, in attempts to induce measurable effects via magnon stirring. The significantly reduced γ-anisotropy of −10% for the axial detector was first thought to be due to excess eddy current heating in the larger copper mass of the top loader configuration used in this particular assembly. Later, it was found that the presence of remarkably small RF fields (−12 dBm), even in bottom loading configuration containing far less Cu in the vicinity of the RF coil, could non thermally degrade the γ-anisotropy to precisely this level whenever the applied field was zero, and over a very wide range of non resonant frequencies. Applying dc fields of a few tenths of a tesla markedly suppressed the effect as was shown in figure 8 of [6]. This has far reaching negative consequences on the sensitivity for weak NMRON signals in the mixed halide system in zero field, but in the context of this paper the spin flop transition remains robust.

Acknowledgements

This work was supported by the Australian Research Council. The authors wish to warmly acknowledge the early assistance of Drs Tobias Funk and Mark Prandolini. Thanks are also due to Dr Vernon Edge for carrying out density determinations on the single crystals.

References

1. Westphal C. H. and Becerra C. C., *J. Phys. C. Solid State Phys.* **13** (1980), L527.
2. Kotlicki A., McLeod B. A., Shott M. and Turrell B. G., *Phys. Rev., B* **29** (1984), 26.
3. Allsop A. L., de Araujo M., Bowden G. J., Clark R. G. and Stone N. J., *J. Phys. C. Solid State Phys.* **17** (1984), 915.
4. Rives J. E. and Benedict V., *Phys. Rev., B* **12** (1975), 1908.
5. Prandolini M. J., Hutchison W. D., Leib J., Chaplin D. H. and Bowden G. J., *Hyperfine Interact. C* **1** (1996), 84.
6. Chaplin D. H. and Hutchison W. D., *Hyperfine Interact.* **136** (2001), 239.
7. Hutchison W. D., Harker S. J., Chaplin D. H. and Bowden G. J., this conference.
8. Le Gros M., Kotlicki A. and Turrell B. G., *Hyperfine Interact.* **77** (1993), 131.

Hyperfine Interactions (2004) 158:181–187
DOI 10.1007/s10751-005-9027-1

Local Magnetic Fields in Some Bismuth-Based Diamagnets. A Survey of NQR Data

E. A. KRAVCHENKO[1,*], V. G. ORLOV[2], V. G. MORGUNOV[1], YU. F. KARGIN[1], A. V. EGORYSHEVA[1] and M. P. SHLIKOV[2]
[1]*Institute of General and Inorganic Chemistry, Russian Academy of Sciences, Moscow, 119991 Russia; e-mail: ekravchenko@igic.ras.ru*
[2]*Russian Research Center, Kurchatov Institute, Moscow, 123182 Russia*

Abstract. Evidence that local ordered magnetic fields from 30 to 250 G exist in bismuth-based diamagnetic compounds comprising neither *d*- nor *f*-elements was given by ^{209}Bi NQR spectroscopy and supported by SQUID measurements of α-Bi_2O_3. The NQR experiments involved a study of the zero-field line shapes, analysis of the Zeeman-perturbed patterns, and examination of the zero-field spin-echo envelopes in single crystals and powders. The results of the experiments followed by computer modeling of the observed spectra were interpreted assuming that ordered magnetic fields are located at the bismuth sites in α-Bi_2O_3, Bi_3O_4Br, $Bi_2Al_4O_9$, $Bi_4Ge_3O_{12}$, $Bi_2Ge_3O_9$ and perhaps in $Bi_3B_5O_{12}$. A survey of the related ^{209}Bi NQR data is here presented.

Key Words: computer modeling, local magnetic fields, nuclear quadrupole resonance.

1. Introduction

Many bismuth-based compounds find diverse applications as diamagnetic materials having valuable physical properties, and α-Bi_2O_3 is a parent reagent for the preparation of HTSC's. However, splittings that looked like typical Zeeman patterns were observed in the zero-field ^{209}Bi NQR spectra of α-Bi_2O_3 and Bi_3O_4Br [1, 2]. Either splittings or line shape asymmetry were also found, in conflict with the X-ray data, in the zero-field ^{209}Bi NQR spectra of $Bi_2Al_4O_9$, $Bi_2Ge_3O_9$, $Bi_4Ge_3O_{12}$. This was understood as an indication that local ordered magnetic fields (H_{loc}) exist in these compounds, the zero-field splittings being consistent with values of H_{loc} from 25 to 250 G [3]. The extended ^{209}Bi NQR Zeeman studies on $Bi_4Ge_3O_{12}$ oriented single crys1tals, followed by a computer modeling of the spectral patterns, were carried out to conclude that an anti-ferromagnetically ordered H_{loc} of the order of 30 G exists in $Bi_4Ge_3O_{12}$ [4, 5].

An examination of powdered $Bi_2Ge_3O_9$ revealed distinctive zero-field modulations displayed by the ^{209}Bi spin–echo envelopes (SEE) [6]. The effect

* Author for correspondence.

was explained, based on the results of modeling the SEE within the density matrix formalism, by the presence of H_{loc} of the order of 65 G at the Bi sites. Evidence was also obtained for the presence of H_{loc} in $Bi_3B_5O_{12}$ [6].

The SQUID-measurements of α-Bi_2O_3 confirmed that it is not diamagnetic in the conventional sense [7, 8]. The magnetization of a single crystal was positive at 4.2 K and highly anisotropic in magnetic fields H_e up to 3−5 kOe [7]. Upon cooling the crystal in a magnetic field (FC), a dramatic increase of the magnetic moment of α-Bi_2O_3, compared to that upon cooling in zero field (ZFC), was observed as H_e increased [8].

The existence of ordered magnetic fields in the compounds comprising neither transition nor rare earth elements is a new phenomenon, and this article surveys the related ^{209}Bi NQR data.

2. Experimental

The preparation of powder samples, crystal growth procedures, results of purity control as well as the details of NQR experiments were described in [2–6]. The procedure of modeling the ^{209}Bi line splittings was described in [3–5]. The main steps of modeling the SEE based on the matrix density formalism were described in [6].

3. Results

In the zero-field ^{209}Bi NQR spectrum, four transition frequencies, ν_i, between five doubly degenerate energy levels $|\pm m\rangle$ are known to be observed (the ^{209}Bi nuclear spin is $I = 9/2$). The zero-field recording made on ~0.5 cm^3 single crystals of α-Bi_2O_3 (monoclinic, $P2_1/c$), however, presented typical Zeeman perturbed patterns: a quadruplet for ν_1 Bi(1) and doublets for all the rest of the lines assigned to both crystallographic Bi(1) and Bi(2) positions (Fig. 1 in [1]). Notable asymmetry was also observed for the zero-field ^{209}Bi lineshapes in powdered Bi_3O_4Br (orthorhombic, *Pnan*) (Fig. 4 in [2]). In $Bi_2Al_4O_9$ (orthorhombic, *Pbam*), $Bi_4Ge_3O_{12}$ (eulitine structure, $I\bar{4}3d$) and $Bi_2Ge_3O_9$ (benitoite-type structure, $P6_3/m$), the zero-field ^{209}Bi NQR spectra were also either split or asymmetric, in conflict with the crystallographic equivalence of the Bi-sites in the unit cells (Fig. 1 in [3]). Assuming that internal magnetic fields of unknown nature cause the splittings of the ^{209}Bi lines, the values and orientations of H_{loc} with respect to the EFG q_{zz} axes were estimated by computer simulation of the line shapes (Table I) [2, 3].

For a more detailed characterization of H_{loc}, an extended Zeeman ^{209}Bi NQR study ($0 \le H_e \le 500$ Oe) of a high-quality single crystal $Bi_4Ge_3O_{12}$, followed by a computer modeling of the spectral patterns, was made assuming that the ^{209}Bi spin system is simultaneously perturbed by H_e and H_{loc} [4, 5]. The symmetry properties of the EFG tensor were analyzed prior to the modeling to

Table I. ^{209}Bi NQR spectra (MHz) and local magnetic fields (H_{loc}, θ, φ) determined from zero-field splittings of the spectra

Compound		T, K	Transition frequencies, MHz[a]				e^2Qq/h, MHz	η	H_{loc}, G	θ^o	φ^o
			ν_1	ν_2	ν_3	ν_4					
α-Bi_2O_3	(1)	77	25.2	45.8	69.4	92.8	556.7	0.13	171	38	0
	(2)		39.3	37.2	58.3	79.4	482.6	0.40	136	0	–
Bi_3O_4Br	(1)	77	22.6	25.1	39.6	53.3	322.6	0.31	250±10	52±2	6±2
	(2)		39.3	37.2	58.3	79.4	482.6	0.40			
$Bi_2Al_4O_9$		77	33.15	50.5	78.0	104.1	626.4	0.19	150±10	0±10	0±5
$Bi_2Ge_3O_9$		77	25.7	51.6	77.3	103.2	618.8	0.0	60±10	25±5	–
$Bi_4Ge_3O_{12}$		300	20.4	40.9	61.3	81.7	490.3	0.0	25±5	45±5	–
$Bi_3B_5O_{12}$	(1)	77	37.40	53.6	83.2	111.6	671.7	0.21			
	(2)		72.16	52.7	73.25	103.0	640.9	0.6			

[a] Averaged over splittings.

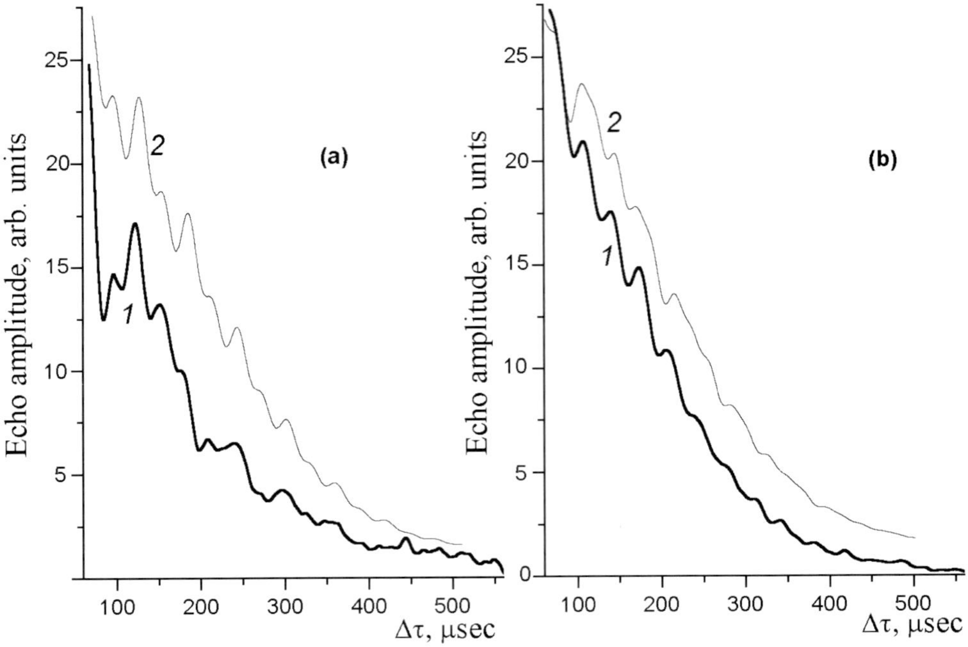

Figure 1. The observed (1) and modeled (2) zero-field ^{209}Bi spin–echo envelopes assigned to the transitions ν_1 ($\Delta m = 1/2 - 3/2$) (a) and ν_2($\Delta m = 3/2 - 5/2$) (b) in powdered $Bi_2Ge_3O_9$ [6].

find that H_{loc} of the order of 30 G is present at each of the 16 Bi sites, being pairwise opposite in direction, i.e., ordered antiferromagnetically [5]. The observed spectra were represented by a superposition of four constituent doublets exchanging their relative positions upon rotation of the crystal around the H_{rf} direction (Fig. 2 in [5]).

No line splitting was found in the zero-field ^{209}Bi NQR spectrum of $Bi_3B_5O_{12}$ (orthorhombic, *P*nma), although the resonance intensity ratio measured in the single crystal for the ν_2 Bi(1) and ν_2 Bi(2) transitions (Table I) deviated considerably from that in a pure NQR spectrum [6]. In addition, the SEE for the ν_2 Bi(1) line was modulated in zero field revealing the presence of the internal source of line splitting. Both findings were understood as an indication that a weak H_{loc} exists in $Bi_3B_5O_{12}$ [6].

Unambiguous evidence for the internal source of line splitting was given by the zero-field modulations of the ^{209}Bi SEE assigned to the ν_1 and ν_2 transitions in powder $Bi_2Ge_3O_9$ (Table I; Figure 1). The program for modeling the SEE, based on the matrix density formalism, was extended to the multilevel system of non-equidistant energy levels and involved the analysis of the ^{209}Bi EFG symmetry in the $Bi_2Ge_3O_9$ structure [6]. The features of the simulated curves were in good agreement with the experiment implying that $H_{loc} = (65 \pm 5)$ G exists in $Bi_2Ge_3O_9$ and directed at an angle $(83 \pm 1)°$ to the ^{209}Bi EFG q_{zz} axis [6].

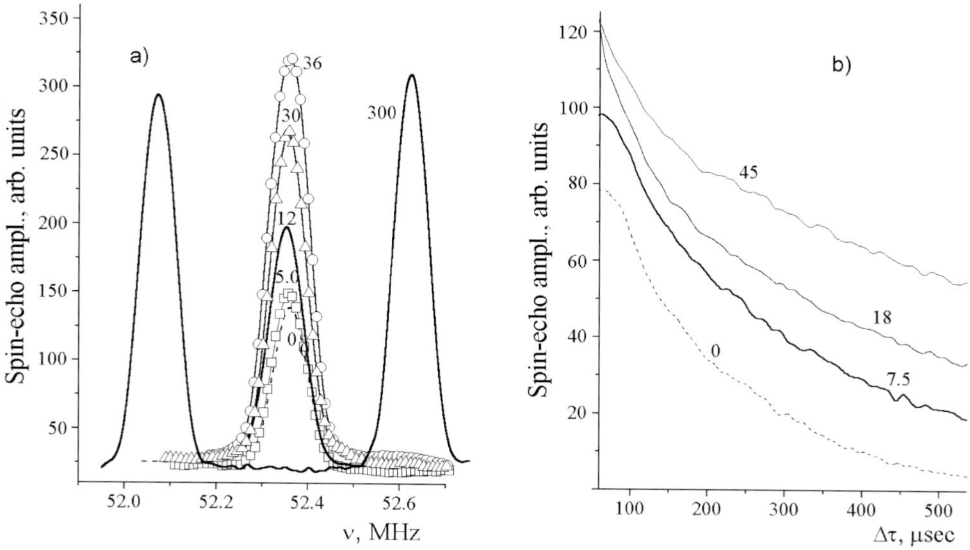

Figure 2. ^{209}Bi NQR line assigned to the ν_2 Bi(2) transition in $Bi_3B_5O_{12}$ single crystal in magnetic fields (a) and spin–echo envelopes in powdered $Bi_3B_5O_{12}$ in magnetic fields (b). The field strengths (Oe) are shown by *numbers*.

4. Discussion

The ^{209}Bi NQR data [1−6] provide reasons to believe that H_{loc} of the order of 30−250 G exist in α-Bi_2O_3, Bi_3O_4Br, $Bi_2Al_4O_9$, $Bi_2Ge_3O_9$, and $Bi_4Ge_3O_{12}$ which are traditionally considered as diamagnets. This on-site effect is much smaller in antimony compounds. Although the explanation implying H_{loc} is consistent with the observed spectra, no unambiguous proof of either the origin of the magnetism or the mechanism of ordering the magnetic moments over a wide temperature range (in α-Bi_2O_3, it exceeds 400°) is available at the moment. The bismuth 6*sp*-valence electrons (most probably, an unshared electron pair) bearing the effective magnetic moment $\mu_{eff} \approx 0.1\mu_B$ [7, 8] are capable of producing the hyperfine filed (~50 G) of the same order as that estimated from the NQR spectra, which suggests the hyperfine nature of H_{loc}. It is reasonable to think that the ordering of so small a μ_{eff} over such a wide temperature range results from the ordering of the electric dipole moments via strong coupling between the magnetic and electronic systems of the compounds. High-temperature anomalies of the dielectric permeability, electric conductivity, etc., accompanied by no change in the crystal symmetry were observed in α-Bi_2O_3 admitting the existence of antiferroelectric correlations [9]. Moreover the magnetoelectric effect (the dielectric polarization of the crystal induced by an applied magnetic field) as well as increased magnetization of α-Bi_2O_3 under the field-cooling (FC) conditions are indicative of the above mentioned coupling [7, 8]. All the discussed compounds featured extraordinary sensitivity of the ^{209}Bi NQR signals to the magnetic fields. A dramatic increase in the resonance intensities depending on the ^{209}Bi EFG symmetry and mutual orientation of H_e and the EFG axes was observed in Bi_3O_4Br [1], α-Bi_2O_3 [1], $Bi_4Ge_3O_{12}$ [5], $Bi_3B_5O_{12}$ [6] upon applying H_e (see, for instance, Figure 2a). A similar phenomenon could explain the deviation of the ν_2Bi(1) and ν_2Bi(2) realtive intensities in $Bi_3B_5O_{12}$ from that in a pure NQR spectrum: H_{loc} unequally influences the appropriate ^{209}Bi EFG tensors having different symmetries and space orientations [6]. In $Bi_4Ge_3O_{12}$, a strong increase in the ^{209}Bi line intensity was suggested to result from an influence of H_e on the spatial orientation of the EFG axes [5]. Recent experiments showed that the intensity increase may be caused by the influence of H_e on the ^{209}Bi spin–spin relaxation time (Figure 2b). A study of this problem for multispin systems with nonequidistant energy levels is, however, ahead.

No intensity increase was observed in related compounds of other Main Group elements. Thus, the ^{115}In NQR (the ^{115}In spin is $I = 9/2$) experiments in magnetic fields on compounds In_xB_y (B = O, Se, Te), where the valence electron shell of the In atom has no unshared pair of electrons, revealed no effect of this kind.

5. Conclusions

The ^{209}Bi NQR spectroscopy in zero and weak (below 500 Oe) constant magnetic fields was found to be a sensitive tool which permitted revealing previously unknown magnetic properties of several bismuth-based compounds, traditionally classified as diamagnets. The results of the ^{209}Bi NQR zero-field, Zeeman experiments and examination of the SEE supported by the SQUID measurements of α-Bi_2O_3 revealed a new phenomenon: local ordered magnetic fields from 30 to 250 G exist in the compounds containing neither transition nor rare earth elements. The key to understanding the phenomenon seems to lie in the strong coupling between the magnetic and electronic systems of the substances.

Acknowledgements

The authors are grateful to the Russian Academy of Sciences (The Fundamental Research Program of the Department of Chemistry and Materials Sciences) and Russian Foundation for Basic Research (project 02-03-33280) for the support. V.G. O. and M.P. S. are grateful to the Council for the support of the Leading Scientific Schools of Russia (grant NS-1572.2003.2).

References

1. Kravchenko E. A. and Orlov V. G., *Z. Naturforsch.* **49a** (1994), 418.
2. Ainbinder N. E., Volgina G. A., Kravchenko E. A., Osipenko A. N., Gippius A. A., Fam S. H. and Bush A. A., *Z. Naturforsch.*, **49a** (1994), 425.

3. Kravchenko E. A., Orlov V. G., Fam S. H. and Kargin Y. F., *Z. Naturforsch.* **53a** (1998), 504.
4. Orlov V. G. and Kravchenko E. A., *Physica B* **259–261** (1999), 564.
5. Kravchenko E. A., Kargin Y. F., Orlov V. G., Okuda T. and Yamada K., *J. Magn. Magn. Mater.* **224** (2001), 249.
6. Kravchenko E. A., Morgunov V. G., Kargin Yu. F., Egorysheva A. V., Orlov V. G. and Shlikov M. P., *Appl. Magn. Reson.* **27**(1–2) (2004), 65.
7. Kharkovskii A. I., Nizhankovskii V. I., Kravchenko E. A. and Orlov V. G., *Z. Naturforsch.* **51a** (1996), 665.
8. Nizhankovskii V. I., Kharkovskii A. I. and Orlov V. G., *Ferroelectrics* **279** (2002), 175.
9. Orlov V. G., Bush A. A., Ivanov S. A. and Zhurov V. V., *J. Low-Temp. Phys.* **105** (1996), 1541.

Hyperfine Interactions (2004) 158:189–193
DOI 10.1007/s10751-005-9028-0

Investigation of the Magnetic Hyperfine Field at ^{140}Ce on Gd Sites in $GdCo_2$ Compound

J. MESTNIK-FILHO[1,*], A. W. CARBONARI[1], H. SAITOVITCH[2] and P. R. J. SILVA[2]
[1]*Instituto de Pesquisas Energéticas e Nucleares, IPEN, São Paulo, SP, Brazil; e-mail: jmestnik@ipen.br*
[2]*Centro Brasileiro de Pesquisas Físicas, CBPF, Rio de Janeiro, RJ, Brazil*

Abstract. Perturbed angular correlation experiments on ^{140}La–^{140}Ce probes substituting for the Gd positions on $GdCo_2$ demonstrate that the magnetic hyperfine field (MHF) on Ce atoms follows an expected Brillouin function but with a substantially reduced saturation value as compared with the MHF acting on free Ce^{+3} ions. The results were interpreted with the aid of first principles electronic structure calculations showing that the reduced value of the MHF is a consequence of a small orbital contribution to the MHF which was attributed to the de-localization of the Ce 4f electronic state.

Key Words: electronic structure, hyperfine field, rare-earth compounds.

1. Introduction

The $GdCo_2$ belongs to a series of rare earth (R) intermetallic compounds where both the rare-earth and Co atoms present magnetic moments exhibiting ferrimagnetic order. In rare earth–transition–metal binary alloys the magnetic properties are determined by the interaction of the magnetic moments of the rare earth 4f electrons with the itinerant 3d electrons of the transition metals. In RCo_2 compounds the 3d magnetization is reached by the local f–d exchange mechanism, i.e., the 4f spins polarize the R 5d band which is hybridized with the Co 3d band. The result is an antiparallel ordering of Co atoms relative to the 4f R spins.

In this work we present the results of first principles calculations of the magnetic hyperfine field (MHF) acting on cerium probes substituting for Gd in $GdCo_2$. It it is seen that the Ce 4f electrons in this system are de-localized and sense the crystalline field effects that substantially quench the orbital moment [1, 2].

* Author for correspondence.

2. Details of the experiment

The samples were prepared by arc melting the constituent high purity elements under argon atmosphere along with neutron irradiated ^{140}La substituting for 0.1 atom percent of Gd. The TDPAC experiments were performed utilizing the 329–487 keV γ–γ cascade of ^{140}Ce populated from the β decay of ^{140}La. The measurement of the magnetic interaction was performed at the ^{140}Ce 2083 keV level, 4+ spin state. Since the quadrupole moment of this state is known to be very small, no quadrupole interactions are expected. The measurements of the magnetic interactions were performed at the temperature range of 8–400 K with a setup consisting of four BaF2 detectors. The resolution of the detector system was 0.6 μs.

3. Electronic structure calculations

The calculations were performed within the scheme of the density functional theory (DFT). In this formalism [3], all the observables are given as a functional of the electronic density (in our case, the spin densities). In particular, the total energy of the system is given by:

$$E_{tot}\left(\sigma^{\uparrow},\sigma^{\downarrow}\right) = T_s + E_{ee} + E_{Ne} + E_{xc} + E_{NN}. \tag{1}$$

Here, $\sigma^{\uparrow}$ and $\sigma^{\downarrow}$ are the spin densities, T_s the single particle kinetic energy, E_{ee} the Hartree component of the electron–electron interactions, E_{Ne} the nuclei-electron Coulomb interaction, E_{xc} the exchange and correlation energy and E_{NN} the Coulomb nuclei-nuclei interaction energy. The aim in these calculations is to find the spin densities which minimize the energy or, equivalently, to solve the one particle Kohn–Sham equation in a self-consistent way [4]. The method employed here utilizes a basis set consisting of augmented plane waves plus local orbitals (APW + lo) as embodied in the WIEN2k [5] computer package. The main limitation of this method lies in the description of the exchange and correlation energy E_{xc}. The usual approach is the local spin density approximation (LSDA) or the improvements with the generalized gradient approximation (GGA), but both suffer for the improper description of the correlation energy and then especially fails to describe the open d or f electronic shells. The calculations were performed with the experimental value for the $GdCo_2$ cell parameter, namely a = 7.28 Å. The number of plane waves was limited to $k_{\max}$ = 7/RMT, with the muffin–tin radius RMT ~ 2.4 a.u. The charge density was Fourier expanded up to $G_{\max}$ = 14 and, and for the Brillouin zone integrations, a tetrahedral mesh of 1200 k points was utilized. Exchange and correlations effects were treated with generalized gradient corrections [6]. The valence states were treated within the scalar–relativistic approach, taking into account the spin–orbit coupling, while the core states were relaxed in a fully relativistic manner. The calculation of the magnetic hyperfine field, also implemented into the WIEN2k code, was performed following the formulas due to Blügel et al. [7], which include relativistic corrections. The calculations were extended to the Ta and Cd impurities, in order to test for the reliability of the calculations to reproduce the MHF on different electronic configurations of the probes. It is expected that the correlation effects would be small for the f states due to the half-filled shell of Gd, just one f electron in case of Ce and the completely populated shell of Ta.

4. Results and discussion

In Figure 1 it is shown the measured Larmor frequency as a function of temperature for the Ce probes in $GdCo_2$. It can be seen that two distinct sites are occupied by the probes with different values of the MHF. The curve with higher values of MHF is attributed to Ce probes occupying Gd sites in $GdCo_2$ since it follows the Brillouin function with the expected Curie temperature of the compound, $T_C = 395$ K. The saturation value of the MHF, 24.6 T at $T = 8$ K, is substantially smaller than the MHF of isolated Ce^{+3} ions in insulators, namely, H_{hf} (Ce^{+3}) = 192 T. There are two possibilities to interpret the observed small MHF: 1) In $GdCo_2$ the Ce probes sense a large transferred MHF highly compensated by a large orbital contribution to the MHF, since they have opposite signals, or 2) The Ce $4f^1$ state senses the crystalline field and partially hybridize with neighboring electronic states and then loses the strictly local character found in insulators. In this case a smaller orbital contribution to the MHF is expected.

In Table I we present the results of the calculations of MHF for Ce, Ta and Cd probes substituting for Gd and for the Gd itself in $GdCo_2$, together with the experimental results taken from reference [8]. It is seen that for Ce probes, the calculated MHF agree quite well with the experiment if it is accepted that both the measured and calculated MHF have the same sign. The result then points to the second possibility stated above, namely, that the orbital contribution to the MHF is small, and thus, the Ce $4f^1$ electrons are partially de-localized. The orbital contribution to the Ce MHF resulted to be -10.6 T.

We can see that the Fermi contact contribution is dominant in all the probes. It can also be seen that the model reproduces quite well the MHF in the case of Ce and Ta but not for Cd. Since in this case we expect a very small core polarization, as it is indeed seen in Table I, the failure would be due to the improper description of the Cd 5s electrons. A possible reason for this result is that the super-cell utilized in the calculations is not sufficiently large, enhancing the Cd–Cd interactions in the simulations even though they were performed with eight atomic layers between two Cd atoms. No experimental value for Gd was found in the literature. Comparing Gd and Ce, which differ only on the number of f electrons, a larger MHF results in the case of Gd, with larger valence and core contributions, due to the larger Gd magnetic moment, as expected. On the other hand, these contributions to the MHF have opposite signs resulting in a relatively small total Fermi contact MHF especially in the case of Gd. In this

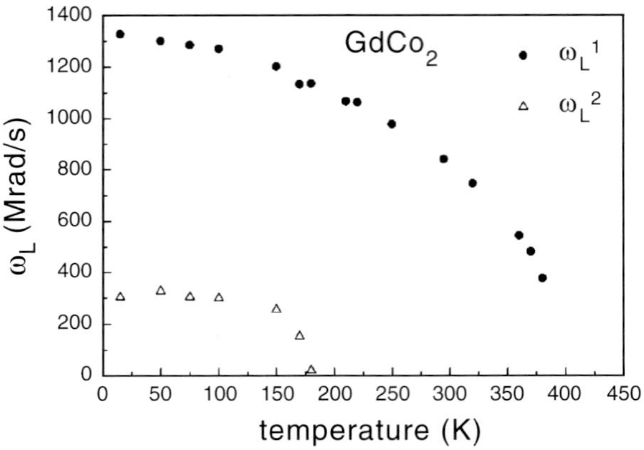

Figure 1. Temperature dependence of the Larmor frequency at ^{140}Ce.

Table I. Calculated and experimental magnetic hyperfine fields at the indicated probes and the Gd site in $GdCo_2$

Probe	Contact	Valence	Core	Orbital	Dipolar	Total	Experiment
Gd	61.0	346.7	−285.7	−3.3	0.38	58.1	–
Ce	33.1	52.3	−19.2	−10.6	0.145	22.4	24.6
Ta	−15.9	−16.6	0.72	−0.27	−0.039	−16.2	19.7
Cd	−34.6	−34.7	0.11		0.0041	−34.6	21.8

Values given in T. Only the magnitude of MHF is given for the experimental value. For the Fermi contact field the valence and core contributions are given in addition.

way, if one imagines a large orbital contribution at the Ce probes, the polarization of valence and core electrons would increase accordingly but not the total contact field. In this case we would expect a large total MHF, being the orbital part the main contribution. This seems not to be the present case, enforcing the argument that the main effect on Ce $4f^1$ electrons in this material is de-localization.

Acknowledgement

Partial support of this research was provided by the Fundação de Amparo à Pesquisa do Estado de São Paulo (FAPESP).

References

1. Eriksson O., Nordstrm L., Brooks M. S. S. and Johansson B., *Phys. Rev. Lett.* **60** (1988), 2523.
2. Delin A., Oppeneer P. M., Brooks M. S. S., Kraft T., Wills J. M., Johansson B. and Eriksson O., *Phys. Rev., B* **55** (1997), R10173.
3. Hohenberg P. and Kohn W., *Phys. Rev.* **136** (1964), B864.

4. Kohn W. and Sham L. J., *Phys. Rev.* **140** (1965), A1133.
5. Blaha P., Schwarz K., Madsen G. K. H., Kvasnicka D. and Luitz J., In: K. Schearz (ed.), *Wien2k, An Augmented Plane Wave + Local Orbitals Program for Calculating Crystal Properties*. Techn. Universitt Wien, Austria, 2001, ISBN 3-9501031-1-2.
6. Perdew J. P., Chevary J. A., Vosko S. H., Jackson K. A., Pederson M. R., Singh D. J. and Fiolhais C., *Phys. Rev., B* **46** (1992), 6671.
7. Blügel S., Akai H., Zeller R. and Dederichs P. H., *Phys. Rev., B* **35** (1987), 3271.
8. Müller S., de la Presa P., Saitovitch H., Silva P. R. J. and Forker M., *Solid State Commun.* **122** (2002), 155.

Hyperfine Interactions (2004) 158:195–198
DOI 10.1007/s10751-005-9032-4

Nuclear Magnetic Resonance on Oriented Nuclei in ^{175}Hf*Fe*

S. MUTO[1,*], T. OHTSUBO[2], S. OHYA[2] and K. NISHIMURA[3]
[1]*Neutron Science Laboratory, KEK, 305-0801, Tsukuba, Japan; e-mail: suguru.muto@kek.jp*
[2]*Department of Physics, Niigata University, 950-2181, Niigata, Japan*
[3]*Faculty of Engineering, Toyama University, 930-8555, Toyama, Japan*

Abstract. Nuclear magnetic resonance on oriented ^{175}Hf in iron host has been measured. Samples of ^{175}Hf*Fe* were made by recoil implantation of the precursor ^{175}Ta isotope. The resonance frequency and the resonance line width have been determined to be 139.0 (1) MHz and FWHM = 2.7 (2) MHz, respectively, in an external magnetic field of 0.1 T. The resonance width was very narrow compared with the previously reported value of 11.0 (1.1) MHz. With the known value of the magnetic moment of $\mu(^{175}\mathrm{Hf}) = -0.62$ (3) μ_N, the hyperfine field has been deduced as $B_{\mathrm{HF}} = -73.6$ (3.5) T.

Key Words: hyperfine field, NMR-ON.

1. Introduction

The nuclear magnetic resonance on oriented nuclei (NMR-ON) is a powerful method for investigating the hyperfine interactions of probe nuclei in ferromagnetic metals. A systematic study of the hyperfine fields in iron gives us very important information on the electronic structures of the elements in iron. Up to now, many samples for NMR-ON experiments have been made by recoil implantation method. Since the group 4 elements of Hf and Zr are little soluble in iron, it is suitable to prepare the dilute alloy samples by recoil implantation. Herzog *et al.* reported the hyperfine field of ^{175}Hf*Fe* to be −64.9 (9.3) T [1]. They prepared the ^{175}Hf*Fe* sample by the mass-separator implantation of ^{175}Hf with an implantation energy of 80 keV. Although they succeeded in the NMR-ON measurement, the observed line width was very large (FWHM = 11.0 (1.1) MHz). The origin of the large line width of the resonance is not clear. Such a metallurgical problem in ^{89}Zr was resolved by recoil implantation of the precursor ^{89}Nb isotope [2].

In this report, we have applied NMR-ON measurements for ^{175}Hf*Fe* using the recoil implantation of precursor ^{175}Ta isotope. It is also important to determine

* Author for correspondence.

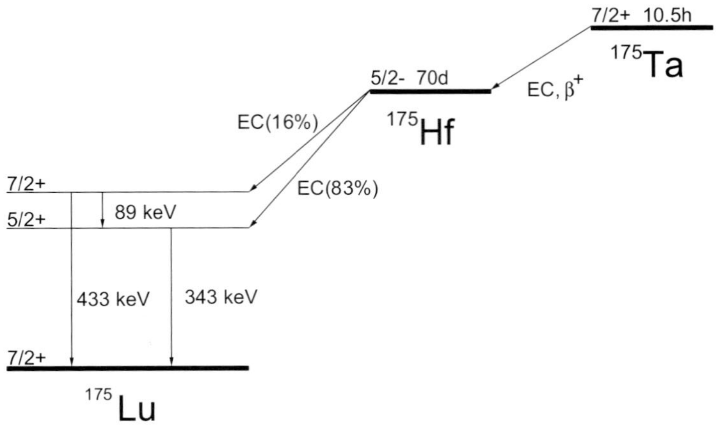

Figure 1. Partial decay scheme of ^{175}Ta and ^{175}Hf.

the hyperfine field precisely for the study of parity mixing and time reversal invariance using Hf*Fe* samples.

2. Experimental procedure

Samples of ^{175}Hf*Fe* were prepared by means of the recoil implantation using ^{175}Lu(α, 4n)^{175}Ta reaction. ^{175}Ta decays to ^{175}Hf with a half-life of 10.5 h. The target of Lu was made by sputtering of natural Lu onto a thin Cu foil. Stacks of alternating targets and pure Fe foils (thickness ~1.5 μm) were irradiated with a 60 MeV α-beam (~1.5 μA) for 35 h from the SF cyclotron at the Institute for Nuclear Study (INS), University of Tokyo. After irradiation, the samples were treated in two ways; one was given no heat treatment, and the other was annealed at 800°C for 1 h in a vacuum. The activated part of the sample foil was soft-soldered to the copper cold finger of a ^{3}He/^{4}He dilution refrigerator, and cooled down to a temperature of about 8 mK. The γ-rays were detected with four pure Ge detectors placed at 0°, 90°, 180° and 270° with respect to the external magnetic field B_0, which defined the orientation axis. The temperature of the samples was monitored using ^{56}Co source produced in the foil by the irradiation. The details of the apparatus have been described in ref. [3].

3. Results and discussion

The partial decay schemes of ^{175}Ta and ^{175}Hf are shown in Figure 1. First, the NMR-ON measurement was done with the unannealed sample. The anisotropy of W(0°) for the 343-keV γ transition in ^{175}Lu was about 6%. The γ-ray was used for detecting NMR-ON resonance. The resonance spectrum was measured at an external magnetic field of B_0 = 0.1 T as shown in Figure 2. The resonance frequency was obtained as ν = 139.0 (1) MHz from the least-squares fit of a

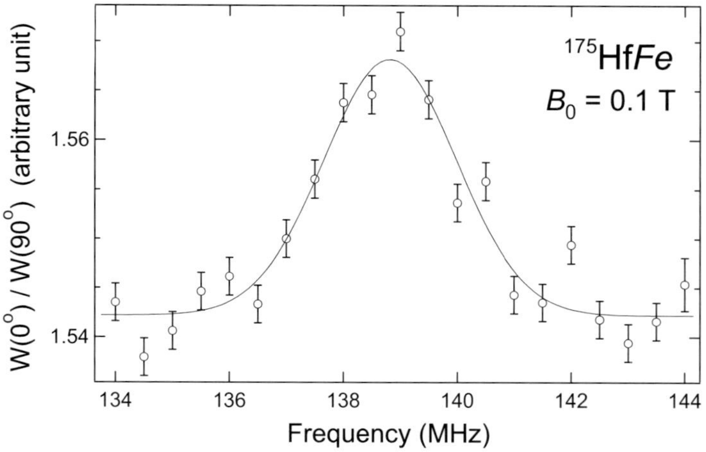

Figure 2. NMR-ON spectrum of ^{175}Hf*Fe*.

Gaussian shape with linear background to the data. The solid line shown in Figure 2. represents the result of the fit. The line width was determined as FWHM = 2.7 (2) MHz with a frequency-modulation bandwidth of ±1.0 MHz. The line width is much narrow than that of 11.0 (1.1) MHz reported by Herzog *et al.* [1]. They made the sample by the implantation of ^{175}Hf. Recently, the magnetic moment of ^{175}Hf was reported to be $\mu(^{175}\text{Hf}) = -0.62$ (3) μ_N [4]. With this magnetic moment, we deduced the hyperfine field value of $B_{HF} = -73.6$ (3.5) T. The value of the hyperfine field agrees well with the previous value of −73.3 (3.5) T [1], which is recalculated with the new value of the magnetic moment. The accuracy of the hyperfine field is entirely limited by the uncertainty of the magnetic moment.

Second, the NMR-ON measurement was done with the annealed sample. No γ-ray anisotropy was observed with the annealed sample. In this sample, the precursor isotopes were considered to replace Fe ions in the substitutional lattice sites, and then decayed to the intended isotopes which remained in the substitutional sites even if they were insoluble in iron. The binding of insoluble isotope in substitutional lattice site was possibly weaker than that of precursor isotope, so that during the heat treatment, if the thermal energy was sufficient, the insoluble isotope could leave substitutional lattice site, and end up in sites with no orienting hyperfine interaction. Therefore, no anisotropy or resonance would be observed. Up to now, recoil implantation of precursor isotope has been used for the NMR-ON measurements of insoluble elements in host metal [2, 5–8]. Similar phenomena were also observed in Y and Zr in iron host [2].

Acknowledgements

The authors would like to thank the crew of the INS cyclotron for their operation of the machine.

References

1. Herzog P., Dämmrich U., Freitag K., Hermann C.-D., Schlösser K. and Soares J. C., *Z. Phys., B* **63** (1986), 241.
2. Ohya S., Ohtsubo T., Komatsuzaki K., Cho D. J. and Muto S., *Phys. Rev., C* **54** (1996), 1129.
3. Nishimura K., Ohya S. and Mutsuro N., *J. Phys. Soc. Jpn.* **56** (1987), 3512.
4. Jin W. G., Wakasugi M., Hies M. G., Inamura T. T., Murayama T., Ariga T., Yamashita A., Wakui T., Katsuragawa H., Ishizuka T., Ruan J. Z. and Sugai I., *Phys. Rev., C* **55** (1997), 1545.
5. Eder R., Hage E. and Zech E., *Nucl. Phys. A* **468** (1987), 348.
6. Hinfurtner B., Hagn E., Zech E. and Eder R., *Phys. Rev. Lett.* **66** (1991), 96.
7. Hinfurtner B., Hagn E. and Zech E., *Phys. Rev. Lett.* **71** (1993), 1736.
8. Hinfurtner B., Seewald G., Hagn E., Zech E. and Towner I. S., *Nucl. Phys. A* **620** (1997), 317.

Hyperfine Interactions (2004) 158:199–203
DOI 10.1007/s10751-005-9033-3

Low-Temperature Nuclear Orientation of ^{144}Pm in Metamagnetic (RE)$NiAl_4$ Single Crystals

K. NISHIMURA[1,*], K. MORI[1], S. TERAOKA[1], W. D. HUTCHISON[2], D. H. CHAPLIN[2], S. OHYA[3], T. OHTSUBO[3], S. MUTO[4] and T. SHINOZUKA[5]
[1]*Faculty of Engineering, Toyama University, Toyama 930-8555, Japan; e-mail: nishi@eng.toyama-u.ac.jp*
[2]*School of Physical, Environmental and Mathematical Science, UNSW@ADFA, ACT2600, Canberra, Australia*
[3]*Department of Physics, Niigata University, Niigata 950-2181, Japan*
[4]*Neutron Science Laboratory, KEK, Tsukuba 305-0801, Japan*
[5]*CYRIC, Tohoku University, Sendai 980-8578, Japan*

Abstract. Magnetic properties of Pm ions in $NdNiAl_4$ were investigated by low-temperature nuclear orientation of ^{144}Pm. The observed γ-ray anisotropy as a function of external fields revealed a change in the magnetic structure of Pm ions at the metamagnetic phase transition of $NdNiAl_4$. The extracted hyperfine field was 50(5) T which is substantially smaller than those previously reported.

Key Words: CEF effect, hyperfine field, magnetic anisotropy, nuclear orientation, Pm ion.

Comprehensive studies of $PrNiAl_4$ and $NdNiAl_4$ ($YNiAl_4$-type orthorhombic crystal structure, *Cmcm*) using single crystals revealed antiferromagnetism and metamagnetic phase transition in these compounds [1, 2]. Magnetization and susceptibility measurements showed that Nd moments ordered along the *b*-axis whilst Pr moments along the *a*-axis. In a more recent work of the pseudo ternary Pr_xNd_{1-x} $NiAl_4$ series, the lowest Neel temperature T_N was found at $x = 0.7$ due to the competing magnetic anisotropies between the Pr and Nd ions [3]. Since low-temperature nuclear orientation (LTNO) is a radioactive detection technique, it can provide some information complementary to those from magnetization measurements and some unique information which is not accessible by the conventional technique. For instance, LTNO experiments, carried out for ^{142}Pr and ^{147}Nd in $Pr_{0.7}Nd_{0.3}NiAl_4$, elucidated that the Pr moments were kept aligned along the *a*-axis due to the large magnetic anisotropy; the Nd moments ordered along the *a*-axis at low field via the exchange interactions with the Pr moments,

* Author for correspondence.

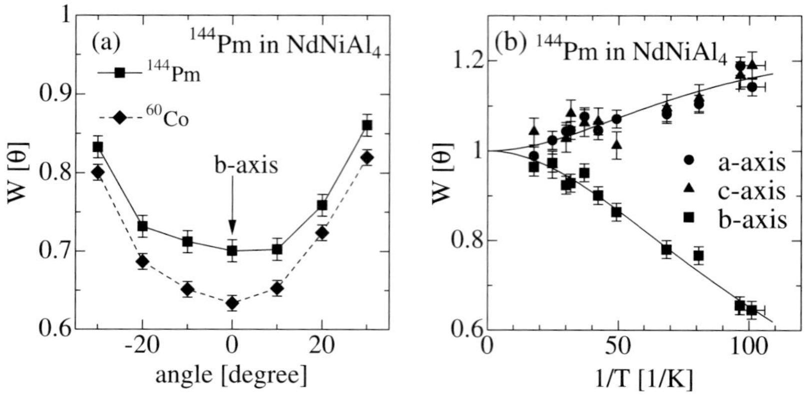

Figure 1. (a) Angular distribution of γ-ray anisotropy of ^{144}Pm in $NdNiAl_4$ at about 9.5 mK. (b) Temperature dependence of γ-ray anisotropies of ^{144}Pm at the *a*-, *b*-, and *c*-axes of $NdNiAl_4$.

but at the metamagnetic transition the Nd moments rotated toward the *b*-axis, as a result of lessening Pr−Nd exchange interaction [4]. It is worth exploring how other RE ions behave at the metamagnetic phase transitions in these compounds. Furthermore the magnetism of Pm ions, in any circumstance, is little studied since this element lacks a stable isotope. In this paper we will report magnetic properties of Pm ions in $NdNiAl_4$ single crystal from LTNO results of dilute ^{144}Pm nuclei. The observed magnetic anisotropy of Pm ions will be discussed in terms of the crystal electric field effect (CEF) based on a simple point-charge model calculation.

Single crystals of $NdNiAl_4$ were made by the Czochralski pulling method. Small pieces of the crystals (approximate dimensions: $3 \times 2 \times 0.2$ mm long along the crystallographic *a*-, *b*-, and *c*-axes) were irradiated with a deuteron beam from a cyclotron at CYRIC, Tohoku University. The radioactive sample was soft-soldered on the cold finger of a ^{3}He/^{4}He dilution refrigerator at UNSW@ADFA or Niigata University. For LTNO measurements, a ^{60}Co in hcp Co nuclear thermometer was soldered near the sample. Both the *b*-axis of $NdNiAl_4$ and the *c*-axis of hcp Co were aligned with the direction of an externally applied field B up to 8 T, provided by a solenoid magnet. Three HP Ge γ-ray detectors were placed along the crystallographic *a*-, *b*-, and *c*-axes.

The observed γ-ray anisotropy, W, was analyzed using the standard formula for uniaxial symmetry, $W(\theta) = 1 + \Sigma_{\lambda = \text{even}} U_\lambda A_\lambda B_\lambda Q_\lambda P_\lambda(\cos\theta)$ [5]. The angular distribution of γ-rays from oriented ^{144}Pm were measured via 618 keV and 697 keV transition with a field of 0.5 T along the *b*-axis in the *a*–*b* plane of the sample, as shown in Figure 1(a). The base temperature was about 9.5 mK and W from the 618 keV and 697 keV γ-rays have been combined at each angle, since both values have been found experimentally to be almost equal [6]. In

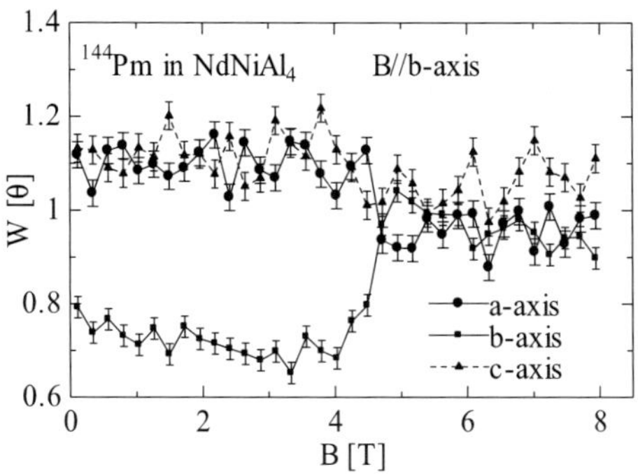

Figure 2. Field dependence of γ-ray anisotropy of ^{144}Pm observed at the *a*-, *b*-, and *c*-axes of $NdNiAl_4$ at about 10 mK with the field along the *b*-axis.

Figure 1(a), the minimum W from ^{144}Pm is shown to be at the *b*-axis, which means the Pm moments were ordered along the *b*-axis via the Nd–Pm exchange interactions.

The temperature dependence of W of ^{144}Pm was measured at the *a*-, *b*-, and *c*-axes directions to estimate the hyperfine field B_{HF}. The fit of the above W equation in Figure 1(b) to the observed W data yielded $B_{HF} = 50(5)$ T, using the known nuclear magnetic moment (1.69 μ_N) [7], and the parameters: $U_2A_2 = -0.4056$, $U_4A_4 = -0.2281$ [8]. This B_{HF} value is considerably smaller than those in PrNi (185 T) [8] and in NdNi (395 T) [6], and the theoretical value, $B_{HF} = 426$ T, for Pm^{3+} free ion [9]. In fact, the γ-ray anisotropies in Figure 1(b) appear substantially smaller than those in the previous work. Antiferromagnetic structure of $NdNiAl_4$ may be a major cause of this small B_{HF}. $NdNiAl_4$ undergoes a metamagnetic phase transition at about $B_C = 4.6$ T at 2 K. The applied field dependence of γ-ray anisotropies at the *a*-, *b*-, and *c*-axes also reflect the change in the magnetic structure of Nd ions at B_C, as can be seen in Figure 2. At lower field than B_C, the Pm moments lay along the *b*-axis. Above B_C, however, the γ-ray anisotropy almost disappear, and the Pm ions seem to have close to random orientation.

For a single 4f magnetic ion interacting with surrounding trivalent rare-earth ions in an orthorhombic crystal structure, in a simple model, the total Hamiltonian can be given as $H^R = H^R_{\mathrm{CEF}} + H^R_{\mathrm{ex}} + H^R_{\mathrm{appl.}}$, where H^R_{CEF} represents the CEF Hamiltonian, H^R_{ex} the exchange interactions, and $H^R_{\mathrm{appl.}}$ the Zeeman interactions. Using the CEF coefficients, B^m_n, and the irreducible spherical tensors, C^m_n, the CEF Hamiltonian is expressed as $H^R_{CEF} = \sum_{n,m} B^m_n C^M_n$ [10]. The B^m_n values can be estimated from the measured susceptibility of $NdNiAl_4$ in Figure 3(a). The experimental data are marked by the discrete symbols: circle (*a*-axis), square (*b*-axis) and triangle (*c*-axis). The results of calculations are displayed with the lines. The best fit between the experimental values and calculated ones was

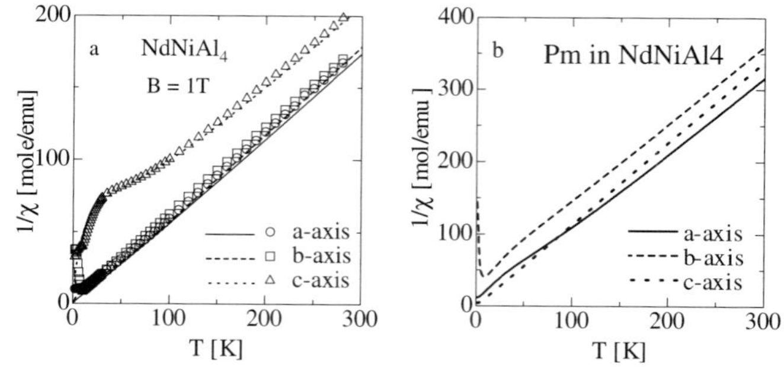

Figure 3. a) Susceptibility of $NdNiAl_4$ with $B = 1$ T at the three crystallographic axis directions. b) Calculated susceptibility for Pm^{3+} ions in $NdNiAl_4$.

achieved with $B_2^0 = -425$ K, $B_2^2 = 95$ K, $B_4^0 = -110$ K, $B_4^2 = 900$ K, and $B_4^4 = 1500$ K (other B_n^m values were neglected). Using these B_n^m values and the C_n^m values calculated for Pm^{3+} ($J = 4$), the expected susceptibility of Pm ions in $NdNiAl_4$ are shown in Figure 3(b). The calculation predicts for Pm to have the magnetically easy axis along the *c*-axis, and the magnetic anisotropy in the *a*–*c* plane is small, but the *b*-axis is the hard axis. The present LTNO results, however, showed that the Pm moments ordered along the *b*-axis of $NdNiAl_4$ in antiferromagnetic state. This fact implies the Nd–Pm exchange interactions dominantly affect the magnetic structure of Pm ions in the antiferromagnetic state. After the metamagnetic transition, the Nd–Pm interactions, the CEF effect and the external field had comparable contributions to the magnetic structure, leading to the random distribution of the Pm moments. To enhance the understanding of Pm ions in this case, it will be of interest to compare the present results with the LTNO of ^{144}Pm in $Pr_{0.5}Nd_{0.5}NiAl_4$ work in progress.

Acknowledgement

K. Nishimura would like to thank for a grant from IKETANI Science and Technology Foundation.

References

1. Mizushima T. and Isikawa Y. *et al., J. Phys. Soc. Jpn.* **66** (1997), 3970.
2. Mizushima T. and Isikawa Y. *et al., J. Phys. Soc. Jpn.* **68** (1999), 637.
3. Nishimura K. and Hutchison W. D. *et al., Jpn. J. Appl. Phys.* **42** (2003), 5565.
4. Nishimura K. and Hutchison W. D. *et al., J. Magn. Magn. Mater.* **226–230** (2001), 1126.

5. Krane K. S., In: Stone N. J. and Postma H. (eds.), *Low-Temperature Nuclear Orientation*, Chap. 2, North-Holland, Amsterdam, 1986.
6. Nishimura K., Mori K. and Ohya S. *et al., Phys. Rev.,* **B53** (1996), 15010.
7. Firestone R. B. and Shirley V. S. (eds.), *Table of Isotopes*, 8th Edition, Wiley, New York, 1996, Appendix E.
8. Nishimura K. and Ohya S. *et al., Hyperfine Interact.* **78** (1993), 475.
9. Bleaney B. In: Elliott R. J. (ed.), *Magnetic Properties of Rare Earth Metals*, Chap. 8, Plenum Press, New York, 1972.
10. Morrison C. A. and Leavitt R. P. In: Gschneidner Jr. K. A. and Eyring L. (eds.), *Handbook on the Physics and Chemistry of Rare Earth*, Vol. 5, North-Holland, Amsterdam, 1982, p. 461.

Hyperfine Interactions (2004) 158:205–209
DOI 10.1007/s10751-005-9034-2

Temperature Dependence of the Hyperfine Magnetic Field at ^{140}Ce in Orthorhombic Tb_3In_5

M. OLZON-DIONYSIO[1,*], S. D. DE SOUZA[1], A. J. A. DE OLIVEIRA[1] and A. W. CARBONARI[2]
[1]*Departamento de Fíisica, UFSCAR, São Carlos, Brazil; e-mail: dmod@df.ufscar.br*
[2]*Instituto de Pesquisas Energéticas e Nucleares-IPEN-CNEN, São Paulo, Brazil*

Abstract. Time differential perturbed γ-γ angular correlation technique was used to measure the magnetic hyperfine field (MHF) at Tb sites in the intermetallic compound Tb_3In_5 using the $^{140}La \rightarrow {}^{140}Ce$ nuclear probe. The measurements were carried out in the temperature range of 8 to 295 K. Two different temperature dependent magnetic frequencies were observed below 30 K, which were assigned as ^{140}Ce substituting the two inequivalent Tb sites in the orthorhombic structure of Tb_3In_5. The temperature dependence of MHF also shows a possible deviation from an expected Brillouin-like behavior for temperatures below 18 K. A Néel transition at 27 K was observed from magnetization measurements in the samples. The magnetization as a function of the applied magnetic field was measured at two temperatures, 5 and 40 K, and the results show anti-ferromagnetic and a typical paramagnetic behavior, respectively. In both cases it was not observed saturation under high magnetic field.

Key Words: ^{140}Ce probe, magnetic hyperfine field, PAC spectroscopy, rare earth magnetism.

1. Introduction

The investigation of the local magnetism in rare earth compounds is very important to understand the rich variety of complex phenomena, which occur in such compounds like Kondo effect, heavy fermion behavior, mixed valence, etc. The origin of the magnetism in these compounds is due to the strongly localized magnetic moment, which is associated to 4f electrons of rare earth atoms. In many compounds, the 4f electrons can interact with other electrons via hybridization with, for instance, s electrons, and lose part of their localized magnetic moment. The investigation of the magnetic hyperfine field (MHF) in rare earth compounds is, therefore important because it is a measurement of the magnetic field in the neighborhood of a particular site and can be associated to magnetic moments in this site or at nearest neighbor sites. The magnetic and electric hyperfine interaction at ^{111}Cd impurity using the Time differential perturbed γ-γ angular correlation (TDPAC) technique has been systematically

* Author for correspondence.

used to investigate intermetallic compounds R_xA_y (R = rare earth element, and A = non-magnetic metal), in the particular those with A = In [1–3], and A = Al [4]. The purpose of such a systematic investigation is to better understand the coupling mechanism that is responsible for the rare earth magnetism. As a part of this study, the intermetallic compound Tb_3In_5 has been investigated ^{111}Cd and the results for previous TDPAC measurements using as probe nuclei were reported [5].

The Tb_3In_5 compound crystallizes in the orthorhombic Pu_3Pd_5-type structure [6, 7], where Tb and In atoms can occupy two and three structurally different sites, respectively. This complex structure makes the local investigation of Tb_3In_5 be quite difficult due to the presence of a complex quadrupole interaction. In the previous measurements [5] with ^{111}Cd, the TDPAC spectrum at 10 K showed a time dependent attenuation of the anisotropy when compared with that obtained at room temperature. This fact was interpreted as due to the presence of the magnetic interaction. However, It was not possible to fit a function that takes into account a combined magnetic and electric interaction. The computer program used to perform this fitting is based on the calculation of Böstrom *et al.* [8], and it is adequate when the electric field gradient (EFG) is axially symmetric ($\eta = 0$). When the asymmetry parameter is small, the program is still adequate if the Larmor frequency (ω_L) of the magnetic interaction is greater than the quadrupole frequency (ω_Q) of the electric interaction. In the case of measurements using ^{111}Cd, the asymmetric electric interaction was very strong and, possibly much larger than the magnetic interaction. Consequently, it was not possible to obtain any information about the magnetic hyperfine field for the Tb_3In_5 compound. The quadrupole moment of the 2083 keV 4^+ state of ^{140}Ce, used as probe nuclei in the present work, is known to be very small [9], and one expects to observe an almost pure magnetic dipole interaction in the TDPAC measurements, which opens the possibility to investigate the local magnetic field in the Tb_3In_5 compound.

2. Experimental

The samples were prepared by repeated arc melting the constituent elements in arc furnace under argon atmosphere purified with a hot titanium getterer. Radioactive ^{140}La nuclei (obtained by neutron irradiation of lanthanum metal in the IEA-R1 research reactor at IPEN) substituting about 0.1% of Tb atoms were melted along with the constituent elements. Samples were annealed at 900°C in a pressure of 1 atm of ultra-pure Ar. Crystal structures of the samples could not be checked by powder X-ray-diffraction measurements due to the plastic deformation during the powdering process [5].

As Tb_3In_5 compound oxidizes very quickly, a second sample, without the radioactive ^{140}La, was prepared for the magnetization measurements. These measurements were carried out in a superconductor quantum interference device

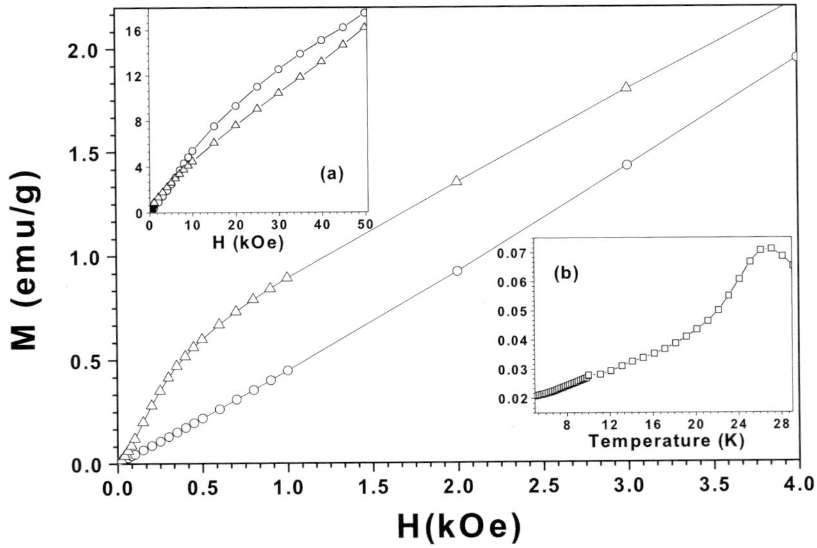

Figure 1. Magnetization curves as a function of applied magnetic field performed at $T = 5$ K (○) and 40 K (△). The insets show (a) in the full interval of H, and (b) the temperature dependence of the magnetization at $H = 1.0$ kOe.

(SQUID) in the temperature range of 5 to 30 K with a rate of 1 K/min and an applied magnetic field of 1 kOe. In addition, the magnetization was also measured as a function of the applied field for 5 and 40 K.

The γ-γ cascade of 329,487 keV populated from the β^- decay of ^{140}La was used for the measurement of magnetic interaction at the 2083 keV intermediate state (4^+ spin) of ^{140}Ce. TDPAC spectra were recorded at several temperatures, in the range from 8 to 295 K, using a standard setup with four BaF_2 detectors arranged in a planar 90°–180° geometry. Twelve delayed coincidence spectra are generated simultaneously with a time resolution of 600 ps. For low-temperature measurements the sample was attached to the cold finger of a closed-cycle-helium refrigerator with temperature control (better than 0.1 K).

3. Results and discussion

Figure 1 shows the results for the magnetization measurements as a function of the applied magnetic field at 5 and 40 K. The magnetization at low field and low temperature shows the characteristic linear behavior expected from a antiferromagnetic system, and the results at 40 K exhibit the typical behavior of a paramagnetic phase. However, it was observed that the magnetization does not saturate at high magnetic fields (see inset a). The antiferromagnetic phase is better characterized by the measurement of the magnetization as a function of temperature for 1.0 kOe (inset b), where the magnetization increases when the temperature increases. This behavior is a characteristic of an antiferromagnetic

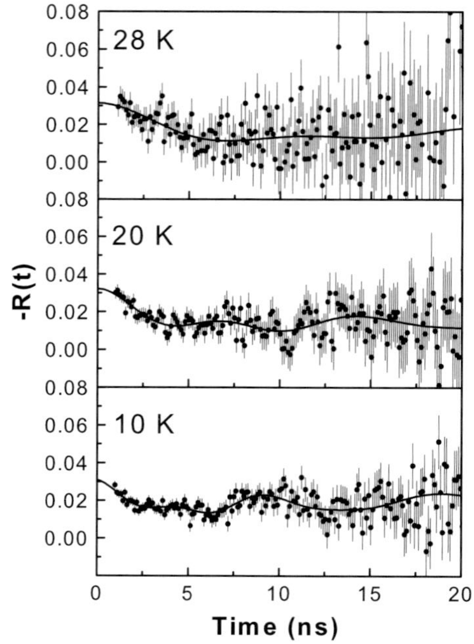

Figure 2. Perturbation functions $R(t)$ for ^{140}Ce in Tb_3In_5 measured at indicated temperature.

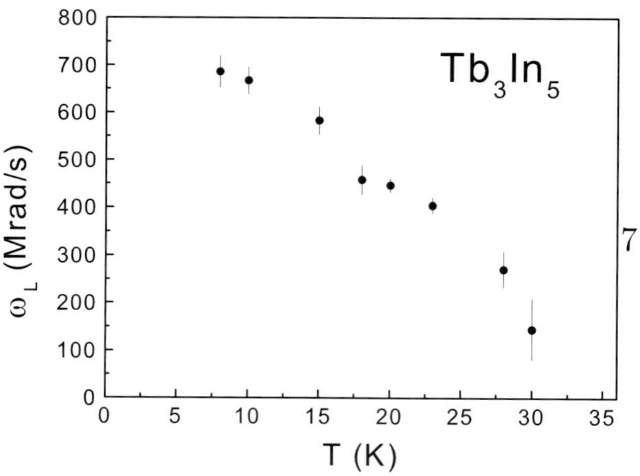

Figure 3. The temperature dependence of fitted Larmor frequency for ^{140}Ce at Tb site in Tb_3In_5.

phase. The Néel temperature (T_N ~27 K) was determined at the peak of the magnetization curve, above of which the magnetization decreases, as expected for a paramagnetic phase.

The spin rotation spectra $R(t)$ is well described by a superposition of two fractions or sites with pure temperature dependent magnetic interactions. The proportion between these fractions is approximately 1:3, as expected for the two Tb sites in the orthorhombic structure of Tb_3In_5. The major fraction (~65%),with distribution $\delta \sim 13\%$, shows higher frequencies than the minor fraction ($\delta \sim 6\%$). Some of the TD-PAC spectra measured with ^{140}Ce probe nuclei are shown in Figure 2. The solid curves are the least squares fit of the experimental data to the appropriate function.

The temperature dependence of the Larmor frequency ω_L corresponding to the major fraction is plotted in Figure 3. The observed magnetic interaction corresponds to the antiferromagnetic ordering of the Tb moments. If the points for 18 and 20 K are not taken into account, one could consider that the data follow a Brillouin function. But, if all data are considered true, they indicate that instead of approaching a saturation value around 500 Mrad/s, ω_L values increases at lower temperatures below around 18 K, and point to a saturation value of ~720 K. This behavior was already observed in $CeMn_2Ge_2$ compound [10] and was attributed to the spin polarization of Ce ions of the probe nuclei at low temperatures induced by the magnetic field from rare earth moments.

Acknowledgements

Partial support for this research was provided by the Fundação de Amparo á Pesquisa do Estado de São Paulo (FAPESP). AWC thankfully acknowledges the support provided by CNP_q in the form of a research fellowship.

References

1. Olzon-Dionysio M., de Souza S. D., Musseler R. and Forker M., *J. Phys., Condens. Matter* (1992), 4307.
2. Musseler R., PhD thesis, ISKP, Bonn University, 1993.
3. Carbonari A. W., Mestnik-Filho J., Saxena R. N. and Saitovitch H., *Hyperfine Interact.* **133** (2001), 77.
4. Forker M., de la Presa P., Olzon-Dionysio M. and de Souza S. D., *J. Magn. Magn. Mater.* **226–230** (2001), 1156.
5. Olzon-Dionysio M., de Souza S. D., Rodriguez A. M., Magon C. J. and de Souza R. R., *Hyperfine Interact.* **136–137** (2001), 367.
6. Villars P. and Calvert L. D. (eds.), *Pearsons Handbook of Crystallographic Data for Intermetallic Phases*, Vol. I., American Society for Metals, USA, 1985.
7. Delfino S., Saccone A. and Ferro R., *J. Less-Common Met.* **102** (1984), 289.
8. Böstrom L., Karlsson E. and Zetterlund S., *Phys. Scr.* **2** (1970), 65.
9. Królas K. and Wodniecka B. H., *Niewodniczanski Institute of Nuclear Physics*, Kraków, Poland, Report No. 1644/OS-1993 (unpublished).
10. Carbonari A. W., Mestnik-Filho J., Saxena R. N. and Lalic M. V., *Phys. Rev., B* **69** (2004), 1444251.

Hyperfine Interactions (2004) 158:211–215
DOI 10.1007/s10751-005-9035-1

Mössbauer Effect Study of $Eu_{0.88}Fe_4Sb_{12}$ Skutterudite

MICHAEL REISSNER[1,*], ERNST BAUER[1], WALTER STEINER[1] and PETER ROGL[2]
[1]*Institute of Solid State Physics, Vienna University of Technology, Austria; e-mail: reissner@ifp.tuwein.ac.at*
[2]*Institute of Physical Chemistry, University of Vienna, Vienna, Austria*

Abstract. The partly filled skutterudites $Eu_{0.88}Fe_4Sb_{12}$ were investigated by ^{57}Fe Mössbauer spectroscopy in the temperature range 4.2 K $\leq T \leq$ 295 K and in external fields up to 13.5 T. The results favour a statistical distribution of Eu and voids in the Fe near neighbour shell. Below 82 ± 1 K magnetic order is present. Debye temperatures Θ_D = 460 ± 20 K and 165 ± 30 K were obtained for Fe with completely occupied Eu sites and for Fe with vacancies in the *R*-neighbour shell, respectively. The temperature dependence of the quadrupole splitting reflects the thermal expansion of the lattice. The induced hyperfine fields at 4.2 K are negative and differ by roughly a factor of two for the two Fe surroundings.

In partly filled skutterudites RFe_4X_{12} (R = rare earth element, X = Sb, P) the R atoms, situated in oversized cages, influence by their local anharmonic vibrations the thermal conductivity [1]. Voids in the R sublattice offer thus the possibility to tailor the thermoelectric figure of merit. $EuFe_4Sb_{12}$ was reported to order ferromagnetically around 82 K with Eu nearly in a divalent state [2, 3]. The saturation moment of 4.26 μ_B obtained at 2 K and 6 T, however, remains significantly below the one expected for completely aligned Eu moments [2]. We started therefore with systematic Mössbauer studies in the temperature range 4.2 K $\leq T \leq$ 295 K and in external fields up to 13.5 T on these skutterudites with P and Sb as pnictides and report in the present paper on the results obtained for $Eu_{0.88}Fe_4Sb_{12}$.

Sample preparation was described in [2]. ^{57}Fe Mössbauer spectra were recorded in transmission geometry with a ^{57}Co**Rh** source relative to which all centre shift (CS) values are given. The spectra were analysed taking into account magnetic and electrostatic hyperfine interactions as well as sample thickness.

The unfilled skutterudites belong to the space group $Im\bar{3}$, where Eu occupies the 2*a*, Fe the 8*c* and Sb the 24*g* positions. Whereas from X-ray investigations the respective lattice sites were found to be fully occupied by Fe and Sb, the occupation number for Eu was only 0.88. Thus Fe may be surrounded by either

* Author for correspondence.

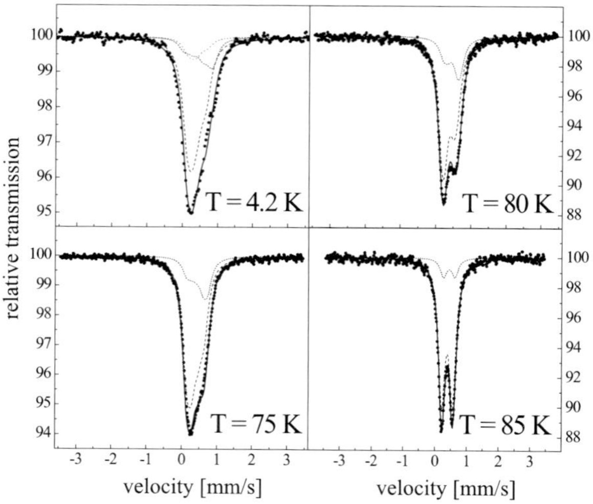

Figure 1. ^{57}Fe Mössbauer spectra recorded at various temperatures.

two, one or zero Eu atoms assuming a statistical distribution in the nearest neighbour shell. These different surroundings can influence the hyperfine parameters leading to the asymmetry of the spectra recorded in the temperature interval between 295 and 85 K (Figure 1) for which the analyses were only possible with a superposition of at least two subspectra (line widths 0.22 mm/s) with different CS and QS (= $eQV_{zz}/4$) values (Figure 2). The relative intensities of these subspectra (0.85 and 0.15) are in fair agreement with the X-ray investigations. Thus the spectrum with the higher intensity may be allocated to Fe with two surrounding Eu atoms and the other to Fe with one or two vacancies in the neighbourhood. A separation of the latter two is out of the experimental resolution, because from an estimation based on a statistical distribution of voids and Eu atoms a relative intensity of only 1.5% follows for Fe with an empty R neighbour shell. We obtained similar results for other RFe_4Sb_{12} (R = La, Pr, Nd) samples [4]. This favours thus for the Fe skutterudites a statistical distribution of voids and Eu atoms on the 2a sites, whereas for $Ce_yFe_{4-x}Co_xSb_{12}$ a solid solution of a completely filled Fe compound in a practically empty Co compound seems more likely to be realised [5, 6].

In agreement with the results of bulk magnetic investigations [2, 3], the appearance of magnetic order is indicated by the change of the overall spectral shape at 82 ± 1 K (Figure 1). For $T > 20$ K a reasonable fit with only two subspectra remains possible. For lower temperatures at least one further subspectrum was necessary (Figure 1), which can be attributed to an impurity phase with magnetic order around 25 K [2].

The measured hyperfine field ($B_{\rm hf}$) for Fe with completely filled Eu neighbour shell is much smaller than the one where voids in the cages are present (Figure 3). Both hyperfine fields exhibit similar temperature dependence and

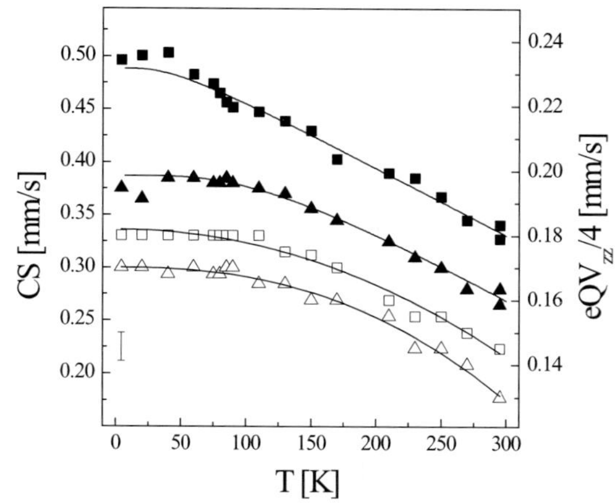

Figure 2. Temperature dependence of the quadrupole splitting $eQV_{zz}/4$ (*open symbols*, details of the fits are described in the text), and the centre shift *CS* (*full symbols*, fit according to the Debye model). Fe completely (△) and partly (□) surrounded by Eu atoms.

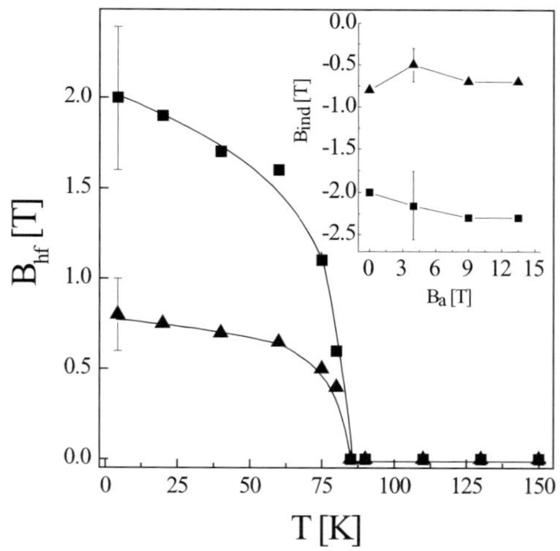

Figure 3. Temperature dependence of the hyperfine field. Inset: dependence of the induced hyperfine field B_{ind} on the external field B_a at 4.2 K. Symbols as in Figure 2. The lines are guides for the eyes.

point to the presence of a magnetic moment on the Fe atoms. From the measurements at 4.2 K in external fields (B_a) induced hyperfine fields were calculated by $B_{ind} = B_{hf} - B_a$. As one would expect for magnetically ordered Fe compounds B_{ind} is negative (inset Figure 3). It differs for the two sites by roughly a factor of two and is not in accordance with our results obtained for samples with R = La,

Pr, and Nd, for which very small and positive values were deduced for both Fe environments. The values are much too small to explain directly the large effective Fe moments measured at higher temperatures. An estimation of valence and core contribution to the hyperfine field by theoretical calculations is necessary to allow final conclusions about the Fe moment.

The temperature dependence of QS is similar for both Fe surroundings (Figure 2) and reflects the dependences found for the Ce compound [5], as well as those for $LaFe_4P_{12}$ [7] and $PrFe_4P_{12}$ [8]. Following Verma and Rao [9], the temperature dependence of QS in a metallic lattice of non-cubic symmetry can be expressed by (1) $QS(T) = a(1 - bT^{\gamma})$, or by (2) $QS(T) = a(1 + bT + cT^2)$. In these equations a equals QS at zero temperature, b and c are free parameters, and γ is either fixed at 3/2 or also a free parameter. Both equations yield equally good fits for the two Fe surroundings, if γ is taken as a free parameter (Figure 2). The resulting γ-values (2.43 and 2.06 for fully occupied Eu sites and with vacancies, respectively) deviate from the expected value for a dominance of thermal vibrations, $\gamma = 3/2$, which was found for $Ce_yFe_{4-x}Co_xSb_{12}$ [5]. It seems that QS(T) in these series, independent of the pnictogen atom and the valence of the R atom, reflects the thermal expansion of the lattice, although the value of QS for the divalent Eu compound is smaller than the one of the compounds where the R atom is in the trivalent state (Pr, Nd) or in a mixed valence state (Yb [10]).

Both Fe environments exhibit similar temperature dependences of CS (Figure 2). The absolute values at room temperature are smaller than the one reported for the La, Ce [5] and Yb [10] compounds reflecting the Eu^{2+} state. Assuming that the main influence of the temperature variation of CS is caused by the second order Doppler shift and using the Debye model yield $\Theta_D = 460 \pm 20$ K for Fe with completely occupied Eu sites (Figure 2). A considerably smaller value of $\Theta_D = 165 \pm 30$ K results for Fe with vacancies in the R–neighbour shell (most probably the one with only one vacancy). The former value is in good agreement with $\Theta_D = 440 - 540$ K obtained for $Ce_{0.98}Fe_4Sb_{12}$ by the same method [5]. For the compounds with P on the pnictide site the reported Debye temperatures scatter between 450 and 600 K [7, 11] indicating that the influence of both the kind and valence of the R atoms and the different mass of the pnictides on the vibrational modes of Fe is small for filled cells. Voids on the R sites have, however, large influences.

References

1. Sales B. C., Mandrus D., Chakoumakos B. C., Keppens V. and Thompson J. R., *Phys. Rev., B* **56** (1997), 15081.
2. Bauer E., Berger S., Galatanu A., Gali M., Michor H., Hilscher G., Paul C., Ni B., Abd-Elmeguid M. M., Tran V. H., Grytsiv A. and Rogl P., *Phys. Rev., B* **63** (2001), 224414.
3. Danebrock M. E., Evers C. B. and Jeitschko W., *J. Phys. Chem. Solids* **57** (1996), 381.

4. Reissner M., Bauer E., Steiner W. and Rogl P., *J. Magn. Magn. Mater.* **272–276** (2004), 813.
5. Long G. J., Hautot D., Grandjean F., Morelli D. T. and Meissner G. P., *Phys. Rev., B* **60** (1999), 7410.
6. Meissner G. P., Morelli D. T., Hu S., Yang J. and Uher C., *Phys. Rev. Lett.* **80** (1998), 3551.
7. Shenoy G. K., Noakes R. and Meissner G. P., *J. Appl. Phys.* **53** (1982), 2628.
8. Tsutsui S., Kuzushita K., Tazaki T., Morimoto S., Nasu S., Matsuda T. D., Sugawara H. and Sato H., *Physica, B* **329–333** (2003), 469.
9. Verma H. C. and Rao G. N., *Hyperfine Interact.* **15–16** (1983), 207.
10. Leithe-Jasper A., Kaczorowski D., Rogl P., Bogner J., Reissner M., Steiner W., Wiesinger G. and Godart C., *Solid State Commun.* **109** (1999), 395.
11. Torikachvili M. S., Chen J. W., Dalichaouch Y., Guertin R. P., McElfresh M. W., Rossel C., Maple M. B. and Meissner G. P., *Phys. Rev., B* **36** (1987), 8660.

Hyperfine Interactions (2004) 158:217–221
DOI 10.1007/s10751-005-9036-0

PAC Investigation of Amorphous Ferromagnets

V. SAMOKHVALOV[1], F. SCHNEIDER[1], S. UNTERRICKER[1,*], M. DIETRICH[2] and THE ISOLDE COLLABORATION[3]

[1]*Institut für Angewandte Physik, TU Bergakademie Freiberg, D-09596 Freiberg (Sachsen), Germany; e-mail: unterr@physik.tu-freiberg.de*
[2]*Technische Physik, Universität des Saarlandes, D-66041 Saarbrücken, Germany*
[3]*CERN, CH-1211 Geneva 23, Switzerland*

Abstract. The amorphous ferromagnetic alloy Fe79/B16/Si5 was investigated by PAC. For that the probe atoms ^{111}In(^{111}Cd) and ^{111m}Cd were implanted at the ISOLDE on-line separator. The distributions of magnetic hyperfine fields were determined. Below the crystallization temperature α-Fe clusters could be observed. For comparison Mössbauer transmission experiments were performed.

Key Words: amorphous ferromagnets, magnetic hyperfine field distributions, PAC.

1. Introduction

Amorphous solids get increasing interest both from technological and scientific point of view because one expects and has already realized favourable technical applications. But there is also a need for clarification of their complicated structure. With nuclear probe methods like Mössbauer spectroscopy (MS) and especially perturbed angular correlations (PAC) the gain of information using the quadrupole interaction is limited. But the situation is much more promising in the case of amorphous ferromagnets were the magnetic hyperfine field distribution reflects the microscopic structure in the probe environments.

Amorphous ferromagnets were intensively investigated by ^{57}Fe MS in the past decades. The spectra were interpreted by a dominating magnetic hyperfine interaction and superimposed quadrupole frequency distributions [1, 2]. But an unambiguous analysis of the spectra is not a simple task and it was an advantage when it was possible to separate the quadrupole interaction by a radio frequency field [3].

PAC investigations of amorphous ferromagnets can hardly be found in the literature. Therefore the potential of this method was an open question for us.

---

* Author for correspondence.

The PAC method can utilize non host probe atoms and the results can be compared with a lot of MS investigations with the host probe ^{57}Fe.

2. PAC experiments

Samples of Goodfellow Metglas FE 820250 (composition Fe79/B16/Si5, thickness 25 μm) were used for the investigations. The material is an amorphous ferromagnetic alloy with a Curie temperature of 405°C and a crystallization temperature of 515°C. The PAC probes ^{111}In(^{111}Cd) and ^{111m}Cd were implanted at the ISOLDE-CERN on-line mass separator [4] with an energy of 50 keV in each case. Both probes make use of the same excited level of ^{111}Cd(5/2+, 84 ns). The PAC apparatus consisted of four BaF_2 detectors with a 90°/180° detector arrangement and the samples were magnetically polarized perpendicular to the detector plane (B_{ext} = 0.5 T).

In Figure 1(a) the result of the as-implanted ^{111}In(^{111}Cd) measurements is presented. The time-dependent coincidence ratio $R(t)$ exhibits the expected strong damping as consequence of the disorder. But the magnetic hyperfine field distribution is simple to evaluate by the Fourier transform $F(\omega)$ though a small quadrupole coupling is superimposed. The mean field of $\langle B_{hf}\rangle = -29(1)$ T is smaller as compared to pure α-Fe ($B_{hf} = -38.4(8)$ T). The reason is the dilution of magnetic moments by the non-magnetic atoms B and Si. Measurements with the sign sensitive ± 135° detector geometry and the Raghavan geometry (polarizing field in the detector plane) gave the same results. All measurements were carried out at R.T. After an annealing at 300°C/10 min. in vacuum the spectra proved to be unchanged.

A heat treatment at 410°C/10 min. caused a considerable variation in the probe environments. The damping is smaller and the mean frequency is higher than before, Figure 1(b). Now the mean field is −38.0(8) T, which is exactly the value of ^{111}Cd substituted in pure α-Fe. Obviously, the In probe atoms are caught in Fe clusters. 410°C is below the crystallization temperature (515°C).

Figure 2 presents a measurement with the probe ^{111m}Cd. The sample was thermally treated (480°C/10 min.) but there are only minor differences to the as-implanted measurement. Though the statistics is not satisfactory it can be recognized that the hyperfine field distribution is practically the same as that for the ^{111}In(^{111}Cd) measurement of Figure 1(a). However, the heating temperature was higher than that of Figure 1(b).

3. Mössbauer investigations

In order to compare our results with that of MS, also such from literature, we made ^{57}Fe Mössbauer transmission measurements on the sample described above. Figure 3 shows two R.T. Mössbauer spectra and the corresponding distributions $P(B_{hf})$ of magnetic fields. The upper picture is a measurement of a

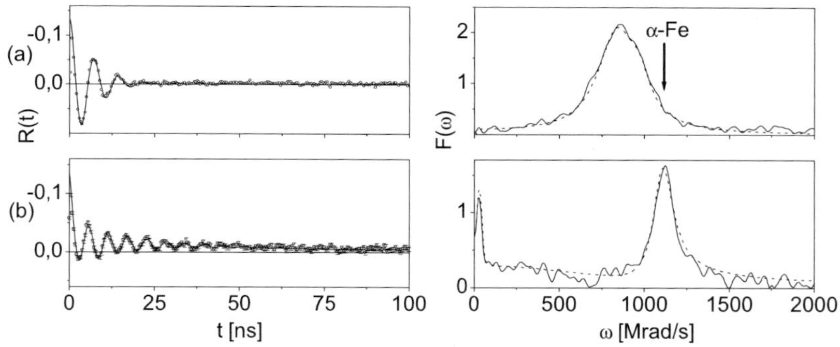

Figure 1. PAC spectra of $^{111}In(^{111}Cd)$ in the amorphous ferromagnetic alloy Fe79/B16/Si5 measured in a 90°/180° detector arrangement with a polarizing field of 0.5 T perpendicular to the detector plane. The Fourier transformation F(ω) (*right*) reflects directly the hyperfine fields distributions P(B_{hf}). The graph (a) shows the as-implanted sample measured at room temperature (*R.T.*). The bottom graph was obtained following a thermal treatment at 410°C/10 min in vacuum.

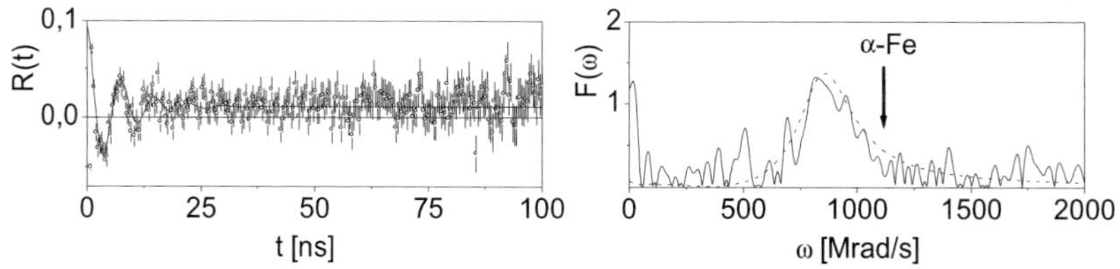

Figure 2. PAC spectrum of ^{111m}Cd in the material of Figure 1. The sample is heat treated at 480°C/10 min. Measuring conditions like that of Figure 1. The poor statistics is a consequence of the ^{111m}Cd half life of only 49 min.

sample without any treatment. Both the spectrum and the magnetic hyperfine field distribution are similar to those, which are well-known from measurements on amorphous $Fe_{80}B_{20}$ and similar materials in the literature [5]. But it is almost impossible to get detailed predictions to the distributions of the other parameters like quadrupole coupling, isomer shift and texture which influence the spectra additionally. Of course, the mean magnetic field is with $\langle B_{hf} \rangle = (-)22(1)$ T smaller than the α-Fe field of −33 T.

By a thermal treatment of the sample a complicated process of structural relaxation starts. Thereby it is interesting that a heating at 500°C/10 min. (crystallization temperature 515°C) produces α-Fe clusters. Figure 3(b) shows the Mössbauer spectrum. About 15% of the ^{57}Fe probes sit in these environments. This behaviour is described in other papers [2, 5, 6] too, in which the spectra were analysed more thoroughly. In these investigations also other crystallization products like Fe_3B and Fe_2B could be identified.

4. Conclusions

The amorphous state is not stable thermodynamically. It is produced by quenching the liquid alloy with a very high cooling rate. Therefore, the results

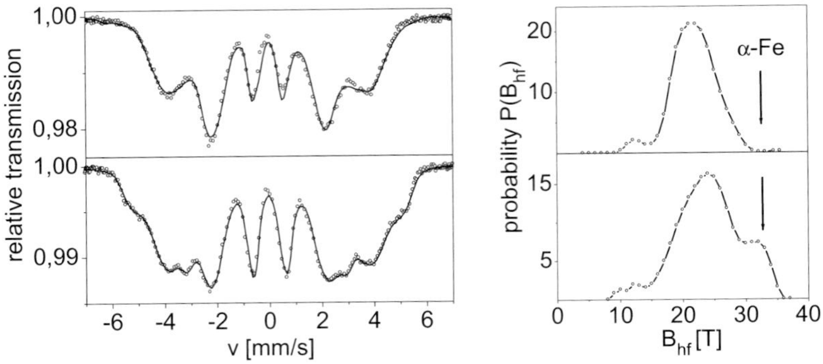

Figure 3. ^{57}Fe Mössbauer transmission spectra of the amorphous ferromagnetic alloy. At the right the distribution of magnetic hyperfine fields $P(B_{hf})$ is shown. The *upper curves* are for the sample without a heat treatment. The *lower curves* for the sample after a thermal treatment at 500°C/10 min.

from different materials can vary depending on the quenching conditions, the aging of the material and the given composition. Moreover, structure differences between surface layers and the bulk material develop as a consequence of varying cooling rates. In this paper only general properties are discussed. If one compares the results of Mössbauer transmission measurements with those of the PAC method one has to consider apart from the different probe atoms also the fact that after implantation only very thin surface layers (mean range of the probes 15 nm) are investigated by PAC.

The observed mean fields and the widths of the field distributions are comparable for the MS and PAC measurements if one considers the different probes in the untreated and the as-implanted sample, respectively. But the In-probes have the peculiarity that they are completely caught by the α-Fe clusters at a temperature distinctly below 500°C. May be this is promoted by surface properties but there should also be an affinity of In atoms to the Fe clusters. The field distribution of Figure 1(b) indicates that the clusters should be nanocrystals of nearly pure α-Fe. Such Fe clusters could not be found in the ^{111m}Cd-PAC measurements though the temperature of heat treatment was near the crystallization temperature. Obviously the Cd probes are not attracted by the Fe clusters. The reason for this different behaviour of In and Cd could be that, according to the corresponding phase diagrams, In has a small solubility in Fe whereas Cd is practically insoluble in Fe. For more detailed findings the statistics of the ^{111m}Cd measurements has to be improved.

Acknowledgements

We thank Prof. D. Rafaja, Institut für Metallkunde, TU Bergakademie Freiberg, Germany, for the small-angle X-ray scattering characterization of the samples. The financial support by the German Bundesministerium für Bildung und Forschung is gratefully acknowledged (BMBF-contract 05 KK1OFA/5).

References

1. Le Caer G., Dubois J. M., Fischer H., Gonser U. and Wagner H. G., *Nucl. Instrum. Methods Phys. Res., B* **5** (1984), 25.
2. Pankhurst Q. A., In: Long G. L. and Grandjean F. (eds.), *Mössbauer Spectroscopy Applied to Magnetism and Materials Science*, Vol. 2, Plenum Press, N. Y. and London, 1996, p. 59.
3. Kopcewicz M., Wagner H. G. and Gonser U., *J. Phys. F. Met. Phys.* **16** (1986), 929.
4. Kugler E., *Hyp. Interact.* **129** (2000), 23.
5. Franke H. and Rosenberg M., *J. Magn. Magn. Mater.* **7** (1978), 168.
6. Ge S. H., Chen G. L., Mao M. X., Xue D. S., Li C. X., Zhang Y. D., Hines W. A. and Budnik J. I., *J. Magn. Magn. Mater.* **129** (1994), 207.

Hyperfine Interactions (2004) 158:223–227
DOI 10.1007/s10751-005-9037-z

Implantation of ^{111}In-probe Nuclei with Nuclear Reactions ^{108}Pd(^{6,7}Li, xn)^{111}In using Pelletron Tandem Accelerator: Study of Local Magnetism in Heusler Alloys

G. A. CABRERA-PASCA[1], M. N. RAO[1], J. R. B. OLIVEIRA[1], M. A. RIZZUTTO[1], N. ADDED[1], W. A. SEALE[1], R. V. RIBAS[1], N. H. MEDINA[1], R. N. SAXENA[2,]* and A. W. CARBONARI[2]
[1]*Instituto de Fisica da Universidade de São Paulo, USP, São Paulo, Brazil*
[2]*Instituto de Pesquisas energéticas e Nucleares – IPEN-CNEN/SP, São Paulo, Brazil; e-mail: rnsaxena@ipen.br*

Abstract. Perturbed Angular Correlation method has been used to study the hyperfine magnetic field in the Heusler alloy Pd_2MnSb(Sn). Ion implantation of the recoil ^{111}In nuclei following heavy ion nuclear reactions ^{108}Pd(^{7}Li, 4n)^{111}In and ^{108}Pd(^{6}Li, 3n)^{111}In has been used to great advantage in the present case resulting in large implantation efficiency. Only a few hours of irradiation time with moderate beam current of the order of 400–500 nA resulted in sufficient implanted ^{111}In activity on the sample for good quality measurements. The hyperfine field was measured at ^{111}Cd probe nuclei substituting Mn and Sb(Sn) sites as a function of temperature. The fraction of ^{111}Cd nuclei occupying Mn atom sites was found to increases with the annealing of sample at higher temperatures.

Key Words: Heusler alloys, magnetic hyperfine field, PAC, probe implantation.

1. Introduction

There are several different ways to introduce radioactive probe nuclei into samples to be measured by Perturbed Angular Correlation (PAC) spectroscopy, each one having its own advantages and disadvantages. The ion implantation process is particularly advantageous because it introduces radioactive probes into an already prepared sample, which avoids extensive manipulation of the radioactive material. What is still a subject of investigations is how the method of introducing probes influences its site location. In the present work, besides reporting a new and efficient way to introduce ^{111}In probe nuclei into samples for PAC measurements, we also show that the method of introducing the probe can give different results. In order to test the method of implantation we have used Pd-

* Author for correspondence.

based Heusler alloy as sample. The Heusler alloys $Pd_2MnSb(Sn)$ have a cubic $L2_1$ structure and order ferromagnetically with a magnetic moment of about 4.3 μB localized on Mn. These alloys have been investigated in the past with PAC spectroscopy [1, 2]. The radioactive ^{111}In probe, introduced in the samples during its preparation by induction melting of component elements, was found to substitute only the Sn and Sb atom sites. On the other hand when ^{111}Ag was introduced in Pd_2MnSn sample through thermal diffusion it occupied the Mn site [3]. In the present experiment heavy ion nuclear reactions ^{108}Pd(^{7}Li, 4n)^{111}In and ^{108}Pd(^{6}Li, 3n)^{111}In, in which $Pd_2MnSb(Sn)$ Heusler alloys themselves served as the reaction target, was used to implant the recoiling ^{111}In nuclei in to the sample.

2. Experimental

The Heusler alloys were prepared by arc-melting the stoichiometric amounts of the constituent elements (99.99% purity) in argon atmosphere purified with a hot titanium getterer. After annealing at 800°C during 24 h in the argon atmosphere the samples were analyzed by X-ray powder difraction, and were found to be essentially single phase with the correct $L2_1$ structure. The ^{111}In nuclei were implanted in the samples by means of nuclear reactions ^{108}Pd(^{7}Li, 4n)^{111}In or ^{108}Pd(^{6}Li, 3n)^{111}In, with a beam energy of 32 MeV using eight UD Pelletron Tandem Accelerator at the Physics Institute of the University of São Paulo. Since Pd is a constituent element of the alloys in the present experiment, the samples themselves served as the reaction targets during irradiation and practically all the ^{111}In nuclei produced in the reaction stopped in the sample. The Heusler alloy samples were cut in the form of a 5 mm diameter disc with a thickness of ~1 mm and mounted in a special reaction chamber [4]. The irradiation times varied from 8–10 h with an average beam current of the order of 400–500 nA on the target. A few hours after the end of irradiation, the samples were examined with HPGe detector for the presence of other radioactive nuclei produced through side reactions with other elements of the sample in addition to the desired ^{111}In. Most of the radioactive products found were either short lived or had little interference in the measurement of coincidence spectra of the gamma–gamma cascade in the decay of ^{111}In.

The TDPAC measurements were carried out with a conventional fast–slow coincidence set-up using four conical BaF_2 detectors. The gamma cascade of 172–245 keV populated from the electron capture decay of ^{111}In, and having an intermediate level at 245 keV with spin $I = 5/2^+$ and $T_{1/2} = 85$ ns, in ^{111}Cd was used to measure the magnetic hyperfine field. The samples were measured in the temperature range of 20–295 K using a closed-cycle helium cryogenic device. The time resolution of the system was about 600 ps.

The TDPAC method is based on the observation of hyperfine interaction of nuclear moments with extra-nuclear magnetic field or electric field gradient. A detailed description of the method can be found elsewhere [5, 6]. The per-

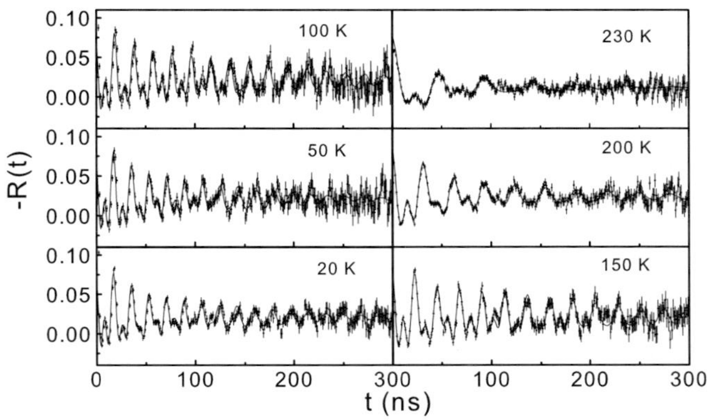

Figure 1. TDPAC spectra of ^{111}Cd implanted in Pd_2MnSb at different temperatures.

turbation factor $G_{22}(t)$ of the correlation function contains detailed information about the hyperfine interaction. Measurement of $G_{22}(t)$ allows the determination of Larmor frequency $\omega L = \mu_N g B_{hf} / \hbar$. From the known g-factor, $g = 0.306$ of the 245 keV state of ^{111}Cd it is thus possible to determine the magnetic hyperfine field B_{hf}.

3. Results and discussion

Some of the TDPAC spectra measured at different temperatures for the alloy Pd_2MnSb are shown in Figure 1. The samples were annealed at 400°C in vacuum for about 4 h, after implantation, before starting the PAC measurements to eliminate the effects of radiation damage. Slight attenuation of the amplitude seen in the spectra results from a low frequency quadrupole interaction, most probably due to some impurities or defects in the crystal structure. The experimental data were analyzed by using a model which included combined magnetic dipole and electric quadrupole interactions. The results show two distinct magnetic interactions at all temperatures in both cases. For Pd_2MnSn the characterstic Larmor frequencies measured at 10 K were found to be 319 Mrad/s for the component with major fraction and 119 Mrad/s for the minor fraction. The corresponding hyperfine fields (B_{hf}) are 21.2(1) T and 7.9(1) T, respectively. The higher frequency component was assigned to probe nuclei occupying Sn site and the lower frequency component to the Mn site in conformity with the previous results [3]. The PAC spectra for the Pd_2MnSb alloy were analyzed in a similar manner. The Larmor frequencies measured at 20 K are 350 Mrad/s (major fraction) and 320 Mrad/s (minor fraction). These were assigned to Sb and Mn sites, respectively, corresponding to hyperfine fields of 23.8(1) T and 21.3(1) T. The values of the hyperfine fields are in good agreement with those obtained in the previous studies [1, 2]. The temperature dependence of hyperfine fields is shown in Figure 2. After an additional annealing of the samples at 800°C for

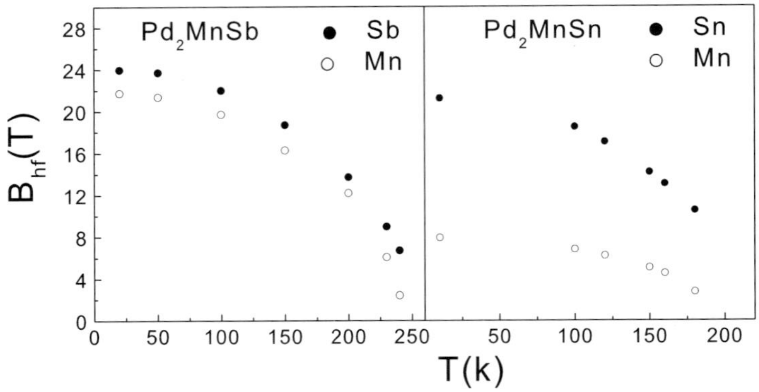

Figure 2. Temperature dependence of magnetic hyperfine fields in Pd_2MnSb and Pd_2MnSn as a function of temperature.

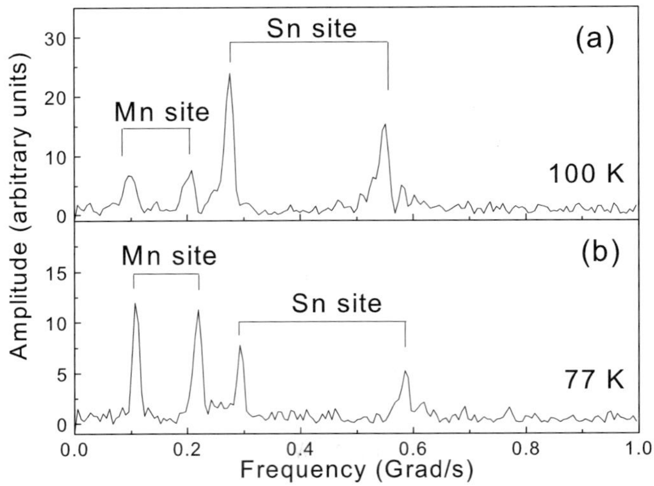

Figure 3. Fast Foureir spectra for the alloy Pd_2MnSn before (a) and after (b) annealing at 800°C for 24 h.

24 h in argon, the site occupation showed significant increase in the fraction of Mn site. This can be seen in the fast Fourier spectra before and after annealing at 800°C of Pd2MnSn (Figure 3).

We have now initiated a more systematic study of the site dependence of the hyperfine fields in the Heusler alloys Pd_2MnZ(Z = Sn, Ge, Sb) with the PAC technique. First measurements on the ternary alloys Pd_2MnSn and Pd_2MnSb with ^{111}In probes showed two Larmor frequencies corresponding to the Mn and Sn sites. This is already an interesting result as it was possible to simultaneously measure the hyperfine fields at more than one site in the same implanted sample. The reason why the implanted ^{111}In nuclei show little preference for Pd sites is not clear at the moment. It is possible, however, that the probe does enter the Pd site during implantation but migrates to other sites on annealing at lower temperatures. Further measurements to better understand the annealing temperature dependence of the ^{111}In-site occupation as well as to obtain the hyperfine field at the Pd site if possible are planned. Present experiment has demonstrated that for samples where Pd is one of the components the process of implantation of ^{111}In using the present nuclear reaction is quite efficient compared to the conventional methods of introducing the probe in the sample. For the implantation of ^{111}In on samples that do not contain Pd the method will require some modifications. The ^{108}Pd(^{7}Li, 4n)^{111}In or ^{108}Pd(^{6}Li, 3n)^{111}In reactions could be produced in a thin foil of Pd (preferably enriched in ^{108}Pd) and swift ^{111}In ions recoiling out of the foil may be stopped in the substrate placed behind the target at a suitable distance and geometry. The reaction chamber for such experiments is under test.

Acknowledgements

Partial financial support for this research was provided by the Fundaçaõ de Amparo á Pesquisa do Estado de São Paulo (FAPESP). GAC-P thankfully acknowledges the student fellowship granted by CAPES.

References

1. Schaf J., Fraga E. R. and Zawislak F. C., *J. Mag. Mag. Mat.* **8** (1978), 297.
2. Schaf J., Pasquevich A. F., Schreiner W. H., Campbell C. C. M., Fraga E. R. and Zawislak F. C., *J. Mag. Mag. Mat.* **21** (1979), 24.
3. Carbonari A. W. and Haas H., *Hyperfine Interact.* **133** (2001), 71.
4. Rao M. N., Oliveira J. R. B., Seale W. A., Rizzutto M. A., Ribas R. V., Alcántara Núñez J. A., Pereira D., Added N., Cybulska E. W., Medina N. H., Saxena R. N. and Carbonari A. W., *Braz. J. Phys.* **33** (2003), 291.
5. Pendl W. Jr., Saxena R. N., Carbonari A. W., Mestnik-Filho J. and Schaft J., *J. Phys., Condens. Matter* **8** (1996), 11317.
6. Attili R. N., Saxena R. N., Carbonari A. W., Mestnik-Filho J., Uhrmacher M. and Lieb K. P., *Phys. Rev. B* **58** (1998), 2563.

Hyperfine Interactions (2004) 158:229–233
DOI 10.1007/s10751-005-9038-y

Magnetic Hyperfine Interaction of a Cubic Defect in α-Iron

S. UNTERRICKER[1,*], V. SAMOKHVALOV[1], F. SCHNEIDER[1], M. DIETRICH[2] and THE ISOLDE COLLABORATION[3]
[1]*Institut für Angewandte Physik, TU Bergakademie Freiberg, D-09596, Freiberg (Sachsen), Germany; e-mail: unterr@physik.tu-freilberg.de*
[2]*Technische Physik, Universität des Saarlandes, D-66041 Saarbrücken, Germany*
[3]*CERN, CH-1211, Geneva 23, Switzerland*

Abstract. PAC measurements on ^{111}In(^{111}Cd) implanted and thermally treated α-Fe have shown an indication for a cubic defect with the ^{111}Cd probe in the centre of it. The measured room temperature (R. T.) magnetic hyperfine fields are $B_{hf1} = -38.4(8)$ T for substitution and $B_{hf2} = +11.5(3)$ T for the cubic defect. Additionally, probes with pure quadrupole frequency distributions were observed, which are incorporated in surface contaminations.

Key Words: α-Fe, cubic defect, implantation, magnetic hyperfine field, PAC.

1. Introduction

With perturbed angular correlations (PAC) point defects in metals have been investigated for a long time and in the last 20 years very successfully also in semiconductors. This method is well suited to obtain a microscopic picture of the defect structure around proper probe nuclei.

Several groups have investigated the behaviour of defects in Ni by PAC. In connection with a certain lattice damage and with ^{111}In(^{111}Cd) as probe, a defect with cubic symmetry was identified and intensively investigated. The structure of this defect is controversially discussed because of its fundamental importance for the interpretation of recovery stage III [1, 2]. Magnetic hyperfine fields at ^{111}Cd probes on different sites of single crystalline Ni surfaces were thoroughly investigated in [6]. Compared to Ni the number of ^{111}In(^{111}Cd) PAC experiments in Fe is small. Reasons may be: the very high frequency of magnetic interaction, the low solubility of In in Fe and the contamination of Fe with several light elements.

* Author for correspondence.

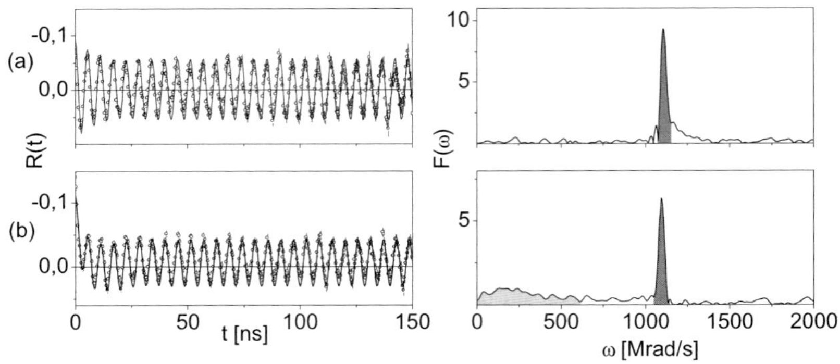

Figure 1. (a) ^{111}In(^{111}Cd) in α-Fe as-implanted (160 keV). Left, time dependent coincidence ratio $R(t)$, right, Fourier transformation $F(\omega)$. (b) Measurement after a heat treatment at 600°C/10 min in vacuum. Probes in non-magnetic environments (frequency distribution) are marked *light grey* in $F(\omega)$. Detector arrangement 90°/180°. All measurements of this contribution were carried out in a polarizing field of 0.5 T perpendicular to the detector plane and at R.T.

2. Behaviour of ^{111}In(^{111}Cd) implanted into α-Fe

An α-Fe foil of 5 μm thickness was implanted at the Institut für Strahlen – und Kernphysik (I. S. K. P.), University of Bonn, with 160 keV ^{111}In probes. The commercial material had a purity of better than 99.99%. The In ions (mean range of about 30 nm) are stopped below a surface layer in which by Auger electron spectroscopy contaminations with oxygen and carbon could be found. The thickness of the contaminated layer (>1% impurity atoms) is approximately 10 nm. In Figure 1(a) a R.T. PAC spectrum of the as-implanted sample is shown, where the magnetic interaction of ^{111}Cd in α-Fe is clearly visible. The PAC apparatus consisted of a four BaF_2 detector 90°/180° arrangement and the sample was magnetically polarized perpendicular to the detector plane. Therefore the measured frequency is twice the Larmor frequency ω_{L1}. A careful check of the $R(t)$ dependence shows two contributions. The larger one is without any damping and corresponds to pure α-Fe environments. Its frequency ω_{L1} = 553(4) Mrad/s is known from diffusion experiments. For the other probes a damping is observed but the mean frequency is near by ω_{L1}. These probes are influenced by non-correlated defects as a consequence of implantation damage (field distributions). A thermal treatment at 300°C for 10 min. in vacuum anneals the environments with frequency distribution completely. Such a behaviour was also described for implanted (120 keV) samples in [3].

If the temperature of thermal treatment is increased to 500°C a substantial out-diffusion of the In probes starts. Now besides the probes with pure Fe environments also those with non-magnetic surroundings appear in the Fourier spectrum, Figure 1(b). They are characterized by broad electric field gradient distributions and correspond to probes which sit in the polluted surface.

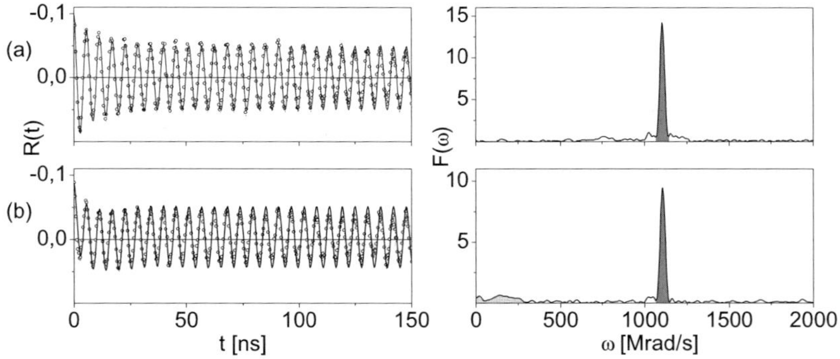

Figure 2. (a) $^{111}In(^{111}Cd)$ in α-Fe as-implanted (50 keV, ISOLDE-CERN). $R(t)$ and $F(\omega)$ correspond to those of Figure 1(a). (b) Measurement after a thermal treatment at 300°C/10 min. Non-magnetic environments are visible here at a lower annealing temperature.

With rising temperature of heat treatment the fraction of probes in non-magnetic environments increases.

3. ISOLDE experiment with ^{111}In(^{111}Cd) implanted into α-Fe

The same α-Fe foil as used in the investigations of Section 2 was utilized for experiments at the ISOLDE on-line separator, CERN-Geneva. Here the implantation energy was only 50 keV but the ^{111}In beam is very pure. The mean range of the In probes amounts to 10 nm, which means that now a large part of probes are influenced by pollutions (mostly O and C) but also other surface defects.

Figure 2(a) shows a PAC-measurement of the as-implanted sample (experimental conditions see figure caption). Again we see only probes in Fe environments, about 50% of them in pure α-Fe, the remainder in surroundings with non-correlated defects. Following a heat treatment at only 300°C in addition to probes in pure α-Fe also such in non-magnetic surroundings are indicated in the Fourier spectrum. The latter sit within surface contaminations, see Figure 1(b). The next step was a thermal treatment at 500°C. Now, Figure 3(a), we have to distinguish between three different probe environments: (1) such with the magnetic interaction of pure α-Fe, (2) probes with a well defined magnetic interaction of ω_{L2} = 176(3) Mrad/s and (3) probes in non-magnetic environments with a broad quadrupole frequency distribution. The result of the sign sensitive ± 135° detector geometry (Figure 3(b)) proves clearly that the probes (3) belong to pure quadrupole interactions which are not visible in this arrangement. The two Larmor frequencies ω_{L1} and ω_{L2} correspond to the magnetic hyperfine fields $B_{hf1} = -38.4(8)$ T and $B_{hf2} = +11.5(3)$ T at R.T. They have opposite signs. B_{hf1} is the substitutional field and B_{hf2} the field at probes within a defect of cubic symmetry and therefore vanishing quadrupole interaction.

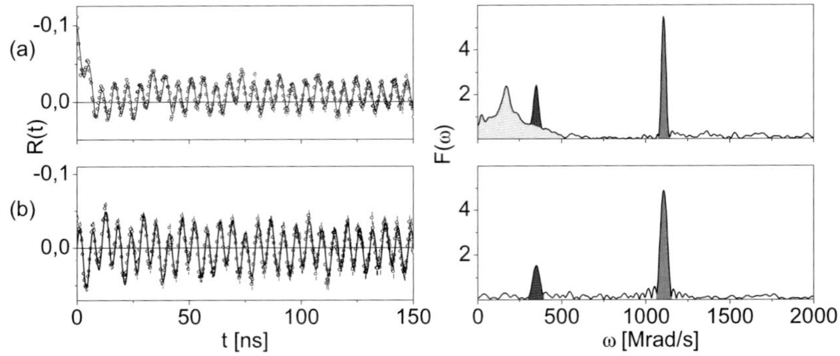

Figure 3. (a) The same sample as in Figure 2 after a heat treatment at 500°C/10 min. $F(\omega)$ shows now peaks for substitution (field B_{hf1}, *grey*), a second defined magnetic interaction (field B_{hf2}, *black*) and a pronounced quadrupole frequency distribution (*light grey*). (b) A measurement with ±135° arrangement proves that the peak of the lower field B_{hf2} corresponds to a defect with cubic symmetry and that the frequency distribution is caused by pure quadrupole interactions.

A heat treatment at 700°C reduces the part of probes in pure α-Fe to less than 10%. The remainder is subjected to the quadrupole interaction mentioned before.

4. Discussion and conclusions

Obviously, the magnetic hyperfine field B_{hf2}, which corresponds to a defined defect of cubic symmetry, develops if the probes become mobile and if they are near surface regions. Such a cubic defect has been observed and investigated with PAC for ^{111}In(^{111}Cd) in Ni since 1974 as well. It can be produced by implantation, by particle irradiation but also by cold-working of Ni. The structure of the cubic defect in fcc Ni is discussed on the basis of two models: (1) the In probes are at an interstitial site which is the centre of a tetrahedron consisting of four nearest neighbour vacancies [2] or (2) the cubic defects are In probes on a tetrahedrally coordinated interstitial position in the fcc lattice [1]. For both configurations mobile vacancies or interstitials in the vicinity of specimen surfaces are necessary.

α-Fe has a bcc lattice structure in which the octahedral interstitial positions have a non-cubic point symmetry. Only the tetrahedral interstitial sites have a perfect tetrahedral coordination of next neighbours which is the requirement for a vanishing or very weak quadrupole interaction, as observed. The unrelaxed nn distances for the tetrahedral interstitial site are 0.161 nm in the Fe cell and 0.151 nm in the Ni cell.

A peculiarity of the Fe experiment is the poor In solubility in comparison with that in Ni and other metals [4]. As a consequence the In probes migrate to outer and inner surfaces and are caught in non-magnetic positions with quadrupole frequency distributions.

It is a realistic goal for the near future to test different defect models for their correctness by modern theoretical methods like WIEN97 [5].

Acknowledgements

We thank Dr. R. Vianden and Dr. P. D. Eversheim, Institut für Strahlen – und Kernphysik, Universität Bonn, Germany, for the implantation of an α-Fe sample and Dr. H. Reuther, Forschungszentrum Rossendorf, Germany, for the AES analysis of the α-Fe foil. The financial support by the German Bundesministerium für Bildung und Forschung is gratefully acknowledged (BMBF-contract 05 KK1OFA/5).

References

1. Aspacher B., Frank W., Kizler P. and Maier K., *Appl. Phys.* **A59** (1994), 339.
2. Pleiter F. and Hohenemser C., *Phys. Rev.* **B25** (1982), 106.
3. Pleiter F., Hohenemser C. and Arends A. R., *Hyperfine Interact.* **10** (1981), 691.
4. Golcewski J. and Maier K., *Nucl. Instrum., Methods Phys. Res.* **A244** (1986), 509.
5. Blaha P., Schwarz K. and Luitz J., WIEN97, 1999, ISBN 3-9501031-0-4.
6. Potzger K., Weber A., Bertschat H. H., Zeitz W.-D. and Dietrich M., *Phys. Rev. Lett.* **88** (2002), 247201.

Hyperfine Interactions (2004) 158:235–243
DOI 10.1007/s10751-005-9039-x

Non-Markovian Dynamics of a Localized Electron Spin Due to the Hyperfine Interaction

W. A. COISH* and DANIEL LOSS
University of Basel, Klingelbergstrasse 82, CH-4056 Basel, Switzerland;
e-mail: bill.coish@unibas.ch

Abstract. We review our theoretical work on the dynamics of a localized electron spin interacting with an environment of nuclear spins. Our perturbative calculation is valid for arbitrary polarization p of the nuclear spin system and arbitrary nuclear spin I in a sufficiently large magnetic field. In general, the electron spin shows rich dynamics, described by a sum of contributions with exponential decay, nonexponential decay, and undamped oscillations. We have found an abrupt crossover in the long-time spin dynamics at a critical shape and dimensionality of the electron envelope wave function. We conclude with a discussion of our proposed scheme to measure the relevant dynamics using a standard spin–echo technique.

Key Words: ESR, hyperfine interaction, nuclear spins, quantum computing, quantum dots, spin-echo, spintronics.

1. Introduction

Experiments with trapped ions [1] and nuclear magnetic resonance (NMR) [2] have proven that the basic elements of a quantum computer can be realized in practice. To progress beyond proof-of-principle experiments, the next generation of quantum information processors must overcome significant obstacles regarding scalability and decoherence. The scalability issue is largely solved by proposals for a solid-state implementation for quantum computing, where established fabrication techniques can be used to multiply the qubits and interface them with existing electronic devices. Due to their relative isolation from the surrounding environment, single electron spins in semiconductor quantum dots are expected to be exceptionally robust against decoherence [3]. Indeed, longitudinal relaxation times in these systems have been measured to be $T_1 = 0.85$ ms in a magnetic field of 8 T [4] and in GaAs quantum wells, the transverse dephasing time T_2^* for an ensemble of electron spins (which typically provides a lower bound for the intrinsic decoherence time T_2 of an isolated spin) has been measured to be in excess of 100 ns [5].

* Author for correspondence.

Possible sources of decoherence for a single electron spin confined to a quantum dot are spin–orbit coupling and the contact hyperfine interaction with the surrounding nuclear spins [6]. The relaxation rate due to spin–orbit coupling $1/T_1$ is suppressed for localized electrons at low temperatures [7] and recent work has shown that T_2, due to spin–orbit coupling, can be as long as T_1 under realistic conditions [8]. However, since spin-carrying isotopes are common in the semiconductor industry, the contact hyperfine interaction (in contrast to the spin–orbit interaction) is likely an unavoidable source of decoherence, which does not vanish with decreasing temperature or carefully chosen quantum dot geometry [9].

In the last few years, a great deal of effort has been focused on a theoretical description of interesting effects arising from the contact hyperfine interaction for a localized electron. The predicted effects include a dramatic variation of T_1 with gate voltage in a quantum dot near the Coulomb blockade peaks or valleys [10], all-optical polarization of the nuclear spins [11], use of the nuclear spin system as a quantum memory [12, 13], and several studies of spin dynamics [14–21]. Here, our system of interest is an electron confined to a single GaAs quantum dot, but this work applies quite generally to other systems, such as electrons trapped at shallow donor impurities in Si:P [9].

An exact solution for the electron spin dynamics has been found in the special case of a fully polarized initial state of the nuclear spin system [14, 22]. This solution shows that the electron spin only decays by a fraction $\propto \frac{1}{N}$ of its initial value, where N is the number of nuclear spins within the extent of the electron wave function. The decaying fraction was shown to have a nonexponential tail for long times, which suggests non-Markovian (history dependent) behavior. For an initial nuclear spin configuration that is not fully polarized, no exact solution is available and standard time-dependent perturbation theory fails [14]. Subsequent exact diagonalization studies on small spin systems [23] have shown that the electron spin dynamics are highly dependent on the type of initial nuclear spin configuration. The unusual (nonexponential) form of decay, and the fraction of the electron spin that undergoes decay may be of interest in quantum error correction (QEC) since QEC schemes typically assume exponential decay to zero.

In the following section we describe a systematic perturbative theory of electron spin dynamics under the action of the Fermi contact hyperfine interaction. Further details of this work can be found in reference [24].

2. Model and perturbative expansion

We investigate electron spin dynamics at times shorter than the nuclear dipole–dipole correlation time τ_{dd} ($\tau_{dd} \approx 10^{-4}$ s in GaAs is given directly by the inverse width of the nuclear magnetic resonance (NMR) line [25]). At these time scales, the relevant Hamiltonian for a description of the electron and nuclear spin dynamics is that for the Fermi contact hyperfine interaction:

$$H = H_0 + V_{ff}, \tag{1}$$

$$H_0 = (b + h_z)S_z; \quad V_{ff} = \frac{1}{2}(S_+h_- + S_-h_+), \tag{2}$$

where $\mathbf{S}$ is the electron spin operator, b gives the electron Zeeman splitting and $\mathbf{h} = \sum_k A_k \mathbf{I}_k$ is the quantum nuclear field. $S_\pm = S_x \pm iS_y$ and $h_\pm = h_x \pm ih_y$ are defined in the usual way. The hyperfine coupling constants are $A_k = Av_0 |\psi(\mathbf{r}_k)|^2$ where v_0 is the volume of a crystal unit cell containing one nuclear spin, $\psi(\mathbf{r})$ is the electron envelope wave function, and A is the strength of the hyperfine coupling. While the total number of nuclear spins in the system, N_{tot}, may be very large, there are fewer spins ($N < N_{\text{tot}}$) within the extent of the electron wave function. In GaAs, all naturally occurring isotopes carry spin $I = \frac{3}{2}$. In bulk GaAs, A has been estimated [25] to be $A = 90\ \mu eV$. This estimate is based on an average over the hyperfine coupling constants for the three nuclear isotopes ^{69}Ga, ^{71}Ga, and ^{75}As, weighted by their relative abundances. Natural silicon contains 4.7% ^{29}Si, which carries $I = \frac{1}{2}$, and 95% ^{28}Si, with $I = 0$. An electron bound to a phosphorus donor impurity in natural Si:P interacts with $N \approx 10^2$ surrounding ^{29}Si nuclear spins, in which case the hyperfine coupling constant is on the order of $A \approx 0.1\ \mu eV$ [9].

For large magnetic fields b, the flip–flop processes due to V_{ff} are suppressed by the electron Zeeman splitting, and it is reasonable to take $H \approx H_0$ (zeroth order in V_{ff}). The zeroth-order problem is algebraically simple, and leads to some interesting conclusions regarding the initial conditions. In this limit the longitudinal spin is time-independent, since $[S_z, H_0] = 0$, but the transverse spin can undergo nontrivial evolution. Assuming uniform hyperfine coupling constants ($A_k = A/N$) and nuclear spin $I = 1/2$ we evaluate the transverse electron spin dynamics for a nuclear spin bath of polarization p along the z-axis and two different nuclear spin initial states. When the A_k are uniform, the transverse spin exhibits periodic dynamics. However, at times much less than the period, given by the inverse level spacing ($t \ll N/A$, setting $\hbar = 1$), we find

$$\langle S_+\rangle_t^{\text{unprep.}} = \langle S_+\rangle_0 e^{i\omega_n t} e^{-t^2/2t_c^2}, \quad t_c = \frac{2}{A}\sqrt{\frac{N}{1-p^2}}, \tag{3}$$

$$\langle S_+\rangle_t^{\text{prep.}} = \langle S_+\rangle_0 e^{i\omega_n t}, \quad \omega_n = b + pA/2. \tag{4}$$

In Equation (3), the nuclear spin system is "unprepared." This state corresponds to either a translationally invariant direct-product state of the form $\Pi_k(\sqrt{f_\uparrow}\,|\uparrow\rangle_k + \sqrt{1-f_\uparrow}\,|\downarrow\rangle_k)$, where $p = 2f_\uparrow + 1$, or to a statistical mixture of product states of the form $|\uparrow\downarrow\uparrow\uparrow\cdots\rangle$, with average polarization p. In contrast, Equation (4) corresponds to the case when the nuclear system has been "prepared" in an eigenstate $|n\rangle$ of the operator $h_z : h_z \mid n\rangle = [h_z]_{nn} \mid n\rangle$, ($[h_z]_{nn} = pA/2$). It is important to note that the decay of the 'unprepared' state is reversible. This decay can be recovered by performing a spin–echo measurement on the electron spin, since a π-rotation of the electron spin $S_z \to -S_z$ reverses the sign of H_0: $H_0 \to -H_0$, and results in time-reversed dynamics. When the hyperfine coupling constants are allowed to vary in space, higher-order corrections in V_{ff} will, however, result in irreversible decay. In the following, we generalize to nonuniform A_k and arbitrary nuclear spin I to evaluate this decay for an initial nuclear state that is an eigenstate of the h_z-operator.

An exact generalized master equation (GME) can be derived for the electron spin operators beginning from the von Neumann equation for the full density operator ($\dot\rho = -i[H,\rho]$) [26]. The resulting GME is expanded in terms of V_{ff} through direct application of the Dyson identity. We find, quite remarkably, that the equations of motion for the longitudinal ($\langle S_z\rangle_t$) and transverse ($\langle S_+\rangle_t = \langle S_x\rangle_t + i\langle S_y\rangle_t$) spin are decoupled to *all orders* in V_{ff} and take the form

$$\langle \dot S_z\rangle_t = N_z(t) - i\int_0^t dt' \Sigma_{zz}(t-t')\langle S_z\rangle_{t'}, \tag{5}$$

$$\langle \dot S_+\rangle_t = i\omega_n\langle S_+\rangle_t - i\int_0^t dt' \Sigma_{++}(t-t')\langle S_+\rangle_{t'}. \tag{6}$$

These integro-differential equations can be converted to a pair of algebraic equations by Laplace transformation $f(s) = \int_0^\infty dse^{-st}f(t)$, Re[s] > 0. The algebraic equations yield

$$S_z(s) = \frac{\langle S_z\rangle_0 + N_z(s)}{s + i\Sigma_{zz}(s)};\quad S_+(s) = \frac{\langle S_+\rangle_0}{s - i\omega_n + i\Sigma_{++}(s)}. \tag{7}$$

The numerator term $N_z(s)$ and self-energy $\Sigma_{zz}(s)$ are related to the self-energy matrix elements for spin-up and spin-down by $N_z(s) = -\frac{i}{2s}\left(\Sigma_{\uparrow\uparrow}(s) + \Sigma_{\uparrow\downarrow}(s)\right)$ and $\Sigma_{zz}(s) = \Sigma_{\uparrow\uparrow}(s) - \Sigma_{\uparrow\downarrow}(s)$. All self-energy matrix elements are written in terms of an expansion in powers of $V_{ff} : \Sigma_{\alpha\beta}(s) = \Sigma^{(2)}_{\alpha\beta}(s) + \Sigma^{(4)}_{\alpha\beta}(s) + \cdots, \alpha,\ \beta = (+,\uparrow,\downarrow)$. For a sufficiently large Zeeman splitting b we find that all higher-order self-energy matrix elements are suppressed by a smallness parameter Δ:

$$\Sigma^{(2(k+1))}_{\alpha\beta}(s) \propto \Delta^k;\ \Delta = \frac{A}{2(b + pIA)}. \tag{8}$$

In Born approximation (second order in the flip–flop terms V_{ff}), and for an initial nuclear spin system with uniform polarization, we find

$$\sum_{\uparrow\uparrow}(s) \approx \sum_{\uparrow\uparrow}^{(2)}(s) = -iNc_+[I_+(s - i\omega_n) + I_-(s + i\omega_n)], \tag{9}$$

$$\sum_{\uparrow\downarrow}(s) \approx \sum_{\uparrow\downarrow}^{(2)}(s) = iNc_-[I_-(s - i\omega_n) + I_+(s + i\omega_n)], \tag{10}$$

$$\sum_{++}(s) \approx \sum_{++}^{(2)}(s) = -iN[c_-I_+(s) + c_+I_-(s)], \tag{11}$$

$$c_\pm = I(I+1) - \frac{1}{N_{\text{tot}}}\sum_k \langle I_k^z(I_k^z \pm 1)\rangle_0, \tag{12}$$

$$I_\pm(s) = \frac{1}{4N}\sum_k \frac{A_k^2}{s \mp i\frac{A_k}{2}} \approx \frac{d}{m}\int_0^1 dx \frac{x|\ln x|^v}{s \mp ix}, \quad v = \frac{d}{m} - 1. \tag{13}$$

In evaluating Equation (13), we have assumed the electron is in its orbital ground state, described by an isotropic envelope wave function of the form $\psi(r_k) = \psi(0)\exp\left[-\frac{1}{2}\left(\frac{r_k}{l_0}\right)^m\right]$. The index k gives the number of nuclear spins within radius r_k of the origin and N is the number of nuclear spins within radius l_0, so in d dimensions $\left(\frac{r_k}{l_0}\right)^d = \frac{k}{N}$. From the definition $A_k \propto |\psi(r_k)|^2$ this gives the coupling constants $A_k = 2\exp\left[-\left(\frac{k}{N}\right)^{\frac{m}{d}}\right]$ in units where $A_0/2 = 1$. For times $t \lesssim N^{3/2}/A$, it is strictly valid to take $N_{\text{tot}} \to \infty$ and change the sum to an integral in Equation (13), [24], which gives the final result for $I_\pm(s)$, above.

The time-dependent spin expectation values can now be recovered from the Laplace transform expressions by evaluating the Bromwich inversion integral:

$$\langle S_z\rangle_t = \frac{1}{2\pi i}\int_{C_B} ds e^{st} S_z(s); \ \langle S_+\rangle_t = \frac{1}{2\pi i}\int_{C_B} ds e^{st} S_+(s). \tag{14}$$

To evaluate these integrals we close the contour in the negative real half-plane, avoiding all branch cuts of the functions $S_z(s)$; $S_+(s)$, and apply the residue theorem. The relevant contour is shown in Figure 1 for $S_+(s)$ or $S_z(s-i\omega_n)$ within Born approximation when $d = m = 2$ (this applies to a two-dimensional quantum dot with parabolic confinement). The spin expectation values that result from this procedure are the sum of contributions with undamped oscillations (from poles on the imaginary axis), exponential decay (from poles with finite negative real part) and nonexponential long-time tails (from branch cut integrals). Since the contributions from poles on the imaginary axis do not decay, the spin expectation

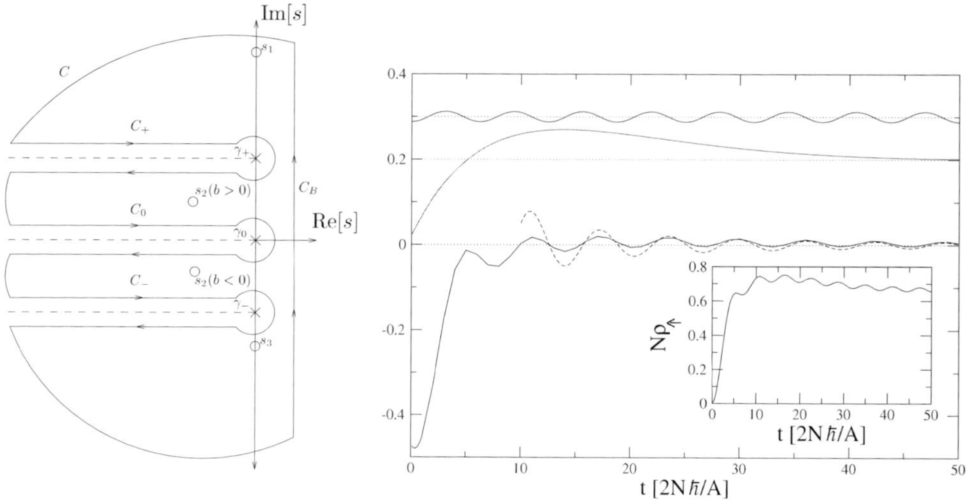

Figure 1. Contour integral (*left*) that determines all contributions (*right*) to the spin expectation values, including undamped oscillations (*right*, *top*), exponential decay (*right*, *middle*) and nonexponential decay (*right*, *bottom*; *solid line*: numerical integration, *dashed line*: analytic asymptotic expression). All three contributions are added to obtain $\rho_\uparrow(t) = 1/2 + \langle S_z \rangle_t$ (*right*, inset). For these calculations we have chosen $d = m = 2$, $I = 1/2$, the initial condition $\rho_\uparrow(0) = 0$, and values of b and p that correspond to $\Delta = 10/11$.

values (in a suitable co-rotating frame) are given generically in terms of a constant piece, and a remainder with nontrivial dynamics:

$$\langle S_X \rangle_t = \langle S_X \rangle_0 + R_X(t), \quad X = +, z. \tag{15}$$

We show the different contributions to $R_z(t)$ in Figure 1 for $d = m = 2$ in the weakly perturbative regime (where $\Delta \lesssim 1$).

In the strongly perturbative regime (when $\Delta \ll 1$), the time dependence of $R_X(t)$ is given exclusively in terms of the functions $I_\pm(t) = \frac{1}{2\pi i}\int_{C_B} ds e^{st} I_\pm(s)$ that appear in Equation (13), above. There is an abrupt crossover in the long-time behavior of these functions at a critical value of d/m. For $d/m < 2$, the major contributions to $I_\pm(t \gg N/A)$ come from the upper limit of the integral in Equation (13), corresponding to nuclear spins near the center of the quantum dot. This leads to a modulation of the spin dynamics at a frequency $A/2N$. When $d/m \geq 2$, the major contributions come from the lower limit, corresponding to nuclear spins far from the dot center. In this case, the long-time dynamics are smoothly varying, with no modulation:

$$R_X(t) \sim I_\pm(t \gg N/A) \propto \begin{cases} \left(\frac{1}{t}\right)^{d/m} e^{\pm i \frac{A}{2N} t}, \nu = \frac{d}{m} - 1 < 1 \\ \frac{\ln^\nu t}{t^2}, \nu = \frac{d}{m} - 1 \geq 1 \end{cases}. \tag{16}$$

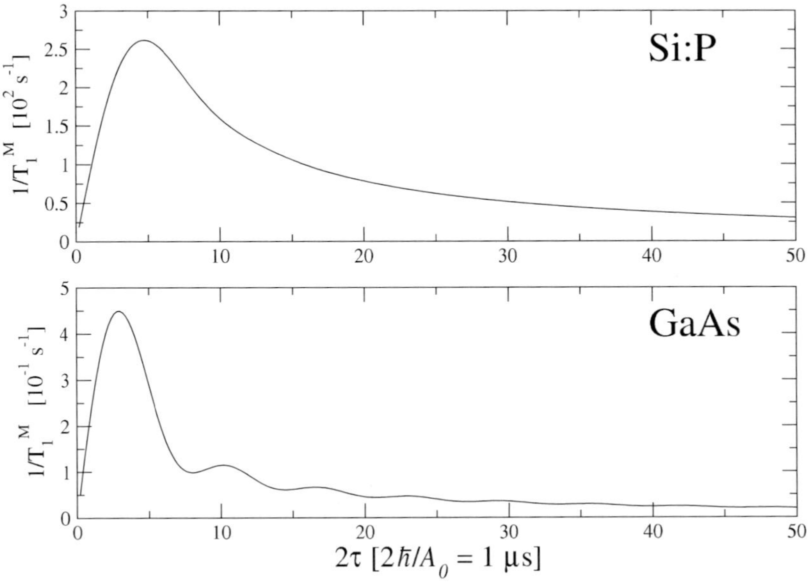

Figure 2. Longitudinal-spin decay rate of the CPMG echo envelope ($1/T_1^{\mathrm{M}} \propto R_z(2\tau)/2\tau$) as a function of the free evolution time 2τ between π-pulses. We plot the results for an electron bound to a phosphorus donor, where $N = 10^2$ (*top*) and a two-dimensional GaAs quantum dot with $N = 10^5$ (*bottom*).

The difference in these two cases should be visible in a spin–echo experiment that uses a Carr–Purcell–Meiboom–Gill (CPMG) spin–echo sequence: $\frac{\pi}{2}-(\tau-\pi_x-\tau-\mathrm{ECHO}-\tau-\pi_{-x}-\tau-\mathrm{ECHO})_{\mathrm{repeat}}$. We consider the strongly perturbative limit ($\Delta \ll 1$), and to resolve the relevant dynamics, the time between π-pulses must satisfy $\tau \ll \sqrt{\delta}\tau_{dd}$, where $\delta = \Delta^2/N$, and τ_{dd} is the nuclear spin dipolar correlation time [24]. Under these conditions the CPMG echo envelope decay rate as a function of τ is determined exclusively by the remainder term according to $1/T_1^{\mathrm{M}} \propto R_z(2\tau)/2\tau$ for the longitudinal component and $1/T_2^M \propto Re[R_+(2\tau)]/2\tau$ for the transverse components. We plot the CPMG decay rate as a function of 2τ in Figure 2 for two systems of interest. For an electron trapped at a donor impurity in bulk silicon, $d = 3$ and the orbital wave function is exponential ($m = 1$). This corresponds to $\nu = 2$ in Equation (16). In a two-dimensional GaAs quantum dot ($d = 2$) with parabolic confinement, the ground-state orbital electron wave function is a Gaussian ($m = 2$), which corresponds to $\nu = 0$ in Equation (16).

3. Conclusions

We have reviewed our theoretical description for the dynamics of a localized electron spin interacting with a nuclear spin environment. We have predicted a sharp crossover in the relevant dynamics at a critical value of the dimensionality and form of the electron envelope wave function, and have described a standard method that could be used to reveal the relevant dynamics. We stress that the electron spin dynamics are in general very rich, described by contributions with exponential decay, nonexponential decay and undamped oscillations. Furthermore, this work may have profound implications for the future of spin-based solid-state quantum information processing and quantum error correction, where previous studies have assumed exponential decay to zero.

Acknowledgements

We thank B. L. Altshuler, O. Chalaev, H.-A. Engel, S. Erlingsson, H. Gassmann, V. Golovach, A. V. Khaetskii, F. Meier, D. S. Saraga, J. Schliemann, N. Shenvi, L. Vandersypen, and E. A. Yuzbashyan for useful discussions. We acknowledge financial support from the Swiss NSF, the NCCR nanoscience, EU RTN Spintronics, DARPA, ARO, ONR and NSERC of Canada.

References

1. Kielpinski D., Meyer V., Rowe M. A., Sackett C. A., Itano W. M., Monroe C. and Wineland D. J., *Science* **291** (2001), 1013.
2. Vandersypen L. M. K., Steffen M., Breyta G., Yannoni C. S., Sherwood M. H. and Chuang I. L., *Nature* **414** (2001), 883.
3. Loss D. and DiVincenzo D. P., *Phys. Rev. A* **57** (1998), 120.
4. Elzerman J. M., Hanson R., van Beveren L. H. W., Witkamp B., Vandersypen L. M. K. and Kouwenhoven L. P., *Nature* **430** (2004), 431.
5. Kikkawa J. M. and Awschalom D. D., *Phys. Rev. Lett.* **80** (1998), 4313.
6. Burkard G., Loss D. and DiVincenzo D. P., *Phys. Rev. B* **59** (1999), 2070.
7. Khaetskii A. V. and Nazarov Y. V., *Phys. Rev. B* **61** (2000), 12639.
8. Golovach V. N., Khaetskii A. and Loss D., *Phys. Rev. Lett.* **93** (2004), 016601.
9. Schliemann J., Khaetskii A. and Loss D., *J. Phys.: Condens. Matter* **15** (2003), R1809.
10. Lyanda-Geller Y. B., Aleiner I. L. and Altshuler B. L., *Phys. Rev. Lett.* **89** (2002), 107602.
11. Imamoḡlu A., Knill E., Tian L. and Zoller P., *Phys. Rev. Lett.* **91** (2003), 017402.
12. Taylor J. M., Marcus C. M. and Lukin M. D., *Phys. Rev. Lett.* **90** (2003), 206803.
13. Taylor J. M., Imamoḡlu A. and Lukin M. D., *Phys. Rev. Lett.* **91** (2003), 246802.
14. Khaetskii A. V., Loss D. and Glazman L., *Phys. Rev. Lett.* **88** (2002), 186802.
15. Erlingsson S. I., Nazarov Y. V. and Fal'ko V. I., *Phys. Rev. B* **64** (2001), 195306.
16. Erlingsson S. I. and Nazarov Y. V., *Phys. Rev. B* **66** (2002), 155327.
17. Semenov Y. G. and Kim K. W., *Phys. Rev. Lett.* **92** (2004), 026601.
18. de Sousa R. and Das Sarma S., *Phys. Rev. B* **67** (2003), 033301.
19. Yuzbashyan E. A., Altshuler B. L., Kuznetsov V. B. and Enolskii V. Z., *cond-mat/0407501* (2004).
20. de Sousa R., Shenvi N. and Whaley K. B., *Phys. Rev. B* **72** (2005), 045330.
21. Shenvi N., de Sousa R. and Whaley K. B., *Phys. Rev. B* **71** (2005), 144419.

22. Khaetskii A., Loss D. and Glazman L., *Phys. Rev. B* **67** (2003), 195329.
23. Schliemann J., Khaetskii A. V. and Loss D., *Phys. Rev. B* **66** (2002), 245303.
24. Coish W. A. and Loss D., *Phys. Rev. B* **70** (2004), 195340.
25. Paget D., Lampel G., Sapoval B. and Safarov V. I., *Phys. Rev. B* **15** (1977), 5780.
26. Fick E. and Sauermann G., *The Quantum Statistics of Dynamic Processes*, Springer, Berlin Heidelberg, New York, 1990, p. 287.

Hyperfine Interactions (2004) 158:245–254
DOI 10.1007/s10751-005-9040-4

Comparative Studies Using EXAFS and PAC of Lattice Damage in Semiconductors

A. P. BYRNE[1,2,*], M. C. RIDGWAY[3], C. J. GLOVER[3] and E. BEZAKOVA[1,†]
[1]*Department of Nuclear Physics, Research School of Physical Sciences and Engineering, The Australian National University, Canberra, Australia; e-mail: aidan.byrne@anu.edu.au*
[2]*Department of Physics, Faculty of Science, ANU, Canberra, Australia*
[3]*Department of Electronic Materials Engineering, RSPhysSE, ANU, Canberra, Australia*

Abstract. We have used the perturbed angular correlation (PAC) method and extended X-ray absorption fine structure spectroscopy (EXAFS), along with microscopic methods to investigate the implantation induced disorder and characterize the ion-induced amorphisation of elemental and compound semiconductors.

Key Words: amorphisation, extended X-ray absorption fine structure spectroscopy (EXAFS), ion-beam induced damage, Perturbed Angular Correlation (PAC), semiconductor materials.

1. Introduction

Ion implantation of semiconductor materials provides a critical processing step in the fabrication of semiconductor devices. Associated with implantation is the introduction of implantation-induced disorder, which can be either desirable or detrimental to device performance. Consequently, the understanding of this disorder is of considerable interest and importance. Many techniques have been applied to the study of semiconductor materials, however no single method can provide a full characterization and a detailed understanding of the physical processes relies on the application of a diverse range of complimentary techniques. A current survey of measurements and calculations is presented by Hobler and Otto [1]. While many methods characterise properties on the macroscopic scale, techniques based on hyperfine methods allow the ability to determine the local atomic environment surrounding probe atoms.

In the present paper we illustrate how two highly local methods, perturbed angular correlations (PAC) and extended X-ray absorption fine structure spectroscopy (EXAFS), can be used to study implantation induced disorder and characterise the ion-induced amorphisation of elemental (Ge) and compound

* Author for correspondence.
† Present Address: Royal Adelaide Hospital, Adelaide, South Australia, Australia.

semiconductors (InP). In both cases the techniques are sensitive to the atomic environment only few atoms away from the probe and provide measures of the short range order of the system.

1.1. PERTURBED ANGULAR CORRELATION MEASUREMENTS

The use of the ^{111}In PAC probe in InP material has provided an ideal system in which to explore the mechanism and dose dependency of ion-induced amorphisation. (A good description of the fundamental principles of the PAC method and its application to semiconductors is provided by the review of Wichert [2].) In our work ^{111}In is introduced as an integral part of the system prior to the modification by ion-beam irradiation and is consequently very sensitive to damage at relatively small doses. A direct production–implantation process has been employed, using a ^{12}C beam from the ANU Heavy-Ion Accelerator Facility. Thin (2.5 μm) rhodium foils were used so that the product nuclei recoil out from the back of the target with relatively wide angular distribution. Up to 60% of all ^{111}In nuclei that leave the target foil can be collected on samples positioned behind the target. TRIM calculations indicate that most of the ^{111}In ions come to rest within 1 to 2.5 microns [3]. While this method has the disadvantage that the radioactive ions are distributed over a relatively large thickness (compared to low energy implanters) it does ensure that the probes remain as very dilute impurities.

After the introduction of ^{111}In, the samples were annealed using rapid thermal annealing (800°C, 10 s) before subsequent processing. Ion beam amorphisation was performed using ^{74}Ge beams from the ANU 1.7 MeV ion implanter with the samples kept at 77 K. As the direct production–implantation process distributes probe nuclei over a significant range, multiple implants at different ion beam energies were used for each implantation dose in order to produce uniform depth profile of damage from just below the surface to 2.5 μm depth.

After implantation, PAC measurements were performed using four BaF_2 detectors arranged in a plane, in close geometry at angles 0°, 90°, 180° and 270°. The samples were positioned perpendicular to the detector plane and at 45° with respect to the detectors.

1.2. EXAFS MEASUREMENTS

The EXAFS method is particularly sensitive to the coordination number, bond length and variations in bond lengths around the probe and a good introduction to the method can be found in the in the work of Sayers and Bunker [4]. The Z dependence of the binding energies means that the process is element specific and in component materials the environments around each constituent can be probed. Although the measurement is intrinsically straightforward, care must be taken in sample preparation and subsequent data analysis. The application to the study of ion beam induced amorphisation is facilitated by the ability to use unimplanted and undamaged samples as reference materials, considerably improving the quality of the data able to be extracted. Ion implantation allows careful control of the physical properties of the material such as density and planarity. For compound semiconductors the implantation process maintains constant stoichiometry.

Transmission and fluorescence EXAFS measurements were performed at a low temperatures ($\approx$10 K) at the Stanford Synchrotron Radiation Laboratory, USA and the Photon Factory, Japan. Sample preparation typically involved the implantation of 2 μm thick crystalline films, their removal from substrates by selective etching and subsequent mounting on thin kapton film. For the transmission measurements the samples were layered so that $\mu x \approx 1$. Details of the EXAFS measurements can be found in the publications [5–11].

2. Results

2.1. AMORPHISATION MEASUREMENTS IN InP AND Ge

Doses between 2×10^{12} and 150×10^{12} ions/cm^2 were implanted into crystalline InP to give a range of samples (11 in total) from undamaged to completely amorphous material. Ratio spectra for some of these samples are shown in Figure 1a. The lowest plot in this series is for the unimplanted sample and shows no time dependance, indicating that the damage caused by the introduction of the radioisotope into the sample has been removed by the annealing and that the probe nuclei are now situated a substitutional sites where they experience no net electric field gradient. At low implantation dose a very gradual loss of alignment is observed consistent with the presence of long range disorder, similar to previous measurements of residual disorder in annealed heavily doped InP samples [12]. For the highest dose indicted here (50×10^{12} ions/cm^2) the spectrum shows a very rapid loss of alignment before the ratio function returns to the 'hard-core' value. Intermediate spectra show both effects, with the data showing a smooth transition from crystalline to amorphous behaviour as a function of implantation dose. Two regimes can be delineated, one where the ^{111}In sits on a weakly disturbed site and the other related to a very damaged (amorphous) environment [3].

A similar methodology has been applied for the PAC study of amorphisation in Ge [5]. Ratio functions for variously damaged Ge crystals are shown in Figure 1b. While similar differentiation into fractions associated with highly damaged amorphous and disordered environments can be identified in this case, two well defined frequencies (49 and 390 MHz), associated with point defects can also distinguished for the implant doses around 10^{12} ions/cm^2 [13]. These can be associated with specific defect configurations trapped at the In impurity and the results are similar to those seen in the PAC study by Haesslein *et al.* [15]

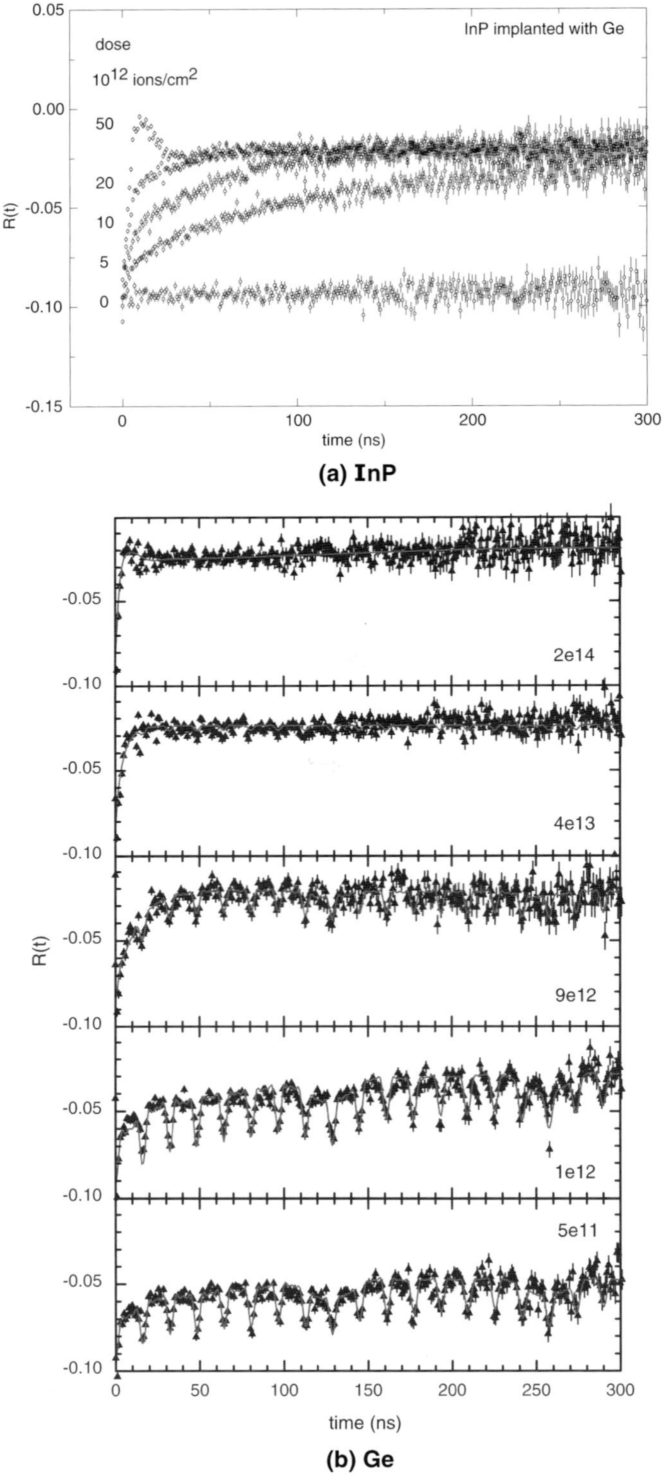

Figure 1. PAC ratio functions using ^{111}In probes in ion-beam damaged semiconductors.

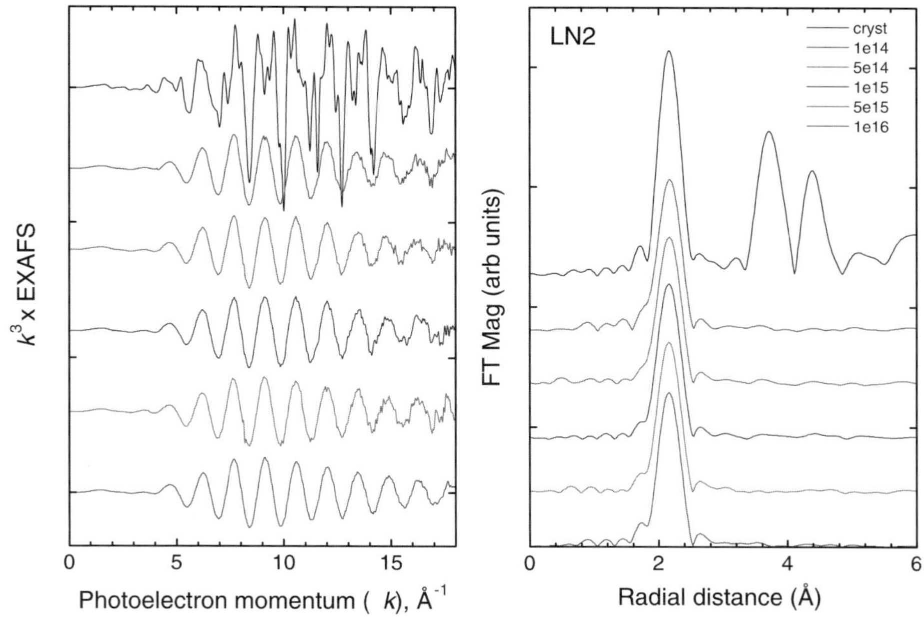

a. EXAFS and Fourier Transforms

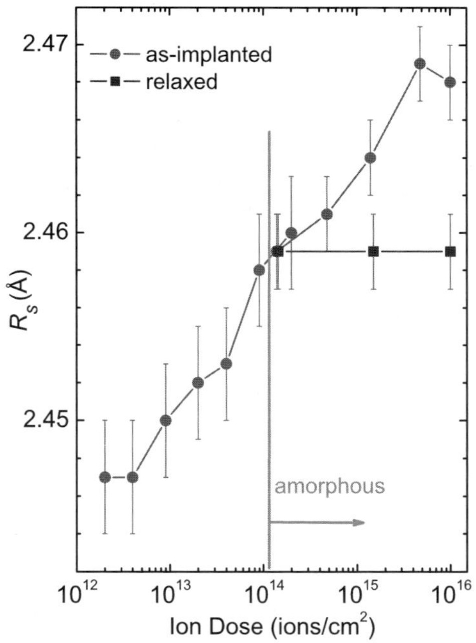

b. Radial distributions

Figure 2. k^3 weighted EXAFS Spectra, corresponding fourier transforms and radial distributions for Ge crystals implanted with Ge.

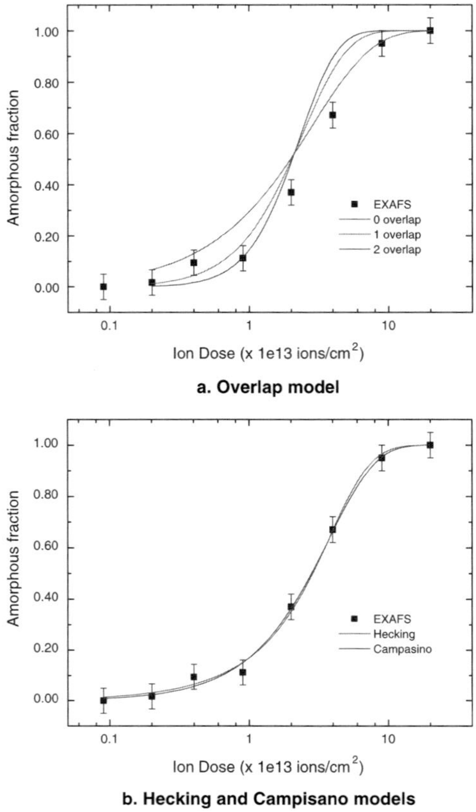

Figure 3. Amorphous fraction as a function of dose deduced from EXAFS measurements in Ge.

of point defects introduced into Ge by electron irradiation and have been observed previously by Feuser *et al.* [14]. Defects in Ge have also been the subject of calculations [16, 17] with the latter work suggesting a split vacancy and a self interstitial for these sites.

2.2. EXAFS RESULTS

EXAFS spectra matching the dose range covered by the PAC studies have been measured for both InP and Ge samples (and for other elemental and compound semiconductors) [5–11]. Figure 2 shows the results of measurements as a function of dose for Ge amorphised at doses higher than that used for the PAC study. The top spectrum shows the data obtained for a crystalline sample while the remainder are for amorphised samples. Three peaks are identifiable in the crystalline material corresponding to the first, second and third nearest neighbour peaks whereas, due to the loss of long range order, only the nearest neighbour peak is observed for the amorphised material. For doses lower than shown in this figure a gradual transition between the amorphous and crystalline material has been observed and, as with the PAC measurements, a two phase distribution as a

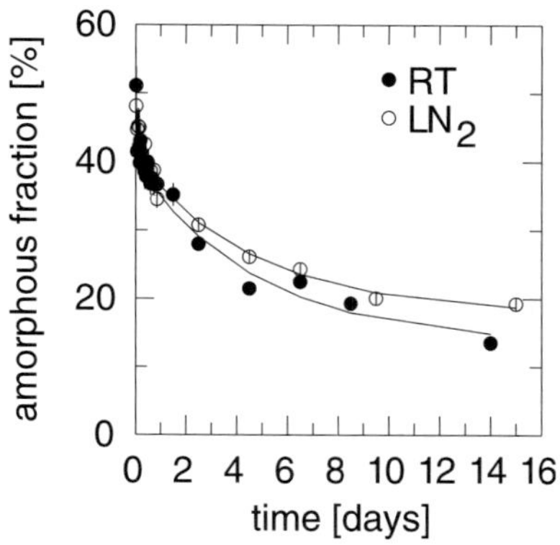

Figure 4. Room temperature relaxation of the amorphous fraction, measured as a function of elapsed time after room temperature (RT) and liquid nitrogen (LN_2) temperature implantation of InP samples with Ge.

function of dose can be determined [6]. For the spectra shown in this figure quantifiable differences in bond lengths and radial distribution functions can be distinguished even though these doses are well above the amorphisation threshold [18].

3. Discussion

3.1. AMORPHISATION

A variety of models have been applied to the study of ion-beam induced amorphisation. Morehead and Crowder [19] present an out-diffusion picture where, following the dissipation of the thermal spike, simple primary defects may diffuse from the boundary of a cylindrical cascade volume before the cascade relaxes leaving an amorphous core. Gibbons' overlap model [20] considers the number of cascade overlaps required to produce a critical defect density for amorphisation. In this model each individual cascade increases the defect density in the material until the free energy is sufficient to allow a spontaneous collapse to the amorphous cascade. Direct amorphisation corresponds to amorphisation induced by a single collision cascade. More realistic models allow for both direct amorphisation and defect production and the nucleation and growth model of Campisano *et al.* [21] assumes that amorphous clusters are nucleated in the irradiated volume. These clusters have large crystalline–amorphous interfaces and act as sinks for ion beam generated defects that, depending of the implantation temperature can cause the volumes to grow or shrink. Hecking *et al.* [22] describe the growth of amorphisation volumes in the case where the incident ion produces both amorphous clusters and the defects that can stimulate further amorphisation. Stimulated amorphisation occurs when the energy spike occurs in regions which already contain amorphous zones. Thus, sub-threshold cascades, which normally would not be able to create amorphous zones, are able to cause growth of existing amorphous regions.

The fraction of amorphised material as a function of ion dose can be used as means of distinguishing between different amorphisation models. For direct amorphisation a simple linear dependance with dose as observed for amorphisation of InP [3]. The data for Ge, deduced from the EXAFS measurements, is shown in Figure 3. Here the data is not well reproduced by assuming direct amorphisation (0 overlap), nor by any higher overlap models. The Campisano model can be adjusted to reproduce the data well (see Figure 3b), however unlike the results for Si for which Campisano model was derived, the data for both Ge (and from the PAC results for InP) imply that the nucleation and growth is columnar rather than 3-dimensional. It is not clear whether this represents a failure of the model or a consequence of ion-implantation into more readily amorphised materials. The Hecking model fit for Ge indicates that the cross-section for stimulated amorphisation is approximately twice that for direct amorphisation. A similar quality fit to the InP data derived from the PAC measurements indicates that in that case the stimulated amorphisation is about one quarter that of the direct process, consistent with the relative ease of amorphisation of the material.

3.2. RELAXATION

Structural relaxation effects resulting from room temperature annealing have also been measured in InP with the measurement of PAC spectra in damaged samples as a function of time.

Decay curves, shown in Figure 4, display two lifetimes components, one of the order of 6 h and the second of the order of five days. The exact origins of these lifetimes are not known however the short lifetime is probably the result of local recombination of point defects and the longer lifetime associated with the migration of vacancies and other defects from further into the lattice. More than 50% of the amorphized clusters experienced relaxation, associated with a dramatic decrease in the mean interaction frequency, suggesting that the local charge distribution has changed significantly. Nevertheless all the indium probes remain in a perturbed lattice environment, i.e., the complete removal of defects was not observed. During ‘topological’ relaxation the atoms change their positions slightly to decrease the bond angle distortion and the defects become embedded in a more ordered network. The continuous presence of defects during the short range ordering could still result in large local EFGs. It is therefore more likely (taking into account the similarity between frequency distributions of the weakly damaged crystalline and relaxed sites) that the relaxation is associated with the annihilation of defects. EXAFS measurements [11] appear to support this mechanism.

4. Conclusions

The highly local methods of PAC and EXAFS can be used effectively to map the transition from the crystalline to amorphous state. Although detailed mechanisms controlling the behaviour are not always evident, the methods display a high sensitivity to structural changes. Clearly, further theoretical calculations such as the MD simulations of Nord *et al.* [23] and the ART results of Mousseau and Barkema [24] are required to understand the complete dynamics of the amorphisation process. PAC measurements have been shown to provide valuable information even in situations where it is not possible identify specific defect configurations. EXAFS measurements can provide high precision measurements of bond length parameters and radial distributions. Both methods are capable of determining structural changes in amorphised materials beyond that available using macroscopic methods. The EXAFS method in particular is capable of showing quantifiable differences in bond length and bond distributions for doses well beyond the amorphisation threshold.

Acknowledgements

We would like to thank the staff of the ANU Heavy-ion Facility for their support in this work and Reiner Vianden for many useful discussions.

References

1. Hobler G. and Otto G., Status and open problems in modeling of as-implanted damange in Si, *Mater. Sci. Semicond. Process.* **6** (2003), 1.
2. Wichert T., *Hyperfine Interact.* **97–98** (1996), 135.
3. Bezakova E. *et al., Appl. Phys. Lett.* **75** (1999), 1923.
4. Sayers D. E. and Bunker B. A., In: Koningsberger D. C. and Prins R. (eds.), *X-ray Absorption: Principles, Applications and Techniques of EXAFS, SEXAFS and XANES*, Wiley, New York, 1988, p. 211.
5. Glover C. J. *et al., Appl. Phys. Lett.* **74** (1999), 1713.
6. Glover C. J. *et al., Nucl. Instrum. Methods* **B161–163** (2000), 1033.
7. Ridgway M. C., Glover C. J., Foran G. J. and Yu K. M., *J. Appl. Phys.* **83** (1998), 4610.
8. Glover C. J. *et al., Phys. Rev., B* **63** (2001), 073294.
9. de M. Azevedo G. *et al., Nucl. Instrum. Methods* **B190** (2002), 851.
10. Glover C. J., Foran G. J. and Ridgway M. C., *Nucl. Instrum. Methods* **B199** (2003), 199.
11. de Azevedo G. M. *et al., Phys. Rev., B* **B68** (2003), 115204.
12. Ridgway M. C., Byrne A. P., Bezakova E., Wehner M. and Vianden R., In: Jagadish C. (ed.), *Proc. 1996 OEMMD Conf.*, (IEEE, Piscataway), 1997.
13. Glover C. J., Byrne A. P. and Ridgway M. C., *Nucl. Instrum. Methods* **B175–177** (2001), 51.
14. Feuser U., Vianden R. and Pasquevich A. F., *Hyperfine Interact.* **60** (1990), 829.
15. Haesslein H., Sielemann R. and Zistl C., *Phys. Rev. Lett.* **80** (1998), 262.
16. da Silva A. J. R. *et al., Phys. Rev., B* **B62** (2000), 9903.

17. Höhler H., Atodiresei N., Schroeder K., Zeller R. and Dederichs P. H., *Phys. Rev., B* **B** (2004) cond-mat/0406616 and *Hyperfine Interact.* This meeting.
18. Glover C. J. *et al.*, *J. Synchrotron Radiat.* **8** 4610 (2001), 773.
19. Morehead F. F. and Crowder R. L., *Radiat. Eff.* **6** (1970), 27.
20. Gibbons J. F., *Proc. IEEE* **60** (1972), 1062.
21. Campisano S. U. *et al.*, *Nucl. Instrum. Methods* **B80/81** (1993), 514.
22. Hecking N., Heidemann K. F., and Te Kaat E., *Nucl. Instrum. Methods* **B15** (1986), 760.
23. Nord J., Nordland K., and Keinonen J., *Nucl. Instrum. Methods* **B193** (2002), 294.
24. Mousseau N. and Barkema G. T., *Phys. Rev., B* **61** (2000), 1898.

Hyperfine Interactions (2004) 158:255–260
DOI 10.1007/s10751-005-9041-3

The Systematics of Muonium Hyperfine Constants

S. F. J. COX[1,2,*] and C. JOHNSON[1]
[1]*ISIS Facility, Rutherford Appleton Laboratory, Chilton, OX11 0QX, UK*
[2]*Condensed Matter and Materials Physics, University College London, London, UK*

Abstract. The hyperfine constants for muonium in elemental and binary inorganic solids suggest formation of three different families of defect centre, with distinct electronic structures. The overall range of values, spanning nearly five orders of magnitude, and their correlation with host properties such as band gap and electron affinity, reveal a deep-to-shallow instability which has profound implications for the electrical properties of hydrogen impurity in electronic materials, both semiconducting and dielectric.

Key Words: electronic materials, hydrogen, muonium.

Abbreviations: μSR – muon spin rotation, relaxation and resonance.

Nomenclature:
Mu – chemical symbol for muonium, light pseudo-isotope of hydrogen.

This paper draws attention to the wealth of hyperfine data which now exists for muonium defect centres in solids, and to the consequent value of muonium as a model for hydrogen in materials where ESR data for hydrogen itself are not available. It highlights important and highly topical new additions to this data base, otherwise accumulated over three decades of μSR spectroscopy, and re–examines which material parameters control the variation of hyperfine constant. The first point to make is that this variation is huge, spanning almost five powers of ten. Thus muonium in quartz or silica has a hyperfine constant close to the free–atom value of 4.45 GHz [1] and its spectrum is visible up to at least 1000 K; in striking contrast, muonium in zinc oxide has a contact term of just 0.5 MHz and disappears around 35 K with an ionization energy of a few tens of meV only [2]. Hydrogen counterparts for both states are known to ESR spectroscopy [3, 4]. The one extreme corresponds to trapped atoms in which the 1s(Mu) or 1s(H) orbitals are confined to interstitial cages in the host lattice, the other to delocalization of the shallow-donor type, in which the orbital is dilated by the the bulk dielectric constant of the medium and, where appropriate, by a low effective mass of the electron.

The question arises whether all intermediate degrees of delocalization of the unpaired electron are permitted, in a continuum between these extremes. It

* Author for correspondence.

appears not. Instead, three categories emerge, which we call quasi–atomic, molecular–radical and extended. There is a limited quantitative variation of hyperfine parameters within each, but the three types are qualitatively or chemically different. They also exhibit different electrical activities, principally as regards their respective ionization behaviours. In other words, there are differences of kind as well as differences of degree.

For the quasi-atomic states, early descriptions in the ESR literature favoured trapped-atom or cavity models in which compression of the 1s(H) orbital by the Pauli exclusion principle competes with its dilation by Van der Waals interactions. The one leads to a slight increase of hyperfine constant, the other to a slight reduction (as for atomic hydrogen in rare gas solids, in quartz or in the alkali fluorides [3, 5–7]) but the overall variation from these effects does not exceed a few percent. Further reduction of spin density on the proton implies a degree of covalency or bonding between the hydrogen atom and the host atoms. The alkali halides provide a good example, where the variation of hyperfine constants for H is closely mimicked by values for Mu – albeit with an almost uniform scaling factor of 98% that can be traced to the greater zero point energy of muonium, as can the similar isotope effect between H and Mu in quartz [3, 7, 8]. The isotope and other dynamical effects are treated in detail by Roduner *et al.* [9].

Moving to the tetrahedrally coordinated semiconductors, ESR data for hydrogen centres are sparse but μSR studies of muonium reveal substantial reductions of hyperfine constant – down to around 30% of the free–atom value [10, 11]. Quantitative computational modelling is largely lacking, apart from some attention to the particular cases of silicon and diamond, but preliminary results from a systematic study by one of us (CJ) for a wider range of materials are given in the insert to Figure 1. These calculations model hydrogen as an interstitial defect in $2 \times 2 \times 2$ supercells, using B3LYP hybrid density functionals within the CRYSTAL2003 code [17–19]. Double-valence basis sets with pseudopotentials to represent core electrons were used in the case of As, Ge, C, Ga, P [20], K, Rb [21] (Hay and Wadt small core), Cl, Br, I [21] (Hay and Wadt large core), while an all–electron 6–21 G basis was used for Si and H [22]. With the curious exception of diamond, a reasonable systematic agreement with the measured values begins to emerge. The broad correlation of hyperfine constant with band-gap, apparent from the extensive experimental data in Figure 1, may also be understood qualitatively: H and Mu retain the greatest atomic character when the admixed states have the greatest disparity in energy, i.e., when there is the widest separation between bonding and antibonding orbitals [14]. The closer energy levels in narrow-gap materials promote greater admixture and loss of atomic character.

Exceptions to this simple correlation are the cuprous compounds, where the unusual proximity of 3d(Cu) orbitals may switch the transfer of spin density from anion to cation. The band-gap correlation also fails when spin transfer from the

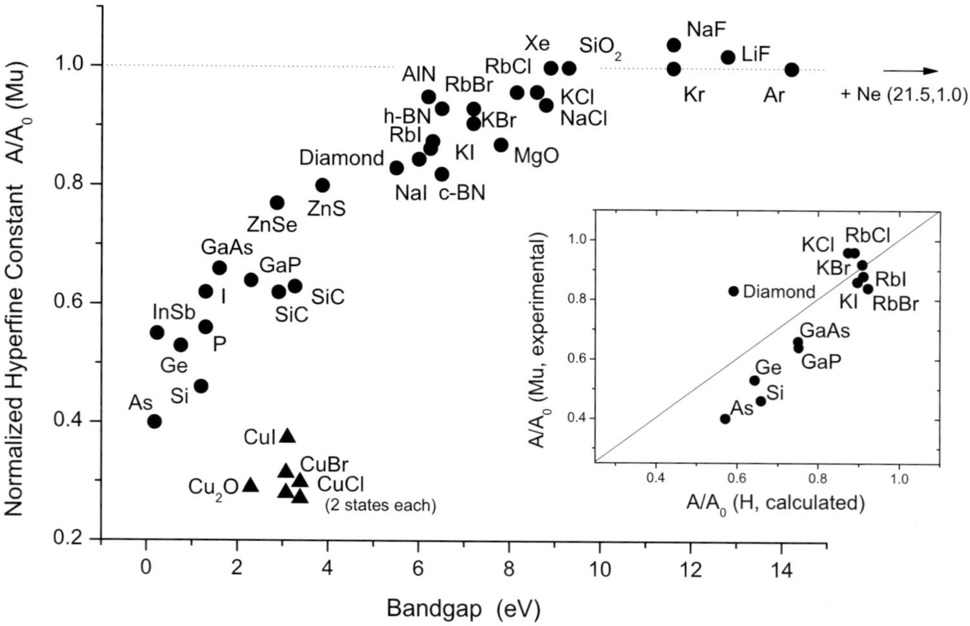

Figure 1. Correlation of muonium hyperfine constant (or spin density on the muon, normalized to the free-atom value) with band gap of the host material (after Cox [12, 13]). The data points for InSb [15] and Cu_2O [16] are new. In the insert, newly calculated values for some of the materials are compared with the experimental data.

muon is more complete. For the centres with molecular–radical character, muonium makes a directional bond with one or more host atoms: their formation may be expressed symbolically by solid-state reactions such as $^{\bullet}\text{Mu} + \text{X}-\text{Y} \rightarrow \text{Mu}-\text{X} + {}^{\bullet}\text{Y}$, in which major spin density (represented by the dot) is transferred to a neighbouring atom or atoms, accounting for the anisotropy of these centres.⋆ The archetypal example is so–called anomalous muonium or Mu* in Si [1]. The extreme uniaxial anisotropy of the hyperfine tensor could only be explained if the muon is located at the centre of a stretched Si—Si bond and the electron occupies the Si—Si antibonding orbital, detached from the conduction band by the local distortion [14]. There is no atomic character whatsoever in this case: the local symmetry forbids any admixture of 1s(Mu) and the contact interaction is in fact negative, representing spin polarization of the Si valence orbitals. Detection of its hydrogen counterpart, originally dubbed the AA9 centre, entirely validated the concept of muonium as a model for hydrogen behaviour and became central to the understanding of the electrical activity of H in Si, where it provides a deep-donor level a few tenths of an eV below the conduction-band miniumum [23]. Bond-centred muonium with deep-donor character is likewise identified in Ge, diamond, GaAs and GaP [10]. A

⋆ We restrict the discussion to muonium defect centres in simple inorganic solids, to the exclusion of the vast number of muoniated organic radicals known to muonium chemistry!

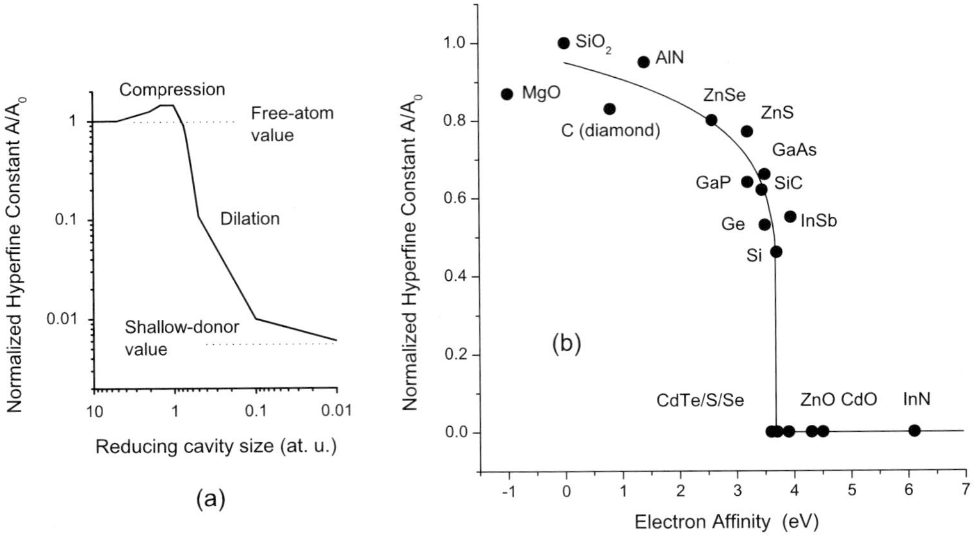

Figure 2. The deep-to-shallow transition. The cavity-model simulation in (a) is due to Claxton and Hall [25]. Correlation of muonium hyperfine constants with electron affinity of the host medium in (b) show a particularly sudden transition (after Cox [13]). A gradual dilation of the atom is implied, as the threshold electron affinity for complete delocalization is approached. The hyperfine constants are expressed as a fraction of the free–atom value; they are not zero to the right of the transition, but typically between 10^{-4} and 10^{-5}.

remarkable electrical activity or fast spin-lattice relaxation is responsible for the difficulty in identifying the elusive muonium centres in the Group-IV chalcogens: these are thought to resemble the hydroxyl radical, OH, the most recent example being identified in tellurium [24].

Although all possible degrees of dilation of the muonium atom appear to be described by cavity models in a dielectric continuum, as depicted in Figure 2(a), these fail to account for the systematics between particular materials [25]. Cavity models also fail to describe any substantial anisotropy, or indeed the seemingly exclusive competition between localization and extension found experimentally. Instead, the electron affinity of the host seems to be the controlling parameter. Figure 2(b) expresses an apparently critical transition between states in which the electron is tightly bound to the muon or proton and those where it is essentially lost to the conduction band. This is the deep-to-shallow instability, described in certain theoretical treatments in terms of an equilibrium between protons and hydride ions [27, 28] but represented more simply by Equation (1). Thus for muonium in CdS, CdSe and CdTe, the unpaired electron is only weakly bound to the interstitial muon at cryogenic temperatures in a wave-packet of conduction band states, as for the shallow-donor state in ZnO [26].

$$H^0 \leftrightarrow H^+ + e_c; \quad Mu^0 \leftrightarrow Mu^+ + e_c. \tag{1}$$

Evidently the molecular–radical states do, in a sense, represent intermediate behaviour between the extremes of Equation (1). The three types of centre have different electrical activity or ionization behaviour, however: the relationship between them may be summarised as follows. Major spin density (of the unpaired electron) is centred on the muon in the quasi-atomic states; their electrical activity is probably that of deep acceptors [13], although conversion to the negative ion has only been explicitly demonstrated in Si and GaAs. For the molecular–radical states, the unpaired electron is localised not on the muon but on one or two nearby atoms; these defect centres have deep-donor character. For the extended states the unpaired electron is spread more widely over many atoms via conduction-band states in the manner of classic shallow donors: their ionization or binding energy is correctly described by effective-mass theory. Again with the curious exception of diamond (for which the electron affinity may be in error) it is interesting that the materials Si, Se, GaAs and GaP exhibiting metastability, i.e., coexistence of quasi-atomic and deep-donor bond-centred muonium, fall near the deep-to-shallow transition of Figure 2(b), i.e., have electron affinities near the threshold value.

In the particular case of oxides, stabilization of the interstitial proton or muon is achieved by formation of the hydroxide ion or its muonium counterpart, as in reaction 2. Evidently no such reaction occurs in quartz, where oxygen shows no affinity for neutral atomic hydrogen and indeed repulsion of the 1s(H) electron by the closed-shell oxygen is responsible for a hyperfine constant higher than the free-atom value [3]. The electron must be picked off by the cation, not the anion. In ZnO this releases the electron into the conduction band but some transition metal cations may trap the electron more locally in deep states.

$$H^0 + O^{2-} \leftrightarrow (OH)^- + e_c; \quad Mu^0 + O^{2-} \leftrightarrow (OMu)^- + e_c. \tag{2}$$

An accompanying paper illustrates this with new examples of each type of centre in three semiconducting oxides: quasi-atomic in Cu_2O, deep donor in Ag_2O and a candidate shallow donor in CdO [16]. The new data points for Cu_2O and CdO are entered in Figures 1 and 2.

We note finally that knowledge of the local electronic structure gained from low-temperature muonium spectroscopy allows atomistic pictures of the electrical activity to be built up that, for hydrogen itself, have largely been lacking. The discoveries of shallow-donor muonium states are particularly significant since they imply that hydrogen impurity could act as an inadvertent dopant in the relevant materials, inducing electronic conductivity. (This is also a remarkable contrast with the possibility of protonic conductivity in more complex oxides). As a method which does not rely on favourable hydrogen solubility, muonium spectroscopy looks set to play an important role in modelling the electrical activity of hydrogen impurity in the new electronic materials, especially in the widegap semiconductors and oxide dielectrics now under development for device applications.

References

1. Brewer J. H. *et al.*, *Phys. Rev. Lett.* **31** (1973), 143.
2. Cox S. F. J. *et al.*, *Phys. Rev. Lett.* **86** (2001), 2601.
3. Weil J. A., *Hyperfine Interact.* **8** (1981), 317.
4. Hofmann D. M. *et al.*, *Phys. Rev. Lett.* **88** (2002), 045504.
5. Foner S. N. *et al.*, *J. Chem. Phys.* **32** (1960), 963.
6. Adrian F. J., *J. Chem. Phys.* **32** (1960), 972.
7. Spaeth J.-M., *Hyperfine Interact.* **32** (1986), 641.
8. Baumeler H. *et al.*, *Hyperfine Interact.* **32** (1986), 659.
9. Roduner E., Percival P., Han P. and Bartels D. M., *J. Chem. Phys.* **102** (1995), 5989.
10. Patterson B. D., *Rev. Mod. Phys.* **60** (1987), 69.
11. Kiefl. R. F., *Phys. Rev. B* **34** (1986), 1474.
12. Cox S. F. J., *J. Phys. C. Solid State Phys.* **20** (1987), 3187.
13. Cox S. F. J., *J. Phys. Condens. Matter.* **15** (2003) R1727.
14. Cox S. F. J. and Symons M. C. R., *Chem. Phys. Lett.* **126** (1986), 516.
15. Storchak V. G., Eshchenko D. G. and Brewer J. H. *et al.*, unpublished.
16. Cox S. F. J., Lord J. S., Cottrell S. P., Alberto H. V., Gil J. M., Piroto Duarte J., Vilão, R. C., Keren A. and Prabhakaran D., Hyperfine Parameters for Muonium in Copper (I), Silver (I) and Cadmium Oxides, in Proceedings for the HFI/NQI 2004, Bonn, Germany, R. Vianden, ed., unpublished.
17. Becke A. D., *J. Chem. Phys.* **98** (1993), 5648.
18. Lee C., Yang W. and Parr R. G., *Phys. Rev. B* **37** (1988), 785.
19. Saunders V. R. *et al.*, CRYSTAL2003 User's Manual, Univerity of Torino, 2003, plus refs therein; see also http://www.crystal.unito.it.
20. Durand P. and Barthelat J. C., *Theor. Chim. Acta* **38** (1975), 283; Barthelat J. C. and Durand P., *Gazz. Chim. Ital.* **108** (1978), 225; Barthelat J. C., Durand P. and Serafini A., *Mol. Phys.* **33** (1977), 159.
21. Hay P. J. and Wadt W. R., *J. Chem. Phys.* **82** (1985), 270 and 299.
22. Binkley J. S., Pople J. A. and Hehre W. J., *J. Am. Chem. Soc.* **102** (1980), 939; Gordon M. S., Binkley J. S., Pople J. A., Pietro W. J. and Hehre W. J., *J. Am. Chem. Soc.* **104** (1983), 2797.
23. Bonde Nielsen K. *et al.*, *Phys. Rev. B* **60** (1999), 1716.
24. Cox S. F. J., Lord S. J., Suleimanov N., Zimmermann U. and Reid I. D., Static and Intermittent Hyperfine Coupling for the Muoniated Radical in Tellurium, in Proceedings for the HFI/NQI 2004, Bonn, Germany, R. Vianden, ed., unpublished.
25. Claxton T. A. and Hall G. G., *Theor. Chim. Acta* **74** (1988), 5.
26. Gil J. M. *et al.*, *Phys. Rev. B* **64** (2001), 075205.
27. Kiliç Ç. and Zunger A., *Appl. Phys. Lett.* **81** (2002), 73.
28. Van de Walle C. G. and Neugebauer J., *Nature* **423** (2003), 626.

Hyperfine Interactions (2004) 158:261–266
DOI 10.1007/s10751-005-9042-2

^{111}Cd PAC Study of Gd–Ni Intermetallic Compounds

P. DE LA PRESA and M. FORKER*
Helmholtz Institut für Strahlen- und Kernphysik der Universität Bonn, Nussallee 14–16, D-53115 Bonn, Germany; e-mail: forker@iskp.uni-bonn.de

Abstract. This paper presents a perturbed angular correlation study of the magnetic and electric hyperfine interactions of ^{111}Cd on Gd sites of the Gd–Ni intermetallic compounds GdNi, $GdNi_2$, $GdNi_3$, Gd_2Ni_7, $GdNi_5$ and Gd_2Ni_{17}.

1. Introduction

The strong axially symmetric electric field gradient experienced by ^{111}Cd nuclei on the cubic Gd site of the C15 Laves phase $GdNi_2$ is now understood to be caused by vacancies trapped by the probe atoms [1]. In the course of the perturbed angular correlation (PAC) study leading to this conclusion, the electric quadrupole interaction (QI) of ^{111}Cd in the paramagnetic phases of the other intermetallic compounds of the Gd–Ni phase diagram had to be determined. The fact that most of these compounds present spontaneous magnetic order at low temperatures suggested an extension of the PAC study to the ferromagnetic phases. In the present contribution, we report measurements of the magnetic hyperfine field at nuclei ^{111}Cd in ferromagnetic GdNi, $GdNi_2$, $GdNi_3$, Gd_2Ni_7, $GdNi_5$ and Gd_2Ni_{17}. The hyperfine field at the Gd nuclei of these compounds has been investigated by the NMR technique [2].

GdNi crystallizes in the orthorhombic CrB-structure with one Gd and one Ni site, $GdNi_2$ in the $MgCu_2$-structure with one Gd (8a) and one Ni (16d) site. The rhombohedral $NbBe_3$-stucture of $GdNi_3$ has two Gd (3a,6c) and three Ni (3b,6c,18h) sites. $GdNi_5$ crystallizes in the hexagonal $CaCu_5$-structure with one Gd (1a) and two Ni (2c,3g) sites. Gd_2Ni_7 exists in a hexagonal and a rhombohedral form with two Gd sites and three Ni sites. The hexagonal Th_2Ni_{17}-structure of Gd_2Ni_{17} has two Gd (2b,2d) and four Ni (6g, 12i, 12k, 4f) sites.

* Author for correspondence.

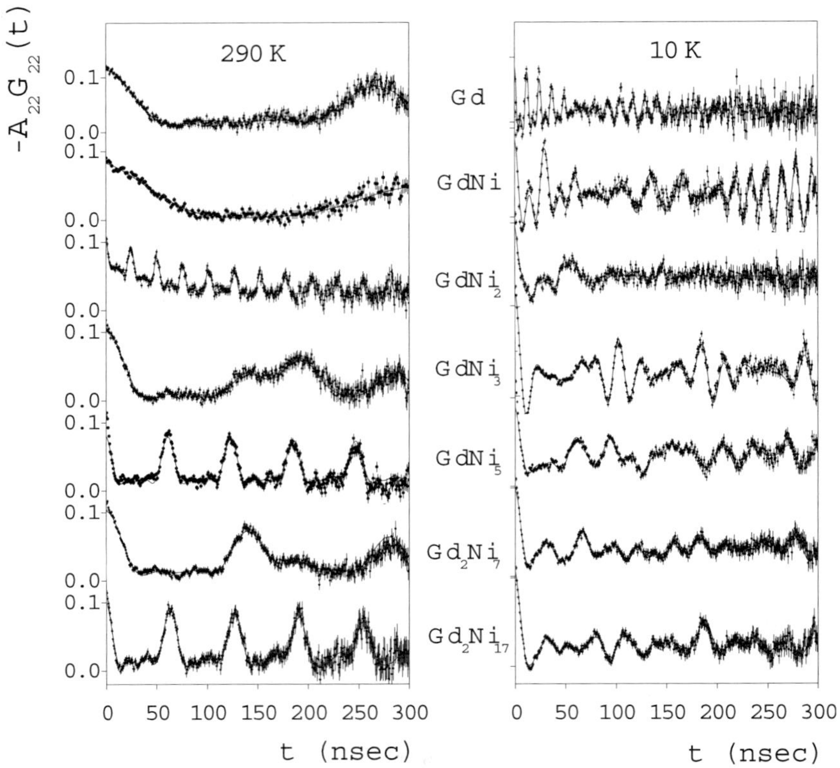

Figure 1. PAC spectra of ^{111}Cd in Gd and in Gd-Ni intermetallic compounds at 290 K and 10 K.

2. Measurements

The PAC measurements were carried out with the 171–245 keV cascade of ^{111}Cd, which is populated by the electron capture decay of the 2.8-d isotope ^{111}In. The compounds, prepared by arc melting of the metallic components in an argon atmosphere, were doped with the PAC probe by diffusion (800°C, 12 h) of carrier-free radioactive ^{111}In into the host lattice. PAC measurements were carried out between 10 and 290 K.

Figure 1 shows typical spectra of ^{111}Cd in Gd metal [3] and in the Gd–Ni intermetallic compounds at 290 and 10 K. In the paramagnetic phase of $GdNi_5$ and Gd_2Ni_{17}, one finds only one strong QI which in both cases has axial symmetry ($\eta = 0$) and very similar quadrupole frequencies (see Table I). The modulation of the angular correlation in paramagnetic GdNi is too slow for a determination of the asymmetry parameter of this weak QI. The frequency given in Table I has been derived with the assumption $\eta = 0$. In $GdNi_2$ one observes at 290 K a superposition of two fractions with relative intensities of 35% and 65%. The rapid periodic oscillation reflects the axially symmetric QI caused by a vacancy trapped at ^{111}Cd, the slowly decaying background corresponds to the

Table 1. Summary of the hyperfine interaction parameters of ^{111}Cd in Gd-Ni intermetallic compounds: The quadrupole frequencies ν_q(290K) and the magnetic interaction frequencies ν_M(10K) in the paramagnetic and the ferromagnetic phases, respectively. The Curie temperatures T_C have been derived from the temperature dependence of the magnetic interaction frequency. The last two columns compare the EFG values of Cd (present study) and Gd nuclei (ref. [2]) in Gd-Ni compounds

Compound	T_C (K)	Fraction	ν_q(290K) (MHz)	ν_M(10K) (MHz)	V_{zz}(CD) (10^{17}V/cm^2)	V_{zz}(Gd) (10^{17}V/cm^2)
Gd	293	1.0	24.9	80.8	1.28	3.4
GdNi	64	1.0	12 (1)	33.1 (1)	0.6	2.2
$GdNi_2$	78	0.35	261 (8a+V)	16-18	13.4	-
		0.65	≤10 (8a)	18.2	0.5	1.9
$GdNi_3$	114	0.20	45.4	25.8	2.34	1.03
		0.80	33.5	21.75	1.73	11.3
Gd_2Ni_7	113	0.25	35.9	22.4 (2)	1.85	
		0.75	47.5	18.2 (1)	2.45	
$GdNi_5$	31	1.0	107.7	1.91 (2)	6.56	8.46
Gd_2Ni_{17}	175	0.4	104.0 (5)	5.45 (5)	5.36	6.25
		0.6	104.0 (5)	2.7 (3), δ=0.25		5.1

majority fraction of ^{111}Cd on site 8a without a nearest neighbour vacancy (for details see [1]). The spectra of paramagnetic $GdNi_3$ and Gd_2Ni_{17} contain two fractions, characterized by axially symmetric QIs. In both compounds the corresponding frequencies are rather similar (ν_q ~ 33–36 and ~ 45–48 MHz, respectively; see Table I), the relative intensities, however, differ strongly. Moreover, the intensities were found to differ considerably between different samples of the same compound.

The presence of a QI in the paramagnetic phase leads to a combined magnetic and electric hyperfine interaction at ^{111}Cd in the ferromagnetic phase of all Gd–Ni intermetallics investigated here (with the exception of the pure magnetic interaction of the majority fraction of $GdNi_2$). Because of the axial symmetry of the QI, the combined interaction is completely described by the magnetic and electric interaction frequencies $\nu_M = g\mu_N B_{hf}/h$ and $\nu_q = eQV_{zz}/h$ and the angle β between the directions of the hyperfine field B_{hf} and the EFG component V_{zz}. The frequency ν_M at 10 K derived by fitting the perturbation function of a combined interaction to the spectra in Figure 1 are listed in Table I. Our result for $GdNi_5$ is in fair agreement with the values previously reported by Pillay *et al.* [4]. In the case of several fractions, the relative intensities were fixed to the room temperature values. In GdNi, the QI is weak compared to the magnetic perturbation and the spectrum at 10 K is therefore insensitive to an eventual asymmetry of the QI. The spectrum of paramagnetic Gd_2Ni_{17} could be well described by a single fraction. For the ferromagnetic phase, however, two

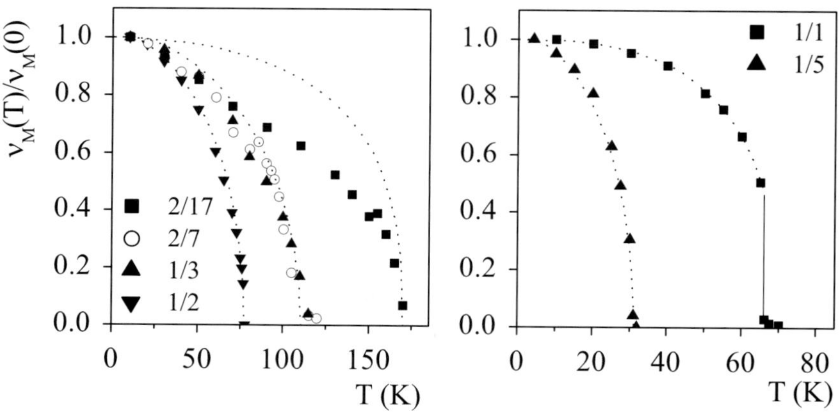

Figure 2. Temperature dependence of the magnetic hyperfine frequency of ^{111}Cd in different Gd-Ni compounds labelled with their Gd/Ni ratio. The dotted lines represent a molecular-field calculation of the temperature dependence of the magnetization for spin $S = 7/2$.

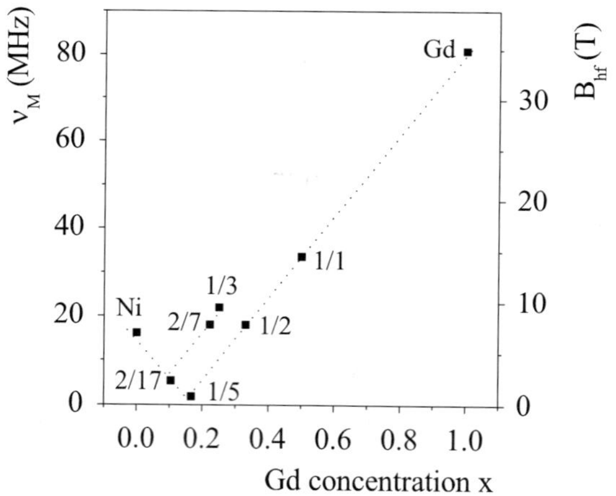

Figure 3. The site-averaged saturation value of the magnetic hyperfine frequency of ^{111}Cd in Gd_xNi_{1-x} compounds, labelled with their Gd/Ni ratio, as a function of the Gd concentration x.

fractions were required, one with a sharply defined frequency ($\nu_M = 5.45$ MHz, rel. intensity ~ 0.4), the other one subject to a distribution (relative width $\delta \sim 0.25$) of magnetic frequencies centred at 2.7 MHz. The origin of this distribution is unclear. Furthermore, Gd_2Ni_{17} presents an unusual temperature dependence of the magnetic frequencies (see below) with a Curie temperature of ~ 175 K, 10–15 K below the values in the literature. These aspects require further studies of Gd_2Ni_{17}. In all compounds, the magnetic hyperfine field was found to be parallel to the EFG direction ($\beta \leq 10°$).

3. Results

The main observations of this study can be summarized as follows:

(i) In a ^{111}Cd PAC study of paramagnetic RNi_5, Devare *et al.* [5] arrived at the conclusion that ^{111}Cd preferentially occupies the rare earth site. From the number of fractions and the symmetry of the QI observed in the present study in the paramagnetic phases, we can generalize this conclusion to all Gd–Ni intermetallic compounds. For ^{111}Cd on the Ni sites one should find up to four fractions and axially asymmetric QIs in a number of cases.

(ii) In all compounds except Gd_2Ni_{17}, the temperature dependence of the magnetic interaction frequency is close to the prediction of a molecular field calculation for spin $S = 7/2$ (see Figure 2). The slight difference between experiment and theory can be attributed to the reduction of the local magnetization caused by the substitution of a magnetic Gd atom by nonmagnetic ^{111}Cd. In GdNi the hyperfine field vanishes discontinuously at $T_C = 66$ K, indicating a first-order magnetic phase transition. The magnetization of GdNi, measured in an external magnetic field of ≤ 8 T, however, shows second-order behaviour [6]. The Curie temperatures listed in Table I agree well with the values in the literature, except for Gd_2Ni_{17}.

(iii) The ^{111}Cd hyperfine field B_{hf} in Gd–Ni compounds depends sensitively on the Gd concentration x (see Figure 3). In the sequence Gd, GdNi, $GdNi_2$, $GdNi_5$, B_{hf} decreases linearly with decreasing Gd contents with a considerable offset at concentration $x = 0$. In these compounds, the 3d-sublattice carries no magnetic moment. The magnetic hyperfine field at nonmagnetic Cd can therefore be attributed to the spin polarization of the s-conduction electrons (CEP) which according to the RKKY theory of indirect 4f–4f exchange is directly proportional to the number of polarizing 4f spins: $\rho_+ - \rho_- \propto J_{sf} \sum_i F(2k_F R_i) \langle S_i^z \rangle$.

Here J_{sf} is the s–f exchange parameter and $F(2k_F R_i)$ the oscillating RKKY function. Considering that J_{sf} and Σ_I $F(2k_F R_i)$ may show some variation between different compounds, the strict linearity of B_{hf} vs. x is probably fortuitous. The overall trend of B_{hf} vs. x, however, is clearly consistent with CEP as main source of the ^{111}Cd-hyperfine field. In contrast to B_{hf} of ^{111}Cd, the hyperfine field at ^{155,157}Gd nuclei in the same compounds shows no systematic concentration dependence [2] because here a major contribution to B_{hf} comes from the core polarization of the Gd shell by its own 4f spin.

In $GdNi_3$, Gd_2Ni_7 and Gd_2Ni_{17}, Ni carries a magnetic moment (0.16, 0.21 and 0.3 μ_B, respectively) and one therefore expects both a 4f- and a 3d-contribution to the magnetic hyperfine field. In this sequence, B_{hf} is again a linear function of the Gd concentration, but – relative to the Gd–$GdNi_5$ sequence – the line is shifted towards higher fields. If this shift can be attributed to the 3d-contribution, is an open question because the present data are not sufficient to disentangle the 3d- and the 4f-part of B_{hf}.

Acknowledgement

This work has been supported by Deutsche Akademischer Austauschdienst (DAAD) and Deutsche Forschungsgemeinschaft (DFG).

References

1. Forker M., de la Presa P., Müller S., Lindbaum A. and Gratz E., *Phys. Rev., B* **70** (2004), 14302.
2. de Jesus V. L. B., Oliveira I. S., Riedi P. C. and Guimarães A. P., *J. Magn. Magn. Mater.* **212** (2000), 125.
3. de la Presa P., Forker M. and Cavalcante L. T., *J. Magn. Magn. Mater.* **272–276**(Supplement) (2004), E401.
4. Pillay R. G., Devare S. H., Pleiter F. and Devare H. G., *Phys. Status Solidi, B* **121** (1984), K141.
5. Devare S. H., Pillay R. G., Malik S. K., Dhar S. K. and Devare H. G., *Phys. Lett.* **79A** (1980), 237.
6. Blanco J. A., Gómez Sal J. C., Rodríguez Fernandez J., Gignoux D., Schmitt D. and Rodríguez-Carvajal J., *J. Phys. Condens. Matter* **4** (1992), 8233.

Hyperfine Interactions (2004) 158:267–271
DOI 10.1007/s10751-005-9043-1

Perturbed Angular Correlation Study of Naturally Occurring Zircon with Very Small Impurity Concentrations

H. JAEGER*, K. S. PLETZKE and S. P. MCBRIDE
Department of Physics, Miami University, Oxford, OH 45056, USA; e-mail: jaegerh@muohio.edu

Abstract. We used ^{181}Ta-PAC spectroscopy to characterize the electric field gradient at Zr sites of zircon samples from the *Harts Range* in Australia. Zircons from this region contain very low levels of radioactive impurities. The PAC spectra between room temperature and 1100°C are axially symmetric and show little damping. The quadrupole frequency decreases linearly with increasing temperature, and the slope of the ν_Q vs. T graph increases above 900°C. This is attributed to a rearrangement of the Si–O coordination, which affects the Zr–O coordination. The spectra of all zircons show a distinct decrease of the anisotropy between 800 and 600°C. This behavior is also seen with samples that do not exhibit the change in slope in the ν_Q vs. T data. We believe that the decrease in the anisotropy is due to an aftereffect following the β-decay of ^{181}Hf.

1. Introduction

Zircon ($ZrSiO_4$) is the primary mineral resource for production of zirconium metal and advanced zirconia-ceramics. Zircon often contains radioactive U and Th impurities and is important for geochronology. Recently zircon attracted attention as model substance for the immobilization of actinide-bearing radioactive waste [1]. Most zircons are found with varying degrees of α-decay radiation damage and are referred to as *metamict*. Metamict zircon differs in a number of properties from crystalline zircon [2]. Metamict zircons recrystallize above 800°C, a process that has been studied in detail over the years [3]. Synthetic zircon exhibits a displacive phase transition between 800 and 1100°C, and it has been suggested that this transition is related to the recrystallization of metamict zircon [4]. During the transition the Si–O bond lengths increase which in turn affects the Zr–O coordination. We have used time-differential perturbed angular correlation spectroscopy (PAC) to characterize zircons with varying degree of metamictization [5], however none of this work shows any evidence for a change in the Zr–O coordination. Recently we began working with zircons from the *Mud Tank* carbonatite in central Australia. Zircons from this area

* Author for correspondence.

contain very few impurities; in particular the concentration of radioactive U and Th is very small, so that *Mud Tank* zircon is essentially free of radiation damage. The purpose of this work is to characterize the electric field gradient (EFG) at a Zr lattice site of *Mud Tank* zircon and compare it to metamict zircon that has been recrystallized.

2. Experimental procedure

For PAC experiments a zircon crystal of size $10 \times 8 \times 5$ mm^3 was ground to a fine powder in an agate mortar. One hundred to 200 mg powder was sealed in an evacuated fused silica capsule and irradiated for 3–5 h at *Ohio State University's Nuclear Reactor Laboratory*. Typically this produced 20 μCi of ^{181}Hf activity and some ^{97}Zr, but this isotope is sufficiently short-lived ($T_{1/2} = 0.7$ days) that it did not interfere with the measurements. This activity is adequate to perform experiments for three months. Typical spectra accumulation times were 25–50 h. During experiments the sample was heated in a furnace in the center of the detector arrangement; the temperature was controlled to ±1°C using a type K thermocouple. Measurements were performed between room temperature and 1100°C using a standard slow-fast PAC spectrometer with four BaF_2 detectors. The perturbation function was analyzed using the expression for static quadrupole interactions in a polycrystalline sample

$$G_2(t) = s_{20}(\eta) + \sum_{i=1}^{3} s_{2i}(\eta)\cos\big(g_i(\eta)\nu_Q t\big)\exp\big(-g_i(\eta)\delta\nu_Q t\big) \tag{1}$$

with η (asymmetry), ν_Q (quadrupole frequency), and δ (damping) characterizing the EFG. The coefficients s_{2i} and g_i depend weakly on η but are independent of ν_Q.

3. Results and discussion

Most naturally occurring zircons have Hf-concentrations of 1–3 mol% residing on Zr lattice sites. Thus ^{181}Ta, a decay product of ^{181}Hf, is an excellent nuclear probe to study the hyperfine interaction on a Zr site. Zircon has a tetragonal unit cell; Zr lattice sites are eightfold coordinated by O forming triangular dodecahedra, which in turn are separated by Si–O tetrahedra [6]. The Zr lattice site has axial symmetry and the EFG of crystalline zircon is expected to have zero asymmetry ($\eta = 0$). Our PAC spectra of *Mud Tank* zircon show three harmonic frequencies characteristic of an electric quadrupole interaction with nearly axially symmetric probe environment.

Figure 1a shows the temperature dependence of the EFG parameters of the *Mud Tank* zircon specimen. The asymmetry parameter η is constant over the temperature range, and the small deviation from zero is attributed to impurities

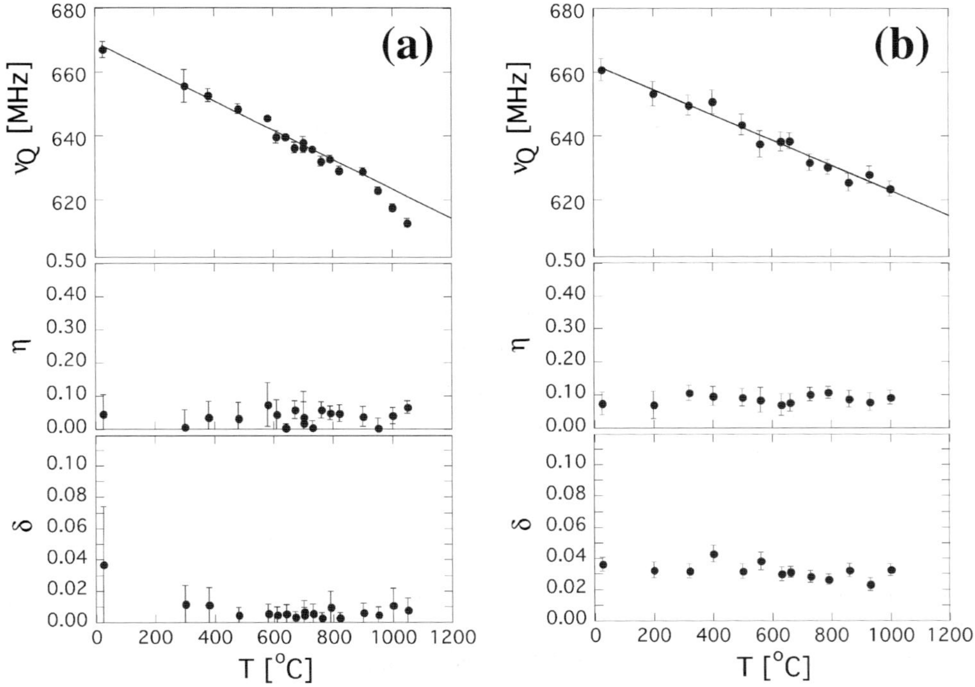

Figure 1. EFG parameters for a *Mud Tank zircon* (a) in comparison with a commercial zircon specimen from *Aldrich* (b). The lines are guides to the eyes.

and lattice imperfections. The damping δ is less than 1% for most temperatures. The first data point at $T = 25°C$ was taken directly after neutron irradiation, so that the higher damping most likely is due to a small amount of neutron-induced radiation damage. This radiation damage annealed after taking the next data point at 300°C. The quadrupole frequency ν_Q decreases linearly with increasing temperature up to 900°C (slope = 46 ± 3 kHz/°C); at higher temperatures ν_Q deviates from this slope and decreases at a higher rate. This temperature dependence, already observed in other *Mud Tank* zircons [7], differs from that of other zircons. For comparison the EFG of a zircon obtained from *Aldrich Corp.* is shown in Figure 1b. ν_Q decreases with a similar slope, but η and δ are significantly larger than the corresponding values for *Mud Tank* zircon. The reason for this is the extraordinary purity of *Mud Tank* zircons [8]. Moreover, ν_Q vs. T of the *Aldrich* zircon does not show the deviation from linearity. We suggest that the change in slope indicates a rearrangement of the Zr–O coordination caused by an increase of the Si–O bond length. The O atoms in a Zr–O dodecahedron can be divided in two groups. Four O are at a distance of 2.13 Å (group I) and four at 2.27 Å (group II) [4]; the group-I O-atoms share a corner with an adjacent Si–O tetrahedron, while the group-II O-atoms share an edge with a Si–O tetrahedron. As the Si–O bond lengths begin to increase the O-atoms in the two Zr-coordination groups show a different thermal behavior. While the

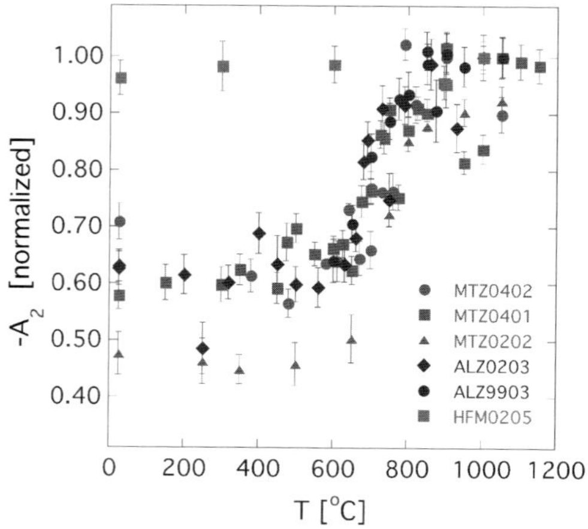

Figure 2. Measured anisotropies for *Mud Tank zircon* (MTZ) and *Aldrich* (ALZ) zircons in comparison with an Hf metal sample (HFM). Data points are normalized to A_2 at $T = 1000°C$.

Zr–O_{II} bond lengths increase steadily with increasing temperature, the Zr–O_I bonds remain nearly constant between 800 and 1100°C. The EFG will reflect this subtle change in the Zr–O coordination and show a different thermal behavior then before the transition. Unfortunately our current furnace does not allow for measurements above 1100°C, and we are working on an alternative design that does not have this limitation.

The primary difference between the *Mud Tank* and other zircons is the much lower impurity concentration that results in a factor of two lower asymmetry and a more than three times lower damping for the *Mud Tank* zircon. It is conceivable that the higher impurity concentration suppresses the subtle change in the bond length of the Zr–O_I group. Moreover, *Mud Tank* zircon never experienced substantial α-decay induced radiation damage, while *Aldrich* zircon has sufficiently high U and Th content that it was partially metamict before it was recrystallized. Possibly the lack in rearrangement of the Zr–O coordination is connected to the recrystallization process.

In our PAC experiments with zircon we observed a consistent drop of the measured anisotropy values with temperature. This was observed with all zircon samples, not just the *Mud Tank* zircons. Figure 2 shows anisotropy data for three *Mud Tank* zircons and two *Aldrich* zircons. Also shown is the anisotropy for a Hf metal sample. While the measured anisotropies for the Hf sample remain at the expected value, the A_2-values of the zircon samples show a 40 to 50% drop between 800 and 600°C. Since sample-detector distance and detector geometry remained the same throughout our experiments this drop is attributed to a rapid relaxation of a fraction of the probe nuclei. Generally we begin our computer fits of the PAC spectrum after the first nanosecond because of a prompt coincidence contribution within that time period due to the limited energy resolution of the BaF_2 detectors. Therefore this relaxation must be sufficiently fast that part of the perturbation function decays to zero within approx. 1 ns. Similar behavior was reported for pure ZrO_2, and it was suggested that this behavior is due to an aftereffect of the ^{181}Hf β-decay [9]. Hence we see the decrease of A_2 in zircon as a further manifestation of a β-decay aftereffect. Following the decay of the ^{181}Hf parent the electronic shell of the ^{181}Ta daughter nucleus is in a metastable configuration. In metals this metastable configuration decays rapidly and all probes see the same static quadrupole interaction at all temperatures. In insulators, such as ZrO_2 and $ZrSiO_4$, this metastable state may last a great deal longer. Just as with ZrO_2 we suggest the existence of an electron trap in the $ZrSiO_4$ band gap. While at high temperatures the trap is depopulated sufficiently fast that by the time the $\gamma\gamma$-cascade takes place the electronic shell of the probe is in thermal equilibrium; however, at low temperatures about 40 to 50% of the probes are still under the influence of a trapped electron and experience a large hyperfine interaction resulting in rapid decay of the perturbation function.

Acknowledgements

The authors wish to thank Dr. John Hanchar of *George Washington University* for providing the zircon sample for this study. Support of the *DOE Reactor Sharing Program* is gratefully acknowledged.

References

1. Ewing R. C., Lutze W. and Weber W. J., *J. Mater. Res.* **10** (1995), 243.
2. Ewing R. C., Haaker R. F. and Lutze W., In: W. Lutze (ed.), *Scientific Basis for Radioactive Waste Management V*, Materials Research Society Proceedings, North Holland, New York, 1983.
3. Weber W. J., Ewing R. C. and Wang L., *J. Mater. Res.* **9** (1994), 688.
4. Mursic Z., Vogt T. and Frey F., *Acta Cryst. B* **48** (1992), 584.
5. Jaeger H., Rambo M. P. and Klueg R. E., *Hyperfine Interact.* **136/137** (2001), 515.
6. Robinson K., Gibbs G. V. and Ribbe P. H., *Am. Mineral.* **56** (1971), 782.
7. Jaeger H., Pletzke K. and Hanchar J. M., In: J. Vienna (ed.), Ceramic Transactions, Vol. 155, Am. Ceram. Soc., Westerville, Ohio, 2004, pp. 31–40.
8. Jaeger H., Rambo M. P., Uhrmacher M. and Lieb K.-P., *Recent Res. Devel. Phys.* **2** (2001), 81.
9. Jaeger H., Su H. T., Gardner J. A., Chen I.-W., Haygarth J. A., Sommers J. A. and Rasera R. L., *Hyperfine Interact.* **60** (1990), 615–618.

Hyperfine Interactions (2004) 158:273–279
DOI 10.1007/s10751-005-9044-0

Anomalous Temperature Dependence of the EFG in AlN Measured with the PAC-Probes ^{181}Hf and ^{111}In

K. LORENZ*,† and R. VIANDEN
HISKP, University of Bonn, Bonn, Germany; e-mail: lorenz@itn.pt

Abstract. ^{181}Hf and ^{111}In ions were implanted into AlN-layers in order to investigate their immediate lattice site environment and its temperature dependence by means of the Perturbed Angular Correlation (PAC) technique. After rapid thermal annealing at 1273 K up to 50% of the probe atoms were incorporated on undisturbed lattice sites defined by an electric field gradient (EFG) of 33 MHz for In and 572 MHz for Hf for measurement at room temperature. PAC-spectra taken at temperatures between 25 and 1200 K show that the EFG measured at the site of the undisturbed probes changes with temperature. While for Hf it decreases by 3%, for In it increases by 25% within the measured temperature range. Thus, the change cannot be due only to the thermal lattice expansion. In the case of In the fraction of probe atoms on substitutional sites increases with temperature until it reaches nearly 100% at 973 K. These effects are fully reversible. For the Hf probe, an additional EFG was detected at temperatures above 300 K.

Key Words: AlN, GaN, implantation, PAC.

Abbreviations: EFG – Electric Field Gradient; QIF – Quadrupole Interaction Frequency; PAC – Perturbed Angular Correlation; RT – Room Temperature.

1. Introduction

Aluminium nitride (AlN) with its wide direct band gap (~6.2 eV) and its alloys with GaN ($Al_xGa_{1-x}N$) are very promising candidates for optoelectronic devices such as detectors and light emitters in the ultraviolet spectral region [1, 2].

Ion-implantation represents an attractive tool for selective area doping, dry-etching or electrical isolation and is a key technology in semiconductor industry [1]. Despite a strong research effort, implantation into group-III-nitrides still suffers from the poor recovery of the crystal during the post-implant annealing process. With the Perturbed Angular Correlation technique (PAC), it is possible to monitor the incorporation of implanted ions into the lattice and probe their immediate lattice environment. Recent PAC-studies of In-implanted GaN

* Author for correspondence.
† Present address: ITN, Lisbon, Portugal. e-mail: lorenz@itn.pt

indicated that after annealing In forms a complex with an unknown defect that is strongly dependent on the temperature at which the measurement is performed [3]. In this work we present temperature dependent PAC studies with the probes ^{181}Hf and ^{111}In implanted into AlN. ^{111}In is an ideal probe to investigate group-III nitrides, being itself a group-III element. Additionally, implanted in GaN and AlN it gives valuable information about the ternary systems InGaN and AlInN, which are relevant materials for various device applications.

2. Experimental details

The radioactive probe atoms ^{181}Hf and ^{111}In were implanted at room temperature (RT) into commercial, 150 nm thick AlN films grown by HVPE on sapphire. The implantation was performed along the surface normal with an energy of 80 keV for ^{181}Hf and 60 keV for ^{111}In and with typical fluences of $\sim 10^{13}$ at/cm^2. SRIM2003 simulations [4] yield an ion range of 28 and 27 nm for Hf and In, respectively. After implantation, the samples were annealed for 120 s at 1273 K in a rapid thermal annealing apparatus between graphite strips in flowing nitrogen gas. This process had previously been optimized for annealing of GaN [5].

A standard PAC set-up with four BaF_2 detectors in a cross-shaped arrangement and conventional fast-slow electronics was used. For all measurements the $\hat{c}$-axis of the AlN samples was aligned with the angle bisector of two detectors under 90°.

PAC-spectra were taken at different sample temperatures between 25 and 1193 K. Measurements below 293 K were carried out under vacuum in a closed cycle He refrigerator. For measurements at elevated temperatures the samples were placed in a graphite heater under low N_2 pressure ($\leq$1 mbar).

3. Results and discussion

Figure 1(a) shows a typical RT PAC-spectrum with ^{181}Hf in AlN after annealing. The results of the temperature dependent PAC-measurements are summarized in Figure 2. The spectra were fitted with the computer code NNfit [6]. Three fractions are necessary to describe the data: an undisturbed fraction f_u, a disturbed fraction f_d with a wide frequency distribution and for temperatures above RT a third fraction f_x. For all three fractions the EFG was assumed to be axially symmetric and oriented along the $\hat{c}$-axis of the wurtzite AlN crystal. The undisturbed fraction f_u is attributed to probes on substitutional lattice sites, as RBS/ Channelling measurements showed that Hf is incorporated on Al-sites after implantation [7]. The quadrupole interaction frequency (QIF) for the undisturbed probes decreases with increasing temperature by 3% from $\nu_u = 576$ MHz at 25 K to $\nu_u = 560$ MHz at 900 K (Figure 2a) with a Lorentzian frequency distribution of

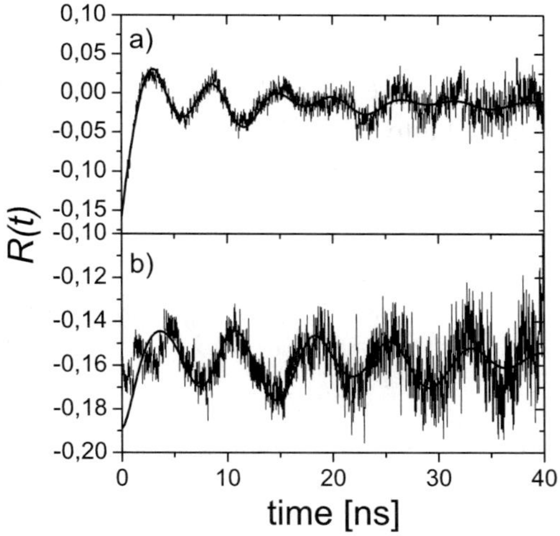

Figure 1. PAC-spectra from [181]Hf-implanted AlN after annealing at 1273 K. a) Spectrum measured at RT. b) Spectrum showing the defect frequency at 873 K. The data shown was derived by correcting the lattice QIF in the fit of the RT-spectrum to the value measured at 873 K and subtracting it from the spectrum at 873 K.

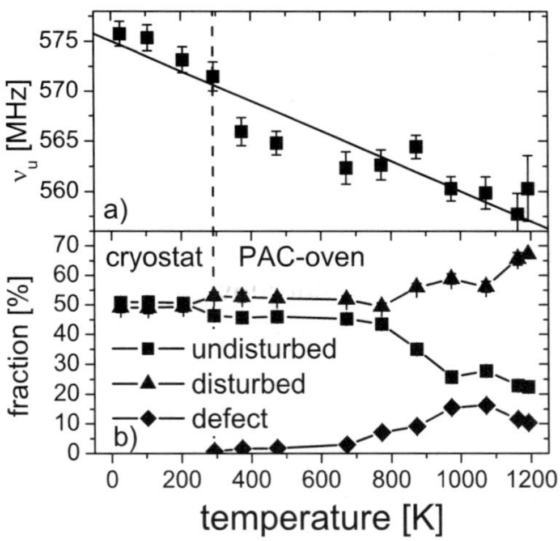

Figure 2. Temperature dependent PAC-measurements with the probe [181]Hf in AlN. a) Change of the QIF for probes on undisturbed lattice sites with temperature. b) Development of the three fractions used to fit the data with temperature.

$\sim$ 7%. Below 700 K f_u and f_d stay constant within the errors at about 50% (Figure 2b). Above RT the third fraction f_x appears with a QIF of $\nu_x = 459$ MHz with a small Lorentzian frequency distribution of $\delta \sim 2\%$. f_x becomes significant for temperatures above 700 K. Figure 1b shows this defect frequency at 873 K. The data shown was obtained by subtracting the fit of the RT spectrum, where only ν_u was changed to the value at 873 K, from the spectrum at 873 K. The

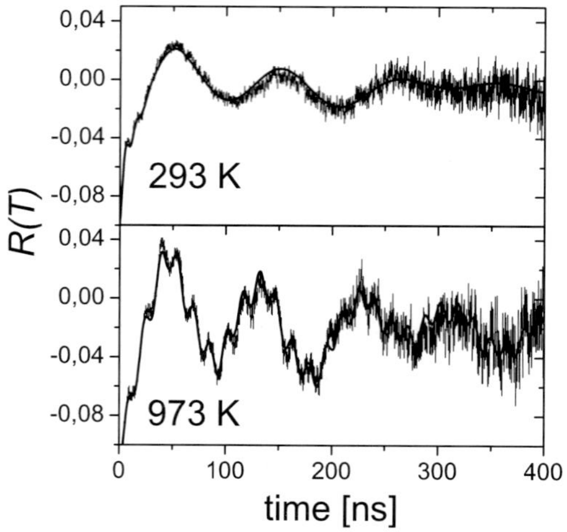

Figure 3. PAC-spectra from ^{111}In-implanted AlN after annealing at 1273 K measured at RT and 973 K.

derived spectrum was then shifted by the anisotropy. Fraction f_x increases with temperature causing a decrease of f_u; above 800 K also f_d increases. Up to 873 K this behaviour is fully reversible, as it was shown with an additional measurement at RT after the one at 873 K; for 1073 K, however, irreversible changes are present in the spectra.

The low frequency distribution of ν_x points to a defined point defect in the neighbourhood of the probe. It is possible that nitrogen vacancies are created at elevated temperature and then are trapped by the probes and cause the additional EFG. When the sample is cooled down again they recombine with interstitials. For temperatures above 1073 K, however, irreversible damage is caused probably due to the out-diffusion of nitrogen.

Figure 3 shows typical PAC spectra taken with the probe ^{111}In at 293 and 973 K after annealing at 1273 K. The results of the entire measuring program are summarized in Figure 4. Three fractions were used for the fit, all with axially symmetric EFG oriented along the $\hat{c}$-axis. A fraction of about 20% of the probes crossed the AlN layer and stopped in the sapphire substrate; we attribute this to channelling effects. This fraction shows a characteristic fast frequency around 215 MHz with a high frequency distribution at low temperatures and very low damping above room temperature (Figure 3). The temperature behaviour of ^{111}In in sapphire has been discussed elsewhere [7, 8]. For measurements below RT, 50% of the probes are incorporated on undisturbed lattice sites with a QIF ν_u = 31 MHz and a Lorentzian frequency distribution of δ = 13%. This undisturbed fraction f_u is attributed to probes on substitutional lattice sites. Emission channelling measurements showed that In is mainly incorporated on substitutional lattice sites [9]. The Al-site is the most probable as In is also a group-III element and the size of the ions is comparable. A third, disturbed fraction f_d

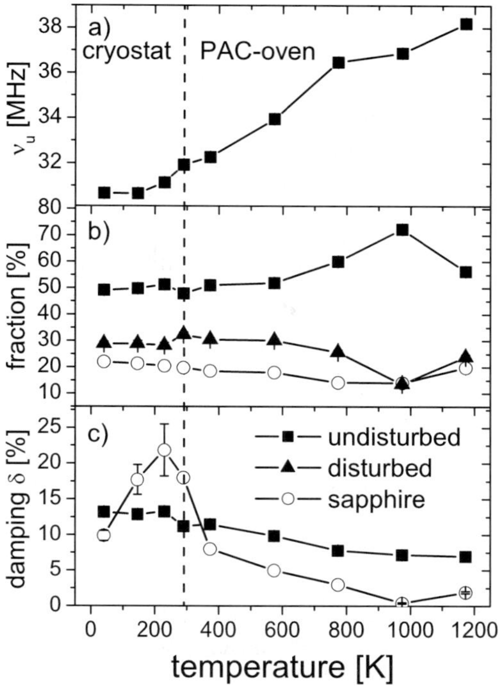

Figure 4. Temperature dependent PAC-measurements with the probe ^{111}In in AlN. a) Change of the QIF for probes on undisturbed lattice sites with temperature. b) Development of the three fractions used to fit the data with temperature. c) The Lorentzian frequency distribution δ of the QIF for undisturbed probes in AlN and sapphire.

shows a frequency of ν_d = 100 MHz with a high Lorentzian frequency distribution $\delta = 25\%$.

Fraction f_u shows a strong dependence on the temperature. The frequency ν_u increases by 25% within the measured temperature range from ν_u = 30.7 MHz at 40 K to ν_u = 38.2 MHz at 1200 K. For comparison, in the same temperature range the QIF for the probes in sapphire decreases by only 3%. The fact, that the EFG is increasing in the case of ^{111}In in AlN but decreasing for ^{181}Hf shows that the change cannot only be due to thermal expansion of the crystal.

For ν_u, δ decreases with temperature from 13% to 7% (Figure 4c) and f_u increases up to nearly 80% at 973 K (Figure 4b). At this temperature, a good fit can also be achieved assuming that all probes that stop in the AlN layer are situated on undisturbed lattice sites. This behaviour described above is fully reversible when going back to lower temperatures, as it was shown with an additional measurement at RT after the one at 973 K. Only for the highest temperature at 1173 K irreversible changes are induced due to the dissociation of the crystal when kept at this temperature for more than 8 h, the time needed for the measurement. This is also seen in the drop of the undisturbed fraction. A similar behaviour was previously reported for ^{111}In implanted into GaN [3]. Also here the undisturbed fraction increases with temperature and reaches 100% above ~500

K and the lattice QIF increases by 30% in the temperature range from 16 to 923 K. The disturbed frequency at low temperature is fast and with a large δ; with increasing temperature, the frequency and δ decrease until they coincide with the undisturbed lattice frequency.

In the case of AlN the disturbed frequency and its distribution are kept constant in the fits as a detailed analysis is not possible because of the superposition of the signal from probes in sapphire.

The results with ^{111}In-implanted AlN as well as GaN show that the low substitutional fractions usually reported for PAC-measurements with ^{111}In at RT [9–13] are not due to an insufficient annealing of the samples and the trapping of probes in highly damaged regions; they point rather to an interaction of the probes with defined defects, exhibiting the discussed temperature dependence. Comparing the results with ^{111}In and ^{181}Hf in AlN it is noticed that the increase of f_u measured with ^{111}In starts at the same temperature when the defect fraction f_x becomes significant in the ^{181}Hf-measurements. A possible explanation is that N-vacancies, produced during the implantation or later when heating the sample, get mobile at temperatures around 600 K and then get trapped by the Hf probe and lead to a well defined EFG. In the case of In, however, they act as a dynamic defect. At low temperatures, they are randomly distributed, causing a distribution of static EFG near some of the probes; at higher temperatures, they move so fast that only the average lattice EFG is visible.

4. Conclusions

We studied the temperature dependence of the EFG in AlN measured with the PAC-probes ^{181}Hf and ^{111}In. In the case of ^{181}Hf the EFG decreases within the measured temperature range by 3% while for ^{111}In the EFG increases by 25%. The fraction of ^{111}In-atoms on undisturbed substitutional lattice sites increases with temperature and at 973 K close to 100% of ^{111}In atoms are found on undisturbed lattice sites in AlN. This behaviour, which has also been observed for In-implanted GaN, shows that the disturbed fraction observed at RT is not due to insufficient annealing of the samples but due to defined defects that show a strong temperature dependence. Comparing the results of the measurements with ^{111}In and ^{181}Hf in AlN suggests that the nitrogen vacancy might be involved in this temperature dependent behaviour.

References

1. Pearton S. J., Zolper J. C., Shul R. J. and Ren F., *J. Appl. Phys.* **86** (1999), 1.
2. Ambacher O., *J. Phys. D* **31** (1998), 2653.
3. Lorenz K., Ruske F. and Vianden R., *Appl. Phys. Lett.* **80** (2002), 4531.

4. Ziegler J. F., Biersack J. P. and Littmark U., *The Stopping and Range of Ions in Solids*, Pergamon, New York, 1985.
5. Bartels J., Freitag K., Marques J. G., Soares J. C. and Vianden R., *Hyperfine Interact.* **120–121** (1999), 397.
6. Barradas N. P., Rots M., Melo A. A. and Soares J. C., *Phys. Rev. B* **47** (1993), 8763.
7. Lorenz K., PhD thesis, University of Bonn, 2002.
8. Penner J., diploma thesis, University of Bonn, 2003.
9. Ronning C., Dalmer M., Uhrmacher M., Restle M., Vetter U., Ziegeler L., Hofsäss H., Gehrke T., Järrendahl K. and Davies R. F., *J. Appl. Phys.* **87** (2000), 2149.
10. Ronning C., Dalmer M., Deicher M., Restle M., Bremser M. D., Davis R. F. and Hofsäss H., *Mat. Res. Soc. Symp. Proc.* **468** (1997), 407.
11. Burchard A., Deicher M., Forkel-Wirth D., Haller E. E., Magerle R., Prospero A. and Stötzler A., *Mater. Sci. Forum* **258–263** (1997), 1099.
12. Lorenz K., Ruske F. and Vianden R., *Phys. Status Solidi, B* **228** (2001), 331.
13. Lorenz K. and Vianden R., *Phys. Status Solidi, C* **1** (2002), 413.

Hyperfine Interactions (2004) 158:281–284
DOI 10.1007/s10751-005-9045-z

The Rare Earth PAC Probe ^{172}Lu in Wide Band-Gap Semiconductors

R. NÉDÉLEC[1,*], R. VIANDEN[1] and THE ISOLDE COLLABORATION[2]
[1] *Universität Bonn, Bonn, Germany; e-mail: nedelec@iskp.uni-bonn.de*
[2] *ISOLDE, CERN, Switzerland*

Abstract. Group III nitrides and other wide band-gap semiconductors like ZnO are very promising materials for application in optoelectronics. However, little is known about the recovery of lattice damage caused by ion implantation necessary to achieve lateral structuring. We use the PAC technique to study the behaviour of the Rare Earth isotope ^{172}Lu(^{172}Yb) after implantation. After annealing, a large fraction of the probes is found on unique lattice sites with axial symmetry and a low damping of the interaction frequency. In addition the corresponding electric field gradient is aligned along the $\langle 0001 \rangle$ axis. A substitutional incorporation of the Rare Earth on regular metal lattice sites is therefore probable.

Key Words: GaN, PAC, perturbed angular correlation, rare earth, ZnO.

1. Introduction

Wide band-gap semiconductors (WBGS) like the group III nitrides or ZnO have recently emerged as important base materials for applications in optoelectronics and in high power, high temperature electronics. An interesting possibility to vary the wavelength of the light emitted by GaN LEDs has been demonstrated by doping it with Rare Earth during growth. In these systems electroluminescence, due to intra-atomic transitions of the RE dopants, has been observed in the visible spectrum with high efficiency at room temperature. Due to their electronic properties, the emitted wavelength depends only on the RE used.

For integrating such optoelectronic devices into circuits an adequate lateral structuring technique is necessary. Ion implantation is such a technique that is commonly used for standard semiconductors like Silicon, but which still needs some development to apply it to wide band-gap semiconductors.

In the presented work we have studied this process with the perturbed angular correlation technique (PAC). This technique is particularly suited for the study of

* Author for correspondence.

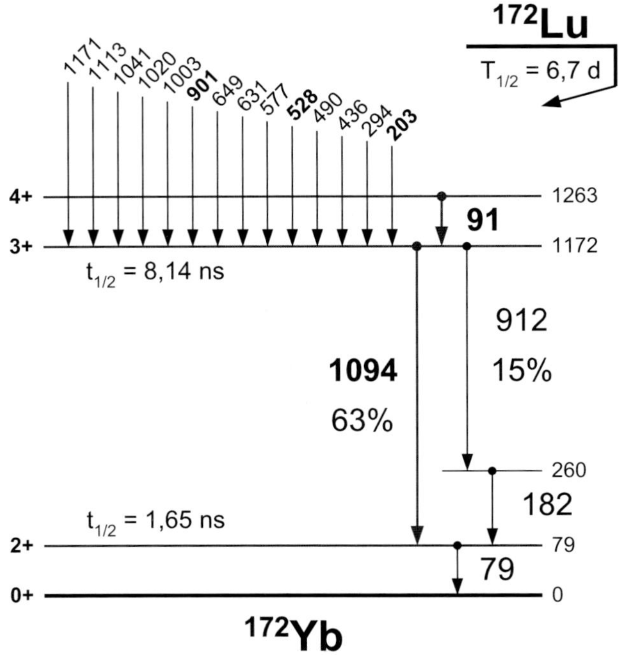

Figure 1. Decay scheme of the nucleus ^{172}Lu. The 91–1,094 keV cascade is suited for PAC experiments [1, 2].

the annealing behaviour of implantation damaged lattice structures. We have used the 91–1,094 keV $\gamma\gamma$-cascade of the PAC Probe ^{172}Lu(^{172}Yb) (Figure 1). The intermediate state has a rather unusual spin of 3^+ and with 8.33(3) ns its half-life is comparatively short. However, the quadrupole moment $Q = 2.87(41)$ and the anisotropy coefficient $A_{22} = 0.387(28)$ of the cascade are fairly large [1, 2].

2. Measurements

The GaN [3] and ZnO [4] samples were implanted at the ISOLDE facility at CERN. The implantation took place at an energy of typically 60 keV. According to the depth implantation profile of a SRIM2003 simulation [5] under these conditions, the implanted atoms come to rest well below the surface. This is mandatory in order to observe the behaviour in the bulk material. After implantation, the recovery process of the samples was examined in an isochronal annealing programme. In order to be able to observe the progress in damage recovery the sample was annealed under flowing nitrogen at different temperature steps for 120 s each in a rapid thermal annealing furnace (RTA) [6]. Between two temperature steps a PAC spectrum was recorded.

Before the first annealing step was carried out a so-called *as implanted* spectrum was recorded (Figure 2a). It shows for both materials a steep drop of the anisotropy function towards the hardcore value. This is characteristic for an

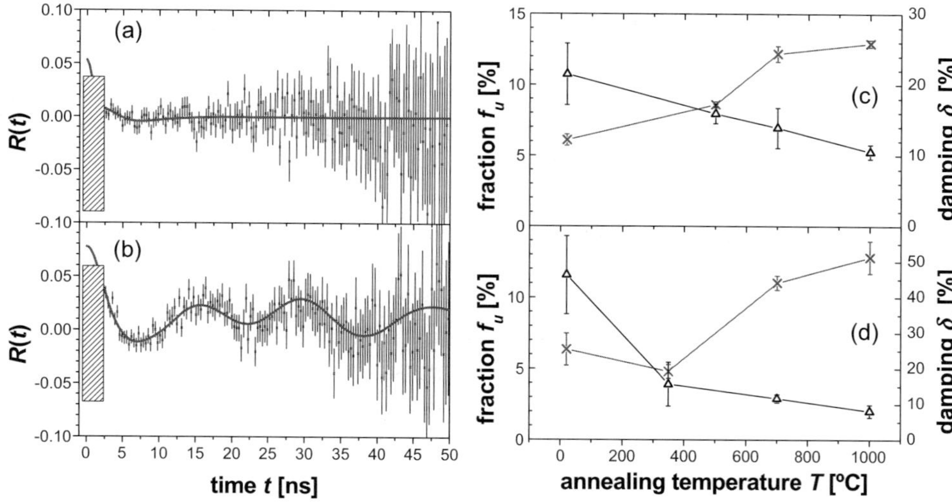

Figure 2. (a) and (b) show the PAC spectra for ^{172}Lu in GaN as implanted and after annealing in a RTA furnace at 1,000°C. The observed frequency in (b) corresponds to an EFG of 3.76(4) $\times$ 10^{17} V cm^{-2} that is attributed to substitutional Lu/Yb on a Ga site. The results of the annealing programme for ZnO and GaN are shown in (c) and (d), respectively. The observed frequency (not shown) remains nearly constant. The undisturbed fraction (crosses) increases strongly from 5% to about 13% for both materials, whereas the damping of the undisturbed fraction (open triangles) is low after annealing at 1,000 °C.

ensemble of probe nuclei in a strongly disturbed environment. The PAC spectra were evaluated with a very simple model assuming two fractions of probe atoms. On the one hand a polycrystalline fraction was used to describe probe nuclei in a statically disturbed environment. Defects are expected to be located randomly around the nucleus. On the other hand, a single crystalline fraction describes the probes sitting on regular lattice sites. Since both GaN and ZnO have wurtzite structure a well-defined electric field gradient (EFG) along the axis of axial symmetry is expected. Slight defects in the probes' surroundings will broaden the observed frequency distribution, leading to a damping of the observed anisotropy over time.

In GaN, already from the beginning a small fraction of probe nuclei is seen on regular lattice sites. This is mainly due to dynamic annealing effects during implantation. However, a strong damping suggests many defects in their environment. After the first annealing step at 350°C this damping δ_u drops from about 45% to about 15%. The single crystalline fraction f_u remains constant. At 700°C the interaction frequency of the single crystalline fraction becomes clearly visible and the fraction rises from about 5% to 10%. From that, one can determine that an annealing temperature of 350°C is high enough to significantly reduce lattice defects in the surroundings of a substitutional probe, but not high enough to bring dislocated probe atoms to regular lattice sites. At an annealing temperature of 1,000°C the single crystalline fraction increases only slightly. In the same time its damping drops to δ = 7%. A clear quadrupole interaction frequency can now be determined for the single crystalline fraction. This frequency of $\nu_Q = 261(3)$ MHz belongs to an EFG of $3.76(4) \times 10^{17}$ V cm^{-2} that is oriented along the $\hat{c}$-axis of the sample. The single crystalline fraction can therefore be attributed to substitutional Lu/Yb on regular lattice sites. The low damping confirms that the neighbouring shells of atoms around the probe atom are in perfect order.

In the case of ZnO the observed behaviour is very similar. Throughout the different temperature steps the single crystalline fraction rises from about 6% to 13% while the damping δ drops from about 20% to 10%. An interaction frequency of $\nu_Q = 146(3)$ MHz, corresponding to an EFG of $2.11(4) \times 10^{17}$ V cm^{-2}, is also attributed to Lu/Yb on regular lattice sites. This interpretation is supported by the fact that the single crystalline fraction shows no asymmetry parameter and that orientation measurements show a clear alignment of the EFG along the $\hat{c}$-axis of the ZnO material (not shown).

3. Summary

In the present work we have shown that it is possible to recover a considerable amount of the damage created in the semiconductors GaN and ZnO during the implantation process. The results of PAC measurements for both materials show that a large fraction of probe atoms is found on sites with a well-defined EFG oriented along $\langle 0001 \rangle$ direction. Moreover, for these sites we observe an axial symmetry of the EFG. Combined with the rather low damping of the interaction, the results presented here favour a substitutional incorporation of the probe atoms into the host lattice.

From the shown PAC spectra (Figure 2) it can be seen that the experimentally observed anisotropy is significantly lower than expected. This is believed to be due to the intermixing of different $\gamma\gamma$-cascades with different anisotropy coefficients in the intermediate state (see Figure 1), which via the Compton effect are registred in the 91 –1,094 keV energy windows. Since an estimation of the expected anisotropy is difficult, the fractions obtained by fitting have been scaled to the literature value [1] with an appropriate correction for the detectors' solid angles applied.

References

1. Tuli J. K. and Martin M. J., *Nucl. Data Sheets* **75/2** (June 1995), 214.
2. Firestone R. B. and Shirley V. S., *Table of Isotopes*, CD-Rom Edition, Version 1.0, March 1996.
3. GaN film (1.3 μm) on sapphire backing, Cree, http://www.cree.com/.
4. ZnO, bulk material, Surface Prep Lab, http://www.surface-prep-lab.com/.
5. Ziegler J. F. and Biersack J. P., SRIM Homepage: http://www.srim.org/.
6. Marx G., diploma thesis, HISKP, University of Bonn, October 1990.

Hyperfine Interactions (2004) 158:285–291
DOI 10.1007/s10751-005-9046-y

TDPAC Study of the Intermetallic Compound $HfCo_3B_2$

I. YAAR*, I. HALEVY, S. SALHOV, E. N. CASPI, M. DUBMAN, S. KAHANE and Z. BERANT
Nuclear Research Center Negev (NRCN), POB 9001, Beer-Sheva, 84190, Israel; e-mail: iyaar@bgu.ac.il

Abstract. The electronic properties of the intermetallic compound $HfCo_3B_2$ were investigated using combined TDPAC measurements and first principles LAPW calculations. The V_{zz} value at the hafnium site is determined from dominant positive *p–p* contribution, with less than 20%, negative *s–d* and *d–d* contributions. Based on the calculated density of state (DOS) at 0 K, a band contribution (γ_{band}) of 7.26 (mJ/mol/K^2) to the value of the electronic specific heat coefficient (γ) was obtained. This relatively low γ_{band} value is attributed to the hybridization between hafnium *d*-states, boron 2*p*-states and cobalt 3*d*-states, formed at the energy interval below E_{Fermi}. This hybridization, together with the dip in the DOS around E_{Fermi}, implies a possible reduction of the low temperature magnetic moment in this compound.

Key Words: hyperfine interaction, LAPW, TDPAC, WIEN97.

1. Introduction

The hexagonal (P6/mmm) $CaCu_5$-type structure is adopted by a large number of binary and ternary intermetallic compounds [1], having a large variety of applications as permanent magnets [2–4] or hydrogen storage materials [5]. Most of the RCo_3B_2 ternary compounds in this family, like $HfCo_3B_2$, are reported to crystallize in the variant $CeCo_3B_2$-type structure, depicted in Figure 1, where the Hf, Co and B atoms occupy the special Wyckoff positions 1a (0, 0, 0), 3g (1/2, 0, 1/2) and 2c (1/3, 2/3, 0), respectively [6, 7].

The magnetic and electronic properties of these compounds have been studied in the past using high-pressure x-ray diffraction technique [8], neutron diffraction, ac and dc-susceptibility measurements [9], Mössbauer studies [10, 11], specific heat measurements [12], TDPAC measurements [13] and full potential Linearized Augmented Plane Wave (LAPW) calculations [14–16].

In the present work, the electronic configuration of the ternary compound $HfCo_3B_2$ was studied, using combined Time Differential Perturbed Angular

* Author for correspondence.

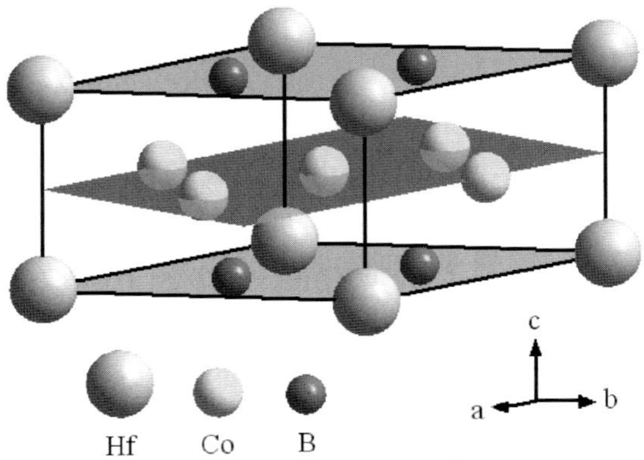

Figure 1. Schematic representation of the $HfCo_3B_2$, $CeCo_3B_2$-type structure [6, 7].

Correlation (TDPAC) technique and full potential linearized augmented plane wave (LAPW) calculations. This combination of both experimental and calculated data validates the further use of the calculations to evaluate the origin of the electric field gradient (efg), and to calculate the expected contribution of electrons to the electronic specific heat coefficient (γ_{band}).

2. Experimental

The sample was prepared by arc melting of cobalt and boron elements together with pre-irradiated hafnium with ≈50 μCi activity of ^{181}Hf ($T_{1/2}$ = 42 *d*), in an argon atmosphere. The obtained sample was annealed at 950°C in a vacuum for 120 h. X-ray diffraction measurements of the sample, taken with Cu–K_α monochromator, indicated two phases; about 50% of the sample crystallized in the expected hexagonal $CeCo_3B_2$-type (P6/mmm) structure with the lattice constants: a = 3.0344 Å and c = 4.8306 Å, in good agreement with previously published results [6]. The second phase was identified as the hexagonal AlB_2-type compound HfB_2 with lattice constants: a = 3.1373 Å and c = 3.4682 Å, in good agreement with JCPDS card #75-1049. HfB_2 is a unique material which has a high melting point, high hardness and high electric conductivity [17].

TDPAC measurements of the above sample were carried out at room temperature using the ^{181}Ta 133–482 keV cascade with an intermediate state half-life of 10.8 ns. This cascade is obtained via the decay of the 18 μs 615 keV state in the ^{181}Ta probe, populated in the β decay of ^{181}Hf → ^{181}Ta [18]. The 18 μs delay before the excited level under observation is reached, gives sufficient time for electronic reorganization around the probe according to the valence change from ^{181}Hf to ^{181}Ta. The time spectra at π and $\pi/2$ were recorded simultaneously using a 4-BaF_2 detector system with a time resolution of 600 ps and a conventional multi-parameter acquisition system MPA-2 (Fast ComTec).

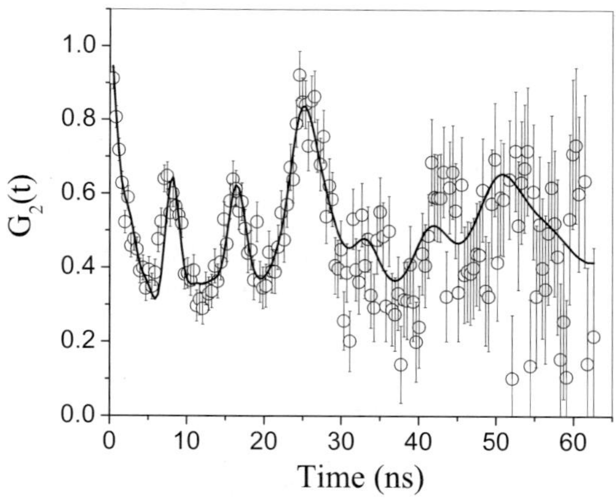

Figure 2. TDPAC spectrum of the sample. The *solid line* is the best fit obtained using the TDPAC experimental efg parameters listed in Table I.

Table I. TDPAC measured and LAPW calculated efg parameters in $HfCo_3B_2$ and HfB_2

Phase	Site	TDPAC measured			LAPW calculated	
		V_{zz} (10^{21} V/m^2)	η	$\delta \times 10^{-2}$	V_{zz} (10^{21} V/m^2)	η
$HfCo_3B_2$	Hf	4.45 ± 0.07	0.1 ± 0.1	1.0 ± 0.2	4.33	0
	Co	–	–	–	−2.08	0.68
	B	–	–	–	−1.05	0
HfB_2	Hf	13.76 ± 0.14	0.13 ± 0.1	9.2 ± 1.6	9.86[a]	0[a]
	B	–	–	–	0.10[a]	–

[a] LAPW calculated results taken from the work of Haas [19].

3. Results

The experimental TDPAC spectrum of the sample is depicted in Figure 2. As expected from a two phase sample, the experimental spectrum was fitted using two sets of efgs, with a relative occupation of ≈50% for each one. The experimental TDPAC parameters of these sites are compared in Table I with LAPW calculated results made with the lattice constants evaluated from the experimental X-ray diffraction data.

As seen in Table I, the TDPAC experimental values of the efg main component (V_{zz}) and asymmetry parameter (η) at the Hf-site for both phases are in reasonable agreement with the LAPW calculated efg parameters of the investigated sample. The origin for the 40% deviation in the measured V_{zz} value at the Hf-site in HfB_2 will be further investigated in the future, using a single phase sample.

4. Calculations

The calculations were done using the full potential linearized plane wave (LAPW) method, as embodied in the WIEN97 code [20]. The unit cell in these calculations is divided into two parts, the atomic spheres and the interstitial region. The wave functions inside the atomic spheres are described by atomic like functions, while in the interstitial region plane waves are used [21]. Exchange and correlation effects are treated within the density functional theory, using the generalized gradient approximation (GGA) [22], where not only the local density, but also its gradient determines the magnitude of the effect. In most cases, the charge asymmetry inside the atomic sphere, where the V_{zz} value is being calculated, determines more than 90% of the V_{zz} value. Inside this sphere, the contribution to V_{zz} can be further divided into *s–d*, *p–p* and *d–d* components, so that the physical origin of the efg can be analyzed [21].

The atomic muffin-tin sphere radius in all of the calculations was taken as 1.50, 1.10 and 0.93 Å for hafnium, cobalt and boron atoms, respectively, and the value of the cutoff parameter $R_{MT}K_{max}$, where R_{MT} is the smallest atomic sphere radius in the unit cell and K_{max} is the magnitude of the largest K vector in the LAPW basis set, was set to 8. The calculated values of V_{zz} and η, obtained in these calculations, are compared in Table I with the experimental TDPAC values. The small discrepancy between the experimental and calculated η value, is indicative of the disorder of the examined sample, as demonstrated in the X-ray results. In a single phase and well ordered $HfCo_3B_2$ sample, the principle axis of the efg will be parallel to the *c*-axis, and a zero η value is expected [16], as demonstrated in the calculated efg parameters given in Table I.

The partial valence *s–d*, *p–p* and *d–d* and the lattice contribution to the efg at the hafnium, cobalt and boron sites are listed in Table II. From these results, it is evident that the major contribution to the V_{zz} value at the hafnium and boron sites comes from a *p–p* contribution next to the probe nucleus, with ≈10% *d–d* negative contribution. Where, at the cobalt site a large and positive *d–d* contribution to the V_{zz} value is observed. This unusual large *d–d* contribution arises from the wide distribution of cobalt 3*d*-electrons, shown in Figure 3.

The partial *s*, *p* and *d*-electrons density of states (DOS) at the hafnium, cobalt and boron sites in $HfCo_3B_2$ are plotted in Figure 3 as a function of energy. The main features of the DOS is the cobalt 3*d*-electrons broad distribution, observed in the energy interval from −7 to 2 eV, relative to E_{Fermi}, were hybridization between cobalt 3*d*-electrons, boron 2*p*-electrons and hafnium 5*d*-electrons, can be seen. This hybridization, together with the low DOS at E_{Fermi}, indicates a possible reduction of the magnetic moment [23] and of the electronic specific heat at low temperature, in this compound [24].

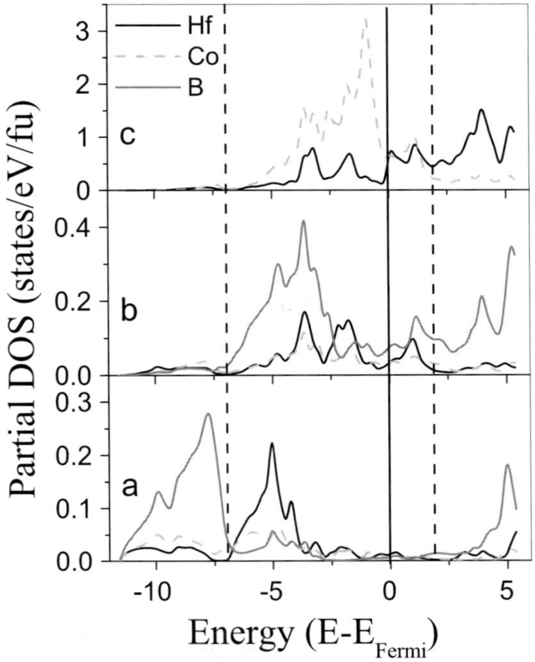

Figure 3. The partial (a) *s*, (b) *p* and (c) *d* electrons DOS at the hafnium, cobalt and boron sites, in $HfCo_3B_2$.

Table II. The partial valence *s–d*, *p–p* and *d–d* and the lattice contributions to the efg at the hafnium, cobalt and boron sites

Atomic site	Contribution to the V_{zz} value at the hafnium 1a site (10^{21} V/m^2)				
	s–d	*p–p*	*d–d*	Lattice	V_{zz}
Hf	−0.412	5.306	−0.564	−0.301	4.33
Co	0.011	−7.679	5.626	0.037	−2.043
B	−0.111	−0.933	−0.012	−0.006	−1.056

Based on the total electronic DOS at the Fermi level calculated for 0 K ($N(E^0_{Fermi})$), the band contribution (γ_{band}) to the value of the electronic specific heat coefficient (γ) can be evaluated using the equation,

$$\gamma_{\text{band}} = \frac{\pi^2 k_B^2}{3} N\left(E^{\,0}_{\text{Fermi}}\right) \tag{1}$$

where k_B is the Boltzmann constant. This equation serves as a link between electronic specific heat measurements and band structure calculations [24].

From the DOS calculated results, depicted in Figure 3, a value of 41.9 (states/Ry/f.u.) is calculated for $N(E^0_{\mathrm{Fermi}})$. This gives, according to Equation (1), a γ_{band} coefficient value of 7.26 (mJ/mol/K^2), a value that is a factor two lower than the previously reported calculated values for the isostructural compounds YCo_3B_2 and $LaCo_3B_2$ [12, 15].

5. Summary

The electronic properties of the intermetallic compound $HfCo_3B_2$ were investigated using combined TDPAC measurements and first principles LAPW calculations. The main features of the DOS is the cobalt 3*d*-electrons broad distribution, observed in the energy interval from -7 to 2 eV, relative to E_{Fermi}, were hybridization between cobalt 3*d*-electrons, boron 2*p*-electrons and hafnium 5*d*-electrons, can be seen. The major contribution to the V_{zz} value at the hafnium and boron sites comes from *p*–*p* contribution, with $\approx$10% negative *d*–*d* contribution. Where, at the cobalt site a large and positive *d*–*d* contribution to the V_{zz} value is observed. This unusual large *d*–*d* contribution at the cobalt site arises from the wide distribution of cobalt 3*d*-electrons.

Based on the calculated DOS at 0 K, a value of 7.26 (mJ/mol/K^2) for the band contribution to the value of the electronic specific heat coefficient was calculated. This relatively low electronic contribution to the specific heat coefficient is mainly attributed to the hybridization between hafnium *d*-states boron *p*-states and cobalt *d*-states, formed at the energy interval from -7 to 2 eV, relative to E_{Fermi}. This hybridization, together with the dip in the DOS around E_{Fermi}, indicates a possible reduction of the magnetic moment in this compound. Further neutron diffraction and specific heat measurements of this compound are planned, as soon as a single $HfCo_3B_2$ phase sample is obtained.

References

1. Guénée L. and Yvon K., *J. Alloys Compd.* **356–357** (2003), 114.
2. Oesterreicher H., Parker F. T. and Misroch M., *J. Appl. Phys.* **12** (1977), 287.
3. Smit H. H. A., Thiel R. C. and Buschow K. H. J., *J. Phys. F. Met. Phys.* **18** (1988), 295.
4. Ido H., Nanjo M. and Yamada M. J., *J. Appl. Phys.* **75** (1994), 7140.
5. Spada F. and Oesterreicher H., *J. Alloys Compd.* **107** (1985), 310.
6. Stadelmaier H. H. and Schöbel J. D., *Monatsh. Chem.* **100** (1969), 224.
7. Perricone A. and Noël H., *J. Alloys Compd.* **367** (2004), 152.
8. Sterer E., Halevy I., Ettedgui H. and Caspi E. N., *J. Phys., Condens. Matter* **14** (2002), 10619.
9. Caspi E. N., Pinto H., Kuznietz M., Ettedgui H., Melamud M., Felner I. and Shaked H., *J. Appl. Phys.* **83** (1998), 6733.
10. Malik S. K., Umarji A. M. and Shenoy G. K., *J. Appl. Phys.* **57** (1985), 2352.
11. Buschow K. H. J., Coehoorn R., Mulder F. M. and Thiel R. C. J., *J. Magn. Magn. Mater.* **118** (1993), 347.
12. Perthold W., Hong N. M., Michor H., Hilscher G., Ido H. and Asano H., *J. Magn. Magn. Mater.* **157/158** (1996), 649.

13. Yaar I., Gavra Z., Cohen D., Levitin Y., Feuerlicht J., Mintz M. H. and Berant Z., *J. Alloys Compd.* **260** (1997), 1.
14. Sandratskii L. M., Kübler J., Zahn P. and Mertig I., *Phys. Rev. B* **50** (1994), 15834.
15. Yamada H., Terao K., Nakazawa H., Kitagawa I., Suzuki N. and Ido H., *J. Magn. Magn. Mater.* **183** (1998), 94.
16. Coehoorn R., Buschow K. H. J., Dirken M. W. and Thiel R. C., *Phys. Rev. B* **42** (1990), 4645.
17. Wang X. B., Tian D. C. and Wang L. L., *J. Phys., Condens Matter* **6** (1994), 10185–10192.
18. Leaderer C. M. and Shirley V. (eds.), *Tables of Isotops*, Wiley, NY, 1978, p. 1135.
19. Haas H., *Hyperfine Interact.* **136/137** (2001), 731.
20. Blaha P., Schwarz K. and Luitz J., Wien97-Code (K. Schwarz, TU Wien, ISBN 3-9501031-0-4) 1999.
21. Blaha P., Dufek P., Schwarz K. and Haas H., *Hyperfine Interact.* **97/98** (1996), 3.
22. Perdew J. P., Burke K. and Ernzerhof M., *Phys. Rev. Lett.* **77** (1996), 3865.
23. White R. M., *Quantum Theory of Magnetism*, Springer, Berlin, Germany, 1983.
24. Kokko K., Ojala E. and Mansikka K., *J. Phys., Condens. Matter* **2** (1990), 4587.

Hyperfine Interactions (2004) 158:293–297
DOI 10.1007/s10751-005-9047-x

Investigations on the Diffusion of Boron in SiGe Mixed Crystals

W.-D. ZEITZ[1,*], J. HATTENDORF[1], W. BOHNE[1], J. RÖHRICH[1], E. STRUB[1] and N. V. ABROSIMOV[2]
[1]*Hahn-Meitner-Institut, Bereich Strukturforschung, Glienicker Strasse 100, D-14109 Berlin, Germany; e-mail: zeitz@hmi.de*
[2]*Institut für Kristallzüchtung, Max-Born-Strasse 2, D-12489 Berlin, Germany*

Abstract. By studying the quadrupolar interaction of ^{12}B in silicon–germanium mixed crystals with the β-NMR method, the boron–germanium pair was identified and the saturation amplitudes for boron in differently composed crystals were measured. The relative saturation amplitudes agree with statistical predictions. At low temperatures boron is preferentially implanted into stable interstitial sites. These sites are converted into substitutional sites by diffusion processes which take advantage of reorientation jumps.

Key Words: β-NMR, boron implantation, diffusion, silicon–germanium crystals.

PACS: 33.25.+k, 76.60.Gv, 61.72.Tt, 31.15.Ar, 33.15.Dj.

For most elements which diffuse by interstitial and vacancy related mechanisms in silicon–germanium mixed crystals, the diffusion coefficient increases with increasing Ge-content. The diffusion of boron, however, shows an unusual behaviour [1–3]: The activation enthalpy is increased in Ge-containing samples in comparison to pure silicon. In addition, boron atoms segregate at the interface between silicon and mixed crystals [4]. The retarded diffusion of boron in the mixed crystals was explained by assuming a boron–germanium pair with a strong bond [5].

By studying the quadrupolar interaction of radioactive ^{12}B in silicon–germanium crystals with the β-NMR method, we have demonstrated that the boron–germanium pair exists [6]. At the site of the boron nucleus, a cylindrical field gradient of $V_{zz} = +3.8$ V/Å^2 was found, which is aligned parallel to the $\langle 111 \rangle$ axis of the crystal. The pair was unambiguously identified on the basis of first principle calculations [7] which also delivered the bond length between the atoms. The positive field gradient which is present at the boron site was mainly deduced from the surplus of charge in the p_x- and p_y-orbitals compared to the p_z-orbital.

---

* Author for correspondence.

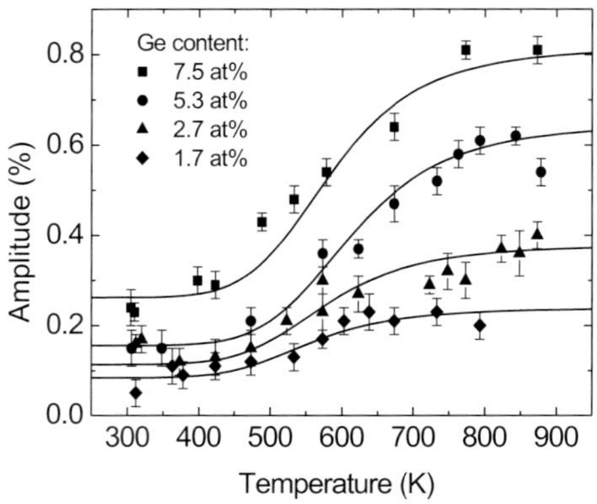

Figure 1. Amplitudes of the $\nu_{-1\leftrightarrow1}$-resonance as function of temperature and composition. This resonance belongs to the boron–germanium pair in $Si_{1-x}Ge_x$ mixed crystals.

In order to get some information on the diffusion process, we measured the amplitude of the $\nu_{-1\leftrightarrow1}$-resonance which belongs to the boron–germanium pair as a function of temperature and composition of the samples. The $\nu_{-1\leftrightarrow1}$-resonance only appears if a quadrupolar interaction is present. We used crystals with a Ge-content of 1.7(1)%, 2.7(2)%, 5.3(3)%, and 7.5(5)%. The compositions were determined with the ERDA-method [8].

The compilation of the results is shown in Figure 1. With unchanged radio frequency power, the resonance amplitudes were taken as relative measures for the number of boron atoms forming this complex in the $Si_{1-x}Ge_x$ mixed crystals. In all cases the amplitudes were low after implantation at room temperature, but increased with rising temperature reaching saturation at about 900 K. The saturation values are different for differently composed crystals. In Figure 2 we compare the relative saturation amplitudes for the different samples with the statistical predictions. The probability $p_n^z(x)$ to find a certain configuration around boron in $Si_{1-x}Ge_x$ mixed crystals can be calculated according to: $p_n^z(x) = (z!/(n!(z-n)!))\, x^n\, (1-x)^{z-n}$ [9]. Here $z = 4$ represents the number of the nearest neighbours in the T_d-symmetry of the lattice and $n = 1$ as there is only one Ge-atom in the complex. The relative amplitudes agree well with the statistical probabilities and do not show any enhancement at low Ge-contents. From this, we conclude that long range attractive potentials between these impurities, which usually lead to enhanced formation of impurity complexes [10], do not play a role here. Therefore, we propose that the retarded diffusion of boron in the mixed crystals should rather be explained by the reduction of lattice strain.

The increase of the amplitude of the $\nu_{-1\leftrightarrow1}$-resonance of the boron–germanium pair resembles the annealing curves of ^{12}B in silicon [11] and may be explained by processes which are proposed to be the basis for boron diffusion in silicon [12]. In Figure 3 this concept is illustrated for boron in *n*-silicon (doping: 1.5×10^{18} cm^{-3} Sb). The resonance signal of the substitutional fraction is small at room temperature, but gradually increases with temperature to reach

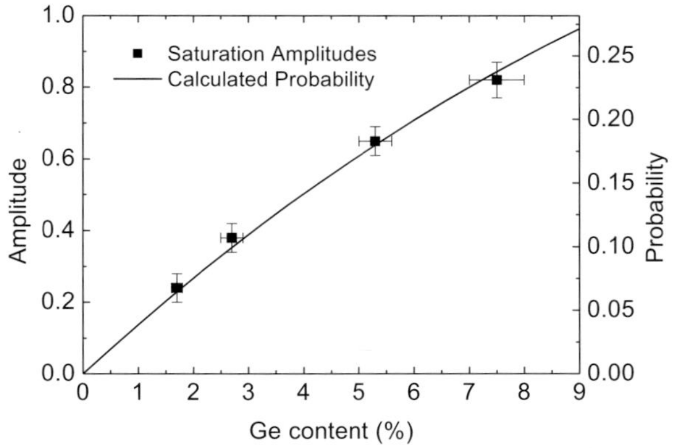

Figure 2. Measured saturation amplitudes of the boron–germanium pair in differently composed samples in comparison to the statistical probabilities (figure similar to [8]). The comparison was made possible by adjusting both scales at the crystal containing 7.5% Ge.

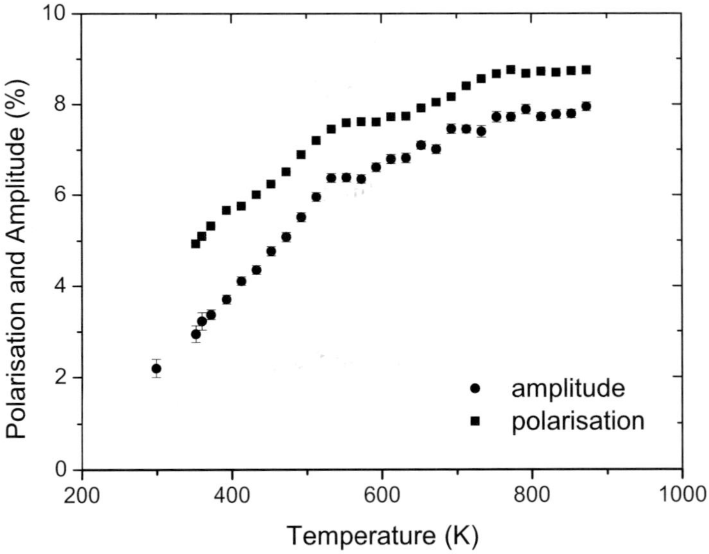

Figure 3. Temperature behaviour of the amplitude of the Larmor resonance and the β-decay asymmetry of boron after implantation in *n*-silicon.

the saturation value of about 8% at $T = 900$ K. A similar temperature behaviour is seen for the total asymmetry of the β-decay of ^{12}B. In the diffusion model, the loss of asymmetry is attributed to fluctuating quadrupolar fields. As the temperature is raised, the dwindling influence of the fluctuating fields gives rise to the increase in asymmetry and in resonance amplitude.

As for the diffusion process itself, boron is regarded to be an interstitial diffuser [5]. Interstitial boron has been identified in earlier β-NMR measurements [13, 14]. Here the electric field gradient of $V_{zz} = -11.2(8)$ V/Å^2, which points along the ⟨111⟩ direction, was correlated to the configuration which was first proposed by Tarnow [15] and strongly supported by DFT calculations [16]. This interstitial B_i^+-complex consisted of a negatively charged boron and a doubly positively charged interstitial silicon attached to it. The interstitial boron defines a negative-U system giving room to geometrically varied configurations after charge transfers [17]. Boron takes positions off the ⟨111⟩-axis in the B_i^0-complex and the negatively charged interstitial boron B_i^- -complex. In consequence of this, boron is expected to have field gradients with non-cylindrical symmetry in both configurations and, in addition, is prone to reorientation jumps which will cause the fluctuating quadrupolar fields.

The experimental group of H. Ackermann in Marburg/Germany found evidence for two kinds of reorientation jumps, when they examined interstitial boron in differently doped silicon crystals as function of temperature and impact of light [12, 18–20]. According to the analysis, the slow reorientation jumps of the B_i^+-configurations between the possible four ⟨111⟩-directions, which start at about room temperature, do not initiate the broadening of resonances and the loss of β-decay asymmetry which is seen in the spectra. But, the reorientation jumps of the B_i^0- and B_i^--configurations were found to be capable to cause the rapidly fluctuating quadrupolar fields. As a consequence, the authors developed a model of diffusion which includes reorientation jumps and charge transfers between positive, neutral and negative boron interstitials as the steps on boron's way to the final occupation of substitutional sites.

The resonance of the interstitial boron complex (Tarnow configuration) has been found in all of our silicon-rich mixed crystals but not in pure germanium. The amplitudes of the resonances were lower in samples with higher Ge-contents. In the spectra for boron in the mixed crystals with relatively low Ge-content, we also see a loss of β-asymmetry and broadening of resonances at certain temperatures. This finding promotes confidence to extend the above model to boron diffusion in the silicon–germanium mixed crystals. Concluding this contribution we like to point out that, to our knowledge, this specific diffusion process has been detected only by the β-NMR technique.

References

1. Kuo P., Hoyt J. L., Gibbons J. F., Turner J. E. and Lefforge D., *Mater. Res. Soc. Symp. Proc.* **379** (1995), 373.
2. Nylandsted Larsen A., *Mater. Res. Symp. Proc.* **532** (1998), 187.
3. Rajendran K. and Schoenmaker W., *J. Appl. Phys.* **89** (2001), 980.
4. Lever R. F., Bonar J. M. and Willoughby A. F. W., *J. Appl. Phys.* **83** (1998), 1998.
5. Zangenberg N., PhD thesis, University of Aarhus, Aarhus (Denmark), 2002.
6. Hattendorf J., Zeitz W.-D., Schröder W. and Abrosimov N. V., *Physica, B* **340-342** (2003), 858.
7. Blaha P., Schwarz K. and Luitz J., WIEN97, Techn. Universität Wien, Austria, 1999, ISBN 3-9501031-0-4.

8. Bohne W., Röhrich J. and Röschert G., *Nucl. Instrum. Methods Phys. Res., B* **136-138** (1998), 633.
9. Jaccarino V. and Walker L. R., *Phys. Rev. Lett.* **15** (1965), 25.
10. Wichert T., Identification of defects in semiconductors, In: Stavola M. (ed), *Semicondutors and Semimetals*, Vol. 51A, Academic, San Diego, 1999, p. 297.
11. Metzner H., Sulzer G., Seelinger W., Ittermann B., Frank H.-P., Fischer B., Ergezinger K.-H., Dippel R., Diehl E., Stöckmann H.-J. and Ackermann H., *Phys. Rev.* **B42** (1990), 11419.
12. Peters D., PhD thesis, University of Marburg, Marburg (Germany), 2001.
13. Fischer B., Seelinger W., Diehl E., Ergezinger K.-H., Frank H.-P., Ittermann B., Mai F., Welker G., Ackermann H. and Stöckmann H.-J., *Mat. Sci. Forum* **83-87** (1992), 269.
14. Hattendorf J., Zeitz W.-D., Abrosimov N. V. and Schröder W., *Physica* **308-310** (2001), 535.
15. Tarnow E., *Europhys. Lett.* **16** (1991), 449.
16. Hakala M., Puska M. J. and Nieminen R. M., *Phys. Rev.* **B61** (2000), 8155.
17. Harris R. D., Newton J. L. and Watkins G. D., *Phys. Rev.* **B36** (1987), 1094.
18. Fischer B., Seelinger W., Frank H.-P., Diehl E., Ergezinger K.-H., Ittermann B., Mai F., Marbach K., Weissenmayer S., Welker G., Stöckmann H.-J. and Ackermann H., *Nucl. Instrum. Methods* **B80/81** (1993), 201.
19. Frank H.-P., Almeida T., Diehl E., Ergezinger K.-H., Fischer B., Ittermann B., Mai F., Seelinger W., Weissenmayer S., Welker G., Ackermann H. and Stöckmann H.-J., *Hyp. Int.* **79** (1993), 655.
20. Frank H.-P., Diehl E., Ergezinger K.-H., Fischer B., Ittermann B., Mai F., Marbach K., Weissenmayer S., Welker G. and Ackermann H., *Mat. Sci. Forum* **143-147** (1994), 135.

Hyperfine Interactions (2004) 158:299–303
DOI 10.1007/s10751-005-9048-9

Local Structure of Implanted Pd in Si Using PAC

D. A. BRETT[1], R. DOGRA[1,2,*], A. P. BYRNE[2,3], M. C. RIDGWAY[1], J. BARTELS[4] and R. VIANDEN[4]

[1]*Department of Electronic Materials Engineering, Research School of Physical Sciences and Engineering, The Australian National University, Canberra, Australia; e-mail: rak109@rsphysse.anu.edu.au*

[2]*Department of Nuclear Physics, RSPhysSE, ANU, Canberra, Australia*

[3]*Department of Physics, Faculty of Science, ANU, Canberra, Australia*

[4]*Institut für Strahlen- und Kernphysik, Universität Bonn, Bonn, Germany*

Abstract. TDPAC has been employed to study the local structure of implanted palladium in silicon utilizing 87–75 keV γ–γ cascade of probe nucleus ^{100}Pd. The observed hyperfine parameters revealed the presence of Pd–V defect pair only in highly doped n-type silicon. A dumbbell structure with substitutional palladium and silicon vacancy as nearest neigbor is suggested for this defect.

Key Words: ion implantation, PAC, transition metals in silicon.

1. Introduction

The transition element impurities such as Pd, Ag, Pt, Au, etc., in semiconductors, present as contaminants or introduced intentionally for technological importance, have attracted a great deal of attention because of their deep levels in the band gap that may trap charge carriers and change the electrical conductivity. These impurities in silicon form an interesting class of defects because of their d-shell electrons. The present understanding of these impurities in silicon is mainly based on a variety of different measurements including EPR, DLTS, PL, RBS/C, Mössbauer measurements and some theoretical calculations. EPR has proven to give detailed information about the structure of deep centers and several isolated impurity-related defects. The ionic and vacancy models [1, 2] were proposed to explain the electronic structure corresponding to substitutional Pd^- or Pt^- in silicon lattices, but only the vacancy model could predict successfully the deep levels associated with transition metals at the end of the d-series. Recent DLTS measurements [3] have revealed that the substitutional palladium impurity in n-type silicon exhibits a single acceptor state (Pd^-) whereas in p-type it occurs in the single as well as in the double donor state ($Pd^{+/++}$). To understand the

* Author for correspondence. On leave from College of Engineering & Technology, Gurdaspur, INDIA–143521.

mechanism that gives rise to the origin of these properties of Pd in n- and p-type silicon, especially their local structure and symmetry, we have used time differential perturbed angular correlation (TDPAC) technique using the radioactive probe nucleus ^{100}Pd.

2. Experimental details

The probe atoms ^{100}Pd $\rightarrow ^{100}$Rh are produced with the 14UD heavy-ion accelerator facility at the ANU via the nuclear reaction ^{92}Zr(^{12}C, 4n)^{100}Pd using a 12C ion beam with an energy of 70 MeV. The reaction subsequently recoil implants the Pd nuclei with an energy up to 8 MeV up to 3 μm deep into the single crystals of silicon (50-μm thick). For present measurements we have used the following single crystals grown along $\langle 100 \rangle$ axis: standard n-type ($\times 10^{15}$ P/cm^3) and p-type ($\times 10^{15}$ B/cm^3), B-doped Si ($\times 10^{19}$/cm^3), P-doped Si ($\times 10^{19}$/cm^3), As-doped Si ($\times 10^{19}$/cm^3) and Sb-doped Si ($\times 10^{18}$/cm^3). In order to remove the radiation damage caused by the implantation, the samples were annealed under nitrogen atmosphere for 30 min in the temperature range 100–600°C. The coincident counts were recorded at room temperature in a conventional setup using the 84–75 keV cascade. The $R(t)$ spectra were fitted with a three-site model: (a) the probe atoms corresponding to unperturbed lattice sites (b) the probe atoms belonging to damaged zones resulting in an exponential fast decay of amplitude, and (c) probe atoms with well-defined unique interaction parameters having various distant defects in the first coordination shells.

3. Results and discussion

The perturbation functions and least squares fits, along with their respective Fourier transforms are shown in Figure 1(a) for the highly P-doped samples. These indicate the presence of a defect in proximity to the probe with well resolved interaction frequency. Figure 1(b) also shows a gradual increase of modulation amplitude with annealing temperature that reaches maximum in the vicinity of 250°C. In all the highly doped n-type samples studied here, we have observed the same quadrupole interaction frequency $\nu_Q \approx 13.1(2)$ MHz at ^{100}Rh probe nuclei whereas highly perturbed spectra have been observed in standard n-, p-type and highly B-doped silicon. The observed zero value of asymmetry parameter, η in highly doped n-type silicon at all annealing temperatures implies that the EFG principal axis at observed site is aligned with the crystallographic axes. Moreover, the <111> orientation of the EFG in all highly doped n-type silicon is observed. The strong similarity of PAC spectra in highly P-, As- and Sb-doped silicon indicates the same defect complex formation in the nearest neighborhood of ^{100}Rh probe nuclei. With increasing annealing temperature, the amplitude of modulation starts decreasing and ultimately diminishes around

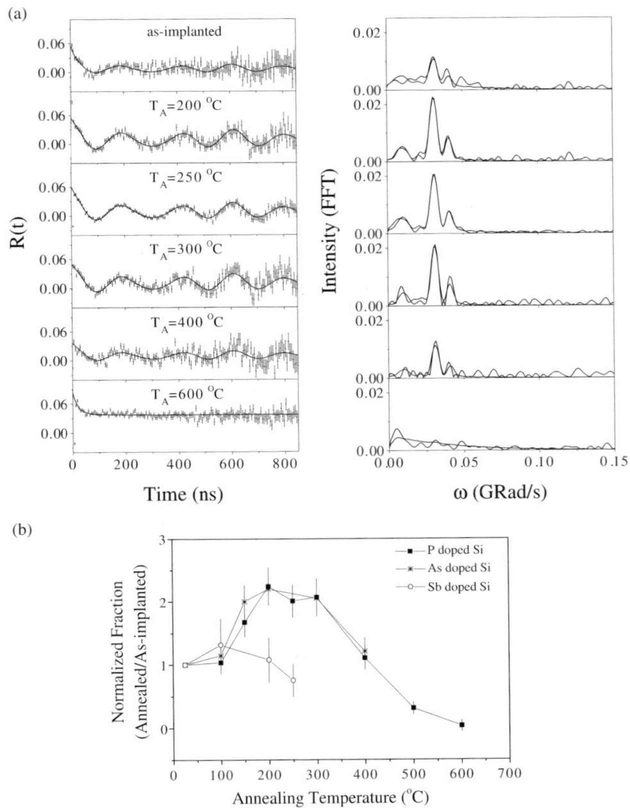

Figure 1. (a) Fitted *R*(*t*) spectra and Fourier transforms for ^{100}Pd in highly P-doped Si. (b) Normalized site fraction as a function of annealing temperature for highly doped n-type Si.

600°C, This corresponds to dissociation of defect complex which is accompanied by an increase of undisturbed fraction of the probes sitting in perfect cubic symmetry. Figure 1(b) shows the normalized population of probe atoms experiencing a unique interaction frequency as a function of annealing temperature in highly doped n-type silicon. In P- and As-doped silicon, the fraction of probe atoms almost doubles around an annealing temperature of 200–300°C, whereas in Sb-doped silicon it decreased rapidly in comparison to as-implanted one. Using first-order kinetics [4], as used for copper in silicon, the activation energy for dissociation of defect complex with Pd impurity in highly doped n-type silicon is estimated to be 2.5(7) eV. The broad range of annealing suggests the retrapping of defects at the probe site.

The lattice locations of the implanted atoms depend on the dose, energy, temperature and the impurity-host atom combinations. Arguments based exclusively on the consideration of charges and radii of host and probe atoms do not always provide unambiguous information about the probe site location. In the past, it has been realized that there are some properties and processes which determine the final location of the implanted atoms in semiconducting materials: size effects, chemical effects, impurity–vacancy complex formation and annealing effects. The observed unique interaction frequency could be assigned to the Pd-defect or Pd-dopant pairs. We have ruled out the possibility of Pd-dopant pairs because of different hybridization between d-shell states of Pd and 3p-, 4p- and 5p-shell states respectively of P, As and Sb. The increased hybridization of d–p shells would have resulted into different lattice relaxation as we move from P–As–Sb and hence the increased trend of interaction frequencies, the observed frequency is same in all highly doped n-type silicon. We attribute our results in highly n-type silicon to the formation of Pd–V complex. The observed symmetric EFG is compatible with the structure of dumbbell consisting of substitutional palladium probe and silicon vacancy on a nearest neighbor lattice site. In such a complex the orientation of principal component of EFG is along the bond center, i.e. <111>, and asymmetry is zero, consistent with the observations. Figure 1(a and b) reveal that at annealing temperature of 600°C, the defect complex Pd–V dissociates and Pd occupies substitutional site with four silicon atoms as nearest neighbors. These results are supported by DLTS measurements in n-type silicon [5] where transfer of an off-center Pd atom to a substitutional site in the silicon lattice on annealing has been observed. The lower probe fraction in Sb-doped sample may be due to the fact that the Sb concentration is lower than P- and As-doped silicon that potentially yields a lesser fraction of charged vacancies and/or palladium ions. The highly disturbed PAC spectra in standard (n- and p-type) and highly p-type Si may be due to randomly distributed ^{100}Pd atoms in the silicon lattice. This kind of effect has recently been observed for Cu and Ag implantation in doped silicon by emission channeling [6, 7].

4. Conclusions

The observed behavior of implanted palladium in silicon is attributed to Pd–V complex in highly doped n-type silicon which is characterized by unique quadrupole interaction frequency of $\nu_Q = 13.1(2)$ MHz and a <111> orientation of the principal component of the associated axially symmetric EFG. The present results are explained by the formation of a dumbbell consisting of substitutional palladium and silicon vacancy as nearest neighbor in the silicon lattice.

References

1. Ludwig G. W. and Woodbury H. H., *Phy. Rev. Lett.* **5** (1960), 98.
2. Watkins G. D., *Physica B* **117–118** (1983), 9. Watkins G. D. and Williams P. M., *Phys. Rev. B* **52** (1995), 16575.
3. Sachse J.-U., Jost W., Weber J. and Lemke L., *Appl. Phys. Lett.* **71** (1997), 1379 and references therein.

4. Wahl U., Vantomme A., Langouche G., Correia J. G. and ISOLDE Collaboration, *Phys. Rev. Lett.* **84** (2000), 1495.
5. Wang L., Yao X. C., Zhou J. and Qin G. G., *Phys. Rev. B* **38** (1988), 13494.
6. Wahl U., Vantomme A., Langouche G., Araujo J. P., Peralta L. and Correia J. G., *Appl. Phys. Lett.* **77** (2000), 2142.
7. Wahl U., Correia J. G., Vantomme A. and ISOLDE Collaboration, *Nucl. Instr. Meth. Phys. Res. B* **190** (2002), 543.

Hyperfine Interactions (2004) 158:305–308
DOI 10.1007/s10751-005-9056-9

Polymorphic Phase Transformation in In_2La and $CeIn_2$

EGBERT R. NIEUWENHUIS, AURÉLIE FAVROT, LI KANG, MATTHEW O. ZACATE[a] and GARY S. COLLINS*
Department of Physics, Washington State University, Pullman, WA 99164, USA; e-mail: collins@wsu.edu

Abstract. Nuclear quadrupole interactions of ^{111}Cd probes in In_2La and $CeIn_2$ were measured using perturbed angular correlation of gamma rays (PAC). Near room temperature, a single non-axially symmetric quadrupole interaction was observed in each compound, with $\omega_0 = eQV_{zz}/\hbar =$ 78.8(2) Mrad/s and $\eta = 0.312(1)$ for In_2La at 11 °C and $\omega_0 = 80(1)$ Mrad/s and $\eta = 0.29(2)$ for $CeIn_2$ at 34 °C. The observed non-axial symmetry is consistent with the reported $CeCu_2$ structure of the phases. The non-axially symmetric interactions were replaced completely by axially symmetric interactions ($\eta = 0$) at 524 °C, with $\omega_0 = 101.0(1)$ Mrad/s in In_2La and $\omega_0 = 96.9(4)$ Mrad/s in $CeIn_2$. The change in symmetry is attributed to a polymorphic phase transformation. Based on symmetry information from the quadrupole interaction and on chemical arguments, it is proposed that high-temperature phases of In_2La and $CeIn_2$ both have the AlB_2 (C32) structure.

Key Words: PAC, phase transformation, polymorph, quadrupole interaction.

1. Introduction

In_2La and $CeIn_2$ compounds have the $CeCu_2$ structure at room temperature [1] and reportedly maintain this structure up to their melting points [2]. Using the method of perturbed angular correlation of gamma rays (PAC), the quadrupole interaction at nuclei of ^{111}Cd probes on the In sublattice has been measured in each compound. Quadrupole interactions measured near room temperature were found to be non-axially symmetric, as expected for the reported $CeCu_2$ structure. However, quadrupole interactions measured at 524 °C were found to be axially symmetric in both compounds and, moreover, to have very similar parameters. This is taken to indicate the existence of a polymorphic phase transformation in both systems to a previously unknown high temperature structure of $CeIn_2$ and In_2La.

---

[a] Present address: Physics and Geology Department, Northern Kentucky University, Highland Heights, KY 41099, USA

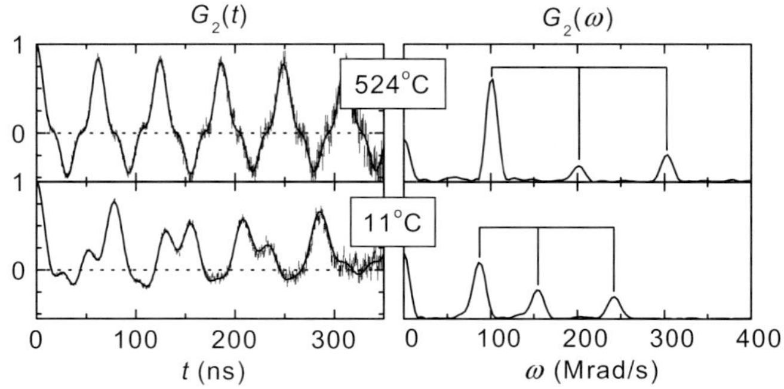

Figure 1. PAC spectra and Fourier transforms of ^{111}Cd in In_2La at indicated temperatures.

A previous report of measurements on In_2La and experimental methods can be found in [3]. Representative PAC spectra of In_2La from that work are shown in Figure 1 for measurements at 11 °C (bottom) and at 524 °C (top). A single quadrupole interaction is observed at 11 °C with $\omega_0 = eQV_{zz}/\hbar = 78.8(2)$ Mrad/s and $\eta = 0.312(1)$, in which ω_0 is the quadrupole interaction frequency and η is the asymmetry parameter of the electric field gradient tensor. At 524 °C, this signal is completely replaced by an axially symmetric signal with $\omega_0 = 101.0(1)$ Mrad/s. At intermediate temperatures, both signals were observed, and site fractions exhibited hysteresis between 200 and 450 °C characteristic of a first order phase transformation. Changes in the quadrupole interactions were reversible when cycling up and down in temperature. More detailed information about site fractions and temperature dependencies of hyperfine parameters can be found in [3].

2. Experiment

For the present work, a Ce–In sample was prepared having a composition of $28.0^{+0.9}_{-1.8}$ at.% Ce, which is in the two-phase field of the Ce–In phase diagram between $CeIn_2$ and $CeIn_3$. Representative spectra are shown in Figure 2 for measurements at 34 °C (bottom) and at 524 °C (top). Two quadrupole interaction signals were observed at 34 °C. One signal had a site fraction of 0.73(4) with $\omega_0 = 73.4(1)$ Mrad/s and was axially symmetric ($\eta = 0$), which is in excellent agreement with the signal previously observed for ^{111}Cd in $CeIn_3$ [4]. The second signal had quadrupole interaction parameters $\omega_0 = 80(1)$ Mrad/s and $\eta = 0.29(2)$, and is attributed to ^{111}Cd on the In site in $CeIn_2$. It can be seen that these latter values are very similar to those observed for In_2La near room temperature cited above. At 524 °C, the signal for $CeIn_3$ is still present, with a frequency slightly lower than that observed near room temperature, which is attributed to thermal expansion, while the signal for $CeIn_2$ is completely replaced by an axially

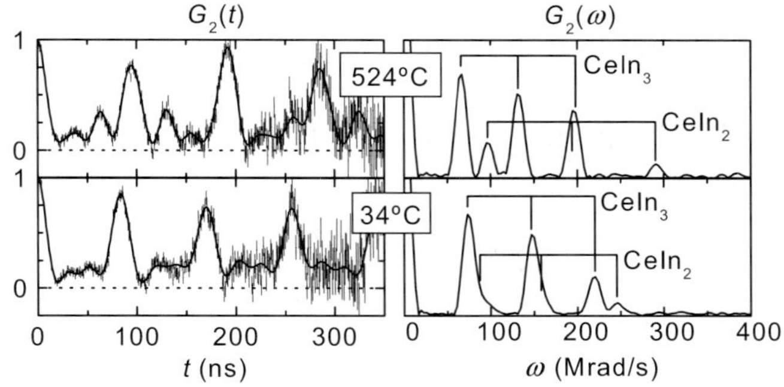

Figure 2. PAC spectra and Fourier transforms at indicated temperatures of ^{111}Cd in a sample containing a mixture of $CeIn_2$ and $CeIn_3$ phases.

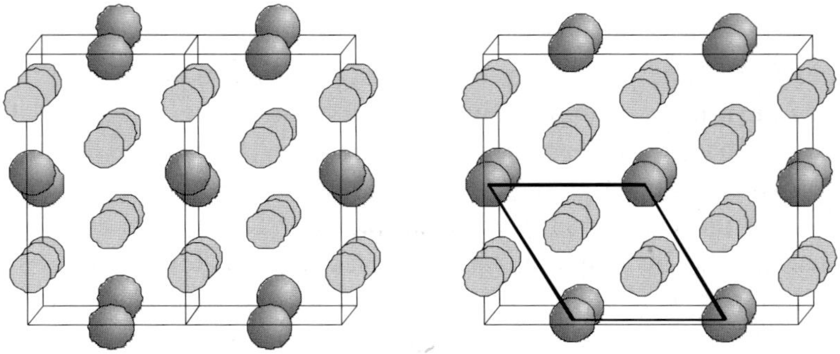

Figure 3. Crystal structures of $CeCu_2$ (*left*) and AlB_2 (*right*), with Ce and Al atoms shown in *light grey* and Cu and B in *dark grey*. On the left, some atoms lying outside the unit cells have been drawn to make comparison easier (adapted from [5]).

symmetric signal having $\omega_0 = 96.9(4)$ Mrad/s. All signals exhibited negligible inhomogeneous broadening.

3. Discussion and conclusions

Observation of an axially symmetric quadrupole interaction at elevated temperature can only come from probes in a structure that differs from $CeCu_2$. In principle, point defects formed at elevated temperature might disturb quadrupole interactions at probe nuclei in the $CeCu_2$ structure, leading to new signals. In such a case, one would observe one or more of the following: inhomogeneous broadening, multiple signals due to different local arrangements of defects, and signals with $\eta > 0$. The PAC signals observed at high temperature are inconsistent with any of these characteristics, and therefore are attributed to probes in a new polymorph with symmetry higher than in the $CeCu_2$ structure. More specifically, observation of $\eta = 0$ requires that there be a single 3-, 4- or 6-fold axis of symmetry at the lattice location of the probe. Because quadrupole interactions of ^{111}Cd probes are observed immediately after decay of parent ^{111}In activity, the probes are located at In lattice locations.

A survey of all common MX_2 compounds (M = metal; X = group IB, IIB, IIIB, IVB elements) listed in Pearson's handbook [1] showed that only four structures have symmetries at the X site that give $\eta = 0$: AlB_2, Cd_2Cd, $CaIn_2$ and $MoSi_2$ [3]. Of these four structures only the AlB_2 structure has been observed for MX_2 compounds in which M and X are both trivalent elements. Therefore, the high-temperature polymorph is believed to have the AlB_2 structure [3]. The $CeCu_2$ and AlB_2 crystal structures are compared in Figure 3. As can be seen, the $CeCu_2$ structure (left) arises through slight distortions of the hexagonal AlB_2 structure (right) that destroy the 3-fold rotational symmetry around the B-site.

The very close values of quadrupole interaction parameters in $CeIn_2$ and In_2La in both LT and HT phases suggest that the phase transformations are of the same type in both systems. Attempts to determine the structure of the high temperature phase of In_2La by X-ray powder diffraction failed due to extremely rapid out-diffusion and oxidation of La [3]. No similar attempt was made for $CeIn_2$ because the same result was expected. However, PAC measurements were readily carried out over weeks using samples in ingot form under vacuum in a pressure less than 5 μPa.

Acknowledgements

This work was supported in part by the National Science Foundation under grant DMR 00-91681 (Metals Program). ERN is grateful to the Marco Polo Funds for support.

References

1. Villars P. and Calvert L. D. (eds.), *Pearson's Handbook of Crystallographic Data for Intermetallic Phases*, 2nd edition, ASM International, Materials Park, Ohio, 1991.
2. Massalski T. B. (ed.), *Binary Alloy Phase Diagrams*, ASM International, Materials Park, Ohio, 1990.
3. Nieuwenhuis E. R., Collins G. S., Favrot A. and Zacate M. O., *Journal of Alloys and Compounds* **387** (2005), 20.
4. Schwartz G. P. and Shirley D. A., *Hyperfine Interact.* **3** (1977), 67.
5. Svoboda P., Doerr M., Loewenhaupt M., Rotter M., Reif T., Bourdarot F. and Burlet P., *Europhys. Lett.* **48**(4) (1999), 410–414.

Hyperfine Interactions (2004) 158:309–312
DOI 10.1007/s10751-005-9057-8

Static and Intermittent Hyperfine Coupling for the Muoniated Radical in Tellurium

S. F. J. COX[1,2,*], J. S. LORD[1], N. SULEIMANOV[3,4], U. ZIMMERMANN[3] and I. D. REID[3,5]
[1]*ISIS Facility, Rutherford Appleton Laboratory, Chilton, OX11 0QX, UK*
[2]*Condensed Matter and Materials Physics, University College London, London, UK*
[3]*Swiss Muon Source, Paul Scherrer Institute, Villigen, Switzerland*
[4]*Zavoisky Physical – Technical Institute, Kazan, Russia*
[5]*Brunel University, London, UK*

Abstract. The hyperfine constant for muonium defect centres in elemental tellurium, measured spectroscopically at low temperature, corresponds to 7% of the free-atom value. The centres are tentatively identified as the diatomic species TeMu, analogous to the hydroxyl radical, OH. Their spectrum disappears as the centres ionize around 80 K but the muon spin response indicates that the same centres form and reionise, rapidly and repeatedly, above 200 K. The capture and loss of charge carriers provide a model for the electrical activity of hydrogen impurity in this low-gap semiconductor.

Key Words: hydrogen, muonium, tellurium.

Nomenclature:
Mu – Chemical symbol for muonium, light pseudo-isotope of hydrogen;
$\gamma_\mu = 2\pi \times 136\text{MHz/T}$ – Muon gyromagnetic ratio

Level crossing resonance reveals the formation of muonium centres following implantation of positive muons into elemental tellurium. This spectroscopic detection of a neutral paramagnetic state is remarkable in a material with such a low band gap, often classed amongst the semimetals. Figure 1 shows the resonant loss of muon polarization as magnetic field is scanned through $B_{\text{res}} \approx 1.2$ T, defining the hyperfine constant as $A \approx (\gamma_\mu/\pi)B_{res} \approx 330\text{MHz}$. The spectrum was recorded on the ALC instrument at PSI; decoupling measurements made at lower longitudinal fields on the EMU instrument at ISIS corroborate this value of the hyperfine constant (defining indeed the search range for the resonance of Figure 1) and indicate that just over 50% of the incoming muons pick up electrons to form such a state in Te at low temperatures.

* Author for correspondence.

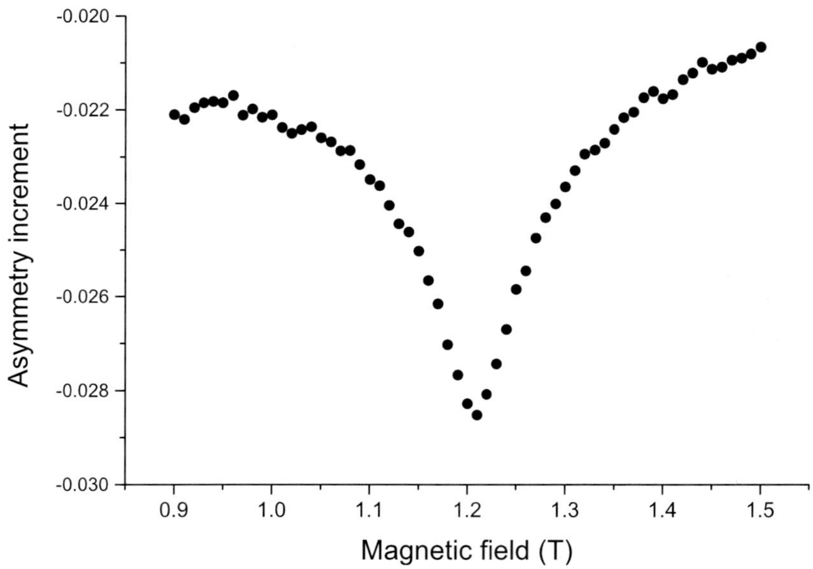

Figure 1. Level crossing resonance for the muonium radical in Te, recorded at 30 K via time-integral measurements of the muon decay asymmetry (PSI data).

Although this value is just 7% of the hyperfine constant for atomic muonium in its free state (A_0 = 4.5 GHz), it is nonetheless surprisingly high for a defect centre in a semiconductor with such a low electron effective mass ($m_n \approx 0.05$) and high dielectric constant ($\varepsilon \approx 35$) as tellurium. The expectations of effective mass theory are that electrons would scarcely be bound to unit charge defects such as the interstitial proton or muon represent: hydrogenic or shallow-donor orbits are dilated in radius by a factor $\varepsilon/m_n \approx 700$ and the atomic volume by the cube of this factor, so that spin density at the central proton or muon would be reduced from the free-atom (H or Mu) values by many orders of magnitude. The observed hyperfine constant is indicative instead of a compact orbital, i.e., a deep-level defect state.

Despite the order of magnitude mass ratio, $m_{\text{Mu}}/m_{\text{H}} \approx 1/9$, muonium serves as a well-behaved isotope and mimics hydrogen reliably in its solid state chemistry. The muonium centre in Te is evidently neither hydrogen-like (atomic) nor hydrogenic (shallow donor) in character but it does have counterparts in sulphur and selenium, i.e., in the other Group-VI elements which are also semiconductors, albeit with considerably higher bandgaps. These were likewise detected via level crossing resonance and assigned, somewhat controversially, to the diatomic radicals SMu and SeMu, formed by an abstraction reaction with the host chalcogen [1, 2]. By analogy, the muonium centre in tellurium would correspond to TeMu. All are envisaged as isolectronic with the hydroxyl radical, OH.

The resonance in tellurium is visible up to about 80 K, whereupon it disappears quite abruptly. Either electrons are then no longer captured or the neutral centres form on implantation but quickly ionize. Above about 200 K, significant spin

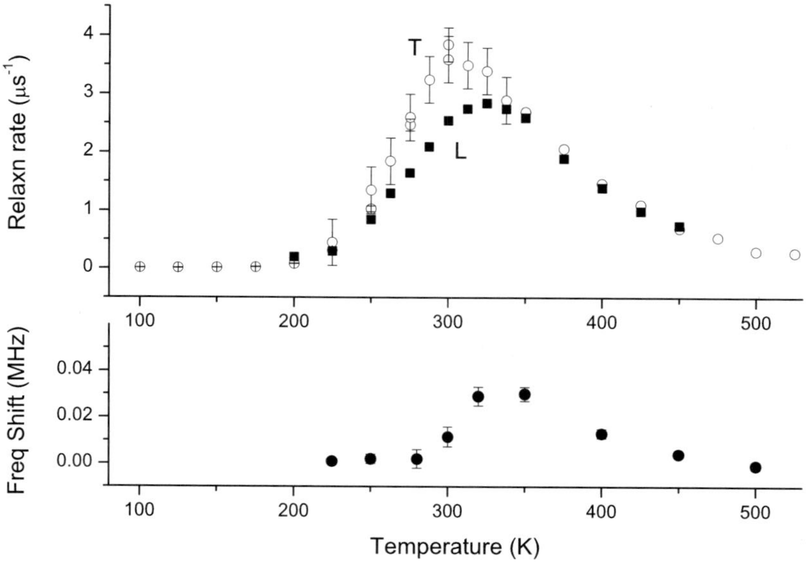

Figure 2. Transverse (T) and longitudinal (L) relaxation rates, both measured at 10 mT (combined ISIS and PSI data), together with the precession frequency shift, measured with respect to the muon Larmor frequency at 55 mT (PSI data).

relaxation sets in, visible both as exponential damping of the precession signal in transverse fields and as a simple exponential decay of the (non-oscillatory) signal in longitudinal field. Both types of relaxation were noted in the early literature [3] but not explored above room temperature. In the present work, we have measured these to sufficient precision to note a significant difference between the respective relaxation rates in this onset régime, the transverse rate being the higher. We find also that both rates peak just above room temperature; thereafter they become essentially identical as they decrease to higher temperatures. A compendium of data to above 500 K, from cryostat and furnace measurements made at both ISIS (EMU) and PSI (GPD) is given in Figure 2.

Although the frequency of muon spin rotation in tellurium lies close to the muon Larmor frequency (appropriate to muons in electronically diamagnetic environments) a significant shift or increase is noticeable, as also shown in Figure 2. This combination of the two types of spin relaxation with an accompanying paramagnetic shift of the precession frequency is characteristic of intermittent hyperfine coupling. It indicates a charge-exchange régime in which short-lived paramagnetic centres are reformed and reionized repeatedly, via capture and loss of charge carriers (whether electrons or holes). Such a régime is well known for muonium in silicon, but only at rather higher temperatures consistent with the higher band gap [4]. Field-dependences of the data are invaluable for estimating the instantaneous hyperfine constant involved here, as well as transition rates into and out of the paramagnetic state. Figure 3 shows these for tellurium, fitted numerically with a density-matrix model due to one of

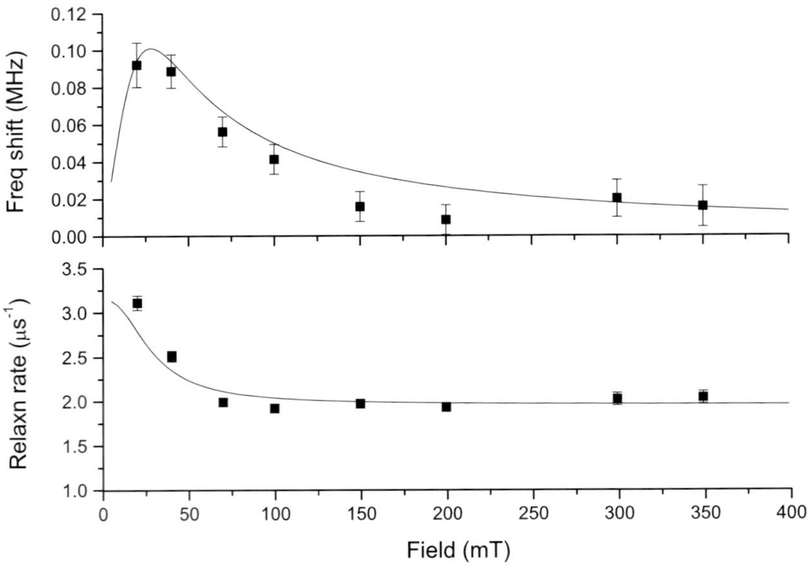

Figure 3. Fitted field dependences of transverse relaxation rate and frequency shift, at 330 K (PSI data).

us (JSL), as described elsewhere [4]. For the conditions of Figure 3, i.e., at 330 K, the lifetime of the paramagnetic state is determined as 280 ps; the average occupancy or *neutral fraction* is 0.7% and there are roughly 50 charge-exchange cycles per muon lifetime $(\tau_\mu = 2.2\mu s)$. Most significantly, the hyperfine constant involved is fitted as 340 ± 90 MHz – entirely consistent with the low-temperature spectroscopic value deduced from Figure 1.

It remains to analyse the data of Figure 2 for the variation of these transition rates with temperature. Roughly speaking, the paramagnetic lifetime is long compared with a hyperfine period to the left of the peaks, short to the right. Preliminary results indicate that activation energies of at least half the bandgap are involved. It may prove possible to distinguish electron and hole processes, i.e. deep-level donor and acceptor functions, or a sequence such as the successive capture of electrons and holes, in which case hydrogen would by inference act as a recombination centre in tellurium.

References

1. Cox S. F. J., Reid I. D., McCormack K. L. and Webster B. C., *Chem. Phys. Lett.* **273** (1992), 179.
2. Reid I. D., Cox S. F. J., Jayasooriya U. A. and Zimmermann U., *Physica B* **326** (2003), 89.
3. Gurevitch I. I. and Ivanter I. G. *et al.*, *Hyperfine Interact.* **6** (1979), 175.
4. Lord J. S. and Cox S. F. J. *et al.*, *J. Phys., Condens. Matter* **16** (2004), S4739, plus references therein.

Hyperfine Interactions (2004) 158:313–316
DOI 10.1007/s10751-005-9058-7

Hyperfine Parameters for Muonium in Copper (I), Silver (I) and Cadmium Oxides

S. F. J. COX[1,2,*], J. S. LORD[1], S. P. COTTRELL[1], H. V. ALBERTO[3], J. M. GIL[3], J. PIROTO DUARTE[3], R. C. VILÃO[3], A. KEREN[4] and D. PRABHAKARAN[5]
[1]*ISIS Facility, Rutherford Appleton Laboratory, Chilton, OX11 0QX, UK*
[2]*Condensed Matter and Materials Physics, University College London, London, UK*
[3]*Department of Physics, University of Coimbra, Coimbra, Portugal*
[4]*Technion, Haifa, Israel*
[5]*Department of Physics, University of Oxford, Oxford, UK*

Abstract. Muonium centres in Cu_2O, Ag_2O and CdO show hyperfine parameters spanning four orders of magnitude. They exemplify the three different categories of hydrogen defect centre in semiconducting and dielectric solids, with very different electronic structure and electrical activity, namely quasi-atomic (possibly deep acceptor), deep donor and shallow donor.

Key Words: hydrogen, muonium, oxides.

Following the success of muonium spectroscopy in modelling the structure and electrical activity of hydrogen defect centres in semiconductors, including the elemental group-IV and compound III–V semiconductors as well as the wider-gap II–VI materials, it is timely to apply the technique to oxides, as these materials are being reappraised and developed as novel electronic materials. The overall systematics show three distinct categories of defect centre, as described in an accompanying paper [1], and we present here the latest examples of each type to be discovered, in Cu_2O, Ag_2O and CdO. Our measurements were made using the EMU and DEVA instruments on the ISIS pulsed muon source.

For Cu_2O and Ag_2O, the character of the centres is best displayed using the longitudinal-field techniques known as repolarization and level-crossing resonance, as in Figure 1. These involve measuring time-average muon polarization as the magnetic field (applied parallel to the polarization of the incoming muon beam) is scanned through regions where the electron, muon and nuclear spin states are mixed or perturbed by the hyperfine and superhyperfine interactions. For Cu_2O, the repolarization is monotonic, representing a simple decoupling of these interactions. The muon-electron hyperfine interaction is fitted from this

* Author for correspondence.

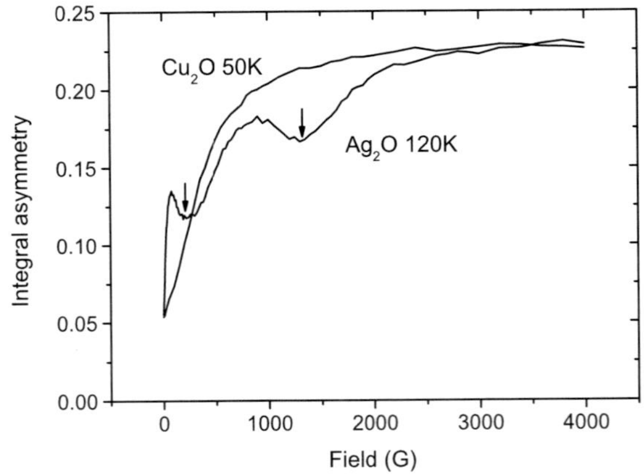

Figure 1. Repolarization curves for muonium in Cu_2O and Ag_2O. The decoupling field for Cu_2O is roughly one third that of atomic muonium; for Ag_2O it is lower still, with level crossing resonances (*arrows*) defining the hyperfine parameters precisely.

curve as 1276 ± 20 MHz, which is just 29% of the free-muonium value (4463 MHz). This is amongst the lowest values for the trapped-atom or quasi-atomic defect centres but is close to those for muonium in the cuprous halides, which span 27% to 37%. The low value suggests a considerable delocalization of spin density onto the surrounding ions, possibly involving transfer to the copper 3d orbitals.

For Ag_2O, the steep initial repolarization indicates a much smaller muon hyperfine constant, but the curve is modified by the appearance of two striking resonances. The higher-field feature we take to represent a resonance of the type denoted $\Delta M = 1$, in which the external field may be said to tune out parallel components of the hyperfine field, leaving the muon spin free to precess in transverse components [2]. The contact interaction is determined rather precisely as 37 ± 0.2 MHz, though its sign is ambiguous: it may be that such a low value represents spin polarization of bonding orbitals. Detection of the resonance implies a degree of anisotropy that may be taken as evidence of directional bonding. The lower-field resonance could conceivably represent a second muonium centre with correspondingly smaller hyperfine constant. Alternatively, it is a resonance of type $\Delta M = 0$ from the same centre. In this case, it represents a flip-flop exchange of polarization between the muon and neighbouring silver nuclei, mediated by their separate interactions with the unpaired electron.

Whereas the centre in Cu_2O appears to be of the quasi-atomic type and is stable up to at least room temperature, that in Ag_2O has hyperfine parameters reminiscent of bond-centred muonium in Si, Ge, GaAs and GaP, where it behaves as a deep donor. Supporting this analogy, the repolarization and resonance features disappear between 150 and 250 K as the unpaired electron is lost by ionization. Figure 2 shows the growth of the muon Larmor precession signal in this régime as neutral muonium converts to the positive ion (the interstitial

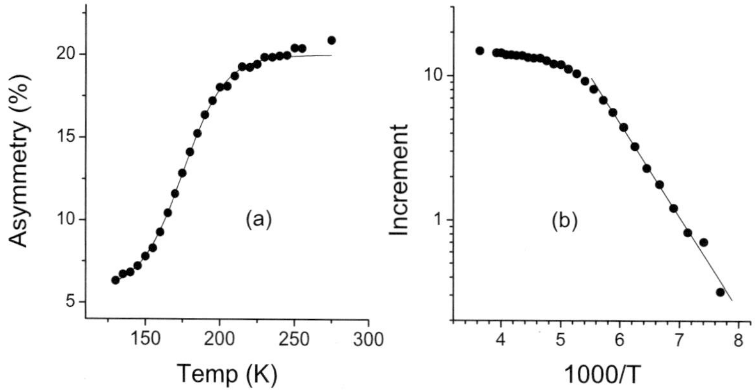

Figure 2. Variation of the diamagnetic or ionic fraction through the ionization régime, for muonium in Ag_2O. Fitted with an equilibrium model in (a), this yields a donor depth of 250 meV; an Arrhenius plot (b), implying direct loss of the electron to the conduction band, gives an ionization energy of half this value.

muon): this signal is measured by the more usual technique of muon spin rotation in transverse magnetic field. Analysis for the donor depth is model-dependent, but lies between 125 and 250 meV.

For CdO, repolarization is complete at much lower fields and the hyperfine parameters are more conveniently determined from the muon spin rotation spectrum, recorded in transverse field. The low-temperature spectrum, obtained by a maximum-entropy transform of the time-domain signal, is shown in Figure 3(a). Satellite lines can be discerned, symmetrically placed about a larger central line. The satellites represent the small proportion of muons which pick up an electron to form paramagnetic muonium, with a hyperfine splitting of about 140 kHz. The central line is the Larmor-precession frequency for muons which fail to pick up an electron on implantation, almost certainly thermalizing as the positive ion. This central line grows at the expense of the satellites in an ionization régime between 50 and 150 K, as shown in Figure 3(b). The evolution is fitted here in the equilibrium model, to give a donor depth of about 90 meV.

In the shallow-donor model, the hydrogenic orbital is dilated both by the dielectric constant of the medium and by the low effective mass of the electron (the latter is weakly bound as a wave packet of conduction- band states). The effective mass is low (0.11) in CdO and the variation of dielectric constant unusually high ($\varepsilon(0) = 21.9$ to $\varepsilon(\infty) = 5.9$). It is normal in effective-mass theory to use the low-frequency value but our results, both for contact interaction and ionization energy, only fit with this model if we use the high-frequency value of dielectric constant. This seems reasonable when ions as heavy as Cd are involved, although other possibilities must be considered, such as association of the muonium with a vacancy or other defect.

It seems likely that the quasi-atomic muonium centre in Cu_2O represents the neutral state of a deep acceptor. However, the acceptor function itself, i.e.,

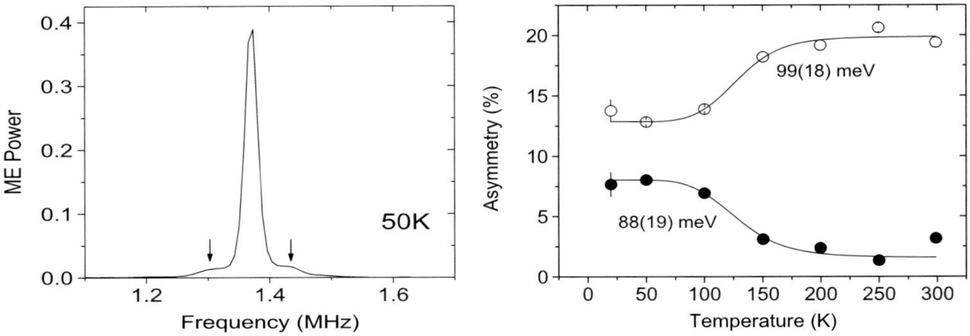

Figure 3. Muon spin rotation spectrum for CdO at 20 K, with hyperfine satellites representing paramagnetic muonium (a) and analysis of the paramagnetic (*lower*) and diamagnetic (*upper*) amplitudes through the ionization régime (b).

conversion to the negative ion, whether by hole ionization or second electron capture, has yet to be demonstrated. The negative ion would be the muonium analogue of the hydride ion, whose importance, or even existence, in oxides is itself a matter of some topical interest. The muonium centre in Ag_2O undoubtedly represents a deep donor whereas that in CdO is a reasonable candidate for an effective-mass shallow donor of the type established in ZnO. (A similar contrast between deep and shallow character has been noted for muonium in HgO and ZnO [3].) The deep donor in Ag_2O and the shallower centre in CdO are both fully ionized by room temperature so that, by implication, hydrogen impurity may act as an inadvertent dopant in these materials, inducing n-type conductivity.

References

1. Cox S. F. J. and Johnson C., (this issue, plus refs therein).
2. Roduner E., in *Muon Science* (Lee S. L., Kilcoyne S. H. and Cywinski R., eds), *Scottish Summer Schools in Physics,* vol. 51 (I.o.P. Bristol, 1998) p. 173.
3. Cox S. F. J., Davis E. A., King P. J. C., Gil J. M. *et al., J. Phys. Cond. Matter* **13** (2001), 9000 (plus references therein): Gil J. M. *et al., ibid*, L613.

Note added in proof. Both resonances for Ag_2O in Figure 1 have since been established as having $\Delta M=1$ character: the low-field resonance is now assigned to a distinct muonium state with shallow-donor character, coexisting with the deep state responsible for the higher-field resonance.

Hyperfine Interactions (2004) 158:317–322
DOI 10.1007/s10751-005-9059-6

PAC Studies on Zr-Based Intermetallic Compounds

L. C. DAMONTE* and L. A. MENDOZA-ZÉLIS
Physics Department, U.N.L.P. and IFLP, CONICET, La Plata, Argentina;
e-mail: damonte@fisica.unlp.edu.ar

Abstract. The Zr_2Al, Zr_3Al_2 and Zr_6NiAl_2 intermetallic compounds were characterized by means of time differential perturbed angular correlation (TDPAC) and X-ray diffraction. Our interest in these Zr(Hf) aluminides comes from crystallization studies of Zr(Hf)-based bulk metallic glasses which have a wide supercooled liquid region.

Key Words: electric field gradient, intermetallic compounds, perturbed angular correlation, transition metal alloys.

1. Introduction

Hafnium and zirconium aluminides have gained attention because of their high melting points, good mechanical properties and high oxidation resistance. The Zr–Al and Hf–Al phase diagrams are ones of the more complicated binary systems since many intermetallic compounds can be formed at different concentrations. Many of these compounds have been studied by a variety of techniques. In particular, electric-field-gradient (EFG) sensitive techniques, such as Perturbed Angular Correlation (PAC), which give information about the atomic distribution around a given atom and may sense the local order. Moreover, the Al rich zone of these phase diagrams were investigated by Wodniecki et al. using the ^{181}Ta and ^{111}Cd probes [1–3].

Our interest in these Zr(Hf) aluminides comes from crystallization studies of Zr(Hf)-based bulk metallic glasses [4, 5]. It is known that crystallization of these amorphous alloys proceeds by nucleation and growth of metastable (even quasicrystalline) phases that evolve to the equilibrium compounds. The formation of a given metastable phase depends on composition, heat treatment, oxygen content, preparation method and the addition of other elements. The possible final crystallization products in Zr–Cu–Al–Ni bulk metallic glasses: Zr_2Al, Zr_3Al_2 and Zr_6NiAl_2, have not been previously investigated by PAC.

In this work, the above-mentioned intermetallic compounds were characterized by means of time differential perturbed angular correlation (TDPAC) and X-ray diffraction. TDPAC measurements were done on as-irradiated and heat

* Author for correspondence.

treated samples. The analyzed systems display more than one quadrupole interaction, in agreement with the existence of different structural sites.

2. Experimental

The samples were prepared by arc melting, under ultra high vacuum, adequate amounts of high-purity elemental powders. A 2 at.% of Hf, required for PAC measurements, was added to each other. With the aim of homogenize all samples, annealing treatments of 24 h at 1000 °C for Zr_2Al and Zr_3Al_2 and at 800 °C for Zr_6NiAl_2 were done.

All the samples for PAC measurements were irradiated with thermal neutrons in order to obtain the required activity of ^{181}Hf by means of the nuclear reaction ^{180}Hf(n, γ)^{181}Hf. The β decay of the ^{181}Hf isotope populates the 133–482 keV γ-γ cascade in ^{181}Ta. A conventional planar setup of two CsF scintillators providing a time resolution (Full Width at Half Maximum) of 0.8 ns was used to collect the coincidences between the two gamma rays. After subtraction of chance coincidence background, time spectra corresponding to angles 90° and 180° between detectors were combined to get the ratio.

$$R(t) = 2\frac{N(180°, t) - N(90°, t)}{N(180°, t) + 2N(90°, t)} = A_{22}^{\exp t} G_2(t) \tag{1}$$

Theoretical functions of the form $A_{22}G_2(t)$, folded with the measured time resolution curve, were fitted to the experimental ratio $R(t)$.

Annealing treatments at 900 °C during 2 and 15 h under vacuum at pressures of 2×10^{-2} atm. and 10^{-5} atm, respectively, were done in order to eliminate radiation damaged.

X-ray diffraction patterns were taken using a Phillips PW 1050 equipment with Cu K_α radiation ($\lambda = 1.5418$ Å).

3. Results and discussion

Figure 1 displays the X-ray diffraction patterns for the three analyzed samples. For Zr_2Al, the characteristic lines of hexagonal Ni_2In type structure (hP6) are observed. In this compound there are two nonequivalent crystalline sites for the Zr atoms with a relative fraction ratio of 1:1. Some impurities lines are also observed but it cannot be decided to which compound corresponds, i.e., ZrAl or Zr_3Al_2 which principal peaks are also displayed.

The characteristic lines for tetragonal Zr_3Al_2 (tP20) are clearly distinguished in the XRD results. In this structure three nonequivalent sites are present with relative fraction 1:1:1. However some Al segregation cannot be discarded since the principal peak of this metal is also observed.

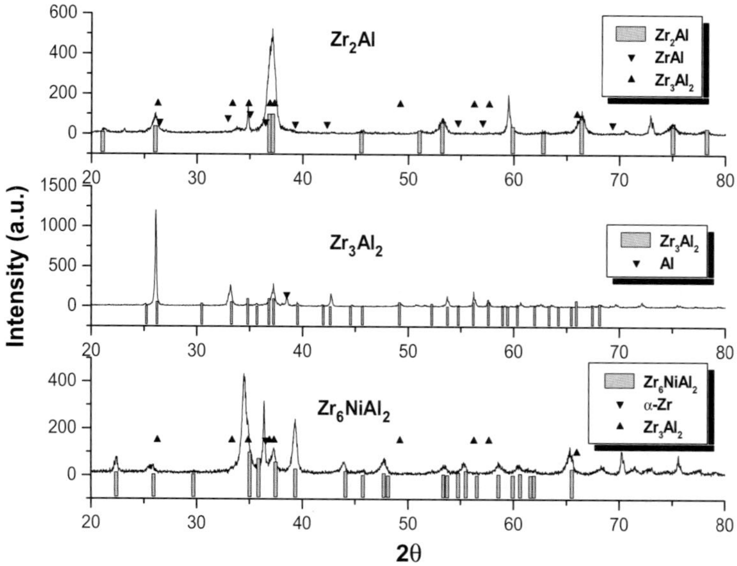

Figure 1. X-Ray diffraction patterns for the Zr crystalline alloys.

For the third compound, Zr_6NiAl_2, the XRD diffraction pattern shows all the characteristic lines corresponding to the Mg_2In hexagonal structure (hP9) with two nonequivalent crystalline sites for the Zr atom in relation 1:1 [6]. Also the most intense diffraction line of α-Zr is distinguished.

The above results show more than a single-phase sample, due to the fact of narrow region for single phases in the complicated equilibrium phase diagram, at least for Zr–Al. In addition, Zr_3Al_2 occurs above 1020 °C and in a composition region near to Zr_2Al. The as-irradiated and annealed samples were measured by PAC at room temperature. For all as-irradiated samples two wide distributions of quadrupolar frequencies are observed. These became better defined with progressive heat treatments [7], which remove radiation damage induced by thermal neutrons.

Figure 2 show the room temperature PAC spectra for all the samples after a heat treatment at 900°C for 15 h. The spectra were analyzed using the function $G_2(t) = \sum_i f_i G_i^2(t, V_{zz}, \eta, \delta)$, where

$$G_i^2(t; V_{zz}, \eta, \delta) = \sum_{i=1}^{3} s_i(\eta)\cos[\omega_i(V_{zz}, \eta)t] \times \exp(-\omega_i^2\delta^2 t^2/2), \qquad (2)$$

the perturbation factor for polycrystalline samples, describes a Lorentzian distribution of quadrupolar frequencies with population f_i. The quadrupole frequency is determined by $\omega_Q = eQV_{zz}/4I(2I-1)\eta$ (V_{zz} is the principal component of the EFG and $Q = 2.51_{15}$ barn, the nuclear quadrupole moment) and the asymmetry parameter, η, defined by $\eta = \left|(V_{xx} - V_{yy})/V_{zz}\right|$.

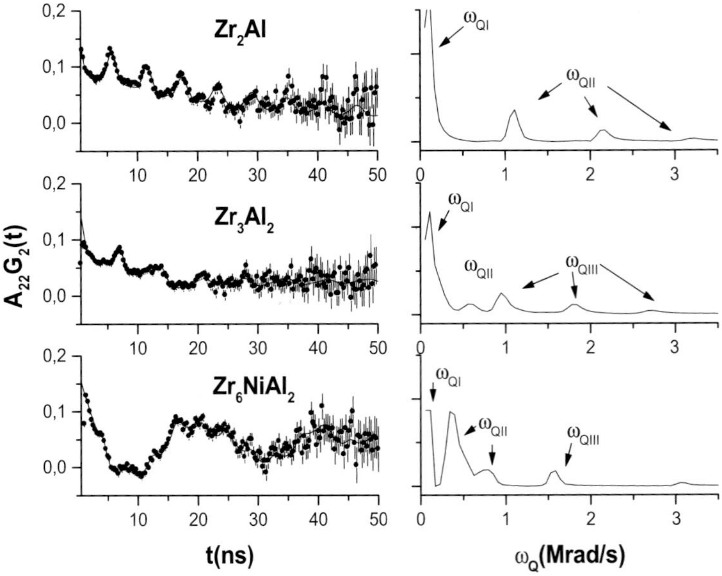

Figure 2. Room temperature PAC spectra (*left side*) and their corresponding Fourier transform (*right side*) for the three crystalline alloys.

Table I. Resulting quadrupolar parameters for the studied alloys

Alloy	f%	ω_Q (Mrad/s)	η	δ	V_{zz} (10^{17} V/cm^2)
Zr_2Al	62_2	0.0	0.0	26_3	0.0
	38_1	180_1	0.03_1	2.3_4	18.9_9
Zr_3Al_2	41_3	12.6_4	0.0	16_6	1.32_9
	27_2	88_2	0.25_7	14_3	9.2_6
	31_2	152_1	0.12_3	4_1	15.9_7
Zr_6NiAl_2	63_8	33_1	0.73_2	7_2	3.5_2
	24_7	41.5_5	0.43_2	2_1	4.4_2
	13_4	144_1	1.0	1.1_6	15.1_7

f, relative populations; ω_Q, quadrupole frequency; η asymmetry parameter and δ distribution. V_{zz} is the major EFG component.

Table I shows the resulting quadrupolar parameters for the three compounds. For Zr_2Al two quadrupolar interactions are observed, but the relative populations are not in a 1:1 relation. These two interactions correspond to the two Zr crystalline sites in this structure, being both sites axially symmetric. Moreover for site I, it has been proposed a distribution around zero frequency value consistent with a highly symmetric environment. Similar results were recently obtained by Wodniecki *et al.* [8]. The difference in their relative fraction is due to the presence of a second phase, which we may attribute to Zr_3Al_2 and which is attributed to ZrAl in [8]. These both have also one EFG component with a low value, present work and [3], but the asymmetry parameter obtained for Zr_3Al_2 is close to zero, as the corresponding one to the extra observed interaction.

Three quadrupole frequencies characterizing the three structural sites are observed for the second intermetallic alloy, Zr_3Al_2 (see Table I). Two of them are symmetric environments ($\eta = 0.0$ and 0.12) in agreement with structural data. A slightly difference in their relative population may be due to the Al segregation reported by XRD results. It is worthwhile to mention that similar results are reported in this Conference Proceedings by Wodniecki *et al.*

For the ternary alloy, Zr_6NiAl_2, three quadrupole interactions are observed. One of them corresponds to α-Zr ($f = 24\%$) and the other two are attributed to the two crystallographic sites of Zr_6NiAl_2. Both sites for Zr in this compound are highly asymmetric, as is seen on the η-values obtained. All the resulting quadrupolar parameters are display in Table I. In this case as for Zr_2Al some other non-desired impurities are present changing the relative fraction for each non-equivalent crystallographic site. Although a definitive assignment cannot be done, the characterizations by PAC of these three compounds allow us to get insight in the crystallization process of bulk metallic glasses [4, 5].

4. Conclusions

The three intermetallic compounds, Zr_2Al, Zr_3Al_2 y Zr_6NiAl_2, were characterized by PAC using the ^{181}Ta probe. Our results for the former one are in agreement with values recently reported. The other two intermetallic compounds, Zr_3Al_2 and Zr_6NiAl_2, to our knowledge, have not been reported before. These data were of great importance for the analysis of short range order thermal evolution in bulk metallic glasses.

Electric field gradient calculations are in progress to make a definite assignment between the lattice site and the observed quadrupole interaction.

Acknowledgements

The authors gratefully acknowledge J. Eckert and S. Deledda from IFW Dresden, Institut fur Metallische Werkstoffe, Dresden, Alemania, for sample preparation. This work was supported by CONICET, Argentina.

References

1. Wodniecki P., Wodniecka B., Kulinska A., Uhrmacher M. and Lieb K. P., *J. Alloys Comp.* **312** (2000), 17–24.
2. Wodniecki P., Wodniecka B., Kulinska A., Uhrmacher M. and Lieb K. P., *J. Alloys Comp.* **335** (2002), 20–25.

3. Wodniecki P., Wodniecka B., Kulinska A., Uhrmacher M. and Lieb K. P., *J. Alloys Comp.* **351** (2003), 1–6.
4. Damonte L. C., Mendoza-Zélis L. and Eckert J., *Mater. Sci. Eng. A* **278** (2000), 16.
5. Damonte L. C., *Ann. Chim. Sci. Mat.* **27** (2002), 61–67.
6. Eckert J., Mattern N., Zinkevitch M. and Seidel M., *Mater. Trans., JIM* **39** (1998), 623.
7. Rebón L., Tielas D. A. and Damonte L. C., *Anales AFA* **13** (2001), 213–217.
8. Wodniecki P., Wodniecka B., Kulinska A., Uhrmacher M. and Lieb K. P., *J. Alloys Comp.* **365** (2004), 52–57.

Hyperfine Interactions (2004) 158:323–328
DOI 10.1007/s10751-005-9060-0

The ^{181}Hf/^{181}Ta Probe in the Li and Nb Sites of Congruent $LiNbO_3$ Co-doped with Mg and Cr Ions Studied by γ–γ PAC

S. A. DIAS[1,*], J. G. MARQUES[1,2], J. G. CORREIA[1,2], J. A. SANZ[3] and J. C. SOARES[1]
[1]*CFNUL, Universidade de Lisboa, Av. Prof. Gama Pinto 2, P-1649-003 Lisboa, Portugal; e-mail: sdias@cii.fc.ul.pt, SOARESJC@cii.fc.ul.pt*
[2]*Instituto Tecnológico e Nuclear, Estrada Nacional 10, P-2685-953 Sacavém, Portugal*
[3]*Departamento de Física de Materiales, C-IV, Universidade Autónoma, E-28049 Madrid, Spain*

Abstract. The results reported in this work show the relevance of the lithium deficiency for the Li lattice site occupation of dopants in congruent lithium niobate crystals co-doped with 0.1 mol% Cr_2O_3 and different concentrations of HfO_2, up to 3 mol%. The same behavior is observed in a crystal doped with 1 mol% of HfO_2 and co-doped with 4 mol% MgO where after annealing most of the Hf/Ta probes occupy the Nb lattice site.

PACS: 23.20. En, 31.30. G, 61.80.H, 77.84.B

Key Words: $LiNbO_3$, neutron-irradiation, perturbed angular correlation spectroscopy.

1. Introduction

Although lithium niobate ($LiNbO_3$) has been studied for over 30 years, there is still today much interest in the characterization of the defects and in the lattice site location of dopants, due to the many applications of this crystal in optoelectronics.

Congruent crystals doped with HfO_2 and as well co-doped with MgO were studied by ion beam channelling and perturbed angular correlation (PAC) experiments [1, 2]. Pure $LiNbO_3$ crystals were also studied by PAC after implantation with the radioactive isotopes ^{181}Hf, ^{111m}Cd and ^{111}In [3, 4]. $LiNbO_3$ doped with Cr_2O_3 and co-doped with HfO_2 and MgO has been studied by optical methods [5].

The main aim of the present work is to compare results obtained for crystals doped with increased concentrations of HfO_2 during growth after neutron

* Author for correspondence.

activation without further treatments with the results of the same crystals after annealing at 800 and 950°C.

It is shown that the Li deficiency in the congruent crystals remains as the most relevant factor for the lattice site occupation of the Hf/Ta probe. The results presented in this work give support for the population of the Li lattice site by the impurity as long as a Li deficiency exists. As soon there is no Li deficiency, some of the impurities are displaced to Nb sites.

2. Experimental details

Two types of crystals were used in this work: a $LiNbO_3$ congruent crystal co-doped with 1 mol% HfO_2 and 4% MgO (MG4), which was transparent, and crystals co-doped with 0.1 mol% of Cr_2O_3 plus 2 mol% HfO_2 (HF2), and 0.1 mol% of Cr_2O_3 plus 3 mol% HfO_2 (HF3), which were green, due to the incorporation of Cr^{3+} in the lattice.

The $LiNbO_3$ crystals were grown at the Autónoma University of Madrid using the Czochralski method from a congruent melt ([Li] / [Nb] = 0.945) with grade I Johnson–Mathey powder. Two samples were cut from the HF2 crystal, one from the upper part and another from the lower part, in order to check for the homogeneous distribution of the Hf/Ta probes during growth.

The ^{181}Hf activity was produced by neutron irradiation of the samples in the Portuguese Research Reactor through the reaction ^{180}Hf (n, γ) ^{181}Hf using a neutron flux of 4×10^{12} $n \cdot cm^{-2} \cdot s^{-1}$ for 3 h. After the decay of the short-lived isotopes, mainly ^{180m}Hf, the PAC measurements were performed with the samples at room temperature and in air. After annealing at 800 and 950°C for 5 h in air, new measurements were performed with the crystals again at room temperature and in air. After the irradiation, the Cr_2O_3 doped crystals lost part of the intense green colour and the crystals doped with MgO lost their transparency. These properties were recovered with annealing.

The γ–γ PAC measurements were done using a 4-detector spectrometer set-up equipped with conical BaF_2 scintillators with a time resolution of 0.6 ns (FWHM) for the 133–482 keV cascade of ^{181}Ta. Two complementary geometries were used, with the c-axis of the single crystals in the detectors plane, at 45° with two detectors (G1) and perpendicular to the detectors' plane (G2). Some measurements were also performed with the c-axis aligned with two detectors, in order to confirm the orientation of the electric field gradients (EFG). The usual $R(t)$ function was determined from the 12 recorded coincidence spectra.

3. Results and discussions

Figure 1 shows the spectra and respective Fourier analysis obtained for the HF2 (upper part of the crystal), HF3 and MG4 crystals after irradiation, in

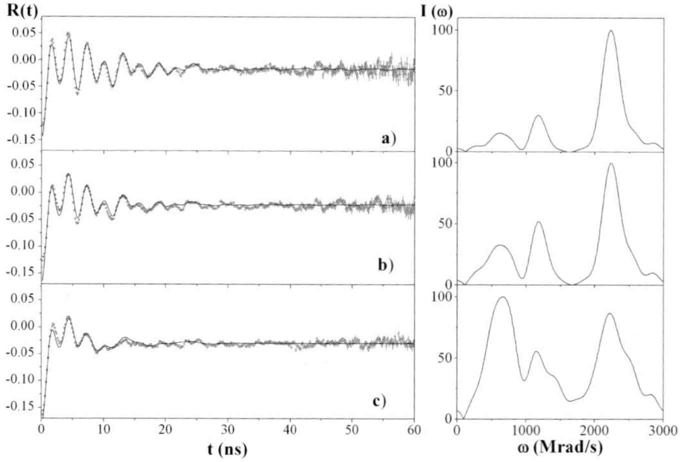

Figure 1. PAC spectra and respective Fourier analysis obtained after irradiation for (a) the HF2 sample, (b) the HF3 sample, and (c) the MG4 sample. The measurements were performed with the *c*-axis in the detectors' plane, at 45° with two detectors.

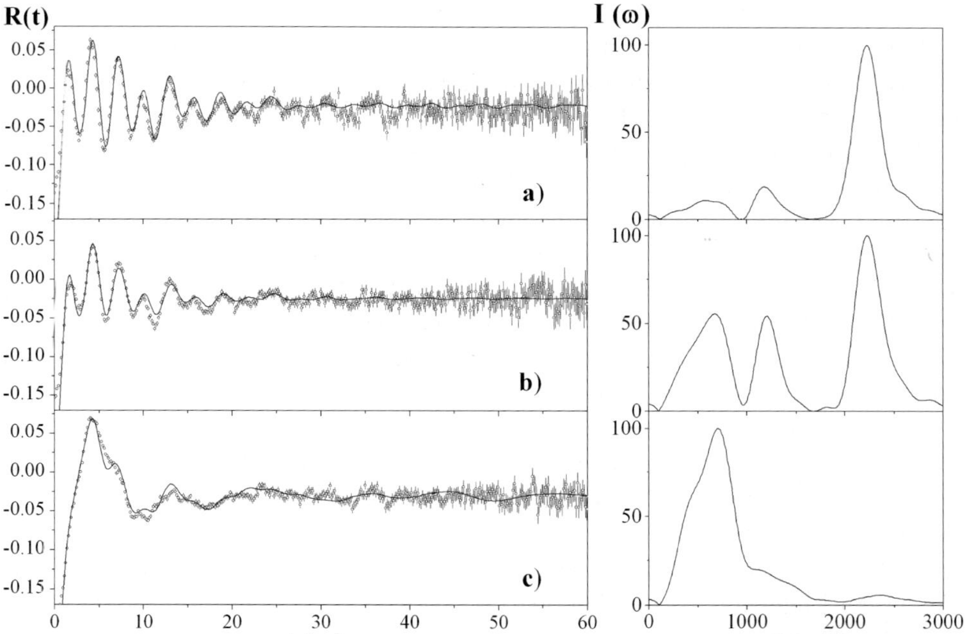

Figure 2. PAC spectra and respective Fourier analysis obtained after annealing for (a) the HF2 sample, (b) the HF3 sample, and (c) the MG4 sample. The measurements were performed with the *c*-axis in the detectors' plane, at 45° with two detectors.

geometry G1. The frequencies shown in the Fourier analysis correspond to the lattice site occupation of the Hf/Ta probes in Nb sites (fundamental frequency ca 350 Mrad/s) and Li sites (ca 1200 Mrad/s), as known from previous studies combining PAC and ion channeling [1, 2].

Table I. Best fit parameters for the spectra obtained after annealing

Sample	Site	ν_Q [MHz]	η	δ_L	Fraction [%]
HF2	Li I	1154 (12)	0.21 (1)	0.032 (5)	35 (5)
	Li II	1214 (12)	0.27 (1)	0.043 (5)	35 (5)
	Nb I	366 (10)	0.51 (5)	0.19 (1)	30 (5)
HF3	Li I	1154 (12)	0.21 (1)	0.052 (5)	33 (5)
	Li II	1214 (12)	0.27 (1)	0.072 (5)	22 (5)
	Nb I	366 (10)	0.41 (2)	0.28 (1)	45 (5)
MG4	Li I	1154 (12)	0.21 (1)	0.054 (5)	15 (5)
	Li II	1214 (12)	0.27 (1)	0.079 (5)	6 (5)
	Nb I	347 (10)	0.42 (2)	0.11 (1)	40 (5)
	Nb II	422 (10)	0.48 (2)	0.15 (1)	39 (5)

Figure 2 shows the corresponding spectra and Fourier analysis obtained after annealing at 950°C. For crystals HF2 and HF3, the main effect of the annealing was an increase of the amplitude of the oscillations. In contrast, for the MG4 crystal, the fraction of probes in Li sites has strongly increased after annealing. Comparing the Fourier analysis of spectra obtained after annealing for the HF2 and HF3 crystals, it is seen that the increase in the Hf concentration leads to a larger fraction of probes in Nb sites.

Corresponding spectra for the lower part of the HF2 crystal (not shown here) indicate a smaller fraction of probes in Nb sites. Thus, a small increase in the Hf concentration from the lower to the upper part of the crystal is derived from our measurements.

Table I summarizes the relevant parameters of the best fits with the NNfit code [6] to the data obtained in geometries G1 and G2 after annealing at 950°C.

For probes in the Li sites of the crystal it was necessary to consider two quadrupole interaction frequencies (QIF), ν_{Q1} = 1154 MHz and ν_{Q1} = 1214 MHz, with non-zero asymmetry parameters (η). Two QIF, ν_{Q1} = 191(2) MHz and ν_{Q2} = 205(2) MHz, with non-zero η were previously found in pure congruent crystals implanted with ^{111}In [4], while for near-stoichiometric crystals only one QIF, ν_{Q1} = 191(2) MHz with $\eta = 0$ was observed [7]. Thus, both the existence of ν_{Q2} and the nonzero η values were attributed to a stoichiometry-related defect. Congruent $LiNbO_3$ is Li deficient and this deficiency must be compensated by a departure from the ideal structure in order to maintain charge neutrality [8].

For probes in Nb sites it was necessary to consider two QIF, ν_{Q3} = 347 MHz and ν_{Q4} = 422 MHz in the case of the MG4 crystal, in agreement with a previous work [2]. The spectra obtained after irradiation and after annealing in the HF2 and HF3 spectra were well fitted with a single QIF for probes replacing Nb, ν_{Q3} = 366 MHz. In all cases relatively large Lorentzian distributions were considered in addition to nonzero asymmetry parameters. No sharp frequencies were observed so far for PAC probes replacing Nb in $LiNbO_3$. The measured EFG are in excellent agreement with NMR measurements [9] after division by the corresponding Sternheimer [10] factors: $V_{zz}(^{181}\text{Ta : Nb}) = 0.104(4) \times 10^{17}$ V/cm^2, for $\nu_Q = 366$ MHz, and $V_{zz}(^{93}\text{Nb}) = 0.104(5) \times 10^{17}$ V/cm^2.

In the MG4 crystal the fraction of Hf/Ta probes in Nb sites after irradiation is only 40(5)%, while after annealing it is 80(5)%, which is in agreement with ion channeling measurements in unirradiated samples [1]. Such decreases after irradiation were observed for other $LiNbO_3$ crystals and were attributed to the recoil to one of the many Li vacancies of part of the probe atoms after the ^{180}Hf(n, γ)^{181}Hf activation reaction [11]. It is not clear why this is not seen in the HF2 and HF3 crystals.

Comparing the HF2 and HF3 crystals after annealing, it is seen that the fraction of Hf/Ta probes in Nb sites is already 30% in the HF2 crystal and increases to 45% in the HF3 crystal. In congruent crystals doped with 1 mol% HfO_2, Hf occupies only Li sites. Thus, the concentration threshold for Hf to start the occupation of Nb sites by Hf must be in the range of 1 to 2 mol% HfO_2. In Mg co-doped crystals, Hf is gradually pushed into occupying Nb sites as the Mg concentration increases [1, 2]. In this work we show that an increase in the Hf concentration also leads to an increased occupation of Nb sites. These results can be explained by the gradual occupation of vacant Li sites in the crystal lattice; as soon as there is no Li deficiency, the impurities are displaced to Nb sites.

4. Conclusions

For the first time results have been obtained for $LiNbO_3$ congruent crystals doped with 0.1% Cr_2O_3 and co-doped with increasing concentrations of HfO_2 (2 and 3 mol%). It is confirmed that increasing the concentrations of the dopants, as soon the Li deficiency disappears, the Hf/Ta probes start to occupy Nb sites. New measurements are planned for crystals co-doped with 0.1% of Cr_2O_3 and higher amounts of HfO_2.

Acknowledgements

One of us (S.A. Dias) acknowledges the Physics Department, the Faculty for Sciences of the University of Lisbon, PRODEP and CFNUL for financial support.

References

1. Rebouta L., da Silva M. F., Soares J. C., Santos M. T., Diéguez E. and Agulló-López F., *Opt. Mat.* **4** (1995), 174.
2. Marques J. G., de Jesus C. M., Melo A. A., Soares J. C., Diéguez E. and Agulló-López F., *Hyperfine Interact. C* **1** (1996), 348.
3. Rebouta L., Soares J. C., da Silva M. F., Sanz-Garcia J. A., Diéguez E. and Agulló-López F., *Nucl. Instrum. Methods B* **45** (1990), 495.

4. Hauer B., Vianden R., Marques J. G., Barradas N. P., Correia J. G., Melo A. A., Soares J. C., Agulló-López F. and Diéguez E., *Phys. Rev., B* **51** (1995) 6208.
5. Cantelar E., Torchis G. A., Bermúdez V., Sanz J. A., Lifante G., Han T. P. J., Jaque F., (private communication) Jaque F., Han T. P. J., Bermúdez V. and Diéguez E., *J. Lumin.* **102/103** (2003), 253.
6. Barradas N. P., Rots M., Melo A. A. and Soares J. C., *Phys. Rev. B* **47** (1993), 8763.
7. Marques J. G., Kling A, Soares J. C., da Silva M. F., Vianden R, Polgár K, Diéguez E and Agulló-López F., *Rad. Eff. Def. Sol.* **150** (1999), 233.
8. Donnerberg H., Tomlinson S. M., Catlow C. R. A. and Schirmer O. F., *Phys. Rev. B* **40** (1989), 11909.
9. Peterson G. E. and Bridenbaugh P. M., *J. Chem. Phys.* **48** (1968), 3402.
10. Feiock F. D. and Johnson W. R., *Phys. Rev.* **187** (1969), 39.
11. Marques J. G., Kling A., de Jesus C. M., Soares J. C., da Silva M. F., Diéguez E. and Agulló-López F., *Nucl. Instrum. Methods B* **141** (1998), 326.

Hyperfine Interactions (2004) 158:329–332
DOI 10.1007/10751-005-9049-8

Comparison of *XYZ* Model Fitting Functions for ^{111}Cd in In_3La

MATTHEW O. ZACATE[1,3,*,†] and WILLIAM E. EVENSON[2]
[1]*Department of Physics, Washington State University, Pullman, WA, 99164, USA*
[2]*School of Science and Health, Utah Valley State College, Orem, UT, 84058, USA; e-mail: bill.evenson@uvsc.edu*
[3]*Department of Physics and Geology, Northern Kentucky University, Highland Heights KY, 41099, USA; e-mail: zacatem1@nku.edu*

Abstract. The *XYZ* model describes the interaction between nuclear probes and an electric field gradient that fluctuates among three orthogonal directions. The model presents a means to calculate the perturbation function that represents spectra obtained using perturbed angular correlation spectroscopy. Three analytic approximations of the perturbation function have been developed previously, and they are evaluated in the present paper in the context of Cd jumping among In-lattice sites in In_3La.

Key Words: PAC, relaxation, *XYZ* model.

Recently, the jump frequency of ^{111}Cd tracers on the In sublattice in In_3La was meaured via nuclear quadrupole relaxation using the method of perturbed angular correlation of gamma rays (PAC) [1]. In_3La has the Cu_3Au, or $L1_2$, structure, for which an In-lattice site has an axially symmetric electric field gradient (EFG). Principal axes of EFGs at neighboring In sites are orthogonal, so that the EFG at the nucleus of a PAC probe reorients by 90° in each jump. A model for such a fluctuating field interacting with a PAC probe has been termed the *XYZ* model [2].

When EFGs reorient in the *XYZ* model, there is a relaxation of the perturbation function $G_2(t)$ due to a loss of coherence in gamma radiation emitted by the ensemble of probes. Measurements of the relaxation can be used to determine the frequency of reorientation w_{EFG} of an EFG from one orientation to one other orientation. For the *XYZ* model, approximate perturbation functions [2–4] of various degrees of sophistication have been developed for spin-5/2 PAC probes. The present work compares the effectiveness of fitting using approximate functions to the results of full numerical fits as were done in [1].

* Author for correspondence.

† Present address: Department of Physics and Geology, Northern Kentucky University, Highland Heights, KY 41099, USA.

The approximate perturbation functions considered here are those by Baudry and Boyer [3], Evenson *et al.* [2], and Guan [4]. In each work, two functions were given: one for the slow fluctuation regime ($w_{EFG} \ll \omega_Q$) and one for the fast fluctuation regime ($w_{EFG} \gg \omega_Q$), where ω_Q is the quadrupole interaction frequency.

Baudry and Boyer found

$$G_2(t) \approx e^{-\lambda t}\left[\frac{1}{5} + \frac{13}{35}\cos(6\omega_Q t) + \frac{2}{7}\cos(12\omega_Q t) + \frac{1}{7}\cos(18\omega_Q t)\right], \tag{1}$$

with relaxation parameter $\lambda = 2w_{EFG}$, for $w_{EFG} \ll \omega_Q$, and

$$G_2(t) \approx e^{-\lambda t}, \tag{2}$$

with relaxation parameter $\lambda = 100\omega_Q{}^2/3w_{EFG}$, for $w_{EFG} \gg \omega_Q$ [3].

In the approximation by Evenson *et al.*,

$$\begin{aligned} G_2(t) &\approx (1/5)e^{-\lambda_0 t} + (13/35)e^{-\lambda_1 t}\cos(6\omega_Q t) + (2/7)e^{-\lambda_2 t}\cos(12\omega_Q t) \\ &\quad + (1/7)e^{-\lambda_3 t}\cos(18\omega_Q t), \end{aligned} \tag{3}$$

with relaxation parameters $\lambda_0 = 1.4896w_{EFG}$, $\lambda_1 = 1.6237w_{EFG}$, $\lambda_2 = 1.8173w_{EFG}$, and $\lambda_3 = 2.1301w_{EFG}$, for $w_{EFG} \ll \omega_Q$ [5], and

$$G_2(t) \approx (4/7)e^{-\lambda_0 t} + 0.3032e^{-\lambda_1 t} + 0.0968e^{-\lambda_2 t} + (1/35)e^{-\lambda_3 t}, \tag{4}$$

with relaxation parameters $\lambda_0 = 38\omega_Q^2/w_{EFG}$, $\lambda_1 = 10.15\omega_Q^2 t/w_{EFG}$, $\lambda_2 = 63.85\omega_Q^2/w_{EFG}$, and $\lambda_3 = 92\omega_Q^2/w_{EFG}$, for $w_{EFG} \gg \omega_Q$ [2].

In the approximation by Guan,[1]

$$\begin{aligned} G_2(t) &\approx (1/5)e^{-\lambda_0 t} + (13/35)e^{-\lambda_1 t}\cos(6\omega_Q t - \phi_1)/\cos(\phi_1) \\ &\quad + (2/7)e^{-\lambda_2 t}\cos(12\omega_Q t - \phi_2)/\cos(\phi_2) \\ &\quad + (1/7)e^{-\lambda_3 t}\cos(18\omega_Q t - \phi_3)/cos(\phi_3), \end{aligned} \tag{5}$$

with relaxation parameters $\lambda_0 = 1.5w_{EFG}$, $\lambda_1 = 1.62w_{EFG}$, $\lambda_2 = 1.8w_{EFG}$, and $\lambda_3 = 2.05w_{EFG}$ and phase factors $\phi_1 = 0.18w/\omega_Q$, $\phi_2 = 0.27w/\omega_Q$, and $\phi_3 = 0.154w/\omega_Q$, for $w_{EFG} \ll \omega_Q$, and

$$G_2(t) \approx (1/5)e^{-\lambda_0 t} + (13/35)e^{-\lambda_1 t} + (2/7)e^{-\lambda_2 t} + (1/7)e^{-\lambda_3 t}, \tag{6}$$

with relaxation parameters $\lambda_0 = 207\omega_Q^2/7w_{EFG}$, $\lambda_1 = 29\omega_Q^2 t/w_{EFG}$, $\lambda_2 = 497\omega_Q^2/10w_{EFG}$, and $\lambda_3 = 19\omega_Q^2/w_{EFG}$, for $w_{EFG} \gg \omega_Q$ [4]. Note that λ_n and w_{EFG} are in MHz when ω_Q is in Mrad/s.

PAC spectra obtained for ^{111}Cd in In_3La at different measurement temperatures can be found in Ref. [1]. w_{EFG}, reported in [1], and ω_Q, reported in [6],

[1] Our Equation (5) has the correct signs for the phase shifts, and has corrected the expressions for ϕ_n in terms of ω_Q.

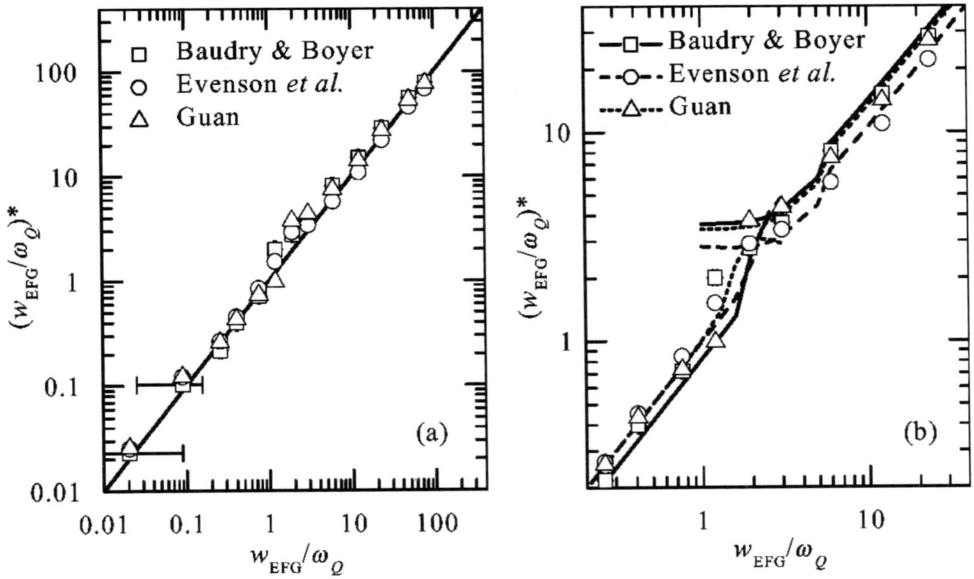

Figure 1. (a) Comparison of results of approximate fits $(w_{EFG}/\omega_Q)^*$ plotted vs. results of full numerical fits w_{EFG}/ω_Q. (b) Enlargement of central portion of (a).

were determined by full numerical fits, i.e. minimizing the difference between spectra and test-$G_2(t)$ functions generated numerically from the *XYZ* model as described in [2] by adjusting w_{EFG} and ω_Q. Spectra were refitted in the present study to Equations (1–6). Results $(w_{EFG}/\omega_Q)^*$ obtained from fits here are plotted vs. results w_{EFG}/ω_Q obtained from full numerical fits as data points in Figure 1. Note that ω_Q can only be fitted in the slow fluctuation regime, so that values of ω_Q used in the fast fluctuation regime were determined by extrapolating results from the slow fluctuation regime.

Figure 1(a) shows results over the range of w_{EFG}/ω_Q that is accessible to measurement. The line indicates values for which $(w_{EFG}/\omega_Q)^*$ would agree exactly with w_{EFG}/ω_Q. As can be seen, results from approximate fits agree well with those of the full numerical fits across the whole range, with the most notable deviation located at the intermediate fluctuation regime ($w_{EFG}/\omega_Q \approx 1$). The deviation arises because the discrepancy between the approximate functions and the *XYZ* model and correlation between w_{EFG} and ω_Q are greatest in this range.

Figure 1(b) focuses on a narrower range of w_{EFG}/ω_Q in order to better compare the different fitting methods. Here, curves indicate results of approximate fits to $G_2(t)$ generated by simulations of the *XYZ* model. These results suggest that one should use the approximations for the slow fluctuation limit up to $w_{EFG}/\omega_Q \approx 2$ and approximations for the fast fluctuation regime above. Some important points regarding fitting PAC spectra using approximate functions for $G_2(t)$ are as follows.

- *Best approximate form.* At least in the case of Cd in In_3La, fits using the approximate forms by Evenson *et al.* (our Equations (3) and (4)) work best.
- *Range of available data affects results.* Fitting with approximate functions is expected to work well when data are available across the full range of slow and fast fluctuation regimes. When data are fitted only for $w_{EFG}/\omega_Q \leq 2$, systematic errors can be expected in the determination of the dependence of fluctuation frequency on temperature or other experimental parameters. The degree of error is expected to increase as the lower limit of the range of fitted data increases.
- *Experimental effects on results.* Measured PAC spectra include data uncertainties that depend on time as well as experimental artifacts coming from the experimental geometry and other aspects of the experimental setup. These experimental factors affect the results of fitting with approximate functions, as can be seen in Figure 1(b) by the deviations between data points and curves.
- *Cubic offset problem in fast fluctuation limit.* If a cubic offset is fitted for spectra, the relaxation parameters and the cubic offset are highly correlated in the fast fluctuation regime. So, the cubic offset must be held fixed at a known value in order to get reliable values of the fluctuation frequencies in the fast regime. This problem applies to full numerical fits as well as approximate fits.

It is now computationally practicable to fit PAC spectra by generating perturbation functions numerically from an appropriate dynamical model, providing the model is sufficiently simple. We advocate fitting spectra with such full numerical fits in order to avoid uncertainties in interpretation of data caused by the above issues.

Acknowledgements

This work was funded in part by the National Science Foundation under grant DMR 00-91681 (Metals Program) at Washington State University. MOZ appreciates the support of Gary S. Collins. WEE acknowledges the support of the Brigham Young University Department of Physics & Astronomy.

References

1. Zacate M. O., Favrot A. and Collins G. S., *Phys. Rev. Lett.* **92** (2004), 225901; Erratum, *Phys. Rev. Lett.* **93** (2004), 049903.
2. Evenson W. E., Gardner J. A., Wang R., Su H.-T. and McKale A. G., *Hyperfine Interact.* **62** (1990), 283.
3. Baudry A. and Boyer P., *Hyperfine Interact.* **34** (1987), 803.
4. Guan H., PhD dissertation, Brigham Young University, 1994.
5. Wang R., PhD thesis, Oregon State University, 1991.
6. Collins G. S., Favort A., Kang L., Nieuwenhuis E. R., Solodovnikov D., Wang J., Zacate M. O., PAC Probes as Diffusion Tracers in Solids, in Proceedings for the HFI/NQI 2004, Bonn, Germany, R. Vianden, ed., unpublished.

Hyperfine Interactions (2004) 158:333–338
DOI 10.1007/s10751-005-9050-2

Site Occupation of In in RAg_6In_6 Studied Using PAC Spectroscopy

M. KRUŻEL[1], K. KRÓLAS[1,*], L. MUSZYŃSKI[1], V. I. ZAREMBA[2] and W. SUSKI[3]
[1]*Institute of Physics, Jagiellonian University, Reymonta 4, PL-30-059 Kraków, Poland; e-mail: krolas@if.uj.edu.pl*
[2]*I. Franko National University of Lviv, Lviv, Ukraine*
[3]*Polish Academy of Sciences, Wroclaw, Poland*

Abstract. PAC measurements were performed for ^{111}In in $LaAg_6In_6$, $PrAg_6In_6$ and $NdAg_6In_6$. It appears that the probe atom occupy three available sites in the crystal lattice with quite a similar probability. There is a small preference for In atoms to occupy a selected lattice site which is not the same in different compounds.

Key Words: PAC, preferential site occupation, ternary alloys.

The tetragonal $ThMn_{12}$ structure has three different Mn positions, denoted in Wyckoff notation by 8*i*, 8*j* and 8*f*. Such a structure is formed by a wide family of ternary compounds in which the Th positions are occupied by rare earth atoms and the Mn positions are occupied by atoms of two elements with specific compositions. For example it has been shown [1] that RAg_6In_6 alloys (R = La, Pr, Nd) crystallize in the $ThMn_{12}$ structure only for the exact 1 : 6 : 6 stoichiometry. The PAC technique offers the possibility to study the relative occupation of the available sites for In atoms.

The PAC measurements were performed for ^{111}In probe in $LaAg_6In_6$, $PrAg_6In_6$ and $NdAg_6In_6$ in a wide temperature range. R–Ag–In alloys have been prepared by melting the elements in stoichiometric quantities in an arc furnace under a protective argon atmosphere. Then the alloys were annealed at 670 K for two weeks. The ^{111}In radioactivity was introduced into the samples by alpha particle irradiations of each compound itself. The ^{111}In isotope was produced in the targets as a result of the $^{109}Ag(\alpha, 2n)^{111}In$ nuclear reaction. Before the PAC measurements the samples were annealed again at 670 K in order to remove radiation damages. Examples of the PAC spectra are shown in Figure 1.

All spectra which were taken below 400 K could be fitted assuming three fractions of ^{111}In atoms exposed to the EFG of three different magnitudes. For

* Author for correspondence.

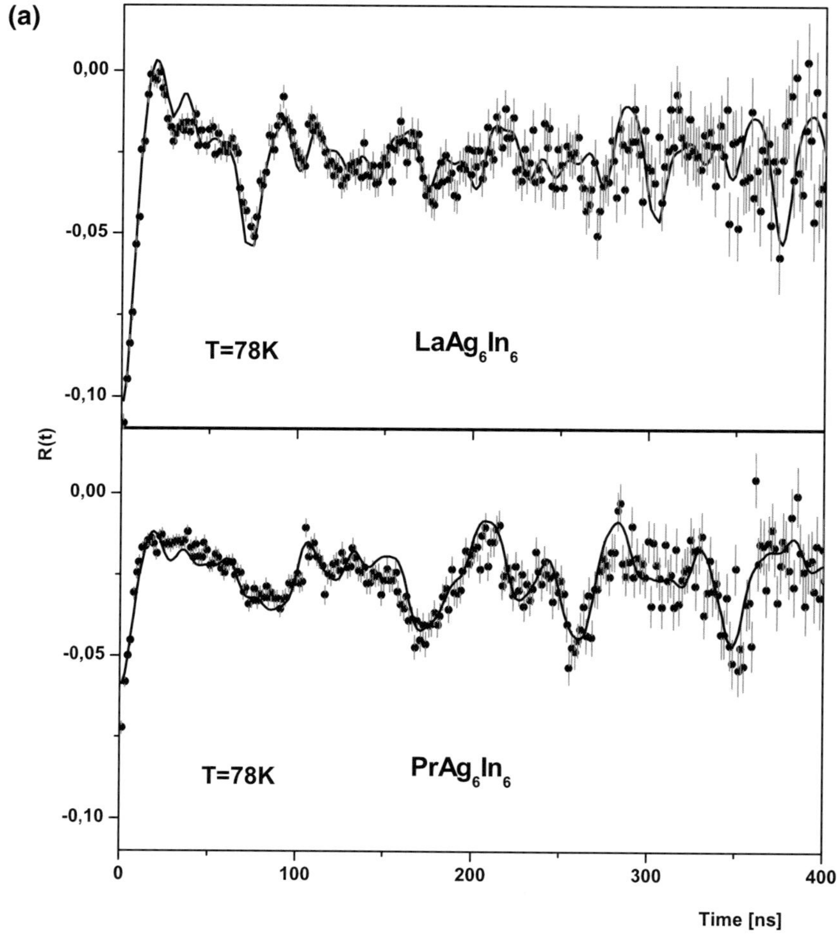

Figure 1. (a) PAC spectra for $LaAg_6In_6$ and $PrAg_6In_6$ measured at 78 K after annealing at 670 K; (b) PAC spectra for $NdAg_6In_6$ measured at 78 and 663 K.

each fraction, f_1, f_2 and f_3, the quadrupole interaction is well defined (no broadening is necessary to fit the spectra) and the EFG is not axially symmetric. Above 400 K an additional fraction, f_0, appears in the spectra. For this fraction of probe atoms the angular correlation is unperturbed. It is well seen in Figure 1b where the fraction f_0 manifest itself in the spectrum taken at 663 K as a distinct increase of the hard-core level. Table I and Figure 2 summarizes the EFG parameters and Table II summarizes the site occupancy fractions for all three compounds.

$LaAg_6In_6$: Figure 3 presents the site-occupation fractions, which were found for ^{111}In probe in $LaAg_6In_6$. The fractions are numbered starting from the site which is characterized by the EFG of the signal having the highest quadrupole interaction constant and highest asymmetry parameter. Obviously, the fractions remain constant at low temperatures up to 400 K. The fraction f_2, which is related to the intermediate frequency of about 90 MHz, is about 0.4 and it is slightly

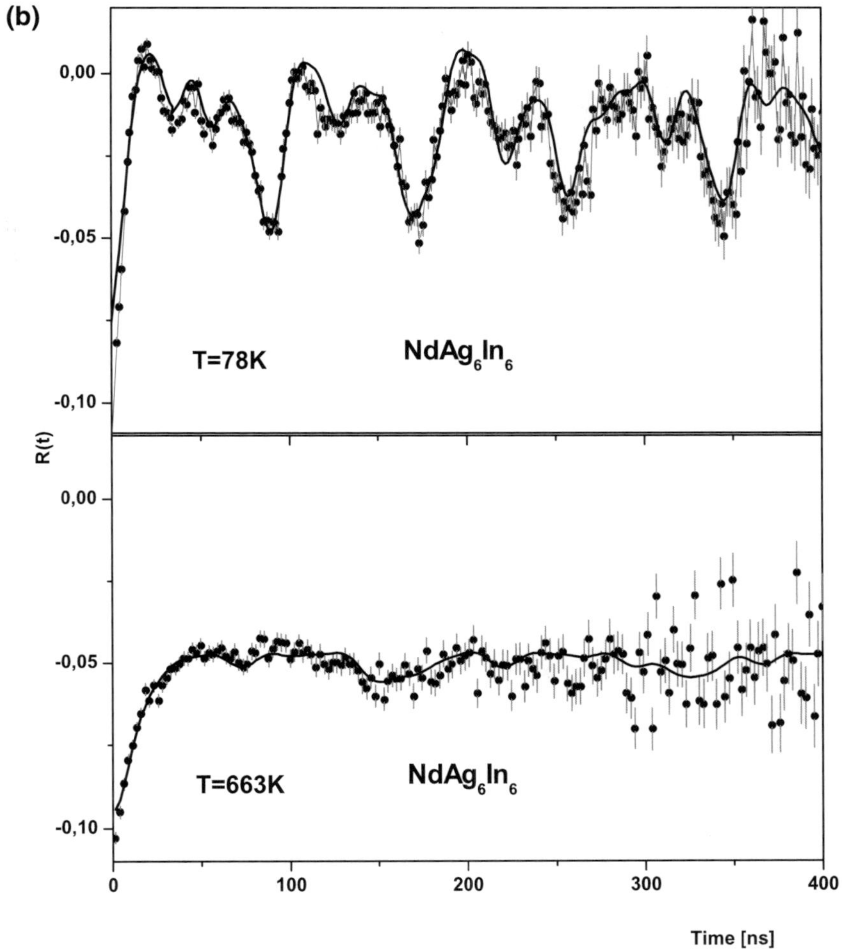

Figure 1. (Continued).

larger than two other fractions. At about 400 K all three fractions tend to decrease. Instead, a new fraction f_0 appears and increases up to 0.2 at 650 K.

$PrAg_6In_6$: As it can be seen from the Table I the EFG parameters for all three sites of $PrAg_6In_6$ lattice are very similar to that of $LaAg_6In_6$ lattice. It is a strong reason to believe that the fractions f_1, f_2 and f_3 represent corresponding lattice sites in both compounds. In the case of $PrAg_6In_6$ the dominant fraction however is the fraction f_3 while in the case of $LaAg_6In_6$ the fraction f_2. In the case of $PrAg_6In_6$ the PAC measurements were performed at rather high temperatures above 650 K up to 990 K. At this temperature range the unperturbed component in the PAC spectra becomes pronounced and finally the angular correlation is fully unperturbed at 990 K. What is important is that the process is reversible. When temperature drops down the fractions f_1, f_2 and f_3 are seen again.

Table I. Electric quadrupole frequencies, ν_Q, and the asymmetry parameters η obtained from the fits to the PAC spectra of ^{111}In in $LaAg_6In_6$, $PrAg_6In_6$ and $NdAg_6In_6$ at 78 K

Compound	Site 1		Site 2		Site 3	
	ν_Q [MHz]	η	ν_Q [MHz]	η	ν_Q [MHz]	η
$LaAg_6In_6$	142(1)	0.65(2)	91(1)	0.40(2)	79(1)	0.28(3)
$PrAg_6In_6$	146(1)	0.68(2)	105(2)	0.45(3)	77(2)	0.16(5)
$NdAg_6In_6$	114(1)	0.64(2)	69(2)	0.88(2)	76(1)	0.22(4)

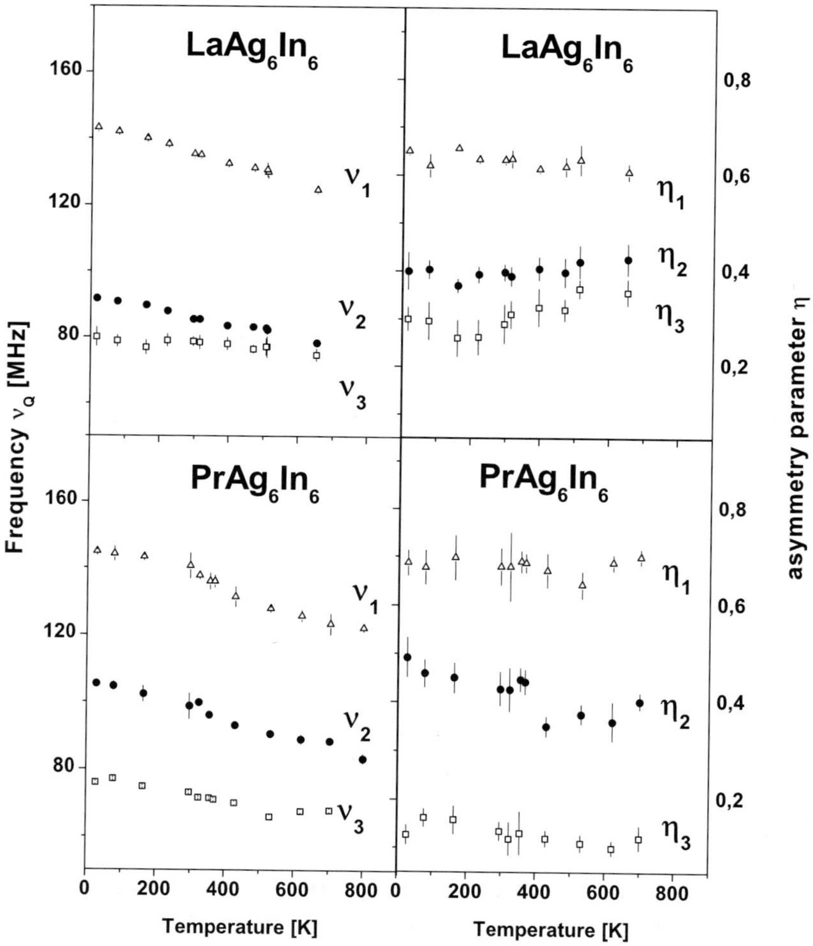

Figure 2. Electric quadrupole frequencies, ν_Q, and the asymmetry parameters η for $LaAg_6In_6$ and $PrAg_6In_6$ *versus* the temperature.

Table II. Site occupancy fractions *f* derived from the spectra measured in the temperature range from 17 to 400 K

Compound	Site 1	Site 2	Site 3
$LaAg_6In_6$	0.27–0.30	0.40–0.43	0.28–0.31
$PrAg_6In_6$	0.27–0.30	0.27–0.31	0.36–0.43
$NdAg_6In_6$	0.16–0.27	0.23–0.26	0.44–0.61

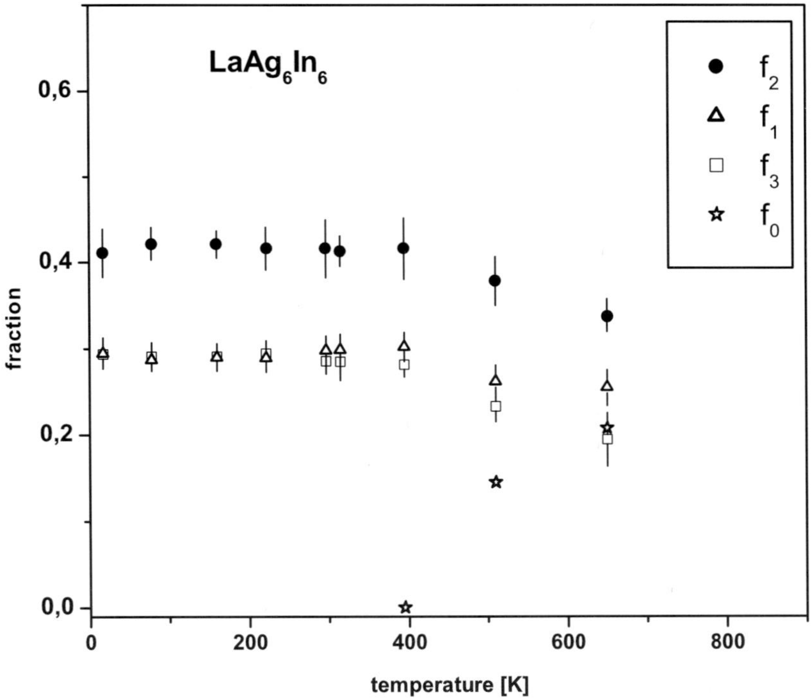

Figure 3. Site occupancy fractions f_1, f_2, f_3 and f_0 for $LaAg_6In_6$ *versus* the temperature.

$NdAg_6In_6$: The quadrupole interaction parameters for $NdAg_6In_6$ are a little bit different than that for two other compounds. Still three lattice sites are occupied by In probe atoms but there is a pronounced preference to occupy the site number 3. The fraction f_0 appears again in the spectra measured above 500 K i.e., at slightly higher temperatures that in the case of La and Pr compounds.

The three fractions in the PAC spectra are attributed to In atoms that occupy all three available sites in the crystal lattice. Since the fractions do not differ too much from the value of $\frac{1}{3}$ it may be concluded that the 8i, 8j and 8f sites are occupied with quite a similar probability. There is a small preference however for In atoms to occupy selected lattice sites. In $LaAg_6In_6$ the preferred site is site number 2 which is characterized by the intermediate values of ν_Q and η. In

$PrAg_6In_6$ compound In atoms tend to locate in the site number 3 with the lowest values of ν_Q and η. Even stronger preferential site occupation is observed in the case of $NdAg_6In_6$. The site of preference is again the site number 3. Unfortunately, it cannot be concluded however which of three 8i, 8j or 8f site is the site of preference without detailed calculation of EFG parameters for the considered sites.

Preferential site occupation was already observed in ternary compounds with $ThMn_{12}$ structure. For example it was shown [2] that Fe atoms occupy almost exclusively the 8f site in YFe_4Al_8. Here the distribution of Ag and In atoms among available sites has more equal probability due to similar chemical properties of these two elements.

There is no final conclusion on the origin of fraction f_0 which develops at high temperatures. The zero-EFG site can be explained either by structural changes in crystal lattice or by the dynamical relaxation of the quadrupole interaction. Further investigations are in progress.

Acknowledgement

This work was supported in part by the funds of the Faculty of Physics, Astronomy and Applied Computer Science.

References

1. Zaremba V. I., Kalychak Y. M., Galadzhun Y. V., Suski W. and Wochowski K., *J. Solid State Chem.* **145** (1999), 216–219.
2. Moze O., Ibberson R. M. and Buschow K. H. J., *J. Phys.: Condens. Matter* **2** (1990), 1677.

Hyperfine Interactions (2004) 158:339–345
DOI 10.1007/s10751-005-9051-1

Lattice Location of ^{181}Ta and ^{111}Cd Probes in Hafnium and Zirconium Aluminides Studied by Perturbed Angular Correlation

P. WODNIECKI[1,2,*], A. KULIŃSKA[1,2], B. WODNIECKA[1], M. UHRMACHER[2] and K. P. LIEB[2]
[1]*IFJ PAN, 31-342 Kraków, Poland*
[2]*Zweites Physikalisches Institut der Universität Göttingen, 37073 Göttingen, Germany*

Abstract. The perturbed angular correlation technique was applied to study the lattice location of the ^{111}In/^{111}Cd and ^{181}Hf/^{181}Ta probe atoms in hafnium and zirconium aluminides. Compounds of different stoichiometries and crystallographic structures were the subject of investigation. According to the expectation, in all investigated compounds ^{181}Hf/^{181}Ta probes occupy the Hf(Zr) crystallographic sites. The ^{111}In/^{111}Cd probes are placed at the sites of all constituent metals – aluminum, hafnium and zirconium, depending on the crystallographic structure of compound, concentration of the constituent metals and temperature of the sample.

Key Words: hyperfine interactions, site preference, Zr(Hf)–Al compounds.

1. Introduction

Aluminum forms with hafnium and zirconium a large number of high-melting intermetallic compounds of different crystallographic structures. This variety of hafnium and zirconium aluminides ensures many technical applications of these materials, attractive for the advanced high temperature application because of their high melting temperature, low density, good oxidation resistance and good thermal stability. However, their applications are limited because they are brittle at low temperatures. Recently, many efforts have been made to improve the ductility of XAl_3 intermetallic compounds. Locations of solutes in binary compounds are of general interest because they affect material properties.

The presented PAC investigations in Hf–Al and Zr–Al systems, employing the most common hyperfine probes ^{181}Hf(β^-)^{181}Ta and ^{111}In(EC)^{111}Cd, aims at determining their location in compounds of different structures. Frequently, probe positions are attributed to particular sites basing on simple comparison of atomic volumes and charges of solute and host atoms. However, such predictions are simplistic because site preferences can change with composition, especially near stoichiometry [1, 2]. Thus, the site of impurity atoms is hard to predict.

* Author for correspondence.

Since the parent nuclei of ^{181}Ta is the hafnium isotope, one may assume that the ^{181}Ta probes substitute hafnium atoms in Hf–Al phases and most probably chemically similar zirconium in Zr – Al phases. The situation is clearly different in the case of ^{111}Cd probes representing dilute impurities in the systems studied and therefore one can observe that some of the crystallographic sites are preferred by the probes.

In the PAC technique each position of the probe atom relative to the neighboring atoms usually gives its own characteristic and distinguishable signal. Number of the observed electric field gradients (EFG) and values of the corresponding hyperfine interaction parameters reflect the positions of the probes in the lattice. In particular, value of the asymmetry parameters, which depends on local point symmetries, can give clear and direct information on probe location. Sometimes a drastic temperature-driven change of the measured EFG can indicate either the probe site switching or a structural phase transition in the host. In such situation an experiment with other probes, preferably identical with the constituent atoms, can be helpful.

2. Experimental details

The investigated compounds were produced by multiple arc melting, under argon atmosphere, followed by a proper annealing. Powder X-ray diffraction analysis was applied to check the structures of the obtained samples. Doping of the investigated compounds with ^{181}Hf activity was performed via the neutron irradiation. In order to introduce ^{111}In/^{111}Cd probes, slices of the same samples (ca. 0.5 mm thick) were implanted with the radioactive ^{111}In ions. Annealing of the samples for few hours at elevated temperature was necessary to remove the irradiation defects and to diffuse the probe atoms to the substitutional lattices sites. PAC measurements were performed in the broad temperature range with a standard fast-slow set-ups equipped with four BaF_2 or NaJ(Tl) detectors.

3. PAC results and discussion

3.1. THE Zr_2Al_3 and Hf_2Al_3 PHASES – ^{111}In/^{111}Cd PROBE ATOMS AT ALUMINIUM SITES

The octahedral unit cell of Hf_2Al_3 and Zr_2Al_3 contains eight molecules. All atoms occupy lattice sites of low symmetry: the Hf(Zr) atoms are located on 16(*b*) sites, while the Al atoms populate the 8(*a*) and 16(*b*) sites [3].

The PAC spectra for ^{181}Ta probes in Zr_2Al_3 and Hf_2Al_3 phases exhibit a single, well defined EFGs with rather high asymmetry parameter values [4, 5]. This unique EFGs correspond to the unique Hf(Zr) 16(*b*) crystallographic site occupied by the ^{181}Ta probes. Very close values of the hyperfine interaction parameters noticed for Zr_2Al_3 and Hf_2Al_3 phases reflect the chemical and structural similarities of these compounds.

In the experiments with ^{111}Cd probes the quadrupole interaction parameters indicate two asymmetric EFG's, nearly identical for both compounds. The measured ratio of the probe fractions amounts ca. 1.5 at room temperature and rises with temperature to ca. 1.8 at 970 K. On that basis we assign the ^{111}Cd probes to 16(*b*) and 8(*a*) sites of the Al sublattice.

3.2. THE $ZrAl_3$ AND $HfAl_3$ PHASES – ^{111}In/^{111}Cd PROBE ATOMS AT HAFNIUM SITES

The tetragonal unit cell of $D0_{23}$ type contains Hf(Zr) atoms in the axially symmetric 4(*e*) sites, while 12 Al atoms placed in the axially symmetric 4(*e*) and 4(*d*) sites and in the non-axially symmetric 4(*c*) sites [3].

In PAC spectra of ^{181}Ta and ^{111}Cd probes in $HfAl_3$ and $ZrAl_3$ phases [4] a single frequency triplet pointing to a unique, axially symmetric EFG was evidenced for each compound. The assignment of ^{181}Ta probes to the unique Hf(Zr) 4(*e*) site is obvious.

In the case of ^{111}In/^{111}Cd probes, since there is only a single Hf(Zr) site in the $D0_{23}$ structure of (Hf/Zr)Al_3, but three non-equivalent Al sites, we are tempted to assign the only EFG found to the ^{111}Cd probes at 4(*e*) Hf(Zr) site.

3.3. THE Zr_3Al PHASE – ^{111}In /^{111}Cd PROBE ATOMS DISTRIBUTED BETWEEN THE SITES OF THE BOTH CONSTITUENTS

A single, axially symmetric EFG measured on ^{181}Ta in a Zr_3Al sample [6] corresponds to the probe location at the unique 3(*c*) Zr site of $L1_2$ lattice. The very narrow frequency distribution reflects a well established crystallographic structure of the sample and a well defined substitutional probe position.

The PAC spectrum for ^{111}Cd probes is quite different. More than 80% of the perturbation factor is described by a vanishing EFG, which is in agreement with the cubic symmetry of the 1(*a*) Al site, while the remaining probes are exposed to a small EFG with $\eta \approx 0$. Hence, we must conclude that some 20% of the ^{111}In probes are incorporated into the Zr sublattice. The occupation of the three times less populated aluminum sites by 80% of ^{111}In probe atoms point to a strong site preference of the indium atoms in favor of the 1(*a*)Al sites.

3.4. THE $HfAl_2$ PHASE – TEMPERATURE-DRIVEN SWITCHING OF ^{111}In/^{111}Cd PROBE ATOMS BETWEEN ALUMINUM AND HAFNIUM SITES

The hexagonal C14 structure of $HfAl_2$ and $ZrAl_2$ Laves phases contains Hf(Zr) atoms in the axially symmetric position 4(*f*), two Al atoms in position (*a*) (also axially symmetric) and six Al atoms in position (*h*) [3].

PAC spectra for ^{181}Ta probes in $ZrAl_2$ and $HfAl_2$ exhibit a single frequency triplet pointing to a unique axially symmetric EFG related to the probes at 4(*f*) Hf(Zr)-site [7].

All PAC spectra obtained with ^{111}Cd probes in $ZrAl_2$ clearly demonstrate that probes occupy two nonequivalent lattice sites (one of them axially symmetric). The number of the observed EFGs and their symmetry point to the ^{111}Cd location in Al-sublattice of $ZrAl_2$ phase.

PAC measurements in the isostructural $HfAl_2$ phase [7] exhibit a very surprising temperature behavior of the ion-implanted ^{111}In atoms. In the $Hf_{33.5}Al_{66.5}$ sample below 370 K PAC spectra are described by a unique well-defined axially symmetric EFG corresponding to probe location at the unique hafnium site. Above 400 K they occupy, similarly as in $ZrAl_2$, the two aluminum sites, however, with a clear preference of the 2(*a*) site. This process, fully reversible when heating and cooling the sample, occurring in a rather narrow temperature interval and accompanying by a broadening of the frequency distributions [8], can be attributed to the atomic size effects.

Surprisingly, in an additional PAC experiment, with $Hf_{34}Al_{66}$ sample of slightly different Hf content, the ^{111}In probes were placed in the whole temperature range only in Al-sublattice and no switching of the probes was evidenced.

3.5. THE ZrAl PHASE – STRONG PREFERENCE OF ^{111}In/^{111}Cd PROBES FOR ALUMINUM SITES IN Zr_2Al_3 CONTAMINATION PHASE

HfAl and ZrAl occur in the orthorhombic B_f type structure with the two 4(*c*) sites of low symmetry occupied by a definite species [3].

The XRD patterns for HfAl and ZrAl samples indicated a single-phase HfAl sample but a small admixture of Zr_2Al_3 phase in ZrAl sample.

The PAC spectra for ^{181}Ta probes in HfAl and ZrAl phases [8] reveal, according to the crystalline B_f structure, the existence of a unique probe site with non-axial environment connected with 100% of probe location in 4(*c*) Hf/Zr site of *m2m* symmetry. It should be pointed out that no evidence of EFG characteristic for the Zr_2Al_3 phase [4] was found in PAC spectra of ZrAl sample, indicating the Zr_2Al_3 phase contamination below few percent.

The PAC experiment for ^{111}Cd in HfAl sample [8] reveal one probe site, populated with most of cadmium probes, characterized by a well defined, non-axial EFG. Because of the same number and symmetry of substitutional lattice sites in Hf- and Al-sublattices in the B_f structure, the question of In impurity localization in HfAl compound cannot be solved basing on the PAC results.

The experiments with ^{111}Cd in ZrAl sample [8] revealed a strong preference of In impurities for the crystallographic sites of Zr_2Al_3 admixture. About 50% of probes were located in Zr_2Al_3 contamination (characterized by two nonaxial EFGs linearly decreasing with temperature [4]). The rest of probes experienced the broad EFG distribution. Experiment with ^{111}Cd probes in ZrAl, repeated for different sample concentrations and thermal treatments, each time resulted in very similar PAC spectra. Therefore, the strong segregation of indium in ZrAl and Zr_2Al_3 mixture to Zr_2Al_3 phase was concluded.

Table I. Lattice site description [3] of the selected hafnium and zirconium aluminides, where different ^{111}In/^{111}Cd probe location was observed [4, 5, 7]

$ZrAl_2$ (C14)				$HfAl_3$ ($D0_{23}$)				Zr_3Al ($L1_2$)				$HfAl_2$ 33.5 at % Hf (C14)			
^{111}In/^{111}Cd in Al-sublattice				^{111}In/^{111}Cd in Hf-sublattice				^{111}In/^{111}Cd in Al and Zr sublattices				Temperature driven switching of ^{111}In/^{111}Cd between Hf and Al sublattices			
Site description				Site description				Site description				Site description			
Lattice site	Coordination			Lattice site	Coordination			Lattice site	Coordination			Lattice site	Coordination		
	n	CN	Distance [Å]		*n*	CN	Distance [Å]		*n*	CN	Distance [Å]		*n*	CN	Distance [Å]
4(*f*) (Zr)	Al	3	3.103	4(*e*) (Hf)	Al	4	2.834	3(*c*) (Zr)	Al	4	3.092	4(*f*) (Hf)	Al	3	3.079
	Al	6	3.1172		Al	4	2.890		Zr	8	3.092		Al	3	3.090
	Al	3	3.1174		Al	4	2.965		Zr	6	4.373		Al	6	3.091
	Zr	3	3.243		Hf	4	3.987						Hf	3	3.217
2(*a*) (Al)	Al	6	2.593	4(*c*) (Al)	Al	4	2.819	1(*a*) (Al)	Al	12	3.092	2(*a*) (Al)	Al	6	2.665
	Zr	6	3.103		Hf	4	2.890		Zr	6	4.373		Hf	6	3.079
6(*h*) (Al)	Al	2	2.573	4(*d*) (Al)	Al	4	2.756					6(*h*) (Al)	Al	2	2.573
	Al	2	2.689		Al	4	2.819						Al	2	2.665
	Al	2	2.699		Al	4	3.987						Al	2	2.678
	Zr	4	3.1172		Hf	4	2.965						Hf	4	3.091
	Zr	2	3.1174	4(*e*) (Al)	Al	4	2.756						Hf	2	3.090
					Al	4	3.108								
					Al	4	3.987								
					Hf	4	2.834								

4. Summary

According to the expectation, in all investigated hafnium and zirconium aluminides the $^{181}Hf/^{181}Ta$ probes occupy hafnium or zirconium crystallographic sites. The situation for the $^{111}In/^{111}Cd$ impurities is more complex. Location of In and Al in the same group of the Periodic Table and the small difference in electronegativity would clearly favor substitution of Al by In. On the other hand, the experimental atomic radii of In, Al, Hf and Zr (equal to 1.55 for Hf, Zr and In and 1.25 Å for Al [9]) favor In atoms to substitute the Hf or Zr site instead of Al. Thus, we could expect a kind of 'competition' between the size effect and electronegativity influence. For Zr_2Al_3, Hf_2Al_3 and $ZrAl_2$ compounds the 'chemical' argument wins over the size argument and all the ^{111}In probe atoms end up at Al sites. On the other hand, in the case of aluminides with $D0_{23}$ structure ($ZrAl_3$ and $HfAl_3$), the substitution of Zr(Hf) atoms for $^{111}In/^{111}Cd$ probes gives evidence for the size argument. In Zr_3Al compound indium impurities were found to occupy both Zr- and Al-sublattices with a strong preference to Al-sites. Finally, our detailed PAC study of $HfAl_2$ phase demonstrated a temperature-driven reversible change of the ^{111}In probes between substitutional sites of different elements. In addition, lack of this switching in $HfAl_2$ sample of slightly different Hf content indicates that solute location can depend also on the components concentration. Similar observation made by Collins *et al.*, [2] for Ni_2Al_3 and Ga_2Al_3 compounds demonstrates that site preferences for an impurity atom can change with composition, especially near stoichiometry. Experiments for ZrAl have shown that site preference can lead to a segregation of $^{111}In/^{111}Cd$ into a trace Zr_2Al_3 phase. Segregation of In solutes in two-phase mixtures was also observed by Collins and Zacate [1, 2] in case of Pd–Ga, Ni–Al and Fe–Al systems.

Different sublattices occupied by $^{111}In/^{111}Cd$ impurity probes in Hf(Zr)–Al compounds, listed in Table I, confirm difficulties connected with the prediction of impurity sites. The theoretical *ab initio* calculations, for example with the WIEN2k code, could dispel some doubts still existing in interpretation of the experimental data.

References

1. Collins G. and Zacate M., *Hyperfine Interact.* **136/137** (2001), 641.
2. Collins G. and Zacate M., *Hyperfine Interact.* **136/137** (2001), 647.
3. Villards P. and Calvert L. D. (eds.), *Pearson's Handbook of Crystallographic Data for Intermetallic Phases*, ASM, Material Park, Ohio, 1991.
4. Wodniecki P., Wodniecka B., Kulińska A., Uhrmacher M. and Lieb K. P., *Hyperfine Interact.* **136/137** (2001), 535.
5. Wodniecki P., Wodniecka B., Kulińska A., Uhrmacher M. and Lieb K. P., *J. Alloys Compd.* **312** (2000), 17.
6. Wodniecki P., Wodniecka B., Kulińska A., Uhrmacher M. and Lieb K. P., *J. Alloys Compd.* **365** (2004), 52.

7. Wodniecki P., Wodniecka B., Kulińska A., Uhrmacher M. and Lieb K. P., *Phys. Lett., A* **288** (2001), 227: *J Alloys Comp* **335** (2002), 20.
8. Wodniecki P., Wodniecka B., Kulińska A., Uhrmacher M. and Lieb K. P., *J. Alloys Compd.* **351** (2003), 1.
9. Slater J. C., *J. Chem.* **39** (1964), 3199.

Hyperfine Interactions (2004) 158:347–351
DOI 10.1007/s10751-005-9052-0

Electrical Field Gradient Studies on $La_{1-x}Cd_xMnO_{3+\delta}$ System

J. P. ARAÚJO[1,*], A. M. L. LOPES[2,4], T. M. MENDONÇA[1], E. RITA[3,4], J. G. CORREIA[3,4], V. S. AMARAL[2] and THE ISOLDE COLLABORATION[4]
[1]*Departamento de Física and IFIMUP, Universidade do Porto, 4169-007 Porto, Portugal; e-mail: jearaujo@fc.up.pt*
[2]*Departamento de Física and CICECO, Universidade de Aveiro, 3810-193 Aveiro, Portugal*
[3]*ITN, E.N. 10, 2686-953 Sacavém, Portugal*
[4]*CERN EP, CH-1211 Geneva 23, Switzerland*

Abstract. A Perturbed Angular Correlation (PAC) study was performed on the $La_{1-x}Cd_xMnO_{3+\delta}$ system that presents a rich variety of structural and magnetic phase transitions as a function of the oxygen content (δ) or as a function of temperature (T). The PAC signal at room temperature allowed the determination of the Electrical Field Gradient (EFG) parameters and to correlate them with the lattice average symmetry namely the orthorhombic (O' or O^*) Pbnm and rhombohedral (R) R$\bar{3}$C phases measured by X-ray diffraction.

Key Words: CMR, manganites, PAC, phase coexistence.

1. Introduction

In the recent years a lot of attention has been given to compounds exhibiting the so-called colossal magnetoresistance (CMR), i.e., the huge change of electrical resistivity due to the application of a magnetic field, rediscovered in the early nineties by Helmolt *et al.* [1] in manganites (e.g., La–CaMnO_3). The reason of such interest is twofold: their large potential for applications and the richness of their electric, magnetic and structural phase diagrams allowing an ideal playground for solid state physics studies in highly correlated electron systems.

The crystalline structures of the manganites differ from that of an ideal perovskite due to lattice distortions, arising from ion size mismatch and from the Mn^{3+} Jahn–Teller effect. These distortions always reduce the cubic symmetry of the ideal perovskite with a corresponding reduction of the Mn–O–Mn bond angle from 180°. When the tolerance factor is close to unity the rhombohedral structure is often found. This structure corresponds to a small rotation of the

* Author for correspondence.

oxygen octahedra about the [111] axis of the cubic perovskite. When the tolerance factor deviates more considerably from unity, the cation misfit is accommodated by a collective buckling of the oxygen octahedra corresponding to a rotation about the [110] axis. This structure is commonly called the *O*-type orthorhombic phase.

Since Mn^{3+} is a Jahn–Teller ion, the static Jahn–Teller effect can stabilize a more severely distorted phase, the O'-type orthorhombic structure. In this structure a strong deformation of the oxygen octahedra results in three different Mn–O bond lengths.

A remarkable property of the $LaMnO_{3+\delta}$ compound is its wide range of oxidative non-stoichiometry ($0 < \delta < 0.2$). The oxygen excess cannot be accommodated interstitially in the lattice but results in the creation of equivalent amounts La and Mn vacancies, inducing Mn^{3+} oxidation to Mn^{4+} ($Mn^{4+} = 2\delta$) [2]. Depending on δ, one of the three crystallographic structures stabilizes, namely strong Jahn–Teller distorted orthorhombic O' ($0 < \delta < 0.04$), weak Jahn–Teller distorted O^* ($0.04 < \delta < 0.11$) and rhombohedric R ($\delta > 0.11$) [3]. In addition, reports exist referring an O^*/R structural phase transition upon temperature variation with a phase coexistence in a certain range of temperatures [4].

Most of the studies on the $La_{1-x}Cd_xMnO_3$ system show that for $x \geq 0.03$ the structure is rhombohedric [5], however, a more recent study point to the existence of Orthorhombic–rhombohedric phase coexistence for samples with $x = 0.33$ [6].

Nuclear hyperfine techniques can be used to monitor, at an atomic scale, structural and magnetic phase transitions. In this work we use the Perturbed Angular Correlation (PAC) spectroscopy via the $^{111m}Cd \rightarrow {}^{111}Cd$ probe, which allows the accurate determination of the EFG parameters, at the probe site.

2. Experimental details

Polycrystalline samples were prepared by solid state reaction or sol–gel methods. In the case of the $LaMnO_{3+\delta}$ system different δ ($0 < \delta < 0.12$) were obtained by suitable thermal treatments under a controlled atmosphere. The crystallographic structures and correspondent lattice parameters were determined by Rietveld refinements of X-ray measurements. The oxygen content was determined from lattice parameters systematics obtained by other authors in samples with well-known δ [3, 4, 7]. Moreover, the Curie–Weiss (θ_p), Néel (T_N) and Curie (T_C) temperatures as well as the saturation moment, determined from magnetization measurements, are in close agreement with those presented by [3], as shown in Figure 1 (left). A Similar analysis for the $La_{1-x}Cd_xMnO_3$ system is shown in Figure 1 (right).

In order to perform PAC measurements, samples were implanted at room temperature with ^{111m}Cd ($T_{1/2} = 48$ m) to a dose of 1.0×10^{12} at/cm^2 and 60 keV energy at the ISOLDE/CERN facility [8]. ^{111m}Cd decays to the 5/2 + 245 keV

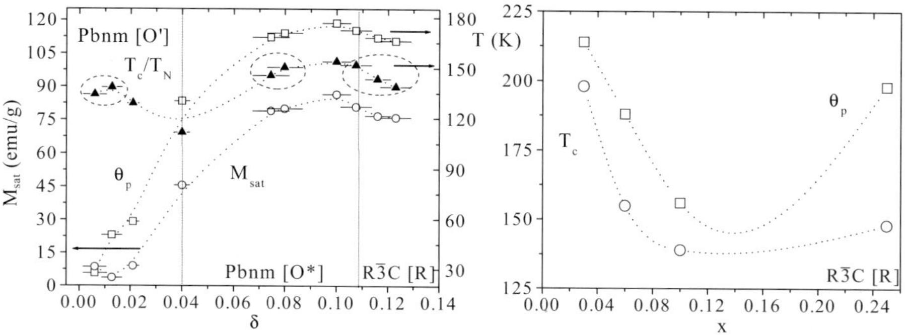

Figure 1. (*Left*) Curie–Weiss (θ_p), Néel (T_N), Curie (T_C) temperatures and saturation magnetization (M_{sat}) as a function of oxygen content (δ) for $LaMnO_{3+\delta}$ samples. *Dotted circles* indicate the oxygen content of samples of the $LaMnO_{3+\delta}$ studied by PAC. (*Right*) Curie–Weiss (θ_p), and Curie (T_C) temperatures as a function of the Cd content (x) for $La_{1-x}Cd_xMnO_3$ samples. *Lines* are guides for the eye.

probing isomeric state of ^{111}Cd. The experimental time perturbation functions, $R(t)$, were measured using a 6-BaF_2-detector spectrometer. After implantation the samples were annealed in a suitable atmosphere during 20 min to recover from implantation damage [9]. The fits to the $R(t)$ experimental PAC functions were calculated numerically by taking into account the full Hamiltonian for the nuclear quadrupole/magnetic combined interactions [10].

3. Experimental results and discussion

In the upper panel of Figure 2a representative $R(t)$ spectra are shown, measured at room temperature for $LaMnO_{3+\delta}$ samples ($x = 0$) with $\delta \sim 0.12$ (top-left) and $\delta \sim 0.08$ (top-right). The fit to each spectrum (continuous lines over the $R(t)$) shows that the Cd probes interact with one main EFG distribution, which was assumed to be Lorentzian-like. This same procedure was applied to the other samples with $x = 0$ and oxygen contents indicated by the dashed circles in Figure 1. The EFG parameters, $\langle V_{zz} \rangle$ and η, found for the main fraction for these seven samples are shown in Figure 2b (top) as a function of 2δ.

As we can see, in the orthorhombic phases ($\delta < 0.11$), only a slight increase of V_{zz} and a decrease of η was found, as function of δ. In contrast, for $\delta > 0.11$, in a small δ window, the EFG parameters change drastically to a much higher value of V_{zz} and a lower η. This is certainly related to the O^*/R phase transition, although, taking into account the higher symmetry of the R phase, the higher V_{zz} is somehow unexpected.

The lower panel of Figure 2a shows representative $R(t)$ spectra also measured at room temperature for $La_{1-x}Cd_xMnO_3$ samples ($\delta \sim 0$) with $x = 0.03$ (bottom-left) and $x \sim 0.25$ (bottom-right). Similarly, the fit to each spectrum shows that the

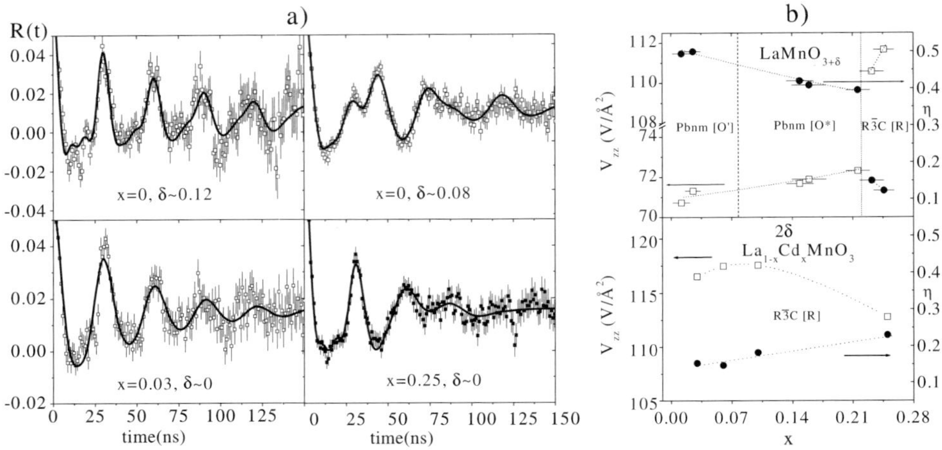

Figure 2. a) *Upper Panel*: representative $R(t)$ experimental functions measured at room temperature for $LaMnO_{3+\delta}$ ($x = 0$) samples with $\delta \sim 0.12$ (*left*) and $\delta \sim 0.08$ (*right*). *Lower panel*: representative $R(t)$ experimental functions measured at room temperature for $La_{1-x}Cd_xMnO_{3+\delta}$ ($\delta \sim 0$) samples with $x \sim 0.03$ (*left*) and $x = 0.25$ (*right*). b) Electrical Field Gradient parameters, V_{ZZ}, η for $LaMnO_{3+\delta}$ samples as a function of 2δ (*top*) and for $La_{1-x}Cd_xMnO_3$ samples as a function of x. *Dotted lines* are guides for the eye.

Cd probes interact with one main EFG distribution, which was also assumed to be Lorentzian-like. The EFG parameters, $\langle V_{zz} \rangle$ and η, found for the main fraction of $La_{1-x}Cd_xMnO_3$ samples are shown in Figure 2b (bottom) as a function of the Cd content (x). Here the V_{zz} decreases slightly, with increasing Cd content and with a corresponding small increase on η. We attribute this change on the EFG parameters to the reduction of the buckling of the MnO_6 octahedra, with corresponding increase of the Mn–O–Mn bond angle.

Finally, it is worth to mention that in all sample with rhombohedral average lattice symmetry ($\delta > 0.11$ and Cd doped samples) a non-negligible fraction of probes (~25%) interact with a second EFG distribution with V_{ZZ} and η values of the same magnitude as those found in samples with orthorhombic average lattice, even in samples that did not evidence any phase coexistence in X-ray diffraction measurements. This suggests the existence of nanoscopic lattice distortions probably related with the formation of Jahn–Teller polarons [11].

4. Conclusions

In summary we performed room temperature PAC measurements on several samples of the system $La_{1-x}Cd_xMnO_{3+\delta}$ as a function of x and δ in order to study the Cd site local vicinity via the EFG parameters. The local structure around the Cd probe was correlated with the average lattice symmetry determined by X-ray diffraction.

Acknowledgements

The authors would like to thank P. B. Tavares and R. Suryanarayanan for providing samples for this study. This work was funded by the FCT, Portugal and FEDER (projects POCTI/CTM/35462/00, CERN/FIS/43725/2001, POCTI/FNU/49509/2002, POCTI/FN/FNU/50183/2003) and the European Union (Large Scale Facility contract HPRI-CT-1999-00018). A. M. L. Lopes and E. Rita acknowledge their fellowships supported by the FCT, Portugal. Professor T. Butz from the Leipzig University is kindly acknowledged for allowing the use of the 6-detector γ-γ PAC spectrometer.

References

1. von Helmolt R., Wecker J., Holzapfel B., Schultz L. and Samwer K., *Phys. Rev. Lett.* **71** (1993), 2331.
2. Van Roosmalen J. A. M., Cordfunke E. H. P., Helmholdt R. B. and Zandbergen H. W., *J. Solid State Chem.* **110** (1994), 100.
3. Prado F., Sanchez R. D., Caneiro A., Causa M. T. and Tovar M., *J. Solid State Chem.* **146** (1999), 418.
4. Ritter C., Ibarra M. R., De Teresa J. M., Algarabel P. A., Marquina C., Blasco J., García J., Oseroff S. and Cheong S-W., *Phys. Rev., B* **56** (1997), 8902.
5. Troyanchuk I. O., Khalyavin D. D., Szymczak H., Khalyavin D. D. and Szymczak H., *Phys. Status Solidi, A* **164** (1997), 821.
6. Peña A., Gutiérrez J., Barandiarán J. M., Chapman J. P., Insausti M. and Rojo T., *J. Solid State Chem.* **174** (2003), 52.
7. Dabrowski B., Dybzinski R., Bukowski Z., Chmaissem O. and Jorgensen J. D., *J. Solid State Chem.* **146** (1999), 448.
8. Kugler E., Fiander D., Jonson B., Haas H., Przewloka A., Ravn H. L., Simon D. J. and Zimmer K., *Nucl. Instrum. Methods Phys. Res., B* **70** (1992), 41.
9. Araújo J. P., Correia J. G., Amaral V. S., Tavares P. B., Lencart-Silva F., Lourenco A. A. C. S., Sousa J. B., Vieira J. M. and Soares J. C., *Hyperfine Interact.* **133** (2001), 89.
10. Barradas N. P., Rots M., Melo A. A. and Soares J. C., *Phys. Rev., B* **47** (1993), 8763.
11. Lopes A. M. L., Araujo J. P., Ramasco J. J., Rita E., Amaral V. S., Correia J. G. and Suryanarayanan R., ISOLDE Collaboration, preprint available at the hyperlink http://arxiv.org/abs/cond-mat?0408150.

Hyperfine Interactions (2004) 158:353–359
DOI 10.1007/s10751-005-9053-z

Experimental Verification of Calculated Lattice Relaxations Around Impurities in CdTe

H.-E. MAHNKE[1,*], H. HAAS[1], V. KOTESKI[1,2], N. NOVAKOVIC[1,2], P. FOCHUK[3] and O. PANCHUK[3]
[1]*Hahn-Meitner-Institut Berlin GmbH, Bereich Strukturforschung, Glienicker Str. 100, 14109 Berlin, Germany; e-mail: mahnke@hmi.de*
[2]*VINČA, POB 522, 11001 Belgrade, Serbia and Montenegro*
[3]*University of Chernivtsi, 274012 Chernivtsi, Ukraine*

Abstract. We have measured the lattice distortion around As (acceptor) and Br (donor) in CdTe with fluorescence detected X-ray absorption spectroscopy. We could experimentally verify the lattice relaxation with a bond length reduction of 8% around the As atom as inferred indirectly from *ab initio* calculations of the electric field gradient performed with the WIEN97 package in comparison with the measured value in a Perturbed Angular Correlation experiment as recently reported. We have complemented our own calculations of relaxation with WIEN97 with calculations using the FHI96md pseudo-potential program, which allows the use of larger super-cell sizes. Encouraged by the good agreement between experiment and model calculation for As in CdTe as well as similarly for the isovalent Se in CdTe, we extended our investigation to Br in CdTe, where the electric field gradient has also been measured, and could not only verify the derived lattice expansion around Br with our EXAFS analysis but additionally observe fractions of Br in the A-center as well as in a DX-center configuration.

Key Words: calculations with DFT theories with LAPW and pseudo-potential methods, dopants in CdTe, fluorescence detected X-ray absorption, lattice relaxation, local structure.

PACS codes: 61.72.-y, 61.72.Vv, 61.10.Ht, 71.15.Ap, 71.15.Mb, 71.55.Gs.

1. Introduction

The electrical and optical properties of semiconductors are determined by the local environment of impurities. Local strain fields around the impurity atom are decisive for the stability of the defect both structurally and functionally as dopant. In the case of As and Br in CdTe acting as acceptor and donor, resp., when incorporated substitutionally on the Te sublattice, information on lattice relaxation around the impurity atom has already been obtained by hyperfine interaction studies. Making use of the donor–acceptor-pair formation between In–As and Ag–Br the electric field gradient (EFG) was determined for the non-

* Author for correspondence.

cubic configuration with the resulting pairs of ^{111}Cd–As [1] and ^{111}Cd–Br [2] after the decay of ^{111}In or ^{111}Ag, resp., employing the perturbed angular correlation (PAC) technique. In *ab initio* calculations using density functional theory (DFT) as implemented in the WIEN97 program the EFGs were reproduced when the lattice distances around the impurity atom were first allowed to relax to minimize the forces within the super-cell [1, 3]. Since the EFG has a high sensitivity to the actual size of atomic relaxations around the probing atom, it is most desirable to experimentally strengthen this approach by determining the bond lengths around the impurity atom directly, which can be achieved with fluorescence detected X-ray Absorption Fine Structure (XAFS with X-ray Absorption Near-Edge Spectroscopy XANES and Extended X-ray Absorption Fine-Structure EXAFS).

The challenge in such experiments lies in the necessary compromise between a concentration high enough for a XAFS detection, but low enough to avoid compensation and clustering of the dopants in the semiconductors. Since a site sensitive XAFS detection has not yet been successfully developed for these systems, one has to be sure that the signal from other possible configurations does not obscure the signal from the assumed purely substitutional one. While for As in CdTe, other conflicting configurations might be various cluster configurations, the most likely alternative configurations for Br in CdTe are the well-known A-center [4] or various forms of DX-centers [5].

2. Experiment

2.1. EXPERIMENTAL DETAILS

While the incorporation of dopants like Br is accomplished in crystal growing from the melt, the incorporation of arsenic as the acceptor dopant in CdTe was done by ion implantation with proper treatment. At the tandetron of FZ Rossendorf, a total dose of 7×10^{15} cm^{-2}-As atoms was implanted almost uniformly distributed up to a depth of 3 μm. The samples, originally from Crystec GmbH,* were thermally treated before and after the implantation following the approaches [6] in which As was predominantly incorporated as acceptor at the Te site as checked by photoluminescence (PL) measurements. To reduce background in the fluorescence-detected EXAFS from undoped sample material, the crystals were thinned from the back to a total thickness of about 30 μm. In the case of Br in CdTe, we used powdered samples, mixed with graphite and polyethylene, and pressed into pellets. Concerning the Br concentration we can only estimate from the concentration in the melt of 5×10^{19} cm^{-3} that the final Br concentration is in the range of a few 10^{18} cm^{-3}.

* Crystec GmbH, D-12555 Berlin.

To determine the local structure around the impurity elements we have measured the K-edge absorption at the X1-beamline of HASYLAB at DESY. In the case of As we measured the absorption in the fluorescence mode with the 5-segment Ge-detector, the fluorescence being detected in line with the polarization vector of the incoming synchrotron radiation. To reduce the influence from elastic and inelastic scattering of the incoming radiation, critical absorption was employed using a foil made out of Ge powder mixed with polyethylene. In the case of Br, a 7-segment Ge-detector could be used with an improved resolution yielding a better sensitivity for lower concentrations. For the measurements, the samples were mounted on a He gas flow cryostat, keeping the temperature at either approximately 18 or 300 K.

The analysis of the observed absorption spectra was done following the standard FEFF procedure [7, 8] including the background treatment with AUTOBK [9].

2.2. THEORETICAL DETAILS

Complementing our experimental work we have performed *ab initio* calculations based on the density functional theory (DFT) with the WIEN97 package [10] which uses the linearized augmented plane wave (LAPW) method, confirming and extending the calculation by Lany *et al.* [1], and with the FHI96md program [11] which uses first-principles pseudo-potentials (PP) and a plane-wave basis set. The latter was used to investigate the size dependence of the super-cells constructed around one substitutional As atom in CdTe. The calculations yielded good agreement with our EXAFS experiment (see Table I) so that the determined relaxations can be taken as a solid basis for further interpretations of derived parameters such as hyperfine interaction parameters of defect complexes (for further details see [1, 12]).

3. Results and discussion

In the case of As in CdTe our EXAFS result yielded a strong inward relaxation of 8% in perfect agreement with the values calculated using the LAPW method which has been applied in the EFG calculation as well as with the pseudo-potential method with cell sizes up to 216 atoms. For comparison we have also determined the relaxation around Se, elemental neighbor to As and isovalent to the anion Te, in a mixed-crystal $CdTe_{1-x}Se_x$ with a concentration of $x = 0.04$ which comes close to the super-cell size of 32 atoms. Again we find perfect agreement between experiment and model calculation with an inward relaxation of 6%. The results are summarized in Table I, more details can be found in [13].

The case of Br in CdTe already provides the opportunity for a more sophisticated analysis of a more complex doping situation. A first EXAFS analysis

Table I. Comparison of experimentally determined nearest neighbor and next-nearest neighbor distances around As, Se and Br impurity atoms in CdTe with calculated values obtained with LAPW (super-cell size 32 atoms) and with PP methods (super-cell size up to 216 atoms)

System	R_{nn} (exp)	R_{nn} (theo)		R_{nnn} (exp)	R_{nnn} (theo)	
		LAPW	PP		LAPW	PP
CdTe	2.806			4.583		
As in CdTe	2.58 (2)	2.57[a]	2.57	4.50 (6)	4.52[a]	4.55
Se in CdTe	2.67 (1)	2.66		4.53 (1)	4.55	
Br in CdTe	2.83 (5)[b]	2.90		4.54 (7)[b]	4.57	

The distances in pure CdTe are derived from the standard lattice parameter. (All values are given in units of 0.1 nm (=1 Å).)
[a] Values as presented in [1].
[b] Assuming substitutional Br as the only configuration.

yields a bond length which is in fair accordance with the bond length derived in the WIEN calculation in reproducing the measured EFG at Cd with one Br neighbor in the Te nearest neighbor shell (see Table I) (the preliminary result was already presented at the E-MRS Spring Meeting 2004 [14]). However, compared to the perfect agreement obtained for the cases As and Se (within much less than 1%), the analysis seems to be questionable. We have therefore analyzed the absorption signal on Br, both in the XANES region and in the EXAFS region, more carefully. Indeed, the larger width of the first shell signal (in R-space) compared to As indicates the possibility of a superposition of different configurations. Since our Br concentration is close or even above the maximum for an uncompensated doping concentration, we included the A-center configuration known as a possible compensating configuration in our analysis. In the XANES region, the model absorption spectra are very similar for both, the substitutional Br and the A-center Br, resp., however it would be very different for a differently charged Br, for which an illustration was presented already in [14]. There is a clear trend that the XANES spectrum corresponds mainly to positively charged Br configurations.

In the EXAFS region, including a sizeable fraction of Br in the A-center configuration significantly improves the fit, however with the distances fixed. A fully satisfactory fit is obtained by including a DX-center configuration as a third contribution of the same size as the substitutional fraction (Figure 1), again however with fixed distances. As distances we took the calculated values R_{nn} = 2.90 Å for substitutional BrTe, 2.80 Å for the A-center Br, and for the DX-center following [5] 2.60 Å for Br–Cd (3-fold coordinated) and 3.10 Å for the more relaxed Br–Cd distance. The next-nearest neighbor shell was included with a fixed value taken from the pure system. The occurrence of various configurations which we experience here may also explain the relatively small fraction with which the PAC experiment detected the ^{111}Ag–Br pair [2].

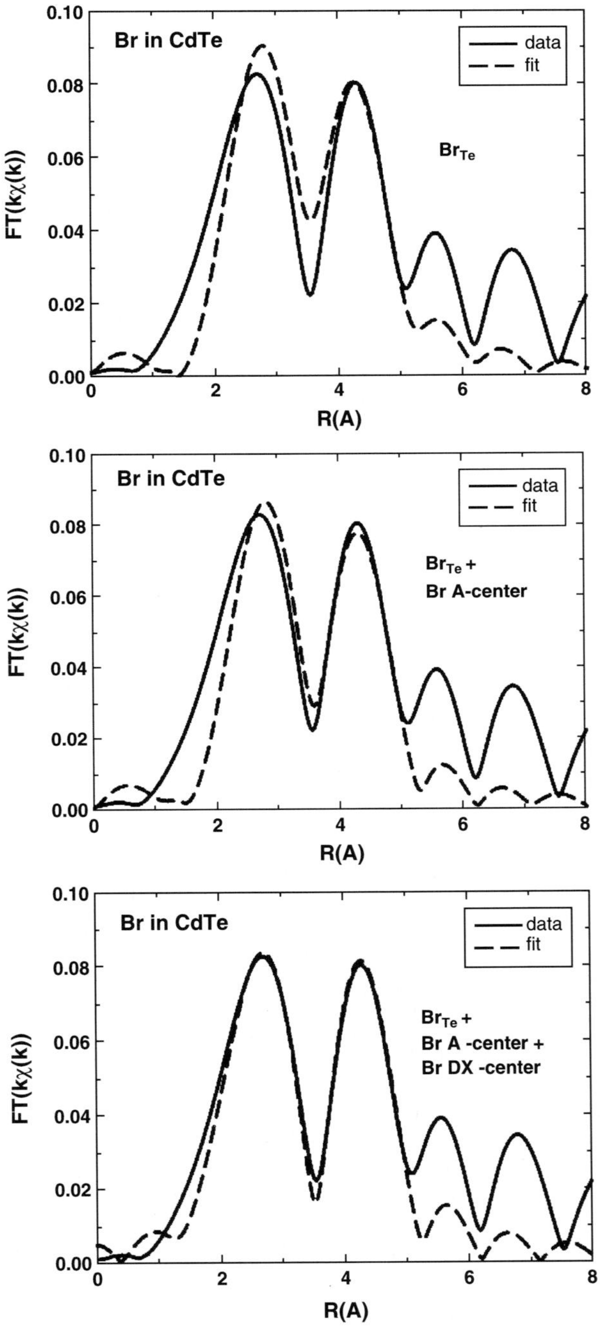

Figure 1. Radial distribution function FT(kχ(k)) of Br in CdTe together with a fit assuming substitutional Br as the only configuration (*top*), with 60% of the Br in the substitutional and 40% in the A-center configuration (*center*), and with Br in three configurations, 30% in the substitutional, 40% in the A-center and 30% in the DX-center configuration (*bottom*).

In summary, in both cases of dopants, the acceptor As as well as the donor Br in CdTe, for which the determination of the EFG on the neighboring Cd atom has pointed towards a sizeable change in atomic distances, these distances have now been measured with EXAFS. Both, the large decrease of 8% for As in CdTe and the remarkable increase of more than 3% for Br in CdTe were found in perfect agreement with the WIEN calculation which reproduced the observed EFGs. In the case of As in CdTe, we additionally tested the theoretical approach by using the pseudo-potential program FHI96md which allowed to use a larger super-cell size finding no significant differences. In the case of Br in CdTe, the absorption spectroscopy revealed the contribution of other configurations to Br substitutionally incorporated which we tentatively identify as the expected A-center and a DX-center.

Acknowledgements

The authors are grateful to the HASYLAB staff at DESY, in particular to N. Haack, E. Welter, and J. Wienold. We thank E. Holub-Krappe for her help in the early phase and H. Rossner for his advice on analysis and fitting of the data and P. Szimkowiak for his help in sample preparation. Help in sample preparation is also acknowledged for the As implantation by M. Friedrich (FZ Rossendorf), and crystal thinning and PL tests by S. Lany, F. Wagner and H. Wolf (University Saarbrücken). We further thank J. Bollmann and J. Weber from the TU Dresden, S. Lany and T. Wichert from the U Saarbrücken, and B. Cekic from VINČA.

References

1. Lany S. *et al.*, *Phys. Rev.*, **B62** (2000), R2259.
2. Ostheimer V. *et al.*, *Phys. Rev.*, **B68** (2003), 235206.
3. Lany S., *Elektronische und strukturelle Eigenschaften von Punktdefekten in II-VI Halbleitern*, PhD thesis, Universität des Saarlandes, Saarbrücken, Shaker, Aachen, 2003, ISBN 3-8322-1302-3.
4. Stadler W. *et al.*, *Phys. Rev.*, **B51** (1995), 10619.
5. Park C. H. and Chadi D. J., *Phys. Rev.*, **B52** (1995), 11884.
6. Lany S. *et al.*, *Physica, B* **302–303** (2001), 114 and ref. therein.
7. Rehr J. J., Mustre de Leon J., Zabinsky S. I. and Albers R. C., *J. Am. Chem. Soc.* **113** (1991), 5135.
8. Stern E. A., Newville M., Ravel B., Yacoby Y. and Haskel D., *Physica* **B208 & 209** (1995), 117.
9. Newville M., Livins P., Yacoby Y., Rehr J. J. and Stern E. A., *Phys. Rev.*, **B47** (1993), 14126.
10. Blaha P., Schwarz K. and Luitz J., *WIEN97, A Full Potential Linearized Augmented Plane Wave Package for Calculating Crystal Properties*, Karlheinz Schwarz, TU Wien, Austria, 1999, ISBN 3-9501031-0-4.

11. Bockstedte M., Kley A., Neugebauer J. and Scheffler M., *Comput. Phys. Comm.* **107** (1997), 187.
12. Koteski V., Ivanovic N., Haas H., Holub-Krappe E. and Mahnke H.-E., *NIM,* **B200** (2003), 60.
13. Koteski V., Haas H., Holub-Krappe E., Ivanovic N. and Mahnke H.-E., *Phys. Scr.*, **T115** (2005), 369.
14. Mahnke H.-E., Haas H., Holub-Krappe E., Koteski V., Novakovic N., Fochuk P. and Panchuk O., *Thin Solid Films*, **480–481** (2005), 279.

Hyperfine Interactions (2004) 158:361–364
DOI 10.1007/s10751-005-9054-y

Hyperfine Interactions of Short-Lived β Emitters in Pd

M. MIHARA[1,*], S. KUMASHIRO[1], K. MATSUTA[1], Y. NAKASHIMA[1], H. FUJIWARA[1], Y. N. ZHENG[2], M. OGURA[1], H. AKAI[1], M. FUKUDA[1] and T. MINAMISONO[1,†]

[1]*Graduate School of Science, Osaka University, Toyonaka, Osaka 560-0043, Japan; e-mail: mihara@vg.phys.sci.osaka-u.ac.jp*

[2]*China Institute of Atomic Energy, Beijing 102413, People's Republic of China*

Abstract. The Knight shifts K and the spin-lattice relaxation time T_1 for the short-lived β-emitters ^{12}B and ^{12}N implanted into Pd have been measured by means of the β-NMR method. The results show that K depends on temperature. The obtained values of experimental K were compared with theoretical values derived with the KKR method.

Key Words: β-NMR, impurity, Knight shift, spin-lattice relaxation time.

1. Introduction

The hyperfine interaction of impurities in metals is of great importance to study the electronic structure of impurities in metals. The β-NMR method using radioactive nuclear probes is one of the most powerful techniques to investigate the hyperfine interaction through the Knight shift K and the spin-lattice relaxation time T_1. The Knight shifts for light impurities in Pt have been studied systematically [1] as a function of valence number of the impurity atoms. Based on the Korringa–Kohn–Rostoker (KKR) band structure calculation [2], the valence number dependence of experimental K in the second row of the periodic system was explained by the systematic change of the position of the anti-bonding state formed by the s–d hybridization [1, 3]. In the case of Pd, the situation is expected to be similar as in the Pt case because of a large density of states in the 4d band. However, the Knight shifts for Pd have not been studied well. In the present study, we have measured, as a function of temperature (100–600 K) by means of the β-NMR technique, K and T_1 for the short-lived β emitters ^{12}B ($I = 1$, $T_{1/2} = 20$ ms) and ^{12}N ($I = 1$, $T_{1/2} = 11$ ms) implanted into Pd samples.

* Author for correspondence.

† Present address: Fukui University of Technology, Fukui 910-8505, Japan.

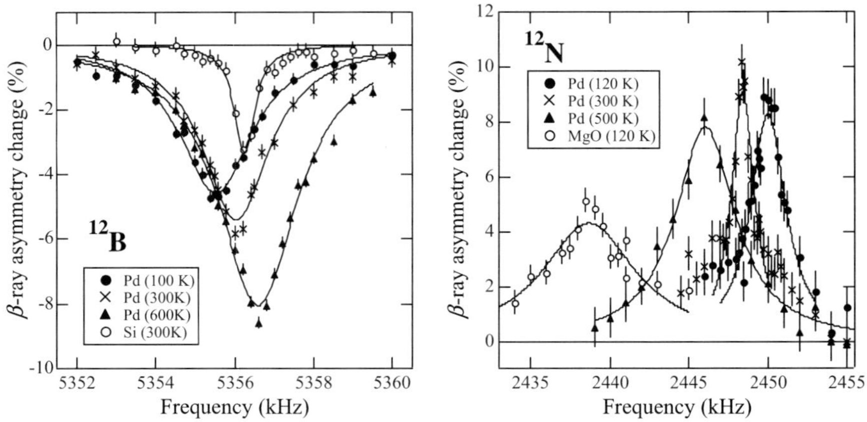

Figure 1. NMR spectra for ^{12}B in Pd and Si, and for ^{12}N in Pd and MgO measured at an external magnetic field of 0.7 T. The *solid curves* are the results of a fit to the data to a Lorenzian distribution.

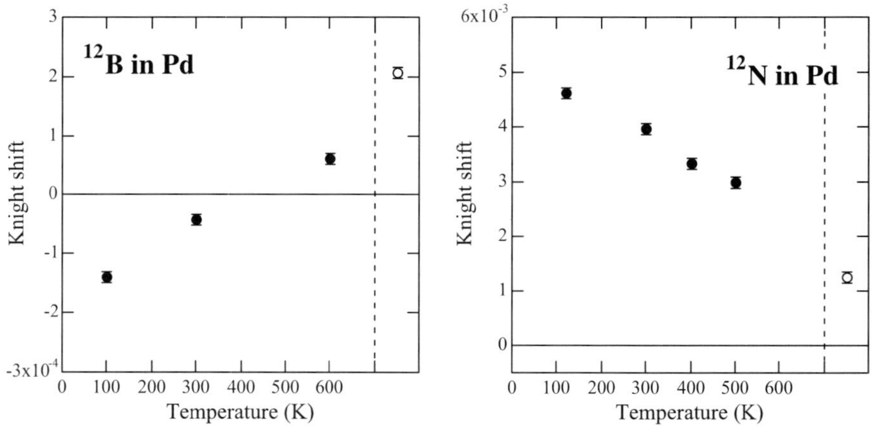

Figure 2. Temperature dependence of the Knight shifts for ^{12}B and ^{12}N in Pd (*solid circle*). The Knight shifts estimated from the T_1T data based on the Korringa relation are indicated with the *open circle*.

2. Experimental

The experimental method and procedure were basically the same as described previously [1, 3]. The experiment was performed at the Van de Graaff accelerator facility in Osaka university. The ^{12}B and ^{12}N nuclei were produced through the nuclear reactions ^{11}B(d, p)^{12}B and ^{10}B(^{3}He, n)^{12}N using 1.5-MeV d and 3.0-MeV ^{3}He beams, respectively. Recoil nuclei were selected with a collimator at 32°–48° for ^{12}B and 12°–28° for ^{12}N to generate spin polarization. They were implanted into a catcher, Pd foil, placed in an external magnetic field $\mu_0 H_0$ of 0.7 T applied parallel to the polarization. Two sets of a plastic scintillation counter telescope were placed above and below the catcher foil along with the

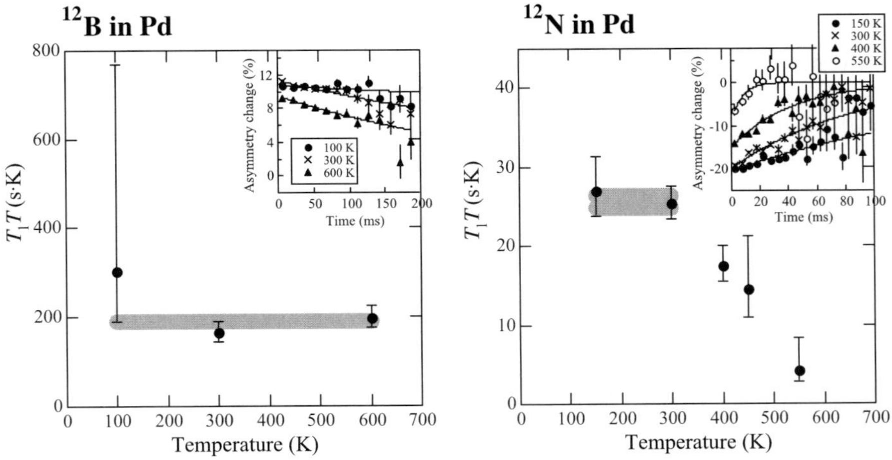

Figure 3. Temperature dependence of T_1T for ^{12}B and ^{12}N in Pd. The inserts are the β-ray asymmetry change measured as a function of time, from which the values of T_1 were deduced by a fit to the data using an exponential function. The *shaded regions* indicate averaged T_1T where the values are constant in temperature.

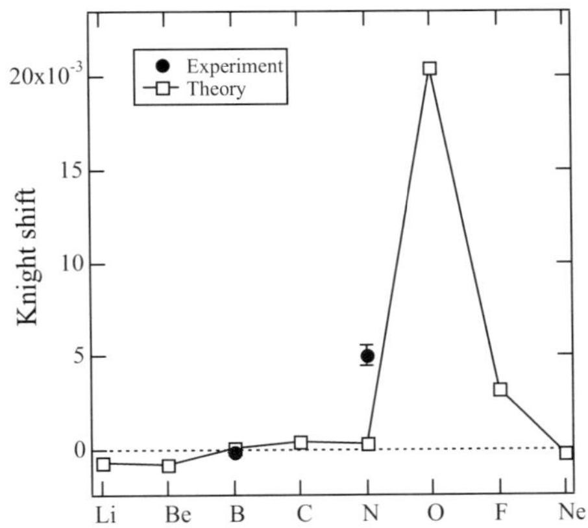

Figure 4. Experimentally determined Knight shifts (*full circles*) and the theoretical ones (*open squares*) for light impurities in Pd at the octahedral interstitial site.

polarization axis in order to detect the β-ray asymmetry. An rf oscillating magnetic field H_1 was applied perpendicular to H_0 to induce transitions. The NMR effect was detected through the β-ray asymmetry change after applying the rf.

3. Results and discussion

Typical NMR spectra observed for ^{12}B and ^{12}N in Pd are shown in Figure 1. It was found that the spectra depend on temperature for both ^{12}B and ^{12}N. NMR spectra for ^{12}B in Si, and for ^{12}N in MgO were observed as reference, which are also shown in Figure 1. These spectra were analyzed using a Lorenzian distribution to extract the resonance frequencies. The Knight shifts were deduced as the shifts between the resonance frequency in Si or MgO and the one in Pd. The deduced values of K are plotted as a function of temperature in Figure 2. A slightly negative value of K for ^{12}B, $-(4.1 \pm 0.7) \times 10^{-5}$ was obtained at 300 K. This value disagrees with the previous ones, $-(32 \pm 15) \times 10^{-5}$ [4] and $+(11 \pm 4) \times 10^{-5}$ [5] measured also at room temperature. The temperature dependence shows that K for ^{12}B crosses zero at little above room temperature and changes to be positive with increasing temperature. On the other hand the results for ^{12}N show a positive K but an opposite slope in the temperature dependence as compared to ^{12}B. These results might be explained mainly by the hybridization of the s state of impurity and the d electrons of Pd, which causes the splitting of the bonding and anti-bonding states. By analogy with the case of Pt [1], the contribution of the bonding state to K is expected to be negative, and as the valence number of the impurity increases, the anti-bonding state becomes lower and passes through the Fermi level, which causes a large positive value of K. The temperature dependence is probably caused by the large and temperature-dependent susceptibility of Pd due to the d-electron polarization [6], which affects the s–d hybridization.

The spin-lattice relaxation times T_1 were extracted from the time dependence of polarization observed in between 100 and 600 K (Figure 3). T_1T for ^{12}B is nearly constant from 100 to 600 K, while for ^{12}N it begins to decrease at above 300 K. The values of T_1T averaged where they are constant in temperature are deduced as (189 ± 17) s·K for ^{12}B and (25.7 ± 1.8) s·K for ^{12}N. The Knight shifts using the Korringa relation calculated from the averaged T_1T are also displayed in Figure 2. The present data for the Knight shifts are compared with the KKR band structure calculation for light impurities in Pd at the octahedral interstitial site as shown in Figure 4. The effect of the local lattice expansion is considered in the calculation [1]. The experimental values of K at 0 K were estimated and plotted in Figure 4 by extrapolating to 0 K from the present temperature dependence with a linear function, because the calculation was performed only at 0 K. The theoretical value of K for ^{12}N is smaller than the experimental one, which seems to be due to incorrect estimation of the local lattice expansion.

One of the authors was partly supported by the Tokui scholarship.

References

1. Matsuta K. *et al.*, *Hyperfine Interact.* **120/121** (1999), 719.
2. Akai H. *et al.*, *Prog. Theor. Phys., Suppl.* **101** (1990), 11.
3. Matsuta K. *et al.*, *Hyperfine Interact.* **97/98** (1996), 501.
4. Williams R. L. Jr., Pfeiffer L., Wells J. C. Jr. and Madansky L., *Phys. Rev., C* **2** (1970), 1219.
5. McDonald R. E. and McNab T. K., *Phys. Rev., C* **10** (1974), 946.
6. Seitchik J. A., Gossard A. C. and Jaccarino V., *Phys. Rev.* **136** (1964), A1119.

Hyperfine Interactions (2004) 158:365–370
DOI 10.1007/s10751-005-9055-x

Hyperfine Interactions in Iron Meteorites: Comparative Study by Mössbauer Spectroscopy

M. I. OSHTRAKH[1,*], O. B. MILDER[2], V. I. GROKHOVSKY[1] and V. A. SEMIONKIN[2]
[1]*Faculty of Physical Techniques and Devices for Quality Control, Ural State Technical University–UPI, Ekaterinburg, 620002, Russian Federation; e-mail: oshtrakh@mail.utnet.ru*
[2]*Faculty of Experimental Physics, Ural State Technical University – UPI, Ekaterinburg, 620002, Russian Federation*

Abstract. The iron meteorites Sikhote–Alin, Bilibino, Chinga and Dronino with different Ni concentration and terrestrial age were studied by Mössbauer spectroscopy. Different Mössbauer hyperfine parameters were determined for studied meteorites and possible Fe–Ni phases were supposed.

Key Words: hyperfine interactions, iron meteorites, Mössbauer spectra.

1. Introduction

Physical characterization of iron meteorites is useful for understanding the origin of extraterrestrial metal formation and its further metamorphism in terrestrial conditions. Usually iron meteorites contain Fe–Ni alloys with various Ni concentrations in one or more phases. Metallic phases of meteorites frequently contain unusual structures which were formed in natural conditions [1]. However, it is not possible to create these structures by modern technologies. Therefore, the study of the features of the phase transformations in meteorites is of interest. Mössbauer spectra of iron meteorites are usually in the form of sextets with non-Lorentzian lines shapes and different line widths and intensities. Therefore, Mössbauer spectra of meteorites may be fitted using several sextets with small differences of the hyperfine field related to various metallic phases. In this work we present the preliminary results of Mössbauer study of the hyperfine interactions in iron meteorites with different Ni concentration and terrestrial age.

* Author for correspondence.

2. Materials and methods

Meteorites Sikhote–Alin and Bilibino were classified as coarse octahedrite (IIB) with about 5.8–6 wt.% of Ni, meteorite Chinga was classified as ataxite (IVB) with about 16.8 wt.% of Ni and meteorite Dronino was classified as anomalous ataxite with about 9.3 wt.% of Ni. The meteorite phases were studied by electron probe microanalysis using scanning electron microscopes JEOL JSM–U3 and Philips 30XL with EDS to determine the phase compositions. The terrestrial ages for these meteorites were evaluated about 200,000 years for Bilibino, about 2,000 years for Chinga, more than 1,000 years for Dronino and about 60 years for Sikhote-Alin. Meteorite samples were prepared as foil and powder.

Mössbauer spectra of meteorite samples were measured at room temperature with a constant acceleration computerized high precision and stable spectrometer that was a part of mult-dimension parametric Mössbauer spectrometer SM–2201 [2]. The noise of velocity signal of spectrometer was 1.5×10^{-3} mm/s, the drift of zero point velocity was $\pm 2.6 \times 10^{-3}$ mm/s, the nonlinearity of velocity signal was 0.01%, the harmonic distortion factor was 0.005% for the frequency band in the range of 0–1120 Hz. A 0.5×10^9 Bq ^{57}Co(Cr) source was used at room temperature. Mössbauer spectra were measured at room temperature in transmission geometry with moving absorber. Mössbauer spectra were computer fitted with the least squares procedure using Lorentzian line shape. Mössbauer parameters isomer shift δ, quadrupole splitting ΔE_Q, hyperfine magnetic field H_{eff}, line width Γ and subspectrum area S were determined. The values of isomer shift are given relative to α–Fe at 295 K.

3. Results and discussion

Mössbauer spectra of meteorite samples are shown in Figure 1. These spectra are similar and look like usual magnetic sextet. However, the results of the better fit of these spectra showed the presence of different quantities of components which could be related to various Fe–Ni phases. Mössbauer parameters obtained from the better fit, possible Fe–Ni phases and Ni content determined by electron probe microanalysis are given in Table I.

Previous study of b.c.c. Fe–Ni alloys prepared in non-equilibrium conditions [3] showed variation of H_{eff} with Ni concentration. Comparison of these results with our data obtained for iron meteorites is shown in Figure 2. The values of H_{eff} for kamacite phases in Bilibino (1), Chinga (2) and Sikhote–Alin (1) were lower than that obtained in [3]. These results for Chinga and Sikhote–Alin meteorites may be related to equilibrium extraterrestrial conditions of kamacite formation. The data for Bilibino meteorite can be explained as a result of the weathering-induced recrystallization process leading to the low temperature equilibrium due to discontinues precipitation reaction in kamacite [4]. The kamacite phase in Bilibino (2) was initially formed in equilibrium extraterrestrial

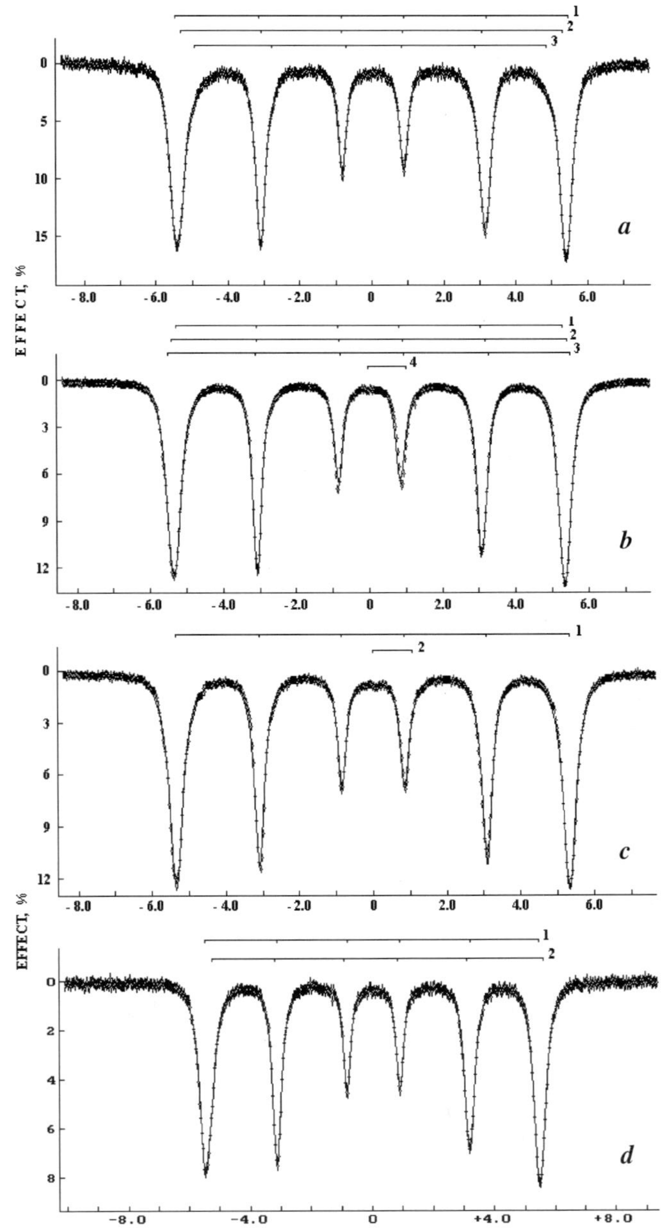

Figure 1. Mössbauer spectra of meteorites Chinga (***a***), Bilibino (***b***), Sikhote–Alin (***c***) and Dronino (***d***) at 295 K. 1–4 are components resulting from the better fit.

conditions, however, its value of H_{eff} coincided with non-equilibrium data from [3]. This fact may be explained as a result of the structural disturbance of residues of the kamacite phase (2) formed in equilibrium conditions due to influence of new kamacite phase (1). The kamacite phase in Dronino (1)

Table I. Mössbauer parameters of iron meteorites at 295 K

Sample	Γ^a, mm/s	δ, mm/s	ΔE_Q, mm/s	H_{eff}, kOe	S, %	Compound[b]	at. % Ni[c]
Chinga	0.355 ± 0.064	0.021 ± 0.032	−0.053 ± 0.032	338.9 ± 1.0	61	α_2–Fe(Ni) (1)	19.6
	0.326 ± 0.064	0.018 ± 0.032	−0.011 ± 0.032	328.9 ± 1.0	34	Kamacite (2)	4.8
	0.350 ± 0.064	0.016 ± 0.032	−0.083 ± 0.032	302.8 ± 1.0	5	Tetrateanite (3)	50.0
Bilibino	0.377 ± 0.064	0.008 ± 0.032	0.012 ± 0.032	328.0 ± 1.0	72	Kamacite (1)	4.3
	0.258 ± 0.064	0.029 ± 0.032	−0.063 ± 0.032	335.5 ± 1.0	18	Kamacite (2)	4.9
	0.336 ± 0.064	0.043 ± 0.032	−0.077 ± 0.032	341.7 ± 1.0	10	α_2–Fe(Ni) (3)	17.2
	0.250 ± 0.064	0.507 ± 0.032	1.036 ± 0.032	–	1	High spin ferric compound (4)	–
Sikhote–Alin	0.430 ± 0.064	0.011 ± 0.032	−0.009 ± 0.032	331.1 ± 1.0	98	Kamacite (1)	5.2
	0.400 ± 0.064	0.545 ± 0.032	1.089 ± 0.032	–	2	High spin ferric compound (2)	–
Dronino	0.407 ± 0.076	0.021 ± 0.038	−0.078 ± 0.038	339.6 ± 1.2	80	Kamacite (1)	6.7
	0.327 ± 0.076	0.066 ± 0.038	−0.218 ± 0.038	337.3 ± 1.2	20	α_2–Fe(Ni) (2)	25.5

[a] The line widths for sextets are given for the 1st and the 6th lines.
[b] The numbers of components in Figure 1 are indicated in parentheses.
[c] Ni concentrations were determined by electron probe microanalysis.

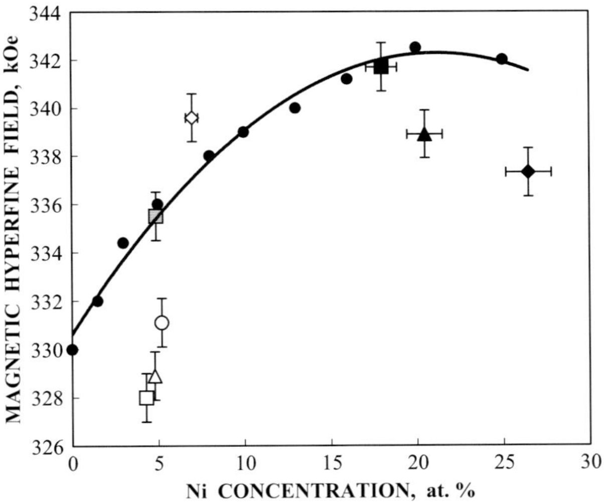

Figure 2. Variation of the magnetic hyperfine field with Ni concentration in b.c.c. Fe–Ni alloys at room temperature: ● – data from [3], □ – Bilibino, kamacite (1), ■ – Bilibino, kamacite (2), ■ – Bilibino, α_2–Fe(Ni) (3), △ – Chinga, kamacite (2), ▲ – Chinga, α_2–Fe(Ni) (1), ○ – Sikhote–Alin, kamacite (1), ◇ – Dronino, kamacite (1), ◆ – Dronino, α_2–Fe(Ni) (2). The numbers in parentheses indicate components in Figure 1.

contained ~0.7 at. % of Co that was higher than that in other meteorite phases. In this case H_{eff} was found larger than for other kamacite phases. The phase α_2–Fe(Ni) found in Bilibino (3), Chinga (1) and Dronino (2) is non-equilibrium ferromagnetic Fe–Ni phase with high Ni concentration which is usually named martensite. The values of H_{eff} for these phases correlated with data from [3] for Bilibino only while those values for Chinga and Dronino were lower. It was determined that α_2–Fe(Ni) phase formation in meteorite Chinga was in extraterrestrial equilibrium conditions. Formation of the α_2–Fe(Ni) phase in meteorite Dronino was considered as a result of the shock-heating process with further cooling.

4. Conclusion

The iron meteorites contained different Fe–Ni phases and were characterized by different values of H_{eff} that were related to the different Ni concentration and the type of phase transformation. Moreover, this Mössbauer study showed that the same Fe–Ni phases (both kamacite and martensite) had small differences of the values of H_{eff} that may be related to the small structural peculiarities arising as a result of different mechanisms of phase transformations in various meteorites. However, this supposition requires further investigations.

References

1. Buchwald V. F., *Handbook of Iron Meteorites*, University of California Press, 1975, p. 1418.
2. Irkaev S. M., Kupriyanov V. V. and Semionkin V. A., British Patent No 10745 (7 May, 1987).
3. Vincze I., Campbell I. A. and Meyer A. J., *Solid State Commun.* **15** (1974), 1495.
4. Grokhovsky V. I., *Meteorit. Planet. Sci.* **32** (1997), A52.

Hyperfine Interactions (2004) 158:371–375
DOI 10.1007/s10751-005-9061-z

A Perturbed-Angular-Correlation Study of Hyperfine Interactions at ^{181}Ta in α-Fe_2O_3

A. F. PASQUEVICH[1,*,a], A. C. JUNQUEIRA[2], A. W. CARBONARI[2] and R. N. SAXENA[2]
[1]*Departamento de Física, Facultad de Ciencias Exactas, Universidad Nacional de La Plata, cc 67, 1900 La Plata, Argentina; e-mail: pasquevi@fisica.unlp.edu.ar*
[2]*Instituto de Pesquisas Energéticas e Nucleares (IPEN), Sao Paulo, Brazil*

Abstract. The hyperfine interactions at ^{181}Ta ions on Fe^{3+} sites in α-Fe_2O_3 (hematite) were studied in the temperature range 11–1100 K by means of the perturbed angular correlation (PAC) technique. The ^{181}Hf(β^-)^{181}Ta probe nuclei were introduced chemically into the sample during the preparation. The hyperfine interaction measurements allow to observe the magnetic phase transition and to characterize the supertransferred hyperfine magnetic field B_{hf} and the electric field gradient (EFG) at the impurity sites. The angles between B_{hf} and the principal axes of the EFG were determined. The Morin transition was also observed. The results are compared with those of similar experiments carried out using ^{111}Cd probe.

Key Words: electric field gradient, hematite, perturbed angular correlation, supertransferred magnetic field.

1. Introduction

The magnetic hyperfine fields in iron oxides have been the subject of several investigations in the past, specially using Mössbauer spectroscopy [1]. Supertransferred hyperfine fields at ^{111}Cd impurity sites in iron oxides were measured by means of the Perturbed Angular Correlation (PAC) technique [2–7]. These studies provided valuable information on the magnetic phase transitions involved in magnetite (Fe_3O_4) and hematite (α-Fe_2O_3). An interesting feature of these experiments is that while aftereffects of the EC decay of ^{111}In to ^{111}Cd appear to play a role in the second case [5–7] they are absent in the first one [2–4, 8]. The aim of the present paper is to investigate the hyperfine fields at ^{181}Ta impurity site in α-Fe_2O_3 by means of the other commonly used PAC radioactive isotope ^{181}Hf, which decays to ^{181}Ta through β^- emission. Aftereffects due to the decay should not occur because the upper level of the 133–482 keV γ–γ cascade has a half life of 12 μs, sufficiently long to allow for electronic adjustment prior to coincidence counting.

*Author for correspondence.
[a]Also at Comisión de Investigaciones Cientificas de la Provincia de Buenos Aires, Argentina.

Among the iron oxides, corundum-type α-Fe_2O_3 (hematite) is the most common on earth. At room temperature, it crystallizes in the rhombohedral corundum structure (space group $R\overline{3}c$). The magnetic moments of Fe^{+3} ions are ordered antiferromagnetically below the Néel temperature T_N = 955 K. As the $^{181}Hf \rightarrow {}^{181}Ta$ probe is expected to substitute the Fe^{+3} in the α-Fe_2O_3 lattice and since the point symmetry of these sites is trigonal with a three-fold axis parallel to [111] axis, ^{181}Ta should experience an axially symmetric Electric Field Gradient (EFG) along this direction. Below the Neél temperature, α-Fe_2O_3 is an antiferromagnetic insulator showing weak ferromagnetism above the Morin temperature, T_M = 260 K due to a slight canting of the two sublattice magnetizations. Below T_M, the direction of the magnetic moments is parallel to the [111]-axis of the hexagonal unit cell while above T_M, the magnetic moments lie in the (111) plane.

2. Experimental

Polycrystalline samples of Fe_2O_3 doped with ^{181}Hf were prepared by adding ~100 μCi of ^{181}Hf as HfF_4 in a dilute hydrofluoric acid solution to a solution of $Fe(NO_3)_3$ obtained by dissolving iron metal in concentrated HNO_3. The radioactive ^{181}Hf was obtained by irradiating approximately 1 mg of Hf metal in the IEA-R1 reactor at IPEN for 64 h with a neutron flux of $\sim 2 \times 10^{13}$ n/cm^2.s. The solution was evaporated to complete dryness and the resulting powder was pressed into a small pellet and sintered for 12 h at 1100 K in air. The PAC spectra were recorded at several temperatures, between 11–1100 K, using a standard setup with four conical BaF_2 detectors arranged in a planar 90–180° geometry, generating simultaneously 12 delayed coincidence spectra. The detector system had a time resolution of ~600 ps. A small tubular furnace was used for the measurements above room temperature and the temperature was controlled to within 1 K. For low-temperature measurements the sample was attached to the cold finger of a closed-cycle helium refrigerator with temperature controlled to better than 0.1 K. An X-ray powder diffractogramme of the sample to make sure that it is α-Fe_2O_3 would have been desirable. But it is known that the calcination of iron salts at high temperature under oxygen always produces hematite. Additionally, the quadrupole interaction data obtained in the paramagnetic phase and the determined phase transition temperature described below are unambiguous proof that we are dealing with hematite.

3. Results and discussion

Typical PAC spectra taken at some of the temperatures are shown in Figure 1. At temperatures above 955 K the spectra are well fitted assuming only one well defined electric quadrupole interaction implying that all the ^{181}Hf probes are located on the Fe^{+3} sites in the oxide and are in the identical environment. The

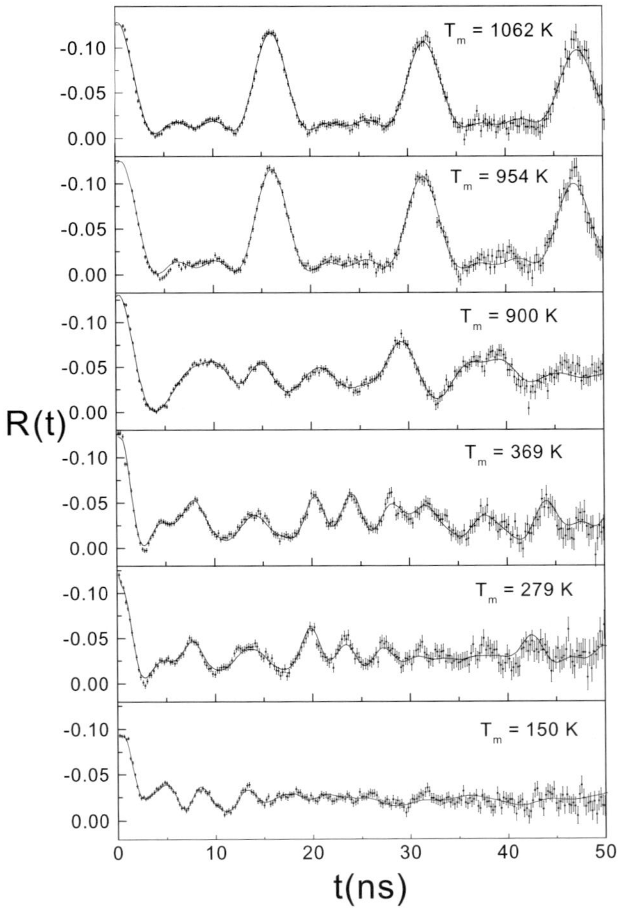

Figure 1. The PAC spectra of ^{181}Ta in α-Fe_2O_3 at different temperatures.

spectrum at 1062 K is a typical one for an axially symmetric EFG as is expected from the fact that α-Fe_2O_3 is paramagnetic at this temperature. The measured hyperfine parameters for the quadrupole interactions at temperatures $T > 955$ K remain nearly constant. The values obtained at 1062 K are: $\nu_Q = 424.1(4)$ Mhz, $\eta = 0.06$ (1) and $\delta = 0.016(2)$. Below T_N, the spectra become complex and temperature dependent indicating combined electric quadrupole and magnetic dipole interactions. The parameters of the electric field gradient show little change. The values of B_{hf} extracted from the Larmor frequency are shown in Figure 2 as a function of temperature. Just below T_N, in the critical region, we used the first eight points to fit the experimental ω_L values to the well-known power-law $\omega_L(T) = \omega_L(0)(1 - T/T_N)^\beta$ for magnetic materials. The resulting parameters are $\omega_L(0) = 842(34)$ Mrads/s, $B_{hf} = 13.3(5)$ T, exponent $\beta = 0.42(3)$, and $T_N = 947(2)$ K. The value of Néel Temperature obtained here (947 K) is somewhat lower than reported earlier (955 K). Above Morin transition temperature (T_M), the angle β between the magnetic field and the Z-axis of the EFG tensor remains close to 90° in agreement with the expected value.

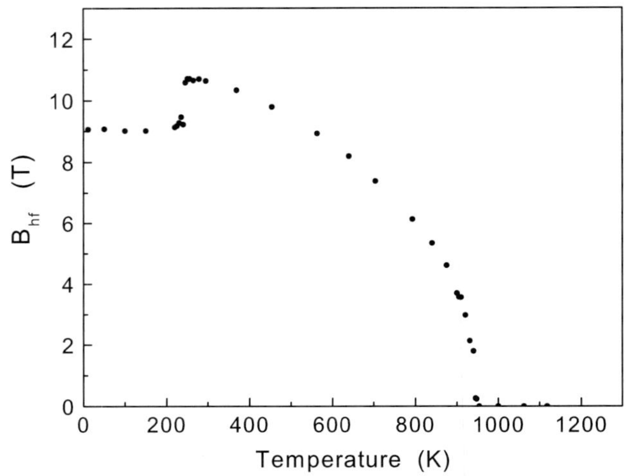

Figure 2. Values of the hyperfine magnetic field as a function of temperature.

Table I. Hyperfine parameters corresponding to the spectra of Figure 1

T_m (K)	B_{hf} (T)	ν_Q (MHz)	η	δ (%)	β (°)
1062	0	424(1)	0.06	0.02	–
954	0	430(1)	0.06	0.02	–
900	3.7(1)	433(1)	0.02	0.03	90
369	10.3(1)	437(4)	0.01	0.03	87
279	10.7(1)	442(7)	0.02	0.03	85
150	9.1(1)	431(7)	0.04	0.11	6

At T_M, the PAC spectra change dramatically, as can be seen in the spectrum at 150 K of Figure 1. Below T_M an additional combined interaction is necessary in order to obtain a good fit. This interaction involves a reduced fraction of probes (~20%). In this contribution we focus our attention in the majority component. The magnetic field shows a discontinuity in which the B_{hf} suddenly drops at around 250 K from a value of 10.7 T to 9.2 T and remains at this value till the lowest temperature. Similar discontinuities have also been observed in the measurement of the hyperfine fields at ^{57}Fe and ^{111}Cd-sites [7, 9]. Below T_M, the hyperfine fields at the ^{181}Ta and ^{111}Cd impurity-sites decrease compared to the values above T_M whereas the field at ^{57}Fe increases. In order to explain the reduced values of the supertransferred hyperfine field below T_M in the case of ^{111}Cd impurity at Fe^{+3} site a model was proposed by Asai et al. [7] in which it was suggested that the directions of the magnetic moments of the nearby Fe^{+3} ions deviate from [111] direction due to the impurity size mismatch when it substitutes a given Fe^{+3} ion. The quadrupole interaction parameters are less sensitive to this transition. The hyperfine parameters corresponding to the spectra of Figure 1 are given in Table I. The angle β between the magnetic hyperfine field and the z-principal axis of the EFG tensor is given. The data for temperatures below T_m correspond to the majority component.

4. Conclusions

We have measured the hyperfine interactions at ^{181}Ta impurities in α-Fe_2O_3 at different temperatures. PAC measurements were performed in the temperature range from 11 to 1060 K to determine the electric field gradient and B_{hf} at the impurity sites. A first principle calculation of EFG and B_{hf} at the ^{181}Ta impurity sites in α-Fe_2O_3 is under way. A model for explaining the discontinuity in the hyperfine field at T_M is under discussion.

Acknowledgements

Partial financial support for this research was provided by the Fundaçaõ de Amparo para Pesquisa do Estado de Saõ Paulo (FAPESP). AFP thankfully acknowledges the financial support provided by FAPESP during the experimental stage in Brazil and the support of CICPBA (Argentine).

References

1. Murad E. and Johnston J. H., In: Long G. J. (ed.), *Mössbauer Spectroscopy Applied to Inorganic Chemistry*, Vol. 2, Plenum, 1987.
2. Inglot Z., Lieb K. P., Uhrmacher M., Wenzel T. and Wiarda D., *Z. Phys. B.* **87** (1992), 323.
3. Inglot Z., Lieb K. P., Uhrmacher M., Wenzel T. and Wiarda D., *Hyperfine Interact.* **120/121** (1999), 237.
4. Inglot Z., Lieb K. P., Uhrmacher M., Wenzel T. and Wiarda D., *J. Phys., Condens. Matter* **3** (1991), 4569.
5. Asai K., Okada T. and Sekizawa H., *J. Phys. Soc. Jpn.* **54** (1985), 4325.
6. Ambe F., Asai K., Ambe S., Okada T. and Sekizawa H., *Hyperfine Interact.* **29** (1984), 1197.
7. Asai K., Ambe F., Ambe S., Okada T. and Sekizawa H., *Phys. Rev. B* **41** (1990), 6124.
8. Pasquevich A. F., Van Eek S. M. and Forker M., *Hyperfine Interact.* **136/137** (2001), 351.
9. van der Woude F., *Phys. Status Solidi* **17** (1966), 417.

Hyperfine Interactions (2004) 158:377–381
DOI 10.1007/s10751-005-9062-y

Perturbed Angular Correlation Study of Order–Disorder Transition in HfW_2O_8

A. F. PASQUEVICH [a]
Departamento de Física, Facultad de Ciencias Exactas, Universidad Nacional de La Plata, cc. 67, 1900 La Plata, Argentina; e-mail: pasquevi@fisica.unlp.edu.ar

Abstract. The perturbed angular correlation (PAC) technique has been used to characterize the degree of atomic order in the neighbourhood of radioactive ^{181}Hf isotopes in HfW_2O_8. PAC measurements were carried out at temperatures between 14 and 723 K. The compound was synthesized starting with the oxides HfO_2 and WO_3, using a method involving ball milling, high temperature annealing and quenching in liquid nitrogen. Fast cooling allows to have the compound at temperatures below 1050 K. The compound has a high degree of stability below such temperature and around 430 K atomic ordering occurs. This transition order–disorder is reversible.

Key Words: oxides, perturbed angular correlation, transition order–disorder.

1. Introduction

The observation that inorganic materials such as ZrW_2O_8 and HfW_2O_8 can undergo significant volume contraction (so-called 'negative thermal expansion') over a wide temperature range has caused significant interest into both the theory of the phenomenon and its potential applications [1, 2] Cubic HfW_2O_8 has been shown to contract continually, isotropically, and reversibly on heating from 2 K to over 1000 K, [3–5] with a coefficient of thermal expansion of $-9.1\ 10^{-6}\ K^{-1}$ between 2 and 350 K The origin of this unusual behavior has been shown to be due to the coupled librations of the semirigid HfO_6 and WO_4 polyhedra that make up the structure [5–8]. These polyhedra are connected via the corner O atoms. An important exception is that one O atom on each WO_4 tetrahedron is not connected to another unit and is therefore considered to be unconstrained. This O (O_4) is located on a WO_4 vertex oriented along one of the [111] directions. Figure 1 shows details of this structure. It is cubic with the space group P213 at temperatures from 0 to 430 K, while the space group changes to $Pa = 3$ as the temperature passes 430 K, with an order-to-disorder phase transition due to the disordering of the orientations of WO_4 tetrahedra [4–6]. This disorder occurs

[a] Also at Comisión de Investigaciones Científicas de la Provincia de Buenos Aires, Argentina.

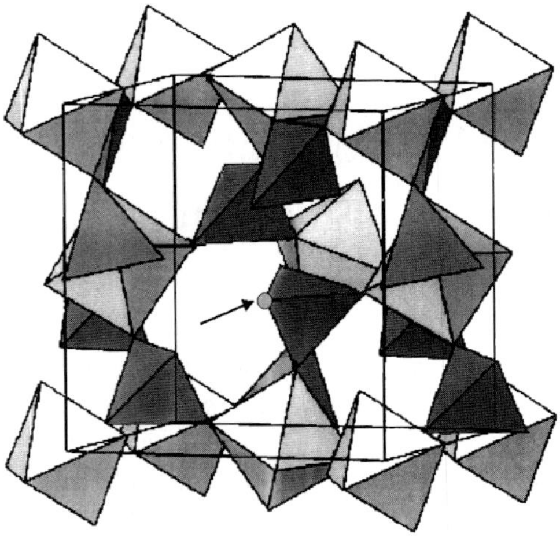

Figure 1. The unit cell for HfW_2O_8 in the low temperature phase. HfO_6 octahedra and WO_4 tetrahedra are shown. The oxygen O_4, which is unshared by other polyhedron, is shown.

due to the hopping of an oxygen (O_4) to an adjacent empty lattice site and a concerted small displacement of W and O_3 ions, resulting in an overall reversal of the direction in which the WO_4 tetrahedra point along the threefold axis [4, 5]. It is an open question whether the two tetrahedra on the [111] diagonals in the unit cells have only two conformations in a correlated manner or they can independently take two orientations [9, 10]. This feature makes this oxide interesting for perturbed angular correlations (PAC) experiments because this technique can give information about the distributions of ions in the neighbourhood of probe atoms located at lattice sites. Additionally, the isotope ^{181}Hf, quite adequate for PAC experiments, can be introduced in the compound as a normal constituent. This feature simplifies the interpretation of the results. A few years ago, Tröger et al. [11] reported PAC experiments in the oxide ZrW_2O_8, which is isomorphous to HfW_2O_8, but they focused on the hyperfine interactions of $^{187}W(\beta^-)^{187}$Re-isotope. The aim of the present investigation is to observe the phase transition and determine the electric field gradient characteristics at both sides of the transition by means of the PAC technique.

2. Experimental

HfW_2O_8 is itself a metastable material but can be made from the binary oxides HfO_2 and WO_3 at high temperature followed by rapid quenching. Once formed, the material is stable to around 1050 K. The samples were prepared by mixing WO_3 and HfO_2 in stoichiometric quantities. The HfO_2 was a mixture of high purity oxide and a couple of milligrams of the same oxide previously irradiated with thermal neutrons to activate the ^{187}Hf isotope. The mixture was ball milled during several hours and a pill was made with the resulting powder. The pill was

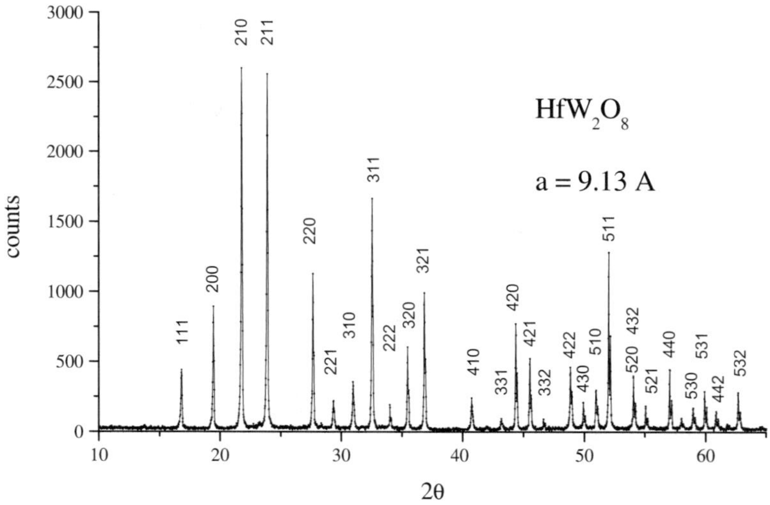

Figure 2. X-ray diffraction pattern of a sample.

annealed at 1473 K during 30 min and after that it was quenched in liquid nitrogen. X-ray diffraction was used to corroborate that this procedure produces a compound with the cubic structure. Figure 2 shows a diffraction pattern.

PAC spectra were recorded at several temperatures using a standard setup with four BaF_2 detectors arranged in a planar 90°–180° geometry, generating simultaneously eight coincidence spectra. The detector system had a time resolution of 700 ps. A small ceramic furnace was used for the measurements above room temperature and the temperature was controlled to within 0.1 K. For low temperature the sample was attached to the cold finger of a closed-cycle helium refrigerator with temperature controlled to better than 0.1 K. Three different samples were measured.

3. Results and discussion

Results of a sequence of measurements are shown in Figure 3A. The spectrum corresponding to room temperature consists of two components characterized by:

$$\begin{array}{llll} f_1 = 74.8(36) & \omega_{Q1} = 52.9(1)\ \text{Mrad/s} & \eta_1 = 0.13(1) & \delta_1 = 1.7(2)\% \\ f_2 = 25.2(12) & \omega_{Q1} = 95.1(20)\ \text{Mrd/s} & \eta_2 = 0.40(3) & \delta_2 = 14.2(2)\%\,. \end{array}$$

As the measurement temperature increases, the first component decreases while the second component increases. From $T_m = 430$ to 723 K (the upper limit of our measurements) the spectra does not change. Cooling down below $T_m = 430$ K recovered the fraction f_1.

In Figure 3B the fractions f_1 measured with three different samples as a function of temperature are shown. The values of each sample were normalized in such a way that the point corresponding to the lowest temperature of

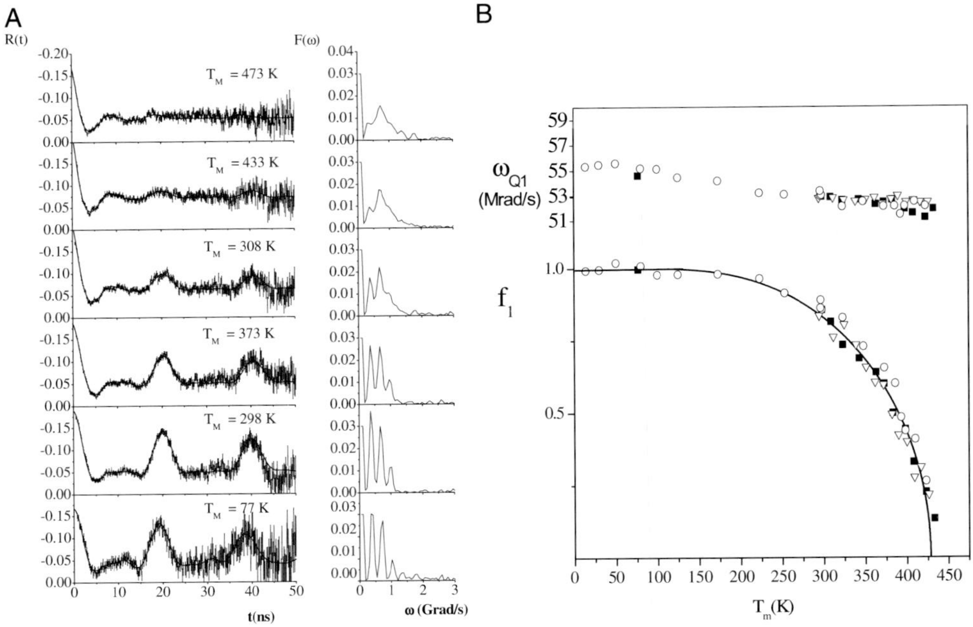

Figure 3. (A) Spectra of ^{181}Hf in HfW_2O_8 taken at different temperatures and the corresponding Fourier transforms. (B) Normalized fraction f_1 and quadrupole frequency ω_{Q1} versus temperature of measurement for three samples. The *solid line* is the order parameter calculated from the approximation of Bragg–Williams for order–disorder transitions in alloys.

measurement fit to the Bragg–Williams approximation curve for the order parameter [12], calculated for an order temperature $T_C = 430$ K. In the same figure the values of the frequency ω_{Q1} for the three samples reported here are displayed. In all the temperature range, η_1 is fairly constant, the values of δ_1 are also constant, except near T_C where it increases. The values of ω_{Q2}, η_2 and δ_2 show more dispersion and depend of the sample history. Typical values, at the high temperature phase are: $\omega_{Q2} = 73$–78 Mrad/s, $\eta_2 = 0.6$–0.7 and $\delta_2 =$ 10%–20%.

At this point we would like to discuss the results. Since Hf is a constituent of the compound and X-ray diffraction confirms the expected structure, the assignment of the interaction labeled 1 to probes atoms at the regular Hf-sites in the α-phase of HfW_2O_8 is straightforward. The second component must be attributed to the β-phase. The frequency distribution of this component could have been anticipated due to the expected disorder of the β-phase. The dispersion in the values of the hyperfine parameters corresponding to this phase can be ascribed to the presence of defects others than the expected oxygens interstitial occupation, produced during the sample preparation. Also, the normalization procedure utilized in relation to f_2-fraction tends to remove from the results the fraction of probes affected by such defects. The agreement between the experimental points and the theoretical curve validates the correction made.

The temperature of the phase transition T_C = 430 K agree quite well with the value reported for Mary *et al.* [4] for HfW_2O_8 and ZrW_2O_8 based on dilatometry data, but the calorimetric studies of Yamamura *et al.* [10] indicate that the transition temperature of HfW_2O_8 was detected at 463 K, 26 K higher than that of ZrW_2O_8. These discrepancies prompted us to carry out PAC experiments in ZrW_2O_8. Electric field gradient calculations using the Point Charge Model and random oxygen distribution are underway trying to relate the measured parameters with the distribution of ions in the lattice.

4. Conclusions

We have characterized the hyperfine interactions of ^{181}Hf in HfW_2O_8 as a function of temperature. The fraction of probes at sites with ordered neighbourhood behaves as the order parameter in the Bragg–Williams approximation and allows to monitor the phase transition. The transition order–disorder was found to occur at T_C = 430 K.

Acknowledgements

The author acknowledges the financial support of CICPBA (Argentine) and AvH - Stiftung (Germany).

References

1. Sleight A. W., *Annu. Rev. Mater. Sci.* **28** (1998), 29.
2. Evans J. S. O., *J. Chem. Soc. Dalton Trans.* (1999), 3317.
3. Evans J. S. O., Mary T. A., Vogt T., Subramanian M. A. and Sleight A. W., *Chem. Mater.* **8** (1996), 2809.
4. Mary T. A., Evans J. S. O., Vogt T. and Sleight A. W., *Science* **272** (1996), 90.
5. Evans J. S. O., David W. I. F. and Sleight A. W., *Acta Crystallogr. B* **55** (1999), 333.
6. Pryde A. K. A., Hammonds K. D., Dove M. T., Heine V., Gale J. D. and Warren M. C., *Phase Transit.* **61** (1997), 141.
7. Mittal R. and Chaplot S. L., *Phys. Rev. B* **60** (1999), 7234.
8. Ernst G., Broholm C., Kowach G. R. and Ramirez A. P., *Nature* **396** (1998), 147.
9. Pryde A. K. A., Hammonds K. D., Dove M. T., Heine V., Gale J. D. and Warren J., *Phys. Condens. Matter.* **8** (1996), 10973.
10. Yamamura Y., Nakajima N. and Tsuji T., *Solid State Commun.* **114** (2000), 453.
11. Tröger W., Ulbrich N. and Butz T., *Hyperfine Interact.* **120/121** (1999), 491.
12. Swalin R. A., *Thermodynamics of Solids*, John Wiley, New York, 1972.

Hyperfine Interactions (2004) 158:383–387
DOI 10.1007/s10751-005-9063-x

Electric Fields Gradients at ^{111}In Sites in $CdIn_2O_4$ Spinel

A. F. PASQUEVICH[1,*,a], A. M. RODRÍGUEZ[1], H. SAITOVITCH[2] and P. R. J. SILVA[2]
[1]*Departamento de Física, Facultad de Ciencias Exactas, Universidad Nacional de La Plata, cc. 67, 1900, La Plata, Argentina; e-mail: pasquevi@fisica.unlp.edu.ar*
[2]*Centro Brasileiro de Pesquisas Fisicas, Rio de Janeiro, Brazil*

Abstract. The perturbed angular correlation (PAC) technique was applied to study the electric field gradients at ^{111}In sites in $CdIn_2O_4$. The measurements were carried out at temperatures in the range 150–1073 K. The aims of the study were both to characterize the quadrupole interactions of the tracers and to determine the distribution of indium ions among the available sites in the spinel structure. Room temperature measurements corresponding to samples cooled at different rates are also reported.

Key Words: PAC technique, quadrupole interactions, oxides, spinels.

1. Introduction

As transparent conductive oxide $CdIn_2O_4$ is a promising material for electronic applications. It is well established that this compound has the spinel structure, but the cation distribution on the lattice is still a subject of interest, mainly due to the role of the cation distribution in the physical properties of the compound. Such distribution, 'inverse' [1] or in contrast 'normal' [2, 3], has been a matter of controversy for many years. The investigation of the inversion degree in this spinel by diffraction methods presents special difficulties, related to the similar X-ray atomic scattering factor of the constituents cations and to the large absorption cross section of ^{113}Cd naturally presented in the samples [4].

This spinel is adequate for PAC experiments using ^{111}Cd($\leftarrow$^{111}In) as probe. Early PAC experiments allow the characterization of the spinel $CdIn_2O_4$ as predominantly normal with an inherent disorder associated with the large cation sizes [5]. In the last years, studies with atom location by channeling enhanced microanalysis technique [4], and combined Rietveld refinements of neutron and X-ray diffraction data [6], gave results which have rekindled the interest of applying PAC technique to study this compound.

* Author for correspondence.
[a] Also at Comisión de Investigaciones Científicas de la Provincia de Buenos Aires, Argentina.

2. The spinel structure and the inversion problem

Spinel are a group of compounds with the general formula AB_2O_4. The spinel has space group Fd3m, cations at the special positions 8a (tetrahedral T-sites) and 16d (octahedral O-sites), and oxygen anions that occupy the general positions 32e [7]. In a normal spinel, the A cation occupies the T-site and the B cations the O-sites. In an inverse spinel, half of B cation occupies T-sites and the others B and A cations randomly occupy the O-sites. In the completely random distribution, a statistically averaged occupation occurs in each site. The cation distribution is conveniently characterized by specifying the inversion parameter, λ, the fraction of B cations at T- sites, i.e., $(A_{1-2\lambda}B_{2\lambda})(A_{2\lambda}B_{2-2\lambda})O_4$. The parameter λ equals 0, 0.33, and 0.5 for normal, random, and inverse spinel, respectively.

3. Experimental

The samples were prepared from stoichiometric amounts of the constituent metals. The metals were dissolved with NO_3H and a few ppm of radiactive ^{111}In in the form of $(NO)_3In$ in aqueous solution was added. The solution was dried and the resultant compound was calcinated at 850 °C for 12 h. Samples without activity prepared in the same way were analyzed by X-ray diffraction. The diffraction pattern corresponded to the spinel structure. In some cases the product consisted in $CdIn_2O_4$ and small amounts of In_2O_3. This is due to partial decomposition of the spinel and sublimation of CdO. The PAC experiments were carried out in both laboratories involved in this investigation. A conventional slow–fast set up of four 2″ × 2″ NaI(Tl) detectors with a resolution time of 3.5 ns was used in Brazil. A fast–fast set of four BaF_2 detectors with a resolution time of 0.7 ns was used in Argentina.

4. Results and discussion

A typical PAC spectrum of high temperature is shown in Figure 1(a). The presence of four hyperfine interactions was observed in this experiment. Two of them, with low population, can be associated to residuals of In_2O_3. Table I displays the parameters of the two other interactions, which correspond to the spinel. One of these interactions (I_1) is very well defined. It gives account of the major characteristics of the Fourier transform. The contribution of the other interaction (I_2) to the transform is shown. These components are also present in the spectrum (Figure 1(b)) taken at room temperature (RT). At this temperature the indium oxide contributions at the spectrum become less important due to the well known damping associated with “after-effects” [8]. In Figure 1(c), a RT spectrum of another sample is shown. In this sample oxide residuals were not found. The spectrum corresponds to a measurement after fast cooling from 1123 K. In this case the amplitude of the component I_2 is greater than before. In

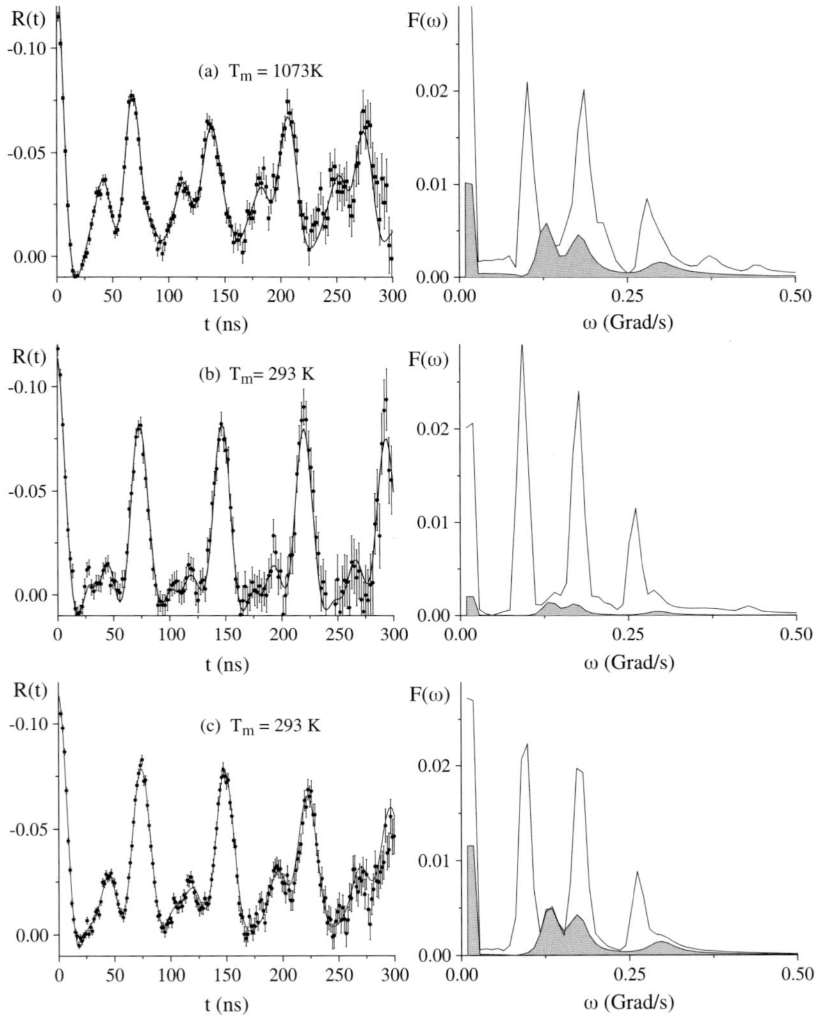

Figure 1. PAC spectra of ^{111}Cd in CdIn$_2$O$_4$ and their Fourier transforms. The contribution of I$_2$ to the transform is indicated by a shadowed area. (a) T_m = 1073 K, (b) obtained at the end of measurements sequence descending from 1073 K, (c) obtained after fast cooling from 1123 K.

Table I. Hyperfine parameters corresponding to the spectra shown in Figure 1

T_M - cooling	I	f (%)	ω (Mrad/s)	η	δ (%)
1073 K	1	62.5 (40)	15.17 (3)	0.18 (1)	1.0 (2)
	2	37.5 (20)	15.23 (15)	0.60 (2)	7.7 (13)
293 K – s.c.	1	91.2 (40)	14.26 (2)	0.11 (1)	0.34 (16)
	2	8.8 (18)	14.86 (42)	0.70 (5)	5.2 (13)
293 K – f.c.	1	69.5 (38)	14.08 (2)	0.16 (1)	1.23 (15)
	2	30.5 (17)	14.77 (16)	0.66 (2)	7.0 (10)

s.c. means “slow cooling”, f.c. “fast cooling.”

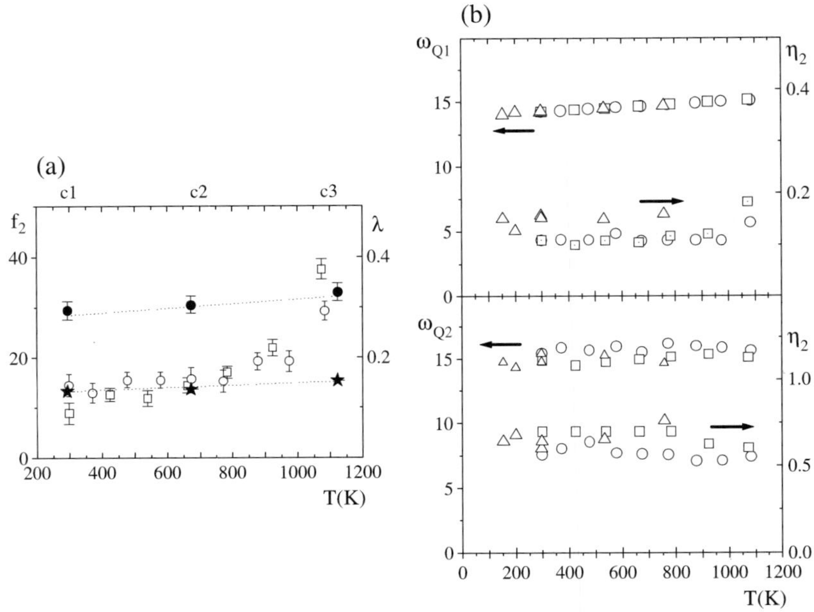

Figure 2. (a) Fractions f_2 (*left and bottom scales*) measured in descendent sequence of temperature (*slow cooling*) with different samples (□, ○). Fractions f_2 (*left scale* and *top labels*) measured at room temperature with samples (●) fast cooled from 1123 K, *c3*: quenching, *c2*: removing from the furnace, *c1*: removing from the furnace after 1 h it was off. Inversion parameter measured at room temperature (*right scale* and *top labels*) ξ (data from ref. 6), *c3*: quenching from 1448 K, *c2*: quenching from 1273 K and *c1*: slow cooling (25 K/h) from 1448 K. (b) Hyperfine parameters for different samples as a function of measurement temperature. □, ○: samples measured in descendent sequence of temperature. Δ: sample fast cooled.

Figure 2(a) the fraction of probes undergoing the I_2 interaction for different samples and temperatures are shown. Also values corresponding to RT measurements following different fast cooling rates are displayed. When the sequence of measurements begins at high temperature, the final f_2-value at RT is much lower.

Whereas the frequencies ω_{Q1} obtained for different samples are in very good agreement, the others hyperfine parameters depend a little on the sample story (Figure 2(b)). For the parameters δ_1 and δ_2, the values given in Table I can be taken as characteristic for both interactions.

As Indium is a constituent of the compound and x-ray diffraction indicates that the compound has the expected structure the well defined component I_1 can be assigned to probes at regular lattice sites in the spinel. Being the f_1-values greater than 0.5, the T-sites can be excluded because the number of such sites in the unit cell is half the number of Indium atoms. Therefore, the interaction I_1 is assigned to probes at O-sites. Concerning to I_2, one could ascribe it to probes at T-sites invoking some degree of inversion. The existence of a non vanishing EFG at these sites would be admissible because the inversion destroys the cubic symmetry otherwise expected. Such assignation would establish an identification of f_2 with λ. Certainly, the f_2-values obtained at high temperatures and those obtained at room temperature after fast cooling, are very similar to the values expected for λ in the case of random distribution at high temperatures. But the results obtained in milled samples with f_2 values well above 0.5 rendered the option under consideration unfeasible [5]. Such results suggested to relate f_2 fraction with probes at sites perturbed by defects. Indeed, computer simulations involving EFG calculations (using the point charge model) and random cation distribution indicated that the main contributions to f_2 come from probes at O-sites with inverted neighbors. In this way the f_2-fraction can be seen as given an upper limit of the degree of inversion λ. In Figure 2(a), experimental λ-values are given for comparison. The small fraction f_2, which appears at room temperature after slow cooling, indicates that the spinel is essentially normal.

5. Conclusions

The hyperfine interactions of ^{111}Cd in the spinel $CdIn_2O_4$ have been characterized as a function of temperature. Two components were observed. One of the components was associated with probes at regular O-sites. The other was associated with probes perturbed by defects and inverted ions.

Acknowledgements

The work was partially supported by CICPBA and CONICET (Argentina) and CNPq (Brazil). One of the authors (AFP) acknowledges the support of AvH – Stiftung (Germany). The authors thank Dr. J. Shitu for the help in performing some of the experiments.

References

1. Skribljak M., Dasgupta S. and Biswas A. B., *Acta Crystallogr.* **12** (1959), 1049.
2. Goodenough J. B. and Loeb A. L., *Phys. Rev.* **98** (1955), 391.
3. Shannon R. D., Gillson J. L. and Bouchard R. J., *J. Phys. Chem. Solids* **38** (1977), 877.
4. Brewer L. N., Kammler D. R., Mason T. O. and Dravid V. P., *J. Appl. Phys.* **89** (2001), 951.
5. Mendoza Zélis L., Pasquevich A. F., Sánchez F. H. and de Virgilis A., *Mat. Sci. Forum* **225** (1996), 401.
6. Ko D., Poeppelmeier K. R., Kammler D. R., Gonzalez G. B., Mason T. O., Williamson D. L., Young D. L. and Coutts T. J., *J. Solid State Chem.* **163** (2002), 259.
7. Wyckoff R. W. G., *Crystal Structures*, Wiley Interscience, New York, 1964.
8. Bibiloni A. G., Desimoni J., Massolo C. P., Mendoza-Zélis L., Pasquevich A. F., Sánchez F. H. and López-García A., *Phys. Rev., B* **29** (1984), 1109.

Hyperfine Interactions (2004) 158:389–394
DOI 10.1007/s10751-005-9064-9

Temperature Dependence of the Quadrupole Interaction for ^{111}In in Sapphire

JAKOB PENNER* and REINER VIANDEN
Helmholtz - Institut für Strahlen- und Kernphysik, Universität Bonn, Bonn, Germany;
e-mail: penner@iskp.uni-bonn.de

Abstract. Perturbed angular correlation measurements of the hyperfine interaction of ^{111}In in sapphire show, that after implantation and annealing at 1000°C, the fraction of undisturbed probe atoms exhibiting a unique quadrupole interaction with $\nu_Q = 219(1)$ MHz ($\eta = 0$) varies between 50% at 4 K, 5% at 100 K and 80% at 973 K in a reversible manner. A possible explanation for this surprising behaviour is the influence of so-called 'after effects' following the EC-decay of ^{111}In to ^{111}Cd. Immediately after the decay the ^{111}Cd is in an ionized state, then collects electrons from its surroundings and reaches the ground state. The different electronic configurations that arise during this relaxation process affect the amplitude (f_u) and the damping (δ_u) of the unique quadrupole interaction.

1. Introduction

Investigations in insulators employing nuclear methods, like the γ–γ perturbed angular correlation (PAC) method [1], have often prompted discussions about the influence of the so-called 'after effects' on the results. Especially an EC decay preceding the γ–γ cascade used for the PAC measurement can lead to highly excited states of the probe atoms electron shells. Some of these states produce strong electromagnetic fields at the site of the nucleus, which can influence or destroy the angular correlation between the γ-rays of the cascade. Here we present a study of the temperature dependence of the quadrupole interaction for ^{111}In in sapphire. This system has been studied in the past [2], but with the increasing importance of sapphire wafers as backing in the wide band gap semiconductor production, the quality of the available material has increased significantly. Thus, new insights into the influence of the 'after effects' are expected.

2. Experimental details

The PAC technique, employed for the experiments presented here, measures the hyperfine interaction of an electric field gradient (EFG) at the site of a radioactive probe with the quadrupole moment Q of the intermediate state of a γ–γ cascade

---

* Author for correspondence.

in the daughter nucleus. This interaction causes a perturbation of the anisotropic emission pattern of the second γ ray with respect to the emission direction of the first. From the time dependence of this emission pattern the parameters of the quadrupole interaction can be derived. It is usually characterized by the quadrupole interaction frequency ν_Q, which is proportional to the product of the principal component of the EFG V_{zz} and the quadrupole moment Q of the intermediate state of the cascade, and the asymmetry parameter η.

The EFG is characteristic for the charge distribution in the lattice surrounding of the probe nucleus. For the present experiments the radioactive probe atom ^{111}In was implanted into sapphire at an implantation energy of 160 keV with typical doses around 5×10^{12} ions/cm^2. Subsequently, the samples were annealed in a rapid thermal annealing apparatus for 120 s at 1273 K with a proximity cap under flowing nitrogen.

A standard PAC set-up of four BaF_2 detectors in a cross-shaped arrangement and conventional fast–slow electronics was used. For all measurements the $\hat{c}$-axis of the Al_2O_3 samples was aligned in the detectors' plane, at 45° between two detectors.

3. Results

Immediately after implantation already a fraction $f_u = 54\%$ of the probe nuclei are found in an undisturbed lattice environment which is characterized by a quadrupole interaction frequency (QI) of $\nu_{Q_1} = (e{\cdot}Q{\cdot}V_{zz})/(h) = 219(2)$ MHz, a relatively small damping of 4.3(1)% or $\delta_2 = (4.3 \pm 0.22)$ MHz and an asymmetry parameter of $\eta = 0$.

The rest of the probe nuclei observe a disturbed lattice environment (ν_{Q1}). The corresponding damping of 40(1)% or $\delta_1 = (68 \pm 1.7)$ MHz is strong and a specific orientation for the EFG tensor cannot be found. This fraction f_g is described best as polycrystalline since defects are distributed randomly around the probe.

After annealing the sample for 120 s at 1000°C we observe the growth of the disturbed fraction f_g and the correlated decrease of the undisturbed fraction f_u. The undisturbed fraction now amounts to less than 10%. At the same time, the damping of this fraction drops below 1%.

Above a sample temperature of about 730 K, and similarly below 100 K a steep increase in the undisturbed fraction is seen. Thereby the damping remains unchanged and the disturbed fraction decreases. The damping of the disturbed fraction remains constant from 100 up to 730 K then drops with higher and lower temperature (see Figures 1 and 2).

4. Discussion

After implantation and annealing at 1000°C for ^{111}In in Al_2O_3 a well-defined QI frequency of ν_{Q1}= 219(2) MHz is observed. This corresponds to an EFG of

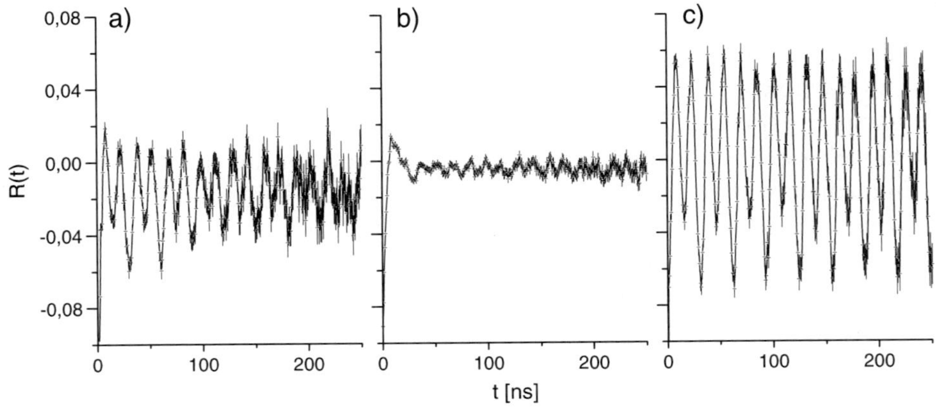

Figure 1. Time dependent anisotropy for ^{111}In(^{111}Cd) in sapphire at different sample temperatures: a) 4 K, b) 293 K, and c) 910 K.

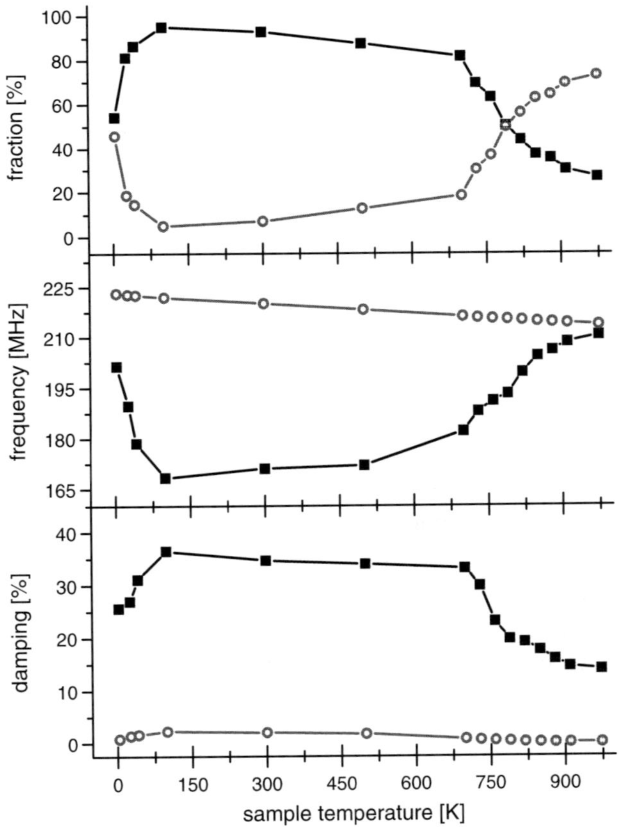

Figure 2. Temperature dependence of the quadrupole interaction frequency, fractions and damping parameter for ^{111}In in sapphire. (*full squares* (■) – disturbed and *open circles* (○) – undisturbed).

$V_{zz} = 1.09 \times 10^{22}$ V/m^2 and has been attributed to substitutional indium on an aluminium lattice site [2].

The fraction f_u of ^{111}In probe atoms underlying this EFG varies in a reversible manner between 50% at 4.2 K, 5% at 100 K and 73% at 973 K. The following model was developed to give a possible explanation for this behaviour.

Immediately after the EC decay of ^{111}In the ^{111}Cd is ionized ('after effect'). Electrons from its surroundings then compensate the charge of the ion. Therefore, the time needed to reach the charge-neutral ground state depends on the speed of this relaxation process. The undisturbed lattice EFG for ^{111}In on Al sites can only be observed in the ground state.

Lupascu *et al.* [2] have shown that it is possible to simulate the influence of this relaxation process on the PAC-perturbation function $G_{22}(t)$ in the following way:

$$G_{22}(t) = \sum_n s_{2n} \cdot \Big\{ [1 - a_n] \cos\left(n\nu_{Q_1} t\right) e^{-(n\delta_1 + \Gamma)t} - b_n \sin\left(n\nu_{Q_1} t\right) e^{-(n\delta_1 + \Gamma)t} + a_n \cos\left(n\nu_{Q_2} t\right) e^{-n\delta_2 t} + b_n \sin\left(n\nu_{Q_2} t\right) e^{-n\delta_2 t} \Big\} \quad (1)$$

with

$$a_n = \frac{\Gamma(n\Delta\delta + \Gamma)}{(\Gamma + n\Delta\delta)^2 + (n\Delta\nu_Q)^2} \qquad b_n = \frac{\Gamma \cdot n\Delta\nu_Q}{(\Gamma + n\Delta\delta)^2 + (n\Delta\nu_Q)^2} \quad (2)$$

$\Delta\nu_Q = \nu_{Q1} - \nu_{Q2}$ describes the difference of the coupling constants and $\Delta = \delta_1 - \delta_2$ the difference of the static distribution widths. This equation describes the perturbation due to an unstable initial EFG$_i$ state (ν_{Q1}, δ_1), which reaches a stable final EFG state (ν_{Q2}, δ_2) with a relaxation rate Γ. The final state's (EFG$_f$) frequency distribution is rather narrow, $\delta_2 \approx 1$ MHz.

It is important to point out that the differences in the static distribution widths $\Delta\delta$, in the coupling constants $\Delta\nu_Q$ and in the relaxation rate Γ contribute quadratically to the amplitudes a_n and b_n. In addition, the final state EFG$_f$ is only damped by its own width δ_2, whereas the initial state is additionally damped by the relaxation rate Γ.

Simulations with constant Γ for ^{111}In in Al_2O_3 have shown, that the PAC spectra can be well described by Equation (1) if one assumes $f_u \approx a_2$ (due to the measurement geometry contributions of a_1 can be neglected).

Values for $\Delta_{\nu Q}$ and $\Delta\delta$ can then be obtained by fitting the amplitude f_u in Figure 3 with

$$f_u = \frac{\Gamma \cdot (\Gamma + \Delta\delta)}{\Delta\nu_Q^2 + (\Gamma + \Delta\delta)^2}. \quad (3)$$

From this fit we derive $\Delta\delta = 90(1)$ MHz and $\Delta\nu_Q = 20(1)$ MHz.

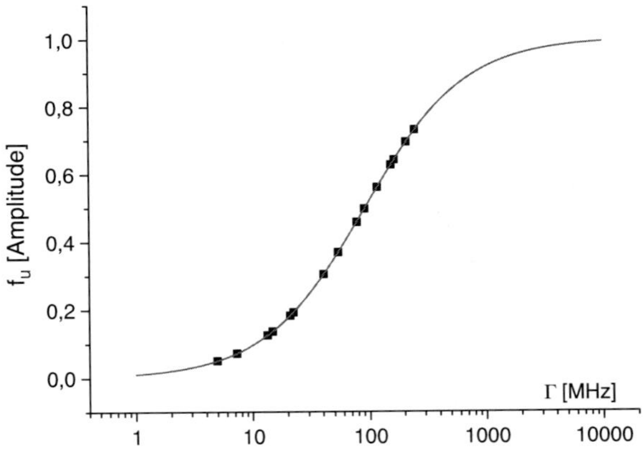

Figure 3. Variation of the amplitude f_u with the relaxation rate Γ.

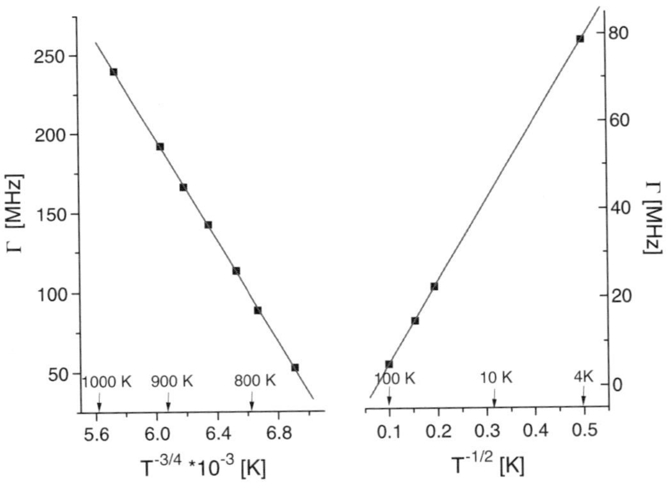

Figure 4. A fit of the $\exp(T^{-3/4})$ bzw. $\exp(T^{1/2})$ dependence predicted by the different models to the relaxation rate at high and low temperatures, respectively.

For temperatures above 650 K Mott and Davis [3] deduce a $\exp(T^{-3/4})$ dependence of Γ. In this regime, transport is assured by electrons hopping over localized states of similar energies and is assisted by phonons. For low temperatures, we applied a model developed by Kehr [4] for Muons. Here, electrons are transported over localized states of the same energy (coherent tunnelling). The conductivity is correlated to the diffusion length and a $\exp(T^{-1/2})$ dependence for Γ is expected.

For the intermediate temperatures, coherent tunnelling is quenched by phonons. At the same time, the variable hopping distance l_{VRH} is still too large to allow the hopping process [3]. Thus, the recombination process is slow, cannot be completed in time and the observed undisturbed fraction remains minimal.

The analysis of the data (Figure 4) shows that these models describe the data very well for low temperatures as well as for the high temperature region. Further, since it can be assumed that the relaxation rate Γ is inversely proportional to the electric conductivity σ of the host, such measurements could contribute to a better understanding of conductive processes in highly resistive materials.

Acknowledgement

This work was partially supported by the DFG under grant no. VI 77/3–1.

References

1. Schatz G., Weidinger A. and Gardner J. A., *Nuclear Condensed Matter Physics*. Wiley, Chichester, 1996.
2. Lupascu D., Habenicht S., Lieb K. P., Neubauer M., Uhrmacher M. and Wenzel T., *Phys. Rev., B* **54** (1996), 871.
3. Mott N. F. and Davis E. A., *Electronic Processes in Non-crystalline Materials*. Oxford University Press, New York, 1979.
4. Kehr K. W., *Hyperfine Interact.* **17-19** (1984), 63.

Hyperfine Interactions (2004) 158:395–400
DOI 10.1007/s10751-005-9065-8

PAC Studies of Implanted ^{111}Ag in Single-Crystalline ZnO

E. RITA[1,2,3,*], J. G. CORREIA[1,2,3], U. WAHL[2,3], E. ALVES[2,3], A. M. L. LOPES[4], J. C. SOARES[3] and THE ISOLDE COLLABORATION[1]
[1]*CERN-PH, C26600, 1211 Geneva 23, Switzerland; e-mail: elisabete.rita@cern.ch*
[2]*Instituto Tecnológico e Nuclear, Sacavém, Portugal*
[3]*Centro de Física Nuclear da Universidade de Lisboa, Lisboa, Portugal*
[4]*Departamento de Física da Universidade de Aveiro, Aveiro, Portugal*

Abstract. The local environment of implanted ^{111}Ag ($t_{1/2}$ = 7.45 d) in single-crystalline [0001] ZnO was evaluated by means of the perturbed angular correlation (PAC) technique. Following the 60 keV low dose (1×10^{13} cm^{-2}) ^{111}Ag implantation, the PAC measurements were performed for the as-implanted state and following 30 min air annealing steps, at temperatures ranging from 200 to 1050 °C. The results revealed that 42% of the probes are located at defect-free S_{Zn} sites ($\nu_Q \sim 32$ MHz, $\eta = 0$) in the as-implanted state and that this fraction did not significantly change with annealing. Moreover, a progressive lattice recovery in the near vicinity of the probes was observed. Different EFGs assigned to point defects were furthermore measured and a general modification of their parameters occurred after 600°C. The 900°C annealing induced the loss of 30% of the ^{111}Ag atoms, 7% of which were located in regions of high defects concentration.

Key Words: Ag, defects, PAC, ZnO.

1. Introduction

The problematic involved with the p-type doping of ZnO is one of the most active trends of investigation in this intrinsically n-type II–VI wurtzite semiconductor. The difficult or non-incorporation of acceptors on appropriate lattice sites and their association with defects are two of the main reasons for the low density of holes achieved in this material [1]. As a *Ib* element, Ag is one of the potential acceptors in ZnO if incorporated on substitutional defect-free Zn (S_{Zn}) sites. In previous studies, the incorporation of Ag at both S_{Zn} and interstitial sites (amphoteric dopant) was suggested [2]. Nevertheless, recent emission channelling investigations for the lattice site location of ^{111}Ag implanted into ZnO single crystals, revealed only one regular site for this element: the S_{Zn} site. In spite that, only 30% of the Ag atoms were found at S_{Zn}, there was no further evidence for Ag in the interstitial form. Moreover, the trapping of Ag by defects was suggested by

* Author for correspondence.

the elevated root mean square displacements from the S_{Zn} site [3], motivating the investigation of ^{111}Ag near neighbourhood.

In this work, the annealing of point defects, in the local environment of implanted ^{111}Ag ($t_{1/2}$ = 7.45 d) into ZnO single crystals was evaluated by means of the Perturbed γ–γ Angular Correlation (PAC) technique. With this method, the Electric Field Gradient (EFG) at the ^{111}Ag site can be measured [4], providing information about the immediate lattice vicinity of the probe. Hence, structural disturbances, such as, dislocations and specific defects located in the ^{111}Ag atoms neighbourhood can be monitored.

2. Experimental details

A commercially available [0001] ZnO single crystal (CrysTec, hydrothermal growth) was homogeneously implanted with 60 keV ^{111}Ag atoms, up to a dose of 1×10^{13} cm^{-2}, at the CERN/ISOLDE facility [5]. These implantation parameters correspond to a projected range, straggling and peak concentration of 195 Å, 74 Å and 5×10^{18} at/cm^3, respectively.

The time dependent perturbation of the angular correlation of the 97–245 keV cascade from ^{111}Cd, populated by the ^{111}Ag β^--decay, was measured with a 4-BaF_2-detector setup. This perturbation results from the interaction between the quadrupole moment (Q) of the cascade's intermediate state (256 keV, $t_{1/2}$ = 85 ns, $I = 5/2^+$) and the EFG at the probe site. For each EFG, the perturbation results in three observable frequencies that will be present in the experimental PAC function $R(t)$ [4]. By fitting $R(t)$ [6], the coupling constant $\nu_Q = 2I(2I - 1)w_0 / (6\pi) = eQV_{zz} / h$ and the asymmetry parameter $\eta = (V_{xx} - V_{yy}) / V_{zz}$ are determined providing, thus, information about the EFG. In this formalism, the EFG is represented by a tensor, V_{ij}, where $|V_{zz}| \geq |V_{yy}| \geq |V_{xx}|$ are the principal components [4].

For the undisturbed crystalline wurtzite structure of [0001] ZnO, a unique EFG is expected with $\eta = 0$ and oriented along the [0001] axis. The PAC measurements were therefore performed in the Raghavan geometry, with the [0001] direction positioned at $\theta = 90°$ and $\varphi = 45°$, relatively to all detectors. In this way, not only ν_Q and η can be extracted for each EFG, but also the orientation relatively to the crystal coordinates. Besides the characteristic lattice EFG, sensed by ^{111}Ag atoms located at undisturbed lattice sites, other EFGs induced by defects are likely to be found. The crystal was measured at room temperature in the as-implanted state and following the 200, 400, 600, 800, 900 and 1050 °C 30 min air annealing steps.

3. Experimental results and discussion

In Figure 1(a)–(f), the $R(t)$ and FFT functions obtained for the as-implanted state and following the 600 and 900 °C air annealings, are presented. For all the

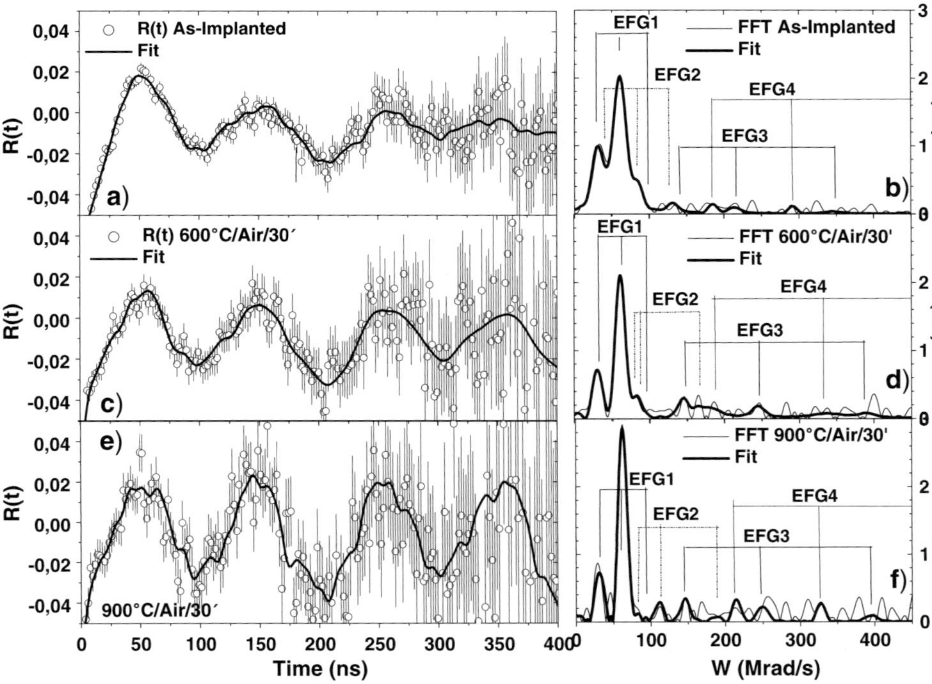

Figure 1. $R(t)$ functions and FFT for the (a, b) as-implanted state and following the (c, d) 600°C and (e, f) 900°C air annealings.

annealing temperatures, approximately 55% of the probes where assigned to four different EFGs in the fitting procedure. The remaining 45% of the ^{111}Ag atoms are interacting with undefined defects configurations, which originate a large EFG distribution that leads to an observed reduced amplitude of the $R(t)$ function. Figure 2(a) and (b) represent the annealing temperature dependence of ν_Q, η, the fractions of ^{111}Ag(f) and attenuation (δ), for each EFG. In the as-implanted state f_1 = 42% of the ^{111}Ag atoms were found at defect-free S_{Zn} sites, experiencing the lattice symmetry EFG$_1$, which is characterized by $\nu_Q(1) \sim 32$ MHz, $\eta(1) = 0$ and orientated along the [0001] axis ($\theta = 90°$ and $\phi = 45°$). In spite that at this stage $\nu_Q(1)$ was still attenuated by $\delta_1 \sim 11\%$, with increasing annealing temperature the lowest value of 0.02% was reached (Figure 2b). This happened already after the 800°C annealing, suggesting a local lattice recovery. Additionally, not only f_1 did not change considerably with the annealings, but also the results are in close agreement to the ^{111}Ag Emission Channelling experiments above referred, where ~30% of Ag was found at regular S_{Zn} sites [3].

The results indicated that three more EFGs with high η, assigned to defects, were present in fractions ranging from 3% to 13%. These values proved to be quite stable throughout the annealing procedure, but no EFG orientation could be

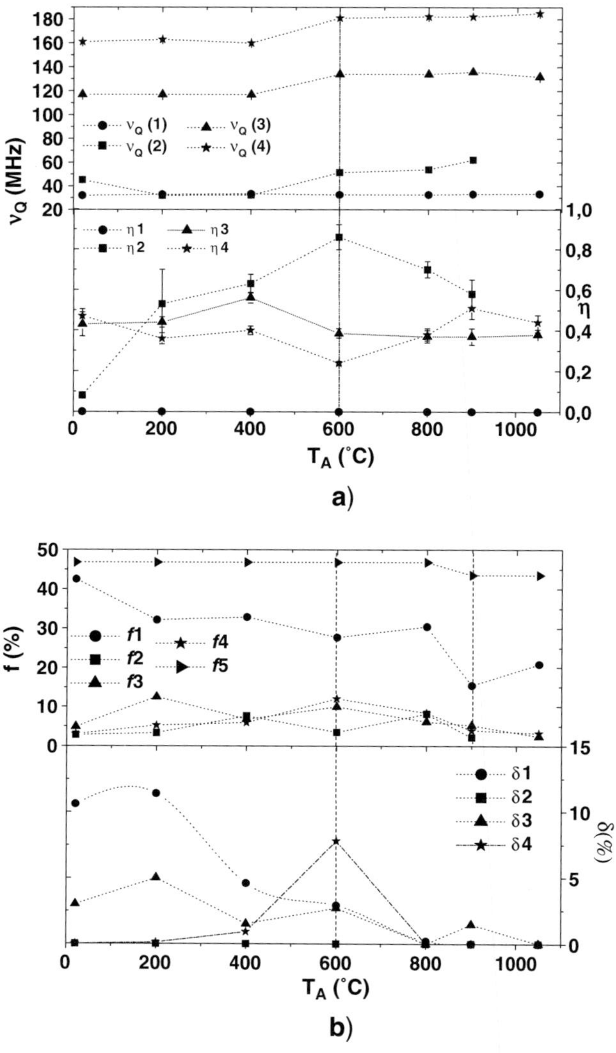

Figure 2. Temperature dependence of (a) ν_Q and η and (b) of the ^{111}Ag fractions f and δ. Above 900°C, the quoted fractions were equally normalized to account for 23% out-diffusion of ^{111}Ag, except for f_5, as from this fraction only 7% out-diffused.

determined. While their asymmetry η was always high, the attenuations δ had a general tendency to decrease (Figure 2(a) and (b)). EFG_2 revealed interesting parameters following the 200 and 400°C annealings. In fact, with $\nu_Q(2) \sim 32$ MHz, close to $\nu_Q(1)$ value, but large $\eta(2) \sim 0.53$–0.61, the presence of ^{111}Ag at S_{Zn} sites *seeing* a specific defect is suggested. EFG_3 and EFG_4 are characterized by $\nu_Q(3) \sim 117$ MHz and $\nu_Q(4) \sim 162$ MHz up to the 400°C annealing. However, after the 600°C annealing the defects configuration has changed, as is seen by their new $\nu_Q(2) \sim 52$ MHz, $\nu_Q(3) \sim 134$ MHz and $\nu_Q(4) \sim 182$ MHz. Moreover, the non-presence of EFG_2 after the 1050°C annealing and the f_1

increase, suggest that the ^{111}Ag atoms formerly associated to EFG_2 where incorporated in defect-free S_{Zn} sites.

It is important to point out that the 900°C air annealing induced the out-diffusion of approximately 30% of the ^{111}Ag atoms. One would be tempted to conclude that the lost ^{111}Ag atoms were the ones in undefined defect configurations (f_5), since they might more easily escape from the crystal. This hypothesis can be pondered by comparing the observable effective amplitude of the R(t) functions for the 800 and 900 °C annealings, since, if an increase of this value is observed, the above-referred assumption could be confirmed. Indeed, such an increase took place, but only by 7%. This hints that from the 30% ^{111}Ag atoms that out-diffused, only 7% belonged to fraction f_5 (undefined defects configuration) and that, therefore, the remaining 23% came from the other ^{111}Ag fractions (defect-free S_{Zn} sites + 3 defect configurations). From our results and fitting procedure, it was not possible to assign these 23% to a specific fraction, for which, in Figure 2b the quoted fractions f_1, f_2, f_3 and f_4 above 900°C were equally normalized to account for this loss. The fact that the higher temperature annealing at 1050°C, did not induce further ^{111}Ag out-diffusion is a suspicious indicator for mainly those atoms in a specific lattice/defect configuration, at a certain activation energy, would have escaped at 900°C. EFG simulations are on their way looking forward to define the origin of some of the defects observed in this study.

In resume, we have shown that in the as-implanted state 42% of the implanted ^{111}Ag atoms occupy defect-free S_{Zn} sites in single crystalline ZnO, experiencing the lattice symmetry EFG ($\nu_Q \sim 32$ MHz, $\eta = 0$), oriented along the [0001] axis. Annealing up to 1050°C in air did not induce a significant variation of this fraction, but a progressive lattice recovery was observed in the near vicinity of the probes. Three other EFGs related with defects where measured, with fractions around 3%–13% and whose configuration changed following the 600°C annealing. The remaining 45% of the probes are considered to be experiencing an EFG distribution in a heavily damaged region. The 900°C air annealing induced the out-diffusion of 30% of the ^{111}Ag atoms, 7% of which were the ones located in the damaged region. From our fitting results it was not possible to disentangle the origin of the remaining 23% lost. Moreover, the higher temperature annealing did not result in further Ag out-diffusion, what suggests Rapid Thermal Annealing treatments to be adequate in this case.

Acknowledgements

This work was funded by the FCT, Portugal (project POCTI-FNU-49503-2002) and by the European Union (Large Scale Facility contract HPRI-CT-1999-00018). E. Rita, U. Wahl and A. M. L. Lopes, acknowledge their fellowships supported by the FCT, Portugal.

References

1. Zhang S. B., *J. Phys., Condens. Matter* **14** (2002), R881–R903.
2. Fan J. and Freer R., *J. Appl. Phys.* **77**(9) (1995), 4795.
3. Rita E. *et al.*, *Physica, B* **340-342** (2003) 240–244.
4. Fraunfelder H. and Steffen R. M., In: Siegbahn K. (ed), *α-, β-, γ-Ray Spectroscopy*, North-Holland, Amsterdam, 1965, p. 997.
5. Kugler E., Fiander D., Jonson B., Haas H., Przewloka A., Ravn H. L., Simon D. J., Zimmer K. and the ISOLDE collaboration, *Nucl. Instrum. Methods Phys. Res., B* **70** (1992), 41.
6. Barradas N. P., Rots M., Melo A. A. and Soares J. C., *Phys. Rev. B* **47** (1993), 8763.

Hyperfine Interactions (2004) 158:401–405
DOI 10.1007/s10751-005-9066-7

Measurement of Quadrupole Interactions in $La_{1-x}Sr_xCoO_3$ Perovskites Using TDPAC Technique

A. C. JUNQUEIRA[1], A. W. CARBONARI[1], R. N. SAXENA[1,*], J. MESTNIK-FILHO[1] and R. DOGRA[2]

[1]*Instituto de Pesquisas energéticas e Nucleares – IPEN-CNEN/SP, São Paulo, Brazil; e-mail: rnsaxena@ipen.br*

[2]*Institute of Advanced Studies, Research School of Physical Sciences and Engineering, The Australian National University, Canberra, Australia*

Abstract. The Perturbed Angular Correlation (PAC) technique was used to study the quadrupole interactions in the $La_{1-x}Sr_xCoO_3$ ($0 \leq x \leq 0.08$) perovskites using ^{111}Cd and ^{181}Ta probes. The radioactive parent nuclei ^{111}In and ^{181}Hf were introduced in the oxide lattice through chemical process during sample preparation and found to occupy only Co sites. The measurements cover a temperature range from 10 to 1150 K except the pure $LaCoO_3$ for which an additional measurement was made at 4.2 K. The measured quadrupole frequencies were found to decrease linearly with increasing temperature as well as with increasing Sr concentration. Temperature dependence of quadrupole frequency in the pure $LaCoO_3$ shows small discontinuities around 70–90 K and 500–600 K which have been atributed to thermally activated spin state transitions, from the low-spin (LS) ground state electronic configuration of Co^{+3} ion to the intermediate-spin (IS) state and from intermediate-spin (IS) state to high-spin (HS) state respectively, observed in some recent studies.

Key Words: electric field gradient, magnetism, metallic oxides, PAC spectroscopy.

1. Introduction

Perovskite oxides of 3d transition metal of the type $LnMO_3$, where Ln is a rare-earth element, are found to exhibit a variety of unusual and interesting transport, magnetic and structural properties. The structure of a perovskite oxide is characterized as a cubic close- packed array of oxygen anions and large rare-earth cations with small transition metal cations occupying the octahedral interstitial sites. The ideal cubic structure is, however, distorted by cation size mismatch and becomes orthorhombic or rhombohedral. The perovskite compound $LaCoO_3$, which crystallizes in a rhombohedrally-distorted structure with

* Author for correspondence.

$R\bar{3}C$ symmetry, has attracted special attention in the past few decades because of the peculiar way its electronic and magnetic properties change with temperature [1].

The ground state electronic spin configuration of the $LaCoO_3$ is low-spin (LS) ($t_{2g}^6e_g^0$) state at low temperatures. The broad transition in the magnetic susceptibility observed at 90–100 K corresponds to the thermal activation of an excited state which could be either a high spin (HS) ($t_{2g}^4e_g^2$) or an intermediate spin (IS) ($t_{2g}^5e_g^1$) configuration. Several studies provided evidence to support one transition type over the other. The majority of new results however point towards the sequence LS → IS → HS with increasing temperature [2, 3] particularly in view of a second transition observed in the magnetic susceptibility at ~500–600 K corresponding to the population of HS state. Substitution of Sr^{2+} for La^{3+} in $LaCoO_3$ brings about significant changes in the system [4, 5]. The introduction of the larger Sr^{+2} cation progressively reduces the rhombohedral distortion present in the parent $LaCoO_3$ compound and suppresses the transition to LS state while stabilizing the IS state. For values of $x \geq 0.18$, $La_{1-x}Sr_xCoO_3$ becomes ferromagnetic. It is expected therefore that a change in the spin state of cobalt as well as the reduced rhombohedral distortion in going from pure to doped compound would result in a corresponding change in the octahedral Co–O bond length. A detailed study of these perovskites with some suitable microscopic technique using an appropriate probe at Co site could reveal important information on the electronic properties including observed LS to IS and IS to HS state transitions observed in $LaCoO_3$.

We have investigated the temperature dependence of quadrupole interaction in a series of perovskite oxides $La_{1-x}Sr_xCoO_3$ ($0 \leq x \leq 0.08$) with TDPAC technique using ^{111}Cd and ^{181}Ta probe nuclei.

2. Experimental

Stoichiometric polycrystalline samples of pure and strontium doped oxides were prepared from a mixture of $La(NO_3)_3$, $Co(NO_3)_3$ and $Sr(NO_3)_2$ solutions to which approximately 20–30 μCi of carrier free ^{111}In in the form of indium chloride solution or ^{181}Hf in the form of hafnium fluoride solution was added. The radioactive ^{181}Hf obtained by irradiating 1 mg of Hf metal with neutrons in the IEA-R1 reactor at IPEN for ~60 h in a flux of 2×10^{13} $n.cm^{-1} \cdot s^{-1}$ was dissolved in dilute HF acid to obtain HfF_4 solution. In each case the mixture was slowly evaporated to dryness and the resulting powder was pressed in to small pellet, which was sintered for about 5 hours at 1000 °C in air. The pellet was ground and sintered again for 5 hours at 1300 °C in air. The samples were analyzed by powder x-ray diffraction method and found to be in single phase.

The TDPAC measurements were carried out in a conventional fast-slow coincidence set-up with four conical BaF_2 detectors generating 12 delayed coincidence spectra simultaneously. Gamma cascade of 133–482 keV populated

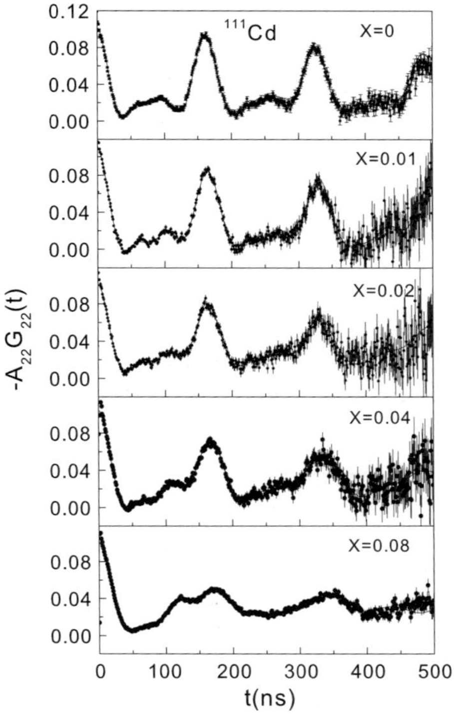

Figure 1. TDPAC spectra of ^{111}Cd for $La_{1-x}Sr_xCoO_3$ at different concentration of Sr.

in the β^- decay of ^{181}Hf and 172–245 keV cascade populated in the electron capture decay of ^{111}In was used to measure the quadrupole interaction of the 482 keV state of ^{181}Ta and 245 keV state of ^{111}Cd respectively. The time resolution of the system was about 700 ps. The samples were measured in the temperature range of 4.2–1150 K. A tubular furnace was used for heating the sample above room temperature. For low temperature measurements the samples were attached to the cold finger of a closed-cycle-helium cryostat. The measurement at 4.2 K was made by immersing the sample in liquid helium.

3. Results and discussion

Room temperature PAC spectra for some of the $La_{1-x}Sr_xCoO_3$ compounds measured at different concentrations of Sr using ^{111}Cd probe are shown in Figure 1 as an example. For ^{111}Cd as well as the ^{181}Ta probe a sharp and well-defined quadrupole frequency was observed for samples with smaller concentrations of Sr and assigned to the probes occupying Co sites. These assignments are essentially based on the results of previous PAC studies with similar perovskites [6, 7]. In both cases a second interaction involving a minor fraction of probe nuclei was observed, which was atributed to some defects in the crystal structure. The quadrupole frequencies in both cases showed increasingly high distribution

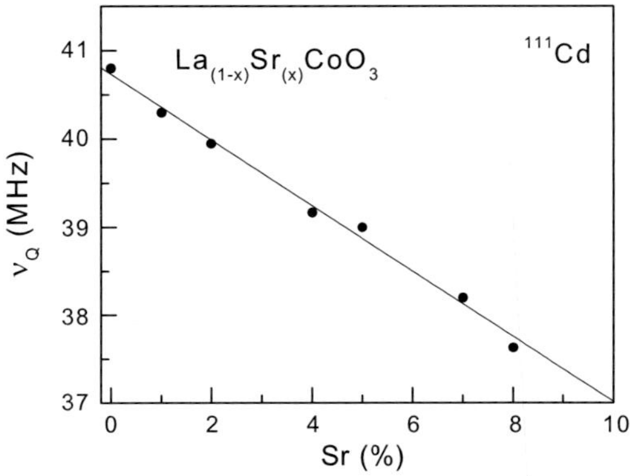

Figure 2. Quadrupole frequencies at room temperature as a function of Sr concentration.

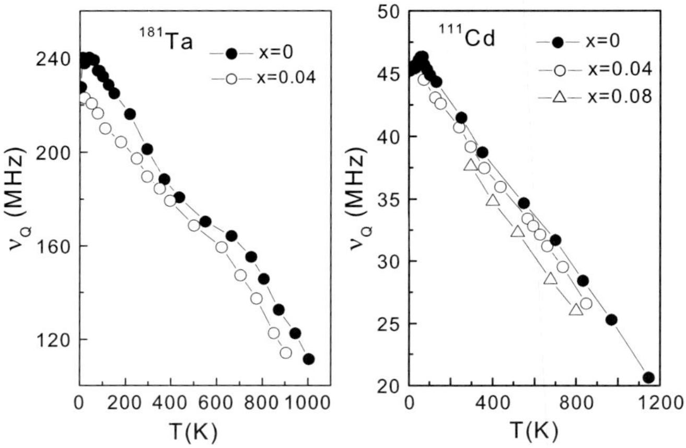

Figure 3. Temperature dependence of quadrupole frequency for pure and Sr doped $LaCoO_3$ determined with ^{111}Cd and ^{181}Ta probes.

at higher concentrations of Sr due to increasing disorder in the compound. As a consequence for Sr concentration greater than 8% it was no longer possible to determine the quadrupole frequencies accurately.

At all temperatures the quadrupole frequencies are found to decrease almost lineary with increasing Sr concentration (Figure 2). The quadrupole frequencies are also found to decrease almost linearly with increasing temperature for the doped as well pure samples. A closer look at the temperature dependence of quadrupole interaction frequencies however shows some important differences between the pure and Sr doped samples, in particular at higher Sr concentration ($x \geq 0.04$) (see Figure 3). The measurements carried out using ^{181}Ta as well as ^{111}Cd on pure $LaCoO_3$ sample show a peak like structure around 70–90 K. A second discontinuity is seen at about 500–600 K, which is barely visible in the case of ^{111}Cd but clearly observed for the data taken with ^{181}Ta probe. Similar discontinuities have been observed earlier in the magnetic susceptibility data as a function of temperature. This behavior is compatible with the idea of a thermally induced spin transition from low-spin (LS) nonmagnetic state ($t_{2g}^6 e_g 0$) at around 70–90 K to an intermediate-spin (IS) state ($t_{2g}^5 e_g^1$) and then from the intermediate-spin (IS) state to high-spin (HS) state ($t_{2g}^4 e_g^2$) at around 500–600 K reported for Co atoms in this compound [3]. With the Sr doping these discontinuities in the linear behavior of efg disappear.

Doping with Sr tends to stabilize the intermediate spin state at lower concentration and therefore the peak at 70–90 K is absent for this compound for $x = 0.04$. The IS to HS transition, however, continues as a function of temperature and the second discontinuity at about 500–600 K is still seen, particularly in the case of measurement with ^{181}Ta (see Figure 3). At still higher Sr concentration ($x \geq 0.08$) the temperature dependence is practically linear and the second peak also disappears. The present experiments with Sr doped $LaCoO_3$ perovskite thus show that the respective spin state transitions are sensitive to Sr concentration and for doping concentration of $x \geq 0.08$ the transition to HS configuration of Co^{+3} ions is almost complete. It is known that for $x \geq 0.18$ $La_{1-x}Sr_xCoO_3$ becomes ferromagnetic [5].

Acknowledgements

Partial financial support for this research was provided by the Fundação de Amparo á Pesquisa do Estado de São Paulo (FAPESP). ACJ thankfully acknowledges the student fellowship granted by FAPESP.

References

1. Senaris Rodriguez M. A. and Goodenough J. B., *J. Solid State Chem.* **116** (1995), 224.
2. Saitoh T., Mizokawa T., Fujimori A., Abbate M., Takeda Y. and Takano M., *Phys. Rev.* **B56** (1997), 1290.
3. Zobel C., Kriener M., Burns D., Baier J., Gruninger M., Lorenz T., Reutler P. and Revcolvschi A., *Phys. Rev.* **B66** (2002), 020402.
4. Yamaguchi S., Okimoto and Tokura Y., *Phys. Rev.* **B55** (1999), R8666.
5. Despina L., Sarro J. L., Thompson J. D., Roder H. and Kwei G. H., *Phys. Rev.* **B60** (1999), 10378.
6. Dogra R., Junqueira A. C., Saxena R. N., Cabonari A. W., Mestnik-Filho J. and Moralles M., *Phys. Rev., B* (2001), 224104.
7. Junqueira A. C., Carbonari A. W., Saxena R. N. and Mestnik-Filho J., *J. Magn. Magn. Mater.* **272–276S** (2004), E1639–E1641.

Coherent Phenomena, Synchrotron Radiation, Quantum Optics

Nuclear Moments, Nuclear Polarization, Nuclear Models

Resonance Methods

Table of Contents Volume II

Hyperfine Interactions (2004) 158:413–416
DOI 10.1007/s10751-005-9068-5

Electric Field Gradients of B in TiO_2

T. SUMIKAMA*,†, M. OGURA, Y. NAKASHIMA, T. IWAKOSHI,
M. MIHARA, M. FUKUDA, K. MATSUTA, T. MINAMISONO and H. AKAI
Department of Physics, Osaka University, Toyonaka, Osaka 560-0043, Japan;
e-mail: sumikama@riken.jp

Abstract. We observed the electric quadrupole interaction of ^{12}B implanted in the interstitial site of TiO_2 using β-NMR method. The electric field gradients including the direction of the principal axes were determined. The direction agreed well with the theoretical calculation.

Key Words: boron, electric field gradient, octahedral interstitial site, TiO_2.

1. Introduction

Hyperfine interactions of short-lived nuclei implanted in crystal lattices have been investigated to obtain new information on the electronic structures of impurities in various crystals as well as on the nuclear properties. We studied the hyperfine interaction of impurities in the TiO_2 (rutile) single crystal using the β-NMR technique, which utilizes the asymmetric β-ray angular distribution from spin polarized nuclei. We found that B [1], N [2], O [3], Na [4] and Sc [5] nuclei maintain at room temperature their full polarization in TiO_2 for several seconds.

In TiO_2, Boron atoms occupy two implantation sites with relative populations 9 : 1. The main site is suggested to be the Ti substitutional site, and the minor one to be the octahedral interstitial site [1]. The direction of the principal axes of electric field gradients (EFGs) have not been determined for the interstitial site. In the present study, we observed the electric quadrupole interaction of the short-lived β emitter ^{12}B ($I^{\pi} = 1^{+}$, $T_{1/2} = 20.2$ ms) implanted into TiO_2 with the β-NMR method. The EFG at the presumably octahedral interstitial site is determined to confirm the identification of this site.

2. Experiment

The experimental setup and procedure are similar to the previous work [1]. ^{12}B nuclei were produced through the nuclear reaction $^{11}B(d, p)\,^{12}B$ using the

* Author for correspondence.

† Present address: RIKEN, Hirosawa 2–1, Wako, Saitama 351-0198, Japan.

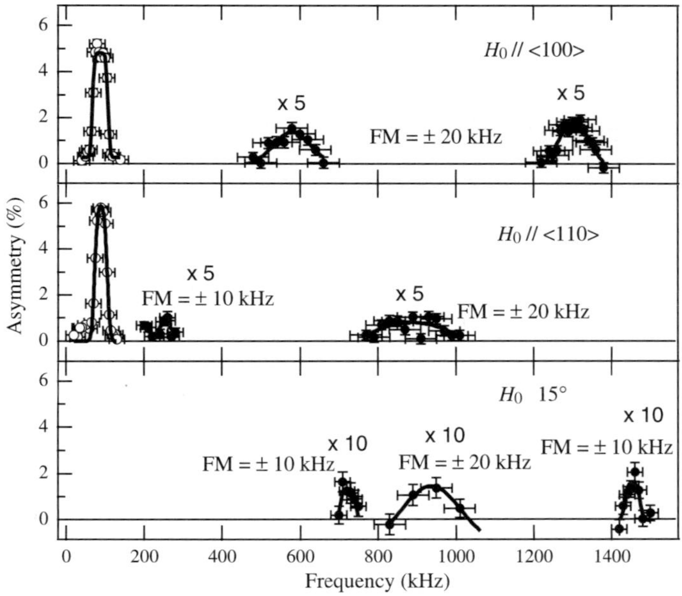

Figure 1. β-NQR spectra of ^{12}B in TiO_2 for three directions as a function of the frequency. The applied Frequency Modulation (FM) was 10 or 20 kHz. The *full circles* are present data for the octahedral interstitial implantation site and the *open ones* are the one for the substitutial implantation site [1].

1.5-MeV deuteron beam provided by the Van de Graaff accelerator at Osaka University. The recoiling ^{12}B nuclei were implanted into a TiO_2 crystal. The recoil angle were selected to be 32–48° so as to create a maximum polarization by the nuclear reaction. Under this condition, the created initial polarization was 10%. An external magnetic field H_0 was applied anti-parallel to the direction of the polarization for maintaining the nuclear polarization and for detecting the NMR signals. To measure the polarization, the β-ray asymmetry was detected by two sets of plastic-scintillation-counter telescopes placed above and bellow the catcher relative to the polarization direction. The RF oscillating magnetic field H_1 was applied perpendicular to the external field H_0 to induce magnetic transitions. The change in β-asymmetry was observed as a function of the frequency difference between the two resonances for $I = 1$ with β-NQR method.

3. Results and discussion

The β-NQR spectra were observed for three directions of H_0 in the (001) plane as shown in Figure 1. They were parallel to the $\langle 100\rangle$ and $\langle 110\rangle$ axes and tilted by 15° relative to the $\langle 100\rangle$ axis. Considering the symmetry of the rutile structure, an observation of an angle θ defined as an angle between the $\langle 100\rangle$ direction and H_0 as shown in Figure 2(b) corresponds to an observation of the four angles, θ, $90° - \theta$, $90° + \theta$ and $180° - \theta$ at the same time. For the special directions $\langle 100\rangle$

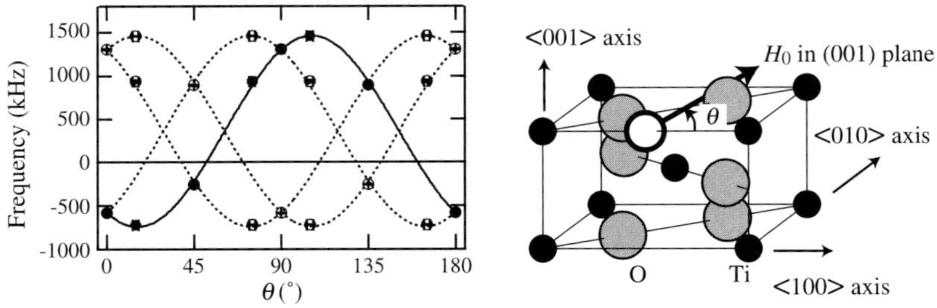

(a) Frequency of ^{12}B in TiO_2 at the octahedral interstitial site.

(b) Angle θ goes from ⟨100⟩ axis to H_0 in (001) plane.

Figure 2. (a) The *lines* show four possible angles. Here, the error of both frequency and angle is smaller than the size of the marks. (b) gives the definition of angle θ. This figure shows the case of the octahedral interstitial site.

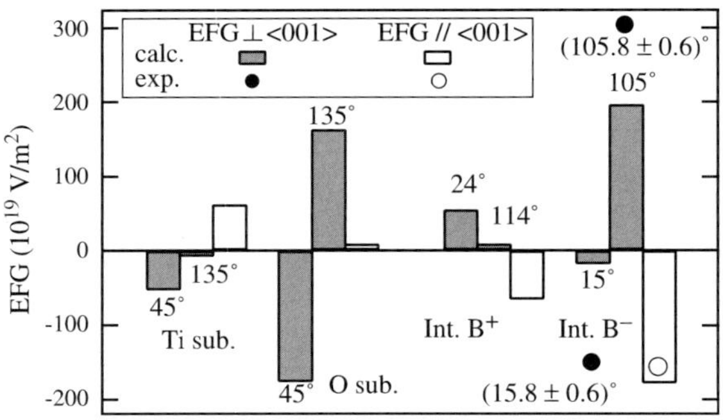

Figure 3. Theoretical and experimental EFGs for B in TiO_2.

and ⟨110⟩, two angles θ and $180°-\theta$, or $90° \pm \theta$ are equivalent, and there is no systematic error caused by the uncertainty of the angle setting since the angular uncertainty separates one frequency f for two equivalent angles to two frequencies $f \pm \delta f$ but will not change their center. For the '15° data,' the uncertainty of an angle setting was ±1.5° and was converted to an error of the frequency. The lowest frequency of the '15° data' couldn't be measured because it is located close to the large resonance from the major site. The obtained frequencies are plotted in Figure 2(a) as a function of the crystal orientation relative to H_0. From the θ dependence, the quadrupole coupling frequency $\nu_Q = 3eqQ/2h = +1457 \pm 5$ kHz, the EFG $q = +(304 \pm 6) \times 1019$ V/m^2 and the asymmetry parameter $\eta = 0.020 \pm 0.006$ were determined with the known quadrupole moment $Q(^{12}\mathrm{B}) = +(13.21 \pm 0.26)$ mb [6].

In order to explain the present results, the EFGs of B in TiO_2 were calculated by use of the Korringa–Kohn–Rostoker (KKR) method within the local density approximation [7]. Three possible implantation sites, the Ti, O substitutional sites and the octahedral like interstitial site were considered. The theoretical EFGs are shown in Figure 3 together with the experimental values. One of the

principal axes of EFGs is parallel to the ⟨001⟩ direction for all sites. The other two axes are parallel to the ⟨110⟩ and $\langle 1\bar{1}0 \rangle$ direction for the substitutional sites. In the case of the interstitial site, the other principal axes are in the (001) plane but are inclined from the ⟨110⟩ or $\langle 1\bar{1}0 \rangle$ axes as shown in Figure 2(b). These results including the EFG principal axes in the (001) plane support the previous identification [1] that the minor site is suggested to be the octahedral interstitial site with B^- state. As a result, the principal axes were determined to be the direction of (15.8 ± 0.6)° for V_{XX} and (105.8 ± 0.6)° for V_{ZZ} relative to the ⟨100⟩ axis in the (001) plane as shown by the solid line in Figure 2(a). It is remarkable that the principal axes are inclined by 29° to the natural symmetry axes ⟨110⟩ and $\langle 1\bar{1}0 \rangle$ of the rutile structure. However the agreement in the magnitudes between the theoretical predictions and experimental EFGs is not satisfactory. The muffin-tin approximation used within the KKR approach may cause these discrepancies and a more realistic treatment of atomic potentials is now in progress.

Acknowledgement

Thanks for the financial assistance from the Special Postdoctoral Researcher Program of RIKEN.

References

1. Ogura M. and Minamisono K. *et al.*, *Hyperfine Interact.* **136/137** (2001), 195.
2. Minamisono T. *et al.*, *Z. Naturforsch.* **53a** (1998), 293.
3. Minamisono T. *et al.*, *Phys. Lett. B* **457** (1999), 9; Matsuta K. *et al.*, *Phys. Lett. B* **459** (1999), 81.
4. Minamisono K. and Matsuta K. *et al.*, KUR Report 88 (2002), 104.
5. Minamisono T. and Fukuda S. *et al.*, *Nucl. Phys., A* **559** (1993), 239.
6. Minamisono T. *et al.*, *Phys. Rev. Lett.* **69** (1992), 2058; Otsubo T. *et al.*, *Hyperfine Interact.* **78** (1993), 185; Yamaguchi T. *et al.*, *Hyperfine Interact.* **120/121** (1999), 689.
7. Akai H. and Akai M. *et al.*, *Prog. Theor. Phys., Suppl.* **101** (1990), 11.

Hyperfine Interactions (2004) 158:417–421
DOI 10.1007/s10751-005-9069-4

Acceleration of Diffusional Jumps of Interstitial Fe with Increasing Ge Concentration in $Si_{1-x}Ge_x$ Alloys Observed by Mössbauer Spectroscopy

G. WEYER[1,2,*], H. P. GUNNLAUGSSON[2], K. BHARUTH-RAM[3], M. DIETRICH[1], R. MANTOVAN[4], V. NAICKER[3], D. NAIDOO[3] and R. SIELEMANN[5]
[1]*ISOLDE Collaboration, EP Division, CERN, Geneva 23, 1211 Switzerland*
[2]*Institute of Physics and Astronomy, Aarhus University, DK-8000 Aarhus C, Denmark*
[3]*School of Pure and Applied Physics, University of Natal, Durban 4041, South Africa*
[4]*Laboratorio MDM-INFM, 20042 Agrate Brianza, Italy*
[5]*Hahn-Meitner Institute, 14109 Berlin, Germany*

Abstract. Radioactive ^{57}Mn isotopes have been implanted into $Si_{1-x}Ge_x$ crystals ($x \leq 0.1$) at elevated temperatures for Mössbauer studies of the diffusion of interstitial ^{57}Fe daughter atoms. The atomic jump frequency is found to increase upon Ge alloying. This is attributed to a lowering of the activation energy, i.e. the saddle point energy at hexagonal interstitial sites with Ge neighbour atoms.

Key Words: diffusion, Fe impurities, Mössbauer spectroscopy, SiGe.

1. Introduction

The diffusion of interstitial iron, Fe_i, in silicon is well investigated experimentally and diffusion coefficients determined, respectively, from long-range diffusion measurements by various techniques [1] and from atomic jump frequencies measured by Mössbauer spectroscopy [2] are in good agreement for both charge states of interstitial $Fe_i^{0/+}$. The diffusion of Fe_i^{+} is faster by about an order of magnitude than that of Fe_i^{0} at low temperatures. The diffusion mechanism is purely interstitial, i.e. atomic jumps between tetrahedral interstitial sites with the hexagonal site as the saddle point, which thus leads to an uncorrelated random walk. It is interesting to note that for the series of interstitial 3d elements in silicon theory predicts a change of the most stable site from a tetrahedral site for the early elements up to Fe_i to a hexagonal site for the late elements Co, Ni and Cu, the fastest diffusing elements in the series [3]. In this contribution we show that alloying of Ge increases the jump frequency of Fe_i in $Si_{1-x}Ge_x$, whereas the

* Author for correspondence.

long-range diffusivity is apparently not affected [4] for $x < 0.1$. Semiconducting $Si_{1-x}Ge_x$, an almost ideal random alloy, may be of interest as a system of particular simplicity to study the alloying effect on interstitially diffusing impurities [4] and interstitial 3d elements and Fe in particular are important, mostly harmful impurities in silicon-based technologies.

2. Experimental

The experiments have been performed at the ISOLDE facility at CERN utilizing the isotopically clean and intense beams of radioactive $^{57}Mn^+$ ions ($T_{1/2}$ = 1.5 min). These were implanted with 60 keV energy to fluences $< 10^{12}/cm^2$ into SiGe single crystals heated to temperatures < 1000 K by means of a halogen lamp. Mössbauer spectra for the 14 keV transition of the ^{57}Fe daughter atoms were recorded by resonance detectors (equipped with ^{57}Fe enriched electrodes) mounted on conventional drive systems outside the implantation chamber. All $Si_{1-x}Ge_x$ ($x \leq 0.1$) samples were n-type material (P or Sb concentrations 10^{16-17} cm^{-3}), epitaxially grown on silicon; a pure silicon n-type reference sample was also employed.

3. Results and discussion

The radiation damage due to the implantation is known to anneal during the lifetime of ^{57}Mn at temperatures > 500 K in silicon and leads to their substitutional incorporation [5]. This was also observed for the SiGe samples. An average recoil energy of 40 eV in the nuclear decay expels the majority of the ^{57}Fe daughter atoms into tetrahedral interstitial sites. Thus diffusional jumps of the interstitial $^{57}Fe_i$ during the lifetime of the Mössbauer state ($T_{1/2}$ = 100 ns) can be detected by the resulting line broadening $\Delta\Gamma$, which is directly proportional to the jump frequency v [6]. In velocity units this is $\Delta\Gamma = 2\hbar c v / E_0$ with E_0 the transition energy, $\hbar$ Planck's constant and c the velocity of light. For a purely random walk the jump frequency is related to the diffusion coefficient D by $v = 6D / l^2$ with l the jump length.

Spectra measured at different temperatures and for two Ge concentrations are compared to those for pure silicon in Figure 1 and the analysis in terms of three spectral components is indicated: two single lines for substitutional Fe_S and interstitial Fe_i^0 on tetrahedral sites, respectively, and a quadrupole-split, broadened line attributed to the formation of Fe_i–V pairs with the vacancy created in the recoil event at high temperatures [7]. As the intensity of the latter line is rather low at the temperatures of interest here, the quadrupole splitting and its $T^{3/2}$ temperature dependence were fixed in the simultaneous analysis of all data for a given sample to the values established for silicon at higher temperatures [7]. Both the line width and the spectral area of all components were free parameters, whereas the isomer shifts (IS) were constrained to follow the second order Doppler shift (SOD). (To ease the comparison of the spectra, in

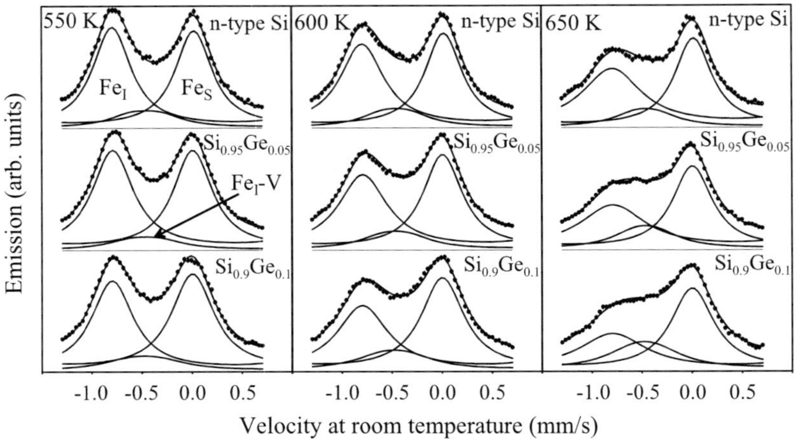

Figure 1. ^{57}Fe Mössbauer spectra obtained after implantation of ^{57}Mn into the Si and SiGe samples held at the temperatures indicated. The *solid lines* show individual fitting components and their sum.

Figure 1 the SOD has been corrected for thus that the velocity scale refers to room temperature and the sextet spectra resulting from the magnetic splitting in the detector have been transformed into a single emission feature). This analysis then gave IS values at room temperature for the Fe_i^0 and Fe_i–V components, which deviated no more than within the error margins from those for pure silicon [7] or germanium [8], which are also very similar for these components. The isomer shift for substitutional Fe_S increased slightly with increasing Ge concentration as would be expected from the difference in the values for pure germanium and silicon; the uncertainties on these values and the restricted Ge concentrations do not allow to deduce a functional dependence. These fits to the data, although not perfect yet as can be seen in Figure 1 by small, apparently systematic deviations, are nevertheless sufficient to extract the line broadening of the interstitial line at high temperatures. In fact, it appears visible directly in the spectra at 600 and 650 K that the broadening increases with increasing Ge concentration for a given temperature. The line broadening extracted from the analysis is plotted as a function of temperature in Figure 2(A) for the samples in this study. Obviously, the broadening is generally more pronounced for the SiGe samples than for pure silicon at a given temperature, however, relatively large error bars and the restricted temperature range, 550–650 K, where significant broadening is detectable and the intensity of the Fe_i–V line is sufficiently small not to affect the results, make difficult the deduction of a functional dependence of the broadening on both temperature and Ge concentration. For a given sample, the jump frequency is assumed to obey the usual diffusion coefficient equation resulting in a temperature dependence of the broadening given by $\Delta\Gamma = \alpha \exp(-E_a / kT)$, where α is a proportionality constant, E_a the (constant) diffusion activation energy and k Boltzmann's constant. As discussed in detail in [2] due to the limitations given above, it is not

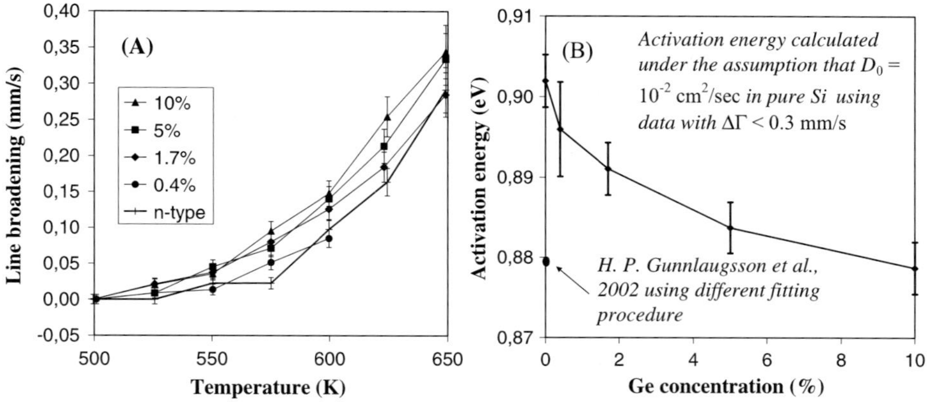

Figure 2. (A) Line broadening of the Fe_i line for the different SiGe samples in this study versus temperature. (B) Activation energy versus Ge concentration obtained from the analysis of the data in (A), assuming a constant pre-exponential factor corresponding to $D_0 = 10^{-2}$ cm^2/s.

possible to determine both α and E_a from a fit to the data with meaningful accuracy. Therefore, in fits the pre-exponential factor was kept constant and equal to that for Fe_i^0 in silicon [7] in order to reveal a possible Ge concentration dependent trend for E_a. As discussed below, thus determined E_a values should be considered as upper limits with unknown deviations from the true values; these deviations should, however, increase with increasing Ge concentration. Figure 2(B) shows the results from this analysis: a clear trend for a decrease of the activation energy with increasing Ge concentration. Note, as indicated in Figure 2(B), that results for Fe_i^0 in silicon from a different analysis model [2] give slightly different absolute values for E_a, however, the trend is robust. Thus the lowering of the activation energy contributes substantially to the increase of the jump frequency with increasing Ge concentration.

As this increase of the jump frequency is much smaller than the difference in jump frequencies for Fe_i^+ and Fe_i^0,we first consider the possibility of a mixture in charge states. The Fe_i donor level is known to approach the valence band with increasing Ge concentration and the band gap shrinks [4]. In view of the illumination of the samples, the band gap shrinkage and the almost constant doping levels of the samples it is therefore likely that a mixture in charge states would tend to more neutral Fe_i with increasing Ge concentration. However, this would result in an effect on the jump frequency opposite to that observed. For positively charged Fe_i^+ no change in the diffusion coefficient has been measured up to $x \sim 0.1$ [4], however, the relative increase detected here for Fe_i^0 if transferred to Fe_i^+ would be within the error margins of those measurements. We thus consider a change in the diffusion mechanism. In the tetrahedral interstitial site in silicon the Fe_i atoms are surrounded by 10 atoms forming four puckered, six-membered rings with the hexagonal site in the center. Depending on its nearest or next-nearest location to a tetrahedral site, a Ge impurity atoms among the 10 atoms would be member of three or two of these rings. A decrease of the

Fe_i energy in such a hexagonal saddle point configuration then corresponds to a decrease of the activation energy and would explain an increase in the jump frequency for such a direction. As a further consequence, the Fe_i atoms can no longer be considered to perform a truly random walk. After an elementary jump via a hexagonal site with one Ge neighbour atom, this same site is available again and thus a return jump more probable than jumps in other directions, giving rise to a (pre-exponential) correlation factor $f < 1$ and such jumps do not contribute to the long-range diffusion for sufficiently small Ge concentrations. Also entropy changes due to alloying tend to decrease the pre-exponential factor. A more detailed discussion of the atomic jumps in such a model will be postponed to a forthcoming publication, however, evidently it offers an attractve, qualitative explanation for the measured effect, which should also show up in the macroscopic diffusion at higher Ge concentrations. In the measurements the recoiling Fe atoms may be assumed to be randomly distributed on tetrahedral sites with a statistical probability of having randomly distributed Ge atoms among the mentioned 10 atoms. As an ensemble average is measured to determine the line broadening, the concluded lowering of the activation energy on an atomic level should be more substantial than is apparent from Figure 2(B).

Acknowledgements

We are grateful to A. Nylandsted Larsen for providing the samples. K. B.-R., V. N. and D. N. acknowledge support from the South African Research Foundation, grant GUN2064730.

References

1. Heiser T. and Mesli A., *Phys. Rev. Lett.* **68** (1992), 978.
2. Gunnlaugsson H. P, Weyer G., Dietrich M., Fanciulli M., Bharuth-Ram K. and Sielemann R., *Appl. Phys. Lett.* **80** (2002), 2657.
3. Kamon Y., Harima H., Yanase A. and Katayama-Yoshida H., *Physica, B* **308–310** (2001), 391.
4. Mesli A., Vileno B., Eckert C., Slaoui A., Pedersen C., Nylandsted Larsen A. and Abrosimov N. V., *Phys. Rev., B* **66** (2002), 045206.
5. Weyer G., Gunnlaugsson H. P., Dietrich M., Fanciulli M., Bharuth-Ram K. and Sielemann R., *Nucl. Instrum. Methods B* **206** (2003), 90.
6. Singwi K. S. and Sjölander A., *Phys. Rev.* **120** (1960), 1093.
7. Gunnlaugsson H. P., Weyer G., Christensen N. E., Dietrich M., Fanciulli M., Bharuth-Ram K., Sielemann R. and Svane A., *Physica B* **340–342** (2003), 532.
8. Gunnlaugsson H. P., Weyer G., Dietrich M., Fanciulli M., Bharuth-Ram K. and Sielemann R., *Physica, B* **340–342** (2003), 537.

Hyperfine Interactions (2004) 158:423–427
DOI 10.1007/s10751-005-9070-y

Hf_2Ni and Zr_2Ni Compounds Studied by PAC with ^{111}Cd Probes

P. WODNIECKI[1,2,*], B. WODNIECKA[1], A. KULIŃSKA[1,2] and M. UHRMACHER[2]
[1]*IFJ PAN, Kraków, Poland*
[2]*II. Physikalisches Institut, Universität Göttingen, D-37077, Göttingen, Germany; e-mail: pawel.wodniecki@ifj.edu.pl*

Abstract. The perturbed angular correlation method (PAC) was applied to investigate the lattice location of implanted ^{111}In probe ions in Hf_2Ni and Zr_2Ni intermetallic compounds. It is concluded that the $^{111}In/^{111}Cd$ probe nuclei experiencing the highly asymmetric electric field gradient (EFG) occupy the unique hafnium or zirconium 8(*h*) sites in the investigated phases. Above room temperature, the EFGs decrease linearly with temperature. The results are compared with that of previous PAC measurements with ^{181}Ta probes.

Key Words: EFG, Hf_2Ni, lattice location, Zr_2Ni.

1. Introduction

The lattice locations of impurity atoms in intermetallic compounds are governed by arguments of size, relative valency and electronegativity [1]. Previous PAC experiments performed in a large variety of Hf/Zr aluminides have shown that $^{111}In/^{111}Cd$ probes, matching well the electronegativity of Al, but oversized, can occupy preferentially Al sites (as in Hf_2Al_3 [2] and $ZrAl_2$ [3]) and/or Hf sites (as in $HfAl_3$ [2]). Moreover, it turned out, that site preferences can change with composition (especially near compound stoichiometry [4, 5]) or temperature [6]. Thus, the sites of the diluted impurity atoms are difficult to predict.

In the former investigation of the Hf_2Ni and Zr_2Ni phases, radioactive ^{181}Hf atoms were used to measure the electric field gradients (EFG) acting on these probes [7, 8]. The tetragonal C16 (Al_2Cu type) structure (Figure 1) is characterized by two sites – the 8(*h*) site with low symmetry at which the EFG with nonzero asymmetry parameter is expected and the 4(*a*) site of high symmetry, at which the EFG should be axially symmetric. Due to this feature it is easy to distinguish between the two possible locations of the impurity probes. The ^{181}Hf probes are, of course, constituent atoms in Hf_2Ni alloy and can be considered as such in Zr_2Ni, due to the chemical similarities of Hf and Zr. Thus,

* Author for correspondence.

site	Coordination n	CN	Distance (Å)
4(a) $\bar{6}m2$	Ni	2	2.634
	Zr	8	2.761
8(h) mm	Ni	4	2.761
	Zr	1	2.989
	Zr	2	3.079
	Zr	4	3.377
	Zr	4	3.432

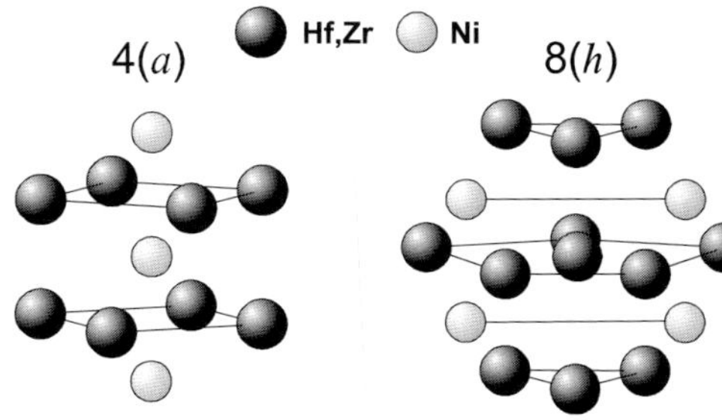

Figure 1. The nearest neighborhoods of 4(a) and 8(h) sites in C16 lattice of Hf_2Ni and Zr_2Ni phases.

as expected, they were found in 8(h) sites of Hf or Zr, on the basis of the measured asymmetry parameters η [7, 8].

Here we have extended the PAC experiments to implanted ^{111}In-ions, which have to be treated as impurities in the Hf_2Ni and Zr_2Ni lattices.

2. Experimental procedure

The Hf_2Ni and Zr_2Ni samples were produced by multiple arc melting, under an argon atmosphere, followed by a few days annealing at 900 K in evacuated and sealed quartz tubes. The powder X-ray diffraction analysis confirmed the C16-type structure of both compounds. Whereas in the case of Zr_2Ni we were dealing with a pure single-phase sample, a small admixture of HfNiO in the Hf_2Ni sample was evidenced.

In order to dope the investigated compounds with ^{111}In/^{111}Cd probes, small pieces of the materials were irradiated at room temperature with radioactive ^{111}In ions at 400 keV with the help of the Göttingen ion implanter IONAS [9]. As directly after implantation the PAC spectra exhibited only a broad distribution of quadrupole frequencies, it was necessary to anneal the samples for 7 h at 1073 K (Hf_2Ni) or 1173 K (Zr_2Ni). These temperatures allowed to remove the irradiation defects and to diffuse the probe atoms to substitutional lattice sites. Higher annealing temperatures caused the outdiffusion of the ^{111}In probes from the samples. The PAC measurements were carried out in the temperature range from 20 to 873 K with a standard fast–slow set-up equipped with four NaI(Tl) detectors.

The analysis of all experimental $R(t)$ spectra was performed applying the expression of the perturbation factor $G_2(t)$ [10] valid for static electric hyperfine interactions:

$$G_2(t) = \sum_{i=1}^{k} f_i \sum_{n=0}^{3} s_{2n}(\eta_i)\cos(g_n(\eta_i)\nu_{Qi}t)\exp(-g_n(\eta_i)\delta_i t), \quad (1)$$

where f_i are fractions of probes exposed to the different EFGs characterized by quadrupole frequencies ν_{Qi}, asymmetry parameters η_i and the width δ_i of the

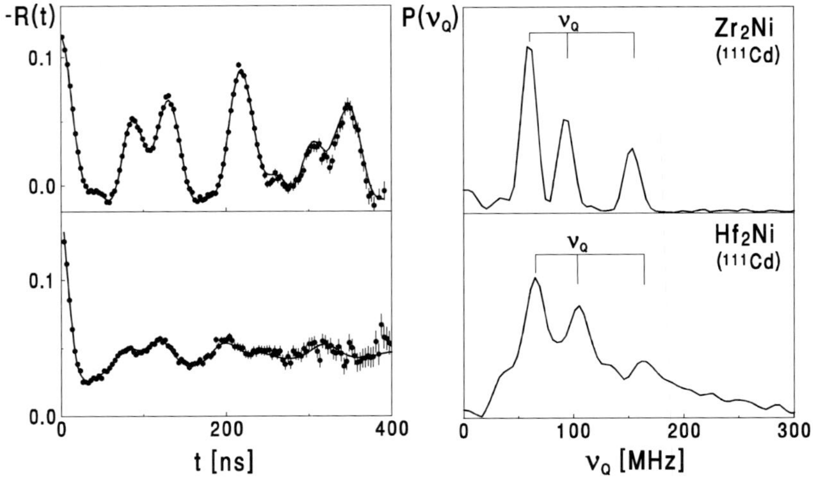

Figure 2. PAC spectra and the Fourier transforms taken at room temperature for ^{111}In/^{111}Cd probes in Zr$_2$Ni and Hf$_2$Ni samples.

Lorentzian ν_{Qi}-distribution. The EFG-values V_{zz} were calculated from the quadrupole frequencies ν_Q according to

$$V_{zz} = h\nu_Q / eQ, \tag{2}$$

adopting the quadrupole moment Q value equal to 0.83(13) b for ^{111}Cd [11], but its error was not taken into consideration in the quoted electric field gradients V_{zz}.

3. Results and discussion

As shown in Figure 2, the ^{111}In/^{111}Cd PAC spectra obtained for both investigated compounds are characterized by the presence of a single EFG of large asymmetry. The very clear PAC signal with a narrow frequency distribution obtained for Zr$_2$Ni corroborates the spectrum measured with ^{181}Ta/^{181}Hf probes in this compound [7, 8] and reflects the well-defined crystallographic site in the single-phase sample. The deduced asymmetry parameter of 0.49(1) excludes the occupation of the axially symmetric 4(a) Ni site and definitely proves substitution of the ^{111}In probes at the Zr 8(h) lattice sites. The PAC spectrum obtained for Hf$_2$Ni revealed a similar, non-axially symmetric EFG ($\eta = 0.44(1)$), indicating that also in this phase the ^{111}In/^{111}Cd probes after implantation and annealing occupy the 8(h) Hf sites. No additional unique frequency was observed in Hf$_2$Ni and the broad quadrupole frequency distribution (as it was also in case of measurements with ^{181}Hf probes [7]) reflects probably the admixture of oxygen in the sample found in the X-ray diffraction analysis.

The temperature dependences of the observed quadrupole frequencies in both compounds exhibit above room temperature a linear decrease with slope

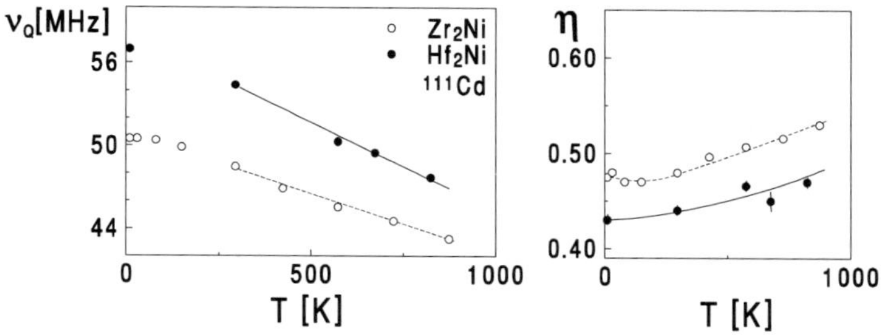

Figure 3. The temperature dependence of the quadrupole frequencies and asymmetry parameters measured for Hf_2Ni and Zr_2Ni phases with ^{111}Cd probes.

Table I. The QI parameters for ^{111}Cd and ^{181}Ta in Zr_2Ni and Hf_2Ni samples

Phase	Probe	Lattice site	υ_Q [MHz]	$\lvert V_{zz}\rvert$ [10^{18}Vcm^{-2}]	η
Zr_2Ni	^{181}Ta [7]	8(*h*) *mm*	628(3)	1.10(1)	0.83(3)
	^{181}Ta [8]	8(*h*) *mm*	625(3)	1.09(1)	0.835(3)
	^{111}Cd	8(*h*) *mm*	48.6(5)	0.24(1)	0.49(1)
Hf_2Ni	^{181}Ta [7]	8(*h*) *mm*	593(2)	1.04(1)	0.74(2)
	^{111}Cd	8(*h*) *mm*	54(1)	0.27(1)	0.44(1)

parameters of $1.8(1) \times 10^{-4}$ K^{-1} and $2.0(3) \times 10^{-4}$ K^{-1} for Zr_2Ni and Hf_2Ni, respectively. At lower temperatures a slight deviation from the linear dependence is observed. The values of the corresponding asymmetry parameters are slightly increasing with temperature over the whole temperature range studied (see Figure 3).

In summary, the deduced hyperfine interaction parameters are presented in Table I and compared with the previous results for ^{181}Hf/^{181}Ta probes. For each of the probes the close values of Hf_2Ni and Zr_2Ni QI parameters can be noticed, reflecting the chemical similarity of Hf and Zr and very close lattice constants of both compounds.

It can be pointed out that the asymmetry parameters found for Cd nuclei are ca. two times smaller than those for Ta probe. Moreover, the experimental EFG values at 8(*h*) lattice site on ^{181}Ta are ca. 4.5 times larger than those measured on ^{111}Cd nuclei, while the ratio of the corresponding antishielding factors for the free ions equals to ca. 2 [12].

The temperature dependencies of V_{zz} and η measured with Cd and Ta probes are also different. For ^{181}Ta in Zr_2Ni lattice the $T^{3/2}$ character of $V_{zz}(T)$ was observed and the decreasing of η values with temperature was found [8], contrary to those observed for ^{111}Cd probe.

Substitution of the Hf/Zr site by In atoms indicates that in this case the size factor wins over the chemical factor. Values of the atomic volumes and electronegativities of the relevant atoms are presented in Figure 4. It is clearly

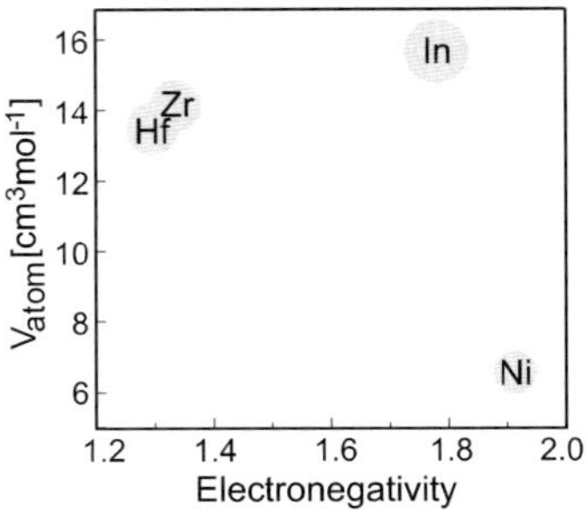

Figure 4. Comparison of atomic volumes and Pauling electronegativities of probe and compound constituent atoms [13, 14].

seen that the large In atoms avoid substituting the much smaller Ni atoms, in spite of the similar electronegativity.

Acknowledgements

The authors gratefully acknowledge the help of D. Purschke and Dr. L. Ziegeler during the ^{111}In implantations at IONAS. This work was funded partially by the Deutsche Forschungsgemeinschaft.

References

1. Miedema A. R., De Chatel P. F. and De Boer F. R., *Physica* **100B** (1980), 1.
2. Wodniecki P., Wodniecka B., Kulińska A., Uhrmacher M. and Lieb K. P., *J. Alloys Compd.* **312** (2000), 17.
3. Wodniecki P., Wodniecka B., Kulińska A., Uhrmacher M. and Lieb K. P., *J. Alloys Compd.* **335** (2002), 20.
4. Collins G. and Zacate M., *Hyperfine Interact.* **136/137** (2001), 647.
5. Wodniecki P., Wodniecka B., Kulińska A., Uhrmacher M., Lieb K. P., In: Görlich E. A., Królas K., and Pędziwiatr A. T. (eds.), *Proc. Int. Conf. on Condensed Matter Studies with Nuclear Methods, Zakopane 2003*, Jagellonian University, Kraków, 2003, p. 155.
6. Wodniecki P., Wodniecka B., Kulińska A., Uhrmacher M. and Lieb K. P., *Phys. Lett. A* **288** (2001), 227.
7. Wodniecka B. Marszalek M., Wodniecki P. and Hrynkiewicz A. Z., *Hyperfine Interact.* **80** (1993), 1039.
8. Poynor A. N., Cumblidge S. E., Rasera R. L., Catchen G. L. and Motta A. T., *Hyperfine Interact.* **136/137** (2001), 549.
9. Uhrmacher M., Neubauer M., Bolse W., Ziegeler L. and Lieb K. P., *Nucl. Instrum. Methods B* **139** (1998), 306.
10. Frauenfelder H. and Steffen R. M., In: Karlsson K., Matthias E. and Siegbahn K. (eds.), *Perturbed Angular Correlations*, North Holland, Amsterdam, 1963.
11. Butz T. and Lerf A., *Phys. Lett., A* **97** (1983), 217.
12. Feiock F. D. and Johnson W. R., *Phys Rev* **187** (1969), 39.
13. Moses A. J. (ed.), *The Practicing Scientist's Handbook*, Van Nostrand Reinhold, London, 1978.
14. Allred A., *J. Inorg. Nucl. Chem.* **17** (1961), 215.

Investigations in Biology, Chemistry, and Medicine

New Directions and Developments in Methodology

Hyperfine Interactions (2004) 159:1–8
DOI 10.1007/s10751-005-9073-8

PAC Probes as Diffusion Tracers in Solids

GARY S. COLLINS*, AURÉLIE FAVROT, LI KANG,
EGBERT REIN NIEUWENHUIS, DENYS SOLODOVNIKOV,
JIPENG WANG and MATTHEW O. ZACATE†
Department of Physics, Washington State University, Pullman Washington 99164, USA;
e-mail: collins@wsu.edu

Abstract. Perturbed angular correlation (PAC) probe atoms have been used as tracers to study diffusion in solids. The method works for diffusion on a sublattice for which the point symmetry is noncubic and the electric field gradient (EFG) at the probe nucleus reorients in each jump. Such motion leads to relaxation of the nuclear quadrupole interaction. Precise values of the tracer jump frequency have been obtained from fits of measured PAC perturbation functions. Results obtained to date are reviewed for Cd tracer atoms in rare-earth indides such as $LaIn_3$ that have the $L1_2$ crystal structure, for which each jump on the In-sublattice reorients the EFG by 90°. New results are presented for $LaSn_3$ and prospects for future studies are outlined.

Key Words: Cu_3Au structure, diffusion, nuclear relaxation, quadrupole interaction.

1. Introduction

Recently, measurements of the motion of PAC probe atoms, used as tracers, were reported by some of us for compounds having the Cu_3Au structure [1, 2]. These measurements differ from previous studies of nuclear quadrupole relaxation using PAC probes, in which probes were generally at fixed locations with defects hopping in the local vicinity such as lattice vacancies (see, e.g., [3–5]), hydrogen atoms (see, e.g., [6, 7]) or electronic defects (see, e.g., [8]). Those studies were mostly made using probe atoms at sites having cubic point symmetry, so that the electric field gradient (EFG) could be attributed entirely to the neighboring defect. In the present approach, one observes motion of the probe atom itself. For this, the probe must jump on a sublattice of sites having non-cubic point symmetry and the EFG must reorient in each jump. In the experiments reported below, no signals were observed that could be attributed to defects such as lattice vacancies, indicating that the residence time of any such defect next to a probe was much less than the mean residence time of the probe on a site. Thus, only the EFG of the perfect lattice was observed. In this paper, results for ^{111}Cd tracer

* Author for correspondence.
† Present address: Physics and Geology Department, Northern Kentucky University, Highland Heights, KY 41099, USA.

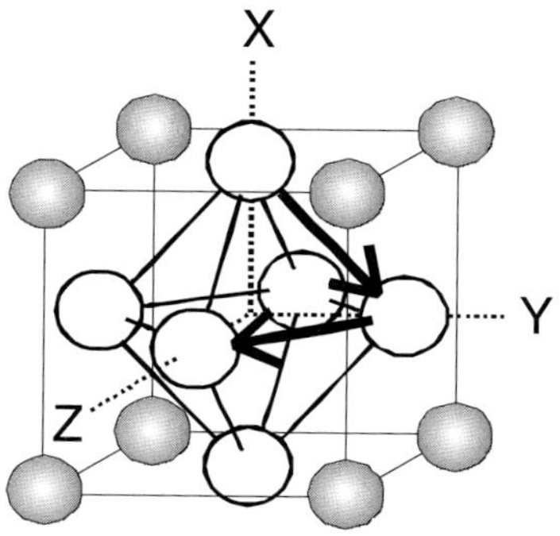

Figure 1. The Cu_3Au ($L1_2$) structure, showing tetragonal Cu sites and reorientation of the axis of the EFG from *X* to *Y* to *Z* directions during successive jumps.

atoms jumping on the Cu-sublattice in the Cu_3Au ($L1_2$) structure obtained to date are surveyed. Figure 1 shows the crystal structure, with Cu sites having tetragonal symmetry, and it can be seen that the EFG reorients along different ⟨100⟩ cube directions in each jump.

We recently reported on atom movement of ^{111}Cd tracers on the In sublattice in $LaIn_3$ [1] and in other tri-indides formed with Ce, Pr, Nd, Gd, Er and Y [2]. While these phases appear as line compounds in binary phase diagrams, their phase fields must have finite widths, however small. Measurements on $LaIn_3$ samples purposely prepared with the two phase boundary (PB) compositions exhibited jump frequencies 10-100 times greater at the more In-rich phase boundary than at the less In-rich boundary [1]. Similar results were obtained for $CeIn_3$ [2]. These results are reviewed and new measurements are presented for Cd tracer motion in $LaSn_3$ at its two PB compositions. In the discussion, general features of the relaxation method are summarized and prospects for further studies are outlined.

2. Experimental methods

Samples of $LaSn_3$ were prepared by melting high purity metal foils with ^{111}In activity under argon in an arc furnace, after which they were given a crystallizing anneal at high temperature under high vacuum prior to the PAC measurements Compositions were 23(1) and 28(1) at.% La. The Sn-rich and Sn-poor compositions were selected to produce two-phase samples having a preponderance of Cu_3Au phase and small amounts of neighboring phases. Accordingly, the Cu_3Au phases had compositions at the more Sn-rich and less Sn-rich PBs of the $LaSn_3$ phase, labeled below as $LaSn_3$(A) and $LaSn_3$(B), respectively. As observed also for the indides, inhomogeneous broadening at low temperature was negligible (<0.2 Mrad/s), suggesting that the phases were highly ordered, had low concentrations of thermal or structural point defects, and had PB compositions differing at most by a few tenths of a percent from the stoichiometric composition.

Direct measurements of the boundary compositions were not attempted because no routine method exists for determination of PB compositions to a precision and accuracy of the order of 0.1 at.%.

PAC measurements showed a predominant signal having a quadrupole interaction frequency equal to that previously attributed to ^{111}Cd on the Sn-site in $LaSn_3$ [9]. (A probe on the cubic La-site would of course have no interaction frequency.) Samples of $LaIn_3$ and $CeIn_3$ having the two boundary compositions were prepared in the same way [1, 2]. The remaining indides were prepared with nominally stoichiometric compositions [2], with their actual compositions expected to be at either one or other of the two PBs. PAC measurements were made using a four-counter spectrometer with BaF_2 scintillators, from which PAC perturbation functions $G_2(t)$ were derived [10]. The $L1_2$ phases exhibit an axially symmetric quadrupole perturbation function which for random polycrystalline texture has the static form

$$G_2^{\text{static}}(t) = \frac{1}{5} + \frac{13}{35}\cos(6\omega_Q t) + \frac{10}{35}\cos(12\omega_Q t) + \frac{5}{35}\cos(18\omega_Q t) \tag{1}$$

at low temperature, in which $\omega_Q \equiv \frac{\pi}{20}|eQV_{zz}|/h$ is the quadrupole interaction frequency. Fitted frequencies were all in excellent agreement with reported values [9]. With increasing temperature, diffusive motion first appears as damping of the static perturbation function. The jump frequency w is defined here as the inverse of the mean residence time of the tracer at a site. Jump frequencies reported in [1, 2] and below were all obtained from direct fits of the measured spectra with perturbation functions computed numerically for a model in which the EFG is axially symmetric and jumps among the three (100) directions. For a description of this fitting method and comparison with various approximate methods, see [11]. Qualitative features of the relaxation are represented well by analytic approximations for PAC perturbation functions given by Baudry and Boyer [12] for a model in which the EFG has constant magnitude, fluctuates among N different orientations, and has an EFG equal to zero when averaged over all orientations. Jumps on the In- or Sn-sublattices fit this model with $N = 3$. In the slow fluctuation regime ($w < \omega_Q$), the perturbation function was approximated by $G_2(t) \cong \exp(-\lambda t)G_2^{\text{static}}(t)$, in which λ is a relaxation frequency. It was found empirically [12] that $\lambda \cong (N-1)w_{\text{EFG}}$, in which w_{EFG} is the relaxation rate caused by fluctuation of the EFG orientation to a *single* other orientation (note that w_{EFG} is called w in [12]). When the EFG can change to any other orientation in a jump, as in the present diffusion model, the total number of orientations differing from the original one is $N-1$ and one sees that the total jump frequency is likewise $w = (N-1)w_{\text{EFG}}$. One thus has the simple result $\lambda = w$. Accordingly, we write for the slow fluctuation regime

$$G_2(t) \cong \exp(-wt)G_2^{\text{static}}(t). \tag{2}$$

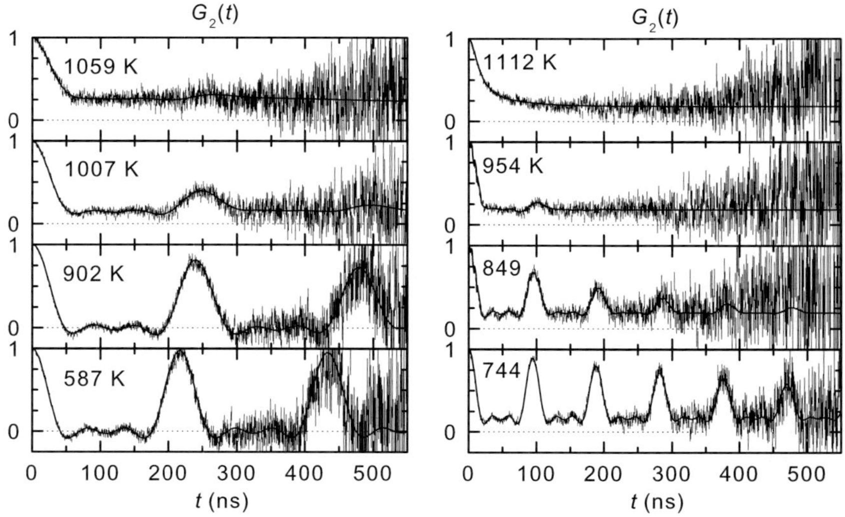

Figure 2. PAC spectra of more Sn-rich $LaSn_3$ (*left*) and more In-rich $CeIn_3$ (*right*) measured at the indicated temperatures. Increasing relaxation at higher temperatures is attributed to faster jumping of ^{111}Cd tracers on the Sn or In sublattices.

For jump frequencies in the fast fluctuation regime ($w > \omega_Q$), similar consideration of the approximate perturbation function given in [12] leads to

$$G_2(t) \cong \exp\left(-200\omega_Q^2 t \big/ 3w\right). \tag{3}$$

While the damping increases with w in the slow fluctuation regime, it decreases with w in the fast fluctuation regime as a consequence of motional averaging of the EFG. At the highest jump frequencies, the EFG approaches its value averaged over all orientations, which is zero. Note that the actual method used to fit the spectra did not involve approximations such as given by Equations (2) and (3) and excellent fits were obtained also in the intermediate fluctuation regime ($w \approx \omega_Q$). For further information, see [1] and [11].

3. Results

Figure 2 shows effects of relaxation in PAC spectra for ^{111}Cd in Sn-rich $LaSn_3$ (left) and In-rich $CeIn_3$ (right). Nearly static perturbation functions observed at low temperatures become increasingly damped at higher temperatures. At high temperatures (1,059 and 954 K), the periodic precessions become completely damped out. Finally, less damping is observed for $CeIn_3$ at 1,112 K in the fast fluctuation regime. Curves drawn show results of the fits to the exact functions and are in excellent agreement with the data.

Quadrupole interaction frequencies observed at room temperature are in excellent agreement with those listed in [9] and attributed to the Cu_3Au structure.

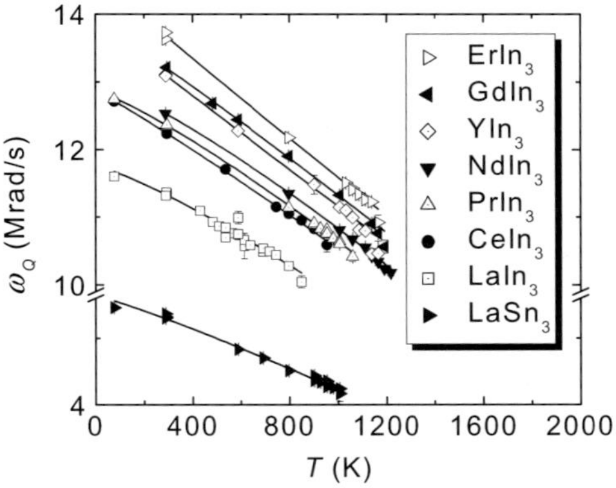

Figure 3. Temperature dependences of quadrupole interaction frequencies for ^{111}Cd probes on the In/Sn sublattices in the indicated compounds. Curves are from fits described in the text.

Table I. Quadrupole interaction parameters and jump frequency activation enthalpies and prefactors for A and B phase boundary compositions

Phase	ω_{Q0} (293 K) (Mrad/s)	B (K^{-1})	ν	PB	Q (eV)	w_0 (THz)
$LaSn_3$	5.64 (5)	2.9 (2) $\times 10^{-4}$	1.13 (8)	A	1.22 (5)	$7._{-3}^{+5}$
				B	1.2 (1)	$8._{-6}^{+21}$
$LaIn_3$	11.76 (5)	2.2 (3) $\times 10^{-4}$	1.21 (9)	A	0.535 (2)	$1.02_{-0.05}^{+0.05}$
				B	0.81 (1)	$1.4_{-0.2}^{+0.2}$
$CeIn_3$	12.86 (3)	2.1 (1) $\times 10^{-4}$	1.08 (4)	A	0.91 (4)	$1.7_{-0.7}^{+1.2}$
				B	1.30 (7)	$11._{-6}^{+13}$
$PrIn_3$	12.87 (4)	2.3 (1) $\times 10^{-4}$	1.20 (5)		1.11 (3)	$3.6_{-1.1}^{+1.5}$
$NdIn_3$	13.00 (7)	2.3 (2) $\times 10^{-4}$	1.23 (9)		1.80 (5)	450_{-180}^{+290}
$GdIn_3$	13.9 (1)	2.1 (2) $\times 10^{-4}$	1.08 (9)		~1.2	~3
$ErIn_3$	14.6 (1)	1.9 (2) $\times 10^{-4}$	0.96 (9)		1.07 (7)	$0.62_{-0.31}^{+0.60}$
YIn_3	13.9 (2)	2.0 (2) $\times 10^{-4}$	1.00 (9)		1.43 (5)	$34._{-15}^{+26}$

Temperature dependences of the quadrupole interaction frequencies are shown in Figure 3. It can be seen that frequencies in some phases appear to decrease linearly with temperature whereas others decrease more rapidly. Temperature dependences were fitted with the empirical expression $\omega_Q = \omega_{Q0}\,(1 - (BT)^{\nu})$ and values obtained for the quadrupole interaction frequency at 293 K, ω_{Q0}, temperature coefficient B and exponent ν are listed in Table I. All exponents are in the range 1.1(1) and show no obvious systematic trend. Coefficients for the indides are in the range 1.9–2.3 $\times 10^{-4}$ K^{-1} whereas a higher value 2.9 $\times 10^{-4}$ K^{-1} was obtained for $LaSn_3$.

Figure 4 shows an Arrhenius plot of jump frequencies for all data sets excepting, for clarity, $CeIn_3$(B) and $LaSn_3$(B). Each data set exhibits a well defined linear Arrhenius dependence on temperature. Despite similar chemistries, the jump frequencies vary by large factors. Jump frequencies were found to be positively correlated with the lattice parameter in the indides [2]. For both $LaIn_3$

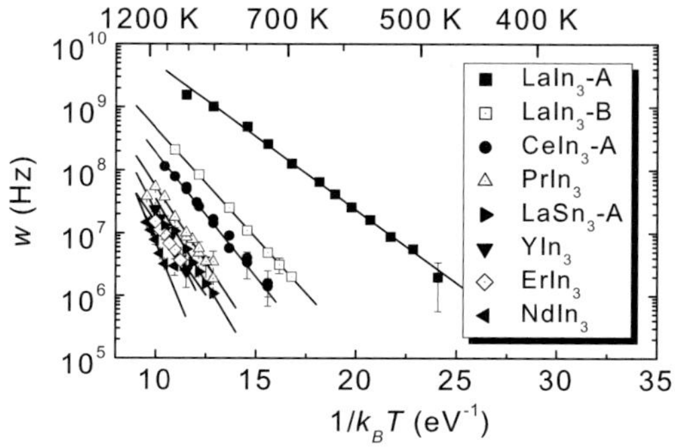

Figure 4. Jump frequencies of ^{111}Cd tracer atoms on the In or Sn sublattices in the indicated compounds.

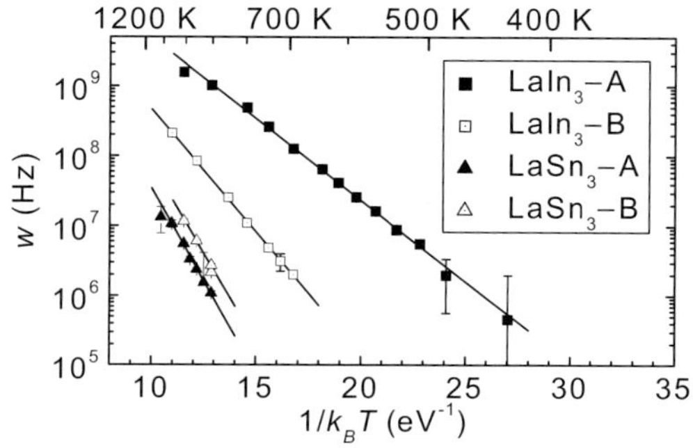

Figure 5. Jump frequencies of ^{111}Cd tracer atoms on the In or Sn sublattices in the indicated compounds.

[1] and $CeIn_3$ [2], the jump frequency is significantly greater at the In-rich boundary composition.

Figure 5 shows Arrhenius plots for the two $LaSn_3$ and two $LaIn_3$ data sets. The jump frequencies can be seen to be greater for $LaIn_3$ at boundary (A) but for $LaSn_3$ at boundary (B). Jump frequencies for $LaSn_3$ are roughly a factor of 10^3 smaller than for $LaIn_3$(A). The small factor-of-two ratio between jump-frequencies in the two $LaSn_3$ samples may indicate that the width of the Cu_3Au phase field in La–Sn is much less than in La–In.

Straight lines drawn in Figures 4 and 5 show results of fits of the jump frequencies for each sample to Arrhenius expressions

$$w = w_0 \exp(-Q/k_B T) \tag{4}$$

in which Q is the jump frequency activation enthalpy and w_0 is a frequency prefactor. Fitted values for Q and w_0 are given in Table I. For $LaIn_3$, $CeIn_3$ and $LaSn_3$, separate results are given for boundary compositions that are more In- or Sn-rich (A) and less In- or Sn-rich (B). The values of Q are quite low in these phases. It was shown in [2] that activation enthalpies along the series of indides formed with La, Ce, Pr and Nd are inversely correlated with the lattice parameter. However, the lattice parameter of $LaSn_3$ is even greater than for $LaIn_3$ while the activation enthalpy is also greater, indicating a systematic difference between indides and stannides.

4. Discussion

4.1. DIFFUSION MECHANISMS

The simplest plausible diffusion mechanism is one in which tracer atoms exchange with vacancies jumping freely on the Sn- or In- sublattice, so that the jump frequency is proportional to the concentration of vacancies. However, it can be shown that concentrations of defects such as lattice vacancies on the In or Sn sublattice can only decrease as the composition of In or Sn increases (e.g., see [13]). The simple sublattice vacancy diffusion mechanism therefore can not explain the observation for La- and Ce- indides that jump frequencies were greater for more In-rich samples. On the other hand, the simple mechanism offers a possible explanation for tracer diffusion in $LaSn_3$ since the jump frequency was smaller for the more Sn-rich sample. Free vacancies in $LaSn_3$ might occur as structural defects in off-stoichiometric compounds or by thermal activation of an equilibrium defect combination that includes vacancies and preserves the composition. Independent information is needed about defect free energies in these systems in order to interpret the activation enthalpy Q in terms of a diffusion mechanism and enthalpies of formation and migration of point defects.

4.2. FEATURES OF THE PAC TRACER METHOD AND FUTURE PROSPECTS

The jump frequency w is related to the diffusivity D in cubic structures via $D = \frac{1}{6} f w \lambda^2$ [14] in which f is the correlation coefficient of diffusion and λ is the jump distance. The range of accessible jump frequencies depends on the meanlife of the PAC level and for ^{111}Cd is roughly 1–1,000 MHz. This translates to diffusivities in the range 10^{-14}–10^{-10} m^2/s using the relation between D and w given above and taking $f \approx 1$. From the same relation, it can be seen that the correlation coefficient of diffusion can be determined experimentally from the ratio of measurements of diffusivity and jump frequency carried out using the same tracer.

The nuclear quadrupole relaxation method is incapable of detecting atom movement on sublattices of cubic sites (e.g., in B2 phases) or on sublattices for which jumps lead to no reorientation of the EFG (e.g., in $L1_0$ phases) because no relaxation is produced. However, there are many sublattices in structures other than $L1_2$ for which atom movement does lead to nuclear relaxation, for example in the A13 structure of β-Mn [15] and γ-brasses or in the B20 (FeSi) structure. One may also study tracer jumps between inequivalent sublattices for which the magnitude and/or orientation of the EFG changes. Finally, for more complex structures in which tracer atoms jump on inequivalent sublattices, such as in the B20 structure, one may measure jump frequencies on different sites and determine ratios of inter and intra-sublattice jump frequencies.

Acknowledgement

This work was supported in part by the National Science Foundation under grant DMR 00-91681 (Metals Program).

References

1. Zacate M. O., Favrot A. and Collins G. S., *Phys. Rev. Lett.* **92** (2004), 225901; and *erratum, op. cit.* **93** (2004), 49903.
2. Collins G. S., Favrot A., Kang L., Solodovnikov D. and Zacate M. O., submitted to Defect and Diffusion Forum **237–240** (2005), 195.
3. Evenson W. E., Gardner J. A., Wang R. W., Su H.-T. and McKale A. G., *Hyperfine Interact.* **62** (1990), 283.
4. Bai B., Collins G. S., Nieuwenhuis H. T., Wei M. and Evenson W. E., In: Mishin Y., Cowan N. E. B., Catlow C. R. A., Farkas D. and Vogl G. (eds.), Diffusion Mechanisms in Crystalline Materials, Materials Research Society Symposium Proceedings **527** (1998), p. 210.
5. Collins G. S. and Nieuwenhuis H. T., *Defect Diffusion Forum* **194–199** (2001), 375.
6. Weidinger A., In: Schlapbach L. (ed.), *Hydrogen in Intermetallic Compounds II*, Topics in Applied Physics, Vol. 67. Berlin, Heidelberg, New York, Springer (1992), 259.
7. Forker M., Herz W., Simon D., Bedi S. C., *Phys. Rev. B* **51** (1995), 15994.
8. Achtziger N. and Witthuhn W., *Phys. Rev., B* **47** (1993), 6990.
9. Schwartz G. P. and Shirley D. A., *Hyperfine Interact.* **3** (1977), 67.
10. Zacate M. O. and Collins G. S., *Phys. Rev. B* **69** (2004), 174202.
11. Zacate M. O. and Evenson W. E., Comparison of XYZ Model Fitting Functions for ^{111}Cd in In_3La, in Proceedings for the HFI/NQI 2004, Bonn, Germany, R. Vianden, ed., unpublished.
12. Baudry A. and Boyer P., *Hyperfine Interact.* **35** (1987), 803.
13. Zacate M. O. and Collins G. S., *Phys. Rev. B* **70** (2004), 24202.
14. Philibert J., *Atom Movements: Diffusion and Mass Transport in Solids* (Les Éditions de Physique, Les Ulis, 1991).
15. Zacate M. O. and Collins G. S., submitted to Defect and Diffusion Forum **237–240** (2005), 396.

Hyperfine Interactions (2004) 159:9–14
DOI 10.1007/s10751-005-9074-7

Nuclear Spin–Lattice Relaxation of Single Crystal $Sr_{14}Cu_{24}O_{41}$

S. OHSUGI[1,*], S. MATSUMOTO[2], Y. KITAOKA[3], M. MATSUDA[4], M. UEHARA[5], T. NAGATA[6] and J. AKIMITSU[5]
[1]*Department of Electrical Engineering and Electronics, College of Industrial Technology, Nishikoya 1-27-1, Amagasaki, Hyogo 661-0047, Japan; e-mail: ohsugi@cit.sangitan.ac.jp*
[2]*Tsukuba Magnet Laboratory, National Institute for Materials Science, Tsukuba, Ibaraki 305-0003, Japan*
[3]*Department of Physical Science, Graduate School of Engineering Science, Osaka University, Toyonaka, Osaka 560-8531, Japan*
[4]*Advanced Science Research Center, Japan Atomic Energy Research Institute, Tokai, Ibaraki 319-1195, Japan*
[5]*Department of Physics, Aoyama-Gakuin University, Chitosedai, Setagaya-ku, Tokyo 157-0071, Japan*
[6]*Department of Physics, Ochanomizu University, Otsuka, Bunkyo-ku, Tokyo 112-8610, Japan*

Abstract. Nuclear spin–lattice relaxation rate T_1^{-1} has been measured for the ladder sites of two single crystals $Sr_{14}Cu_{24}O_{41}$ (Sr14-A,B) by ^{63}Cu NMR/NQR. The hole localization around 100 K appears as a peak in the T variation of T_1^{-1}(NQR). On the other hand, it is suppressed in the T_1^{-1}(NMR) data under the magnetic field $H \sim 11$ T, and a new peak appears around 20 K. T_1^{-1}(NMR) around the peak is more enlarged for Sr14-B than for Sr14-A. Hence, holes on the ladders of Sr14-B tend to be more localized. This is considered to be an origin for the occurrence of the magnetic order in Sr14-B under $H \sim 11$ T.

Key Words: Cu NMR/NQR, magnetic ordering, nuclear spin–lattice relaxation, spin ladder, $Sr_{14}Cu_{24}O_{41}$.

1. Introduction

Cu NMR spectrum of nonmagnetic impurity Zn-doped spin-1/2 Heisenberg two-leg ladder compound $SrCu_2O_3$ (Sr123) is broadened with Curie-like T dependence [1]. Our Cu-NQR/NMR results have demonstrated that the magnetic order in Zn-doped Sr123 is three-dimensional (3D) antiferromagnetic (AF),

* Author for correspondence.

where an interladder interaction is in a weakly coupled quasi-one-dimensional (WC-Q1D) regime. A WC-Q1D staggered polarization (SP) model has explained the T dependence of the broadened spectrum. Such a SP has been also revealed in Cu NMR measurements of $Sr_{14-x}Ca_xCu_{24}O_{41}$ (Cax) with hole-doped ladders. We have reported Cu NMR/NQR results on the magnetic ordering in Ca11.5 [2]. Origin of the 3D long-range (LR) ordering in Zn-doped Sr123 and Cax around $x = 11.5$ at low T is considered to be similar. Once SPs with an unpaired spin are induced on the ladders and effective weak 3D interlayer interactions occur, the localized spins undergo the LR ordering. Recently, we have reported on the field-induced 3D LR ordering below $T_N \sim 20$ K in a single crystal $Sr_{14}Cu_{24}O_{41}$ (Sr14-B), revealed in the Cu NMR measurements under $H \sim 11$ T [3, 4]. In order to clarify that the magnetic ordering in the Q1D two-leg ladder systems is associated with the SP and the weak 3D interaction originating from an appearance of unpaired spins on the ladders, we measured T_1^{-1} for the ladder-Cu sites of single crystals Sr14-A,B.

2. Results and discussion

Figure 1 shows the ^{63}Cu ($-1/2\leftrightarrow 1/2$)-transition NMR spectra for the ladder sites of Sr14-B under $H\|b \sim 11$ T. The spectra are those after deducting the spin-Knight shift K_S from the raw spectra. Nearly the same magnitude of the spin gap, $\Delta_{A,B} \sim 500$ K [3] has been estimated for Sr14-A,B from the T variation of K_S above 100 K, using the relation [5],

$$K_S \sim \chi_S \sim \frac{1}{\sqrt{T}} \exp\left(-\frac{\Delta}{k_B T}\right). \tag{1}$$

Above 200 K, remarkable difference has not been observed between both the Sr14-A,B spectra. However, apparently, the spectrum for Sr14-B starts to experience a gradual spliting into two spectra with Curie-like broadening as T decreases from 200 K, whereas the spectrum for Sr14-A dose not. The splitting and broadening have been also observed under $H\|a,c$. The anisotropy of the separations is compatible to that of hyperfine form factor $(A_b - 3B)$ / $(A_{a,c} - 3B) \sim 2.8$ [6] above 20 K, but deviates from ~ 2.8 below 20 K. The splitting in the Sr14-B spectrum clearly indicates that a short-range (SR) SP originated from a single-hole localization spreads over the near neighbor ladders below 200 K. Moreover, LR ordering seems to be occurred below $T_N \sim 20$ K. A small average spontaneous moment $\langle\mu\rangle_{\mathrm{ladder}} \sim 2 \times 10^{-2}\mu_B$ [3] on the ladders at 5 K has been estimated from the separations in the spectra under $H\|a,b,c$, using the hyperfine-coupling constants $A_{a,c} = (48\ \mathrm{kOe})/\mu_B$ and $A_b = (-120\ \mathrm{kOe})/\mu_B$ estimated for the ladders in Sr123 [7].

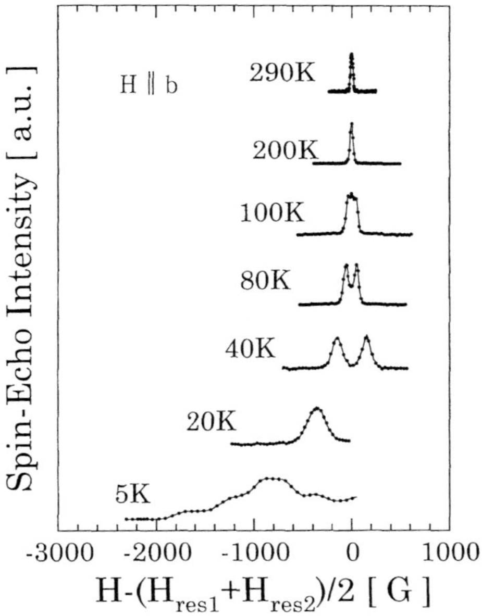

Figure 1. ^{63}Cu ($-1/2\leftrightarrow 1/2$)-transition NMR spectra for the ladder sites of Sr14-B under $H\|b\sim 11$ T at 125.1 MHz. The spectra are those after deducting the spin-Knight shift K_s from the raw spectra.

Not only the ladder-NMR spectrum but also the NMR spectrum for the Zhang-Rice (ZR) singlet sites in the chains starts to separate below 20 K, whereas the spectra above 20 K have only the T dependence of K_S due to the dimer formation with a gap $\Delta_{\rm dimer} = 127$ K [4]. The nucleus at the nonmagnetic ZR site is magnetically coupled with the nearest neighbor dimer-Cu spins. From the maximum interval in the splitting ZR-NMR spectrum at 5 K, a tiny dimer moment $\langle\mu\rangle_{\rm dimer}\sim 1.4\times 10^{-2}\mu_B$ has been estimated, using $A_{a,c} = (-14.8\ {\rm kOe})/\mu_B$ and $A_b = (-18.9\ {\rm kOe})/\mu_B$ as previously reported for Sr14 [8].

Next, we show Cu NQR spectra for the ladder sites of Sr14-A,B in Figure 2. At high T up to 200 K, spectral shape is simple. Full width at half maximum of the ^{63}Cu spectrum FWHM $\sim$ 200 kHz for Sr14-A is as large as that for Sr14-B. However, at low T down to 4.2 K, the NQR spectrum for Sr14-B possesses a distinct multi-peak structure, originated from the different electric field gradient (EFG) at each Cu site due to the single-hole localization. On the other hand, in the Sr14-A spectrum at 4.2 K, the hyperfine peaks are smeared out. The frustration effect of the interladder 90° Cu–O–Cu bond between the ladders is weaker in Sr14-B than in Sr14-A, and the holes are further localized with the growing the interladder coupling. Namely, the degree of hole localization on the ladders associated with some structural disorder is supposed to be higher in Sr14-B than in Sr14-A. In fact, such a characteristic is also revealed in T_1 measurements.

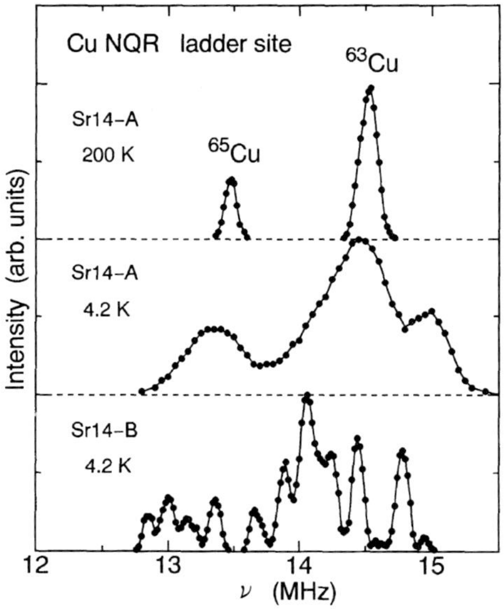

Figure 2. Cu NQR spectra for the ladder sites of Sr14-A,B in a range of frequency $\nu = 12.75$–15.5 MHz at 200 K and 4.2 K.

T_1 was measured with the saturation-recovery method [9]. Relaxation curve of the nuclear magnetization $M(t)$ at a time t after the saturation pulses is expressed for the Cu (nuclear spin $I = 3/2$) ($-1/2 \leftrightarrow 1/2$)-transition NMR as

$$M(t) = M\left[1 - 0.1\exp\left(-\frac{t}{T_1}\right) - 0.9\exp\left(-\frac{6t}{T_1}\right)\right] \tag{2}$$

and for the Cu NQR,

$$M(t) = M\left[1 - \exp\left(-\frac{3t}{T_1}\right)\right]. \tag{3}$$

Figure 3 shows T dependence of T_1^{-1} for the ladder sites of Sr14-A,B by ^{63}Cu NMR/NQR. At low T below 30 K, T_1^{-1}(A, NQR) becomes constant. Since T_1^{-1} is suppressed under $H \sim 11$ T, it might be due to an existence of some magnetic impurities in the sample. Local impurity spins are easily polarized along the H, and which do not contribute to the T_1 process. T_1^{-1}(NMR) above 200 K is uniquely estimated with the Equation (2), while short components appear in the $M(t)$ curves below 200 K, where holes start to localize. T_1^{-1}(NMR) in Figure 3 was estimated from a fit to the relaxation curve in a range of $M(t)/M = 1 - 0.1$.

Solid curves are the fits for the T_1^{-1}(NMR,NQR) data above 200 K, using the equation for the activation energy of a spin gap Δ

$$\frac{1}{T_1} \sim \exp\left(-\frac{\Delta}{\kappa_B T}\right) \tag{4}$$

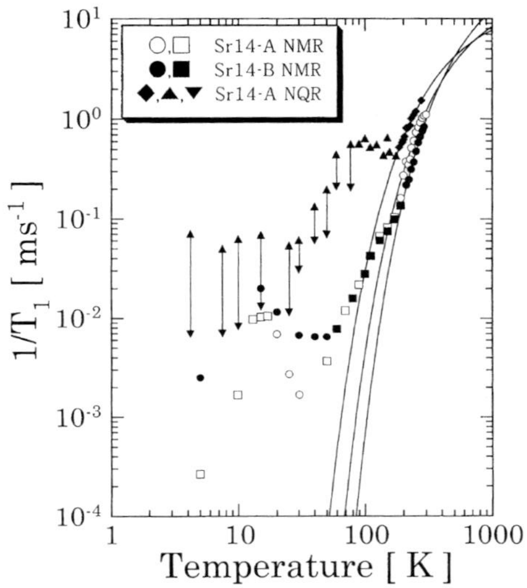

Figure 3. T dependence of T_1^{-1} for the ladder ^{63}Cu sites of Sr14-A,B. *Open* [*solid*] *circles* and *squares* denote T_1^{-1}(A, NMR) [T_1^{-1}(B, NMR)] for Sr14-A [Sr14-B] under the magnetic field parallel to the b-axis, $H\|b \sim 11$ T at 125.1 MHz. *Bars* between the *triangles* below 70 K indicate distribution ranges of T_1^{-1} (A, NQR) components. For *solid curves*, see the text.

at $T \ll \Delta$ [5]. Δ(NMR) ~ 900 K is estimated for Sr14-A,B [Δ(NMR) = 830 (1080) K for Sr14-A (Sr14-B)]. The deference between Δ_{T_i} and Δ_{K_S} has been theoretically explained as to be $\Delta_{T_1} \simeq \sqrt{3}\Delta_{K_S}$ at high T [10] and $\Delta_{T_1} = 1.5\Delta_{K_S}$ at low T [11]. Δ(NQR) = 620 K for Sr14-A is somewhat smaller than Δ(NMR) ~ 900 K. This might be because in the Δ(NQR) estimation for the data above 150 K, T_1^{-1}(A, NQR) contains the quadrupole relaxation rate due to the slow hole motions (EFG fluctuations) starting from 200 K.

An important finding is that the peak in T_1^{-1}(NQR) around 100 K [8] due to the single-hole localization is suppressed to 20 K in the T_1^{-1}(NMR) data under $H \sim 11$ T. On top of that, T_1^{-1}(NMR) around the peak is more enlarged for Sr14-B than for Sr14-A, indicating that the holes on the ladders in Sr14-B tend to be more localized. Thus, this is considered to be an origin for the occurrence of the magnetic order in Sr14-B. SP on the ladders and localized spins on the chains in Sr14-B experience the LR ordering below $T_N \sim 20$ K by growing the interlayer coupling under $H \sim 11$ T.

3. Conclusion

Occurrence of a 3D LR order in Sr14-B under $H \sim 11$ T has been demonstrated from Cu-NMR/NQR measurements. The present results obtained from the NQR spectra and the T dependence of T_1^{-1} suggest that holes on the ladders in Sr14-B tend to be more localized than in Sr14-A. With enough large 3D interaction, SP on the ladders and localized spins on the chains in Sr14-B undergo the 3D LR ordering below $T_N \sim 20$ K in $H \sim 11$ T.

Acknowledgements

This work was supported by CREST (Core Research for Evolutional Science and Technology) of Japan Science and Technology Corporation (JST) and also partly by the COE Research (10CE2004) and (09440138) in a Grant-in-Aid for Scientific Research from the Ministry of Education, Sports, Science, and Culture of Japan.

References

1. Ohsugi S. *et al.*, *Phys. Rev., B* **60** (1999), 4181.
2. Ohsugi S. *et al.*, *Phys. Rev. Lett.* **82** (1999), 4715.
3. Ohsugi S., Matsumoto S., Kitaoka Y., Matsuda M., Uehara M., Nagata T. and Akimitsu J., *Phys. B* **312–313** (2002), 603; *Z. Naturforsch.* **57a** (2002), 509.
4. Ohsugi S., Matsumoto S., Kitaoka Y., Matsuda M., Uehara M., Nagata T. and Akimitsu J., *J. Magn. Magn. Mater.* **272–276** (2004), e683.
5. Troyer M., Tsunetsugu H. and Würtz D., *Phys. Rev., B* **50** (1994), 13515.
6. Magishi K., Matsumoto S., Kitaoka Y., Ishida K., Asayama K., Uehara M., Nagata T. and Akimitsu J., *Phys. Rev., B* **57** (1998), 11533.
7. Ishida K., Kitaoka Y., Asayama K., Azuma M., Hiroi Z. and Takano M., *J. Phys. Soc. Jpn.* **63** (1994), 3222; *Phys. Rev., B* **53** (1996), 2827.
8. Takigawa M., Motoyama N., Eisaki H. and Uchida S., *Phys. Rev., B* **57** (1998), 1124.
9. Narath A., *Phys. Rev., B* **13** (1976), 3724.
10. Kishine J. and Fukuyama H., *J. Phys. Soc. Jpn.* **66** (1997), 26.
11. Sachdev S. and Damle K., *Phys. Rev. Lett.* **78** (1997), 943.

Hyperfine Interactions (2004) 159:15–19
DOI 10.1007/s10751-005-9075-6

Metal Atom Dynamics and Spin-Lattice Relaxation in Multilayer Sandwich Compounds

ISRAEL NOWIK and ROLFE H. HERBER*
Racah Institute of Physics, The Hebrew University of Jerusalem, 91904 Jerusalem, Israel; e-mail: HERBER@VMS.HUJI.AC.IL

Abstract. Temperature-dependent ^{57}Fe Mössbauer spectroscopy has been used to elucidate the hyperfine parameters and dynamical behavior of the metal atom in several organo-iron complexes which have one or more η^5 P_5 ring structures as ligated groups. The spin-lattice relaxation of the (paramagnetic) one-electron oxidation products occurs on a time scale fast compared to $\tau_{1/2}$ (ME) at temperatures in the range $85 < T < 320$ K.

Key Words: ^{57}Fe Mössbauer spectroscopy, η^5 P_5 iron complexes, spin-lattice relaxation.

1. Introduction

Temperature-dependent ^{57}Fe Mössbauer spectroscopy has been widely used to characterize diamagnetic (neutral) organometallic complexes, and the information derived therefrom has served to elucidate key questions concerning structure and bonding in such compounds. Significantly less attention has been paid to the one-electron oxidation products of such compounds, although a number of ferricinium salts were characterized in the early days of this spectroscopy. In particular, attention has focused recently on the relaxation rate of the spin orientation in these paramagnetic compounds, and it has been shown [1] that the principal relaxation mechanism is that of a spin-lattice interaction *via* an Orbach type process [2]. The early pioneering work of Collins [3] showed that the large (~2.0–2.4 mm sec^{-1}) quadrupole splitting in ferrocene and related compounds collapses to ~0.2–0.4 mm sec^{-1} on the removal of one electron, giving rise to an asymmetric broadened resonance line. The methods used to analyze spin relaxation phenomena in Mössbauer spectra have been reviewed by Wickmann and Wertheim [4], by Hoy [5] and others and depend in part on the temperature dependent non-Lorentzian line shape of the absorbance. A number of such studies involving ferrocenoid solids have been reported in the literature [6].

More recently, a detailed study of a number of ferrocene-related compounds, in which one or more of the ligand rings contain one or several P atoms, and

* Author for correspondence.

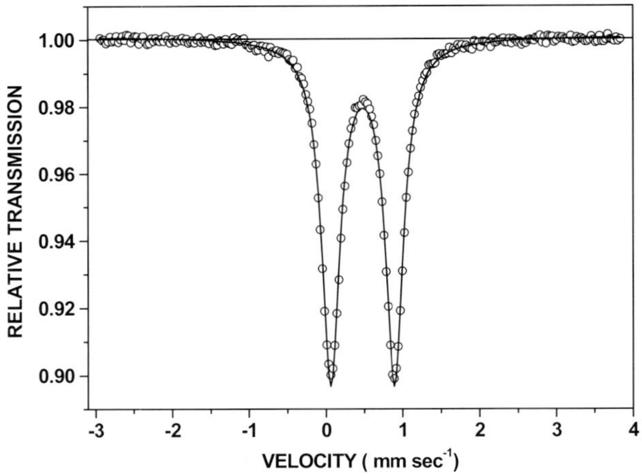

Figure 1. ^{57}Fe Mössbauer spectrum of **I** at 90 K. The velocity scale is with respect to the centroid of an α-Fe absorber spectrum which was also used for spectrometer calibration.

Table I. ^{57}Fe Mössbauer parameters for the three compounds discussed in the text

Param/Compd.	I	II	III	Units
IS(90)	0.477(4)	0.461(3)	0.469(4)	mm sec^{-1}
QS(90)	0.830(4)	0.639(3)	0.880(4)	mm sec^{-1}
−dIS/dT	4.85(11)	5.06(9)	4.84(3)	mm sec^{-1}K^{-1} × 10^{-4}
−d ln A/dT	4.43	10.6(6)	12.89(4)	K^{-1} × 10^{-3}

The IS values are reported with respect to the centroid of a room temperature α-Fe spectrum which was also used for spectrometer calibration. The parenthetical numbers are the errors associated with the last place(s) of the reported values.

which have been subjected to a one-electron oxidation process, have been described in the literature [7], including several multi-layer sandwich compounds in which the iron-containing cationic moieties are accompanied by a balancing anion such as BF_4^- and PF_6^-.

A compound of particular interest in this regard is the paramagnetic complex η^5–Cp*Fe–η^5–P_5 Fe η^5–Cp*$^+$ BF_4^- (**I**) [where Cp* is pentamethyl cyclopentadienyl] which is related both to η^5–Cp*Fe–η^5–P_5 (**II**) (neutral) and η^5–Cp* Fe η^5–P_5 Fe η^5–Cp$^+$$PF_6^-$ (**III**) (paramagnetic). A typical spectrum of **I** is shown in Figure 1. The observed spectra of this paramagnetic systems displays a very narrow line quadrupole doublet, showing that unlike in paramagnetic ferrocene-related cations, here the spin relaxation rates are extremely rapid compared to the characteristic Mössbauer time scale ($\tau_{1/2} \sim 10^{-7}$ sec). The hyperfine parameters at 90 K and related parameters of these three compounds are summarized in Table I. It is noteworthy that the quadrupole splitting (QS) in **I** is very much smaller than that observed in neutral ferrocenes, but approximately the same as in the neutral **II** in which a η^5–P_5 is ligated directly to the metal

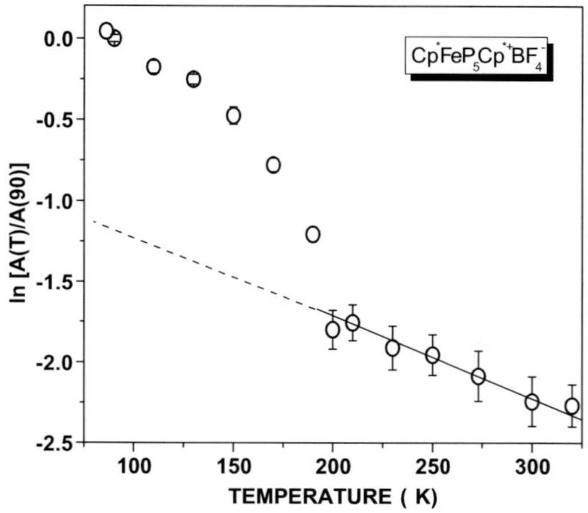

Figure 2. Temperature dependence of the recoil-free fraction of **I**. The high temperature slope is indicated by the straight line segment of the data.

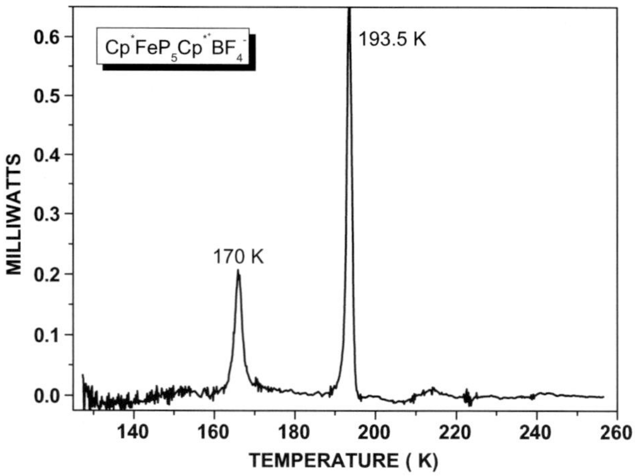

Figure 3. Differential scanning calorimetry data for **I**. The raw data have been base-line corrected. The scan rate (on warming) was 5 degrees per minute.

atom. Moreover, the QS is larger by a factor of ~4 compared to the ferricinium homologue and is nearly temperature independent in the range $85 < T < 200$ K, and then decreases slowly at higher temperatures. The temperature-dependence of the recoil-free fraction (f) for **I**, which for a thin absorber is equal to ln A(T), where A is the area under the resonance curve, is summarized graphically in Figure 2 from which it is noted that there is a marked discontinuity in the slope at ~195 K. The corresponding differential scanning calorimetry data are shown in Figure 3. Thermodynamic parameters associated with the latter are $\Delta H = 1.19$ kJ mol^{-1} and $\Delta S = 6.16$ J mol^{-1}K^{-1}, similar to the corresponding values − 1.88 kJ mol^{-1} and 8.17 J mol^{-1}K^{-1} determined for **II**.

As noted from Table I, quite similar parameters associated with the Mössbauer spectra are observed for **III** which differs from **I** only in the fact that one of the multilayer rings is Cp rather than Cp* and that the counter anion is PF_6^- instead of BF_4^-. Again, the spectra in the range $90 < T < 320$ K consist of a well-resolved doublet pattern with a QS (90 K) which is almost identical to that observed in **I**. Clearly, with respect to the hyperfine parameters, the Cp and Cp* ligands are indistinguishable within the error limits of the measurements. One additional point, however, does deserve mention: the ln A(T) data for **III** are well fitted by a linear regression over the whole temperature range, and do not show any evidence of the discontinuity reported (above) for **I**. This observation may be associated with the onset of ring rotation in the two solids which is expected to reflect the differences in the steric requirements of the Cp and Cp* ligands as well as of the two counter anions.

Both the fast spin-lattice relaxation noted for **I** and related phospholyl complexes [7], as well as the magnitude of the QS interactions in these complexes, as well as in **II** and **III**, compared to values reported earlier for ferrocenoid complexes without ring phospholyl groups, is presumably associated with the difference in the metal atom–ligand bonding obtaining in these compounds. An MO correlation diagram for η^5–N_5 ligands has been discussed by Frenking [8], and for η^5–P_5 by Schleyer et al. [9] and Malar [10]. However, a detailed orbital description of cyclic ligands containing both CH and P groups, and their bonding interaction with Fe central atoms appears to be not yet available. The metal atom–ligand bonding in these complexes is known to involve an overlap between the metal atom d-orbitals and the π-density of the cyclic ligand. Therefore a detailed understanding of the magnitude of the Q.S. in such compounds will require a focused study of the MO levels participating in these interactions, and of the configuration of the one-electron oxidation products here discussed.

Acknowledgements

The authors are indebted to Prof. A. R. Kudinov of the A. N. Nesmeyanov Institute for Organoelement Compounds for making his interesting compounds available for study and to Ms. Janet Zoldan of the Israel Institute of Technology for the careful differential scanning calorimetry data here reported.

References

1. Schottenberger H., Wurst K., Griesser U. J., Jetti R. K. R., Laus G., Herber R. H., Nowik I., *J. Am. Chem. Soc.* **127** (2005), 6795–6801.
2. Abragam A. and Bleany B., In: *Electron Paramagnetic Resonance of Transition Ions*, Clarendon Press, Oxford, 1970, pp. 560–561; Finn C. B. P., Orbach R. and Wolf W. P., *Proc. Phys. Soc.* **77** (1961), 267.

3. Collins R. L., *J. Chem. Phys.* **42** (1965), 1072–1080.
4. Wickman H. H. and Wertheim G. K., In: Gol'danskii V. I. and Herber R. H. (eds.), *Chemical Applications of Mössbauer Spectroscopy,* Chapter 11, Plenum Press, New York, 1968, pp. 548–621.
5. Hoy G. R., In: Long G. J. (ed.), *Mössbauer Spectroscopy Applied to Inorganic Chemistry,* Chapter 8, Plenum Press, New York, 1984, pp. 195–225.
6. Herber R. H., Nowik I. and Rosenblum M., *Organometallics* **21** (2002), 846–851; Schottenberger H., Wurst K. and Herber R. H., *J. Organomet. Chem.* **625** (2001), 200–207; Nowik I. and Herber R. H., *Inorg. Chim. Acta* **310** (2000), 191–195; Herber R. H., Gattinger I. and Köhler F. H., *Inorg. Chem.* **39** (2000), 851–853; and references therein.
7. Herber R. H., Nowik I., Loginov D. A., Starikova Z. A. and Kudinov A. R., *Eur. J. Inorg. Chem.* (2004), 3476–3483.
8. Frenking G., *J. Organomet. Chem.* **635** (2001), 9–23.
9. Urnezius E., Brennessel W. W., Cramer C. J., Ellis J. E. and von Schleyer P. R., *Science* **295** (2002), 832–834.
10. Malar E. J. P., *Eur. J. Inorg. Chem.* (2004), 2723–2732.

Hyperfine Interactions (2004) 159:21–27
DOI 10.1007/s10751-005-9076-6

Comparison of the Impurity Effects on Lattice Dynamics in K_2SnCl_6 between Isomorphic and Nonisomorphic Systems Near the Structural Phase Transition Temperature Revealed by Nuclear Resonance

Y. M. SEO[1,*], B. S. KIM[1], S. K. SONG[1] and J. PELZL[2]
[1]*Department of Physics, Myongji University, Yongin Kyunggi-do, 449-728, South Korea; e-mail: symseoul@unitel.co.kr; sksong@mju.ac.kr*
[2]*Institut fuer Experimentalphysik 3, Ruhr Universitaet, Bochum, Germany*

Abstract. Nuclear resonance studies of the two different types impurity doped potassium hexachloro-stannates, the isomorphic system such as $(K–Rb)_2SnCl_6$ and $K_2(Sn–Re)Cl_6$ and the nonisomorphic system $K_2SnCl_6:Al^{3+}$ in the high temperature cubic phase revealed contrasting features with one the other characterized by static in the former and dynamic feature in the latter case respectively. The resonance spectra of the nonisomorphic system indicate additionally a sign of the local structural transition above the conventional structural phase transition temperature. This seems to be triggered by the ligand-deficient octahedral defects and can be explained in terms of the enhanced activity of the octahedral defects for the hindered rotation.

Key Words: isomorphic/nonisomorphic impurity effects, nuclear resonance, pretransition.

1. Introduction

This contribution concerns the nuclear resonance studies of the potassium hexachloro-stannate, K_2SnCl_6, the structure of which is intentionally disordered by introducing two different types of impurities in the crystal according to whether the starting materials for impurity belong to the same family, the so-called antifluorite A_2BCl_6 as the host K_2SnCl_6 or not. Among the three kinds of impurity doped K_2SnCl_6 considered in this work, $(K–Rb)_2SnCl_6$, $K_2(Sn–Re)Cl_6$ and $K_2SnCl_6:Al^{3+}$, the two compounds Rb_2SnCl_6 and K_2ReCl_6 used as the starting materials for impurity in the first two crystal systems $(K–Rb)_2SnCl_6$ and $K_2(Sn–Re)Cl_6$ are isostructural with K_2SnCl_6 in the high temperature cubic phase. On the other hand, in the crystal $K_2SnCl_6:Al^{3+}$, Al^{3+} metal ions are extracted from the nonisostructural compound $AlCl_3$. Thus the first two crystals

* Author for correspondence.

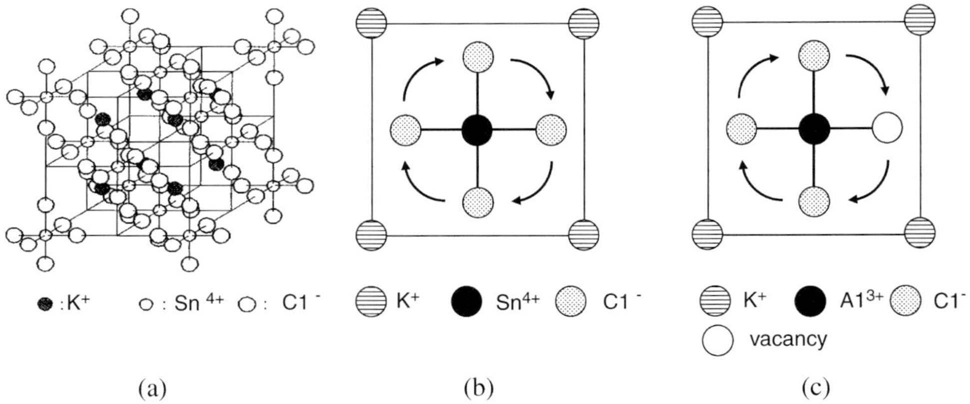

Figure 1. (a) Unit cell of cubic K_2SnCl_6. (b) Intact $SnCl_6^{2-}$ and (c) ligand-deficient $AlCl_5^{2-}$ in the K^+-cage.

are denoted by isomorphic system and the crystals K_2SnCl_6:Al^{3+} by nonisomorphic system respectively. The accumulation of the nuclear resonance studies in both systems indicates markedly distinguishing impurity features between them that can be interpreted in the context of the static compositional disorder in the isomorphic system resulting from the statistical distribution of two sorts of host and impurity atoms and dynamic disorder in the nonisomorphic system produced by the impurity induced lattice defects respectively [1, 2].

Many of the A_2BCl_6 compounds exhibit structural phase transitions. The phase transitions in A_2BCl_6 are driven by the condensation of the rotary lattice mode associated with the small angle rotative vibration of the BCl_6-complex about one of the three B–Cl bond axes at optical frequencies. It changes to the thermally activated rotation of the BCl_6-octahedra by 90°, the hindered rotation, at high temperatures during which a simultaneous reorientation of four halogens about any of the four fold B–Cl bond axes occurs leaving the other two in their position. The interplay of these two types of motion seems to be responsible for the crystal instabilities and consequently for the structural transitions. K_2SnCl_6 undergoes two structural changes at $T_{c1} = 262$ K and $T_{c2} = 256$ K from face centered cubic at high temperatures to monoclinic at low temperatures (Figure 1).

The study of nuclear resonance can provide detailed information on the static and the dynamic properties of the structure on the scale of the interatomic spacings since one of the main hamiltonians of the nuclear spin system, the quadrupole hamiltonian, consists of the local field. To understand the local dynamics in the iso/nonisomorphic K_2SnCl_6, the main concern of which is the temporal fluctuation of the local field, a semiclassical description of the relaxation seems to be appropriate. This leads to two limiting cases; the high frequency local field fluctuation with small amplitudes on the one hand and the infrequent and large amplitude fluctuation of the local field on the other hand. The so-called 'weak collision' approach is relevant to the former case while to the latter case the 'strong collision' approach is applicable [3].

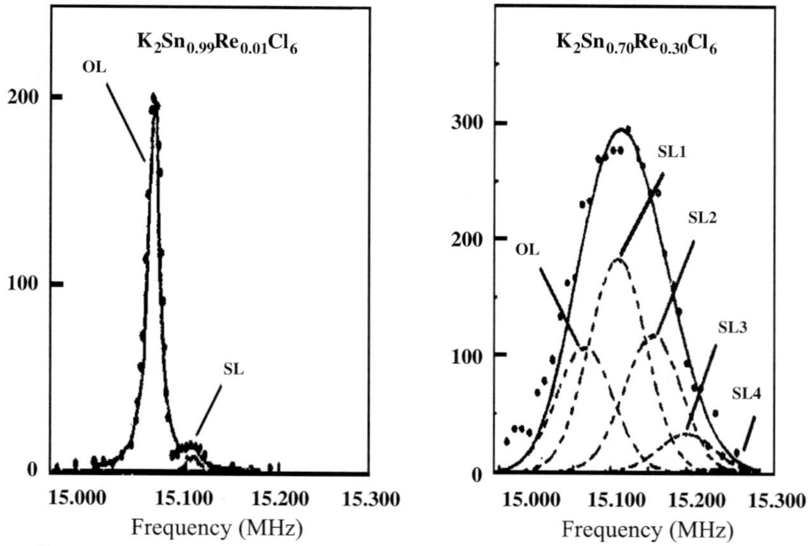

Figure 2. ^{35}Cl NQR line spectra of the K_2SnCl_6-structure in $K_2Sn_{1-x}Re_xCl_6$ for small x and large x. The *dashed lines* denote the individual line (original and satellite lines) intensities obtained from the theoretical model superposed to the resultant total intensity distribution (*solid lines*).

Following the general procedure, the weak collision approach yields the relaxation rates in the fast motion regime

$$T_{1,2} \approx \frac{|H'_{\mathrm{mm'}}|^2}{\eta^2} 2\tau_{\mathrm{c}}$$

with the matrix element of the perturbation hamiltonian $H'_{\mathrm{mm'}}$ and the characteristic time of fluctuation τ_{c}. The characteristic of the strong collision approach is that the order of magnitude of the spin–lattice and the spin–spin relaxation time is the same as the characteristic time of the local field fluctuation τ_c.

$$T_1 \approx T_2 \approx \tau_c.$$

2. Results and discussion

2.1. ISOMORPHIC SYSTEM; K_2(SN–RE)CL_6 AND (K–RB)$_2SNCL_6$

2.1.1. *The ^{35}Cl NQR line spectra in K_2(Sn–Re)Cl_6*

The ^{35}Cl NQR line spectra reflect directly the spatial inhomogeneity of the isomorphic crystals. In addition to the line width broadening, this is demonstrated by the appearance of the satellite lines near the original resonance line (Figure 2). The characteristic of the satellite lines, equidistant line position in the frequency domain and the shift of the center of gravity in line intensity into the high frequency region with increasing impurity concentration, may be explained in terms of the local point symmetry in the crystal and the probability for the impurities to occupate the lattice sites around the symmetry points.

With a strong fourfold feature of the lattice points in K_2SnCl_6 in the cubic phase, the appearance of the total four satellite lines in addition to the original one in the crystal with especially large concentration of impurity is the manifestation of the random distribution of the static impurities at the corresponding lattice sites [2].

2.1.2. ^{87}Rb *NMR relaxation in* $(K\text{–}Rb)_2SnCl_6$

The temperature behaviours of the ^{87}Rb NMR relaxation measured at 65.400 MHz are characterized in the range from 320 K to 140 K by the temperature independence of the spin–lattice and spin–spin relaxation (Figure 3a) and the Gaussianlike decay of the nuclear magnetisation for the spin–spin relaxation. The relaxation time amounts to 20 ms for T_1 and 2 ms for T_2, respectively. The temperature independence of both relaxation times T_1 and T_2 and the Gaussianlike decay of the ^{87}Rb NMR relaxation indicate again a static impurity effects the feature of which coincides with the results of ^{35}Cl NQR in $K_2(Sn\text{–}Re)Cl_6$.

2.2. NONISOMORPHIC SYSTEM; K_2SNCL_6:AL_{3+}

2.2.1. ^{35}Cl *NQR and* ^{27}Al*-NMR*

In the pure K_2SnCl_6 the ^{35}Cl NQR line intensity maintains its strength on appraching T_{c1} from above up to the point just above the first transition temperature T_{c1} and reduces rapidly to zero within the temperature interval $\Delta T =$ 0.5 K around T_{c1} with further decreasing temperature. In contrast to this, the ^{35}Cl NQR line intensity in K_2SnCl_6:Al^{3+} begins to decrease already at temperatures, which are designated by T_I, around room temperature ($T_I \approx 300$ K) with decreasing temperature in the cubic phase that are quite above the first transition temperature ($T_{c1} = 262$ K). At T_{c1} the residual sharp cubic line vanishes abruptly. Below T_{c1} new lines of the low temperature phase evolve gradually with decreasing temperature that possess much broader line width than that of the cubic line from the start (Figure 4a).

The progressive reduction of the line intensity below T_I was found to be attributed to the changeover of the nuclear magnetisation for both spin–spin and spin–lattice relaxation from the single exponential to the two exponential decay with two distinct components of the relaxation times $T_{1,2}^L$, $T_{1,2}^S$, the long and the short component. The decay of the nuclear magnetisation $M(t)$ is for $T_{c1} \leq T \leq T_I$ (Figure 4b)

$$M(t) = A_L \exp\left(-\frac{t}{T_{1,2}^L}\right) + A_s \exp\left(-\frac{t}{T_{1,2}^s}\right) + M(t = \infty).$$

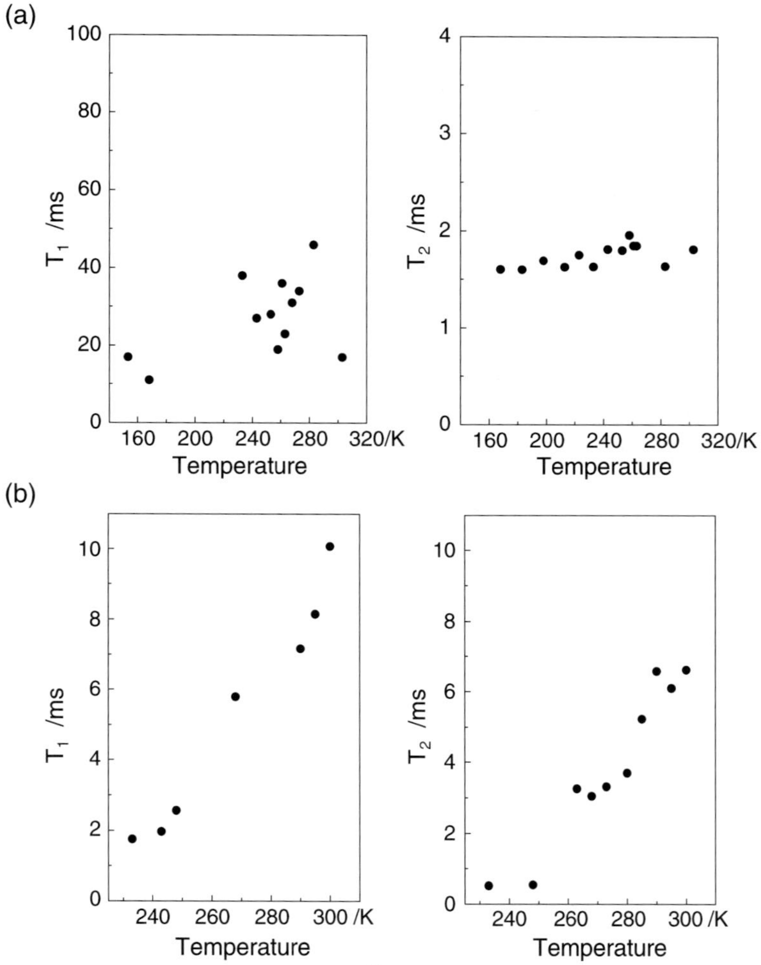

Figure 3. (a) Temperature dependence of the ^{87}Rb NMR in $(K_{1-x}Rb_x)_2SnCl_6$. (b) Temperature dependence of the ^{27}Al NMR in K_2SnCl_6:Al^{3+}.

For both spin–spin and spin–lattice relaxations the relative fraction of the short to long component A_s/A_L increases on approaching T_{c1}. The short components of both relaxation times T_1^S ($\approx$100 μs) and T_2^S ($\approx$70 μs) are the same in the order of magnitude and increases slightly on approaching T_{c1} from above.

The ^{27}Al NMR relaxation measured at 52.114 MHz shows a gradual reduction of T_1 and T_2 over the temperature range from 300 K up to T_{c1} and undergoes a sudden shrinkage below T_{c1} (Figure 3b). The temperature behaviour of T_1 and T_2 in the ^{27}Al NMR and an abrupt increase in the ^{35}Cl NQR line width below T_{c1} indicate that in K_2SnCl_6:Al^{3+} there is a motion associated with the Al^{3+} impurities that is slowed down with decreasing temperature and frozen in below T_{c1} due to the change in the lattice potential.

On the other hand, two components of the ^{35}Cl NQR relaxation time indicate there are two sets of nuclei having different environments. The increase in the fraction of the short component T_1^S and T_2^S on approaching T_{c1} from above

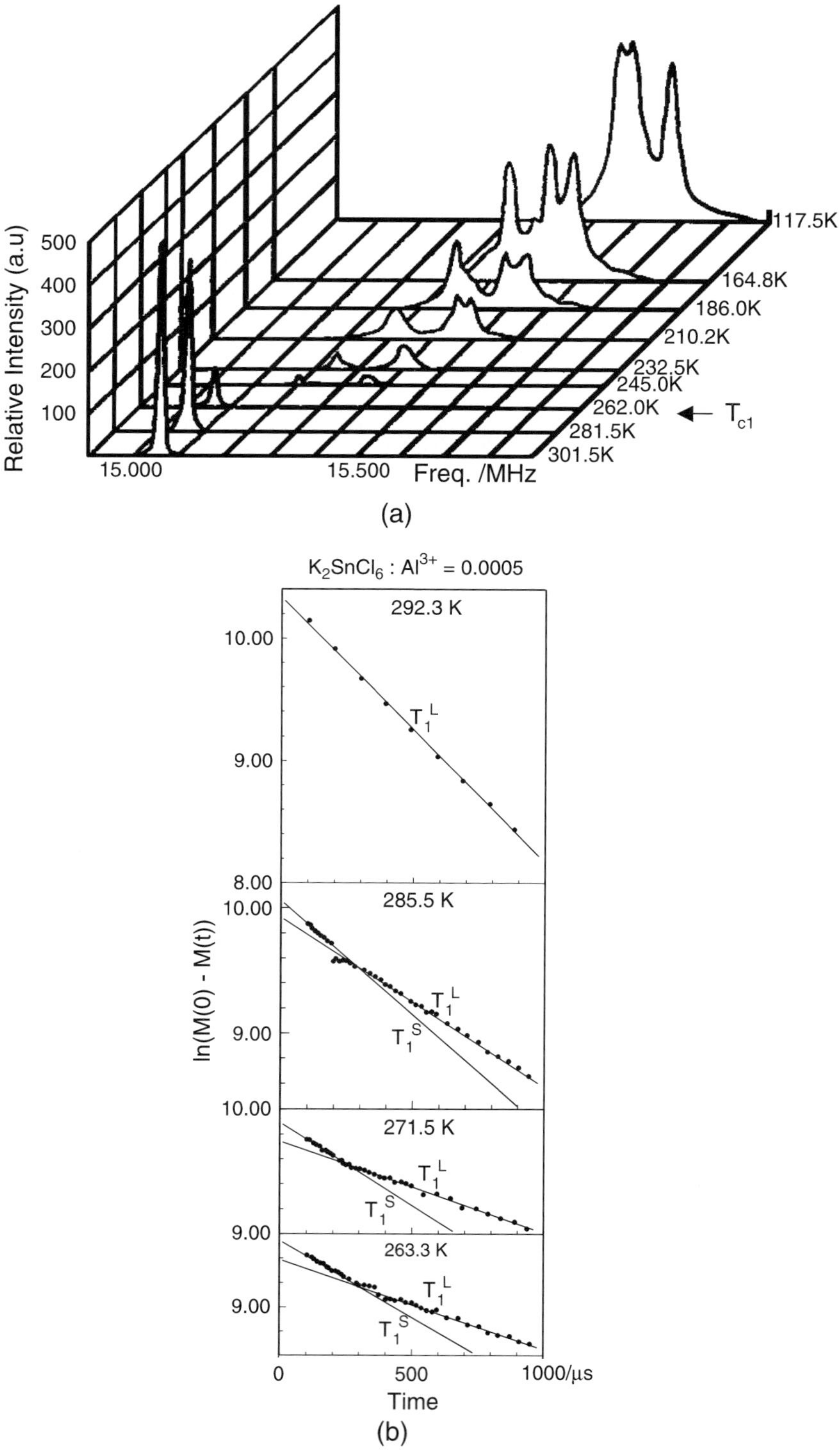

Figure 4. (a) Temperature dependence of the ^{35}Cl NQR line spectra around the structural transition in K_2SnCl_6:Al^{3+}. (b) Temperature dependence of the ^{35}Cl NQR spin–lattice relaxation on approaching T_{c1} from above.

implies this component is associated with the phase transition. The temperature dependence of the long components T_1^L and T_2^L was identified to be the same as that of the T_1 and T_2 in the pure K_2SnCl_6. Considering the characteristic of the short relaxation times and the interrelation between the hindered rotation of $SnCl_6^{2-}$ and the phase transition in the crystal suggest that a special type of impurity-induced defects such as the ligand-deficient octahedral complexes $AlCl_5^{2-}$ are the possible candidate being responsible for the anomaly in the temperature range $T_{c1} \leq T \leq T_I$ (Figure 1b).

Such type of defects may be produced by replacing Sn^{4+} by Al^{3+} in the $SnCl_6^{2-}$ octahedra with a simultaneous formation of a vacancy at one of the six chlorine ligand sites as a charge compensation process. Comparing with the intact $SnCl_6^{2-}$ the spatial arrangement of the ligand-deficient $AlCl_5^{2-}$ octahedra is more favourable to the hindered rotation. This suggests that the ligand-deficient octahedral defects $AlCl_5^{2-}$ may trigger the collective rotation of the local site octahedra driving consequently to the pretransition in the cubic phase near T_{c1}.

3. Conclusion

Nuclear resonance studies of the iso- and nonisomorphically impurity doped K_2SnCl_6 crystals arround the structural transition temperature revealed their own characteristic of impurity effects that are contrasting with one another in the high temperature cubic phase.

The appearance of the satellite lines besides the original resonance line is one of the representative features in the isomorphic system that may be well explained in the context of the static distribution of the unmobile impurituies on the lattice sites.

In contrast to this, the impurity effects of the nonisomorphic system is rather dynamic due to the mobile lattice defects. The resonance spectra suggest that the enhanced activity of the hindered rotationis through the formation of the ligand-deficient octahadral defect $AlCl_5^{2-}$ is responsible for the pretransition near the transition temperature in the cubic phase.

References

1. Seo Y. M., Pelzl J. and Dimitropoulos C., *Z. Naturforsch.* **53a** (1998), 552.
2. Seo Y. M., Song S. K. and Pelzl J., *Z. Naturforsch* **55a** (2000), 207.
3. Rigamonti A., *Adv. Phys.* **33** (1984), 115.

Hyperfine Interactions (2004) 159:29–34
DOI 10.1007/s10751-005-9077-4

Nuclear Spin-Lattice Relaxation of $^{82}Br\underline{Fe}$

C. TRAMM*, P. D. EVERSHEIM and P. HERZOG
Helmholtz-Institut für Strahlen-und Kernphysik der Universität Bonn, D-53115 Bonn, Germany; e-mail: tramm@hiskp.uni-bonn.de

Abstract. The field dependence of the nuclear spin-lattice relaxation (SLR) of cold implanted ^{82}Br ($T \leq 25$ mK) in α-Fe single crystals was investigated with nuclear magnetic resonance of oriented nuclei (NMR/ON) at low temperatures as experimental technique. The SLR at the lattice sites with the hyperfine fields found by earlier NMR/ON experiments was measured as a function of the applied external magnetic field B_{ext} parallel to the three principle axes [100], [110] and [111] of the iron single crystal. The data were evaluated with the full relaxation formalism in the *single impurity limit* and for comparison also with the often employed model of a single exponential function with an effective relaxation time T_1'. With a phenomenological model the high field values of the relaxation rates $r_{\infty,\,[100]'} = 6.6(2) \cdot 10^{-15}\ \mathrm{T^2sK^{-1}}$, $r_{\infty,\,[110]} = 5.4(2) \cdot 10^{-15}\ \mathrm{T^2sK^{-1}}$ and $r_{\infty,\,[111]} = 5.2(1) \cdot 10^{-15}\ \mathrm{T^2sK^{-1}}$ were obtained.

Key Words: field dependence, low temperature nuclear orientation, nuclear magnetic resonance, spin-lattice relaxation.

1. Introduction

In the investigation of the nuclear spin-lattice relaxation (SLR) behaviour at low temperatures care is needed to avoid systematic errors due to the experimental technique or inadequate evaluation of experimental data. Knowledge of the correct relaxation behaviour is e.g., important for on-line nuclear orientation experiments but especially most essential for comparison of experimental values with theoretical relaxation models. We have performed very careful studies of the SLR of ^{82}Br in iron single crystals and its variation with the applied external field and its direction relative to the main crystal axes. The data were evaluated with full relaxation formalism in the *single impurity limit* and for comparison also with the usually employed model of a single exponential function with an effective relaxation time T_1' [1]. From earlier experiments [2, 3] it is known that different hyperfine fields, associated with different lattice sites, contribute to the effective hyperfine field of $^{82}Br\underline{Fe}$. We studied the SLR at the lattice sites with the hyperfine fields ($B_{hf} = +\ 81.\overline{31}(3)$T [4]) found by NMR/ON [2, 3].

* Author for correspondence.

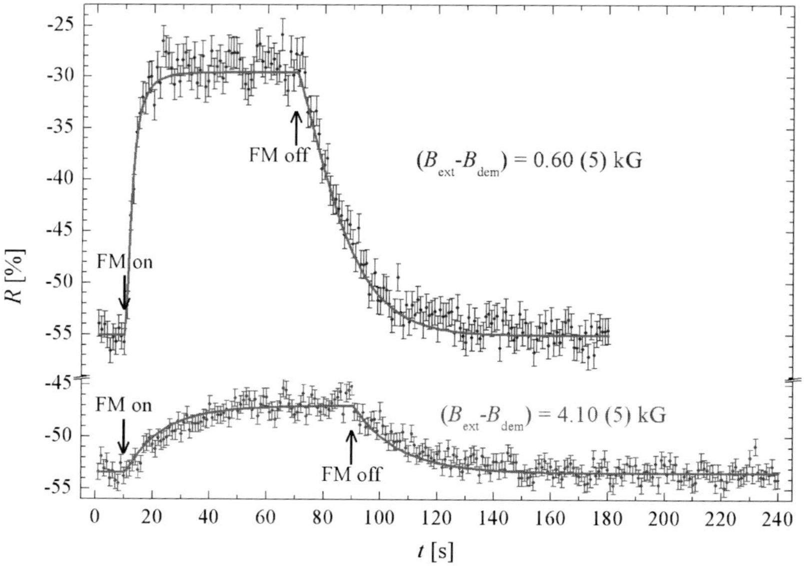

Figure 1. Relaxation curves for ^{82}Br<u>Fe</u> (anisotropy of the 619 keV γ-line at 0°, $[100]' \parallel \overrightarrow{B}_{ext}$) at two different effective external magnetic fields $(B_{ext} - B_{dem})$ where B_{ext} is the applied and B_{dem} is the demagnetisation field. The *solid line* represents a fit with full relaxation formalism in the single impurity limit [1]. The *arrows* indicate the time at which the frequency modulation (FM) is switched on and off, respectively.

2. Experiments and results

The ^{82}Br activity was prepared by neutron irradiation of KBr in the reactor at the Forschungszentrum Geesthacht GmbH (Germany) and implanted with an implantation energy of 80 keV into cold α-iron single crystals (*bcc*, $T \leq 25$ mK, dose $\approx 3.5 \cdot 10^{13}$ cm^{-2}) using a negative surface-ionization ion source [5]. The almost 1 mm thick single crystals were soldered with Woods metal to the cold-finger of the ^{3}He$-^4$He dilution refrigerator of the system FOLBIS [5, 6] and one of the three main crystal axes [100], [110] and [111] was aligned parallel to the applied external magnetic field $\overrightarrow{B}_{ext}$. Two different single crystals were used: no. 1, a 0.8 mm thick (11×9) mm^2 quader, for [100] and $[110] \parallel \overrightarrow{B}_{ext}$ and no. 2, a 0.7 mm thick elliptical disk with the semiaxes 5 mm and 6 mm, for $[111] \parallel \overrightarrow{B}_{ext}$. Due to the special cut of single crystal no. 1 the [100] direction is tilted about 11.2° out of the surface plane (the (001) plane is rotated about 16° relative to the surface plane with the edge as rotation axis which is parallel to the $[1\bar{1}0]$ direction). For that reason it was not possible to attach crystal no. 1 with exactly $[100] \parallel \overrightarrow{B}_{ext}$. Therefore we call this case the $[100]'$ direction. The surface of crystal no. 2 is parallel to the $(0\bar{1}1)$ plane. Before each implantation the surface of crystal no. 1 was chemically polished with an HF-H_2O_2–H_2O mixture [7]. The single crystal no. 2 was prepared by the crystal laboratory of the TU München. A $(1 \times 1 \times 12)$ mm^3 ^{60}Co<u>Co</u>(hcp) nuclear thermometer on the backside of the cold-finger was used for thermometry.

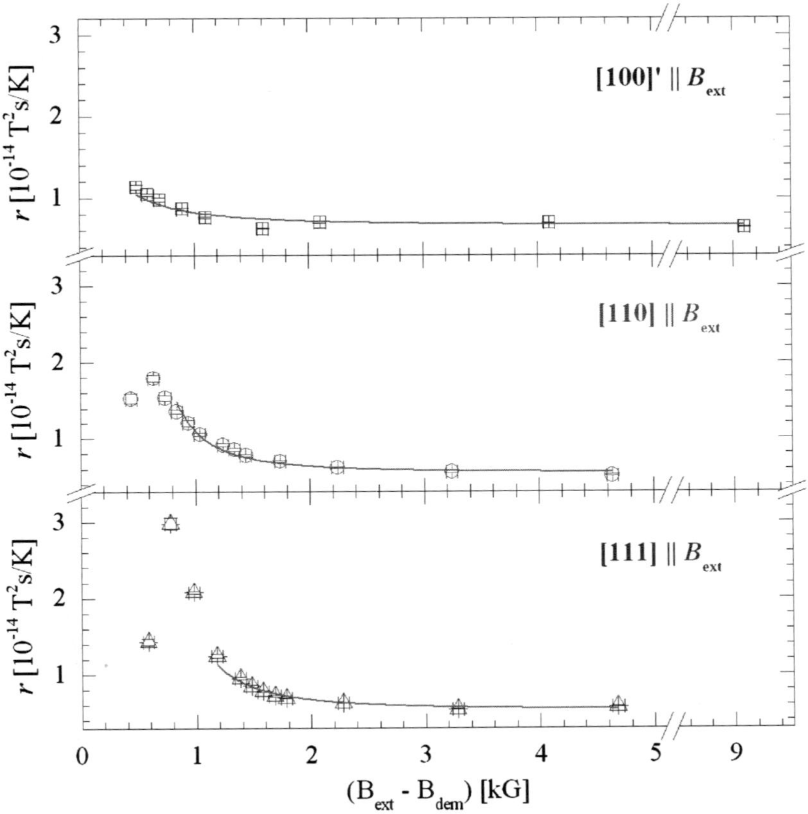

Figure 2. Relaxation rate r versus effective external magnetic field for the three principle axes [100]′, [110] and [111] || $\vec{B}_{ext}$ (for [100]′ || $\vec{B}_{ext}$, see text). The *solid lines* represent least square fits of Equation (1) to the data.

The γ-radiation in the ^{82}Br decay (and for thermometry in the ^{60}Co decay) was detected with Ge detectors at up to three angles 0°, 180° and 90° against the orientation axis defined by $\vec{B}_{ext}$. The anisotropy of the γ-rays $R = n_c/n_w - 1$ was calculated from the line intensities n which were corrected for dead time, life time and measuring time, where n_w corresponds to the warm sample (unpolarized nuclei) and n_c to the cold sample (polarized nuclei). The relaxation curves were measured with the experimental technique nuclear magnetic resonance of oriented nuclei (NMR/ON). The time dependent anisotropy, for example shown in Figure 1 for the 619 keV γ-line, was obtained with continuous input of radio frequency to achieve a constant base temperature. For a short interval the frequency modulation was switched on to resonate the nuclei. In the rest of the time the back relaxation was observed [8]. The temperature of the samples lay between 8 and 10 mK for all measurements. In the full relaxation formalism in the single impurity limit the relaxation behaviour is completely defined by the

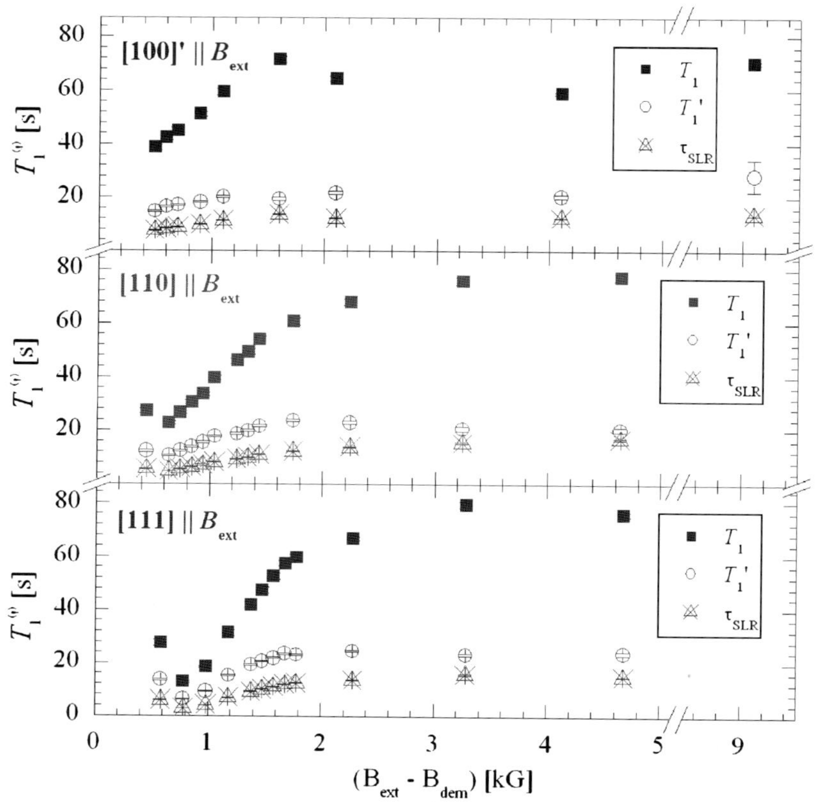

Figure 3. Relaxation time T_1, effective relaxation time T_1' and the low temperature value of τ_{SLR} versus the effective external magnetic field for the three principle axes [100]′, [110] and [111] $\|$ $\vec{B}_{ext}$ of the single crystal (for $[100]' \|\vec{B}_{ext}$, see text).

Korringa constant C_K, the interaction temperature $T_{int} = \Delta E/k_B$ (with ΔE the nuclear level splitting and k_B the Boltzmann constant) and the sample temperature T [1]. Therefore one can define the relaxation time $T_1 = 2C_K/T_{int} \cdot \tanh(T_{int}/(2T))$ [1], which normaly differs from the effective relaxation time T_1' defined by $R(t) = \left(R_0 - R_{eq}\right) \cdot e^{-t/T_1'} + R_{eq}$, with R_0 being the anisotropy just before switching off the frequency modulation and R_{eq} the anisotropy in thermal equilibrium. Because the Korringa law $C_K = T_1 \cdot T$ is no longer valid at low temperatures the time constant $\tau_{SLR} = C_K/(I \cdot T_{int})$ for $T \leq I \cdot T_{int}$ is often used ($\tau_{SLR} = C_K/T$ for $T \geq I \cdot T_{int}$, I is the nuclear spin).

The data evaluation in the single impurity limit yielded the relaxation rates $r = 1/(\gamma^2 \cdot C_K) \cdot$ (with $\gamma = g\mu_N/\hbar$ the gyromagnetic ratio) shown in Figure 2 where the relaxation rate is plotted in dependence of the effective external magnetic field $(B_{ext} - B_{dem})$ (where B_{dem} is the demagnetisation field) for the three different orientations of the single crystals.

3. Discussion

The relaxation rate shows for all three principle axes a smooth decrease at higher fields. For the two directions of hard magnetisation we obtained at low fields a maximum like structure for r in contrast to the direction of easy magnetisation ([100] $\parallel \vec{B}_{ext}$ for α-iron) where the data set does not show such a structure. The maximum for [111] $\parallel \vec{B}_{ext}$ is about five times higher than the high field value whereas the maximum for [110] $\parallel \vec{B}_{ext}$ is only three times higher. In comparison with the magnetisation curves [4] it is noticeable that these maxima correspond to the 'sharp bend' of the magnetisation curves. The turning of the magnetisation seems to enhance the relaxation so that r increases until the magnetisation points in the same direction as $\vec{B}_{ext}$. The case $[100]' \parallel \vec{B}_{ext}$ should also show a maximum for r because $\vec{B}_{ext}$ is not exactly parallel to the easy direction and the magnetisation must be turned away from the [100] direction. However, by comparison with the magnetisation curve (measured, but not shown) we found out that an existing maximum might be at external fields lower than the lowest experimental point. In the region $(B_{ext} - B_{dem}) \approx B_a$ (with B_a being the anisotropy field) the magnetisation turns clearly faster, it can even make a 'jump' [9]. In such unstable position of the magnetisation the relaxation is enhanced. With higher magnetic fields the spins "are held" ever more strongly so that the relaxation rate decreases. The 'jump' is bigger for mounting the single crystal with [111] $\parallel \vec{B}_{ext}$ (110 plane) than with [110] $\parallel \vec{B}_{ext}$ (100 plane) [9], which can explain the more pronounced maximum of the [111] direction. The high field values of the relaxation rates, which principally should be equal, are $r_{\infty,\,[100]'} = 6.6(2) \cdot 10^{-15}\ \mathrm{T^2sK^{-1}}$, $r_{\infty,\,[110]} = 5.4(2) \cdot 10^{-15}\ \mathrm{T^2sK^{-1}}$ and $r_{\infty,\,[111]} = 5.2(1) \cdot 10^{-15}\ \mathrm{T^2sK^{-1}}$. The difference of the saturation values is significant, it is, however, small compared to the differences of r at low fields. An explanation for this difference could not be found. The high field values are determined with the phenomenological model [10]

$$r = r_\infty + c_r \cdot \eta^2. \tag{1}$$

Here $\eta = 1 + B_{hf}\,(B_{ext} - B_{dem} \pm B_a)$ is the enhancement factor of the rf power [9] and c_r is a constant scaling factor. Earlier experiments of H.D. Rüter *et al.* [11] on ^{131}I$\underline{\text{Fe}}$ yielded the same order of magnitude for the relaxation rates and a quite similar field dependence, where it has to be considered that the single crystals of ref. [11] were partly cut in other planes and thus the maxima can be slightly different.

The relaxation time T_1, the low temperature value of τ_{SLR} and the effective relaxation time T_1' are shown in Figure 3 as function of the effective external magnetic field $(B_{ext} - B_{dem})$. The data show corresponding to the relaxation rate a low field minimum for [110] and [111] $\parallel \vec{B}_{ext}$ and saturation behaviour at high fields. The time constant T_1, extracted from the data with the full relaxation formalism, is at all points much bigger than the effective relaxation time T_1'

which was evaluated with a single exponential fit. This confirms that the relaxation time T_1, as F. Bacon *et al.* mentioned in [12], has no direct relation to the relaxation in low temperature nuclear orientation experiments. The interaction temperature of $^{82}\text{Br}\underline{\text{Fe}}$ is $T_{\text{int}} \approx 10$ mK [4] so for all measurements $T \leq I \cdot T_{\text{int}}$ was fulfilled (I (^{82}Br) = 5). Accordingly $\tau_{\text{SLR}} = C_{\text{K}}/(I \cdot T_{\text{int}})$ was used for calculating τ_{SLR}. Often the effective relaxation time T_1' is interpreted as the low temperature value of τ_{SLR}. This is not the case here, because τ_{SLR} is in the entire magnetic field range around the factor 1.2 to 2.3 smaller than the effective relaxation time T_1'. However, the low temperature value of τ_{SLR} represents a clearly better approximation for T_1' in low temperature nuclear orientation experiments than the relaxation time T_1, where τ_{SLR} is a lower and T_1 an upper limit for T_1' for the system $^{82}\text{Br}\underline{\text{Fe}}$.

Acknowledgements

The authors wish to thank the nuclear orientation group of the TU München for one of the single crystals. We also thank K. Maier for helpful advise in preparing the surface of the samples. We acknowledge the support by the Forschungszentrum Geesthacht GmbH where the neutron activations were performed.

References

1. Klein E., In: Stone N. J. and Postma H. (eds.), *Low-Temperature Nuclear Orientation*, North-Holland, Amsterdam, 1986, p. 579.
2. Herzog P., Dämmrich U., Freitag K., Herrmann C.-D. and Schlösser K., *Z. Phys. B* **64** (1986), 353.
3. Dämmrich U., PhD Thesis, Universität Bonn, 1989.
4. Tramm C., PhD Thesis, Universität Bonn, 2004.
5. Herzog P., Dämmrich U., Freitag K., Prillwitz B., Prinz J. and Schlösser K., *Nucl. Instrum. Methods B* **26** (1987), 471.
6. Herzog P., Folle H.-R., Freitag K., Kluge A., Reuschenbach M. and Bodenstedt E., *Nucl. Instrum. Methods* **155** (1978), 421.
7. Maier K., Private communication, Universität Bonn, 1999.
8. Stone N. J., In: Stone N. J. and Postma H. (eds.), *Low-Temperature Nuclear Orientation*, North-Holland, Amsterdam, 1986, p. 641.
9. Seewald G., PhD Thesis, Technische Universität München, 1999.
10. Seewald G., Private communication, Technische Universität München, 2004.
11. Rüter H. D., Haaks W., Duczynski E. W., Gerdau E., Visser D. and Niesen L., *Hyperfine Int.* **9** (1981), 385.
12. Bacon F., Barclay J. A., Brewer W. D., Shirley D. A. and Templeton J. E., *Phys. Rev. B* **5** (1972), 2397.

Hyperfine Interactions (2004) 159:35–42
DOI 10.1007/s10751-005-9078-3

Electronic Relaxation in Indium Oxide Films Studied with Perturbed Angular Correlations

A. LOHSTROH[1], M. UHRMACHER[1,*], P.-J. WILBRANDT[2], H. WULFF[3], L. ZIEGELER[1] and K. P. LIEB[1]

[1]*II. Physikalisches Institut, Universität Göttingen, Friedrich-Hund-Platz 1, D-37077, Göttingen, Germany; e-mail: muhrmac@gwdg.de*

[2]*Institut für Materialphysik, Universität Göttingen, Friedrich-Hund-Platz 1, D-37077, Göttingen, Germany*

[3]*Institut für Chemie und Biochemie, Universität Greifswald, D-17489, Greifswald, Germany*

Abstract. The electronic relaxation due to the ^{111}In → ^{111}Cd electron capture process was studied by measuring perturbed γγ-angular correlations after ^{111}In-ion implantation into indium oxide films of various compositions and thicknesses and deposited on different substrates. X-ray diffraction, Rutherford backscattering and secondary ion mass spectrometry were used to characterize the microstructure and composition of the samples. The film thickness and type of substrate were found to have much less influence on the after effect than the deposition parameters and impurities.

Key Words: decay after effects, electronic relaxation, indium oxide, perturbed angular correlations (PAC), RBS, SIMS, XRD.

1. Introduction

Indium oxide is the base material for producing indium tin oxide (ITO), one of the technologically most important transparent conductive oxides (TCO). By means of perturbed angular correlations (PAC) with ^{111}In tracers in pure or doped In_2O_3 [1–4] powder samples the so-called "after-effects" of the radioactive ^{111}In → ^{111}Cd electron capture have been studied. Generally, in semiconductors or insulators, the relaxation time of the excited electron shell may be longer than the mean life of the hyperfine-sensitive intermediate state in ^{111}Cd ($\tau = 122$ ns), leading to time-dependent (fluctuating) electric field gradients (EFG) and damped PAC perturbation functions. Dopants can strongly influence electronic relaxations as shown e.g., for La_2O_3 [5, 6].

The present investigation was motivated by a previous PAC study of a thin In_2O_3 film on an Al substrate [7], in which no fluctuating EFG was found contrary to the expectations. We investigated this system in detail using ^{111}In

* Author for correspondence.

Table I. Overview of the samples analyzed in the present work

Sample name	Substrate	Thickness [nm]	Processing	$E(^{111}In)$ [keV]	Analysis			
					PAC	RBS	XRD	SIMS
Au 1000	Au	1000	Reactive sputtering	100	X	–	X	X
Au 200	Au	200	Reactive sputtering	100	X	–	–	X
Si 1000	Si	1000	Reactive sputtering	400	X	X	X	X
Si 100	Si	100	Reactive sputtering	100	X	X	–	–
Al 600a	Al	600	E-gun evaporation	400	X	X	X	X
Al 600b	Al	600	E-gun evaporation	400	X	X	X	X
Al_2O_3 500	Al_2O_3	400–600	MO-CVD	400	X	X	X	X

ions implanted into In_2O_3 films of various thicknesses (100–1000 nm), which were deposited onto different substrates (Au, Al, Si, Al_2O_3) and contained different types and concentrations of dopants. In order to characterize the microstructure and composition of the samples, we used X-ray diffraction (XRD), Rutherford backscattering (RBS) and time-of-flight secondary ion mass spectrometry (TOF-SIMS).

2. Experimental

Table I summarizes all the samples discussed, including the deposition and annealing conditions and substrates. The samples on Al (Al 600a, Al 600b) were produced by electron-gun evaporation of indium oxide powder in the presence of oxygen and indium [8]. Most samples were deposited by reactive sputtering of a metallic indium target in oxygen atmosphere [9]. The film named Al_2O_3 500 was grown on a polycrystalline Al_2O_3 substrate using low-pressure metal–organic chemical vapor deposition. In this process a single-source precursor ($^iPr_2InOH\cdot{}^iPr_2InNH_2$) and nitrogen as carrier gas were used, which allowed to grow In_2O_3 without any additional oxygen [10].

The thickness and element composition were measured by means of RBS and for selected samples by TOF-SIMS, while the phases and microstructure was determined via XRD. RBS was carried out with the 900 keV He^{2+} beam of the Göttingen ion implanter *IONAS* [11]. SIMS analyses were performed with a TOF-SIMS IV spectrometer of *IONTOF* either at a pressure of 10^{-10} mbar or under a partial oxygen pressure of 1×10^{-6} mbar. The SIMS depth profiles for the samples Al 600a and Si 1000 are displayed in Figure 1, showing the sputtering yields of $^{115}In^+$, $^{16}O^-$ and $^{27}Al^+$ ions (corresponding to their elemental concentrations after appropriate calibration) as function of the number of sputtering cycles (corresponding to the thickness of the sputtered layer). In Figure 1 we note the striking difference in the element composition at the interface: for the e-gun evaporated film Al 600a the interface is sharp, but it is

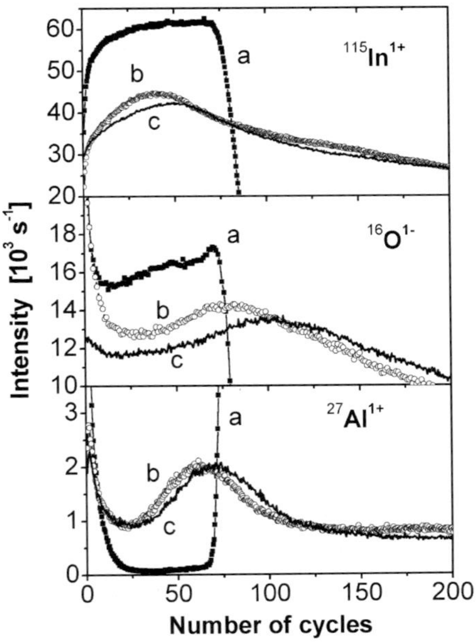

Figure 1. Intensities of $^{115}In^{+}$, $^{16}O^{-}$ and $^{27}Al^{+}$ ions detected during TOF-SIMS analyses: (a) Sample Al 600 as deposited, (b) sample Si 1000 as deposited, and (c) sample Si 1000 after annealing.

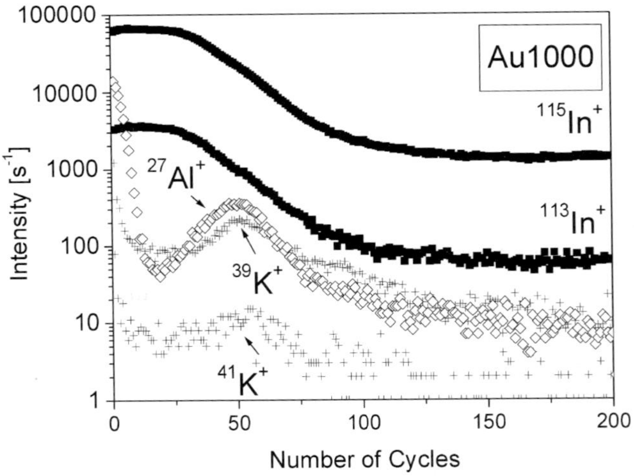

Figure 2. Intensities of the main positive ion species detected during the TOF-SIMS analysis of the sample Au 1000.

much broader in the case of the reactive-sputtered film Si 1000. In fact, for the latter sample SIMS proved the presence of Si in the In_2O_3 film and indicated strong interface diffusion. The SIMS spectra were also used to search for additional impurities in the films. The sample Al_2O_3 500 showed very small amounts of carbon and hydrogen, while the sample Au 1000 contained Al and K

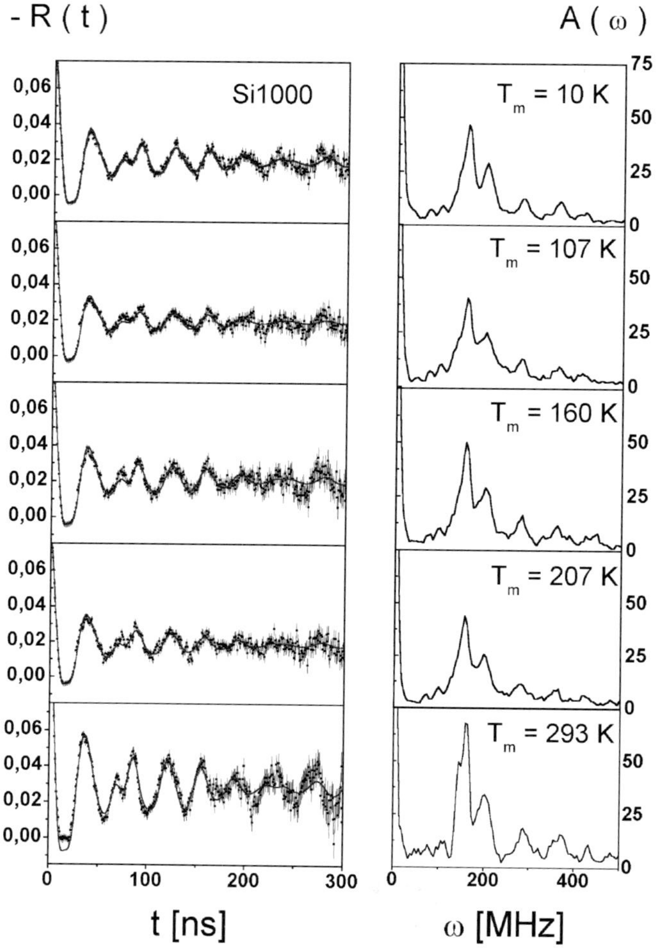

Figure 3. PAC spectra taken for the sample Si 1000 after a 1-h annealing in air at 623 K.

impurities, having their maximum concentration at the film–substrate interface (see Figure 2). Such impurities were also observed in all the other samples produced via reactive sputtering onto Si substrates and in e-gun evaporated films on Al.

The radioactive $^{111}In^{+}$-ions were implanted with *IONAS* [11] at either 400 or 100 keV, corresponding to the mean implantation depth of 70 or 23 nm. The samples were annealed in air at 623 K to remove radiation damage; the samples Si 1000, Al 500b and $Al_2O_3$500 were annealed *in vacuo* at a pressure of less than 10^{-5} mbar. The PAC spectra were measured at 10–673 K in a four-detector, fast-slow coincidence apparatus equipped with NaI(Tl) scintillators. The measurements below room temperature were carried out using a closed-cycle helium cryostate, those above room temperature at 10^{-5} mbar in a chamber equipped with an oven.

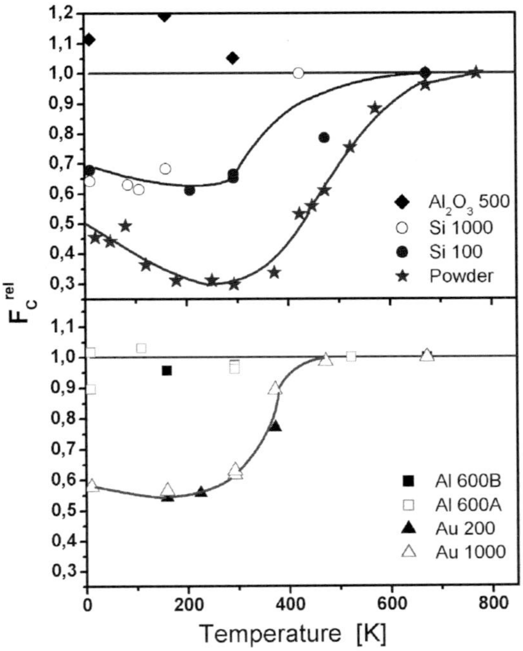

Figure 4. Relaxation factor F_C^{rel} of site C extracted from the PAC data for the various samples. The *lines* are given to guide the eye.

After appropriate annealing the ^{111}In tracers and their daughter atoms ^{111}Cd are found at the well known two cation sites of the In_2O_3 bixbyite lattice, named C (asymmetric EFG, $\nu_{QC} \approx 117$ MHz, $\eta_C \approx 0.73$) and D (symmetric EFG, $\nu_{QD} \approx 155$ MHz) [7, 12–16]. All previous PAC studies indicated that spectra taken at $T_{meas} \geq 623$ K can be fully described by two *static* EFG fractions f_{site} with $f_C + f_D = 1$ [1–4, 17]. In the case of fluctuating EFGs in In_2O_3 [4] and La_2O_3 [17] we modeled the situation after the EC-decay as an *uni-directional* relaxation, where a statistical EFG distribution evolves in time with an average relaxation rate Γ_r and finally reaches a static EFG. Here the $R(t)$ function contains a fast loss of anisotropy before reaching an undamped pattern with reduced amplitudes $f_C(T)$ and $f_D(T)$, but $f_C + f_D < 1$. If we define the ratios between the reduced fractions $f_i(T)$ and their high-temperature (static) values f_i^{max} as the relaxation factors $F_i^{rel}(T) \equiv f_i(T)/f_i^{max}$, then F_i^{rel} are related to the relaxation rate Γ_r [4, 17] via the relation

$$F^{rel} := \frac{f(T)}{f^{max}} = \frac{\Gamma_r(a + \Gamma_r)}{b^2 + (a + \Gamma_r)^2} \tag{1}$$

The parameters a and b for ^{111}Cd at the two lattice sites in In_2O_3 are $a_C = 143(4)$ MHz, $b_C = 86(6)$ MHz for site C and $a_D = 118(4)$ MHz, $b_D = 91(5)$ MHz for site D [4].

3. Results and discussion

As a typical example Figure 3 summarizes the PAC spectra and Fourier transforms for the sample Si 1000. The fitted hyperfine parameters ν_{QD} = 150–156 MHz, ν_{QC} = 114–120 MHz and η_C = 0.71–0.76 indicate a weak temperature dependence. As a first important result we emphasize that at *high temperatures* the hyperfine parameters of all measured In_2O_3 films agree well with the previous data on In_2O_3 powder samples [4]: $F_i^{rel}(T) = 1$ (no relaxation). As shown in Figure 4, the main differences between the various samples occur for the $F_i^{rel}(T)$ values at intermediate temperatures. As the typical error of F_C^{rel} is somewhat smaller (≈0.15) than that of F_D^{rel} (≈0.20), due to the better statistics because of its higher site occupation, we restrict the discussion to the site C. Nevertheless, the temperature evolution at site D was found to follow that at site C.

General: It is obvious that in none of the present thin film samples the static amplitude f_C was as strongly reduced ($F_C^{rel} = 0.55$–0.65) as for the powder sample ($F_C^{rel} \approx 0.30$ [4]). Furthermore, no clear influence of the film thickness was observed neither for the Si-substrates (Si100, Si 1000) nor the Au-substrates (Au 200, Au 1000): the relaxation factors F_C^{rel} for the thin and the thick films and for both substrates overlap rather well. However, we noted that no reduction of f_C was observed at any temperature for the samples Al 600a, Al 600b and $Al_2O_3$500.

Substrate material and film thickness: Most importantly, Figure 4 proves that the relaxation neither depends on the substrate material nor on the film thickness. The independence on the film thickness suggests that long-range electron transport over hundreds of nanometer from the substrate to the ^{111}Cd probes, which are located near the surface, has a very small, if any influence on the relaxation process of the probe ions. On the other hand, an undamped fraction f_C^{max} was observed at all temperatures for the metallic substrates Al 600a and Al 600b. This might possibly support the assumption of long-range electronic transport, but on gold backings the PAC spectra show clear relaxation. Finally, the absence of relaxation for the insulator Al_2O_3 500 clearly rules out a classification by the substrate-type.

Dopants: Previous PAC-experiments in semiconducting oxides proved that small amounts (≈0.5 at%) of electron-donor impurities can suppress the EC-decay after-effect very effectively [2, 4–6, 18]. In fact, the technically interesting ITO is produced by doping indium oxide with Sn, which acts as an electron-donor, similar to Zr and Al [19, 20]; adding Al increases the conductivity of In_2O_3. The SIMS measurements gave no evidence of neither Sn nor Zr in the samples, but of Al and K at the ^{111}In implantation depth (see Figure 2). It is thus most likely that these metallic impurities in the In_2O_3 films are responsible for the different degrees of relaxation in the various samples. One may also argue that dispersed metallic In precipitates in In_2O_3, which cannot be detected by RBS nor SIMS, may provide the electrons needed to accelerate the after-effect. Such precipitates have been identified by XRD in the sample Si 1000. The precipitates would show up in the PAC-spectra by the presence of an EFG typical of ^{111}Cd in metallic In and have previously been found after heating In_2O_3 above 973 K, where it disintegrates to In_2O, which is not stable at RT and further decays to In [21]. However, *none* of the PAC data showed any fraction (≥5%) of metallic In precipitates. This indicates that the precipitates, if they exist, are located at larger depths, far away from the ^{111}In probes. This also explains why the sample Si 1000, which showed the after-effect in the PAC spectra, contained metallic indium as deduced from XRD. In conclusion, no correlation was found between the damping of the PAC spectra and the observation of In precipitates via XRD.

Temperature dependence of the electronic relaxation: All the thin film samples – except for Al 600a, Al 600b and Al_2O_3 500 – showed a minimum of F_C^{rel} at 200–300 K, in agreement with previous results on pure and Sn-doped In_2O_3 [2, 4] and other oxides showing EC after-effects [1, 2, 8, 17, 18]. At a charge carrier density of 10^{19} cm^{-3} (Mott criterion [22]) doped In_2O_3 undergoes a metal–insulator transition. In the case of a lower charge carrier density, a linear increase in conductivity at temperatures between 70 and 300 K is reported [23]. This increase matches the temperature dependence of the PAC spectra. An alternative explanation may be a variable range hopping (VRH) process as discussed for La_2O_3 [17, 24]. Here the model was able to explain the increase in conductivity at low temperatures, which was observed in amorphous semi-conductors [25]. Unfortunately, our data are not sufficient to verify this interpretation. Especially the observed increase in the relaxation rate below 100 K still remains unexplained.

Conclusions: The present study on In_2O_3 films demonstrates that parameters of the film production such as metallic impurities, inhomogeneities, deviations from the stoichiometry and the microstructure appear to influence the electronic relaxation behavior more strongly than the film thickness and/or the type of substrate material. The electronic relaxation rates deduced from the PAC spectra do not depend on long-range electron transport over hundreds of nanometer. The measured temperature dependence of relaxation agrees qualitatively with previous measurements of In_2O_3 powder. The new experiments demonstrate again [1, 2] that PAC-measurements with ^{111}In probes in semiconducting or insulating materials and in particular the analysis of the damping due to the EC after-effect are an extremely sensitive method to detect the availability of electrons at the site of the hyperfine probe atom. It may be used to measure the conductivity in 'non-conducting' materials far below the detection limit of other techniques.

Acknowledgements

The authors gratefully acknowledge the help of D. Purschke during the ^{111}In implantations, S. Dhar, G. A. Müller and M. Schwickert with the RBS analyses and E. Carpene with the XRD measurements. We thank P. Lobinger for preparing the MO-CVD sample and S. Habenicht for useful comments. This work was supported by the Deutsche Forschungsgemeinschaft in the project UH104-1/1.

References

1. Bibiloni A. G. *et al.*, *Phys. Rev.* **B29** (1984), 1109.
2. Bibiloni A. G. *et al.*, *Phys. Rev.* **B32** (1985), 2393.
3. Renteria M. *et al.*, *Phys. Rev.* **B55** (1997), 14200.
4. Habenicht S. *et al.*, *Z. Phys.* **B101** (1996), 187.
5. Lupascu D., Uhrmacher M. and Lieb K. P., *J. Phys., Condens. Matter* **6** (1994), 10445.
6. Lupascu D. *et al.*, *Nucl. Instrum. Methods* **B113** (1996), 507.
7. Bolse W., Uhrmacher M. and Kesten J., *Hyperfine Interact.* **35** (1987), 931.
8. Ovadyahu Z., Ovryn B. and Kraner H. W., *J. Elec. Soc.* **130** (1983), 917.
9. Wulff H., Quaas M. and Steffen H., *Thin Solid Films* **355–356** (1999), 395.
10. Lobinger P., Park H. S., Hohmeister H. and Roesky H. W., *Chem. Vap. Depos.* **3** (2001), 105.
11. Uhrmacher M. *et al.*, *Nucl. Instrum. Methods* **B9** (1985), 234; *Nucl. Instrum. Methods* **B139** (1998), 306.
12. Bartos A. *et al.*, *Phys. Rev.* **B157** (1991), 513.
13. Bartos A., Lieb K. P., Uhrmacher M. and Wiarda D., *Acta Crystallogr.* **B49** (1993), 165.
14. Rogers J. D. and Vasquez A., *Nucl. Instrum. Methods* **130** (1975), 539.
15. Marezio M., *Acta Crystallogr.* **20** (1966), 723.
16. Wiarda D., Uhrmacher M., Bartos A. and Lieb K. P., *J. Phys., Condens. Matter* **5** (1993), 4111.
17. Lupascu D. *et al.*, *Phys. Rev.* **B54** (1996), 871; Tomala K. and Görlich E. A. (eds.), *Proc. XXX Zakopane School of Physics*, Krakow, 1995, p. 196.
18. Uhrmacher M., Neubauer M., Lupascu D. and Lieb K. P., *Proc. Intern. Workshop: 25th Anniversary of Hyperfine Interactions at La Plata,* 1995, p. 82.
19. Frank G. and Köstlin H., *Appl. Phys.* **A27** (1982), 197.
20. Safi I. and Howson R. P., *Thin Solid Films* **343–344** (1999), 115.
21. De Wit J. H., *Solid State Chem.* **8** (1973), 142.
22. Morikawa H. H. and Fujita M., *Thin Solid Films* **359** (2000), 61.
23. Schwartz I., Shaft S., Moalem A. and Ovadyahu Z., *Phil. Mag.* **B50** (1984), 221.
24. Pike G. E., *Phys. Rev.* **B6** (1972), 1572.
25. Tsuda N., Nasu K., Fujimori A. and Siratori K., *Electronic Conduction in Oxides*, Springer Series "Solid State Sciences," Springer, Berlin Heidelberg New York, **94**, 2000.

Hyperfine Interactions (2004) 159:43–48
DOI 10.1007/s10751-005-9079-2

Layered Inclusion Compounds Containing Aniline and Polyaniline Studied by NQR and IR Spectroscopy

T. A. BABUSHKINA*, T. P. KLIMOVA, L. D. KVACHEVA
and S. I. KUZNETSOV
A. N. Nesmeyanov Institute of Organoelement Compounds, Russian Academy of Sciences, Vavilov str., 28, 119991 Moscow, Russia; e-mail: tab@ineos.ac.ru

Abstract. Temperature dependence of the ^{127}I NQR frequencies in the intercalates of PbI_2 with aniline were measured. Unusual temperature coefficients of frequency and the electric field gradient asymmetry parameters for these compounds were discussed, taking into account the structure peculiarities and the amplitude of iodine oscillations.

Key words: aniline, intercalate, IR spectra, NQR, polyaniline.

1. Introduction

Lead diiodide intercalated by aniline (or polyaniline) can be considered as natural analogue of layered composite nanostructures formed by alternation of semi-conducting layers of matrix lattice and dielectric (or conducting) layers.

Prolonged heating of intercalates PbI_2 with aniline at a temperature about 100°C can lead to aniline polymerization. Polymerization in air can result in conducting molecular polymers. We believed that interaction between semi-conducting hosts (PbI_2) and conducting guest (polyaniline) can give composites with interesting physical properties.

2. Experimental

The inclusion of aniline forming sandwiches with layered hosts of PbI_2 was performed by slow precipitation from an excess of aniline after sonication of the solutions for 15−20 min or without it. Rich yellow color of PbI_2 changes to yellowish-green color of intercalate already in process of sonication [1]. The inclusion of aniline molecules can take place with the aniline plane both perpendicular and parallel to the iodine layers.

* Author for correspondence.

The sonication induces strong dispersion of the initial PbI_2 powder and can result in the formation of nanoparticles. It is believed that for PbI_2 the smallest stable nanoparticle is a fine hexagonal crystal comprised of two iodine layers with seven iodine atoms in each layer and two lead layers [2]. Probably the formation of nanoparticles facilitates the inclusion process.

3. Results and discussion

We had three samples of $PbI_2 \cdot an_x$ obtained without (sample 1 and 2) and with sonication (sample 3). X-Ray diffraction patterns of the compounds obtained display the lines from large inter-planar spacing (from 8.27; 10.16 to 11.19 Å). The inter-plane distance in the PbI_2 is equal to 7.08 Å. This is the evidence of aniline inclusion into the PbI_2 structure.

Intercalate of PbI_2 with polyaniline was synthesized in aqueous medium through the interaction of PbI_2 and aniline hydrochloride.The reaction results in new unstable intercalate with the inter-plane distance of 12.6 Å.The following heating at 70–100°C destroys the intercalate giving pure Pb_2 and polyaniline.

Adducts PbI_2 with aniline of the composition 1:1; 1:2.5; 1:3 are known [3]. The latter two have the structures with 5- and 6-layers packing of aniline molecules. Determination of an exact composition of the intercalates studied in this work is precluded by formation of the oxidation products of free aniline if the samples are stored in air. According to thermogravimetric data the sample 3 has the composition 1:3.

NQR spectra of all samples investigated have one common set of lines belonging to the transitions $\Delta m = 1/2–3/2$ and $\Delta m = 3/2–5/2$. The maximum intercalated sample 3 as obtained with sonication has only this set of NQR frequencies. The other two samples have besides the first set another set of NQR lines, individual for every sample (Table I) [4.5]. In the sample 2 the frequencies of two sets (2a and 2b) are very similar. Therefore NQR data point to existence of three so called "phases" of $PbI_2 \cdot an_x$.

IR spectroscopy needs for analysis much smaller quantity of substance in comparison with the NQR method. Hence we tried to separate by hand two powders with different colors mixed in the sample 1 (samples 1a and 1b). IR spectra of the samples 1a and 3 are identical. IR spectra of samples 3 (1a) and 1b have in the range 4000–400 cm^{-1} the bands characteristic for aniline (Figure 1).

So, in the region 3600–3000 cm^{-1} there are the bands of stretching NH_2 modes. The positions of these bands point to the presence of free (3575–3507) and H-bonded (3305–3241) amino groups. The comparison of their intensities leads to conclusion that the number of H - bonded NH_2 groups in the sample 3 (1a) is more, than that in the sample 1b.

The band of the bending NH_2-modes is positioned in the range 1560–1500 cm^{-1} in the spectrum of the sample 1b. In the case of the sample 3 this band

Table I. The NQR frequencies, quadrupole coupling constants (QCC) and asymmetry parameters (η) of electric field gradient (EFG) at 77 K of samples $PbI_2 \cdot an_x$

Sample	$\nu_{1/2-3/2}$, MHz	$\nu_{3/2-5/2}$, MHz	QCC, MHz	η,%
1a	71.85	105.74	370.5	54
1b	35.08	39.85	146.9	85
2a	71.85	105.74	370.5	54
2b	72.27	107.78	377.3	54
3	71.85	105.74	370.5	54

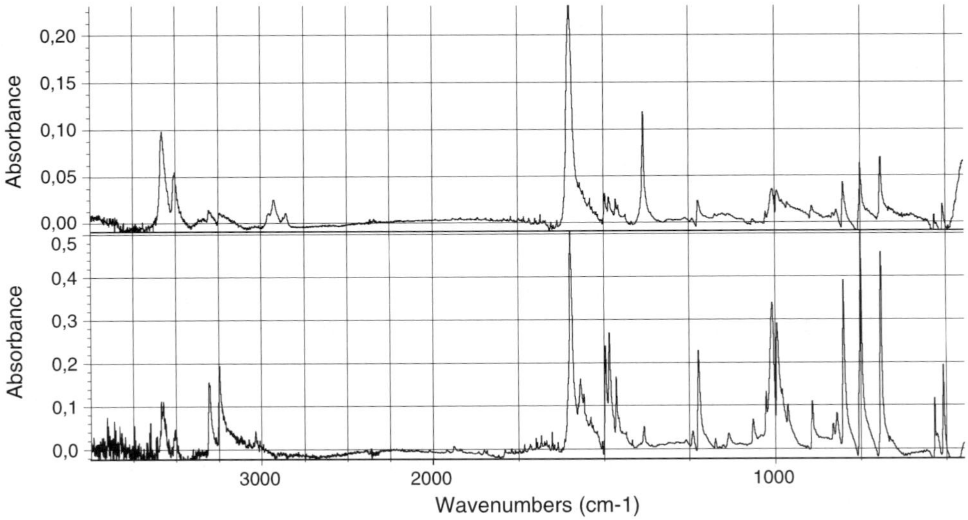

Figure 1. IR spectra of the intercalates samples Pb · an_x. Sample 1b (*top*) and 1a (3) (*bottom*).

shifts to 1600 cm^{-1} and overlaps with the band of the benzene ring. This also points to the difference of NH_2-groups in these samples.

The intensities of the absorption bands of benzene ring in the range 800–700 cm^{-1} in the spectrum of the sample 3 are much higher than in the spectrum of 1b. As the weights of the samples are equal, this difference points to the larger content of PbI_2 in the sample 1b. So the guest/host ratios in these samples are different.

As NQR frequencies of the phase 2a (3) and 2b differ very little we believe that the composition of the phase 2b is equal to 1 : 2.5. Hence the phase 1b has the composition 1 : 1 in accordance with the data of elemental analysis.

We studied the temperature dependence of the ^{127}I NQR signals over a wide temperature range (from 77 K to room temperature) for the intercalation compounds of PbI_2 [6].

The temperature dependence of NQR frequencies for two phases 1b and 3 of the $PbI_2 \cdot an_x$ ($x = 1$, 3 accordingly) is shown in Figure 2.

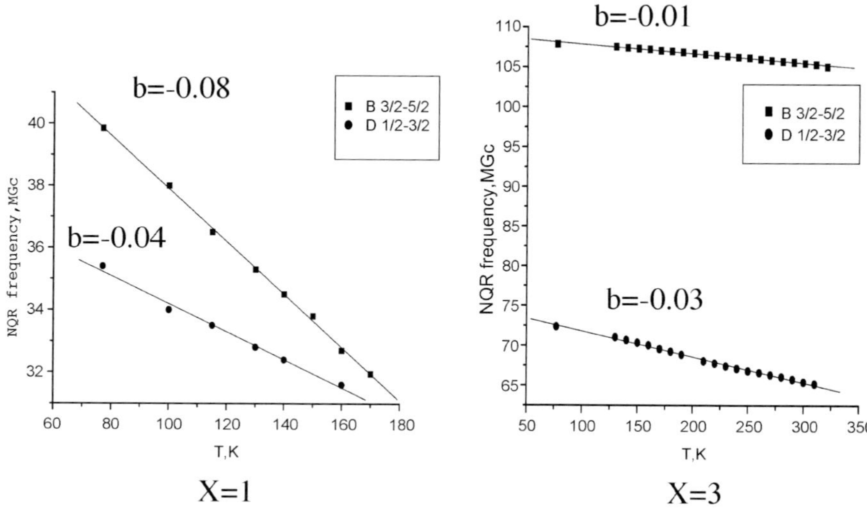

Figure 2. The temperature dependence of the frequencies of NQR lines for the intercalates PbI_2 an_x. The values b are the slopes of linear fits $\nu = \nu_o + bT$.

For phase PbI_2 an (1b), the raising of the temperature leads to continuous rapprochement of the observed NQR lines (supposed intersection point is $T \approx$ 190 K) and, as a consequence, to the increase in the asymmetry parameter of EFG up to 100%. The temperature coefficient of the QCC of this phase is very high. The slope of the linear fit line of the temperature dependence of the frequency of $\Delta m = 3/2$–$5/2$ transition is equal to -0.08 for the phase with $X = 1$, while that for the phase with $X = 3$ is equal to -0.01. Probably the large temperature coefficient of the QCC for this phase with $X = 1$ can be due to high mobility of iodine atoms (or ions) leading to the flat distribution of EFG.

Earlier the anomalous increase of the asymmetry parameter of EFG at the heating was shown for ^{139}La atom in the anion conductor LaF_3 [7]. Authors relate these data to formation of vacancies due to the transport of F^- anions. We believe that this can make a parallel between our results and quoted data.

The significant change of the QCC in the sample 3 in relation to QCC of the initial PbI_2 is the evidence of the transport of the lone pair charge of the nitrogen atom about the ~0.15 e.

The calculation of EFG on the iodine atoms in $PbI_2 \cdot an_x$ by using a model of the point charges and dipoles [8] shows that the maximum EFG change can have place at the increase of the inter-plane distance for about 3 Å because the interaction of the iodine atoms in the adjacent layers decreases after intercalation. Further increase doesn't influence on the EFG value significantly. Besides, the increase of the interplane distance by 1 Å reduces the initial three-dimensional lattice to the system of lower dimensions due to addition of the periodic potential of the guest molecules and changes of the direction of the max EFG component. Therefore the increase of the inter-plane distance in the sample 3 (the com-

Table II. Distribution of the intercalates of $PbI_2(CdI_2)$ with nitrogen heterocycles

1 type	QCC, MHz at 77 K	η, %	2 type	QCC, MHz at 77 K	η, %	3 type	QCC, MHz at 77 K	η, %
CdI_2 an_2	556.3	21						
CdI_2 py_2	666.3	5	CdI_2 py_4	365.9	21	CdI_2 py_6	216.9	99.5
CdI_2 pip_2	640.2	4						
			PbI_2 an_3	370.3	54	PbI_2 an	146.9	85
				377.5	54			
			PbI_2 py_2	335.4	78			
PbI_2 pip	485.2	13				PbI_2 pip_8	72.3	80

position 1:3) found by X-ray diffraction could be one of the causes for high frequencies of NQR in that phase. Lower NQR frequencies and the high asymmetry parameter in the sample 1b (1:1) are explained partly by the increase of the inter-plane distance for about 1 Å.

We tried to discuss all experimental data obtained by many methods and investigators [1, 4–6, 9–11] and considered possible interactions in the types of the intercalates studied.

Solvent molecules (aniline, pyridine (py), piperidine (pip), etc.) orient in the electric field of iodine ions of the PbI_2 or CdI_2 matrices by ion–dipole interaction. External conditions (the rate of the precipitation, a sonication and etc.) can determine the type of substances obtained. The sonication promotes maximum intercalation and therefore leads to an appreciable charge transfer. The ion–dipole interaction can promote realization of the structure that is most advantageous energetically. So, one can discern three types of interaction of matrix atoms and lone pair electron of nitrogen atoms of solvent molecules.

First type of interaction (the charge transfer on the metal ions of matrices) changes, as a rule, the hexagonal structure of matrices so molecular complexes are formed. Really, the QCC values are high – that is characteristic for the compounds with covalent bonds (complex ions). Second type of interaction (the charge transfer to the iodine ions of matrices) and third type (preferential charge–dipole interaction), as a rule don't change the structure of precursor matrices and therefore the intercalates forms. NQR spectra are the most informative in resolving the substances of the second and the third types (Table II, the NQR data are taken from [1, 4–6, 10]). The compounds $CdI_2 \cdot py_6$ и $PbI_2 \cdot$ an (1:1) were assigned to compounds with the third type of interaction because of NQR spectra peculiarities. QCC's in these substances are small in comparison to QCC's of compounds of another composition that is certainly not the case of the prominent charge transfer. The asymmetry parameters (η) of EFG on iodine atom in these adducts are large enough, near to 100%. Then the EFG have only two components that are not equal to zero that is the EFG has only two dimensions. This detail (large value of η) shows that compound $PbI_2 \cdot pip_8$ can be also the substance of the third type. The substances $PbI_2 \cdot an_3$, $PbI_2 \cdot an_{2.5}$ and $CdI_2 \cdot py_4$ have higher values of QCC's and were assigned to the second type.

Thus NQR method helps to resolve the intercalates and molecular complexes, it is very useful for determination of the interaction type of compounds investigated.

Acknowledgement

This work is supported by RFBR (grant 04-02-17303).

References

1. Konopleva K. G., Venskovskii N. U., Tupoleva A. L. and Babushkina T. A., *Zh.Inorg.Khim. (Russ. J. Inorg. Chem.)* **45** (2000), 1283.
2. Sandorf C. J., Hwang D. M. and Chung W. M., *Phys. Rev., B Condens. Matter* **33** (1986), 5953.
3. Zabrodskii Yu., Tolmachev D. L., Pai'vin V. S., Mil'ner A. P., Koshkin V. M., *Fiz. Tverd. Tela* **33** (1986), 5953 (in Russian).
4. Lyfar D. L. and Ryabchenko, S. M., *J. Mol. Struct.* **83** (1982), 353.
5. Babushkina, T. A. and Seryukova, I. V., *Z. Naturforsch.* **53a** (1998), 585.
6. Babushkina T. A., Konopleva K. G., Tupoleva A. L., Venskovskii N. U., Guibe L., Gourdji M. and Peneau A., *Z. Naturforsch.* **55a** (2000), 139.
7. Poeppelmeier K. R. and Hwu S.-J., *Inorg. Chem.* **26** (1987), 3297.
8. Moskalev A. K. and Seryukova I. V., *Vestn. Krasnoarsk. Gos. Univ.* **3** (1998), 29 (in Russian).
9. Koshkin V. M. and Dmitriev Yu. N., *Chem. Rev.* **19**(2) (1994), 139p.
10. Konopleva K. G., Venskovskii N. U., Tupoleva A. L. and Babushkina T. A., *Koordinatsionnaya Khimiya* (Russian) **25** (1999), 505.
11. Mehrotra V., Lombardo S., Thompson M. O. and Giannelis E. P., *Phys. Rev.* **44b** (1991), 5786.

Hyperfine Interactions (2004) 159:49–53
DOI 10.1007/s10751-005-9080-9

High Temperature Diffusion of ^{6}Li Adsorbed on a Ru(001) Single Crystal Surface as seen by Pulse NMR

H. J. JÄNSCH*, H. LÖSER, A. VOSS and D. FICK
Fachbereich Physik und Wissenschaftliches Zentrum für Materialwissenschaften, Philipps-Universität, D-35032 Marburg, Germany;
e-mail: heinz.jaensch@physik.uni-marburg.de

Abstract. Diffusion measurements on lithium atoms adsorbed on a ruthenium single crystal were performed in the high temperature regime (1100–1200 K). Pulsed NMR techniques were utilized to produce and observe the decay of magnetization patterns from which the diffusion coefficient was extracted. The observed temperature dependence could be described by $D = (10 \pm 7)$ cm^2/s · exp ($-$(0.46 ± 0.07) eV/kT). The extremely high diffusion coefficient and prefactor are understood by a gas like adsorbate behavior. The electric field gradient has been measured with ^{7}Li: $V_{zz} = -5.0 \pm 0.1\ 10^{15}$ V/cm^2 with an inhomogeneity of less then 1% as judged by the width of the satellite transitions.

Key Words: adsorption, alkali metal, diffusion, EFG, NMR.

1. Introduction

Diffusion of adatoms on surfaces is a very general phenomenon. It occurs at any temperature and rates vary enormously. Going from low to high temperature the diffusion mechanism changes from dominantly jump diffusion to multi jump or gas like diffusion mechanisms. At high temperatures the thermal desorption of adsorbates is often in competition with diffusion. This competition is what *high* means in the current context. This important regime is typical for catalytic processes or chemical vapor deposition among others. Many methods have been devised to study *microscopic* and also *macroscopic* diffusion although mostly at low temperatures [1–3]. The transition from jump to gas diffusion has been seen in CO on Pt(111) [4, 5].

For Li on Ru(001) diffusion measurements at lower temperatures exist. Terrace jump diffusion was identified with a barrier of 0.18 eV and an additional barrier of 0.46 eV was found and tentatively assigned to step edges or other abundant defects [6–8]. High temperature data are lacking.

---

* Author for correspondence.

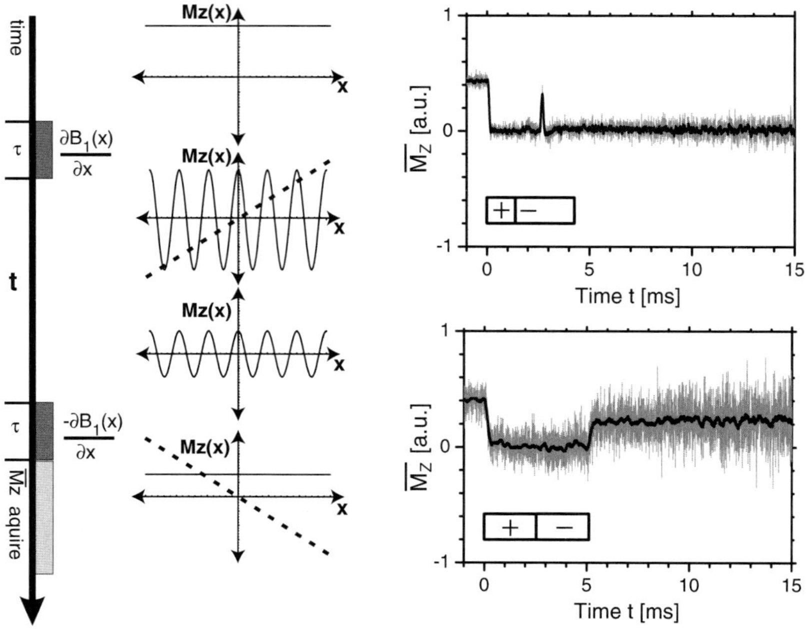

Figure 1. On the *left panel* the timing of the diffusion measurement is schematically shown. (See text.) On the *upper right panel* the $M_z(t)$ is shown. The first pulse (+) reduces the average magnetization to zero. The inverted pulse (−) refocuses the magnetization. The *lower panel* shows $M_z(t)$ when the refocusing pulse is stopped at an optimal time. Introducing a waiting time between the pulses reduces the magnetization in an exponential fashion which is visible in Figure 2.

Here we demonstrate a new method to measure, at high temperature, the macroscopic diffusion coefficient for Li adsorbed on a Ru(001) single crystal surface, using an NMR technique. On zeolites and powders NMR it is widely used to study surfaces [9, 10]. Conventional NMR needs $\approx 10^{18}$ equivalent nuclei [11]. A surface has only $\approx 10^{15}$ sites per cm^2. Highly nuclear spin polarized adsorbates and single photon counting detection techniques are used to increase the sensitivity by at least seven orders of magnitude [12–15]. Like in conventional NMR the transverse magnetization can be measured as a free induction decay (FID) signal but also the longitudinal magnetization (M_z) can be observed directly, both in beam-foil-spectroscopy. The EFG is measured by observing the FID of ^{7}Li and the diffusion coefficient by measuring M_z of ^{6}Li as described below.

2. Experimental

Only a brief description is given. For more detail see references [15–17]. All experiments are performed in a UHV chamber with a base pressure of

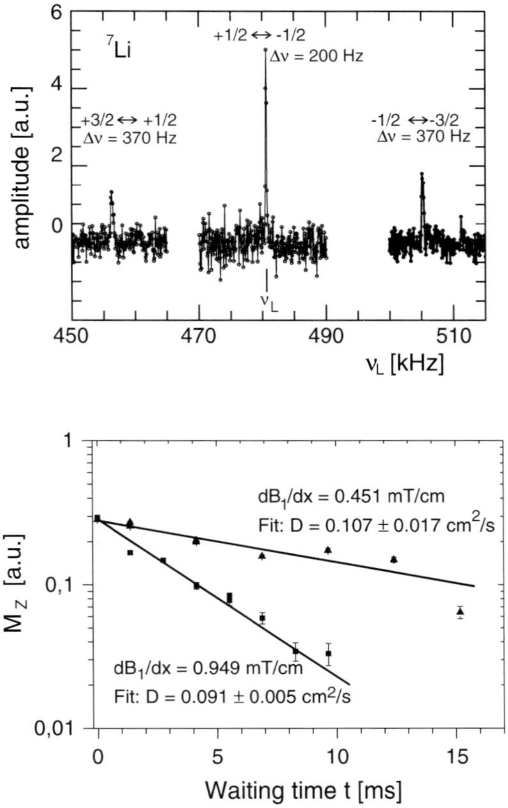

Figure 2. Upper: ^{7}Li NMR spectrum of adsorbed lithium on Ru(001). *Lower:* Average magnetization as function of waiting time between rf pulses.

2×10^{-11} mbar. The Ru(001) surface is cleaned by sputtering and oxygen treatment and during the NMR experiments by flashes to 1650 K every 30 s. Atomic beams of highly nuclear spin polarized ($P_z \approx 0.5$) ^{6}Li or ^{7}Li are directed towards the Ru surface. Where the atoms reside for mean residence times of $1 \ldots 10^{-3}$ s at $T = 1100 \ldots 1200$ K. Desorbing ions ($\approx$1/3 of the particles) are accelerated (10 keV) and shot through a thin carbon foil. These beam-foil excited atoms radiate resonant light; the polarization of which is dependent on the nuclear polarization before the excitation due to hyperfine interaction in the excited atom. Polarization analysis of the photons and the spatial direction in which they are observed determine the type of nuclear polarization observed (1st or 2nd rank, transverse or longitudinal). The FID is obtained by observing in transverse direction the time dependent circular polarization after the atoms on the surface have been excited by a $\pi/2$ rf pulse [15]. For the diffusion measurement spatially inhomogeneous rf radiation is used with a gradient (dB_1/dx) along the surface. Gradient pulses are applied twice: at first to create a spatial magnetization pattern and a second time to regain an average magnetization

($\overline{M_z}$) by applying the inverse rf pulse (Figure 1). $\overline{M_z}$ is reduced exponentially by introduction a waiting time between the pulses. The reduction also depends on the diffusion coefficient (D), the strength and duration of the rf pulses, and the size of T_1 and T_2 [16, 17]. Only D is unknown and thus can be determined.

3. Results and discussion

3.1. EFG AND HOMOGENEITY

The upper panel of Figure 2 shows the NMR spectrum of ^{7}Li (spin I = 3/2) on Ru(001). The lines result from a homogeneous electric field gradient together with a strong magnetic field. The EFG is $V_{\text{eff}} = -5.0 \pm 0.1\ 10^{15}$ V/cm^2. Since the adsorbate is highly mobile and the symmetry axis (surface normal) is along the magnetic field, we assume $V_{\text{eff}} = V_{zz}$ as has been seen in similar cases [18]. The line width of the central transition is 200 Hz and that of the satellites is 370 Hz. The first one is caused by magnetic inhomogeneity and the second one additionally by the inhomogeneity of the EFG. This yields a $\Delta B = 0.12$ G and assuming Gaussian line shape $\Delta V_{\text{eff}} = 0.07\ 10^{15}$ V/cm^2. The quadrupole moment of ^{6}Li is about 50 times smaller then that of ^{7}Li the γ-factors only by 2.6. Therefore, the line width of ^{6}Li is dominated by ΔB.

3.2. DIFFUSION

Figure 1 shows the timing of the pulse experiment (*left*) and two examples of $\overline{M_z}$ (t), namely dephasing and reconstruction of $\overline{M_z}$. The lower panel of Figure 2 shows $\overline{M_z}$, after optimal reconstruction when a waiting time is introduced between the gradient pulses. $\overline{M_z}$ is reduced exponentially as expected [16]. The two sets of data where obtained with different gradient strength. The are fitted together to give a common and very high diffusion constant of $D = 0.092 \pm 0.005$ cm^2/s. A temperature dependence was measured between 1100 and 1200 K. It can be fitted to an Arrhenius law with a diffusion barrier of 0.46 ± 0.07 eV and a prefactor of $D_0 = 10 \pm 7$ cm^2/s.

Next neighbor jump (3 Å) diffusion cannot explain these results. The jump frequency and the average velocity of the adatoms are unphysical. More likely is a 2D adsorbate gaseous state and ballistic flight paths with localization at large distances. An interpretation along these lines gives jump distances of ≈250 Å which is in the vicinity of the typical step edge distances on the crystal used. This maybe accidental but hints at a possible localization mechanism, namely, collision with an abundant defect.

Acknowledgement

The support of the “Deutsche Forschungsgemeinschaft” (DFG) is kindly acknowledged.

References

1. Gomer R., *Rep. Prog. Phys.* **53** (1990), 917–1002.
2. Barth J. V., *Surf. Sci. Rep.* **40** (2000), 75–149.
3. Ala-Nissila T., Ferrando R. and Ying S. C., *Adv. Phys.* **51** (2002), 949–1078.
4. Reutt-Robey J. E., Doren D. J., Chabal Y. J. and Christman S. B., *J. Chem. Phys.* **93** (1990), 9113–9129.
5. Froitzheim H. and Schulze M., *Surf. Sci.* **320** (1994), 85–92.
6. Ebinger H. D., Jänsch H. J., Polenz C., Polivka B., Preyß W., Saier V., Veith R. and Fick D., *Phys. Rev. Lett.* **76** (1996), 656–659.
7. Ebinger H. D., Arnolds H., Polenz C., Polivka B., Preyß W., Veith R., Fick D. and Jänsch H. J., *Surf. Sci.* **412/413** (1998), 586–615.
8. Kirchner G., Czanta M., Dellemann G., Jänsch H. J., Mannstadt W., Paggel J. J., Platzer R., Weindel C., Winnefeld H. and Fick D., *Surf. Sci.* **494** (2001), 281–288.
9. Fraissard J., *Catal. Today* **51** (1999), 481–499.
10. Bonardet J. L., Fraissard J., Gedeon A. and Springuel-Huet M. A., *Catal. Rev.* **41** (1999), 115–225.
11. Slichter C. P., *Annu. Rev. Phys. Chem.* **37** (1986), 25–51.
12. Riehl-Chudoba M., Memmert U. and Fick D., *Surf. Sci.* **245** (1991), 180–190.
13. Kaack M. and Fick D., *Phys. Rev., B* **51** (1995), 17902–17909.
14. Arnolds H. and Jänsch H. J., *Chem. Phys. Lett.* **272** (1997), 13–17.
15. Arnolds H., Fick D., Unterhalt H., Voß A. and Jänsch H. J., *Solid State Nucl. Mag. Reson.* **11** (1998), 87–102.
16. Löser H., *Hochtemperatur Diffusion von Lithium auf Ru(001)*, PhD thesis, Philipps-Universität, Marburg, 2002, http://archiv.ub.uni-marburg.de/diss/z2003/0165.
17. Löser H., Fick D. and Jänsch H. J., *J. Phys. Chem., B* **108** (2004), 14440–14445.
18. Koch E., Horn B. and Fick D., *Surf. Sci.* **173** (1986), 639–664.

Hyperfine Interactions (2004) 159:55–61
DOI 10.1007/s10751-005-9081-8

Structural Properties of the Donor Indium in Nanocrystalline ZnO

T. AGNE[1,*], M. DEICHER[1], V. KOTESKI[2], H.-E. MAHNKE[2], H. WOLF[1] and T. WICHERT[1]
[1]*Technische Physik, Universität des Saarlandes, 66041, Saarbrücken, Germany*
[2]*Bereich Strukturforschung, Hahn-Meitner-Institut Berlin, 14109, Berlin, Germany;*
e-mail: Thomas.Agne@tech-phys.uni-sb.de

Abstract. The structural properties of the nanocrystalline semiconductor ZnO (nano-ZnO) doped with the donor Indium were investigated by perturbed $\gamma\gamma$ angular correlation spectroscopy (PAC) and extended X-ray absorption fine structure measurements (EXAFS). Up to an average concentration of one In atom per nanocrystallite, PAC measurements show that about 12% of the ^{111}In atoms are incorporated on substitutional Zn sites. At higher In concentrations, new In defect complexes are visible in the PAC spectra, which dominate the spectra if the average In concentration exceeds one In atoms per nanocrystallite. In addition, the local environment of Zn and In atoms in In doped nano-ZnO was investigated by EXAFS. The measurements at the K edge of Zn show that the crystal structure of nano-ZnO corresponds to bulk ZnO. In heavily In doped nano-ZnO the EXAFS experiments at the K edge of In exhibit an expansion of the first O shell about the In site. Since about four O atoms are detected in this first shell a substitutional incorporation of the In atoms in the ZnO lattice is suggested. The second shell to be occupied by Zn atoms as well as higher shells are almost invisible, which might have the same microscopic origin as the occurrence of defect complexes observed by PAC.

1. Introduction

In nanometer scale semiconductors, several optical and electrical properties exhibit a strong size dependence [1]. The doping of nanocrystalline semiconductors, which is the basis for the successful application of semiconductors, still poses a severe problem [2]. Most effort for solving this problem has been focused on II–VI semiconductor nanocrystals that were doped with impurities, such as Mn, Cu, or rare earth-elements, such as Tb (Ref. 2 and references therein). The doping with impurities that introduce shallow donor or acceptor levels has been reported by only a few groups [3–5].

In a previous work, we reported on nanocrystalline ZnO (nano-ZnO) with a mean grain size of 11 nm, which was successfully doped with the radioactive donor ^{111}In [5]. The required incorporation of ^{111}In atoms on undisturbed Zn sites after a hydrothermal treatment at 473 K was shown by detecting the site

* Author for correspondence.

specific electric field gradient (EFG) in nanocrystalline ZnO using perturbed $\gamma\gamma$ angular correlation spectroscopy (PAC) [6]. At 473 K, the crystalline quality of nano-ZnO is significantly improved as revealed by different experimental techniques (TEM, XRD, UV–VIS absorption, and photoluminescence spectroscopy). The analytical results indicate that the successful doping of nano-ZnO with In donors at 473 K is accompanied by the onset of crystal growth and might be accompanied by the removal of intrinsic defects in the nanocrystallites. Based on the known experimental conditions for the incorporation of ^{111}In atoms in nano-ZnO, experiments with stable In atoms as a function of In concentrations are possible, now.

In this work, the structural properties of nano-ZnO doped with the donor Indium in a range of the relative In concentration [In/ZnO] of 10^{-5} – 10^{-3} are presented. The environment of the In atoms in In doped nano-ZnO is investigated by PAC using the probe ^{111}In/^{111}Cd. Additional information on the local structure of In and Zn atoms in In doped nano-ZnO is obtained by extended X-ray absorption fine structure measurements (EXAFS). For the latter experiment a relative In concentration of at least 10^{-3} is required.

2. Sample preparation

In doped nano-ZnO samples were produced by electrochemical deposition under oxidizing conditions (EDOC) [7, 8]. The electrolysis process uses dissolved Zn^{2+} ions and takes place in a solution of 0.1 M tetrabutylammonium-bromide (TBA-Br) dissolved in 2-propanol. During the electrolysis a current density of 10 mA/cm^2 was applied at a temperature of about 303 K. The nano-ZnO crystallites were doped *in situ* with In atoms by adding a solution of stable In along with radioactive ^{111}In atoms to the electrolyte. During this procedure the ZnO clusters are coated with TBA, which works as a stabilizer of the colloidal nano-crystallites. In this way it was possible to obtain relative concentrations of In atoms in the range between 10^{-6} and 10^{-3}, which were verified by chemical analysis. For hydrothermal treatment, the ZnO nanocrystals were dissolved in 2-propanol and the solution, encapsulated into a pressure vessel was heated to a temperature of 473 K, yielding a pressure of about 2.5 MPa. After that treatment the ZnO nanocrystallites have a volume-weighted mean grain size of 11 nm.

3. PAC experiments

In ZnO, the incorporation of the ^{111}In atoms on Zn sites can easily be verified by comparing the electric field gradient (EFG), measured at the site of the radioactive dopant ^{111}In/^{111}Cd in a PAC experiment [6], with the well-known EFG caused by the hexagonal ZnO lattice at the probe ^{111}In/^{111}Cd on a Zn site. Choosing V_{zz} as the largest component of the traceless EFG tensor and the asymmetry parameter $\eta = (V_{xx}-V_{yy})/V_{zz}$, the EFG measured by the probe ^{111}In/^{111}Cd in bulk ZnO is characterized by the coupling constant $\nu_Q = eQV_{zz}/$

h = 31.2(1) MHz and $\eta = 0$ [9]; here, Q is the nuclear quadrupole moment of the isomeric 245 keV − $5/2^+$ level of ^{111}Cd used for measuring the nuclear quadrupole interaction.

For In doped nano-ZnO with relative In concentrations of 10^{-5}, 10^{-4}, and 10^{-3}, Figure 1 shows PAC spectra (left) along with their Fourier transforms (right); the corresponding average numbers of In atoms per nanocrystallite n_{In} are (top to bottom) 0.1, 1, and 10, respectively. Up to a concentration of 10^{-4} (n_{In} = 1) an EFG characterized by $\nu_Q = 31(2)$ MHz and $\eta = 0.2(1)$ is observed, which is detected by a fraction of about 12% of the ^{111}In atoms. The coupling constant ν_Q matches well to that of the EFG of ^{111}In on undisturbed Zn sites in bulk ZnO, mentioned above. The difference in the asymmetry parameter η of the observed EFG might indicate distortions of the ZnO lattice in the environment of the ^{111}In atoms. These distortions can be caused by the incorporation of ^{111}In atoms near the surface of the crystallite or by internal strain present in the nanocrystallites. The existence of lattice distortions in nanocrystallites has been confirmed by EXAFS experiments [10]. At the relative concentration of 10^{-4} obviously new perturbations arise in the PAC spectra, which are characterized by two EFGs with the parameters $\nu_Q = 130(10)$ MHz, $\eta = 0.4(1)$ and $\nu_Q = 150(12)$ MHz, η = 0.7(1). These parameters are detected by 17% and 14% of the probe atoms, respectively. The new EFGs are assigned to In defect complexes. At still higher In concentration of 10^{-3} (n_{In} = 10), the PAC spectrum exclusively exhibits the EFGs of the In defect complex, whereas the signal of In atoms located on unperturbed Zn sites has totally vanished. The observation of the various EFGs in the PAC spectra of In doped nano-ZnO is obviously closely connected with the actual In concentrations in the ZnO nanocrystals. Finally, it should be noted that in all experiments the majority of In atoms (70–80%) is exposed to a wide distribution of EFG leading to a strong relaxation in the PAC spectra. These In atoms might by located in defect-rich surroundings, e.g. surface sites or dislocations.

The appearance of additional EFGs at In concentrations of 10^{18} cm^{-3} (corresponding to a relative In concentration of $\sim 10^{-4}$) and above was also observed in the II–VI compounds CdTe, ZnTe, ZnSe, CdSe, CdS, ZnS [11]. In the case of CdTe, these EFGs were caused by Cd vacancies [12].

4. EXAFS experiments

The local structure of In and Zn atoms in In doped nano-ZnO was investigated by EXAFS measurements at a temperature of 20 K in fluorescence mode at the X1 beam line of HASYLAB at DESY. The analysis of the absorption spectra was done following the standard FEFF procedure [13–16].

The EXAFS results are plotted in Figure 2: Panel (b) shows the K edge absorption $\chi(k)$ at the Zn host atoms together with its Fourier transform (FT). The data can be fitted up to the fourth nearest-neighbour shell with parameters known from bulk ZnO (see Figure 2(a)). The fit parameters of the first O shell

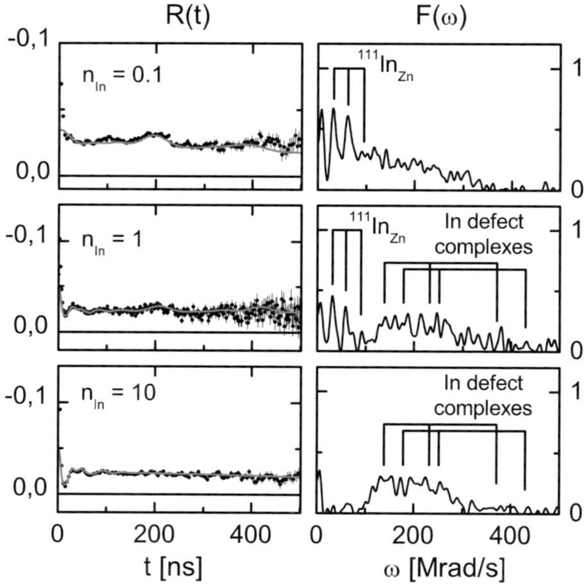

Figure 1. PAC spectra (left) and their Fourier transforms (right) of nanocrystalline ZnO doped with the donor In at different average number of In atoms per crystallite (n_{In}).

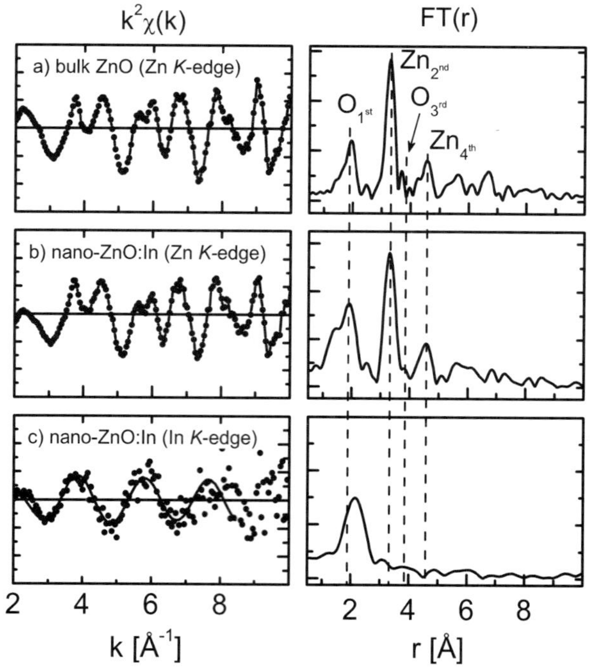

Figure 2. EXAFS spectra (left) and their Fourier transforms (right) of (a) bulk ZnO measured at 20 K at the K edge of Zn and of In doped nano-ZnO ([In/ZnO] = 10^{-3}) measured at the K edges of (b) Zn and (c) In.

Table I. Fit parameters of EXAFS experiments performed at the K edges of Zn and In in bulk ZnO and highly In doped nano-ZnO

Sample	Absorber	$O_{1st\ shell}$			$Zn_{2nd\ shell}$		
		n	R (Å)	σ^2 (10^{-3}Å^{-2})	n	R (Å)	σ^2 (10^{-3}Å^{-2})
Bulk ZnO	Zn	4.1 ± 0.2	1.96 ± 0.01	2.6 ± 0.2	11.9 ± 0.3	3.23 ± 0.01	2.6 ± 0.8
Nano-ZnO:In	Zn	3.5 ± 0.8	1.96 ± 0.02	3.0 ± 1.5	11.3 ± 1.0	3.22 ± 0.02	4.3 ± 0.6
Nano-ZnO:In	In	3.3 ± 1.4	2.17 ± 0.05	4.01 ± 7.9	–	–	–

n number of atoms; R distance to the absorber atom; σ^2 rms fluctuation in the distance.

(O_{1st}) and the second Zn shell (Zn_{2nd}) are listed in Table I together with the corresponding parameters of ZnO bulk material [17]. Figure 2(c) shows the K edge absorption $\chi(k)$ at the In dopant atoms along with its Fourier transform. In the FT spectrum the first O shell (O_{1st}) is clearly visible, whereas the atoms in the second Zn shell (Zn_{2nd}) and higher shells are almost invisible. An analysis of the first O shell about the In atoms yields the radial distance of about 2.17(5) Å (Zn–O_{NN}: 1.96(1) Å), which seems to be caused by the larger covalent radius of the In atoms as compared to the Zn host atoms. The number of atoms in the O shell is 3.3 ± 1.4. Since the number of O atoms in the first shell in bulk ZnO is four and in bulk In_2O_3 is six, the measured value suggests that the In atoms are incorporated on Zn sites in the ZnO lattice. The deficiency of atoms in the Zn_{2nd} shell as observed by EXAFS at the In K edge (see Figure 2(c)) might have the same microscopic origin as the occurrence of the defect complexes observed by PAC (see Figure 1(c)).

5. Conclusion

Up to a relative In concentration of 10^{-4}, PAC investigations performed at heavily In doped nano-ZnO show that about 12% of the ^{111}In atoms are incorporated on substitutional Zn sites. At the relative In concentration of 10^{-4}, however, new In defect complexes are visible in the PAC spectra, which finally at a relative In concentration of 10^{-3}dominate the spectra. It is concluded that the In defect complexes are formed if there are more than one In atom per ZnO nanocrystal.

EXAFS measurements performed at the K edge of the Zn atoms show that the crystal structure of nano-ZnO is very close to the structure of bulk ZnO. EXAFS measurements performed at the K edge of In in heavily In doped nano-ZnO exhibit an expansion of the crystal lattice to the first O shell. Since nearly four O atoms are detected in the first shell it is suggested that the In atoms are incorporated substitutionally in the ZnO host The second Zn shell and higher are almost invisible, which might have the same microscopic origin as the defect complexes observed in the PAC experiments. Further experiments are necessary in order to establish this correlation on an atomic scale.

Acknowledgements

The financial support by the DFG within the SFB 277 is gratefully acknowledged. The authors are grateful to the HASYLAB staff for their assistance in the EXAFS experiments.

References

1. Alivisato A. P., *Science* **271** (1996), 933.
2. Shim M., Wang C., Norris D. J. and Guyot-Sionnest P., *MRS Bull.* **26** (2001), 1005.

3. Rockenberger J., zum Felde U., Tischer M., Tröger L. Haase M. and Weller H., *J. Chem. Phys.* **112** (2000), 4296.
4. Fujii M., Mimura A., Hayashi S., Yamamoto Y. and Murakami K., *Phys. Rev. Lett.* **89** (2002), 206805.
5. Agne T., Guan Z., Hempelmann R., Li X. M., Natter H., Wolf H. and Wichert T., *Appl. Phys. Lett.* **83** (2003), 1204.
6. Wichert T. In: Stavola M. (ed.), *Identification of Defects in Semiconductors*, Academic, London, 1999, p. 297.
7. Agne T., Guan Z., Li X. M., Wolf H. and Wichert T., *Phys. Status Solidi B* **229** (2002), 819.
8. Dierstein A., Natter H., Meyer F., Stephan H.-O., Kropf C. and Hempelmann R., *Scr. Mater.* **44** (2001), 2209.
9. Wolf H., Deubler S., Forkel D., Foettinger H., Iwatschenko-Borho M., Meyer F., Renn M., Witthuhn W. and Helbig R., *Mater. Sci. Forum* **10–12** (1986), 863.
10. Rockenberger J., Tröger L., Kornowski A., Vossmeyer T., Eychmüller A., Feldhaus J. and Weller H., *J. Phys. Chem. B* **101** (1997), 2691.
11. Ostheimer V., Ph.D thesis, Universität des Saarlandes, Saarbrücken, Germany.
12. Lany S., Ostheimer V., Wolf H. and Wichert T., *Physica B* **308–310** (2003), 958.
13. Rehr J. J. and Albers R. C., *Rev. Mod. Phys.* **72** (2000), 621.
14. Stern E. A., Newville M., Ravel B., Yacoby Y. and Haskel D., *Physica B* **208–209** (1995), 154.
15. Newville M., *J. Synchrotron Radiat.* 8 (2001), 322.
16. Rehr J. J., Albers R. C. and Zabinsky S. I., *Phys. Rev. Lett.* **69** (1992), 3397.
17. Landolt/Börnstein. In: Rössler U. (ed), *Numerical Data and Functional Relationships in Science and Technology – New Series III*. Vol. 41B, Berlin Heidelberg, New York, Springer 1999.

Hyperfine Interactions (2004) 159:63–69
DOI 10.1007/s10751-005-9084-5

Grain Size Effect on the Temperature Dependence of the Electric Field Gradient in Nanocrystalline In

X. M. LI, Z. GUAN, T. AGNE, H. WOLF* and T. WICHERT
Technische Physik (Geb. 38), Universität des Saarlandes, Postfach 151150, D-66041 Saarbrücken, Germany; e-mail: herbert@tech-phys.uni-sb.de

Abstract. Nanocrystalline indium (nano-In) was prepared with different particle sizes by electrochemical deposition. The temperature dependence of the local electric field gradient (EFG) of nano-In was investigated in a temperature range of 20–300 K using the probe ^{111}In for perturbed $\gamma\gamma$ angular correlation (PAC) experiments. The temperature dependence of the EFG of nano-In can be described by a $(1-BT^{3/2})$ dependence as in bulk-In. It is shown that the low temperature value of the EFG and the proportionality constant B vary systematically with particle size.

Key Words: electric field gradient, nano-indium, PAC, temperature dependence.

1. Introduction

The electrical field gradient (EFG) in non-cubic metals has been investigated as a function of temperature in the past using e.g., the perturbed $\gamma\gamma$ angular correlation technique (PAC). For the temperature dependence, the relation

$$\nu_Q(T) = \nu_Q(0)\left(1 - BT^{3/2}\right), \tag{1}$$

with B = const, has been found which is valid for many non-cubic metals [1]. This temperature dependence can be explained theoretically on the basis of the interaction of the metal electrons with phonons [2, 3]. For the hexagonal metals Zn and Cd a significant increase of the lattice EFG at temperatures below 260 K has been found after cold working of polycrystalline samples [4]. The authors explained their results by a slight change of the c/a ratio due to intrinsic defects generated by the cold working process leading to the change of the low temperature value of the EFG. However, also changes in the phonon spectrum due to the limited grain size of the cold worked samples were taken into consideration. The grain size dependence of the EFG in nano-In was investigated by Sinha and Collins at a temperature of 77 K [5]. The authors proposed that a decrease of the c/a ratio of the tetragonal lattice of In is responsible for the

* Author for correspondence.

Table I. Preparation conditions and grain sizes of nano-Indium

Sample	Temperature (°C)	Current density (mA/cm^2)	Particle size (nm)	
			TEM	XRD
n-In#1	5	8	11(2)	26
n-In#2	5	15	12(4)	37.5
n-In#3	40	40	27(4)	45

decrease of the EFG with decreasing grain size. In the present work, the temperature dependence of the EFG in nano-In metal is investigated using PAC.

2. Experimental details

Nano-In was prepared by electrochemical deposition which is based on the reduction of metal ions supplied from a sacrificial anode of In metal in a non-aqueous electrolyte [6]. In this process, the particle size is influenced by the temperature of the electrolyte and the current density during deposition. Samples of nano-In with three different particle sizes were prepared by variation of temperature and current density during deposition (see Table I). The In nanocrystals were doped with radioactive ^{111}In probe atoms by diffusing ^{111}In atoms into the sacrificial In-anode before the electrolysis process. The particle sizes were determined by transmission electron microscopy (TEM) and by X-ray diffraction (XRD). The temperature dependence of the lattice EFG in nano-In was investigated by PAC measurements in the range 20 K to 300 K.

3. Results and discussion

3.1. DETERMINATION OF PARTICLE SIZE

Directly after preparation, a small droplet of the solution containing the In nanocrystals was extracted for taking TEM images. The TEM images show almost spherical particles having an average size of 11(2), 12(4), and 27(4) nm for samples n-In#1, n-In#2, and n-In#3, respectively (Figure 1). The numbers in brackets indicate the size distributions (FWHM) obtained from a statistical analysis.

After the radioactive decay of the ^{111}In atoms, i.e., about 6–8 weeks after preparation, for each sample the whole sample volume was used for XRD measurements. The XRD spectra taken from the three samples exhibit the tetragonal lattice structure of bulk In (Figure 2). Using the Scherrer formula, the average particle sizes D were determined to 26, 37.5 and 45 nm for n-In#1, n-In#2 and n-In#3, respectively, but the Scherrer analysis does not yield information of the size distribution. There is an obvious, systematic difference

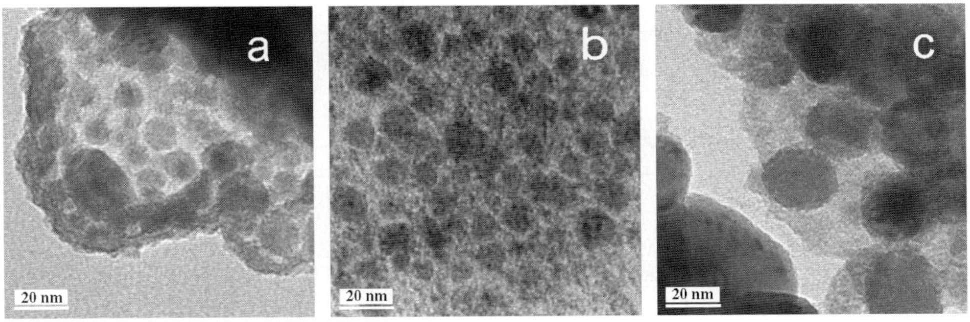

Figure 1. TEM images of the three different nano-Indium samples. The average particle sizes are (a) 11 nm, (b) 12 nm, and (c) 27 nm for n-In#1, n-In#2, and n-In#3, respectively (see Table I).

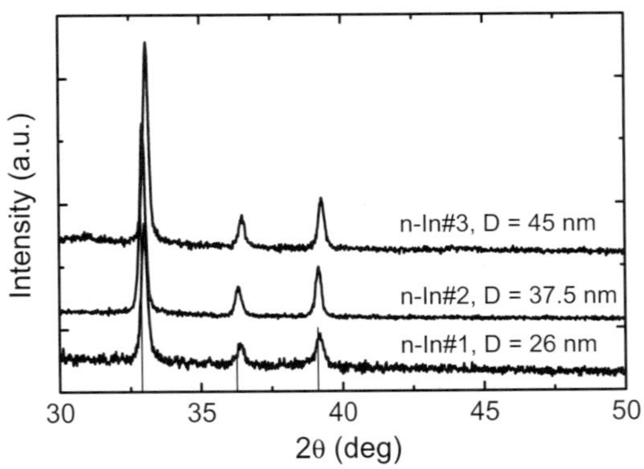

Figure 2. XRD data of the three investigated samples of nano-Indium showing the size dependence of the lattice constants. The *vertical lines* correspond to bulk Indium.

between the particle sizes as determined by TEM and XRD. One reason might be the fact that for each TEM analysis only a small droplet, containing 29, 35, and 26 particles for samples n-In#1, n-In#2, and n-In#3, respectively, was extracted from the solution, which might be not representative for the whole sample. Secondly, the XRD investigations were performed 6–8 weeks after preparation and due to the low melting point of In (430 K) a significant grain growth may have taken place during the storage time at ambient temperature. Both effects would easily explain the systematic difference between TEM and the XRD results. In the following, the smaller sizes, as determined by TEM, are assigned to the PAC experiments.

3.2. PAC INVESTIGATIONS

Figure 3 shows the PAC time spectra (left) along with their Fourier transforms (right) of bulk and nano-Indium samples measured at 20 K. All spectra show the

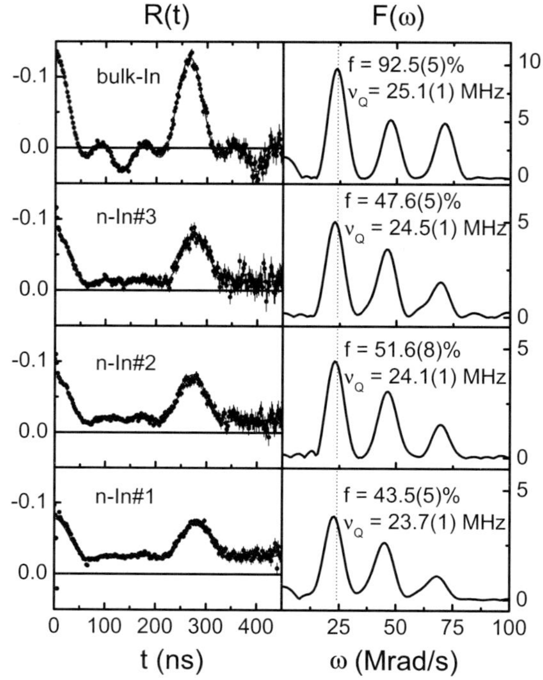

Figure 3. PAC spectra (*left*) and corresponding Fourier transforms (*right*) of bulk and nano-Indium measured at 20 K.

modulation caused by the tetragonal lattice of In metal. The corresponding fit parameters are listed in Table II. The coupling constant ν_Q, i.e., the EFG at the substitutional lattice site, at a fixed temperature below ambient temperature decreases slightly with decreasing particle size, which is consistent with observations made by Sinha and Collins [5]. However, there is no systematic dependence on the particle size observed for the fraction f of ^{111}In atoms on substitutional lattice sites and for the EFG distribution, i.e., the damping factor δ. Although Sinha and Collins did not give exact numbers for the EFG measured for nano-In at 77 K, it can be estimated that the difference of the EFGs between bulk In and nano-In of 12 nm particle size seems to be larger than in the present investigations. This may be an indication that the determination of the particle size in the present investigations is affected by systematic errors. The deviation of the EFG of nano-In from the bulk value becomes smaller with increasing temperature (see Figure 4). While in bulk-In, as expected, almost all ^{111}In probe atoms reside on substitutional lattice sites, in the nano-In samples only a fraction of f $\approx$50% of all probe atoms detects the EFG corresponding to unperturbed substitutional lattice sites of In.

The observed reduced fraction of ^{111}In probe atoms on unperturbed lattice sites in nano-In is qualitatively consistent with the observations made by Sinha and Collins [5]. This reduction can be caused on the one hand by a significant

Table II. PAC parameters (f, ν_Q, δ) measured at 20 K and temperature dependence of the EFG ($\nu_Q(0)$, B) for the investigated samples of nano-In and of bulk In

Sample	Size (nm)	f (%)	ν_Q (MHz)	δ (MHz)	$\nu_Q(0)$ (MHz)	B ($K^{-3/2}$)
n-In#1	11(2)	43.5(5)	23.7(1)	0.65	23.80	$5.76 \cdot 10^{-5}$
n-In#2	12(4)	51.6(8)	24.1(1)	0.2	24.24	$5.82 \cdot 10^{-5}$
n-In#3	27(4)	47.6(5)	24.5(1)	0.3	24.54	$6.03 \cdot 10^{-5}$
bulk		92.5(5)	25.1(1)	0.0	25.35	$6.11 \cdot 10^{-5}$

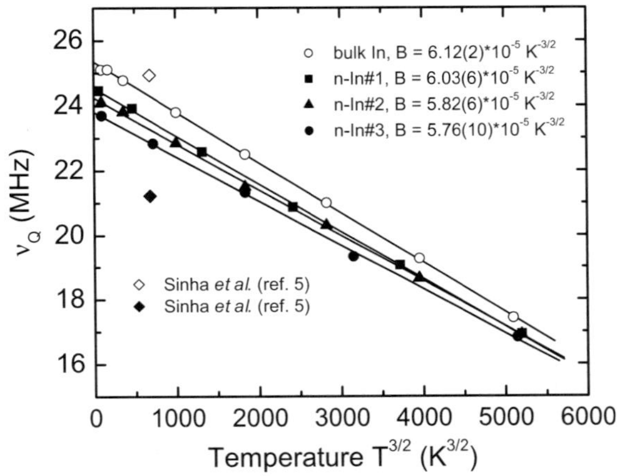

Figure 4. Temperature dependence of EFG for different grain sizes of nano-In compared to bulk In. The *lines* correspond to fits using Equation (1).

number of lattice sites near the surface of the crystallite. For detecting the lattice EFG a minimum distance to the surface is required. For In atoms residing at or near to the surface additional contributions to the EFG become important. On the other hand, there might be impurity atoms or intrinsic defects incorporated during preparation giving rise to non-uniform defect related EFGs. Assuming that this reduced fraction is essentially determined by surface effects, for the ^{111}In probe atoms a minimum distance of about 2 nm to the surface necessary for detecting the lattice EFG is estimated. However, Sinha and Collins estimated a surface layer of only 0.6 nm in which the ^{111}In atoms do not contribute to the observed lattice EFG [5]. In the present investigations, therefore, impurities and/ or lattice defects seem to be of importance for the reduced fraction of unperturbed lattice sites.

3.3. TEMPERATURE DEPENDENCE OF THE EFG

The temperature dependence of the EFG of nano-In and bulk-In has been measured in the temperature range between 20 and 300 K. The data are shown in

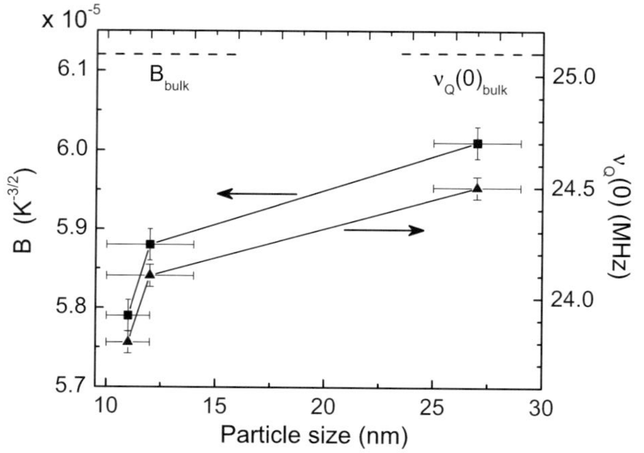

Figure 5. The dependence of the parameters $\nu_Q(0)$ (*triangles*) and B (*squares*) (see Equation (1)) on the particle size.

Figure 4 along with fits using Equation (1), whereby $\nu_Q(0)$ and B are treated as free fit parameters. Figure 5 shows the extracted fit parameters $\nu_Q(0)$ and B as a function of the particle size determined by TEM (see also Table II). The data in Figures 4 and 5 show that the temperature dependence of the EFG is well described by Equation (1) for all particle sizes investigated and that the values of both parameters decrease with decreasing particle size.

Since the zero temperature value $\nu_Q(0)$ obviously depends on the particle size, as is shown in Figure 5, it is concluded that changes in the phonon spectrum due to the reduced particle size cannot be the only reason for the differences observed in nano- and polycrystalline In. At zero temperature, no phonons are excited and, therefore, no differences of the EFG due to differences of the phonon spectrum should be present. On the other hand, slightly changed lattice constants of the In nanocrystals as compared to bulk In might be of importance. The dependence of $\nu_Q(0)$ on the particle size can qualitatively be explained in terms of such slight variations of the lattice constants. The XRD data, however, do not show this effect directly due to their limited accuracy. In order to explain the decrease of the deviation of the EFG from the bulk value with increasing temperature, the values of the lattice constants should approach the bulk values with increasing temperature.

The decrease of the EFG in nano-In with decreasing particle size is consistent with the observations from Sinha and Collins [5], but is in contradiction to the observations made by Meyer *et al.* in the hexagonal metals Zn and Cd [4]. However, there is no reason that changes of the lattice constants alter the EFGs in a unique direction for different materials. By Oshima *et al.* is reported that the structure of nano-In transforms from bct to fcc structure if the particle size becomes less than 5 nm [7]. The decreasing EFG observed here and by Sinha and Collins might indicate the transition from the bct to the fcc structure.

In nanocrystals, it can be expected that due to the particle size the long wavelength part of the phonon spectrum is cut off. Since at low temperatures essentially phonons of long wavelengths are excited, changes of physical properties depending on the excitation of phonons can be expected to take place at these temperatures. A quantitative estimation of the effect of a modified phonon spectrum on the temperature dependence of the lattice EFG in non-cubic metals is difficult to perform at the present time. On the basis of the available experimental data it is not possible to find unique indications for effects of changes of the phonon spectrum on the temperature dependence of the EFG in nanocrystalline In.

4. Summary

The temperature dependence of the EFG was investigated in nano-In. It is found that the relation $\nu_Q(T) = \nu_Q(0)\left(1 - BT^{3/2}\right)$ which is well known to be valid for bulk non-cubic metals, still holds for nanocrystal metal In. At the same time it is observed that $\nu_Q(0)$ and B depend on the particle size. The modifications of the EFG seems to be connected to slightly changed lattice constants of the nanocrystals as compared to the bulk values, but additional effects based on changes of the phonon spectrum due to the reduced size of the crystallites cannot be excluded.

Acknowledgements

We thank Dr. H. Natter for the XRD analyses and Dr. H. Shen for the TEM images. The financial support by the Deutsche Forschungsgemeinschaft (SFB 277) is gratefully acknowledged.

References

1. Christiansen J., Heubes P., Keitel R., Klinger W., Loeffler W., Sandner W. and Witthuhn W., *Z. Phys. B* **24** (1976), 177.
2. Jena P., *Phys. Rev. Lett.* **36** (1976), 418.
3. Nishiyama K., Dimmling F., Kornrumpf Th. and Riegel D., *Phys. Rev. Lett.* **37** (1976), 357.
4. Meyer F., Deubler S., Plank H., Witthuhn W. and Wolf H., *Hyperfine Interact.* **34** (1987), 243.
5. Sinha P. and Collins G. S., *Nanostruct. Mater.* **3** (1993), 217.
6. Hempelmann R. and Natter H., Patent: DE 198 40 842 A 1 (2000).
7. Oshima Y., Nangou T., Hirayama H. and Takayanagi K., *Surf. Sci.* **476** (2001), 107.

Hyperfine Interactions (2004) 159:71–74
DOI 10.1007/s10751-005-9085-4

Split and Compensated Hyperfine Fields in Magnetic Metal Clusters

H. NAKAMURA[1,*], H. CHUDO[2,a], M. SHIGA[2] and T. KOHARA[1]
[1]*Graduate School of Material Science, University of Hyogo, Kamigori, Ako-gun, Hyogo 678-1297, Japan; e-mail: naka6ra@sci.himeji-tech.ac.jp*
[2]*Department of Materials Science and Engineering, Kyoto University, Kyoto 606-8501, Japan*

Abstract. As prominent characteristics of magnetic metal cluster found in vanadium sulfides, we point out marked separation and compensation of the hyperfine field at the nuclear site; these are in somewhat discordance with the common sense for 3d transition-metal magnets, where the on-site isotropic field, scaling the ordered moment magnitude, is dominant.

Key Words: GaV_4S_8, GeV_4S_8, metal cluster, NMR.

The measurement of hyperfine fields $\boldsymbol{H}_{\rm int}$ in magnetically ordered materials is one of classical applications of NMR. $\boldsymbol{H}_{\rm int}$ is often referred as a rough estimate of the ordered moment $\boldsymbol{m}$ with use of the hyperfine coupling tensor $A_{\rm hf}$, which converts $\boldsymbol{H}_{\rm int}$ to $\boldsymbol{m}$. It is generally believed that, for 3*d*-transition-metal magnets, the on-site *isotropic* field $\boldsymbol{H}^{\rm on}_{\rm iso}$, made through the core-electron polarization exchange-induced by magnetic electrons, is the major contribution. As $A_{\rm hf}$, the value estimated from the Knight-shift analysis in the paramagnetic state or from theoretical considerations is often used. However, we sometimes encounter the cases such that the naive analysis does not work at all. An example is found in the case of a quasi-one-dimensional vanadium sulfide $BaVS_3$, which exhibits a metal-insulator transition at $T_{\rm MI} \simeq 70$ K [1] and long-range antiferromagnetic ordering at $T_{\rm X} \simeq 30$ K [2, 3]. The antiferromagnetic $BaVS_3$ exhibits markedly split zero-field resonances [4] at around frequencies much lower than the expectation from the actual moment value [2]. At present we speculate that the puzzling NMR spectrum is related to the metal-atom clustering [5, 6] associated with the metal-insulator transition.

To obtain information on the hyperfine coupling characteristic to such metal clusters, we investigate stable and well-isolated clusters. One is the tetrahedral (tetra-nuclear) V_4 cluster (cubane type V_4S_4 cluster with two unpaired spins) involved in an antiferromagnetic matrix crystal GeV_4S_8 ($T_{\rm N} \simeq 16$ K) with the cubic $GaMo_4S_8$ type crystal structure [7]. The other is the ferromagnetic

* Author for correspondence.
[a] Present address: Department of Chemistry, Kyoto University, Kyoto 606-8502, Japan.

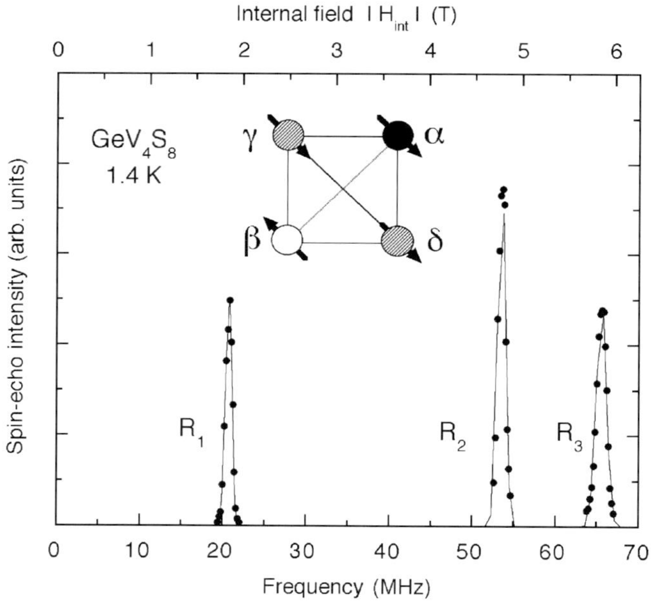

Figure 1. The NMR spectrum measured for GeV_4S_8 at 1.4 K under zero external field. The *inset* is the projection of the tetrahedral V_4 cluster on the (001) plane with taking β-α and γ-δ bonds along [110] and $[1\bar{1}0]$ directions, respectively. *Arrows* represent the proposed directions of spins ($[1\bar{1}0]$ or $[\bar{1}10]$), being common for all V_4 clusters. *Black*, *white* and *hatched sites* sense different hyperfine fields due to different local magnetic environments.

analogue (with an unpaired spin per cluster) in GaV_4S_8 ($T_C \simeq 13$ K) with the saturation magnetization $m \simeq 0.2\ \mu_B$/V [8]. For these clusters, intercluster hyperfine interaction is probably minor comparing with the intracluster interaction. Hence we have $\boldsymbol{H}_{\text{int}} \simeq \boldsymbol{H}_{\text{on}} + \boldsymbol{H}_{\text{trans}}$, where $\boldsymbol{H}_{\text{on}}$ is the field made by on-site electrons and $\boldsymbol{H}_{\text{trans}}$ the transferred fields summed over other three atoms inside the cluster. Note that the V site has trigonal symmetry along the diagonal axis of the tetrahedron, $\langle 111 \rangle$, but the symmetry axes are locally different from each other within the cluster.

Figure 1 shows the zero-field NMR spectrum measured for the V_4 cluster in GeV_4S_8. Three resonances were found in a relatively wide frequency range; $|H_{\text{int}}| \simeq 1.9$, 4.8 and 5.9 T. Neutron diffraction experiment suggested that, as one of most probable candidates, the spin structure of the V_4 cluster in GeV_4S_8 is common as $\uparrow\uparrow\uparrow\downarrow$ with the spin directions as shown in the inset of Figure 1 [6]. No evidence indicating the spin density modulation has been found, and the moment magnitude was estimated to be 0.50 μ_B/V. Believing in the spin arrangement with the unique moment magnitude, the local magnetic environments (magnetic sites) are classified into three kinds: The angle between the moment and the local symmetry axis θ is 90° for α, β and 35° for γ, δ (see the inset of Figure 1 for the labels of atoms). This results in the separation of the anisotropic component in the on-site field $\boldsymbol{H}_{\text{on}}^{\text{aniso}}$. In addition, the arrangement of neighbor

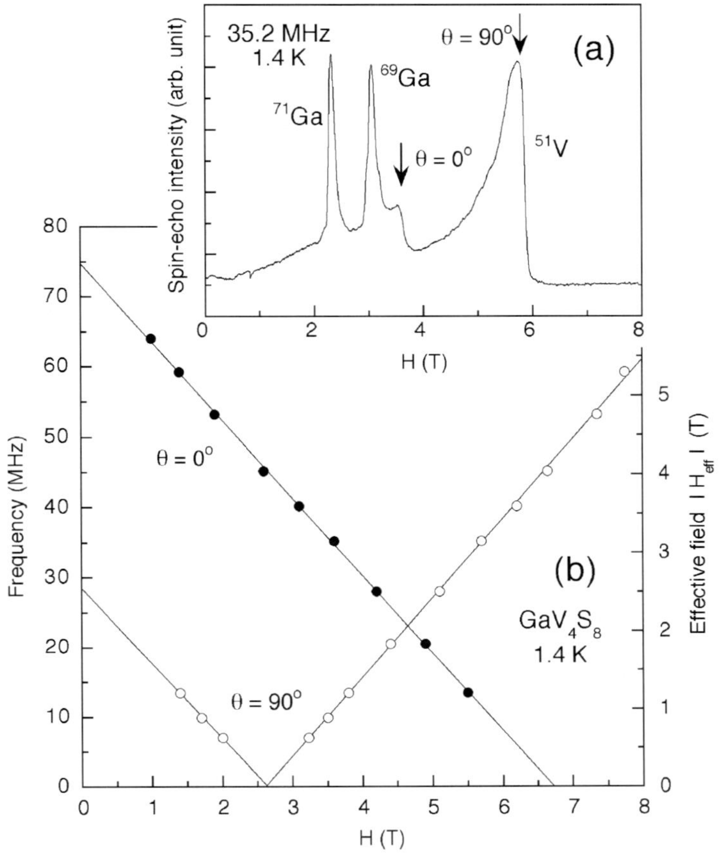

Figure 2. (a) An example of the field-swept ^{51}V NMR spectrum measured for GeV$_4$S$_8$ at 1.4 K and 35.2 MHz. Sharp peaks are ^{71}Ga and ^{69}Ga resonances. *Arrows* indicate the singular points corresponding to $\theta = 0°$ and 90°. (b) The frequency dependence of the fields corresponding to $\theta =$ 0° (*full circles*) and 90° (*open circles*).

spins are ↑↑↓ for α, γ, δ and ↓↓↓ for β, resulting in the difference in $\boldsymbol{H}_{\text{trans}}$. Above all, we expect three types of magnetic sites. Although the site assignment is not possible with only the present experimental data, it is qualitatively clear that the contributions other than $\boldsymbol{H}_{\text{on}}^{\text{iso}}$, namely $\boldsymbol{H}_{\text{on}}^{\text{aniso}}$ and $\boldsymbol{H}_{\text{trans}}$, dominate $\boldsymbol{H}_{\text{int}}$ in the ferrimagnetic cluster.

Experiments in external fields H_{ext} on the ferromagnetic V$_4$ cluster in GaV$_4$S$_8$ are useful to separate the isotropic and anisotropic contributions. Figure 2(a) shows an example of the field-swept ^{51}V NMR spectrum measured for the powder sample, which is markedly broadened. In this condition (under sufficiently high fields), the magnitude of the effective field at the nuclear site, $|H_{\text{eff}}|$, is given by $H_{\text{int}} + H_{\text{ext}}$ since moments orient along the external field. The broadened lineshape is understood as a uniaxial powder pattern associated with the distribution of θ in the powder sample; H_{eff} ranges from negative to positive in the shown spectrum. Singular points in the lineshape correspond to $\theta = 0°$ and 90°. The fields of θ = 0° and 90° are plotted against various experimental frequencies in Figure 2(b). Extrapolation of the lines to the ordinate gives $H_{int}(90°) = \pm 2.6$ and $H_{int}(0°) = \mp 6.7$ T. From these values we have $|H_{iso}| = 0.5$ and $|H_{ansio}| = 6.2$ T with the definition $H_{int}(\theta) = H_{iso} + \frac{1}{2}H_{aniso}(3\cos^2\theta - 1)$. This result indicates the considerable anisotropy in $\boldsymbol{H}_{int}$ and an abnormally small isotropic field. The small isotropic term may be ascribed to the compensation between on-site and transferred fields.

In summary, in the metal clusters in the vanadium sulfides, (1) the hyperfine field is considerably anisotropic, (2) intracluster transferred fields are possibly large and site-dependent due to asymmetric bonds (or the effect of surface) inherent to the small cluster. The compensation between on-site and transferred fields may give rise to an anomalously small isotropic field. We conclude that the internal fields in these clusters are determined by the *direction* of moments rather than the *magnitude*. We also suggest a unique possibility of NMR to test the metal-atom clustering in magnetic materials.

References

1. Gardner R. A., Vlasse M. and Wold A., *Acta Crystallogr., B* **25** (1969), 781.
2. Nakamura H. *et al.*, *J. Phys. Soc. Jpn.* **69** (2000), 2763.
3. Higemoto W. *et al.*, *J. Phys. Soc. Jpn.* **71** (2002), 2361.
4. Nakamura H., Imai H. and Shiga M., *Phys. Rev. Lett.* **79** (1997), 3779.
5. Inami T. *et al.*, *Phys. Rev., B* **66** (2002), 073108.
6. Nakamura H. *et al.*, in preparation.
7. Johrendt D., *Z. Anorg. Allg. Chem.* **624** (1998), 952.
8. Brasen D. *et al.*, *J. Solid State Chem.* **13** (1975), 298.

Hyperfine Interactions (2004) 159:75–80
DOI 10.1007/s10751-005-9082-7

Hyperfine Fields in Nanocrystalline $Fe_{0.48}Al_{0.52}$

K. SZYMAŃSKI[1,*], D. SATUŁA[1], L. DOBRZYŃSKI[1,2], E. VORONINA[3] and E. P. YELSUKOV[3]
[1]*Institute of Experimental Physics, University of Białystok, 15-424 Białystok, Poland; e-mail: kszym@alpha.uwb.edu.pl*
[2]*The Sołtan Institute for Nuclear Studies, 05-400 Otwock-Świerk, Poland*
[3]*Physical–Technical Institute of Ural Division of Russian Academy of Science, Izhevsk, Russia*

Abstract. Mössbauer measurements with circularly polarized radiation were performed on a nanocrystalline, disordered $Fe_{48}Al_{52}$ alloy. The analysis of the data for various polarization states resulted in the characterization of the hyperfine magnetic field distribution and the dependence of the average *z*-component of hyperfine field on the chemical environment. An increasing number of Al in the first coordination shell causes not only a decrease of magnetic moments but also introduces noncollinearity.

Key Words: circularly polarized radiation, Fe–Al alloys, nuclear resonant magnetometry.

1. Introduction

The magnetic ground state of Fe–Al alloys in the equiatomic concentration range depends strongly on the composition and structural order. Alloys of bcc-type structure were suspected to be antiferromagnets [1, 2], however, no antiferromagnetic ordering was detected in neutron scattering experiments. In many works noncollinear or antiparallel arrangements of magnetic moments were considered [3–5], but no experimental evidences was found. Directions of local Fe magnetic moments, including parallel or antiparallel configurations, can be investigated by Mössbauer polarimetry with circularly polarized radiation [6]. The present paper demonstrates how the magnetic moments of iron depend on the local environment in one selected Fe–Al alloy.

2. Sample

Because of the great sensitivity of magnetic properties on crystalline order, we aimed at the preparation of a homogeneous, disordered bcc state. Therefore melting, homogenization and grinding processes were applied as follows.

* Author for correspondence.

Fe–Al ingots were alloyed from high-purity components in an induction furnace in Ar atmosphere. They were homogenized in a vacuum furnace at 1400 K during 6 h. The chemical analysis showed that the Al concentration was (52.0 ± 0.5) at.%. At first, the ingots were milled in a mortar by hand. The powder with a particle size less than 300 μm was used for further mechanical grinding in a planetary ball mill with vials and balls made of tungsten carbide. The milling was performed in an inert gas atmosphere during 10 h, the time sufficient for creating disordered Fe–Al alloys [7]. X-ray studies showed a lattice parameter of 0.2918(4) nm, a mean grain size (4.0 ± 0.3) nm and microstrains (0.5 ± 0.1)%, the last values found from the broadening of the diffraction lines using harmonic analysis [8]. Magnetic measurements showed a Curie temperature T_C = (110 ± 10) K, the presence of a hysteresis, and lack of saturation at $T = 5$ K at fields up to 15 T. The powdered $Fe_{0.48}Al_{0.52}$ sample was mixed with Li_2CO_3 and epoxy glue in order to obtain texture free absorbers for Mössbauer measurements.

3. Principles of resonant polarimetric methods

For the case of measurements with circularly polarized radiation, the relative line intensity i_n of the nth nuclear transition in the Zeeman sextet depends on the average cosine square of the angle between k vector of photon and vector of B_{hf}, parameter which we call c_2, and on the average cosine of the angle between k and B_{hf}, called c_1 [6, 9]:

$$\begin{aligned} 16i_1 &= 48i_4 = 3(1 \pm 2c_1 + c_2), \\ 4i_2 &= 4i_5 = 1 - c_2 \\ 48i_3 &= 16i_6 = 3(1\mu 2c_1 + c_2). \end{aligned} \tag{1}$$

The upper and lower signs in (1) correspond to two opposite circular polarizations. In the case of a magnetic texture with axial symmetry, c_2 is a measure of the perpendicular component of the hyperfine field: when $c_2 = 0$ the perpendicular component achieves a maximum, while when $c_2 = 1$ this component is zero. The line intensities for unpolarized radiation can be obtained from (1) by setting $c_1 = 0$. The quantity $c_1 \times B_{hf}$ thus is an average component of the hyperfine field in the direction of the k vector and we will abbreviate it as a z-component of B_{hf}. It follows from equation (1) that polarized radiation allows to measure the average component of the hyperfine field along magnetization direction.

There are well estabilished methods for evaluation of $p(B_{hf})$ distributions from Mössbauer spectra. We have developed an algorithm for evaluation of the $p(B_{hf})$, $c_1(B_{hf})$ and $c_2(B_{hf})$ from simultaneous fitting of set of a spectra measured for different polarization states [9]. Polarized, monochromatic radiation is obtained by resonant filter technique [6].

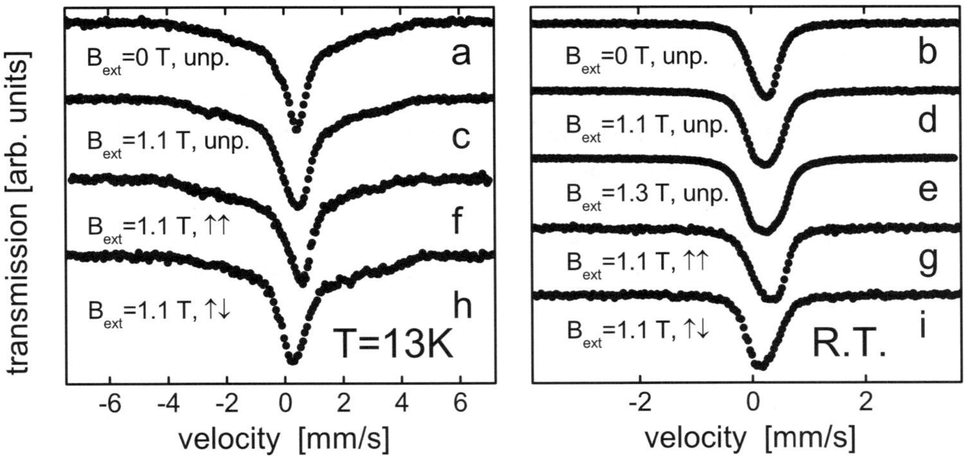

Figure 1. Mössbauer spectra measured with unpolarized radiation (a–e), circularly polarized radiation (f–i). *Arrows* ⇈ and ⇅ indicate two opposite circular polarizations. B_{ext} is applied magnetic field parallel to the photon k vector.

4. Measurements and results

The absorbers were placed inside the permanent magnet. Results of measurements which were performed at $T = 13$ K with two opposite circular polarizations and with unpolarized radiation in an external magnetic field of 1.1 T and in zero field, and spectra obtained at room temperature (well above Curie temperature) are presented in Figure 1.

A reversal of circular polarization state results in clear shift of the position of central absorption line (see Figure 1 f, h). There is also a change in the wide absorption area located at about −3 mm/s and + 3 mm/s in Figure 1f, h. This behaviour shows unambiguously that large Fe hyperfine fields (and related large magnetic moments) are parallel to the net magnetization as in Fe-based ferromagnets.

Detailed analysis based on the algorithm described in [9] allows one to get the distribution of hyperfine field shown in Figure 2a. The broad distribution of hyperfine field does not permit to make definite conclusions about transverse field component. In contrast, a clear-cut result is obtained for the c_1 parameter, see Figure 2b.

The hyperfine field in α-Fe is antiparallel to the direction of the magnetic moment of the iron atom [10] and to the magnetization. In Figure 2b the vertical axis corresponds to the $-c_1$ parameter. It is clear that iron magnetic moments with negative c_1 (large B_{hf}) are oriented parallel to the magnetization and those with positive c_1 (small B_{hf}) are oriented antiparallel.

The results presented in Figure 2a,b can be summarised as follows. Distribution $p(B_{hf})$ has a large value in the vicinity of $B_{hf} = 0$ and a tail ending at $B_{hf} \approx 30$ T. This result is in full agreement with earlier Mössbauer studies performed with unpolarized radiation [7].

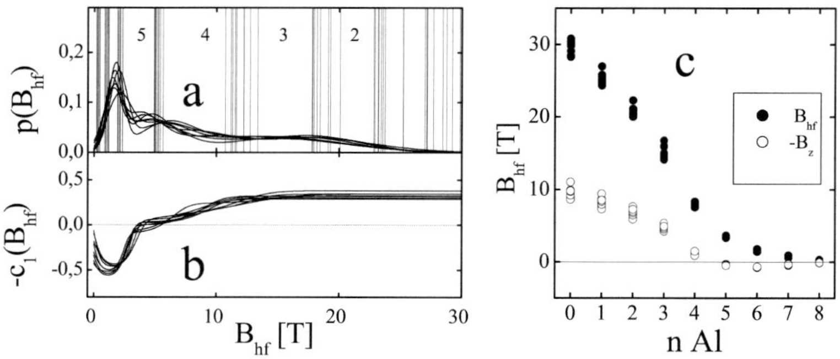

Figure 2. (a) Distribution of $p(B_{hf})$ and (b) dependence of c_1 on B_{hf} obtained from the data presented in Figure 1. *Vertical lines* in (a) show decomposition of the $p(B_{hf})$ distribution into sectors with area proportional to the probability given by binomial distribution $p(N,n,x)$. Numbers shown in rows correspond to n (number of Al atoms in the nearest coordination shell). (c) Average hyperfine field and average z-component of the hyperfine field related to sectors from (a) and (b). Note that there is a minus sign before c_1 in the vertical axis in (b) and before B_z in (c).

The most important result concerns the $c_1(B_{hf})$ dependence. The iron atoms with B_{hf} fields smaller than about 3.4 T develop a pronounced positive B_z component (Figure 2b). Next we observe that in the range of $B_{hf} = 5 \div 10$ T the c_1 parameter decreases and crosses zero. In that region the z-component of B_{hf} has a positive as well as a negative contribution to the hyperfine field, resulting in zero average value. The z-component saturates at B_{hf} larger than 15 T.

5. Discussion

The hyperfine magnetic field brings indirect information about the Fe magnetic moment. The positive values of c_1 seen in Figure 2b at $B_{hf} \approx 2$ T, could be interpreted assuming an antiparallel orientation of some fraction of the iron atoms, but could also be attributed to atoms with zero magnetic moment. To check what is the case we have to analyse the room temperature measurements, which are carried out in the paramagnetic state. They show a peak in the $c_1(B_{hf})$ parameter (not shown), the shape of which is similar to the one observed at T = 13 K. If the peak at $B_{hf} = 1 \div 2$ T in Figure 2b corresponded to iron atoms loosely bound to the rest of magnetic system, as is the case of a paramagnetic state, then one should expect to see the narrow peak precisely at 1.1 T. The most likely reason for seeing the broad peak at about 3 T is presence of small quadrupole interactions and a distribution of the isomer shift arising from variety of chemical environments [3]. Experimental confirmation of this interpretation is formed in the value of FWHM of the room temperature spectrum in zero applied field, equal to 0.62 ± 0.01 mm/s, see Figure 1b., while the FWHM for calibration spectra is 0.24 ± 0.01 mm/s only. We also see a clear asymmetry of the spectra measured at field $B_{ext} = 1.3$ T (Figure 1e). The asymmetry results from mixed magnetic dipole and electric quadrupole interactions. To obtain a correct description of the spectra one has to use a full Hamiltonian treatment, which is beyond the scope of this paper. Small value of the parameter c_1 in the range of B_{hf} in range of 5 ÷ 10 T clearly shows either noncollinearity or antiparallel orientation of a fraction of Fe magnetic moments.

It is reasonable to assume that (1) the smaller B_{hf} values are observed, when there is a larger number n of Al which surround Fe atom, and that (2) the distribution of Fe and Al atoms is random. These two assumptions allow to divide the $p(B_{hf})$ distribution into a few sections, each one having the area proportional to the probability $p(N,n,x)$ given by binomial distribution:

$$p(N,n,x) = \binom{N}{n}(1-x)^n x^{N-n}. \tag{2}$$

Here $x = 0.48$, $N = 8$ and $n = 0 \ldots N$. The sectioning is displayed in Figure 2a. For every section one can find an average $\overline{B}_{hf}(n)$. Similarly, using the same sectioning one can find from Figure 2b average z-component, which is displayed in Figure 2c. It is remarkable that the value $\overline{B}_{hf}(n)$ for $n = 0$ (Figure 2c), determined from the high field tail of the distribution (Figure 2a), coincides with the value for Fe surrounded by 0 Al in ordered Fe_3Al at low temperatures 31.8 T [11].

It follows from Figure 2b that $c_1(B_{hf})$ is constant for $B_{hf} > 15$ T, which strongly suggests (see Figure 2c) that Fe atoms surrounded by $n < 4$ Al exhibit the same type of magnetic structure. A reasonable assumption is that these moments are locally collinear.

Substantial changes are observed when n varies from $n = 3$ to $n = 4$. The hyperfine field is reduced to about 8 T, the value much smaller than 24.2 T found in ordered Fe_3Al [11]. Moreover, the c_1 (or B_z in Figure 2c for $n = 4$) is close to zero indicating that Fe moments become disordered. Fe atoms with $n = 5$ Al neighbours have small but nonzero total magnetic moments and the c_1 parameters indicate that, on the average, these moments are oriented antiparallel, see Figure 2b,c for $n = 5$.

When the number of Al nearest neighbors exceed $n = 5$ the Fe total magnetic moments are further reduced, see full points in Figure 2c. However, $n > 5$ is located in the region of B_{hf} equal or smaller than applied field 1.1T (see Figure 2a). Because the value of the applied field is comparable to the hyperfine field itself, and, as was already discussed, the quadrupole splitting and isomer shift interaction influence small magnetic interaction, information about the values and arrangements of these magnetic moments are not available at this stage of analysis.

In the presented analysis, random Fe distribution was assumed. If, however, short range order is present, sectioning shown in Figure 2a may not be valid. If the probabilities of the Al-rich environments are larger than given by 2, vertical lines with small n will shifts towards higher B_{hf} values.

In summary, we present results of investigations of an magnetic moment arrangements in disordered $Fe_{0.48}Al_{0.52}$ by Mössbauer polarimetric technique.

We have demonstrated that an increase of the number of Al in the first coordination shell causes a decrease of magnetic moments and introduces noncollinearity or antiparallel orientation when B_{hf} is in the range of $5 \div 10$ T. These noncollinear moments have positive as well as negative contributions to the net magnetization. The noncollinearity, present at low temperatures, explains the large coercivity, the the lack of saturation and reduction of the ratio of magnetization and average hyperfine field.

References

1. Arrott A. and Sato H., *Phys. Rev.* **114** (1959), 1420.
2. Sato H. and Arrott A., *Phys. Rev.* **114** (1959), 1427.
3. Bogner J. *et al.*, *Phys. Rev. B* **58** (1998), 149922.
4. Jo T. and Akai H., *J. Phys. Soc. Jpn.* **50** (1981), 70.
5. Jo T., *J. Phys. Soc. Jpn.* **40** (1976), 715.
6. Szymański K., Dobrzyński L., Prus B. and Cooper M. J., *Nucl. Instrum. Methods B* **119** (1996), 438.
7. Yelsukov E. P., Voronina E. V. and Barinov V. A., *JMMM* **115** (1992), 271.
8. Warren B. E. and Averbach J., *J. Appl. Phys.* **21** (1950), 595.
9. Szymański K., *Nucl. Instrum. Methods B* **134** (1998), 405;
   Szymański K., Satuła D. and Dobrzyński L., *Hyperfine Interact.* **156/157** (2004), 21.
10. Hanna S. S. *et al.*, *Phys. Rev. Lett.* **4** (1960), 513.
11. Johnson C. E., Ridout M. S. and Cranshaw T. E., *Proc. Phys. Soc* **81** (1963), 1079.

Hyperfine Interactions (2004) 159:81–86
DOI 10.1007/s10751-005-9083-6

Mössbauer Investigation of Highly Dispersed Iron Particles in Crazed Porous Polymers

E. S. TROFIMCHUK*, N. I. NIKONOROVA, S. K. DEDUSHENKO and Y. D. PERFILIEV
Department of Chemistry, Moscow State University, Lenin Hills, Moscow, 119992, Russia; e-mail: elena_trofimchuk@mail.ru

Abstract. Formation and stability of highly dispersed iron particles in crazed porous polymer matrices were studied. The iron–polymer composites obtained were characterized by different morphologies and dimensions of iron particles. The phase content of the iron constituent in a composite studied by Mössbauer spectroscopy was shown to depend on the type of the iron salt and the method of introduction of the initial reagents into a polymer.

Key Words: crazing, iron–polymer composites, Mössbauer spectroscopy.

1. Introduction

Materials containing iron nanoparticles are of interest for high-density magnetic recording. In view of that, the preparation and stabilization of iron particles with controlled dimensions is an important problem. One way of nanoparticle stabilization is the synthesis of iron particles inside polymer matrices. There are several approaches for the synthesis of iron particles in a polymer, e.g. thermo-chemical decomposition of iron carbonyl in polymer melts [1], electrochemical reduction inside solvent swollen polymer matrix [2], and ion implantation [3]. The main disadvantages of the known approaches are oxidation and destruction of the polymer matrix under the influence of temperature or of the Fe^+ ion beam [1, 3], a relatively high cost of polymers which are capable of swelling [2], and poor mechanical properties of such composites.

In this paper, an alternative approach to the preparation of iron–polymer composites based on commercial polymers is proposed. Here, the reduction of iron ions is carried out *in situ* within nanometre-sized pores of a polymer obtained via solvent crazing [4]. Solvent crazing is the transition of glassy and semicrystalline polymers in a highly dispersed and oriented state upon uniaxial stretching in the presence of adsorptionally active media. Earlier, highly dispersed particles of nickel, copper and silver were introduced into porous

* Author for correspondence.

polyethylene, isotactic polypropylene, poly (ethylene terephthalate), and poly (vinyl chloride) films using solvent crazing. It is important that the reduction of metal particles in crazed polymer was carried out without destruction of the polymer film. Those metal–polymer composites were shown in [5, 6] to be characterized by a high level of mutual dispersion of thermodynamically incompatible components and by good mechanical properties.

The objective of this work was to investigate the formation and stability of iron particles inside crazed porous films of high-density polyethylene by Mössbauer spectroscopy, X-ray diffraction and scanning electron microscopy (SEM).

2. Experimental

Commercial films of isotropic high-density polyethylene (HDPE) ($\overline{M}_w = 200000$, thickness 75 μm) were used as initial materials. To obtain porous samples, the films of HDPE were stretched up to 200% in the presence of isopropanol with a rate of 5 mm/min using hand-operated clamps. In this case craze nucleation and growth proceed via the mechanism of delocalized solvent crazing [7], i.e. throughout the bulk of the sample. The as-received porous HDPE matrices were characterized by their effective volume porosity, which was estimated by the changes in geometrical size of the polymer under crazing. Its value was ~40 vol.%. The average diameter of the polymer pores evaluated using the method of pressure-driven liquid permeability was 6 nm.

We studied the development of iron particles as a result of chemical reduction of ammonium iron(II) sulfate hexahydrate, $(NH_4)_2Fe(SO_4)_2 \cdot 6H_2O$, and iron(III) chloride hexahydrate, $FeCl_3 \cdot 6H_2O$, using sodium boron hydride, $NaBH_4$, as a reducing agent. The concentration of water–alcohol solutions of the initial reagents was 0.1 mol/l (the content of alcohol was 20 vol.%).

Low-molecular-weight substances were introduced into polymer matrices using two approaches:

I First, a porous structure of the polymer was obtained by deforming the polymer film in iron-free isopropanol. Then, streams of the metal salt and reducing agent proceeded to the polymer using counter-current diffusion. In this case, driving force of the reagent diffusion to the reaction front was provided by precipitation and removal of the as-formed substance from the reaction sphere. The reduction time was 1 h. After reduction the samples were washed in distilled water and dried in the isometric state at 100°C during 2 h in the atmosphere of argon.

II Pores were created and filled simultaneously during the deformation of the polymer film in the water–isopropanol solution of an iron salt. Then, the filled polymer film was stored in the solution of $NaBH_4$ to reduce iron.

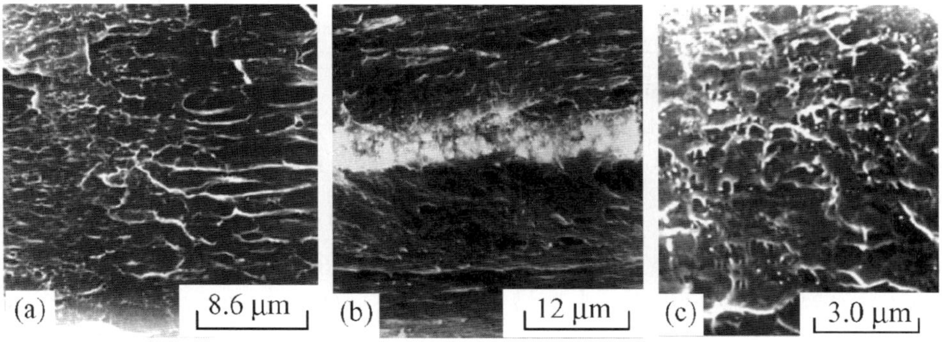

Figure 1. SEM-micrographs of (a) the initial porous HDPE matrix and iron-containing composites obtained by (b) method **I** and (c) method **II**.

The content of the iron introduced into the film was determined gravimetrically. In this work, composites containing up to 40 wt.% of iron were obtained.

The phase composition of the iron–polymer samples was studied by absorption Mössbauer spectroscopy. Mössbauer spectra were measured with a Perseus spectrometer working at constant velocities; the control and adjustment of the transducer velocity were performed by a laser interferometer. A standard γ-source of ^{57}Co in the matrix of metallic rhodium with the activity of 0.6 GBk (a product of Cyclotron Co., Ltd., Obninsk, Russia) was employed. In this work, isomer shift values are related to α-Fe at room temperature.

X-ray diffraction analysis of samples was performed with a DRON-3M diffractometer using MoK_{α} radiation (Zr filter). The average crystallite size was estimated using Scherrer's formula.

The morphology of composites obtained was examined with a Hitachi S-520 scanning electron microscope. Samples were prepared using the procedure of brittle fracture in liquid nitrogen and plating with a platinum–palladium alloy.

3. Results and discussion

After reduction, the iron–polymer composite films become dark-gray. X-ray investigation of samples obtained by method **I** shows that they contain metallic iron. Peaks of α-Fe with the lattice distance of d_{110} = 2.02(1) Å and d_{200} = 1.43(1) Å (2.0268 and 1.4332 Å [8]) are observed in their diffraction patterns. The average diameter of iron crystallites is 40(5) nm for different preparation conditions. Samples obtained by method b do not show any peaks in X-ray patterns.

SEM micrographs of the initial HDPE crazed film (a) and iron–HDPE composite films obtained via methods **I** (b) and **II** (c) are presented in Figure 1. It can be seen that the structure of the initial crazed HDPE film represents a system of co-penetrating pores. In the composite obtained via method **II**, the spherical particles with the dimension of 0.1 μm are discretely and uniformly distributed within the porous polymer matrix. When the counter-current diffusion approach

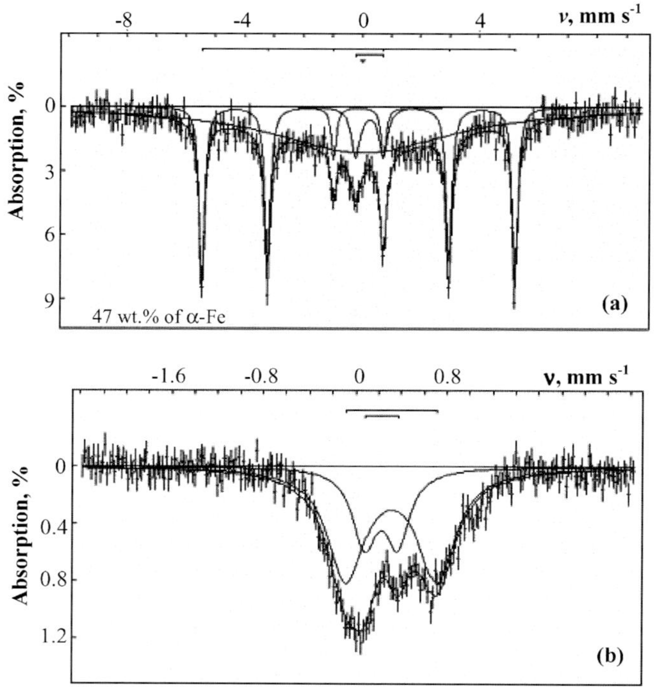

Figure 2. Mössbauer absorption spectra of the iron-containing HDPE composites obtained by (a) method **I** and (b) method **II**.

(method **I**) is used, spherical particles with a diameter of 0.2–0.4 μm form a layer with a thickness of 4–6 μm inside a polymer. It is important to stress that the structure of the original porous matrices is unchanged after the reduction of iron inside a polymer.

It is known [9] that chemical reduction of metals from their salts by $NaBH_4$ gives a mixture of products. However, X-ray diffraction and microscopic investigations of metal–polymer composites do not allow us to determine amorphous admixtures that are produced during the reduction process within a polymer film. For example, reduction of iron by $NaBH_4$ can be described by the following reactions:

$$5(NH_4)_2Fe(SO_4)_2 + 10NaBH_4 = 5Fe + 10B + 20H_2 + 5(NH_4)_2SO_4 + 5Na_2SO_4, \quad (1)$$

$$2FeCl_3 + 6NaBH_4 = 2Fe + 6B + 12H_2 + 6NaCl. \quad (2)$$

From the literature data [9], 3.5 to 10 wt.% of boron is obtained in these processes. In reaction (2), 5 to 10 wt.% of iron(III) hydroxide is also formed owing to the high pH (pH = 11) of an aqueous $NaBH_4$ solution.

The composite film was found to change its color immediately from dark-gray to yellow-brown after the removal of the reducer excess, which is connected with the oxidation of highly dispersed iron particles in air and with the formation of different iron oxo-, hydroxo-derivatives. It was not possible to avoid iron oxidation even by drying and storing the iron–polymer composite films in the inert atmosphere of argon.

Mössbauer spectroscopy is a very informative technique for the analysis of iron–polymer samples. It allows one to confirm the presence of metallic iron in a polymer and to estimate the amount of admixture of other iron-containing phases. Mössbauer spectra of the samples obtained are presented in Figure 2. For a sample received via method a (Figure 2(a)) we found a sextet of α-Fe and a doublet related to Fe(III) compounds. The content of α-Fe depends on the type of iron salt used and varies from 20 to 60 wt.%. The amount of Fe(III) compounds increases when we use $FeCl_3$. There is also an extended absorption background in the spectrum that is difficult to explain. Due to the layer distribution of highly dispersed iron particles within HDPE the chemical composition of samples remains constant even during storage of the composites in the presence of air for 3 months. Apparently, a thin oxide film that prevents penetration of oxygen molecules into the layer is obtained on the surface of the iron layer within a polymer.

Mössbauer absorption spectrum of the iron–polymer sample obtained by method **II** (Figure 2(b)) differs completely from that of the sample obtained by method **I**. This spectrum can be presented as a superposition of two doublets with isomer shifts $\delta_1 = 0.33(1)\ \mathrm{mm \cdot s^{-1}}$ and $\delta_2 = 0.24(1)\ \mathrm{mm \cdot s^{-1}}$ and quadrupole splitting values $\Delta_1 = 0.79(1)\ \mathrm{mm \cdot s^{-1}}$ and $\Delta_2 = 0.29(1)\ \mathrm{mm \cdot s^{-1}}$. These Mössbauer parameters correspond to oxo-compounds of Fe(III) with octahedral and tetrahedral coordination of the metal, respectively [10]. There are no α-Fe lines in the spectrum that can be explained by rapid oxidation of highly dispersed iron particles, discretely distributed within the porous HDPE matrix, by water and alcohol (Figure 1(b)).

Thus, iron nanoparticles are formed in pores of a crazed polymer during reduction of Fe^{2+} or Fe^{3+} ions by sodium boron hydride. The nanoparticles are extremely reactive, so that their oxidation by solvents or oxygen starts immediately after their formation. Aggregation of metallic particles decreases the oxidation rate. In addition, these aggregates are covered by a layer of oxidation products protecting the metallic nuclei from further oxidation. The aggregates are well visible in micrographs and have bigger sizes as compared to the size of iron crystallites. In the case of a low concentration of iron nanoparticles in a polymer (if method **II** is applied), aggregation is not possible and each nanoparticle is oxidized completely. Micrographs show big particles (see Figure 1(c)) consisting of insoluble iron oxidation products.

We were unable to stabilize iron nanoparticle with super-paramagnetic properties–stable particles were too big. Nonetheless, crazed porous polymers seem to be promising as templates for synthesis and stabilization of iron nanoparticles and preparation of composite materials with different distributions of metallic particles inside a polymer. Particle sizes can be diminished by introducing surfactants and other additives during the synthesis, which would cover iron particles immediately after their formation and protect them from oxidation.

Acknowledgements

This work was financially supported by the Russian Foundation for Basic Research (project # 04-03-42744).

References

1. Kozinkin A. V., Vlasenko V. G., Gubin S. P., Shuvaev A. T. and Dubovtsev I. A., *Inorg. Mater.* **32** (1996), 376.
2. Yoon M., Kim Y. M., Kim Y., Volkov V., Song H. J., Park Y. J., Vasilyak S. L. and Park I.-W., *J. Magn. Magn. Mater.* **265** (2003), 357.
3. Petukhov V. Y., Ibragimova M. I., Khabibullina N. R., Shulyndin S. V., Osin Y. N., Zheglov E. P., Vakhonina T. A. and Khaibullin I. B., *Polym. Sci., Ser. A* **43** (2001), 1154.
4. Bakeev N. F. and Volynskii A. L., *Solvent Crazing of Polymers*, Elsevier, Amsterdam, 1995.
5. Stakhanova S. V., Trofimchuk E. S., Nikonorova N. I., Rebrov A. V., Ozerin A. N., Volynskii A. L. and Bakeev N. F., *Polym. Sci. Ser. A* **39** (1997), 229.
6. Volynskii A. L., Arzhakova O. V., Yarysheva L. M. and Bakeev N. F., *Polym. Sci. Ser. C.* **44** (2002), 83.
7. Volynskii A. L., Arzhakova O. V., Yarysheva L. M. and Bakeev N. F., *Polym. Sci. Ser. B* **42** (2000), 70.
8. ICDD-PDF-Database, 2004.
9. Mal'zeva N. N. and Khain V. S., *Sodium Boron Hydride*, Nauka, Moscow 1985 (in Russian).
10. Menil F., *J. Phys. Chem. Solids* **46** (1985), 763.

Hyperfine Interactions (2004) 159:87–93
DOI 10.1007/s10751-005-9086-3

NMR Studies on the Internal Structure of High-T_c Superconductors and Other Anorganic Compounds

K. KUMAGAI[1,*], K. KAKUYANAGI[1], M. SAITOH[1], Y. MATSUDA[2], M. HASEGAWA[3], S. TAKASHIMA[4], M. NOHARA[4] and H. TAKAGI[4]
[1]*Division of Physics, Graduate School of Science, Hokkaido University, Sapporo 060-0810, Japan; e-mail: kumagai@phys.sci.hokudai.ac.jp*
[2]*Institute for Solid State Physics, University of Tokyo, Kashiwa, Chiba 277-8581, Japan*
[3]*Institute of Material Research, Tohoku University, Katahira, Sendai 980-8577, Japan*
[4]*Department of Advanced Materials Science, University of Tokyo, Kashiwa, Chiba 277-8581, Japan*

Abstract. Spatially-resolved NMR is used to probe internal structures in highly correlated superconductors of optimally-doped $Tl_2Ba_2CuO_{6+\delta}$ ($T_c = 85$ K) and a heavy fermion superconductor $CeCoIn_5$ ($T_c = 2.3$ K). The characteristic change of the properties of ^{205}Tl-NMR in the vortex state provides a clear evidence of the antiferromagnetic order in the vortex cores below 20 K in $Tl_2Ba_2CuO_{6+\delta}$. We also obtain anomalous ^{115}In-NMR spectra of $CeCoIn_5$, which provides a microscopic evidence for the occurrence of a spatially-modulated superconducting order parameter expected in a Fulde–Ferrel–Larkin–Ovchinnkov (FFLO) state.

Key Words: antiferromagnetic order, FFLO state, NMR, superconductivity, vortex core.

1. Introduction

The relation between superconductivity (SC) and magnetism has been recognized as a central issue in the physics of highly correlated materials such as high-T_c cuprates (HTSC) and heavy fermion compounds. The microscopic structure of the vortex state in those substances comes out to be a very interesting subject [1]. Recently, the coexistence or competition of SC and magnetic ground states have been intensively studied [2, 3].

Very recent heat capacity measurements of heavy fermion compound $CeCoIn_5$ revealed that a second order phase transition takes place at $T^{\#}(H)$ within the SC state in the vicinity of the upper critical field at low temperatures [4, 5]. The transition line branches from $H_{c2}^{\parallel}$- line and decreases with decreasing T, indicating the presence of a novel SC phase. While recent experimental results make the Fulde–Ferrel–Larkin–Ovchinikov (FFLO) [6, 7] scenario a very

* Author for correspondence.

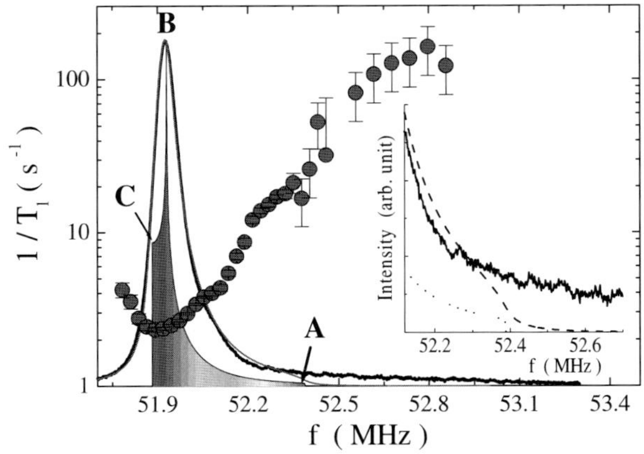

Figure 1. ^{205}Tl-NMR spectrum (*solid line*) at 5 K. The intensity is plotted in a linear scale. The thin solid line depict the histogram at particular local fields of the Readfield pattern. The dotted line represents the spectrum convoluted with Lorentzian broadening function, $f(H_{\rm loc}) = \sigma/(4H_{\rm loc}^2 + \sigma^2)$ using $\sigma = 42$ kHz. The *filled circles* show the frequency dependence of $^{205}T_1^{-1}$ at the T_1 site. A, B and C represent the position of center of vortex core, saddle point and center of vortex lattice for the field distribution in the square vortex lattice, respectively. The inset shows the detail of the spectrum around the vortex core in the enlarged scale.

appealing one for $CeCoIn_5$, there is no direct experimental evidence so far which verifies the spatially nonuniform SC state expected in the FFLO state.

As NMR is particularly suitable for the above purpose because NMR can monitor the local information about low energy quasiparticle excitations and antiferromagnetic fluctuations sensitively. In this paper, we report new findings associated with the microscopic vortex structure, namely, antiferromagnetic vortex core in an optimally-doped HTSC, $Tl_2Ba_2CuO_{6+\delta}$, [8] and the spatially modulated superconducting gap (FFLO) state in a strong magnetic field exceed a Pauli limit in $CeCoIn_5$ [9].

2. Experimental

NMR measurements were carried out on the c-axis oriented polycrystalline powder of $Tl_2Ba_2CuO_{6+\delta}$ ($T_c = 85$ K) and also on a single crystal of $CeCoIn_5$ in the magnetic field H parallel to the [100]-direction. The nuclear spin-lattice relaxation time of ^{205}Tl in $Tl_2Ba_2CuO_{6+\delta}$ and the Knight shift of ^{115}In-NMR at the In(1) site with axial symmetry were intensively investigated.

3. Results and discussion

A clear asymmetric pattern of the ^{205}Tl-NMR spectrum, which originates from the local field distribution associated with the vortex lattice (the Redfield pattern), is observed below the vortex lattice melting temperature (~60 K at

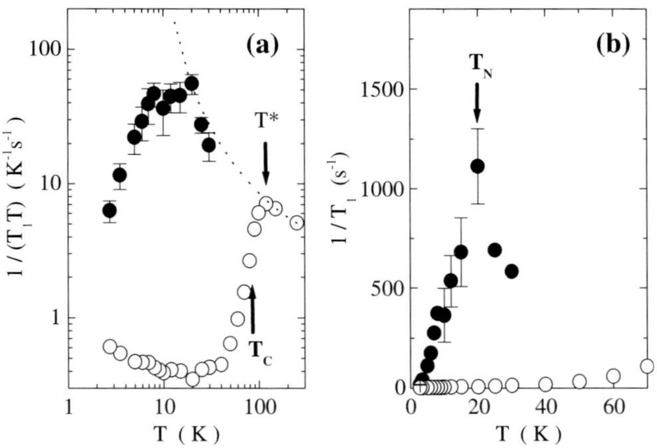

Figure 2. T-dependence of $(^{205}T_1\,T)^{-1}$ (a) and of $^{205}T_1^{-1}$ (b) at low temperatures. The *filled and open circles* represent the data at vortex cores and at the saddle point, respectively. $T^* \backsimeq 120$ K is the pseudogap temperature. The *dotted line* represents the Curie–Weiss law which is determined above T^*. In (b), T_N is the temperature at which $^{205}T_1^{-1}$ at the core exhibits a peak.

$H = 2.1$ T). The solid line in Figure 1 depicts the NMR spectra at $T = 5$ K and $H = 2.1$ T. The local field profile in the vortex state shown in the inset of Figure 1 is given by approximating $H_{\rm loc}(\boldsymbol{r})$ with the London result for square vortex lattice. The thin solid line in Figure 1 depicts the histogram at a particular local field which is given by the local field distribution $f(H_{\rm loc}) = \int \delta[H_{\rm loc}(\boldsymbol{r}) - H_{\rm loc}]d^2\boldsymbol{r}$. In the calculation we used $\xi_{ab} = 18$ Å and $\lambda_{ab} = 1700$ Å, and assumed the square vortex lattice. The theoretical curve reproduces the data well above $T = 20$ K. On the other hand, the spectrum at $T = 5$ K shows significant broadening at high frequency region (core region), while it can be well fitted below 52.1 MHz as seen in the inset of Figure 1. The line broadening occurs only in the high frequency core region. The observed broadening below ~20 K is explained by the appearance of static magnetism within the vortex cores below ~20 K.

The filled circles in Figure 1 show the frequency dependence of $^{205}T_1^{-1}$. On scanning from outside into cores, $^{205}T_1^{-1}$ increases rapidly after showing a minimum near saddle points. The magnitude of $^{205}T_1^{-1}$ in the core region is almost two orders of magnitude larger than that near the saddle point. The remarkable enhancement of $^{205}T_1^{-1}$ provides a direct evidence that the AF correlation is strongly enhanced near the vortex core region. Figures 2(a) and (b) depict the T-dependences of $(^{205}T_1T)^{-1}$ and $^{205}T_1^{-1}$ within cores (*filled circles*) and at the saddle points (*open circles*), respectively. From high temperatures down to about 120 K, $(^{205}T_1T)^{-1}$ obeys the Curie–Weiss law, $(^{205}T_1T)^{-1} \propto 1/(T + \theta)$. The lowest T at which this law holds is conveniently called the pseudogap temperature T^*. Below T^*, $(^{205}T_1T)^{-1}$ decreases rapidly without showing any anomaly associated with the superconducting transition at T_c, similar to other HTC. The important signature is that the T-dependences of T_1^{-1}

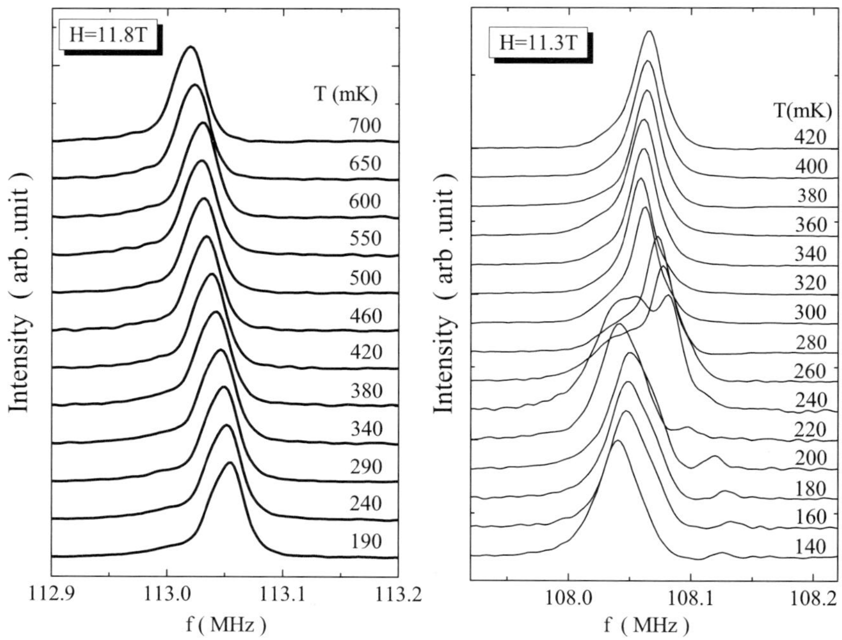

Figure 3. ^{115}In-NMR spectra (central line of ($\pm 1/2 \leftrightarrow \mp 1/2$ transition) in $CeCoIn_5$ as a function of frequency for various temperatures at $H = 11.3$ T.

in the vortex core have a peak around $T = 20$ K, and below $T \sim 20$ K, $(^{205}T_1{}^{\rm core})^{-1}$ decreases rapidly with decreasing T as seen in Figure 2(b).

On basis of the results of the broadening of the NMR spectra below 20 K and the T-dependence of $1/T_1$, we are lead to conclude that the local AF ordering takes place in the core region at T_N=20 K; T_N corresponds to the Néel temperature within the core [8]. The experimental feature is well explained with the model of the AF vortex core. The appearance of the AF ordering is consistent with the prediction of recent theories based on the SO(5) and $t - J$ models [2, 3].

Figure 3 shows ^{115}In-NMR spectra of $CeCoIn_5$ at various temperatures down to 140 mK at $H = 11.8$ T (in the normal state) and $H = 11.3$ T (across the normal to the SC phase at $T \sim 700$ mK and also the second phase transition at $T \sim 300$ mK). The spectrum with single peak is observed down to $T^{\#}$(=300 mK) and the peak position is slightly decrease with decreasing temperature. Then, the spectrum shows complex feature just across the boundary of the high field SC phase at $T^{\#}$. Below $T^{\#}$, the spectrum shows double peaks. The separation of the upper and lower peak position increases with decreasing temperature. The relative ratio of the intensities at each peak changes drastically.

The Knight shift at $H = 11.3$ T shows quite unusual temperature dependence as seen in Figure 4. Here, the Knight shift is evaluated from the peak position by taking into account the shift due to large electric quadrupole interaction with using a second order perturbation calculation for parameters, $\nu_Q = 8.173$ MHz

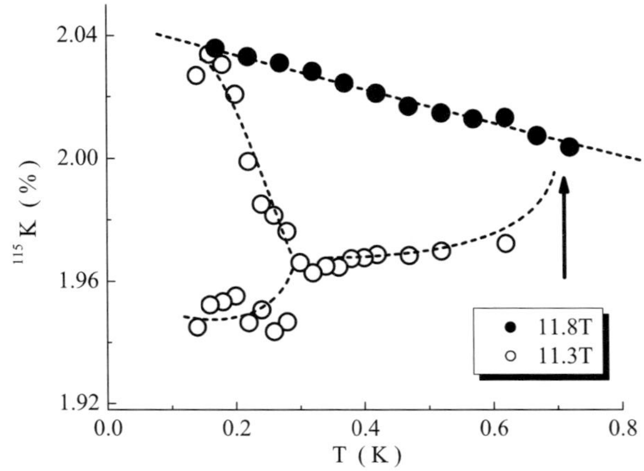

Figure 4. Temperature dependence of Knight shift ^{115}K at H = 11.8 T (•) and 11.3 T (○) in $CeCoIn_5$. The *arrow* shows the SC transition temperature at corresponding field. The *broken lines* are guide for eyes.

and $\eta = 0$ for the ^{115}In(1) site (nuclear spin: $I = 9/2$). Just across the high field SC phase at $T^{\#}$, the Knight shift of the upper one increases largely with lowing temperature. The Knight shift of the upper peak is coincident with the values in the normal state at the lowest temperature.

In the SC state, the penetration depth for NMR rf excitation field, H_1, is reduced largely. In the FFLO state, the SC order parameter is nearly sinusoidal along the vortex direction (parallel to H_0). With the nodal sheet structure perpendicular to H_0 (along the a-axis), the penetration depth of H_1 is expected to be modulated. For following discussions about a simulation of the spectrum, we use a simple sinusoidal relation of the FFLO SC gap with the relation of

$$\Delta(x) = \Delta_0 \sin Qx.$$

Here, a modulation length of the SC order parameter $\Lambda = 2\pi/Q$ [10]. From the London equation, the spatial distribution of H_{rf} in the FFLO state is given by

$$H_1(x) \propto \frac{\sinh\frac{\Lambda/2-x}{\lambda} + \sinh\frac{x}{\lambda}}{\sinh\frac{\Lambda}{2\lambda}}$$

at the boundary condition that H_1 is equal to that of the normal state at $x = 0$ and $x = \Lambda/2$. With using the spatial distribution of H_1 amplitude, the NMR spectrum is given by

$$I(k) = \int_0^{\Lambda/2} \delta\left(K\left(\frac{\Delta(x)}{T}\right) - k\right)[H_1(x)]^3 \mathrm{d}x$$

Here, $K\left(\frac{\Delta(x)}{T}\right)$ is the Yoshida function for the Knight shift in the SC state. Here, adjustable parameters are only two, $\frac{\Lambda}{\lambda}$ and $\frac{T}{\Delta_0}$ [9].

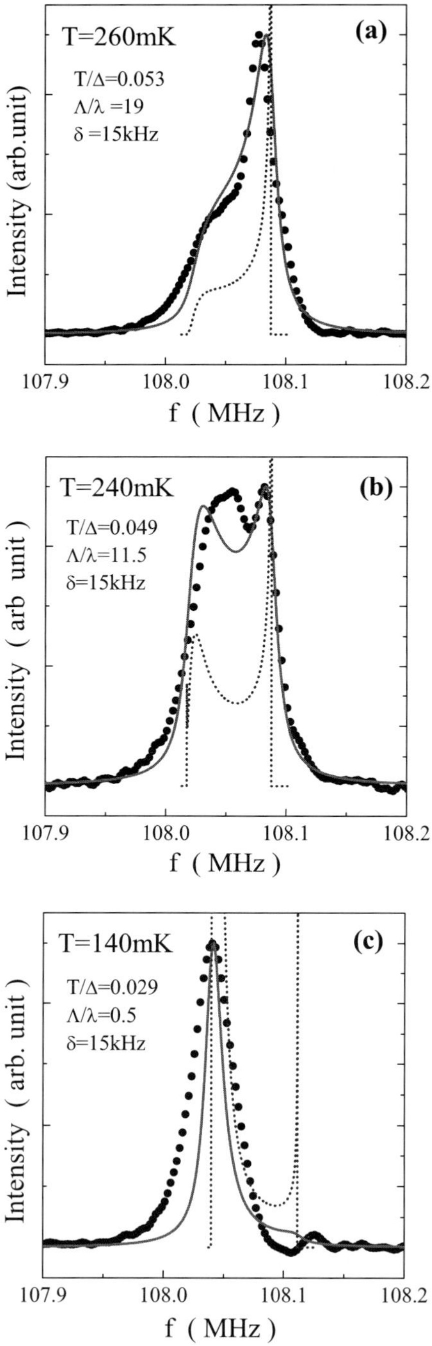

Figure 5. ^{115}In-NMR spectra (•) at $T = 260$ mK, $T = 240$ mK and $T = 140$ mK. The *solid line* and the *dotted line* represent the simulation spectra with and without a convolution of an inhomogeneous broadening with a Lorentian function (σ = 15 kHz), respectively.

Our simulations seem to interpret very well the split spectrum with reasonable change of the Knight shift and the intensity as shown in Figure 5. The important result seen from the simulation analysis is that the wavelength of the spatial oscillation of the SC order parameter in the FFLO state decreases with lowering temperature. Near the transition boundary between non-modulated SC and the FFLO phase at $T^{\#}$ = 300 mK, Λ is 50 times larger than the penetration length, $\lambda \sim 2000$ Å just below $T^{\#}$. At temperature where Λ/λ is comparable with $10 \sim 1$, the NMR line shape changes drastically. At low temperature where Λ is smaller than λ, the NMR intensity from the SC region becomes dominant and is temperature-independent. The simulation in good agreement with the experimental spectra indicates the existence of the spatially modulated SC order gap as expected in the FFLO phase.

In summary, from the spatially-resolved ^{205}Tl-NMR in $Tl_2Ba_2CuO_{6+\delta}$ and ^{115}In(1)-NMR in $CeCoIn_5$, we have obtained novel features for the internal structure of superconductors associated with the vortex state. In the vortex core region in $Tl_2Ba_2CuO_{6+\delta}$, the AF spin correlations are extremely enhanced, and that the paramagnetic–AF order transition of the Cu moments takes place at $T_N \sim$ 20 K. The anomalous change of the ^{115}In(1)-NMR spectra in $CeCoIn_5$ are well characterized by taking into account a spatial modulation of the SC gap and the penetration depth of rf field (H_1). The present study provides the strong evidence from a microscopic point of view that the high field phase within the SC state in $CeCoIn_5$ is the FFLO phase.

References

1. Kivelson S. A. *et al.*, *Rev. Mod. Phys.* **75** (2003), 1201.
2. Zhang S. C., *Science* **275** (1997), 1089.
3. Himeda A. *et al.*, *J. Phys. Soc. Jpn.* **66** (1997), 3367;
   Ogata M., *Int. J. Mod. Phys.* **B13** (1999), 3560.
4. Radovan H. A. *et al.*, *Nature* **425** (2003), 51.
5. Bianchi A. *et al.*, *Phys. Rev. Lett.* **91** (2003), 187004.
6. Fulde P. and Ferrel R. A., *Phys. Rev. A* **135** (1964), 550.
7. Larkin A. I. and Ovchinnikov Y. N., *Sov. Phys. JETP* **20** (1965), 762.
8. Kakuyanagi K., Kumagai K., Matsuda Y. and Hasegawa M., *Phys. Rev. Lett.* **90** (2002), 197003.
9. Kakuyanagi K. *et al.*, *Phys. Rev. Lett.* **94** (2005), 47602.
10. Tachiki M. *et al.*, *Z. Phys.* **B100** (1996), 369.

Hyperfine Interactions (2004) 159:95–102
DOI 10.1007/s10751-005-9198-9

Structure and Ionic Conductivity of Halocomplexes of Main Group Metallic Elements Studied by NMR and NQR

TSUTOMU OKUDA and KOJI YAMADA*
Graduate School of Science, Hiroshima University, Kagamiyama 1-3, 739-8526 Higashi-Hiroshima, Japan; e-mail: kyamada@sci.hiroshima-u.ac.jp

Abstract. Li_3InBr_6 undergoes phase transition to a lithium superionic conductor at $T_{tr} = 314$ K ($\sigma = 5.0 \times 10^{-4}$ S cm^{-1} at 330 K). The Rietveld analysis and the DSC measurement suggested that the positional disorder is introduced at the cationic sites above T_{tr}. The X-ray powder diffraction pattern at the superionic phase changes gradually with temperature and finally shows a simple powder pattern at 420 K which is quite similar to that of LiBr. This rock salt structure contains intrinsic vacancies because one In^{3+} and two vacancies substitute for three Li^+. ^7Li and ^{115}In NMR support the rapid diffusion of the Li^+ and the introduction of the In^{3+} into the rock salt structure. On the other hand, the ionic conductivity for Na_3InCl_6 was 10^{-5} S cm^{-1} even at 500 K. Conduction path for the sodium ions could be proposed by means of the Rietveld analysis and the NMR experiment using a single crystal.

Key Words: ^7Li NMR, ^{115}In NMR, conduction mechanism, quadrupole coupling constant, ternary halide, X-ray diffraction.

1. Introduction

In our previous papers, we have reported extremely high conductivity of Li_3InBr_6 at the high-temperature phase (hereafter abbreviated as HT phase) [1–3]. Its conductivity reaches to 5×10^{-4} S cm^{-1} just above the T_{tr} as shown in Figure 1. This conductivity is much higher than that reported for Li_3MX_6 (M = Lanthanoids, X = Cl and Br) [4, 5] and is comparable to that of B_2S_3–Li_2S glass [6]. In the course of the development new superionic conductor in these ternary halides, Na_3InCl_6 was synthesized and its structure and conductivity were investigated. As Figure 1 shows, however, its conductivity was much lower than that of Li_3InBr_6. In this report, we will discuss the difference between their

* Author for correspondence.

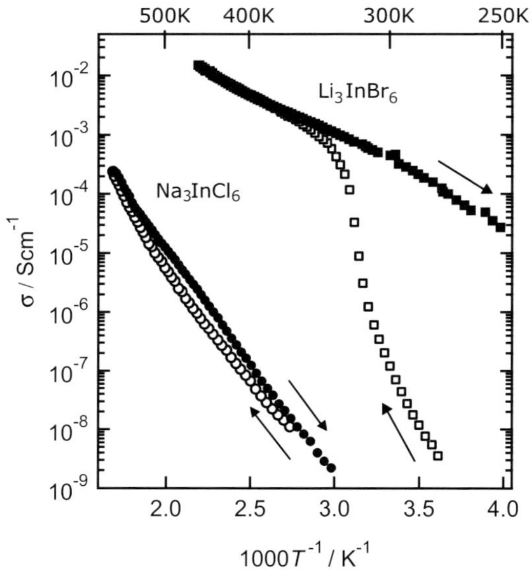

Figure 1. Ionic conductivity for Li_3InBr_6 and Na_3InCl_6 determined by the complex impedance method.

conduction mechanisms on the basis of their structures and dynamical information from the NMR experiments.

2. Experimental section

2.1. SAMPLE SYNTHESIS AND CHARACTERIZATION

Polycrystalline sample of Li_3InBr_6 was prepared by the solid-state reaction at 473 K between stoichiometric mixture of LiBr and $InBr_3$ [1, 2]. Single crystal of Na_3InCl_6 was synthesized from the melt using a Bridgman furnace. Both samples were characterized by the chemical analysis or by the powder X-ray diffraction.

2.2. X-RAY DIFFRACTION AND NMR

A single crystal of Na_3InCl_6 suitable for the X-ray diffraction was selected from the broken pieces of crystals. All measurements were performed at 200 K by a Mac Science DIP2030 imaging-plate diffractometer with a monochromatized Mo Kα radiation. Powder X-ray diffraction was observed using a Rigaku Rint PC diffractometer with a homemade variable temperature attachment. Rietveld refinements on these samples were performed using RIETAN [7]. ^{7}Li, ^{23}Na and ^{115}In NMR were observed by means of a homemade pulsed spectrometer at 6.4 T. A single pulse sequence followed by a Fourier transformation was applied for all measurements.

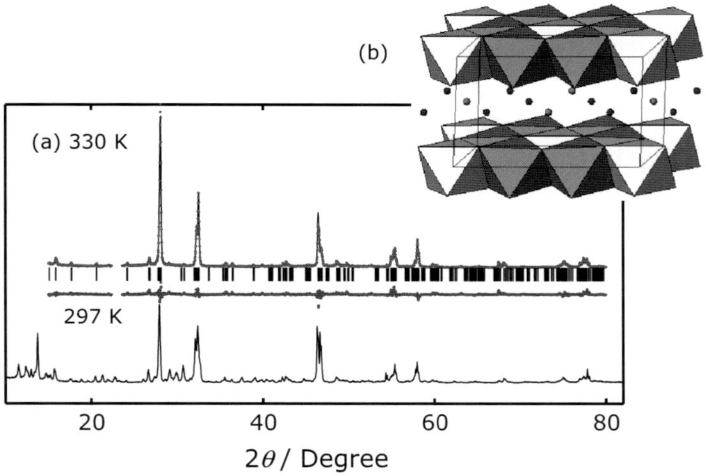

Figure 2. (a) X-ray powder diffraction pattern of Li_3InBr_6 at 297 and 330 K. (b) Structure model at 330 K.

3. Results and discussion

3.1. CRYSTAL STRUCTURE OF Li_3InBr_6 AT 330 K

Figure 2 shows the powder pattern followed by the Rietveld analysis just above T_{tr} together with that of at 297 K. The structure of Li_3InBr_6 at HT phase was isostructural with Li_3InCl_6 as shown in Figure 2(b) [8]. Two crystallographic different $InBr_6$ octahedra are partially occupied with In^{3+}. This structure suggests that the phase transition is accompanied by an order–disorder phenomenon at least at the In^{3+} site. Since we could not get enough structural information about the Li^+, we evaluated the transition entropy from the DSC measurement. The transition entropy was evaluated to be 15(1) J K^{-1} mol^{-1}. This value is larger than $R\ln(3)$ = 9.1 J K^{-1} mol^{-1} which is expected for the structural shown in Figure 2(b). If we assume a positional disorder at the Li^+ sites using all possible octahedral sites, the number of the Li^+ sites increases two times. The experimental transition entropy agrees well with this model where both cations are in the disordered state, $R\ln(3 \times 2)$ = 14.9 J K^{-1} mol^{-1}. This structure is essentially rock salt structure, except the ordering of the In^{3+}. With increasing the temperate, weak reflections disappeared gradually and a simple powder pattern which was quite similar to that of LiBr was appeared. Figure 3(a) shows a typical example observed at 420 K. The weak reflections denoted by arrows could be assigned to the superlattice reflections corresponding to a cubic lattice with $a = 2a_{LiBr}$ where a_{LiBr} is a lattice constant of LiBr. These superlattice reflections suggest some cationic order is kept up to 420 K. The powder patter including these superlattice reflections could be indexed as a trigonal. Figure 3(b) shows the proposed trigonal structure. It is interesting to note that these superlattice reflections tend to disappear when the sample was kept at 420 K. This finding

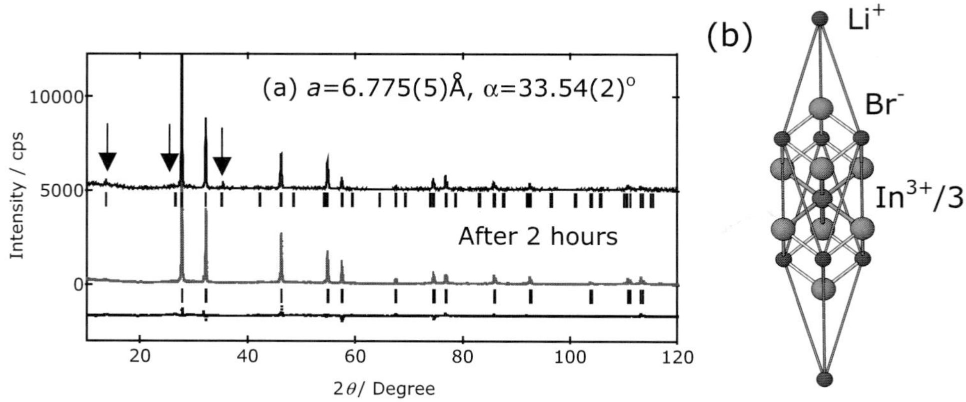

Figure 3. (a) X-ray powder diffraction pattern of Li_3InBr_6 at 420 K (*top*) and after 2 h (*bottom*). (b) Trigonal structural model at 420 K.

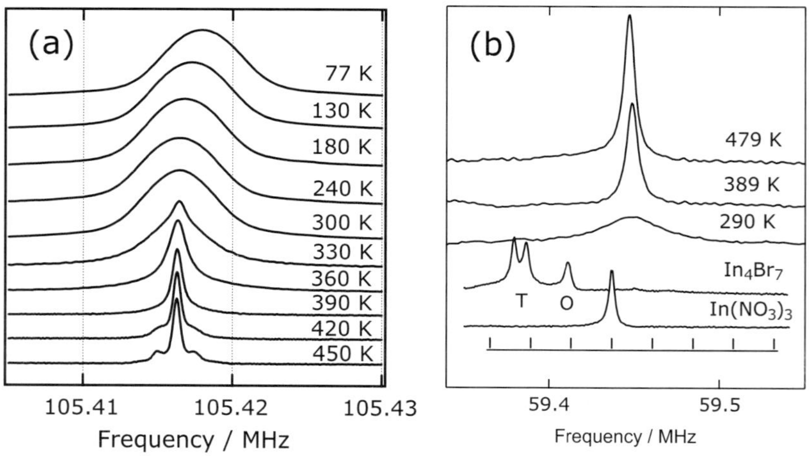

Figure 4. (a) ^{7}Li NMR as a function of temperature. (b) ^{115}In NMR for Li_3InBr_6 together with In_4Br_7.

suggests that the diffusion of In^{3+} as well as Li^{+} takes place throughout the crystal and results in an average structure having rock salt structure, in which 1/3 of the cationic sites are vacant.

3.2. ^{7}Li AND ^{115}In NMR ON Li_3InBr_6

Figure 4(a) and (b) show the ^{7}Li and ^{115}In NMR spectra. As expected from the high ionic conductivity, the motional narrowing was observed for ^{7}Li NMR above T_{tr}. It is interesting to note that the small splitting due to first order quadrupole effect was appeared above 420 K suggesting that the axially

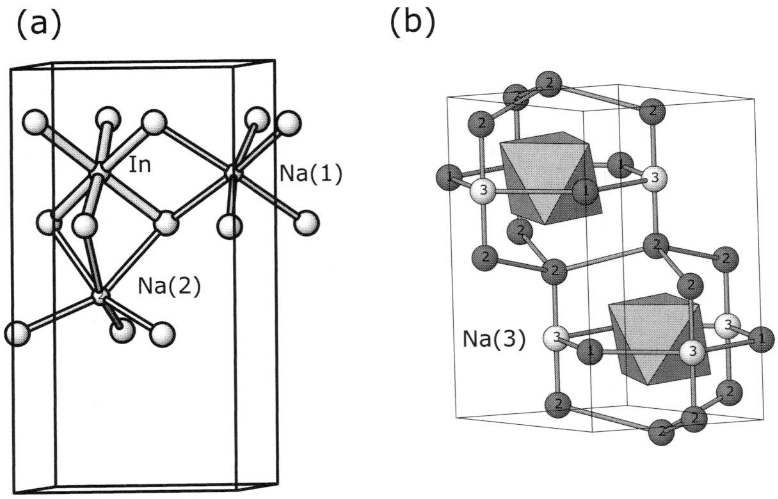

Figure 5. (a) Crystal structure of Na_3InCl_6 at 200 K. (b) Conduction path of the sodium ions at 450 K using a partially occupied Na(3) site.

symmetric environment at the Li^+. This is consistent with the trigonal structure. On the other hand, the rapid Li^+ diffusion leads to the disappearance of the time averaged e^2Qq/h at the ^{115}In site and results in a sharp ^{115}In spectrum above ca 400 K. Furthermore, the chemical shift is different from that of the isolated $InBr_6^{3-}$ or $InBr_4^-$ existing in In_4Br_7 [9].

3.3. SINGLE CRYSTAL X-RAY ANALYSIS ON Na_3InCl_6 AT 200 K

Na_3InCl_6 belongs to a trigonal system (*P*-31*c*) with a = 6.8915(1) Å and c = 12.3820(4) Å. Figure 5(a) shows the structure, in which an isolated $InCl_6^{3-}$ anion and two crystallographic Na^+ exist in the unit cell. All these cations are located on the special high symmetry positions 32 or 3. The chloride ions form a hexagonal close packing (hcp) in contrast to the cubic close packing (ccp) of the bromide ions found in Li_3InBr_6 [3]. Na(1) and Na(2) also form $NaCl_6$ octahedra; however, the distortion at the Na(2) site is remarkable probably due to the electrostatic repulsion between In^{3+} and Na^+.

3.4. ^{115}In NMR AND ^{23}Na NMR

Since all cations are located on the threefold axes, the quadrupole coupling tensors must be cylindrical ($\eta = 0$) with their principal axes parallel to the c-axis. Figure 6 shows the temperature dependence of the ^{115}In NMR using polycrystalline sample. The powder pattern simulation based on the second-order quadruple effect using ν_L = 59.439 MHz and e^2Qq/h = 20.1 MHz is shown at the bottom. ν_L agrees very well to that observed for the cubic elpasolite

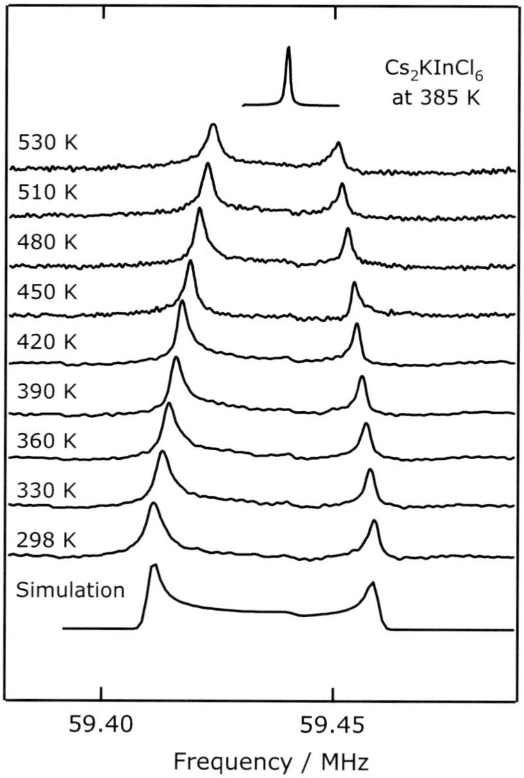

Figure 6. ^{115}In NMR using polycrystalline Na_3InCl_6 as a function of temperature.

Cs_2KInCl_6 which contains a regular octahedral $InCl_6{}^{3-}$ anion. Although the separation between two singularities decreased continuously with increasing temperature, no discontinuity was detected.

In contrast to the ^{115}In NMR, ^{23}Na NMR showed the first-order quadrupole effect. Figure 7 plots the splitting between a pair of satellite transitions ($\Delta m = 1/2 \leftrightarrow 3/2$ and $-1/2 \leftrightarrow -3/2$) as a function of the crystal rotation. Two pairs of satellite transitions with the intensity ratio 2:1 were observed. We could assign the stronger pair to the Na(2). Since e^2Qq/h tensors must be axially symmetric, the splitting ($\Delta\nu$) could be expressed as a function of the crystal rotation (θ) as [10],

$$\begin{aligned}\Delta\nu &= \nu(1/2 \leftrightarrow 3/2) - \nu(-1/2 \leftrightarrow -3/2) \\ &= 2\nu_q\left(3\sin^2\phi\cos^2\theta - 1\right)/2, \end{aligned} \quad (1)$$

where $\nu_q = 3(e^2Qq/h)/2I(2I-1)$ with $I = 3/2$, ϕ is the angle between rotation axis and the principal axis of the e^2Qq/h tensor. Two sets of the numerical solutions were obtained for each site, however, referring to the ϕ value estimated from the

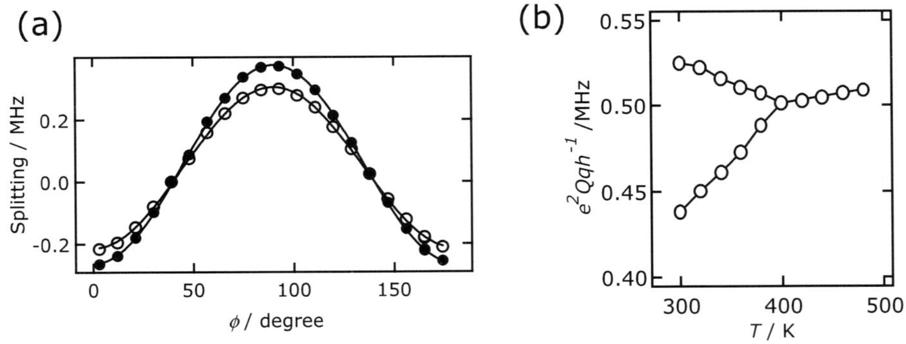

Figure 7. (a) ^{23}Na NMR using single crystal of Na_3InCl_6. Splitting between a pair of satellite transitions is plotted against the crystal rotation. (b) Temperature dependence of the e^2Qq/h at the Na(1) and Na(2) sites.

single crystal ^{115}In NMR, e^2Qq/h were determined uniquely to be 0.426(2) and 0.522(2) MHz for Na(1) and Na(2) site, respectively. Figure 7(b) shows temperature dependence of the e^2Qq/h determined at $\theta = 90$, where $\Delta\nu$ corresponds to $(1/2) \cdot e^2Qq/h$. It is interesting to note that the two crystallographically nonequivalent sites merge into one line at ca 400 K. Since these is no phase transition around this temperature region, this coalescence may suggest the chemical exchange between these two sites faster than the NMR timescale ($\sim 10^{-4}$ s). On the other hand, the motional correlation time of the sodium ion (τ_{NMR}) was estimated from the temperature dependence of the FWHM line width.

$$\tau_{NMR} = 3.5 \times 10^{-11}\,\mathrm{s}\,\exp\left(46(3)\,\mathrm{kJ\,mol^{-1}}/RT\right). \tag{2}$$

Equation (2) indicates that $1/\tau_{NMR}$ reaches 24 kHz at 400 K and this value is reasonable to explain the coalescence phenomenon shown in Figure 7.

3.5. RIETVELD ANALYSIS AT 450 K

In order to confirm the chemical exchange stated above, Rietveld analysis was performed at 450 K. The X-ray powder diffraction at 295 K could be reproduced quite well using the structural parameters at 200 K except thermal parameters. On the other hand, we could confirm a new sodium site Na(3) at 450 K. Figure 5(b) shows Na(3) site together with the possible conduction path. In a typical two-site chemical exchange process, two original peaks do not shift but show only broadening at the beginnings. In this case, however, e^2Qq/h at the Na(1) site increases with temperature probably suggests the rapid translational motion between Na(1) and Na(3) sites before the overall exchange between Na(1) and Na(2).

4. Conclusions

Li_3InBr_6 undergoes phase transition to a lithium superionic conductor at 314 K, where a rock salt structure with the substitution $3Li^+ = In^{3+} + 2(V)$ is formed essentially. Whereas Na_3InCl_6, the chloride ions form hexagonal close packing array in contrast to the cubic close packing in Li_3InBr_6. A conduction path of the sodium ion could be proposed by means of the Rietveld analysis and single crystal ^{23}Na NMR.

Acknowledgement

This work was supported by a Grant-in-Aid for Scientific Research (No. 15550121) from the Ministry of Education, Culture, Sports, Science and Technology of Japan.

References

1. Tomita Y., Fujii A., Ohki H., Yamada K. and Okuda T., *Chem. Lett.* (1998), 223.
2. Tomita Y., Yamada K., Ohki H. and Okuda T., *Z. Naturforsch.* **53a** (1998), 466.
3. Yamada K., Iwaki K., Okuda T. and Tomita, Y. In: Chowdari, B. V. R., Prabaharan, S. R. S., Yahaya M. and Talib I. A. (eds.), *Solid State Ionics: Trends in the New Millennium*, World Scientific, Singapore, 2002, pp. 621–628.
4. Bohnsack A., Stenzel F., Zajonc A., Balzer G., Wickleder M. S. and Meyer G., *Z. Anorg. Allg. Chem.* **623** (1997), 1067.
5. Bohnsack A., Balzer G., Wickleder M. S., Güdel H.-U. and Meyer G., *Z. Anorg. Allg. Chem.* **623** (1997), 1352.
6. Menetrier M., Hojjaji A., Estournes C. and Levasseur A., *Solid State Ionics* **48** (1991), 325.
7. Izumi F. and Ikeda T., *Mat. Sci. Forum* **321–324** (2000), 198.
8. Schmidt M. O., Wickleder M. S. and Meyer G., *Z. Anorg. Allg. Chem.* **625** (1999), 539.
9. Yamada K., Mohara H., Kubo T., Imanaka T., Iwaki K., Ohki H. and Okuda T., *Z. Naturforsch.* **57a** (2002), 375–380.
10. Abragam A., *Principles of Nuclear Magnetism*, Chap. VII, Oxford University Press, London, 1961.

Hyperfine Interactions (2004) 159:103–108
DOI 10.1007/s10751-005-9087-2

Phase Transition and Orientational Disorder of the Cation in [(PyO)(H/D)][$AuCl_4$] (PyO = C_5H_5NO) Crystal

T. ASAJI[1,*], E. AKIYAMA[1], F. TAJIMA[2], K. EDA[2],
M. HASHIMOTO[2] and Y. FURUKAWA[3]
[1]*Department of Chemistry, Graduate School of Integrated Basic Sciences, Nihon University, Sakurajosui, Setagaya-ku, 3-25-40, Tokyo 156-8550, Japan; e-mail: asaji@chs.nihon-u.ac.jp*
[2]*Department of Chemistry, Faculty of Science, Kobe University, Nada-ku Kobe 657-8501, Japan*
[3]*Department of Science Education, Graduate School of Education, Hiroshima University, Kagamiyama, Higashi-Hiroshima 739-8524, Japan*

Abstract. An orientational disorder of the cation in [(PyO)D][$AuCl_4$] crystal was investigated by the ^{35}Cl NQR and ^{1}H NMR measurements. A structural phase transition was found at ca. 70 K from the temperature dependence of the NQR frequencies both in [(PyO)D][$AuCl_4$] and [(PyO)H][$AuCl_4$]. Temperature dependence of the spin-lattice relaxation time T_1 of the NQR of $[AuCl_4]^-$ could be interpreted by an electric field gradient modulation due to the motion of the cation. Characteristics of T_1 of ^{35}Cl NQR as well as that of ^{1}H NMR suggest a dynamic orientational disorder of the cation.

Key Words: hydrogen bond, NQR, order–disorder, phase transition.

1. Introduction

It is known that pyridine *N*-oxide (PyO) forms monomeric hydrogen adduct $[(PyO)H]^+$ in its salts [1–3]. In addition, PyO has a tendency to form monovalent diadduct type cation $[(PyO)_2H]^+$, in which the two PyO units are bridged by a proton [3–5]. Both types of cations are incorporated in $[(PyO)_mH_n][AuCl_4]_n$ having a PyO/$HAuCl_4$ ratio of m/n = 4/3 and 3/2 as evidenced by X-ray structural analysis [6]. It is interesting that an orientational disorder of the monomeric cation is observed for the compounds with m/n = 3/2 and 1/1. In this study, phase transition and orientational disorder of the cation in the 1/1 compound [(PyO)(H/D)][$AuCl_4$] will be discussed. In [6], we reported that the D-salt, [(PyO)D][$AuCl_4$], exhibited phase transitions at ca.70 and 273 K, although the latter one detected only by differential thermal analysis (DTA) and differential

* Author for correspondence.

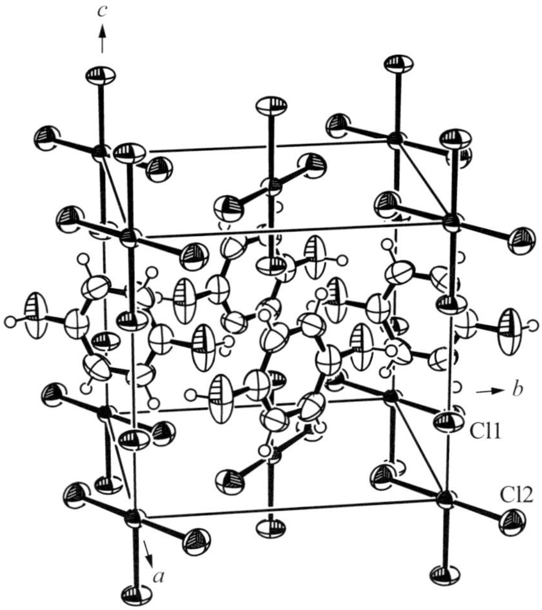

Figure 1. Crystal structure of [(PyO)D][$AuCl_4$].

scanning calorimetry (DSC) measurements, has not yet been well established. In the present work, the very subtle heat anomaly at 273 K was proved to be due to a trace of contamination-water that had persisted even after drying under *vacuo* at room temperature. The phases below and above 70 K will be tentatively referred to as LTP and HTP, respectively.

The crystal structure of HTP of the D-salt at 298 K (Figure 1) consists of an alternative stacking of the two-dimensional array of the cations and that of the anions [6]. In the respective layer, the molecular plane of the cation or anion is perpendicular to the layer plane. In this structure, the symmetry of $[(PyO)D]^+$ ion is C_{2h}, that is, there must exist an orientational disorder of the cation. Then, there is possibility that the transition between LTP and HTP accompanies order–disorder of the cation. In order to clarify the relation between the thermal motion and the disorder of the cation, the measurements of spin-lattice relaxation time T_1 of ^{35}Cl NQR and that of ^{1}H NMR were carried out as a function of temperature. The study of the phase transition in this compound is very interesting from the point of view of order–disorder transition in a crystal having typical layer structure.

2. Results and discussion

Temperature dependences of the NQR frequencies and spin-lattice relaxation times of the D-salt are given in Figures 2 and 3, respectively. Similar results were

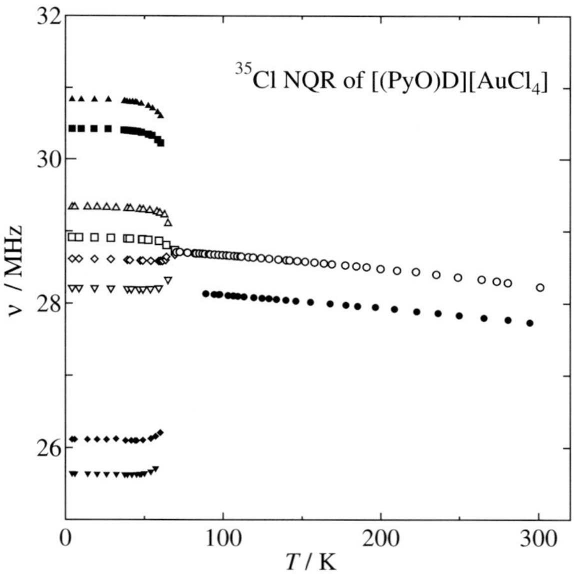

Figure 2. Temperature dependence of ^{35}Cl NQR frequencies of [(PyO)D][$AuCl_4$].

obtained as well for the H-salt, [(PyO)H][$AuCl_4$]. At around 273 K, there is no appreciable anomaly in the NQR frequency *versus* temperature curve. By a single crystal X-ray work as well, no detectable change was observed in the crystal structure across 273 K. Furthermore, recent DTA measurements on a specimen dried under *vacuo* at 423 K showed no heat anomaly at 273 K suggesting that the previous observation is related to contamination-water.

The ^{35}Cl NQR frequencies for the D-salt (the H-salt) at 119 K are 28.634 (28.636) and 28.080 (28.084) MHz. These observations indicate that the D-salt is isostructural with the H-salt. In the crystal structure at 298 K, there exist two crystallographically nonequivalent chlorine atoms (Cl1 and Cl2 in Figure 1). The structural environments of these chlorine atoms are as follows [6]. Au–Cl1 and Au–Cl2 distances are both 2.27_7 Å, Cl1···O and Cl2···O contacts are 3.16_3 and 4.03_4 Å, respectively, and Au-Cl1···O and Au-Cl2···O bond angles are 122.4_1° and 94.5_2°, respectively. Judging from van der Waals radii of Cl, O, and H, which are 1.80, 1.40, and 1.2 Å, respectively, the Cl1···O contact of 3.16 Å suggests an appreciable hydrogen bond. Since this hydrogen bond is expected to be roughly perpendicular to the direction of Au–Cl bond, the low- and high-frequency lines observed at room temperature can be assigned to Cl1 and Cl2, respectively, by applying simple Townes–Dailey theory [7].

Below T_c = ca. 70 K, both in the D- and H-salts, each resonance line splits into quartet indicating a phase transition. The ^{35}Cl NQR frequencies for the D-salt (the H-salt) at 4.2 K are 30.833 (30.815), 30.423 (30.411), 29.331 (29.335), 28.913 (28.915), 28.612 (28.605), 28.212 (28.205), 26.115 (26.134), and 25.644 (25.660) MHz. The lower-frequency line which is assignable to the hydrogen

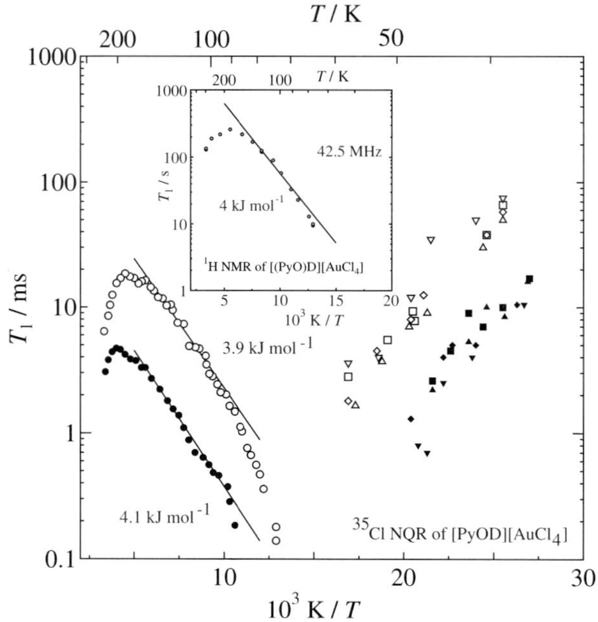

Figure 3. Temperature dependence of ^{35}Cl NQR spin-lattice relaxation time T_1 of [(PyO)D][AuCl$_4$]. The same symbols as Figure 2 are used for the data of the respective NQR lines. The inset shows temperature dependence of ^{1}H NMR spin-lattice relaxation time T_1 of [(PyO)D][AuCl$_4$].

bonded chlorine atom (Cl1) disappeared at ca. 90 K on cooling and showed a very large splitting of ca. 5 MHz in the LTP. From this observation, it is very plausible that an orientational ordering of the cation takes place below T_c.

As shown in Figure 3, the T_1 decreases when temperature approaches to T_c from both the low- and high-temperature side resulting in the characteristic V-shape in the lnT_1 *versus* $1/T$ plot. The D-salt gave a little deeper V-shape in this plot. As a relaxation mechanism responsible to such temperature dependence, electric field gradient modulation due to a motion of nearby ion is highly possible [8–10]. On the assumption of a relaxation mechanism due to the modulation effect, the activation energy of the motion of the cation [(PyO)D]$^+$ was estimated to be ca. 4 kJ mol^{-1} from the Arrhenius-type decrease of T_1 with decreasing temperature down to T_c (see Figure 3).

In order to confirm that the motion of the cation as a whole really takes place with frequency comparable to the NQR frequency, even at 77 K, we measured ^{1}H NMR T_1 of the D-salt. The results are shown in the inset of Figure 3. The ^{1}H T_1 showed a similar Arrhenius-type decrease with decreasing temperature down to T_c. This temperature dependence indicates that the motion of [(PyO)D]$^+$, which is responsible for the NMR relaxation, is fast enough to compared with the Larmor frequency of 42.5 MHz employed in the NMR measurements. The slope of lnT_1 *versus* $1/T$ plot gives an activation energy of ca. 4 kJ mol^{-1} for the cationic motion which value agrees very well with that derived from the temperature dependence of the NQR T_1 assuming the electric field gradient modulation effect.

These experimental results suggest a dynamical orientational disorder of the $[(PyO)D]^+$ above T_c. The fact that no appreciable isotope dependence was observed in T_c can be well understood if this phase transition is mainly related with an orientational order–disorder of the cation as a whole and a motion of the acid hydrogen itself is not so important in the transition mechanism. The activation energy of ca. 4 kJ mol^{-1} obtained for the motion of $[(PyO)D]^+$ seems to be considerably low.

The heat capacity measurements, which are now in progress, revealed that another transition took place at 63 K just below T_c of 70 K. The detail of the experimental results of the heat capacity measurements and the temperature dependence of the NQR frequencies in the vicinity of the successive phase transitions will be reported in a forthcoming paper [11].

3. Conclusion

An orientational disorder of the cation in $[(PyO)D][AuCl_4]$ crystal was concluded to be dynamic one by temperature dependence of the spin-lattice relaxation time of ^{35}Cl NQR as well as that of ^1H NMR. A structural phase transition which can be related to the ordering of the $[(PyO)D]^+$ orientation, was located at ca. 70 K both in $[(PyO)D][AuCl_4]$ and $[(PyO)H][AuCl_4]$.

Acknowledgements

This work was partially funded by "High-Tech Research Center" Project for Private Universities, matching fund subsidy from the Ministry of Education, Culture, Sports, Science and Technology (MEXT) 2000–2004, a Grant for 2004 from the College of Humanities and Sciences, Nihon University, and Nihon University Individual Research Grant for 2004.

References

1. Tsoucaris P. G., *Acta Crystallogr.* **14** (1961), 914–917.
2. Moers O., Wijaya K., Hamann T., Blaschette A. and Jones P. G., *Z. Naturforsch.* **56b** (2001), 1052–1062.
3. Asaji T., Tajima F. and Hashimoto M., *Polyhedron* **21** (2002), 2207–2213.
4. Hussain M. S. and Schlemper E. O., *J. Chem. Soc., Dalton Trans.* (1982), 751–755.
5. Wąsicki J., Jaskólski M., Pająk Z., Szafran M., Dega-Szafran Z., Adams M. A. and Parker S. F., *J. Mol. Struct.* **476** (1999), 81–95.
6. Asaji T., Akiyama E., Tajima F., Eda K., Hashimoto M. and Furukawa Y., *Polyhedron* **23** (2004), 1605–1611.
7. Nakamura D., Ikeda R. and Kubo M., *Coord. Chem. Rev.* **17** (1975), 281–316.

8. Ishikawa A., Ito Y., Horiuchi K., Asaji T. and Nakamura D., *J. Mol. Struct.* **192** (1989), 221–227.
9. Asaji T., Horiuchi K., Chiba T., Shimizu T. and Ikeda R., *Z. Naturforsch.* **53a** (1998), 419–426.
10. Asaji T., Yoza M. and Ishizaka T., *J. Phys., Condens. Matter* **11** (1999), 5219–5223.
11. Asaji T., Watanabe J., Akiyama E., Fujimori H. and Oguni M., to be published.

Hyperfine Interactions (2004) 159:109–119
DOI 10.1007/s10751-005-9088-1

NQR Spin Diffusion in an Inhomogeneous Internal Field

GREGORY B. FURMAN* and SHAUL D. GOREN
Physics Department, Ben Gurion University, Beer Sheva, 84105, Israel;
e-mail: gregoryf@bgu.ac.il

Abstract. The theory of NQR spin diffusion is extended to the case of spin lattice relaxation and spin diffusion in an inhomogeneous field. Two coupled equations describing the mutual relaxation and the spin diffusion of the nuclear magnetization and dipolar energy were obtained by using the method of nonequilibrium state operator. The equations were solved for short and long times approximation corresponding to the direct and diffusion relaxation regimes.

Key Words: NQR, spin diffusion.

1. Introduction

Studies of the NMR [1–3] and NQR of nuclei have demonstrated that spin diffusion plays an important role in the relaxation of nuclei in the presence of paramagnetic impurities (PI). Such type of relaxation originates from the magnetic dipole–dipole interaction of PI with neighboring nuclei, which is inversely proportional to the sixth power of the distance. Thus, near the PI the equilibrium with the lattice is reached at a faster rate [1–3]. The nuclear magnetization during the relaxation process is spatially inhomogeneous over a sample volume, and this induces a spatial diffusion of the nuclear spin energy by, for example, flip–flop transitions due to the dipole–dipole interactions between nuclear spin.

However, first, till now most studies of the nuclear spin diffusion were related to systems with nuclear spin $I = 1/2$, described by the Hamiltonian, whose main term includes just linear functions of spin operators and, correspondingly, forms equidistant energy spectra and, second, most of them deal with the process of a spin diffusion in homogeneous magnetic fields [4, 5].

In many samples spin systems consist of nuclei with $I > 1/2$ and they interact with their environment through the electric quadrupole moment Q, and these interactions are strong enough to observe magnetic resonance of nuclei in the absence of an external magnetic field (pure NQR-case). Unlike the NMR-case, the NQR energy spectrum is non-equidistant and, in many cases, degenerated. These circumstances lead to certain difficulties in obtaining a diffusion equation and a calculation of a diffusion coefficient.

* Author for correspondence.

Besides, the spin diffusion processes no longer exactly conserve nuclear quadrupole energy in an inhomogeneous field [6] because the quadrupole interaction energy is not identical for neighboring nuclear spins. In order for the spin diffusion process to take place, the nuclear quadrupole energy difference must be taken up by another thermodynamic reservoir, for example, by the dipole–dipole energy one.

Recently the theories for spin diffusion of the nuclear dipolar order via PIs [8, 9] in NMR and NQR [10] and spin diffusion in an inhomogeneous field [11] have been developed. It was shown that thermodynamics reservoir of the dipolar order plays an important role in the spin lattice relaxation and spin diffusion in an inhomogeneous field. Nuclear dipolar order is characterized by a state with nuclear spins oriented along an internal local field generated by the dipole–dipole interactions (DDI) and can be described by a dipolar temperature [12–15]. Here we consider the phenomena of spin lattice relaxation and spin diffusion for both the nuclear quadrupole and dipolar energies of the nuclear spins due to their DDI in solids containing PIs in an inhomogeneous field.

2. Theory

2.1. HAMILTONIAN

The evolution of the spin system consisting of nuclear spins with $I > 1/2$ and PI spins may be described by a solution of the equation for density matrix $\rho\ (t)$ (in units of $\hbar = 1$)

$$i\frac{d\rho(t)}{dt} = [\mathcal{H}(t), \rho(t)] \tag{1}$$

with the Hamiltonian

$$\mathcal{H}(t) = \mathcal{H}_Q + \mathcal{H}_{dd} + \mathcal{H}_{PI} + \mathcal{H}_P + \mathcal{H}_{br}(t). \tag{2}$$

Here $\mathcal{H}_Q$ represents the interaction of the I-spin system with the EFG; $\mathcal{H}_{dd}$ and $\mathcal{H}_{PI}$ are the Hamiltonians of the dipole–dipole interaction between nuclear spins and nuclear and PI spins, respectively; $\mathcal{H}_P$ describes the impurity spin system; $\mathcal{H}_{br}(t) = \sum_{q=-2}^{2} E^{(-q)}(t)A^q$ the spin–lattice interaction Hamiltonian, describes spin–lattice relaxation caused by the torsional vibrations (Bayer mechanism) [16], where A^q is a bilinear function of the spin operators and $E^{(-q)}(t)$ is a random function of time [14].

Using the projection operators [17] e^{μ}_{mn} and ε^{j}_{mn} defined by their matrix elements $\langle m'|e^{\mu}_{mn}|n'\rangle = \delta_{m'm}\delta_{n'n}$ and $\langle \nu'|\varepsilon^{\mu}_{\nu\sigma}|\sigma'\rangle = \delta_{\nu'\nu}\delta_{\sigma'\sigma}$ and introducing a projection density operators, $e_{mn}\ (\vec{r})$, for the nuclear spins I, and $\varepsilon_{mn}\ (\vec{r})$ for PI spins

$$e_{mn}(\vec{r}) = \sum_{\mu} \delta(\vec{r} - \vec{r}_{\mu})e^{\mu}_{mn}; \quad \varepsilon_{mn}(\vec{r}) = \sum_{j} \delta(\vec{r} - \vec{r}_{j})\varepsilon^{j}_{mn} \tag{3}$$

the density of the Hamiltonians $\mathcal{H}_Q$, $\mathcal{H}_{dd}$, and $\mathcal{H}_{PI}$ can be written down in the following form:

$$H_Q(\vec{r}) = (2I+1)^{-1}\sum_{mn}\omega_{mn}^0 e_{mm}(\vec{r}), \tag{4}$$

$$\mathcal{H}_{dd}(\vec{r}) = \int d\vec{r}\,' \sum_{mnm'n'} g_{mn}^{m'n'}(\vec{r}-\vec{r}\,')e_{mn}(\vec{r})e_{m'n'}(\vec{r}\,'), \tag{5}$$

$$\mathcal{H}_{PI}(\vec{r}) = \int d\vec{r}\,' \sum_{mnm'n'} f_{mn}^{m'n'}(\vec{r}-\vec{r}\,')e_{mn}(\vec{r})\varepsilon_{m'n'}(\vec{r}\,'), \tag{6}$$

$$\mathcal{H}_{br}(t) = \sum_q\sum_{mn} E^{(-q)}(t)A_{mn}^q \int d\vec{r}\, e_{mn}(\vec{r}), \tag{7}$$

where $\omega_{mn}^0 = \lambda_m - \lambda_n$, λ_m, $|m\rangle$ and $|n\rangle$ are the eigenvalues and eigenvectors of the operator $\mathcal{H}_Q$. $|\nu\rangle$ and $|\sigma\rangle$ are eigenvectors of the operator $\mathcal{H}_P$; matrix Elements $g_{mn}^{m'n'}(\vec{r}-\vec{r}\,')$ and $f_{mn}^{m'n'}(\vec{r}-\vec{r}\,')$ can be presented as

$$g_{mn}^{m'n'}(\vec{r}-\vec{r}\,') = G_{mn}^{m'n'}(\vec{r}-\vec{r}\,')\left[(\delta_{mn}+\delta_{pq})(\delta_{m\bar{n}}+\delta_{p\bar{q}}) + (\delta_{mq}+\delta_{pn})(\delta_{m\bar{q}}+\delta_{p\bar{n}})\right] \tag{8}$$

with $\bar{n} = n$ and

$$f_{mn}^{m'n'}(\vec{r}-\vec{r}\,') = F_{mn}^{m'n'}(\vec{r}-\vec{r}\,')(\delta_{mn}+\delta_{pq})(\delta_{m\bar{n}}+\delta_{p\bar{q}}), \tag{9}$$

$G_{mn}^{m'n'}$ and $F_{mn}^{m'n'}$ are matrix elements of the dipole–dipole Hamiltonians $\mathcal{H}_{dd}$ and $\mathcal{H}_{PI}$ in $\mathcal{H}_Q - representation$ [17].

2.2. DIFFUSION EQUATIONS

To obtain the equation describing the spin diffusion and spin lattice relaxation of both the quadrupole and dipolar orders we will use the method of none-quilibrium state operator [18], which has been applied to obtain the diffusion equation in cases of the Zeeman order spin diffusion [19] and dipolar order [8] spin diffusion.

Using the commutation rules between the components of the projection operators (3) $[e_{mn}(\vec{r}), e_{m'n'}(\vec{r}\,')] = \delta(\vec{r}-\vec{r}\,')(\delta_{nm'}e_{mn'}(\vec{r}) - \delta_{n'm}e_{m'n}(\vec{r}))$ and

$[e_{mn}(\vec{r}), \varepsilon_{m'n'}(\vec{r}')] = 0$, we can obtain the following equations in the form of localized laws of conservation of the spin energy densities

$$\frac{\partial e_{mm}(\vec{r})}{\partial t} + div\left(\vec{j}_{mm}(\vec{r})\right) = K_{mm}(\vec{r}) + L_{mn}(\vec{r}), \tag{10}$$

$$\frac{\partial \mathcal{H}_{dd}(\vec{r})}{\partial t} + div\left(\vec{j}_{dd}(\vec{r})\right) + \sum_{mn} \vec{j}_{mm}(\vec{r}) \frac{\partial \omega^0_{mn}(\vec{r})}{\partial \vec{r}}$$
$$= K_{dP}(\vec{r}) + L_{dP}(\vec{r}), \tag{11}$$

$$\frac{\partial \mathcal{H}_P}{\partial t} = -\int d\vec{r}\left\{(2I+1)^{-1}\frac{\partial}{\partial t}\left[\sum_{mn} \beta_{mn}(\vec{r})\omega^0_{mn}(\vec{r})e_{mm}(\vec{r})\right]\right.$$
$$\left. + \frac{\partial \mathcal{H}_{dd}(\vec{r})}{\partial t}\right\}. \tag{12}$$

The last equation is result of the energy conservation law. In Eq. (10) $\vec{j}_{mm}(\vec{r})$ is the flux of operator $e_{mm}(\vec{r})$,

$$\vec{j}_{mm}(\vec{r}) = -\frac{i}{2}\int d\vec{r}'\sum_{k}(\vec{r} - \vec{r}')G^{km}_{mk}(\vec{r} - \vec{r}')\left\{e_{mk}(\vec{r}')e_{km}(\vec{r})\right.$$
$$\left. - e_{mk}(\vec{r})e_{km}(\vec{r}') - e_{km}(\vec{r}')e_{mk}(\vec{r}) + e_{km}(\vec{r})e_{mk}(\vec{r}')\right\} \tag{13}$$

$K_{mm}(\vec{r}) = i[\mathcal{H}_{PI}, e_{mm}(\vec{r})]$ and $L_{mm}(\vec{r}) = i[\mathcal{H}_{br}, e_{mm}(\vec{r})]$ in Equation (10) are the change of the nuclear quadrupolar energy density due to the interaction with the PI and caused by the torsional vibrations (Bayer mechanism), respectively. In Equation (11), $\vec{j}_{dd}(\vec{r})$ is the operator of the flux of nuclear dipolar energy,

$$\vec{j}_{dd}(\vec{r}) = -2i\sum_{mnpq}\sum_{m'n'p'}\int d\vec{r}'\int d\vec{r}''(r'' - r)G_{mnpq}(r'' - r)e_{mn}(\vec{r}'')e_{m'n'}$$
$$\times(\vec{r}')\left[G_{m'n'qp'}(r' - r)e_{pp'}(\vec{r}) - G_{m'n'p'p}(r' - r)e_{p'q}(\vec{r})\right], \tag{14}$$

$K_{dP}(\vec{r}) = i[\mathcal{H}_{PI}, \mathcal{H}_{dd}(\vec{r})]$ and $L_{dP}(\vec{r}) = i[\mathcal{H}_{br}, \mathcal{H}_{dd}(\vec{r})]$ in Equation (11) are the change of the nuclear dipolar energy density due to the interaction with the PI and caused by the torsional vibrations (Bayer mechanism), respectively. Note, that in the case with a homogenous magnetic field, $\frac{\partial \omega^0_{mn}(\vec{r})}{\partial \vec{r}}$, from the system Equations (10) and (11) we have two separate equations: Equation (10) leads to the localized law of conservation of the quadrupolar energy densities [20] and Equation (11) leads to conservation law of the dipolar energy [10].

In the high temperature approximation we can write the density matrix in the following form [18]

$$\sigma = \left\{1 - \int_0^1 d\lambda[\mathcal{B}(t+i\lambda) - \langle \mathcal{B}(t+i\lambda)\rangle]\right\}\sigma_{eq}, \tag{15}$$

where the thermodynamic average $\langle \ldots \rangle$ corresponds to an average with the quasi-equilibrium operator $\sigma_{eq} = e^{-\mathcal{A}}/Tre^{-\mathcal{A}}$, and

$$\mathcal{A} = \int d\vec{r}\left[(2I+1)^{-1}\sum_{mn}\beta_{mn}(\vec{r})\omega^0_{mn}(\vec{r})e_{mn}(\vec{r}) + \beta_d(\vec{r})\mathcal{H}_{dd}(\vec{r})\right] - \beta_p\mathcal{H}_p, \tag{16}$$

$$\begin{aligned}\mathcal{B}(t+i\lambda) = e^{-\lambda\mathcal{A}}\int_{-\infty}^0 dte^{\varepsilon t}\int d\vec{r}(2I+1)^{-1}\sum_{mn}\Big\{&\omega^0_{mn}(\vec{r})\vec{j}_{mm}(\vec{r},t)\\ \times\nabla\beta_{mm}(\vec{r},t) + \vec{j}_{mm}(\vec{r},t)\times\left[\beta_{mn}(\vec{r},t) - \beta_d(\vec{r},t)\right]&\\ \times\frac{\partial\omega^0_{mn}(\vec{r})}{\partial\vec{r}} + \vec{j}_d(\vec{r},t)\nabla\beta_d(\vec{r},t) + \left[\beta_{mn}(\vec{r},t) - \beta_L\right]&\\ \times K_{ZS}(\vec{r},t) + \left[\beta_d(\vec{r},t) - \beta_L\right]K_{dS}\left(\vec{r},t\right)\Big\}e^{\lambda\mathcal{A}}.&\end{aligned} \tag{17}$$

By using Equations (10)–(16), and taking into account that for single crystal sample of cubic symmetry, the diffusion coefficients, both for quadrupolar and dipolar energies, which in the general case of noncubic symmetry is a symmetrical tensor of second rank [5], reduces to a scalar quantity. By introduction the quantities $\left[\beta_{mn}(\vec{r},t) - \beta_L\right] = \xi_{mn}(\vec{r},t)$ and $\left[\beta_d(\vec{r},t) - \beta_L\right] = \zeta(\vec{r},t)$, the diffusion equations can be obtained

$$\begin{aligned}\frac{\partial\xi_{mn}(\vec{r},t)}{\partial} = \frac{1}{\omega^0_{mn}(\vec{r})}\nabla\{D_{mn}(\vec{r})\left[\omega^0_{mn}(\vec{r})\nabla\xi_{mn}(\vec{r},t) + (\xi_{mn}(\vec{r},t)\right.&\\ \left.-\zeta(\vec{r},t))\nabla\omega^0_{mn}(\vec{r})\right]\} - W_{mn}(\vec{r})\xi(\vec{r},t),&\end{aligned} \tag{18}$$

$$\begin{aligned}\frac{\partial\zeta(\vec{r},t)}{\partial t} = \sum_{mn}\frac{D_{mn}(\vec{r})\nabla\omega^0_{mn}(\vec{r})}{M_{mm}}\left[\omega^0_{mn}(\vec{r})\nabla\xi_{mn}(\vec{r},t) + (\xi_{mn}(\vec{r},t)\right.&\\ \left.-\zeta(\vec{r},t))\nabla\omega^0_{mn}(\vec{r})\right] + \nabla\left[D_d(\vec{r})\nabla\zeta(\vec{r},t)\right] - W_d(\vec{r})\zeta(\vec{r},t).&\end{aligned} \tag{19}$$

where $M_{mm} = \int d\vec{r}'G^2(\vec{r} - \vec{r}')\langle e^2_{mm}(\vec{r})\rangle$.

The first term in the square brackets of the right side of Equation (18) describes the time dependence of the $\xi_{mn}(\overrightarrow{r},t)$ due to the spin diffusion with a diffusion coefficient of

$$D_{mn}(\overrightarrow{r}) = \int_0^1 d\lambda \int_{-\infty}^{\infty} dt e^{\varepsilon t} \left\langle \overrightarrow{j}_{mn}(\overrightarrow{r},\lambda,t)\overrightarrow{j}_{mn}(\overrightarrow{r})\right\rangle \Big/ Tr(e^2_{mm}(\overrightarrow{r})). \tag{20}$$

The second term gives the variation of $\beta_{mn}(\overrightarrow{r},t)$ due to interaction with the dipolar reservoir in the inhomogeneous field. The last term in the right side of Equation (18) gives the relaxation of $\xi_{mn}(\overrightarrow{r},t)$ toward the inverse lattice temperature with density of the transition probability per unit time, $W_{mn}(\overrightarrow{r})$, which for a cubic crystal is given by $(K_{mm}(\overrightarrow{r}) + L_{mm}(\overrightarrow{r}))$, where

$$W_{mn}(\overrightarrow{r}) = \int_0^1 d\lambda \int_{-\infty}^{\infty} dt e^{\varepsilon t} \Big\langle (K_{mm}(\overrightarrow{r},\lambda,t) + L_{mm}(\overrightarrow{r},\lambda,t))(K_{mm}(\overrightarrow{r}) + L_{mm}(\overrightarrow{r}))\Big\rangle \Big/ Tr(e_{mm}e_{mm}(\overrightarrow{r})). \tag{21}$$

The first term in the curly brackets of right side of Equation (19) describes the time variation of the dipolar energy due to the spin diffusion with diffusion coefficient

$$D_d(\overrightarrow{r}) = \int_0^1 d\lambda \int_{-\infty}^{\infty} dt e^{\varepsilon t} \left\langle \overrightarrow{j}_d(\overrightarrow{r},\lambda,t)\overrightarrow{j}_d(\overrightarrow{r})\right\rangle \Big/ Tr(\mathcal{H}_d\mathcal{H}_d(\overrightarrow{r})). \tag{22}$$

The second term gives the variation of $\beta_d(\overrightarrow{r},t)$ due to the interaction with the quadrupole reservoir in an inhomogeneous field. The last term in the right side of Equation (19) gives the relaxation with the density of the transition probability per unit time, $W_d(\overrightarrow{r})$, which for a cubic crystal is given by

$$W_d(\overrightarrow{r}) = \int_0^1 d\lambda \int_{-\infty}^{\infty} dt e^{\varepsilon t} \Big\langle [K_{dP}(\overrightarrow{r},\lambda,t) + L_{dP}(\overrightarrow{r},\lambda,t)][K_{dP}(\overrightarrow{r}) + L_{dP}(\overrightarrow{r})]\Big\rangle \Big/ Tr(\mathcal{H}_d\mathcal{H}_d(\overrightarrow{r})). \tag{23}$$

The boundary conditions can be introduced by defining a sphere with radius l about each PI, called the spin diffusion barrier radius. Inside this sphere the spin diffusion process goes to zero:

$$\nabla\beta_{mn}(\overrightarrow{r},t)\Big|_{|\overrightarrow{r}|=l} = 0 \quad \text{and} \quad \nabla\beta_d(\overrightarrow{r},t)\Big|_{|\overrightarrow{r}|=l_d} = 0. \tag{24}$$

To obtain the radius of spin diffusion barrier, let us emphasize that $D_{mn}(\overrightarrow{r})$ and $D_d(\overrightarrow{r})$ are a function of the distance $\overrightarrow{r}$ from the nearest PI. In the Gaussian limiting case the stochastic theory of magnetic resonance [21] the dimensional dependence of the diffusion coefficients $D_{mn}(\overrightarrow{r})$ and $D_d(\overrightarrow{r})$ can be expressed by the following function [19]

$$D_{mn}(\overrightarrow{r}), D_d(\overrightarrow{r}) \sim \exp\left[-\left(\frac{\nabla\omega_{mn}^0(\overrightarrow{r})\overrightarrow{r}}{\omega_d}\right)^2\right]. \tag{25}$$

Using Equation (26) the diffusion barrier radius [1, 2, 22, 23, 25] for the spin diffusion of quadrupole energy can be found by solving the equation,

$$\frac{3\gamma_S}{l^3}\langle S_Z\rangle\left[\frac{qe\mu}{\langle S_z\rangle\gamma_S}\left|\overrightarrow{r}_0\right| + 1 \pm \left(1 - \frac{\left|\overrightarrow{r}_0\right|}{l}\right)\right] = \frac{6\gamma_I}{\left|\overrightarrow{r}_0\right|^3}, \tag{26}$$

where the first term in square brackets in Equation (26) describes a distortion of the crystal field as a result of the inclusion of the PI. In Equation (26) r_0 is distance between neighboring nuclei, μ is the Sternheimer antishielding factor [24]. It was assumed that the distortion of the electric field is equivalent to the presence of a charge q [25].

The examination of the functional dependence (25) for the dipolar diffusion coefficient results that the main term for $D_d(\overrightarrow{r})$ does not include any dimensional dependence. Thus the radius of the diffusion barrier for dipolar energy is $l_d = \left|\overrightarrow{r}_0\right|$ which corresponds to non-barrier diffusion and to the fastest relaxation of the dipolar energy.

In the case of a homogeneous magnetic field, $\nabla\omega_{mn}^0(\overrightarrow{r}) = 0$, Equations (20) and (21) give the results obtain earlier for the spin diffusion of the quadrupole [5] and of the dipolar energies [8]. From Equations (20) and (21) we get that the dissipation of the density quadrupolar and dipolar energies are driven by: 1) the exchange between them; 2) spin diffusion process: and 3) direct relaxation to the PI.

2.3. DIRECT RELAXATION REGIME

Exact solutions of Equations (20) and (21) are extremely difficult, even for simple a model situations. That is why we consider evolution of the spin system in time by using the next considerations. Immediately after a disturbance of the nu-clear spin system, the gradients of $\xi_{mn}(\overrightarrow{r},t)$ and $\zeta(\overrightarrow{r},t)$ are sufficiently small and diffusion cannot be of importance at the start of the relaxation process [22], this is the so-called diffusion vanishing regime [Lowe]. To describe the relaxation at that time interval we can use Equations (20) and (21) by putting all inverse temperature gradient-terms equal zero, $\nabla\xi_{mn}(\overrightarrow{r},t) = 0$ and $\nabla\zeta(\overrightarrow{r},t) = 0$. We also accept the approximation that at distances larger then the radius of the diffusion barrier the diffusion coefficient is independent of $\overrightarrow{r}$ [4]. Under these approximations Equations (20) and (21) come to

$$\frac{\partial\xi_{mn}(\overrightarrow{r},t)}{\partial t} = \frac{D_{mn}(\overrightarrow{r})\Delta\omega_{mn}^0(\overrightarrow{r})}{\omega_{mn}^0(\overrightarrow{r})}\left\{\left[\xi_{mn}(\overrightarrow{r},t) - \zeta(\overrightarrow{r},t)\right]\right\} - W_{mn}(\overrightarrow{r})\xi(\overrightarrow{r},t), \tag{27}$$

$$\frac{\partial\zeta(\vec{r},t)}{\partial t} = \sum_{mn} \frac{D_{mn}(\vec{r})\left[\nabla\omega_{mn}^{0}(\vec{r})\right]^{2}}{M_{mm}}\left[\xi_{mn}(\vec{r},t) - \zeta(\vec{r},t)\right]$$
$$-W_{d}(\vec{r})\zeta(\vec{r},t). \tag{28}$$

The evolutions of the $\xi_{mn}(\vec{r},t)$ and $\zeta(\vec{r},t)$ toward their steady-state values is a linear combination of $(2I+1)$ exponents with $N = (2I+1)$ relaxation times $\tau_N(\vec{r})$. These relaxation times $\tau_N(\vec{r})$ are the function of the position $\vec{r}$. In order to obtain the experimentally observed signals, the solutions of Equations (27) and (28), must be averaged over the sample. For this averaging procedure one needs the knowledge the field distribution.

As a result of the diffusion vanishing relaxation regime the local inverse temperatures, $\xi_{mn}(\vec{r},t)$ and $\zeta(\vec{r},t)$; become spatially distributed over the sample with a distribution which is not the equilibrium one. In this case we have to take into account also the gradient-terms, $\nabla\xi_{mn}(\vec{r},t)$ and $\nabla\zeta(\vec{r},t)$ in Equations (20) and (21). In the next section we will consider the influence of the spin diffusion process.

2.4. DIFFUSION RELAXATION REGIME

Assuming that at distances larger then the radius of the diffusion barrier the diffusion coefficient is independent of $\vec{r}$ [4]. Multiplying Equation (20) by $\gamma\omega_{mn}^{0}(\vec{r})\langle e_{mm}(\vec{r})\rangle$ and Equation (21) by $\sum_m M_{mn}$ and by integrating Equations (29) and (30) over space variable $\vec{r}$, we obtain the equations describe the evolution of the experimentally observed values, the quantity that connects with the Z component of the total nuclear magnetization $\left(M_z(t) - M_z(0) = \sum_{mn} E_{mn}(t)\right)$, $E_{mn}(t)$

$$\frac{\partial E_{mn}(t)}{\partial t} = -\frac{\gamma}{\sum_m M_{mm}} D_{mm} \int d\vec{r} E_d(\vec{r},t)\nabla\omega_{mn}^{0}(\vec{r})]$$
$$-\int d\vec{r} W_{mn}(\vec{r}) E_{mn}(\vec{r},t), \tag{29}$$

and the total dipolar energy, $E_d\,(t)$

$$\frac{\partial E_d(\vec{r},t)}{\partial t} = -\sum_{mn} D_{mn} \int d\vec{r} E_{mn}(\vec{r},t)\omega_{mn}^{0}(\vec{r}) - \int d\vec{r} W_d(\vec{r}) E_d(\vec{r},t), \tag{30}$$

where $E_{mn}(t) = \int d\vec{r} E_{mn}(\vec{r},t)$, $E_{mn}(\vec{r},t) = \gamma\omega_{mn}^{0}(\vec{r})\xi_{mn}\langle e_{mm}(\vec{r},t)\rangle$, $E_d(t) = \int d\vec{r} E_d(\vec{r},t)$, and $E_d(\vec{r},t) = \sum_m M_{mm}\zeta(\vec{r},t)$.

To obtain the solution of the Equations (29) and (30) and to calculate the relation times, both for the nuclear magnetization and for the dipolar energy, we need to know the internal distribution function of the field, $\omega_{mn}^{0}(\vec{r})$. As it follows from Equations (29) and (30) that for a special distribution of the internal field, $\Delta\omega^{0}_{mn}(\vec{r}) = 0$, the diffusion equations gives two uncoupled equations:

$$\frac{\partial E_{mn}(t)}{\partial t} = -\int d\vec{r}\, W_{mn}(\vec{r}) E_{mn}(\vec{r}, t), \tag{31}$$

and

$$\frac{\partial E_{d}(t)}{\partial t} = -\int d\vec{r}\, W_{d}(\vec{r}) E_{d}(\vec{r}, t). \tag{32}$$

Solving Equations (31) and (32), we obtain the normalized relaxation functions

$$R_{mn}(t) = \frac{E_{mn}(t) - E_{mn}(\infty)}{E_{mn}(0) - E_{mn}(\infty)} = e^{-t/T_{1mn}} \tag{33}$$

and

$$R_{d}(t) = \frac{E_{d}(t) - E_{d}(\infty)}{E_{d}(0) - E_{d}(\infty)} = e^{-t/T_{1d}}, \tag{34}$$

where $T_{1mn} = \frac{0.12}{C_p D_{mn}^{3/4} F}$ and $T_{1d} = \frac{0.12}{C_p D_{d}^{3/4} F}$, C_p is the concentration of the PI and F is the angular average of the coupling dipolar constant of the DDI between the nuclear and PI. It follows from the solutions (33) and (34) that at a long time after the excitation of the spin system, the nuclear magnetization describes by the sum of exponents $M_z \simeq \sum_{mn} a_{mn} e^{-t/T_{1mn}}$, while the dipolar energy decreases to equilibrium exponentially.

3. Results and discussion

We will compare the results obtained here with the relaxation processes of the nuclear magnetization [7] and the dipolar energy [6] in the mixed state conventional superconducting vanadium ($I = 7/2$). In the type II superconductors, an applied magnetic field $\vec{H}_0$, in the range between the lower and upper critical field, $H_{c1} < H_0 < H_{c2}$, penetrates into the bulk sample in the form of vortices, each with a quantum flux of $\Phi_0 = \frac{c\hbar}{2e}$, which form a two-dimensional structure in the plane perpendicular to $\vec{H}_0$ [27]. The distribution of the internal vortex field can be obtained by solving the Landau–Ginzburg equation, which gives

$$H(\rho) = \frac{\phi_0}{2\pi\lambda^2} \ln\frac{\rho}{\lambda} \text{ for } r < \lambda, \tag{35}$$

where λ is the London penetration length and ρ is the distance from the core of vortex in the cylindrical coordinate, $r^2 = \rho^2 + z^2$. Using experimental data [6, 7] we obtain the spin relaxation time for the nuclear magnetization, $T_1 = 43$ s and the spin relaxation time for the dipolar energy, $T_{1d} = 93$ ms. As a consequence, the dipolar energy decreases to the equilibrium state with an anomalously short time as compared to the relaxation time of the nuclear magnetization, $\frac{T_1}{T_{1d}} = 442$. Theoretical estimation of the ratio of the relaxation time, using Equations (33) and (34), results $\frac{T_1}{T_{1d}} = \left(\frac{l}{r_0}\right)^3$. Taking into account that the distance between neighboring vanadium nuclei $r_0 = 2.63 \times 10^{-8} cm$ and the radius $l = 1.96 \times 10^{-7}$ cm, for the ratio $\frac{T_1}{T_{1d}} = 413$, which is in a good agreement with the result obtained from experimental data.

4. Conclusions

In conclusion, we obtained coupled equations describing mutual relaxation and spin diffusion of the quadrupole energy and dipolar energy by using the method of nonequilibrium state operator [18]. The equations were solved at a short and long times approximations corresponding to the direct and diffusion relaxation regimes. We showed that at the beginning of relaxation process, the direct relaxation regime is preferred. The relaxation regime changes both for the nuclear quadrupolar and the dipolar energies, to the diffusion one.

References

1. Bloembergen N., *Physica* **15** (1949), 386.
2. Khutsishvili G. R., *Sov. Phys. JETP* **4** (1957), 382.
3. de Gennes P. G., *J. Phys. Chem. Solids* **3** (1958), 345.
4. Khutsishvili G. R., *Sov. Phys. Uspekhi* **8** (1966), 743.
5. Khutsishvili G. R., *Sov. Phys. Uspekhi* **11** (1969), 802.
6. Genack A. Z. and Redfield A. G., *Phys. Rev. Lett.* **31** (1973), 1204.
7. Redfield A. G. and Yu W. N., *Phys. Rev.* **169** (1968), 443; *Phys. Rev.* **177** (1968), 1018.
8. Furman G. B. and Goren S. D., *J. Phys. Condens. Matter* **11** (1999), 4045.
9. Boutis G. S., Greenbaum D., Cho H., Cory D. G. and Ramanathan C., *Phys. Rev. Lett* **92** (2004), 137201–1.
10. Furman G. B. and Goren S. D., *Sol. St. Nucl. Magn. Res.* **16** (2000), 199.
11. Furman G. B. and Goren S. D., *Phys. Rev. B* **68** (2003), 064402.
12. Jeneer J. and Broekaert P., *Phys. Rev.* **157** (1967), 232.
13. Goldman M., *Spin Temperature and Nuclear Resonance in Solids*, Claredon Press, Oxford, 1970.
14. Abragam A. and Goldman M., *Nuclear Magnetism: Order and Disorder*, Clarendon, Oxford, 1982.
15. Slichter C. P., *Principles of Magnetic Resonance*. Springer Verlag, Berlin Heidelberg, New York, 1980.
16. Bayer H., *Z. Phys.* **130** (1951), 227.
17. Ainbinder N. E. and Furman G. B., *Zh. Eksp. Teor. Fiz.* **85** (1983), 998; *Sov. Phys. JETP* **58** (1983), 575.
18. Zubarev D. N., *Nonequilibrium Statistical Thermodynamics*, Imprint Consultants Bureau, New York, 1974.
19. Buishvili L. L. and Zubarev D. N., *Sov. Phys. Solid State* **3** (1965), 580.
20. Furman G. B., Goren S. D., Panich A. M. and Shames A. I., *Z. Naturforsch.* **55a** (2000), 54–60.

21. Kubo K. and Tomita K., *J. Phys.Soc. Japan* **9** (1954), 888.
22. Blumberg W. E., *Phys. Rev.* **119** (1960), 79.
23. Rorschach H. E. Jr., *Physica* **30** (1964), 38.
24. Das T. P. and Hahn E. L., *Nuclear Quadrupole Resonance Spectroscopy*, Suppl. 1 to Solid State Phys., Academic, New York, 1958.
25. Bukhbinder I. L. and Kessel A. R., *Zh. Eksp. Teor. Fiz.* **65** (1973), 1498; *Sov. Phys. JETP* **38**, (1974), 745.
26. Landau L. D. and Lifshitz E. M., *Quantum Mechanics -Non Relativistic Theory*, Pergamon, Oxford, 1989.
27. Abrikosov A. A., *Zh. Eksp. Teor. Fiz.* **32** (1957), 1442; *Sov. Phys JETP* **5** (1957), 1174.

Hyperfine Interactions (2004) 159:121–125
DOI 10.1007/s10751-005-9089-0

The Nature of Line Broadening in Thermally Detected $^{57}\mathrm{Fe}\underline{\mathrm{Fe}}$ NMR

W. D. HUTCHISON*, G. A. STEWART, S. J. HARKER† and D. H. CHAPLIN
School of Physical, Environmental and Mathematical Sciences, The University of New South Wales at the Australian Defence Force Academy, Canberra, Australia;
e-mail: w.hutchison@adfa.edu.au

Abstract. $^{57}\mathrm{Fe}\underline{\mathrm{Fe}}$ with isotopic concentrations from 15 to 95% is studied using NMR thermally detected by nuclear orientation. Lines are found to be consistently homogeneous. The contrast with previous inhomogeneous $^{57}\mathrm{Fe}\underline{\mathrm{Fe}}$ lines from Mössbauer detected NMR is explained by differences in radio frequency field strength.

Key Words: ferromagnetism, line widths, NMR, nuclear orientation.

1. Introduction

Recently we demonstrated the use of Mössbauer effect spectroscopy, in absorber geometry, to detect nuclear magnetic resonance (MNMR) of ^{57}Fe in 30% isotopically enriched Fe foil at millikelvin temperatures [1]. Cooling the absorber to millikelvin temperatures orients the ^{57}Fe nuclei and 'tilts' the six line Mössbauer spectrum. NMR tends to equalise the $I = 1/2$ populations and therefore reduces the tilt of the Mössbauer spectrum. The NMR line in this case was predominately inhomogeneous. That is, in order to maximise the continuous wave NMR signal it was necessary to modulate the radio frequency (RF) back and forth across the resonance line. Surprisingly, this contrasted with the homogeneous nature of the ^{57}Fe resonance line recorded via NMR that was thermally detected using a nuclear orientation thermometer (NMR-TDNO) [2].

In this current work we explore the basis of this difference in line broadening mechanisms for what were quite similar specimens both subject to NMR but using differing detection techniques. One obvious difference between the two original ^{57}Fe enriched foils was the actual isotopic concentration of the ^{57}Fe, at 30% (MNMR) and 95% (NMR-TDNO). Therefore the starting point in this

* Author for correspondence.
† Present address: School of Physics and Materials Engineering, Monash University, Melbourne, Australia.

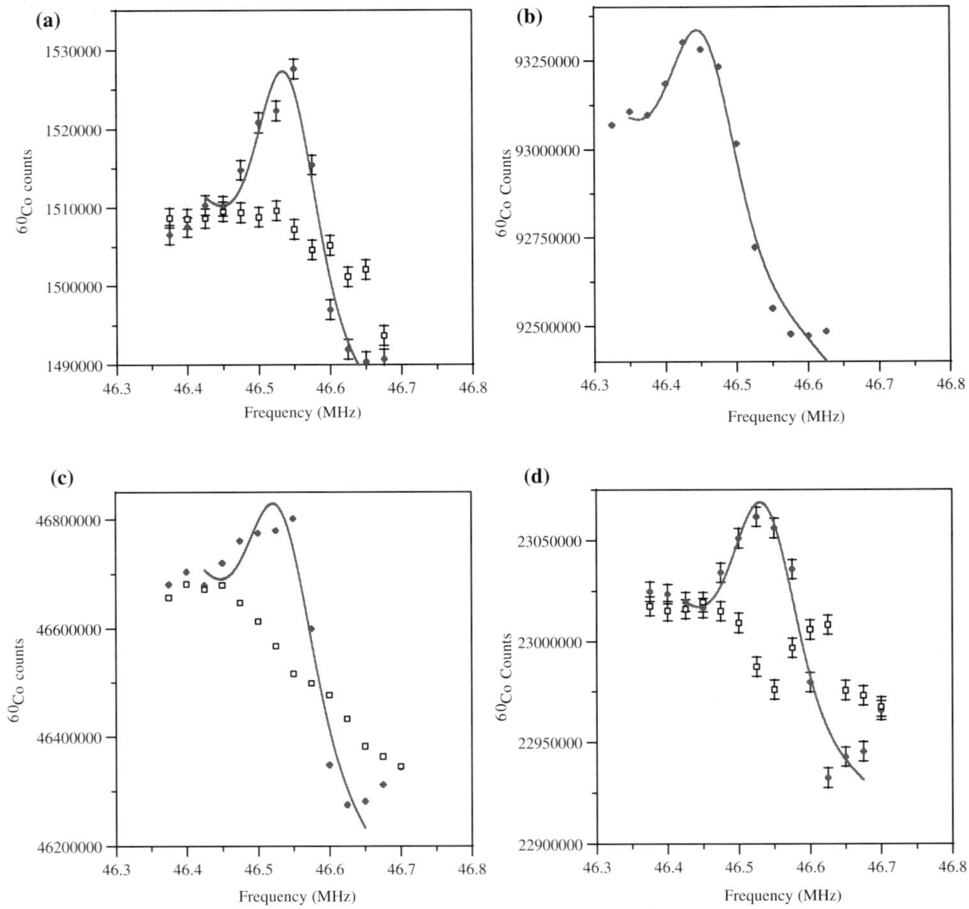

Figure 1. NMR-TDNO of Fe foils with varying ^{57}Fe concentrations (*open symbols* = ±200 kHz FM): (a) 95% ^{57}Fe, B_{app} = 0.05 T, FWHM = 90(14) kHz, (b) 60% ^{57}Fe, B_{app} = 0.10 T, FWHM = 98(14) kHz, (c) 30% ^{57}Fe, B_{app} = 0.05 T, FWHM = 96(22) kHz, and (d) 15% ^{57}Fe, B_{app} = 0.05 T, FWHM = 95(18) kHz.

investigation was further NMR-TDNO measurements for a range (15–95%) of ^{57}Fe concentrations.

2. Experimental details

^{57}Fe enriched Fe foils with isotopic concentrations of 15%, 30%, 60% and 95% were prepared by arc melting and subsequent rolling to around 1 μm thickness. ^{60}Co activity was then diffused into each foil via annealing at 850°C under flowing hydrogen. In turn, each sample was soldered to a 'silver serpentine,' weak link cold finger [3] and cooled to ~8 mK in a dilution refrigerator. Polarising static and RF magnetic fields were provided by a mutually perpendicular

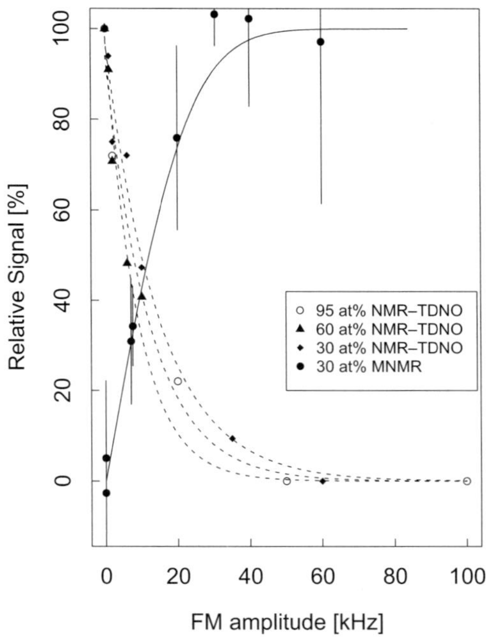

Figure 2. Relative 'on resonance' signal verses FM amplitude (±value) for three Fe foils of differing isotopic concentrations via the NMR-TDNO method. These show the response expected of a homogeneous line. MNMR in contrast has maximum signal for a FM width of order the inhomogeneous line width [1].

superconducting solenoid and twin loop coil set, respectively. Low temperature nuclear orientation (LTNO) was observed via detection of the gamma-rays emitted by the ^{60}Ni daughters of the ^{60}Co thermometric probes. The MNMR [1], by comparison, utilised the same dilution refrigerator fitted with mylar windows and smaller lower cans that minimised the gamma-ray path length. The Fe foil was mounted on a copper 'picture frame' and backed by a very thin copper foil to provide both good thermal contact and acceptable resonant gamma-ray transmission.

3. Results and discussion

NMR-TDNO results for the four Fe foils of different ^{57}Fe concentration are shown in Figure 1. In each case the ^{57}Fe NMR via LTNO of *in situ* ^{60}Co probes shows a homogeneous line, manifest in the observation of maximum resonant signals without frequency modulation (FM), and independent of ^{57}Fe concentration (15% to 95%).

The difference in the nature of the NMR lines between the two detection techniques is highlighted in Figure 2. Ordinarily, NMR lines of concentrated systems are homogeneous through the affect of nuclear spin–spin coupling. In

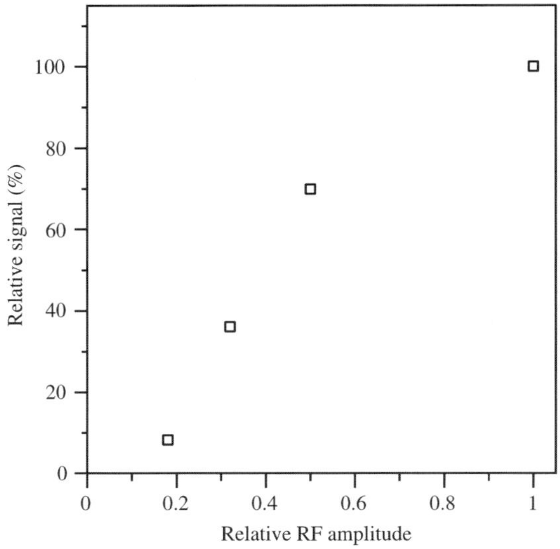

Figure 3. Relative peak NMR-TDNO signal for a 95% ^{57}Fe foil ($B_{app} = 0.05$ T) as a function of input RF amplitude.

ferromagnets at low temperature the dominant coupling mechanism is the electron spin mediated Suhl–Nakamura interaction [4]. For ^{57}FeFe even this term is relatively weak due largely to the small value of the nuclear magnetic moment. Following [5], the Suhl–Nakamura line-width is estimated to be about 200 Hz, too weak to be important c.f. present line-widths (≥70 kHz).

$$\Delta f_{SN} = \frac{1}{h}\left(\frac{I(I+1)}{6\pi}\right)^{1/2} \frac{A^2 S}{g_S \mu_B B_E}\left(\frac{B_E}{B_{pol}+B_A}\right)^{1/4} \approx 200\,\text{Hz}$$

An alternative explanation is required. RF power broadening could be the mechanism for the homogeneous line in the NMR-TDNO case and, moreover, a relatively small difference in rf power between the two experimental set-ups could lead to the change in character from the inhomogeneous MNMR case to the NMR-TDNO case where homogeneous broadening dominates. Due to the technical requirements of the MNMR experiment, the configuration in the vicinity of the cold finger is quite different. We would expect the effective rf field strength B_1 to be somewhat lower in the MNMR case, for a given nominal RF input level. Also we know, via ^{60}Co LTNO, that the average temperature in the Fe foil was higher (~17 mK for MNMR vs. ~8–10 mK for NMR-TDNO). The total line widths were also measurably different, for example at $B_{app} = 0.05$ T, being ~95 kHz for NMR-TDNO and ~70 kHz for MNMR [1]. A total line width Δf_T composed of homogeneous (Δf_h) and inhomogeneous (Δf_i) components is given by $\Delta f_T = \sqrt{\Delta f_h^2 + \Delta f_i^2}$. For the ^{57}FeFe case Δf_h can be equated with the RF power broadened line width Δf_1 by $\Delta f_h \approx \Delta f_1 = \sqrt{f_1^2 T_1/T_2}$ (see for example [6] and $f_1 = \gamma B_1$). Assume that $T_2 = 1/\Delta f_{SN}$ in this case and note that T_1 is quite long at 735 s in 0.1 T at 10 mK [3]. Together with an estimate for the rf field strength from an independent ^{60}Co$\underline{\text{Fe}}$ single pulse NMRON experiment, these conditions lead to an estimate of Δf_h(NMR-TDNO) $\approx$ 60 kHz for the typical RF input level used. Setting $\Delta f_i \approx$ 69 kHz for both cases and choosing Δf_h(MNMR) $\approx$ 10 kHz, ($\times$ 5 smaller f_1(MNMR) together with a shorter T_1 due to the higher temperature), allows calculation of total line widths consistent with the experimental values. Further NMR-TDNO measurements recorded for a 95% ^{57}Fe foil support the rf broadening premise, although the extent of errors in fitting line shapes makes it difficult to draw a definitive conclusion. The unique parameter values in this case however, allow an alternative test. Saturation of an NMR resonance occurs for the condition $f_1^2 T_1 T_2 > 1$ [6]. This relation is satisfied by the f_1 values from our above estimate for the NMR-TDNO case. Indeed, the resonance is saturated at the standard RF levels used and would remain so even if reduced by several orders of magnitude. As a result, we would expect the 'on resonance' signal strength to be reduced according to the ratio of effective line width Δf_1 to the inhomogeneous line width Δf_i alone, in much the same way that MNMR signal would be affected by changing FM amplitude. This suggestion is validated in Figure 3 and it would appear that variation of RF strength is the important difference between the NMR-TDNO and MNMR results. Furthermore we have shown that ^{57}Fe$\underline{\text{Fe}}$ is an interesting test bed for some fundamental principles of NMR.

Acknowledgement

Project funded in part by a University of New South Wales URSP research grant.

References

1. Stewart G. A., Hutchison W. D., Harker S. J. and Chaplin D. H., *Phys. Rev., B* **66** (2002), 134415.
2. Hutchison W. D., Harker S. J., Chaplin D. H., Funk T. and Klein E., *Hyperfine Interact.* **120/121** (1999), 193.
3. Funk T., Klein E. and Brewer W. D., *Phys. Lett., A* **248** (1998), 457.
4. Turov E. A. and Petrov M. P., *Nuclear Magnetic Resonance in Ferro- and Antiferromagnets*, translated by E. Harnik, Halsted Press, 1972.
5. Suhl H., *Phys. Rev.* **109** (1958), 606.
6. Abragam A., *The Principles of Nuclear Magnetism*, Oxford University Press, Oxford, 1961.

Hyperfine Interactions (2004) 159:127–130
DOI 10.1007/s10751-005-9090-7

Pulsed ^{14}N NQR Device Designed to Detect Substances in the Presence of Environmental Noise[†]

GERÓNIMO E. POLETTO[1], TRISTÁN M. OSÁN[2]
and DANIEL J. PUSIOL[1,2,*]
[1]*Spinlock S.R.L. C. Arenal 1020 X5000GZU, Córdoba, Argentina*
[2]*Fa.M.A.F.-U.N.C. Medina Allende s/n X5016LAE, Córdoba, Argentina;*
e-mail: dpusiol@spinlock.com.ar

Abstract. A new device designed for both volumetric and surface NQR detection of substances spatially located in several positions, and in the presence of environmental interference, is described. The device consists of two probe coils, placed on the same detection plane, for excitation and detection of NQR signals. Experimental results obtained using Strong Off Resonance Comb (SORC) pulse sequences, for the excitation of the ν_{-} transition in samples of Sodium Nitrite ($NaNO_2$), are presented. It is shown that, when the total signals induced in each coil are properly combined, the interference commonly detected in both coils is attenuated relative to the NQR signal detected by either one or both probe coils. NQR signal can be detected by either one or both coils, but in both cases the noise induced by distant environmental sources is attenuated.

Key Words: auto-shielding coils, NQR, surface coils.

1. Introduction

The use of an arrangement of two coils for detecting NQR signals in the presence of environmental noise was previously proposed [1]. In that work the NQR signal was mainly detected in one coil while the interference was detected in both coils. The result was that the interference was attenuated relative to the detected NQR signal.

In this work we present a new device designed for both volumetric and surface Nuclear Quadrupole Resonance (NQR) detection of substances spatially located in several positions (like drugs and/or explosive substances carried inside shoes, for example), and in presence of environmental interference. The system probe for the excitation and detection of the NQR signals comprises two coils placed on the same detection plane. It is shown that the NQR signals can be detected when the samples are placed whithin or close to one or both probe coils whereas the environmental noise produced by distant sources is attenuated.

---

* Author for correspondence.

† Argentinian and International patents pending.

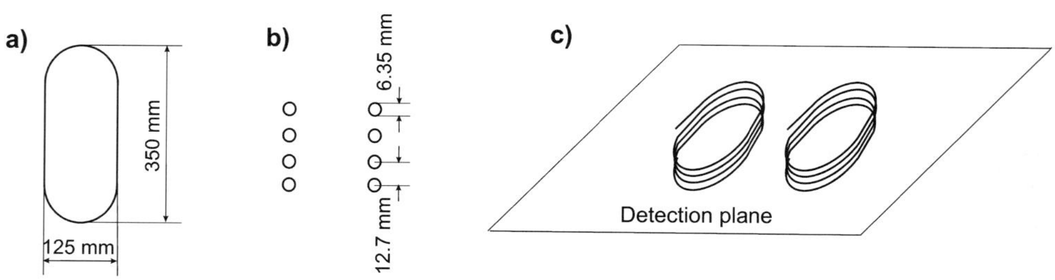

Figure 1. a) View of section and geometric dimensions of probe coils. b) Sketch of turns spacing of probe coils. c) Schematic view of the experimental arrangement of the two probe coils lying on a common detection plane.

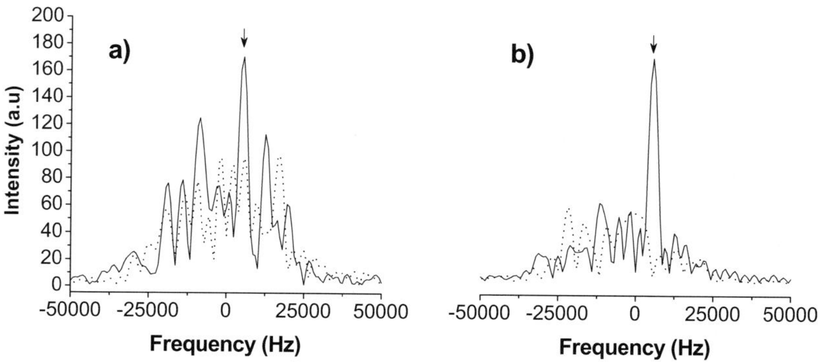

Figure 2. a) FT of the detected signals with the one-coil probe (see Section 3). b) FT of the detected signals with the two-coil probe (see Section 3). The little arrows indicate the peaks corresponding to the ν_- transition in $NaNO_2$. The intensities of the FT are given in arbitrary units (a.u).

2. Experimental

The device comprises two independent tuned circuits. Each circuit has an electrically balanced solenoidal-like probe coil with 4 shoe-like shape turns. The coils were made with a copper pipe of 6.35 mm of external diameter. Coils geometry is presented in Figure 1(a,b).

The two probe coils are intended to excite and detect NQR signals and can operate in two different ways. In the first case, the two coils are wound in the same sense. They are excited by RF pulses 180° out of phase to each other, at a frequency very near of the resonance frequency of the desired substance to be detected. Hence, the two coils produce RF fields which have a phase difference of 180° to each other. During detection, the induced voltage in each coil is composed of the environmental noise and possibly NQR signals. Because the two coils are wound in the same sense, the noise produced by distant environmental sources will be induced with the same phase in both coils but the NQR signals will be 180° out of phase to each other. After a pre-amplification, the phase of the signal coming from one of the coils is shifted by 180°. Therefore, when the

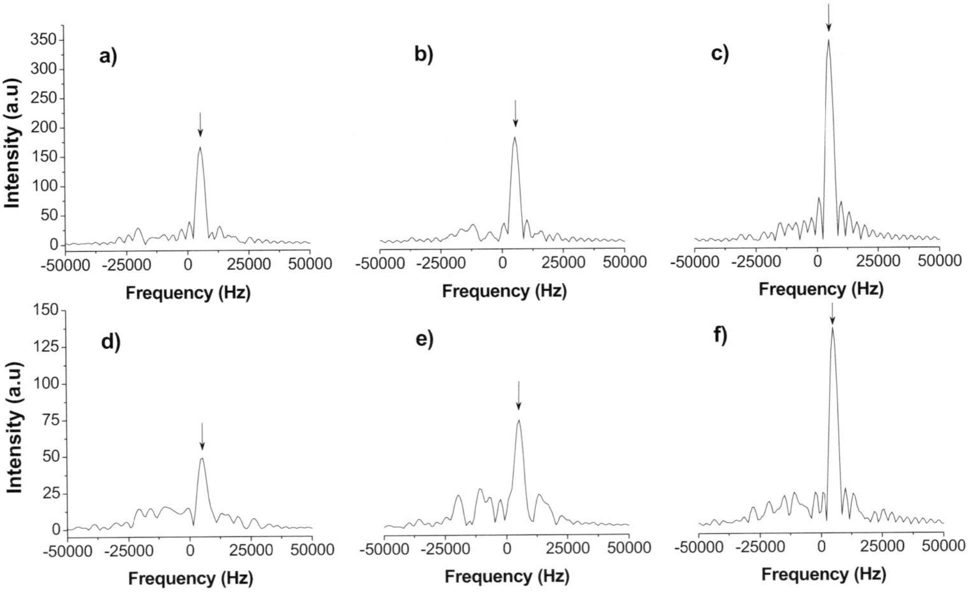

Figure 3. a), b) and c) FT of the detected signals when samples of $NaNO_2$ are placed inside the probe coils (see Section 3). d), e) and f) FT of the detected signals when samples of $NaNO_2$ are placed at 2 cm above the upper plane of the probe coils (see Section 3). The little arrows indicate the peaks corresponding to the $\nu\_$ transition in $NaNO_2$. The intensities of the FT are given in arbitrary units (a.u).

total signals are added, the NQR signals sum up and the environmental noise cancels out.

In the other case the coils are wound in opposite senses to each other. The excitation signal applied to both coils must be now in phase. Then, the NQR signals induced in each coil will be in phase to each other while the environmental noise produced by distant sources will be induced 180° out of phase. Thus, when the two induced signals are added, the NQR signals are summed and the environmental noise is cancelled.

Measurements were performed adopting a probe configuration with two coils lying in the detection plane and wound in the same sense. A schematical view of this experimental arrangement can be seen in Figure 1(c).

To test the device we used SORC pulse sequences [2] to excite the $\nu\_ = 3.600$ MHz NQR transition of Sodium Nitrite ($NaNO_2$) at room temperature. Measurements were performed with samples of 300 g of $NaNO_2$ placed either inside or outside of the probe coils.

3. Results and discussion

In order to test the device introduced in this work, we performed in the first place measurements with just one probe coil (the other coil was properly disconnected). Figure 2a shows the Fourier Transform (FT) of the detected signals. The solid line shows the FT of the detected signal when there is a sample of $NaNO_2$ placed inside the coil. The dotted line shows the FT of the detected signal in the absence of the sample. Figure 2b shows the experimental results when two coils wound in the same sense to each other (as was described in Experimental section) are engaged. The solid line shows the FT of the detected signal when the same sample of $NaNO_2$ is placed inside one of the probe coils. The dotted line shows the FT of the detected signal in the absence of the sample. As can be seen, the configuration with two coils has a much better signal-to-noise (SNR) ratio than the configuration comprising just one probe coil.

The following experimental results were obtained when the two coils are engaged. Figure 3a shows the FT of the detected signal when there is just one sample of $NaNO_2$ placed inside one of the probe coils. Figure 3b shows the FT of the detected signal when the same sample is placed inside the other probe coil. Figure 3c shows the FT of the detected NQR signals when there is a sample of 300 g of $NaNO_2$ placed inside each probe coil. As can be seen, the NQR signal can be detected by either one or both probe coils.

In addition, measurements were performed with samples placed at 2 cm above the upper plane of the probe coils. Figure 3d shows the FT of the detected signal when a sample of $NaNO_2$ is placed at 2 cm above the upper plane of one probe coil. Figure 3e shows the FT of the detected signal when the same sample is placed at 2 cm above the upper plane of the other probe coil. Figure 3f shows the FT of the detected signal when there is a sample of 300 g of $NaNO_2$ placed at 2 cm above the upper plane of each probe coil. In this case, the experimental results show that the sensitivity of the device allows for surface detection of NQR signals by either one or both probe coils.

4. Conclusions

The experimental results show that the configuration with two coils for the excitation and detection of NQR signals, as described in the Experimental section, is less sensitive to environmental interference than the configuration comprising just one probe coil. In addition, the results show that NQR signals can be detected when the samples are placed close to or inside of either one or both probe coils. Thus, the two-coil configuration yields a higher SNR than the single-coil arrangement.

References

1. Smith J. A. S. and Rowe M. D., U. S. Patent Number 6,486,838, November 2002.
2. Kim, S. S., Jayakody J. R. P. and Marino R. A., *Z. Naturforsch.* **47a** (1992), 415.

Hyperfine Interactions (2004) 159:131–136
DOI 10.1007/s10751-005-9091-6

Some Aspects of Dynamics of Nitrogen-14 Quadrupolar Spin-System

T. N. RUDAKOV*, P. A. HAYES and W. P. CHISHOLM
QRSciences Limited, 8-10 Hamilton Street, WA 6107, Cannington, Australia;
e-mail: trudakov@qrsciences.com

Abstract. This is a study of the behaviour of nuclear quadrupole resonance (NQR) signals in the "observation windows" of multi-pulse sequence for a nitrogen-14 spin-system. Obtained results revealed steady state (SS) and spin echo (SE) components of the signal. The results contribute to the understanding the dynamic properties of the quadrupolar spin-system.

Key words: multi-pulse sequence, nuclear quadrupole resonance, pulsed spin locking.

1. Introduction

Understanding dynamics of quadrupolar spin-systems has a direct impact on the successful practical application of nuclear quadrupole resonance (NQR) method, namely the detection of specific dangerous or illicit substances. The multi-pulse NQR techniques based on using steady-state free precession (SSFP) [1, 2] and spin-locking (SL) [3, 4] effects are of particular interest. The SSFP sequences have been used to increase sensitivity both in multi-pulse NMR and NQR experiments [4–7]. It is a mandatory requirement for an SSFP multi-pulse sequence that the RF pulse repetition period τ is less than or equal to the spin–spin relaxation time T_2. Under these conditions a relatively large and continuous steady-state signal can be produced which recurs for as long as the sequence is applied. The spin-locking multi-pulse (SLMP) sequences cause a refocusing of transverse magnetisation for periods much longer than T_2 and produce a train of spin-echoes decaying with the effective relaxation time T_{2e}, which is limited by spin–lattice relaxation time in the rotating frame $T_{1\rho}$, which is less than the spin–lattice relaxation time T_1.

The effect of various multi-pulse sequences on the quadrupolar spin-system was investigated both experimentally and theoretically taking into account the difference between the NMR and NQR spin-systems [8–12]. The common behaviour of a quadrupolar spin-system from the application of an SLMP

* Author for correspondence.

sequence is the establishment of a quasi-steady state for a period of time of around T_2, this quasi-stationary magnetization then decays to equilibrium or steady-state level [10–12]. Previously it was considered that the steady-state signal emerges only at times $t > T_1$. The structure of a quasi-steady state signal has not been studied. Usually this signal is treated simply as an SE signal. Previous investigations by the authors have examined the SL and SSFP effects in quadrupolar spin systems with axial symmetric and non-symmetric electric field gradient (EFG) tensor [13–15]. In this work [15] based on experiments it was proposed that some refinements describing the behaviour of the spin-system should be introduced into the model. The approach taken was to consider a quasi-steady state signal to be a combination of at least the SL and SS components. In reality experiments show that when pulse separation τ is short, an SS component of NQR signal appears together with an SE component in the interval between pulses which usually forms the "observation windows." The SE signals add to the steady-state signals in the observation windows and decay with effective time T_{2e}. Thus after a period of time, in the region of several multiples of T_1, when the effect of the preparatory pulse is "forgotten" by the spin-system, the pulse sequence effect becomes equivalent to the conventional SSFP sequence. The present work seeks to extend this research for a clearer understanding of the behaviour in these transitional regions of time. Following on from this idea, it would also be interesting to experimentally examine if the quasi-stationary and equilibrium magnetisation can be refocused again after a long SSFP sequence. If, for instance, we use a long SSFP sequence instead of a preparatory pulse followed by a train of RF pulses will a train of echoes be observed? This question was also examined in the present work.

2. Experimental details

The experiments were carried out using a pulsed NQR spectrometer based on the TECMAG "Apollo" console, designed to operate in a low-frequency band (0.3–10 MHz) [13]. In these tests the duration of a single so-called "90° pulse" was about 80 μs, this was determined to be the excitation time for maximum FID amplitude. In all experiments the delay between the end of an RF pulse and the beginning of detection was 0.2 ms. The sample used in these experiments was 60 g of polycrystalline sodium nitrite ($NaNO_2$). The transition frequency of $\nu_+ = 4.640$ MHz was used for all experiments. The spin–lattice relaxation time T_1 was measured at room temperature (297 K) and found to be $T_1 \approx 90$ ms. The spin–spin relaxation time T_2 was about 5 ms. The base SLMP sequence used in our experiments can be written as

$$\theta^0_\varphi - \left(\tau - \theta^1_{\varphi+90^\circ} - \tau-\right)_N, \qquad (1)$$

where θ^0 is the flipping angle of the preparatory pulse, θ^1 and is the flipping angle of the other pulses in the sequence and φ is the phase of the preparatory pulse, τ is the time intervals between θ^0 and θ^1 pulses, subsequent pulses are separated by the period 2τ. In order to investigate different components of NQR signal we have employed the phase-cycling multi-pulse-method [13, 15].

3. Results and discussion

The resultant NQR signal components (SE and SS) obtained with pulse sequence (1) at resonance frequency having $\tau = 0.5$ ms are displayed in Figure 1. As expected the SE signals decay with effective time T_{2e} which in this case is about 65 ms. The SE signal intensity dependence with time after the end of preparatory pulse is presented in Figure 2 (curve A). It is noted that this dependence is quite well described by a single exponential function. The behaviour of the SS components is more complicated. It can be seen from Figure 1b that the SS signal quickly appears after the first several pulses and achieves its quasi-steady state value during a period close to T_2. After that time interval the quasi-stationary magnetization rises and reaches a constant value. This dependence is also presented in Figure 2 (curve B). This behaviour can not be described by a single exponential function. In this case the shape of the response with time is described well by a bi-exponential function

$$S = S_\infty \left[1 - a_s \exp\left(-\frac{\mathrm{t}}{\mathrm{T}_{1es}} \right) - (1 - \mathrm{a_s}) \exp\left(-\frac{\mathrm{t}}{\mathrm{T}_{1el}} \right) \right], \tag{2}$$

where t is the time after the end of the preparatory pulse, S_∞ is the signal intensity after a long t and the a_s coefficient varies from 0 to 1 and indicates the weight of the short exponential term. T_{1es} and T_{1el} are respectively defined as the short and long effective decay times. For curve B approximated by the function (2) it is found that $T_{1es} = 3$ ms, $T_{1el} = 90$ ms and $a_s = 0.4$. Note that value of T_{1el} equals that of T_1 and T_{1es} is close to T_2^*. This behaviour implies that the process of recovering SS component progresses more quickly during the initial stage of the pulse sequence with short time constant T_{1es}. Then the response assumes long time constant T_{1el} which is equal or very close to T_1.

Following on from the behaviour at the beginning of the sequence it would be interesting to view the dependence after steady state is established and whether an echo can be recovered. Several experiments were performed to study some aspects of the behaviour from this equilibrium magnetisation. A similar pulse sequence as (1) was employed where a long SSFP sequence is used instead of a preparatory pulse. The length of this sequence was about 450 ms which is five times longer than T_1. The pulse repetition time for SSFP sequence was shorter than 2τ. As in the case of conventional SLMP sequence, a

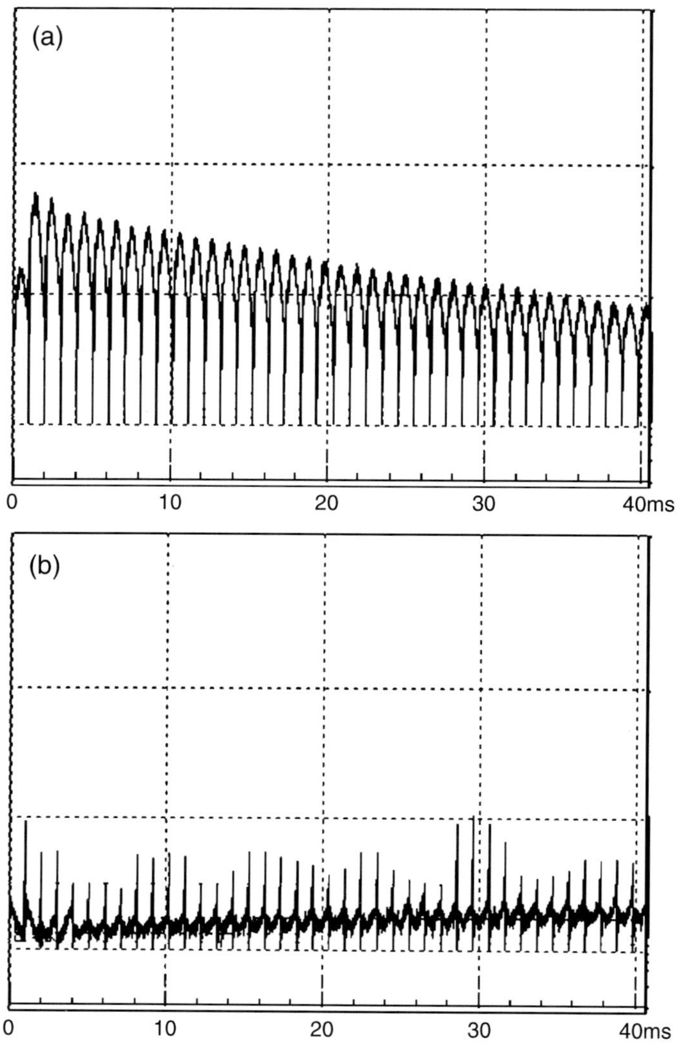

Figure 1. The magnitude of the NQR signal trains in a powdered sample of sodium nitrite ($NaNO_2$) obtained with pulse sequence (1) at resonance frequency using the phase-cycling multi-pulse-method: (a) SE component, (b) SS component; $\tau = 0.5$ ms, $\varphi = 0°$ and $180°$.

train of echoes was observed in the intervals between θ^1 pulses. This fact is clearly demonstrated in Figure 3 for the ν_+ line at resonant frequency. One can see that the SE decay envelope is very similar to that found in the case of a conventional SLMP sequence. Note that as the steady-state signal intensity depends strongly on the frequency offset Δf it is only natural that the SE intensities also depend on Δf. In the present work this dependence is only indicated its further investigation was not pursued at this time and requires additional research.

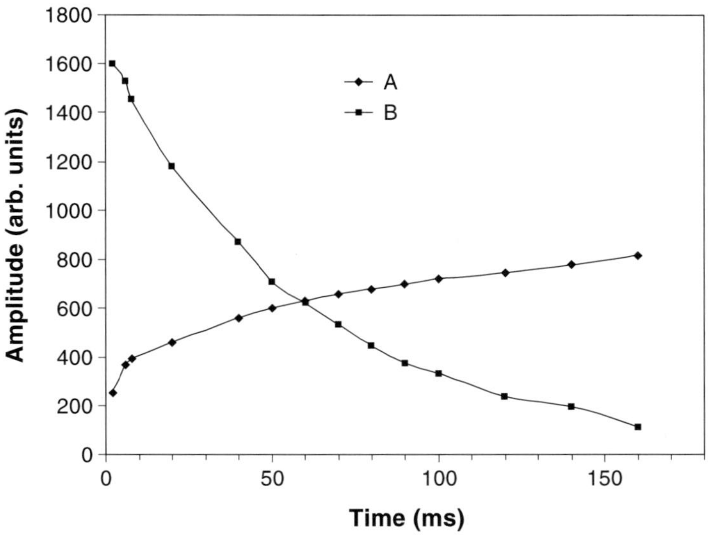

Figure 2. The dependence of the NQR signal intensity in sodium nitrite ($NaNO_2$) with increasing time directly after the preparatory pulse. The pulse sequence (1) was used at resonance frequency. Curve A and B describe SE and SS components, respectively.

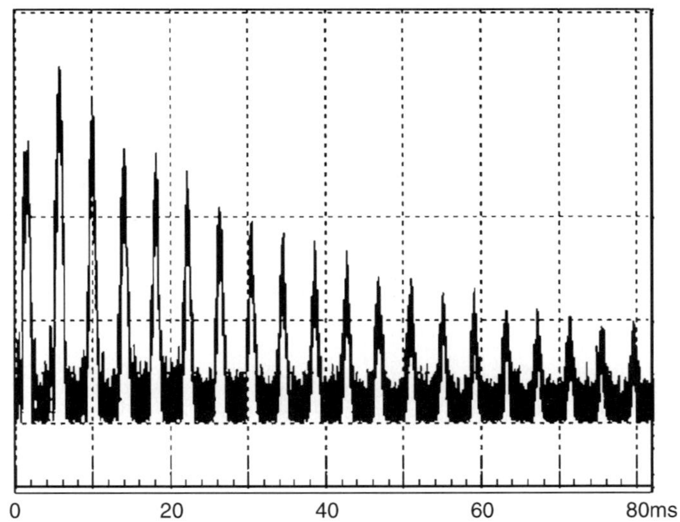

Figure 3. The magnitude of the spin–echo signal train in sodium nitrite ($NaNO_2$) obtained after a long SSFP sequence followed by a train of RF pulses with repetition time $2\tau = 4$ ms. The duration of SSFP sequence is 450 ms and pulse repetition time of 0.5 ms.

4. Conclusion

The quasi-steady-state NQR signal has a complex structure and contains SS and SE components each of them has different dependence on the time after the preparatory pulse. These dependences can explain the behaviour of the total NQR signal.

It was shown that the equilibrium magnetisation can be refocused again after a SSFP sequence even if its duration is much longer than T_1. This resultant behaviour is interesting and may have practical applications in pulsed spin locking experiments.

References

1. Bradford R., Clay C. and Stick E., *Phys. Rev.* **84** (1951), 157.
2. Ostroff E. D. and Waugh J. S., *Phys. Rev. Lett.* **16** (1966), 1097.
3. Mansfield P. and Waugh J. S., *Phys. Lett.* **22** (1966), 133.
4. Carr H. Y., *Phys. Rev.* **112** (1958), 1693.
5. Freeman R. and Hill H. D. W., *J. Magn. Reson.* **4** (1971), 366.
6. Buess M. L., Garroway A. N. and Miller J. B., *J. Magn. Reson.* **92** (1991), 348.
7. Klainer S. M., Hirschfeld T. B. and Marino R. A., In: Marshal A. G. (ed), *Fourier, Hadamard and Hilbert Transforms in Chemistry*, Plenum, NY, 1982, pp 147–181.
8. Cantor R. S. and Waugh J. S., *J. Chem. Phys.* **73** (1980), 1054.
9. Hitrin A. K., Karnaukh G. E. and Provotorov B. N., *J. Mol. Struct.* **83** (1982), 269.
10. Matti Maricq M., *Phys. Rev., B* **33** (1986), 4501.
11. Osokin D. Ya., *Mol. Phys.* **48** (1983), 283.
12. Osokin D. Ya., Ermakov V. L., Kurbanov R. H. and Shagalov V. A., *Z. Naturforsh.* **47a** (1992), 439.
13. Rudakov T. N. and Mikhaltsevich V. T., *Chem. Phys. Lett.* **363** (2002), 1.
14. Rudakov T. N., Mikhaltsevich V. T., Flexman J. H., Hayes P. A. and Chisholm W. P., *Appl. Magn. Reson.* **25** (2004), 467.
15. Rudakov T. N., Mikhaltsevich V. T., Hayes P. A. and Chisholm W. P., *Chem. Phys. Lett.* **387** (2004), 405.

Hyperfine Interactions (2004) 159:137–141
DOI 10.1007/s10751-005-9092-5

^{14}N Study of Aromatic Nitroso Compounds

N. BEEPATH* and D. STEPHENSON
Chemistry Department, University of the West Indies, West Indies, Trinidad and Tobago; e-mail: nadiabeepath@hotmail.com

Abstract. The ^{14}N NQR frequencies of selected aromatic C-nitroso compounds were measured using cross relaxation double resonance. The monomers have high ^{14}N NQR frequencies with χ and η values of 5700 to 6100 kHz and 0.5 ± 0.1, respectively. The dimers and polymer show lower frequencies with values of 2300 to 2600 kHz and 0.8 ± 0.05. Benzofuroxan has values typical of both the monomer and dimer.

Key Words: 1,4-dinitrosobenzene, ^{14}N, nitrosobenzene, NQR.

1. Introduction

^{14}N NQR data on aromatic C-nitroso compounds are limited, although these compounds are theoretically and industrially significant, and have associated health implications. They exist as blue or green monomers or as white, cream, colourless, or yellow dimers (*cis* and/or *trans*) in the solid state (and solution). Principally, the nitroso monomer and dimer differ in the chemical bonding of the nitrogen atom. The monomers are linked via *N*,*N*-azodioxy bonds in the dimer. Exchange between the monomer and dimer is relatively facile, especially in solution. However, the solid lattice stabilizes these compounds thereby facilitating retention of their solid state structures. NQR allows the direct study of ^{14}N in the solid state and is very sensitive to changes in the N orbital population. Therefore, any change in the ^{14}N electronic distribution due to dimerization can be detected by NQR. The ^{14}N NQR frequencies for aromatic C-nitroso compounds with known structures (see Table I) are reported. 1,4-dinitrosobenzene and 1,2-dinitrosobenzene are also included. Electric field gradients (sign and direction) around the N atom in the nitroso monomer and dimer were assigned via ab-initio calculations [1].

2. Experimental

The compounds were either purchased or synthesized using known literature procedures.

Identification was confirmed by ^{1}H NMR, Infrared Spectroscopy and powder X-ray Diffraction. All NQR spectra were recorded using double resonance cross

* Author for correspondence.

Table I. Experimental ^{14}N NQR data of the aromatic C-nitroso compounds

Compound	Form	ν_0 (kHz)	ν_- (kHz)	ν_+ (kHz)	χ (kHz)	η
N,N-dimethyl-4-nitrosoaniline	Monomer	–[a]	3000[a]	3000[a]	4000[a]	–[a]
		1368[b]	3658[b]	5026[b]	5790[b]	0.473[b]
		1474[b]	3568[b]	5132[b]	5860[b]	0.503[b]
4-iodonitrosobenzene	Monomer	1712	3652	5364	6011	0.570
Nitrosobenzene	Cis dimer	969	1391	2360	2500	0.775
2,6-Dimethylnitrosobenzene	Trans dimer	1000	1253	2267	2347	0.852
1,4-dinitrosobenzene	Polymer	1043	1357	2400	2505	0.833
Benzofuroxan (Benzofurazan-1-oxide)	–	895	1737	2632	2912	0.615
		1737[b]	3000[b]	4737[b]	5158[b]	0.674[b]
		1790[b]	2895[b]	4737[b]	5088[b]	0.704[b]

[a]Amine signal.
[b]Two similar ^{14}N electronic environments.

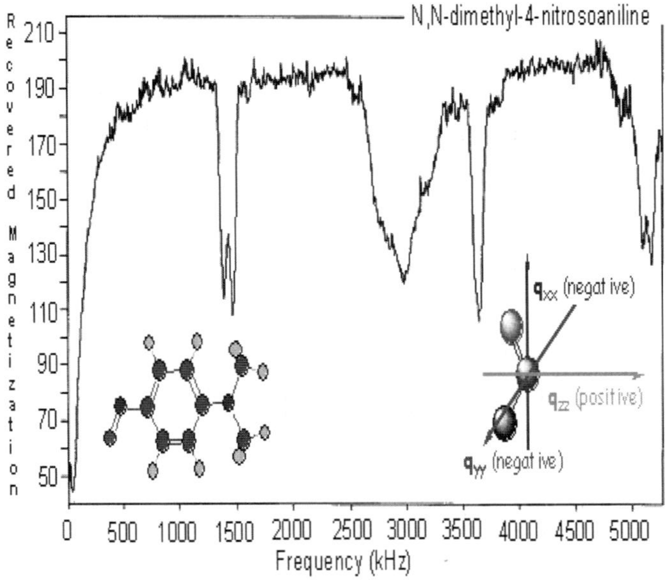

Figure 1. ^{14}N NQR spectrum of *N,N*-dimethyl-4-nitrosoaniline (nitroso monomer).

relaxation [2]. For ^{14}N with nuclear spin I = 1, the technique has many advantages. No radio-frequency radiation needs to be applied to the spin system, the method has high sensitivity and all three transitions can be easily detected. The disadvantage is that the ^{14}N spins are observed in a finite magnetic field often resulting in a broadened and distorted signal. For ^{14}N, however, with a small gyromagnetic ratio, the matching fields are usually less than 0.1 T, and the line broadening is small unless the asymmetry parameter is close to zero. Two basic strategies, adiabatic demagnetization and adiabatic remagnetization, were used. Cartesian coordinates based on X-ray data [3, 4] were input into the program GAMESS [1] to calculate EFG values, from which the quadrupole coupling constant (χ) and asymmetry parameter (η) were deduced.

3. Results and discussion

Table I gives the experimental ^{14}N NQR data. Selected NQR spectra are shown in Figures 1 and 2. For nitrosobenzene and benzofuroxan, the high field proton relaxation times were too long to allow the standard demagnetization experiment to be done in a reasonable time. Instead the quicker (but less sensitive) remagnetization approach [2] was used. Thus, the resonance frequencies for these compounds appear as peaks (Figure 1) instead of dips (Figure 2). The nitroso dimers and monomers show very different quadrupole frequencies. The amino nitrogen of *N,N*-dimethyl-4-nitrosoaniline gave a very broad peak, the broadening caused by a small asymmetry parameter. The remaining three peaks, similar to those of 4-iodonitrosobenzene, were assigned to the monomeric nitroso

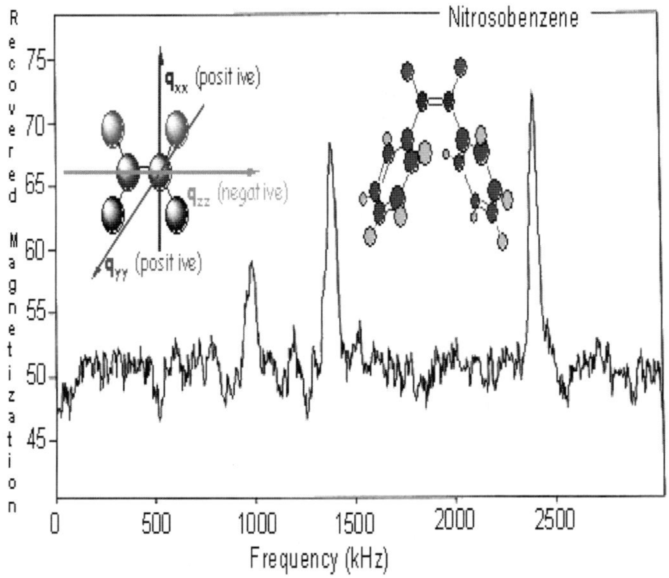

Figure 2. ^{14}N NQR spectrum of nitrosobenzene (nitroso dimer).

Table II. The experimental and theoretical EFG components, χ and η values of *N,N*-dimethyl-4-nitrosoaniline (N_M) and Nitrosobenzene (N_D)

Compound	Electric field gradient (Vm^{-2})			χ (kHz)	η
	q_{zz}	q_{yy}	q_{xx}		
(N_M) (experimental)[a]	(+) 1.196E + 22	(−) 8.810E + 21	(−) 3.155E + 21	5790[a]	0.473[a]
	(+) 1.198E + 22	(−) 9.227E + 21	(−) 2.761E + 21	5860[a]	0.503[a]
(N_M) (theoretical)	+ 1.449E + 22	− 1.204E + 22	− 2.455E + 21	7163	0.661
(N_D) (experimental)[a]	(−)5.167E + 21	(+) 4.587E + 21	(+) 5.813E + 20	2500	0.775
(N_D) (theoretical)	− 4.951E + 21	+ 4.308E + 21	+ 4.848E + 20	2447	0.739

[a] The experimentally determined EFG components were assumed to have the same sign as those obtained by ab-initio calculations.

nitrogen atom (N_M). The appearance of doublets indicated that there was some disorder in the monomer possibly arising from different spatial orientations of the molecule. Only three resonance frequencies were observed for the dimers, indicating that both azodioxy nitrogen atoms (N_D) are in equivalent environments. The characteristic χ and η values of N_M are close to 6000 kHz and 0.5, respectively while N_D has values around 2400 kHz and 0.8. Although the frequencies for both *cis* and *trans* dimers are similar, the η value is slightly higher for the *trans* dimer.

1,4-dinitrosobenzene is a polymer comprised of *N,N*-azodioxy linkages and is used as a rubber vulcanising agent. 1,4-dinitrosobenzene shows three sharp ^{14}N

NQR resonances, suggesting a fairly ordered structure. Its low frequencies, χ and η values are similar to those of N_D. In fact, its η value is closer to that of the *trans* dimer possibly confirming its predominantly *trans* linkages consistent with its observed infrared and ultraviolet spectra. 1,2-dinitrosobenzene only exists as a transient intermediate in the benzofurazan-1-oxide equilibrium. It dimerizes internally to form the stable benzofuroxan at room temperature. The ^{14}N NQR data indicates two different nitrogen atoms present as expected of benzofuroxan. The frequencies of both are comparable to those of N_M and N_D suggesting that the bonding of the nitrogen atoms in benzofuroxan is similar to that in the nitroso compounds. The lower NQR frequencies were assigned to that nitrogen with four (three covalent and one coordinate) bonds similar to that of N_D. The higher frequencies were assigned to second nitrogen atom analogous to N_M. Table II compares the electric field gradients, χ and η values calculated from experimental and ab-initio results.

The calculated χ and η values for N_M were relatively higher than the experimental values as shown in Table II. However, for N_D both values were consistent with the experimental values. This may be explained by the fact that the X-ray data for nitrosobenzene was more accurate. It is also speculated that the quinoid structure contributes to the nitroso monomer and this was not taken into account in the ab-initio calculations. Ab-initio calculations showed that the main EFG component, q_{zz}, in N_M lies along the lone pair on the nitrogen atom. Q_{zz} is positive, indicating a region of depleted electronic charge and associated with closed shell systems. The orientation of all three components is seen in Figure 1. The main EFG component of N_D lies along the N=N bond and is negative, indicating a region of concentrated electronic charge and typical of covalent bonds. All three EFG components can be seen in Figure 2.

References

1. Schmidt M. W., Baldridge K. K., Boatz J. A., Elbert S. T., Gordon M. S., Jensen J. H., Koseki S., Matsunaga N., Nguyen K. A., Su S. J., Windus T. L., Dupuis M. and Montgomery J. A., *J. Comput. Chem.* **14** (1993), 1347–1363.
2. Stephenson D. and Smith J. A. S., *Proc. R. Soc. Lond., A* **416** (1988), 149.
3. Romming C. H. R and Talberg H. J., *Acta Chem. Scand.* **27**(6) (1973), 2246.
4. Dieterich D. A., Paul I. C. and Curtin D. Y., *J. Am. Chem. Soc.* **96**(20) (1974), 6372.

Hyperfine Interactions (2004) 159:143–148
DOI 10.1007/s10751-005-9093-4

NQR, NMR and Crystal Structure Studies of $[C(NH_2)_3]_2HgX_4$ (X = Br, I)

Y. FURUKAWA[1,*], H. TERAO[2], H. ISHIHARA[3], T. M. GESING[4] and J.-C. BUHL[4]

[1]*Department of Science Education, Graduate School of Education, Hiroshima University, 739-8524 Higashi-Hiroshima, Japan; e-mail: yfuruka@hiroshima-u.ac.jp*
[2]*Department of Chemistry, Faculty of Integrated Arts and Sciences, Tokushima University, 770-8502 Tokushima, Japan*
[3]*Faculty of Culture and Education, Saga University, 840-8502 Saga, Japan*
[4]*Institute of Mineralogy, The University of Hannover, 30167 Hannover, Germany*

Abstract. The crystal structure of $[C(NH_2)_3]_2HgBr_4$ has been determined at room temperature: monoclinic, space group $C2/c$, with $a = 10.035(2)$, $b = 11.164(2)$, $c = 13.358(3)$ Å, $\beta = 111.67(3)°$, and $Z = 4$. The crystal consists of planar $[C(NH_2)_3]^+$ and distorted tetrahedral $[HgBr_4]^{2-}$ ions. The Hg atom is located on a two-fold axis such that two sets of inequivalent Br atoms exist in an $[HgBr_4]^{2-}$ ion. In accordance with the crystal structure, two ^{81}Br NQR lines widely separated in frequency were observed between 77 and ca. 380 K. $[C(NH_2)_3]_2HgI_4$ yielded four ^{127}I NQR lines ascribable to $m = \pm 1/2 \leftrightarrow \pm 3/2$ transitions, indicating that its crystal structure is different from the bromide complex. The ^{1}H NMR T_1 measurements showed a single minimum for the bromide but two minima for the iodide. The analyses based on the C_3 reorientations of the planar $[C(NH_2)_3]^+$ ions gave the activation energies of 29.8 kJ mol^{-1} for the bromide, and 30.2 and 40.0 kJ mol^{-1} for the iodide.

Key Words: ^{81}Br and ^{127}I NQR, ^{1}H NMR T_1, crystal structure, molecular motion.

1. Introduction

The guanidinium ion $[C(NH_2)_3]^+$ is interesting in its ability of making hydrogen bonds as well as its unique planar shape. The ions have been found subjected to the hydrogen bonds as many as possible, resulting in the stabilization of crystal structures (see, e.g., [1, 2]). On the other hand, Hg–halogen bonds seem to be especially sensitive to the intermolecular interactions such as hydrogen bonding owing to the soft-acid nature of Hg atom which implies that its bonds are easily sustained to polarization. As an example, through the observation of halogen NQR of a series of piperazinium tet-

* Author for correspondence.

rahalogenometallates(II) of Zn, Cd, and Hg, we found that the halogen NQR resonance lines of the Hg compounds were most widely spread in frequency, in accordance with the order of softness of metals, thus, the polarizability of the bonds Zn–X < Cd–X Hg–X (X = Br, I) [3]. In the dynamical point of view, the guanidinium ions tend interestingly to undergo reorientation motions about their (pseudo) C_3 axes in the crystals [4, 5]. In this study following the previous one for $[C(NH_2)_3]HgX_3$ (X = Br, I) [2], we measured the halogen NQR and ^1H NMR spin-lattice relaxation time (T_1) for $[C(NH_2)_3]_2HgX_4$ (X = Br, I) and also determined the crystal structure of the bromide at room temperature, in order to get information about the molecular motions as well as the structure and bonding.

2. Experimental

Both crystals of $[C(NH_2)_3]_2HgX_4$ (X = Br, I) were obtained by slow concentration of methanol solutions containing HgX_2 and $[C(NH_2)_3]X$ in the molar ratio of 1:2, followed by recrystallization from the same solvent. The results of C, H, and N elemental analyses; found/calc.; wt.%: C: 3.65/3.75; H: 1.72/1.89; N: 13.08/13.12 for $[C(NH_2)_3]_2HgBr_4$; C: 2.91/2.90; H: 1.34/1.46; N: 10.21/10.15 for $[C(NH_2)_3]_2HgI_4$.

The crystal structure was determined at 293 K. The details of the structure determination will be submitted elsewhere.

The NQR spectra were recorded by using a super-regenerative spectrometer with Zeeman modulation. The accuracy of frequency measurements is estimated to be within ± 0.02 MHz. ^1H spin-lattice relaxation times (T_1) were measured at 42.5 MHz by the inversion recovery method on a standard pulse NMR apparatus.

3. Results and discussion

$[C(NH_2)_3]_2HgBr_4$: The crystal structure of $[C(NH_2)_3]_2HgBr_4$ is shown in Figure 1. The crystal belongs to a monoclinic system $C2/c$ with $a = 10.035(2)$, $b = 11.164(2)$, $c = 13.358(3)$ Å, $\beta = 111.67(3)°$, and $Z = 4$. The crystal consists of nearly planar $[C(NH_2)_3]^+$ and distorted tetrahedral $[HgBr_4]^{2-}$ ions. The C–N bond lengths (1.306(1) Å) as well as the N–C–N bond angles (~120°) in the cation are normal. The anion tetrahedron has a twofold symmetry so that two inequivalent Br atoms exist in the anion. The Hg–Br bond lengths are largely different: 2.664(1) Å for Br(1) and 2.559(2) Å for Br(2). The Br–Hg–Br bond angles are also widely spread between 101.6 (Br(1)–Hg–Br(1)′) and 121.7° (Br(2)–Hg–Br(2)′), indicating a severe distortion of the anion. The distortion of anion is the largest among the relevant compounds: $[N(CH_3)_4]_2HgBr_4$ (2.585 − 2.589 Å and 108.0 − 113.7°) [6], $[(CH_3)_2NH_2]_2HgBr_4$ (2.569 − 2.650 Å and 103.0 − 112.9°) [7], and $(CH_3NH_3)_2HgBr_4$ (2.591 − 2.602 Å and 105.3 −

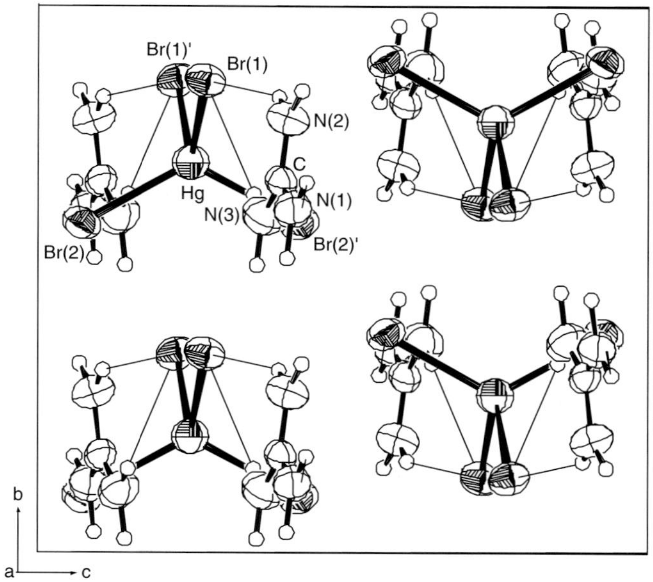

Figure 1. Crystal structure of $[C(NH_2)_3]_2HgBr_4$.

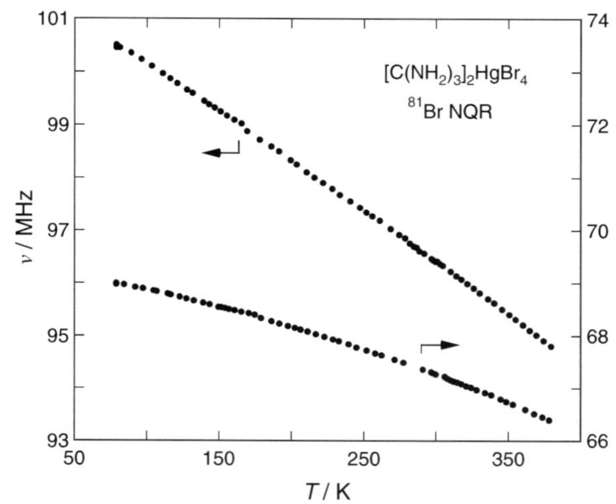

Figure 2. Temperature dependence of ^{81}Br NQR frequencies of $[C(NH_2)_3]_2HgBr_4$.

112.0°) [8]. There are several hydrogen bonds N–H···Br between $[C(NH_2)_3]^+$ and $[HgBr_4]^{2-}$ ions as shown in Figure 1. The shortest H···Br distance is 2.46(10) Å for Br(1) and 2.87(11) Å for Br(2). Thus, the bond of Br(1) atom is expected more ionic than that of Br(2) atom.

In accordance with the structure, two ^{81}Br NQR resonance lines widely separated in frequency were found. The low frequency line (68.98 MHz at 77 K and 67.29 MHz at 298 K) and the high frequency one (100.50 MHz at

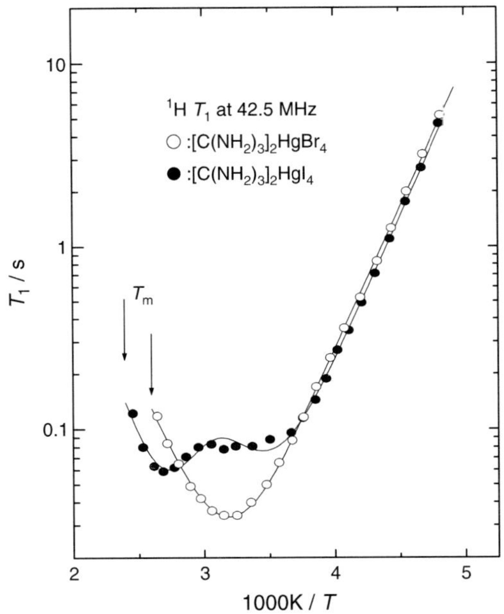

Figure 3. Temperature dependence of ^{1}H NMR spin-lattice relaxation time T_1 at 42.5 MHz of $[C(NH_2)_3]_2HgBr_4$ and $[C(NH_2)_3]_2HgI_4$.

77 K and 96.44 MHz at 298 K) may be assignable to Br(1) with the longer Hg–Br bond and Br(2) with the shorter one, respectively. It is noticed that the frequencies cover a rather wide range compared to those in $[N(CH_3)_4]_2HgBr_4$ (84.02, 88.16, 90.42, 91.91 MHz at 77 K) [9]. As in the latter compound the formation of N–H···Br hydrogen bonds is not possible, this extension in frequency indicates that the hydrogen bonding interactions in addition to ionic ones may affect additional effects on the nature of Hg–Br bonds, followed by electronic redistribution in $[HgBr_4]^{2-}$ ions. The temperature dependence of the NQR frequencies is given in Figure 2. Their frequencies decrease monotonously with increasing temperature from 77 K to ca. 380 K without showing any signs of occurrence of phase transition.

The temperature dependence of ^{1}H NMR T_1 is shown in Figure 3. Only one T_1 minimum with 34 ms was located near 310 K. The planar $[C(NH_2)_3]^+$ ions usually undergo C_3 reorientation in crystalline solids [2, 4, 5], and hence this minimum is assignable to the cationic C_3 reorientation. The temperature dependence of T_1 is then analyzed by using BPP type equation and the Arrhenius relation for the correlation time τ of the reorientation.

$$T_1^{-1} = C\left[\tau/\left(1+\omega^2\tau^2\right) + 4\tau/\left(1+4\omega^2\tau^2\right)\right], \tag{1}$$

and

$$\tau = \tau_0 \exp\left(E_a/RT\right). \tag{2}$$

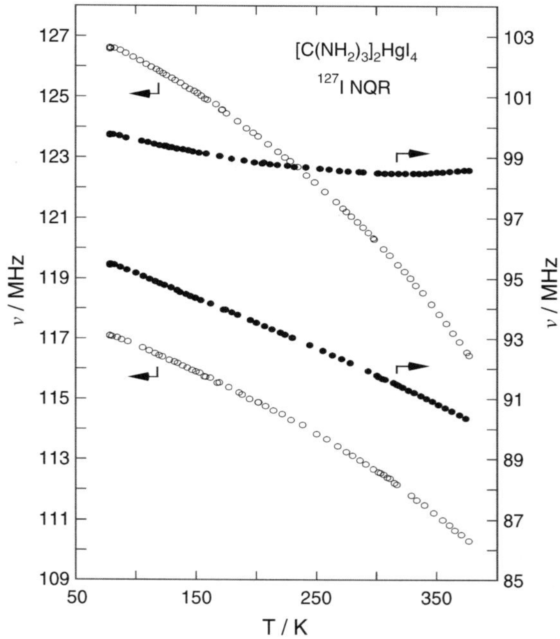

Figure 4. Temperature dependence of ^{127}I NQR frequencies ($m = \pm1/2 \leftrightarrow \pm3/2$) of $[C(NH_2)_3]_2HgI_4$.

By least-squares calculation, the motional constant C, the correlation time at infinite temperature τ_0, and the activation energy E_a were determined as $C = 5.6 \times 10^9$ s^{-2}, $\tau_0 = 2.5 \times 10^{-14}$ s, and $E_a = 29.8$ kJ mol^{-1}. This E_a value is slightly smaller than the value of 35.0 kJmol^{-1} in $[C(NH_2)_3]HgBr_3$ [2].

$[C(NH_2)_3]_2HgI_4$: The crystal structure of $[C(NH_2)_3]_2HgI_4$ has been yet unknown. Four ^{127}I NQR lines ascribable to the $m = \pm1/2$ to $\pm3/2$ transitions were observed: 95.43, 99.73, 117.10, and 126.57 MHz at 77 K (91.81, 98.48, 112.66, and 120.30 MHz at 298 K), indicating the existence of four inequivalent I atoms in the crystal and thus a different structure from the bromide analogue. These frequencies are also widely spread, compared to those in $[N(CH_3)_4]_2HgI_4$ [9] (eight lines in a narrower range between 103.76 and 118.80 MHz at 77 K). This fact may also show that the N–H···I hydrogen bond interactions are reflected on the ^{127}I NQR frequencies, because the N–H···I hydrogen bonds do not exist in $[N(CH_3)_4]_2HgI_4$. As seen in Figure 4, three of the four lines decrease in frequency as normal up to 380 K, but the remaining one has a slightly positive temperature coefficient above ca. 300 K, indicating a dynamical effect on the hydrogen bonds in which the relevant I atoms may be concerned.

Two ^{1}H T_1 minima with different depths near 300 and 370 K (Figure 3) correspond to the existence of at least two kinds of crystallographically and dynamically different $[C(NH_2)_3]^+$ cations. Thus, we postulate the existence of two inequivalent cations in the iodide crystal as in the structures of related compounds, $[N(CH_3)_4]_2HgBr_4$ [6], $[(CH_3)_2NH_2]_2HgBr_4$ [7]; $[CH_3NH_3]_2HgBr_4$

[8], and $[CH_3NH_3]_2HgI_4$ [8], all of which include one type of anions but two independent cations. The T_1 analysis by two sets of Equations (1) and (2) yield motional parameters as $C = 2.3 \times 10^9$ s^{-2}, $\tau_0 = 7.4 \times 10^{-15}$ s, $E_a = 30.2$ kJ mol^{-1}. and $C = 2.8 \times 10^9$ s^{-2}, $\tau_0 = 4.9 \times 10^{-15}$ s, and $E_a = 40.0$ kJmol^{-1}. The τ_0 values of the order 10^{-15} s seem too small as ascribed simply to reorientation motions. Therefore, another approach will be needed for more elaborate analyses. One of such analyses is given by Slotfeldt-Ellingsen and Pedersen [10], where τ_0 is formulated to depend on E_a and include both frequency and entropy factors. As the hindering potential to the C_3 reorientation of cation as a whole is created by intermolecular interactions, the largely different E_a values may reflect their different crystal packing and/or hydrogen bond scheme. From the ^{127}I NQR as well as ^{1}H T_1, no phase transition was evidenced between 77 K and their melting points.

References

1. Krishnan V. G., Shi-qi D. and Weiss Al., *Z. Naturforsch.* **46a** (1991), 1063.
2. Terao H., Hashimoto M., Hashimoto A. and Furukawa Y., *Z. Naturforsch.* **55a** (2000), 230.
3. Ishihara H., Hatano N., Horiuchi K. and Terao H., *Z. Naturforsch.* **57a** (2002), 343.
4. Pajak Z., Grottel M. and Koziol A., *J. Chem. Soc., Faraday Trans.* **2**(78) (1972), 1529.
5. Gima S., Furukawa Y. and Nakamura D., *Ber. Bunsenges. Phys. Chem.* **88** (1984), 939.
6. Kamenar B. and Nagl A., *Acta Crystallogr.* **B32** (1976), 1414.
7. Pabst I., Bats J. W. and Fuess H., *Acta Crystallogr.* **B46** (1990), 503.
8. Körfer M., Fuess H., Bates J. W. and Anorg Z., *Allg. Chem.* **548** (1986), 104.
9. Terao H., Okuda T., Yamada K., Ishihara H. and Weiss A., *Z. Naturforsch.* **51a** (1996), 755.
10. Slotfeldt-Ellingsen D. and Pedersen B., *J. Phys. Chem. Solids* **38** (1977), 65.

Hyperfine Interactions (2004) 159:149–155
DOI 10.1007/s10751-005-9100-9

^{81}Br NQR Study of $[NH_3(CH_2)_nNH_3]CdBr_4$ (n = 4 and 5) and $[NH_3(CH_2)_nNH_3]ZnBr_4$ (n = 5 and 6)

HIDETA ISHIHARA[1,*] and KEIZO HORIUCHI[2]
[1]*Faculty of Culture and Education, Saga University, Honjo-machi 1, Saga 840-8502, Japan*
[2]*Faculty of Science, University of the Ryukyus, Nishihara-cho, Okinawa 903-0213, Japan*
e-mail: horiuchi@sci.u-ryukyu.ac.jp

Abstract. ^{81}Br NQR frequencies and differential scanning calorimetry (DSC) were measured as a function of temperature. $[NH_3(CH_2)_4\ NH_3]CdBr_4$ (**1**) and $[NH_3(CH_2)_5NH_3]CdBr_4$ (**2**) showed a doublet and quartet ^{81}Br NQR spectrum, respectively. $[NH_3(CH_2)_5NH_3]ZnBr_4$ (**3**) and $[NH_3(CH_2)_6NH_3]ZnBr_4$ (**4**) exhibited a four-line ^{81}Br NQR spectrum. From the NQR results, it is inferred that (**1**) and (**2**) consist of infinite two-dimensional sheets of corner-sharing $CdBr_6$ octahedra, whereas (**3**) and (**4**) have isolated $[ZnBr_4]^{2-}$ tetrahedra. All of the crystals except (**1**) showed at least one structural phase transition above 380 K.

Key Words: Br NQR, crystal structure, molecular structure, structural phase transition.

1. Introduction

The compounds of the general formula $(n\text{-}C_nH_{2n+1}NH_3)_2MCl_4$ with M = Mn, Fe, Cu, and Cd crystalizes in the layered-perovskite structure, where layers of an infinite two-dimensional network of corner-sharing divalent-metal-chloride octahedra MCl_6 are attached from both sides by layers of isolated alkylammonium groups. On the other hand, $(n\text{-}C_nH_{2n+1}NH_3)_2MCl_4$ crystals with M = Co, Zn, and Hg are known to have a structure with isolated MCl_4 tetrahedra. These two kinds of compounds, especially the former ones, have attracted much attention of researchers because of their sequential structural phase transitions [1]. The alkylenediammonium compounds $[NH_3(CH_2)_nNH_3]MX_4$ with M = Mn, Fe, Cu, Cd, Co, Zn, Hg, and X = Cl, Br, however, have not been studied extensively so far compared with the normal-alkylammonium compounds $(n\text{-}C_nH_{2n+1}NH_3)_2MX_4$. In order to examine crystal structures and structural phase transitions in the alkylenediammonium compounds $[NH_3(CH_2)_nNH_3]CdBr_4$ (n = 4 and 5) and $[NH_3(CH_2)_nNH_3]ZnBr_4$ (n = 5 and 6), we have measured the temperature dependences of ^{81}Br NQR frequencies and differential scanning calorimetry (DSC). In this paper, we will abbreviate $[NH_3(CH_2)_4NH_3]CdBr_4$,

* Author for correspondence.

Table I. Results of preparation and chemical analysis

Compound	T_m/K	Chemical Analysis (in weight %); found (calculated)		
		C	H	N
$NH_3(CH_2)_4NH_3CdBr_4$	518	9.24(9.20)	3.46(3.35)	4.63(4.65)
$NH_3(CH_2)_5NH_3CdBr_4$	499	11.24(11.19)	2.99(3.00)	5.27(5.22)
$NH_3(CH_2)_5NH_3ZnBr_4$	504	12.30(12.27)	3.32(3.29)	5.76(5.72)
$NH_3(CH_2)_6NH_3ZnBr_4$	475	14.39(14.32)	3.61(3.60)	5.61(5.56)

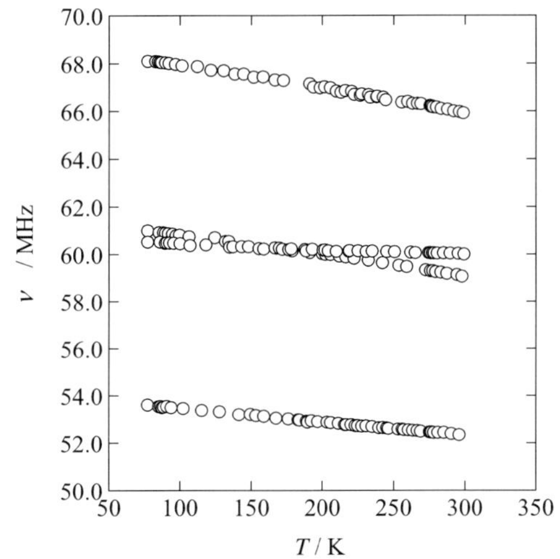

Figure 1. Temperature dependences of ^{81}Br NQR frequencies in 1,5-pentylenediammonium tetrabromozincate(II) $[NH_3(CH_2)_5NH_3]ZnBr_4$.

$[NH_3(CH_2)_5NH_3]CdBr_4$, $[NH_3(CH_2)_5NH_3]ZnBr_4$, and $[NH_3(CH_2)_6NH_3]ZnBr_4$ as buCd, peCd, peZn, and heZn, respectively.

2. Experimental

Samples were prepared as follows. $[NH_3(CH_2)_4NH_3]CdBr_4$: $[NH_3(CH_2)_4NH_3]Br_2$ solution obtained by neutralizing $NH_2(CH_2)_4NH_2$ with HBr, and $CdBr_2$ solution obtained by dissolving $CdCO_3$ in HBr, were mixed with a molar ratio of 1:1; the mixed solution was adjusted at pH < 3 and polycrystalline samples were obtained by a slow evaporation of water in a desiccator over P_2O_5. The other samples were obtained in a similar manner. The results of chemical analysis are given in Table I. All crystals are colourless. buCd and peCd crystallized in form of tabular and rod, respectively. peZn and heZn crystallized in a block form.

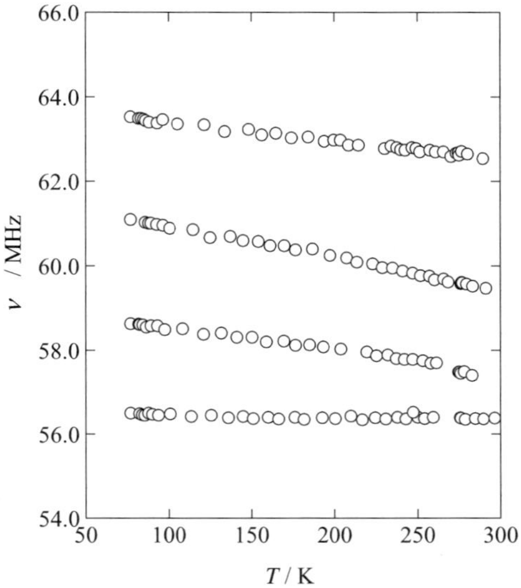

Figure 2. Temperature dependences of ^{81}Br NQR frequencies in 1,6-hexenediammonoim tetrabromozincate(II) $[NH_3(CH_2)_6NH_3]ZnBr_4$.

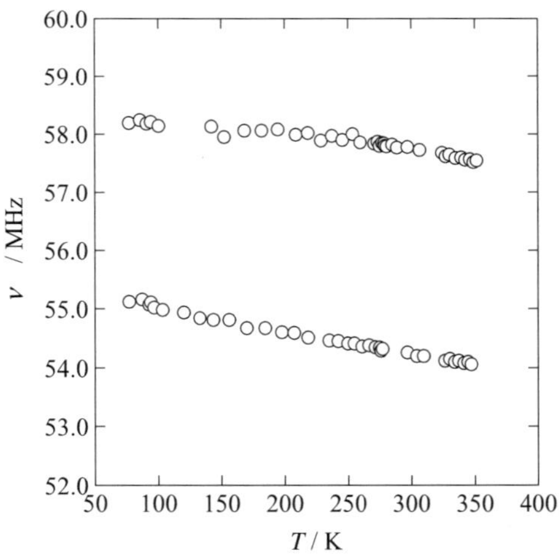

Figure 3. Temperature dependences of ^{81}Br NQR frequencies in 1,4-butylenediammonium tetrabromocadmate(II) $[NH_3(CH_2)_4NH_3]CdBr_4$.

^{81}Br NQR spectra were recorded with an NQR spectrometer working in the super-regenerative mode and resonance frequencies were determined by counting techniques. The sample temperature was measured by a copper-constantan thermocouple. A differential scanning calorimeter DSC220 with a disk-station SSC5200 from Seiko Instruments Inc. was used for thermal measurements. Samples of around 10 mg were employed, and the heating and cooling rates were usually set at 10 and 5 K min^{-1}, respectively. The measurements were carried

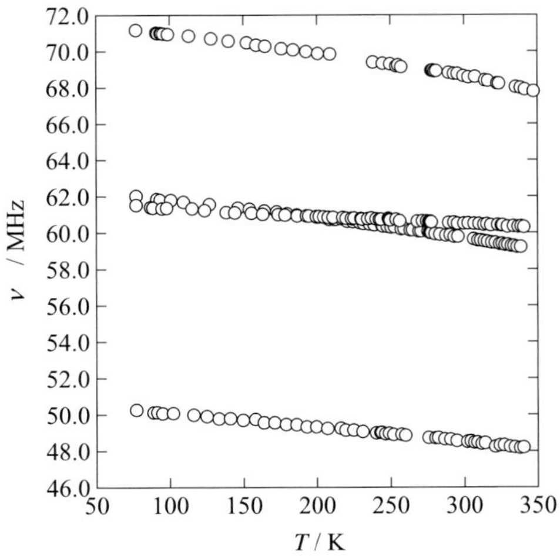

Figure 4. Temperature dependences of [81]Br NQR frequencies in 1,5-pentylenediammonium tetrabromocadmate(II) $[NH_3(CH_2)_5NH_3]CdBr_4$.

Table II. [81]Br NQR frequencies at several temperatures and temperature coefficients $K_\nu = \Delta\nu/\Delta T$ between 77 K and room temperature

Compound	ν/MHz (*T*/K)	ν/MHz (*T*/K)	ν/MHz (*T*/K)	K_ν /kHz K^{-1}
$NH_3(CH_2)_4NH_3CdBr_4$	58.192 (77)	57.78 (297)	57.52 (349)	−1.9
	55.120 (77)	54.26 (297)	54.06 (347)	−3.9
$NH_3(CH_2)_5NH_3CdBr_4$	71.183 (77)	68.76 (296)	67.89 (342)	−11.1
	62.028 (77)	59.79 (296)	59.22 (339)	−10.2
	61.526 (77)	60.50 (296)	60.31 (341)	−4.68
	50.267 (77)	48.56 (295)	48.20 (340)	−7.83
$NH_3(CH_2)_5NH_3ZnBr_4$	68.096 (77)	65.92 (299)		−9.80
	60.998 (77)	59.06 (298)		−8.77
	60.519 (77)	60.00 (299)		−2.3
	53.607 (77)	52.34 (296)		−5.79
$NH_3(CH_2)_6NH_3ZnBr_4$	63.533 (77)	62.54 (290)		−4.7
	61.098 (77)	59.47 (291)		−7.6
	58.635 (77)	57.40 (283)		−6.0
	56.498 (77)	56.38 (297)		−0.54

out under an atmosphere of dry N_2 gas with the flow rate of about 40 ml min^{-1} and repeated more than three times.

3. Results and discussion

[81]Br NQR frequencies were measured above 77 K. The results are shown in Figures 1–4, and the numerical values at some temperatures along with the tem-

perature coefficients $\Delta\nu/\Delta T$ are given in Table II. buCd and the other compounds gave two and four resonance lines with equal intensity, respectively, over the whole temperature region observed. NQR frequencies of all compounds decreased almost proportional to temperature on heating; the temperature coefficients are rather small, especially in buCd, the second-lowest-frequency line in peZn and the lowest line in heZn. These temperature dependences are considered to be ascribed to N–H···Br hydrogen bonds between cations and anions [2].

3.1. $[NH_3(CH_2)_5NH_3]ZnBr_4$ AND $[NH_3(CH_2)_6NH_3]ZnBr_4$

The temperature dependences of ^{81}Br NQR frequencies observed in $[NH_3(CH_2)_3NH_3]ZnBr_4$ (prZn) are similar to those of peZn and heZn [3]. The crystal structure of prZn at room temperature is monoclinic with space group of $P2_1/n$ and $Z = 4$ [3]; there is one crystallographically nonequivalent anion in a unit cell and $[ZnBr_4]^{2-}$ ion adopts a shape of a slightly distorted tetrahedron. From the NQR results in prZn, peZn, and heZn, we can infer that peZn and heZn form an isolated $[ZnBr_4]^{2-}$ tetrahedron, and that the crystal structures of peZn and heZn are very similar to or the same as that of prZn. In this case, four resonance signals should be assigned to four bromine atoms in the same tetrahedron. The NQR frequencies observed for peZn spread over 14.4 MHz at 77 K, while for heZn spread over 7.0 MHz at 77 K. This suggests that a relatively much difference exists in zinc-bromine distances in the $[ZnBr_4]^{2-}$ tetrahedron in peZn.

All of the resonance lines faded out about 300 and 290 K in peZn and heZn, respectively, which can be explained by the fluctuation of the electric field gradients at the bromine nuclei caused by reorientational motions of anions [4]. On the other hand, four ^{81}Br NQR signals in prZn can be observed even at 350 K. This fact can be explained by the difference in interionic interactions in these crystals. prZn has N–H···Br hydrogen-bond networks, where two $[ZnBr_4]^{2-}$ tetrahedra receive nine hydrogen bonds, and its melting point is 577 K [3], which is about 70 and 100 K higher than those of peZn and heZn, respectively. The increase in length of alkylenediammonium cations appears to weaken the cohesive forces in crystals leading to the reduction in the melting points and the activation energies of thermal motions of $[ZnBr_4]^{2-}$ ions.

From the DSC measurements between ca. 130 K and the melting point, a single phase transition was observed at 471 ± 1 K in peZn; moreover, heZn undergoes structural phase transitions at 389 ± 2 and 424 ± 2 K. Although no phase transition was observed below room temperature in peZn and heZn, some normal-alkylammonium compounds $(n\text{-}C_nH_{2n+1}NH_3)_2ZnBr_4$ undergo structural phase transitions below room temperature: $(C_2H_5NH_3)_2ZnBr_4$ (299.5 K); $(n\text{-}C_3H_7NH_3)_2ZnBr_4$ (130 K); $(n\text{-}C_4H_9NH_3)_2ZnBr_4$ (299 K); $(n\text{-}C_5H_{11}NH_3)_2ZnBr_4$ (244 K) [3, 5]. This lattice stability below room temperature in the alkylenediammonium compounds seem to be ascribed to differences in crystal structures between the alkylenediammonium and alkylammonium compounds.

$(n\text{-}C_3H_7NH_3)_2ZnBr_4$ and $(n\text{-}C_4H_9NH_3)_2ZnBr_4$ crystalize monoclinic with space group of $P2_1/c$ and $Z = 4$ at room temperature [3]; the layers made up of $[ZnBr_4]^{2-}$ anions and cations are linked by N–H···Br hydrogen bonds; the cationic layers consist of two kinds of alkylammonium cations; although both cations are connected with the anionic layers by the hydrogen bonds, they are to each other joined by van der Waals interactions. On the other hand, cations and anions in prZn are joined by the N–H···Br hydrogen bonds. This relatively strong cohesive forces seem to make the alkylenediammonium compounds stable below room temperature.

3.2. $[NH_3(CH_2)_4NH_3]CdBr_4$ AND $[NH_3(CH_2)_5NH_3]CdBr_4$

$[NH_3(CH_2)_3NH_3]CdBr_4$ (prCd) gives a doublet ^{81}Br NQR spectrum and crystallizes orthorhombic with space group of *Pnma* and $Z = 4$ at room temperature [6]; Cd is surrounded by six bromine atoms, and a nearly regular octahedron $CdBr_6$ is formed. This structure has two kinds of bromine atoms, i.e., terminal and bridging bromines. The crystal structure of buCd is considered to be very similar to that of prCd, that is, anions in buCd adopt a shape of corner-sharing octahedra. The fact that strong NQR signals can be observed even at 350 K is also explained by this structure. Compared with the isolated tetrahedral structure, the two-dimensional network structure is tightly joined to each other, preventing anions from performing thermal motions. The frequency difference between high- and low-frequency lines at 77 K in prCd is 15.2 MHz, on the other hand that in buCd is only 3.1 MHz, suggesting that in buCd two kinds of Cd–Br bond distances are very close.

The crystal structure of peCd is clearly different from those of buCd and prCd. The temperature dependences of NQR frequencies in peCd are very similar to those in peZn and there are some $A(I)_2CdX_4$ or $A(II)CdX_4$ compounds having isolated tetrahedral anions [7]. Hence it may be difficult to infer the structure of peCd only from the present results. Nevertheless, it seems to be natural to conclude that peCd has octahedral anions.

A single heat anomaly due to a structural phase transition was observed at 419 ± 2 K in thermal measurements for peCd between ca. 130 K and the melting point of 499 K. On the other hand, for buCd no structural phase transition could be detected from NQR and DSC measurements. Hence, we can conclude that the crystal lattices of both compounds are very stable. This lattice stability results from anions being tightly bound to each other, and cations and anions being connected by N–H···Br hydrogen bonds.

Acknowledgements

This work is partly supported by a Grant-in-Aids for Scientific Research No. 06640450 from the Ministry of Education, Science and Culture, Japan.

References

1. For example: (Mn) Murti P., Kind R. and Buehrer W., *Phys. Rev. B* **38** (1988), 666; (Cu) Jahn I. R., Knorr K. and Ihringer J., *J. Phys., Condens. Matter* **1** (1989), 6005; (Cd) Chapuis G., Kind R. and Arend H., *Phys. Status Solidi, A* **36** (1976), 285; (Zn) Sakiyama Y., Horiuchi K. and Ikeda R., *J. Phys., Condens. Matter* **8** (1996), 5345.
2. Nakamura D., Ikeda R. and Kubo M., *Coord. Chem. Rev.* **17** (1965), 281; Horiuchi K., Sasane A., Mori Y., Asaji T. and Nakamura D., *Bull. Chem. Soc. Jpn.* **59** (1986), 2639; Horiuchi K., *J. Chem. Soc., Faraday Trans.* **89** (1993), 3359.
3. Ishihara H., Dou S.-q., Horiuchi K., Paulus H., Fuess H. and Weiss Al., *Z. Naturforsch.* **52a** (1997), 550.
4. Chihara H. and Nakamura N., *Adv. Nucl. Quadrup. Reson.* **4** (1980), 1.
5. Horiuchi K., *J. Phys. Soc. Jpn.* **63** (1994), 363; Horiuchi K. and Weiss Al., *J. Mol. Struc.* **345** (1995), 97.
6. Ishihara H., Dou S.-q., Horiuchi K., Krishnan V. G., Paulus H., Fuess H. and Weiss Al., *Z. Naturforsch.* **51a** (1996), 1216.
7. For example; Ishihara H., Dou S.-q., Horiuchi K., Krishnan V. G., Paulus H., Fuess H. and Weiss Al., *Z. Naturforsch.* **51a** (1996), 1027; Ishihara H., Horiuchi K., Dou S.-q., Gesing T. M., Buhl J.-C., Paulus H. and Fuess H., *Z. Naturforsch.* **53a** (1998), 717.

Hyperfine Interactions (2004) 159:157–172
DOI 10.1007/s10751-005-9101-8

Multi-Frequency Resonances in Pure Multiple-Pulse NQR

G. B. FURMAN[1,*], G. E. KIBRIK[2] and A. YU. POLYAKOV[2]
[1]*Ben-Gurion University, Be'er-Sheva, Israel; e-mail: gregoryf@bgu.ac.il*
[2]*Perm State University, Perm, Russia*

Abstract. We have observed multi-frequency resonances in a system with a spin 3/2 irradiated simultaneously by a multiple-pulse radiofrequency sequence and a low frequency field swept in the range 0 ÷ 80 kHz. The theoretical description of the effect is presented using both the rotating frame approximation and the Floquet theory. Both approaches give indentical results at the calculation of the resonance frequencies, transition probabilities and shifts of resonance frequency. The calculated magnetization vs. the frequency of the low-frequency field agrees with the obtained experimental data.

1. Introduction

One of the most effective and promising high-resolution nuclear magnetic resonance (NMR) and nuclear quadrupole resonance (NQR) techniques for the study of solids is a multiple-pulse radiofrequency (RF) action [1, 2]. The multiple-pulse methods allow one to remove dipolar broadening from a resonance line in solids thus increasing by several orders the sensitivity of the NMR and NQR spectroscopy in the study of weak interaction. These methods are very effective in the study of the spin lattice relaxation processes due to a slow atomic motion. Usually the theoretical description of multiple-pulse experiments both in NMR [3] and NQR [4] is based on the construction of the effective time-independent Hamiltonian by using the conditions for periodicity and cyclicity of the pulsed action. Then the dynamics of a spin system subjected by pulsed RF fields is presented in an equivalent form as the motion of nuclear spins in a constant effective field H_e [5]. The magnitude and direction of this effective field are determined by parameters of the multiple-pulse sequence. An experimental measurement of the value of the effective field is important for the confirmation of this theoretical model.

It is reasonable to suggest that an additional field with an angular frequency Ω close to $\omega_e = \gamma H_e$ should cause resonance absorption of energy (γ is the gyro-magnetic ratio of nuclei). Spin–echo signals observed between RF pulse se-

* Author for correspondence.

quence would allow us to determine H_e as well as to obtain the information on slow atomic motion that is not available from the traditional methods.

With this in mind, we have studied experimentally resonance transitions in the nuclear spin system subjected by a simultaneous action of a multiple-pulse RF sequence and an additional low frequency (LF) field with an angular frequency Ω. The results of our experiments described in the next section have shown that resonance transitions were observed not only at the frequency close to $\Omega_0 = \omega_e$, but also at frequencies close to Ω_n given by the expression:

$$\Omega_n = |\omega_e \pm 2\pi n/t_c|, \quad n = \pm 1, \pm 2 \ldots \tag{1}$$

where t_c is the period of the multiple-pulse RF sequence.

Multiple resonance modes of higher orders have been detected by microwave spectroscopy [6], molecular beam technique [7], optical pumping [8] and observed previously in NMR experiments [9–13]. However the amplitude of these resonances decreased abruptly with the mode order of the resonance. As distinct from this, the amplitude of the resonances observed in our experiments decreased slowly and the resonances of higher orders were well observable.

Because the nuclear spin system possesses a set of the resonance frequencies, the relaxation measurements performed on one resonance frequency Ω_n can give the information on oscillations of atoms on all the frequencies from the spectrum determined by (1). It moves us to comprehensive experimental and theoretical study of this system.

The theoretical treatment of NMR phenomena is usually based on three approaches: i) a semi-classical mathematical approach [14]; ii) a second quantization method [15]; and iii) the Floquet theory [17].

The semi-classical mathematical approach [14], where the field is considered as a classical system and the atomic system as a quantum one, has allowed one to explain of series of experimentally observed phenomena. This approach is quite natural if to take into account, that the average number of photons in a mode of the periodic field is extremely great. The main method used in the framework of the semi-classical approach is the so-called "rotating frame approximation", keeping exactly just the terms that are resonant. The remaining non-resonant terms are considered as a perturbation.

Intrinsic inconsistency of the semi-classical approach is obviated in the framework of the secondary quantization method [15, 16]. Treating the RF field as photons, the evolution of the united system "atom + field" (so-called "dressed" atom) is described by the Hamiltonian which is independent of time, and its investigation turns out simpler than solving the Schrödinger equation with the time-dependent Hamiltonian. With the time-independent Hamiltonian, one can define energy levels of the physical system. Each of these levels corresponds to a stationary state of the system: atom dressed by photons. The dependence of the energy of the united system on the field frequency allows one to construct the diagram of levels. The coupling between RF photons and an atom perturbs the energy levels and allows one to interpret all resonances as "crossings" and "anticrossings" of the conforming levels [18].

The Floquet theory [17] is a power method widely used in NMR spectroscopy for solving time-dependent problems [10, 19]. On the one hand, this approach uses the advantages of the secondary quantization method resulting in the time-independent Hamiltonian. On the other hand, it simplifies calculations as the terms related to the system of free RF photons are not included in the Hamiltonian. The last is justified because an RF field applied in the NMR method contains abundant photons and changes in the state of the photon system may be neglected. In the framework of the Floquet theory, the interaction of the spin system with RF field is considered as completely quantum-mechanical. Therefore, the Floquet theory can be considered to be a bridge between the semi-classical and the quantum methods.

To calculate the transition probabilities and the shift of the center of the resonance line, two theoretical methods are used: the semi-classical approach and the Floquet theory. We extend the theory of the multiphoton resonances [14, 15, 17] to the case of the pure multiple-pulse NQR. The results given by both theoretical approaches are compared with each other and with the experiment.

2. Experimental

The experiments were performed using an automated multiple-pulse NQR spectrometer. Mutually perpendicular continuous LF and pulse RF magnetic fields were generated by crossed coils. Resonances were observed in the effective field of a multiple-pulse RF sequence $(\pi/2)_y - (t_c/2 - \varphi_x - t_c/2)^N$, where φ_x denotes the pulse that rotates the nuclear magnetization about direction of RF field in the rotating frame by an angle φ, and N is the number of pulses in the sequence. This sequence consisted of $N = 256$ pulses and spin locking signal was sampled in the interval between them. The period of the pulse sequence was $t_c = 100$ ms.

The NQR of ^{35}Cl nuclei was observed in polycrystalline $KClO_3$ at 77 K. Without action of the LF field, we recorded the sequence of echo signals in the interval between pulses at 28.9539 MHz with practically constant amplitude (height of the spin echo signals) during the observation time (Figure 1). This amplitude corresponds to a quasi-equilibrium state of the spin system with magnetization M_0.

LF field with the amplitude 2.5 G, nonsynchronized with RF pulses, was swept in the range 1 ÷ 80 KHz. The resonance reduction of the magnetization amplitude was observed (Figure 1) at several different frequencies Ω. The transient process with the spin–spin relaxation time T_2 was followed by establishment of a new quasi-equilibrium state of the spin system with the

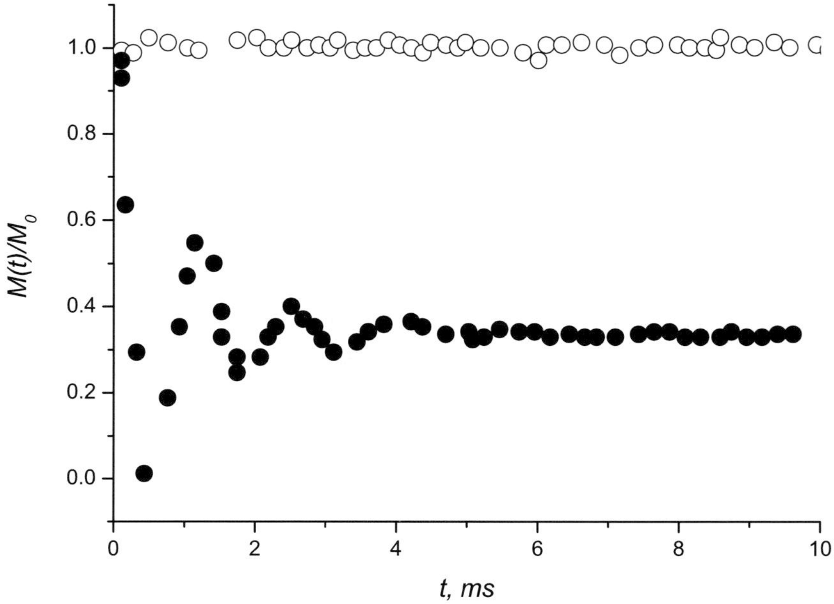

Figure 1. Time dependence of relative magnetization $\frac{M}{M_0}$ in the absence of the LF field (*open circle*) and in the present of the LF field (*solid circle*) at $\varphi = \frac{\pi}{2}$ in polycrystalline $KClO_3$.

reduced amplitude of the magnetization. Our measurements give the value $T_2 =$ 455 ms.

Figure 2 shows the dependence of the relative magnetization M/M_0 on the frequency Ω of the LF field for the $\varphi = \frac{\pi}{2}$ RF pulse (corresponding to the pulse duration $t_w = 15$ μs). The effective frequency of the RF field is determined by $\omega_e = \frac{\varphi}{t_c}$. For example, $\frac{\omega_e}{2\pi} = 2.4$ kHz for $\varphi = \frac{\pi}{2}$.

As follows from Figure 2, the amplitude of the resonances decreases slowly with increasing the mode number n.

3. Theory

3.1. SEMI-CLASSICAL APPROXIMATION: ROTATING FRAME APPROXIMATION

Let us consider a system of nuclear spins $I = 3/2$ placed in an inhomogenous electric field gradiet (EFG) and subjected to a joint action of two time-dependent magnetic fields: a Multiple pulse RF sequence with the angular frequency ω equaled to the resonance frequency ω_0 and a continuous low frequency (LF) field with the angular frequency Ω.

In the rotating frame the equation for the density matrix of the system takes the form:

$$i\frac{d\rho(t)}{dt} = [\mathcal{H}(t), \rho(t)], \tag{2}$$

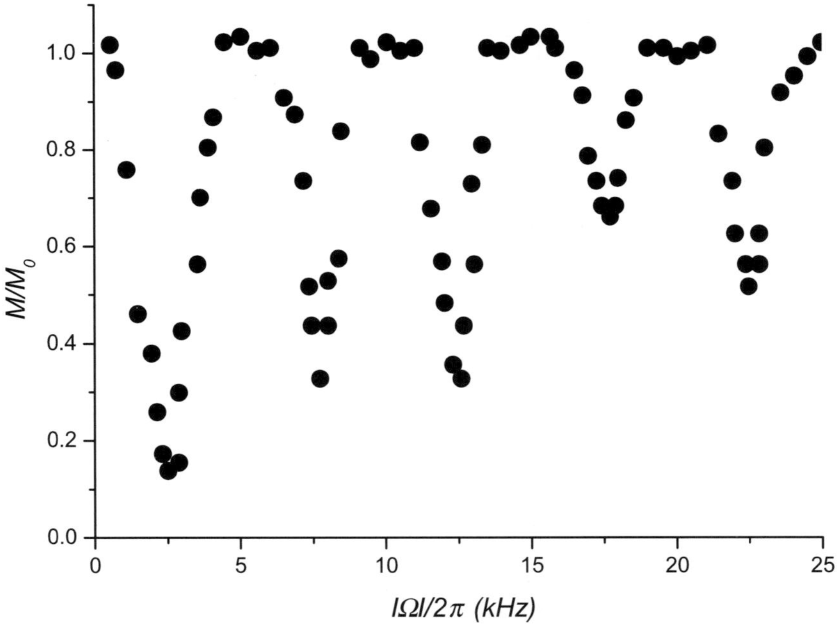

Figure 2. Dependence of relative magnetization $\frac{M}{M_0}$ on the LF field frequency Ω at $\varphi = \frac{\pi}{2}$ in polycrystalline $KClO_3$.

where the Hamiltonian of the system is

$$\mathcal{H}(t) = 2\omega_2 S_z \cos \Omega t + S_x \varphi \sum_{k=1}^{\infty} \delta(t - kt_c - t_c/2), \tag{3}$$

S_x and S_z are *X*- and *Z*-components of the effective spin operator [4]; the pulse angle $\varphi = \gamma H_1 t_w$ and $0 < \varphi < 2\pi$; $\omega_2 = \gamma H_2$; H_1 and H_2 are the amplitudes of the RF pulse and LF fields, respectively. The initial phase of LF field is chosen equal zero.

To solve Equation (2), we apply the unitary transformation

$$\widetilde{\rho}(t) = e^{-i\Omega S_x t} P^+(t)\rho(t)P(t)e^{i\Omega S_x t}, \tag{4}$$

where the unitary operator $P(t)$ is given by the solution of the equation

$$i\frac{dP(t)}{dt} = S_x \varphi \sum_{k=1}^{\infty} \delta(t - kt_c - t_c/2)P(t) - P(t)\mathcal{H}_e \tag{5}$$

with the initial condition

$$P(0) = 1 \tag{6}$$

and the effective time-independent Hamiltonian

$$\mathcal{H}_e = \omega_e S_x. \tag{7}$$

After performing the unitary transformation (4) we obtain the following equation for the transformed density matrix $\widetilde{\rho}(t)$

$$\frac{d\widetilde{\rho}(t)}{dt} = [\mathcal{H}^{tr}(t), \widetilde{\rho}(t)] \tag{8}$$

with

$$\mathcal{H}^{tr}(t) = (\omega_e - \Omega)S_x + \widetilde{S}_z(t), \tag{9}$$

where $\widetilde{S}_z(t) = P^+(t)S_zP(t)$ is the periodic function of time and can be expanded into the Fourier series

$$\widetilde{S}_z(t) = S_+ \sum_{n=-\infty}^{\infty} b_n e^{-in\omega_c t} + S_- \sum_{n=-\infty}^{\infty} b_n e^{in\omega_c t} \tag{10}$$

where $S_\pm = S_z \pm iS_y$, $b_n = \omega_2 \frac{(-1)^n \sin\frac{\varphi}{2}}{\varphi + 2\pi n}$ are the Fourier coefficients, and $\omega_c = \frac{2\pi}{t_c}$.

The resonance terms in the solution of Equation (8) are determined by the lowest values of the differences $\xi = \left|\omega_e - \Omega \pm n\omega_c\right|$. At $\xi = 0$ we obtain the expression (1) for resonance conditions. The rest non-resonant terms determine a frequency shift Δ for the position of the resonance dips shown in Figures 1 and 2. Let us estimate this shift as a function of the frequency Ω using the average Hamiltonian theory [21, 24]. We will consider three various cases depending on the value of the frequency Ω.

The first case, $\Omega = 0$, corresponds to the appearance of an additional constant magnetic field along Z-axes resulting in the difference between ω and ω_0.

To determine the influence of all the non-resonant terms, we will perform the unitary transformation under Equation (8),

$$\rho'(t) = e^{i\varphi S_x \frac{t}{t_c}} \widetilde{\rho}(t) e^{-i\varphi S_x \frac{t}{t_c}}, \tag{11}$$

which leads to

$$\frac{d\rho'(t)}{dt} = [\mathcal{H}'(t), \rho'(t)]$$

$$= \left[S_+ \sum_{n=-\infty}^{\infty} b_n e^{-i(2\pi n + \varphi)\frac{t}{t_c}} + S_- \sum_{n=-\infty}^{\infty} b_n e^{i(2\pi n + \varphi)\frac{t}{t_c}}, \rho'(t)\right]. \tag{12}$$

Let us replace $H'(t)$ by the average Hamiltonian up to second order of the expansion in ω_2/ω_e:

$$\mathcal{H}^{av} = t_c \sum_{n=-\infty}^{\infty} \frac{b_n^2}{2(2\pi n + \varphi)} [S_-, S_+], \tag{13}$$

Note that in the case of non-zero initial phase of the LF field $\overline{H}$ includes $|b_n|^2$. Therefore the initial phase of the LF field does not influence the result. Performing the summation in Equation (13) we obtain

$$\Delta_1 = \frac{\omega_2^2 t_c}{8} \cot \frac{1}{2}\varphi \text{ with } \varphi \neq -2\pi n. \tag{14}$$

In the limit that $t_c \to 0$, a one long pulse, we obtain from Equation (14) a simple expression for the shift $\Delta_1 = \frac{\omega_2^2}{4\omega_e}$, which is similar to the Bloch–Siegert results [22]. However, in contrast to the Bloch–Siegert shift which is produced by the RF field and has the value of the order of $\left(\frac{\omega_1}{\omega_0}\right)^2 \sim 10^{-6}$, the obtained shift is sufficiently higher because it is determined by the ratio of two weak fields with the amplitudes H_2 and H_e. For example, for the angle $\varphi = \frac{\pi}{2}$, the shift Δ_1 is of the order of 10^{-2}. The shift Δ_1 is caused by a deviation of the frequency of the applied RF field from the Larmor frequency and should be taken into account when the results of multiple-pulse experiments are analyzed.

The second case: $\Omega = \omega_e$. The average Hamiltonian $\overline{H}$ is

$$\mathcal{H}^{av} = 2b_0 S_x + \sum_{n=-\infty,\, n\neq 0}^{\infty} \frac{b_n^2}{2n\omega_c}[S_-, S_+]. \tag{15}$$

After the summation in Equation (15) we obtain the shift due to common action of the non-resonant terms

$$\Delta_2 = \frac{\omega_2^2 t_c}{\varphi^3}\left[\sin^2 \frac{1}{2}\varphi - \frac{\varphi}{8}(\varphi + \sin\varphi)\right]. \tag{16}$$

Thus, the resonance frequency differs from ω_e. A shift appears also when we consider $\Omega = \omega_e + n\omega_c$.

In order to determine correctly the resonance conditions, let us consider the third case: $\Omega \neq \omega_e + n\omega_c$. Using the transformation

$$\rho''(t) = e^{i(\varphi-\Phi)S_x \frac{t}{t_c}} \widetilde{\rho}(t) e^{-i(\varphi-\Phi)S_x \frac{t}{t_c}}, \tag{17}$$

where $\Phi = \Omega t_c$, we obtain

$$\frac{d\rho''(t)}{dt} = \left[S_+ \sum_{n=-\infty}^{\infty} b_n e^{-i(\varphi-\Phi+2\pi n)\frac{t}{t_c}} + S_- \sum_{n=-\infty}^{\infty} b_n e^{i(\varphi-\Phi+2\pi n)\frac{t}{t_c}}, \rho''(t)\right]. \tag{18}$$

Let us exclude from the sums in (18) the terms with $n = -\frac{\varphi-\Phi}{2\pi}$:

$$\frac{d\rho''(t)}{dt} = \left[(-1)^{\frac{\varphi-\Phi}{2\pi}} \cdot \frac{w_2}{\Phi} \sin\frac{\varphi}{2} S_x + S_+ \sum_{n=-\infty, n\neq -\frac{\varphi-\Phi}{2\pi}}^{\infty} b_n e^{-i(\varphi-\Phi+2\pi n)\frac{t}{t_c}} \right.$$
$$\left. + S_- \sum_{n=-\infty,\, n\neq -\frac{\varphi-\Phi}{2\pi}}^{\infty} b_n e^{i(\varphi-\Phi+2\pi n)\frac{t}{t_c}}, \rho''(t)\right] \tag{19}$$

The average Hamiltonian up to second order of the expansion in ω_2/ω_e

$$\mathcal{H}^{av} = (-1)^{\frac{\varphi-\Phi}{2\pi}} \frac{\omega_2}{\Phi} \sin\frac{\varphi}{2} S_x + \omega_2^2 t_c \sin^2\frac{\varphi}{2} \sum_{n=-\infty, n\neq -\frac{\varphi-\Phi}{2\pi}}^{\infty} \frac{1}{(\varphi+2\pi n)^2(\varphi-\Phi+2\pi n)} [S_-, S_+]. \quad (20)$$

Because in the considered case $\varphi - \Phi \neq -2\pi n$, the first term in (20) disappears and the summation over all n gives the following equation for the frequency shift:

$$\Delta_3 = \omega_2^2 t_c \left(1 - \frac{2\sin\frac{\varphi}{2}\sin\frac{\Phi}{2}}{\Phi\sin\frac{\varphi-\Phi}{2}}\right). \quad (21)$$

Emphasize that Φ is a function of Δ_3 and Equation (21) has infinite number of roots. Resonance transitions are realized at $\Omega = \widetilde{\Omega}_n$, where

$$\widetilde{\Omega}_n = \omega_e + \Delta_{3n} + n\omega_c, \ n = 0, \pm 1, \pm 2 \ldots \quad (22)$$

Resonance frequencies calculated according (22) using numerical solutions of (21) for the shift are presented in Tables I and II along with the experimental data. The shift is caused by off-resonant component of the LF field.

The time-average transition probability, $\overline{P}$ can be determined using the method developed in [23]:

$$\overline{P} = \frac{1}{2}\sum_n \frac{b_n^2}{(\omega_e - \Omega - n\omega_c + \Delta_3)^2 + b_n^2}, \quad (23)$$

Since $\left(\frac{\omega_2}{\omega_e}\right)^2 << 1$ (for example, $\left(\frac{\omega_2}{\omega_e}\right)^2 = 6.5 \times 10^{-2}$ for $\varphi - \frac{\pi}{2}$), the transition probabilities become significant only in the vicinity of the resonance frequency determined by the condition (22). The expression for the nuclear magnetization M, the quantity needed to compare with experiment, can be found using Equation (23):

$$\frac{M}{M_0} = 1 - \sum_n \frac{b_n^2}{(\omega_e - \Omega - n\omega_c + \Delta_3)^2 + b_n^2}. \quad (24)$$

Expressions (23) and (24), obtained in the framework of the semi-classical approach, determine a multi-line resonance spectrum of the two-energy level system with spin 1/2 irradiated by multiple pulse RF and LF fields. As follows from (24) the amplitudes of the resonances decreases slightly with the number n. There is disparity between a two-level energy model of the spin system and the multi-line resonance spectrum. In order to make a physical interpretation clear, the energy spectrum and transition probabilities will be calculated in the next section by using the Floquet theory [17].

3.2. FLOQUET THEORY

The Floquet theory provides a quite natural description of the multi-line resonance spectrum of the two-energy level spin system. Let us introduce the time-dependent operator

$$U(t, t_0) = F(t)F^{+}(t_0), \tag{25}$$

where the unitary operator $F(t)$ obeys to the evolution equation

$$\frac{dF(t)}{dt} = \mathcal{H}^{tr}(t)F(t) \tag{26}$$

with the periodic time-dependent Hamiltonian (9). The general form of the solution of Equation (26) is given by the Floquet theorem [17] is

$$F(t) = \Psi(t)e^{-i\mathcal{H}^{F}t}, \tag{27}$$

where $\Psi(t)$ is a periodic function of time with the frequency ω_c, and $\mathcal{H}^F$ is the time-independent effective Hamiltonian. Using the periodic character of the Hamiltonian $\widetilde{H}(t)$ and function $\Psi(t)$, the time-dependent Equation (26) can be rewritten as an infinite set of the coupled equations [17]

$$\sum_{\mu k} \left(\mathcal{H}^{n-k}_{\alpha\mu} + n\omega_c \delta_{nk}\delta_{\alpha\mu}\right) F^{k}_{\mu\beta} = h_\beta F^{n}_{\alpha\beta}, \tag{28}$$

where h_β are diagonal elements of the effective Hamiltonian $\mathcal{H}^F$, $\mathcal{H}^{n-k}_{\alpha\mu}$ and $F^{n}_{\alpha\beta}$ are the Fourier components of the matrix elements of $\mathcal{H}^t(t)$ and $F(t)$:

$$F_{\alpha\beta}(t) = \sum_{n} F^{n}_{\alpha\beta} e^{inw_c t} e^{-ih\beta}, \tag{29}$$

$$\mathcal{H}^{t}_{\alpha\beta}(t) = \sum_{n} \mathcal{H}^{n}_{\alpha\beta} e^{in\omega_c t}. \tag{30}$$

Here $\alpha = \pm 1/2$, $\beta = \mp 1/2$ characterize various effective spin states. Using the orthogonal basis $|\alpha n\rangle$, where n denotes the Fourier component, the term in the brackets of the Equation (28) can be presented as the Floquet Hamiltonian [17]

$$\langle \alpha n|\mathcal{H}_F|\beta k\rangle = H^{n-k}_{\alpha\beta} + n\omega_c\delta_{nk}\delta_{\alpha\beta}, \tag{31}$$

Using the operator basis $e_{an,\beta m}$ [24] with matrix elements $\langle \mu p|e_{an,\beta m}|\eta q\rangle = \delta_{\alpha\mu}\delta_{\beta\eta}\delta_{pn}\delta_{mq}$ the Floquet Hamiltonian can be expressed as

$$\mathcal{H}_F = \sum_{\alpha,n} (\varepsilon_\alpha + n\omega_c)e_{\alpha n,\alpha n} + \sum_{\alpha\neq\beta,n,m} \left(b^{\alpha\beta}_{m-n} e_{\alpha n,\beta m} + b^{\beta\alpha}_{n-m} e_{\beta n,\alpha m}\right) \tag{32}$$

where $\varepsilon_\alpha = \alpha(\omega_e - \Omega)$, $b^{\alpha\beta}_{m-n} = b_{m-n}$ are the Fourier coefficients determined in (10). The first term in (32) is diagonal, while the second one is off-diagonal.

As follows from (32), nonzero off-diagonal matrix elements between $|\alpha n\rangle$ and $|\beta m\rangle$ for various n and m appear in first order of $\frac{\omega_2}{\omega_e}$. Therefore all resonance transitions between different states of the Hamiltonian (32) can be excited in the straight way without involving intermediate states as it is used in the scheme of multiphoton excitation proposed in [10, 19, 25].

The intensity of a resonance transition can be estimated using the perturbation method [7] based on the approximation of the Floquet Hamiltonian H_F (32) by the 2 × 2 matrix ($\alpha \neq \beta$)

$$H_F' = \begin{matrix} \varepsilon_\alpha + m\omega_c + c_{\alpha m} & b_{m-n} + d_{n-m}^{\alpha\beta} \\ b_{m-n} + d_{n-m}^{\alpha\beta} & \varepsilon_\beta + n\omega_c + c_{\beta n} \end{matrix}, \tag{33}$$

with matrix elements corrected by taking into account the rest part of the Hamiltonian: the diagonal elements – by introducing the additional terms $c_{\alpha n}$ which are up to second order to within $\left(\frac{\omega_2}{\omega_e}\right)^2 << 1$ is

$$c_{\alpha n} = \sum_{\mu,k} \frac{\langle \alpha n|\mathcal{H}_F|\mu k\rangle\langle\mu k|\mathcal{H}_F|\alpha n\rangle}{E_{\alpha n} - E_{\mu k}}. \tag{34}$$

and the off-diagonal ones – by adding the terms $d_{n-m}^{\alpha\beta}$ of the same order

$$d_{n-m}^{\alpha\beta} = 2\sum_{\mu k} \frac{\langle \beta n|\mathcal{H}_F|\mu k\rangle\langle\mu k|\mathcal{H}_F|\alpha m\rangle.}{E_{\beta n} - E_{\mu k}}, \tag{35}$$

In (34) and (35) $E_{\beta n}$ are eigenvalues of the Floquet Hamiltonian H_F. Diagonalizing the matrix (33) gives the eigenvalues of the Hamiltonian H_F':

$$\begin{aligned} 2E_{m,n}^{\pm(\alpha\beta)} &= \left[\varepsilon_\alpha + \varepsilon_\beta + (m+n)\omega_c\right] \\ &\pm \left\{\left[(\varepsilon_\alpha + m\omega_c + c_{\alpha m}) - (\varepsilon_\beta + n\omega_c + c_{\beta n})\right]^2 + \left(b_{m-n} + d_{n-m}^{\alpha\beta}\right)^2\right\}^{1/2}. \end{aligned} \tag{36}$$

Substituting the matrix elements (31) into (34) we obtain the expressions for the energy level shifts

$$c_{\alpha n} = -c_{\beta n} = \sum_{k=-\infty}^{\infty} \frac{b_{k-n}^2}{(\varphi - \Phi + 2\pi(k-n))}. \tag{37}$$

The summation in Equation (37) gives the same equation for the frequency shift as above obtained using the rotating frame approximation (see Equation (20)):

$$\Delta_3 = c_{\alpha n} - c_{\beta n} = \frac{\omega_2^2 t_c}{2}\left(1 - \frac{2\sin\frac{\varphi}{2}\sin\frac{\Phi}{2}}{\Phi\sin\frac{\varphi-\Phi}{2}}\right). \tag{38}$$

As follows from Equation (35), the terms $d^{\alpha\beta}_{n-m}$ are proportional to $\left(\frac{\omega_2}{\omega_e}\right)^2$ and can be neglected in Equation (36). Therefore the energy levels are given by

$$E^{\pm(\alpha\beta)}_{m,n} = \frac{1}{2}\left\{(m+n)\omega_c \pm \left[(\omega_e - \Omega - (m-n)\omega_c + \Delta_3)^2 + b^2_{n-m}\right]^{1/2}\right\}. \tag{39}$$

The normalized energy $\frac{2E^{\pm(\alpha\beta)}_{m,n}}{\omega_e}$ is presented in Figure 3 as a function of $\frac{\Omega}{\omega_e}$ for $m - n = 0$.

The average probability [17] of the resonance transition is

$$\overline{P}_{\alpha\to\beta} = \frac{1}{2}\sum_{m,n}\frac{b^2_{m-n}}{(\omega_e - \Omega - (m-n)\omega_c + \Delta_3)^2 + b^2_{m-n}}. \tag{40}$$

The probability calculated according to Equation (40) is presented in Figure 4 at $\varphi = \frac{\pi}{2}$ as a function of $x = \frac{\Omega}{\omega_e}$ for the terms in the sum with $m - n = 0, \pm 1, \pm 2$.

Using (40) the expression for the nuclear magnetization M, is the quantity needed to compare with experiment, can be found:

$$\frac{M}{M_0} = 1 - \sum_{m,n}\frac{b^2_{m-n}}{(\omega_e - \Omega - (m-n)\omega_c + \Delta_3)^2 + b^2_{m-n}}. \tag{41}$$

Equations (40) and (41) coincide (by changing the summation index: $k = m - n$) with the equations obtained for the transition probability and the magnetization in the previous section (see Equations (23) and (24)). The magnetization calculated according to Equation (41) is presented in Figure 5.

4. Results and discussion

As follows clearly from the comparison of expressions (23), (24), (40), and (41), the calculations using both the semi-classical and Floquet methods give the identical results for the transition probability and nuclear magnetization. The results for the resonance conditions and for the shift Δ_3 of the resonant frequency obtained within the framework of these methods are also identical. Moreover, the semi-classical approach allows one to obtain automatically the shift Δ_1 under off-resonance conditions $\omega_0 \neq \omega$ ($\Omega = 0$) from RF excitation. However the interpretation of the multi-frequency resonance transitions for two-level energy system in the framework of the semi-classical approach involves diffculties.

The Floquet method gives clear physical description of observed phenomena. The time-independent Hamiltonian obtained using this method leads to the multi-level energy spectrum. Each of these levels corresponds to a stationary state of the system: spin dressed by photons. The coupling between photons and spin allows one to interpret the resonances as "anticrossing" of the levels [10, 17].

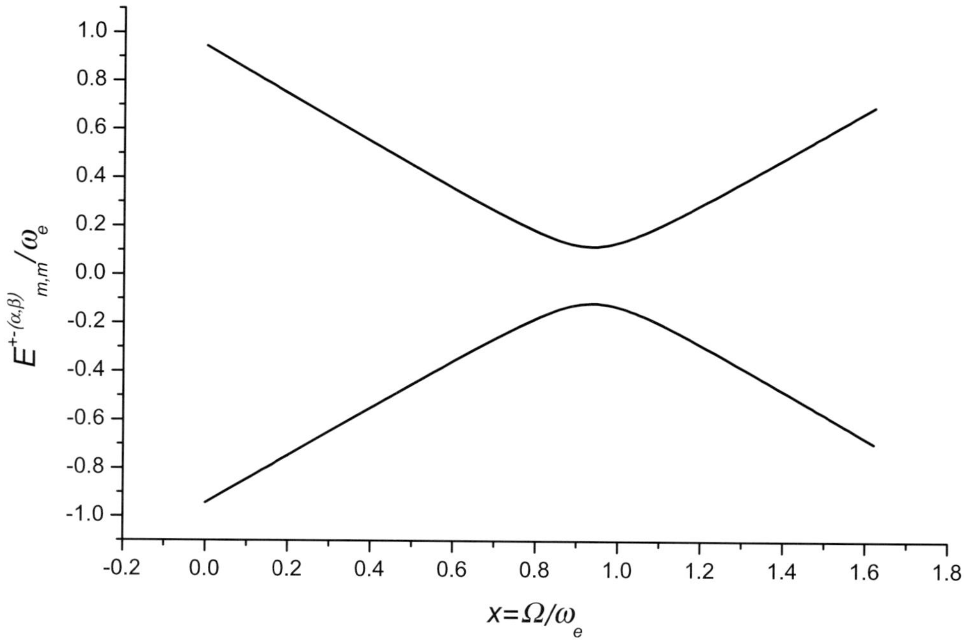

Figure 3. The normalized energy as a function of $x = \frac{\Omega}{w_e}$ for $m - n = 0$.

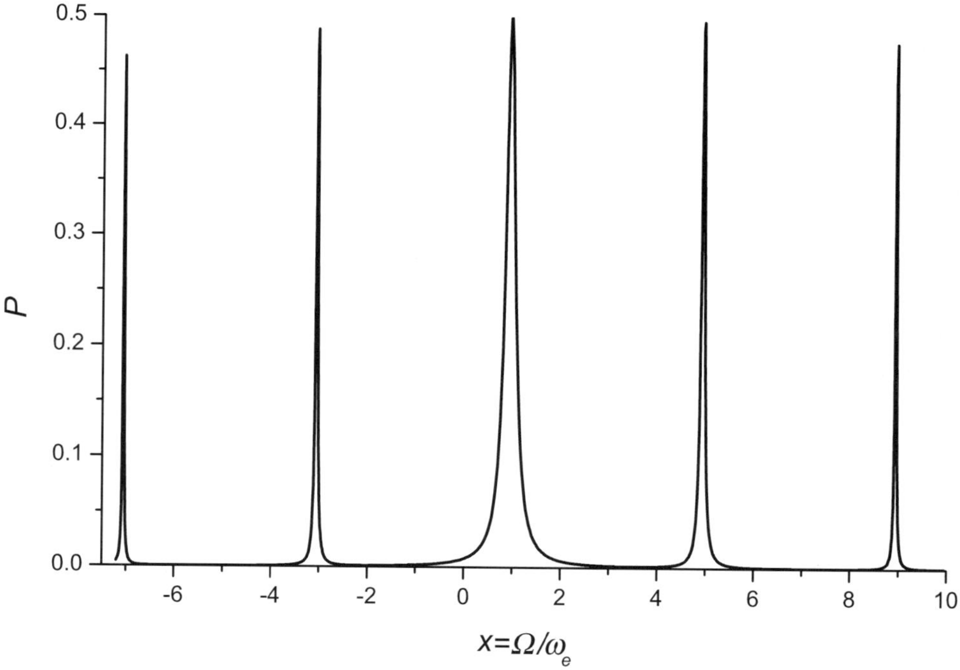

Figure 4. The probabilities as functions of $x = \frac{\Omega}{w_e}$ for $\varphi = \frac{\pi}{2}$ and $m - n = 0, \pm 1, \pm 2$.

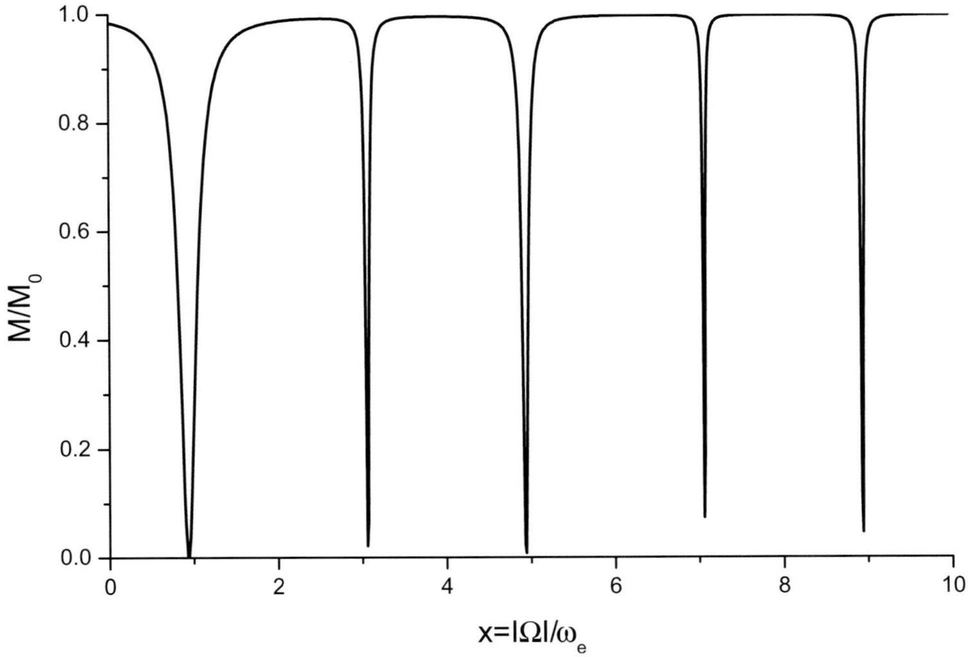

Figure 5. Normalized magnetization $\frac{M}{M_0}$ as function of $x = \left|\Omega/\mathrm{w}_e\right|$ calculated according Equation (41) for $\varphi = \frac{\pi}{2}$ and $m - n = 0, \pm 1, \pm 2$.

Matching of the computed results for the nuclear magnetization with the experimental data displays series of essential differences. One of these differences consists in the fact that the observed NMR signal does not damp up to zero (Figures 1 and 2), while the calculations predict zero magnetization at the resonant frequency (Equation (41) and Figure 5). A possible reason of the observed finite value of $\frac{M}{M_0}$ is an inhomogeneity of the RF field in the bulk of the sample resulting in a distribution of angles φ over the sample. To compare correctly the theoretical result for the nuclear magnetization with the experiment, we have to take into account the distribution of φ over the real resonance line with nonzero line width averaging expression (41) over the line:

$$\left\langle \frac{M}{M_0} \right\rangle = \frac{1}{M_0} \int d\varkappa M(\varkappa) g(\varkappa). \tag{42}$$

Let us consider the Lorentzian distribution of frequencies over a line having width c. It leads to the following distribution of angles φ

$$g(\varkappa) = \frac{1}{\pi} \frac{c}{c^2 + \varkappa^2}, \tag{43}$$

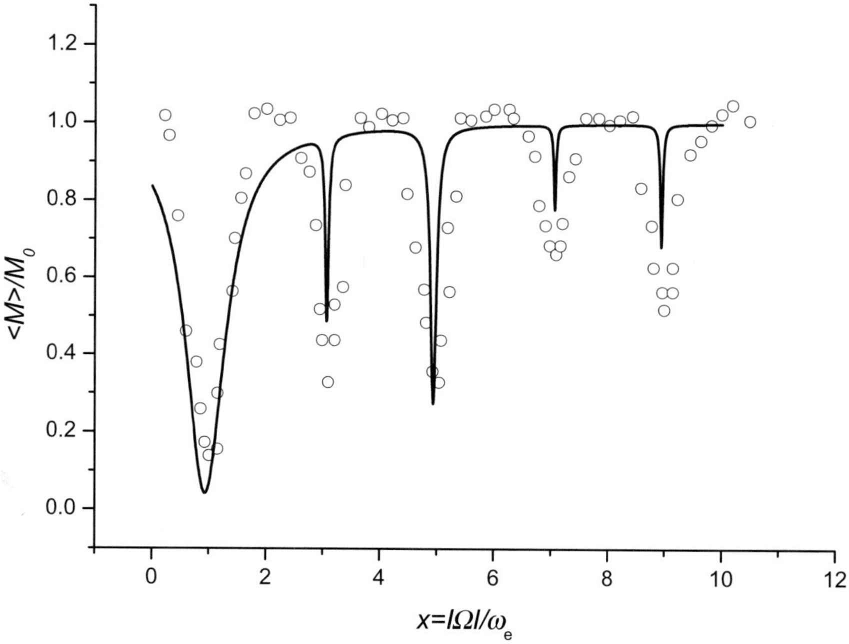

Figure 6. Dependence of averaged magnetization $\left\langle \frac{M}{M_0} \right\rangle$ of $x = |\Omega/\mathrm{w}_e|$ at $\varphi = \frac{\pi}{2}$ in polycrystalline $KClO_3$: *solid curve* – theory (44), *open circle* –experiment.

where $\varkappa = \varphi - \varphi_0$ and $\varphi_0 = \langle \varphi \rangle_V$ is an angle averaged over the sample. After the integration of (42) with the substitution of (43) we obtain

$$\left\langle \frac{M}{M_0} \right\rangle = 1 - \sum_n \frac{1 + a_n c}{a_n^2 \left(1 - x - \frac{2\pi}{\varphi_0} n + \frac{\Delta_3 t_c}{\varphi_0}\right)^2 + (1 + a_n c)^2}, \tag{44}$$

where

$$a_n = \frac{\varphi_0^4}{(\omega_2 t_c)^2 \sin^2 \frac{\varphi_0}{2}} \left(1 + \frac{2\pi}{\varphi_0} n\right)^2, \tag{45}$$

The average magnetization, calculated according Equation (44) with $c = 0.018$, is presented in Figure 6 at $\varphi_0 = \frac{\pi}{2}$ as a function of $x = \left|\frac{\Omega}{w_e}\right|$ along with the experimental results. Using the relationship $c = \langle (\varphi - \varphi_0)/\varphi_0 \rangle$ the dispersion of angles can be estimated as $\langle (\varphi - \varphi_0) \rangle = 2.827\,4 \times 10^{-2}$ rad or 1.62^0. One can see the excellent agreement of the theoretical and experimental results.

The resonance transitions corresponding to negative frequencies Ω (see Figure 5) can be interpreted as ones caused by the rotating in the opposite direction component of the linearly polarized LF field.

The multi-frequency resonances were observed in the experiments with other excitation schemes [10, 19, 25] and explained by multiphoton transitions between different states. Because the multiphoton transitions involves intermediate states and the transition probability decreases with number of photons [10, 19, 25], the intensity of the resonance signal diminishes quickly with number of the mode. A peculiarity of the excitation scheme proposed in the present paper is that, along with multiphoton transitions, there are direct transitions between dressed spin states with $(n - m) > 1$. Therefore, the transition probability is the same for all the differences $n - m$. As a result, the intensity of the resonance signal is slowly decreased with number of the mode (Figures 1 and 2).

5. Conclusions

We have studied both experimentally and theoretically the dynamics of a spin system with the spin 3/2 under simultaneous influence of DC, multiple-pulse RF and LF fields. An important peculiarity of the applied excitation scheme is the possibility of the measurement of the effective field H_e and observation of odd and even resonances with large mode numbers. It was shown that the intensity of the resonance lines decreases slowly.

Both used theoretical approaches, the semi-classical and the Floquet methods, give the identical results at the calculation of the resonant frequencies, magnetization amplitudes and shifts of resonance frequency. These results are in a good agreement with the experimental data for different angles and mode numbers equaled to $n = 0, \pm 1, \pm 2$. Some difference between the theoretical and experimental values of the signal intensity can be explained by the inhomogeneity of the RF field over the sample leading to the dispersion of the angle φ.

The proposed technique can be incorporated for the study of the slow atomic motion in systems involving spin-3/2 nuclei.

Acknowledgement

This research was supported in part by a Grant from the U. S.-Israel Binational Science Foundation (BSF).

References

1. Haeberlen U., *High Resolution NMR Spectroscopy in Solids: Selective Averaging*, Academic, New York, San Francisco, London, 1976.
2. Mehring M., *High Resolution NMR Spectroscopy in Solids*, Springer, Berlin Heidelberg New York, 1983.
3. Ivanov Y. N., Provotorov B. N. and Fel'dman E. B., *Pis'ma v Z. Eksp. Teor. Fiz.* **27** (1978), 164; (*Sov. Phys. JETP* **24** (1978), 135).
4. Ainbinder N. E. and Furman G. B., *Z. Eksp. Teor. Fiz.* **85**, (1983), 988; *Sov. Phys. JETP* **58** (1983), 575.
5. Maricq M. M., *Phys. Rev. Lett.* **56** (1986), 1433.
6. Autler A. and Tones C. H., *Phys. Rev.* **100** (1955), 703.

7. Salwen H., *Phys. Rev.* **99** (1955), 1274.
8. Margerie J. and Brossel J., *Compt. Rend.* **241** (1955), 373.
9. Bucci P. and Santucci S., *Phys. Rev., A* **2** (1970), 1105.
10. Zur Y., Levitt M. H. and Vega S., *J. Chem. Phys.* **78** (1983), 5293.
11. Schmidt A. and Vega S., *J. Chem. Phys.* **96** (1992), 2655.
12. Filip C., Hafner S., Schnell I., Demco D. E. and Spiess H. W., *J. Chem. Phys.* **110** (1999), 423.
13. Hafner S. and Demco D. E., *Solid State Nucl. Magn. Reson.* **22** (2002), 247.
14. Abragam A., *The Principles of Nuclear Magnetism,* Claredon Press, Oxford, 1961.
15. Polonsky P. N. and Cohen-Tannoudji C., *J. de Physique (Paris)* **26** (1965), 409.
16. Cohen-Tannoudji C., *Atom in Electromagnetic Fields*, Word Scientific, London, 1994.
17. Shirley J. H., *Phys. Rev., B* **138** (1965), 979.
18. Cohen-Tannoudji C., Dupont-Roc J. and Fabre C., *J. Phys., B* **6** (1973), L218.
19. Krauss E. M. and Vega S., *Phys. Rev. A* **34** (1986), 333.
20. Allen P. S. and Cowking A., *J. Chem. Phys.* **47** (1967), 4286.
21. Buishivili L. L. and Menabde M., *Sov. Phys. JETP* **50** (1979), 1176.
22. Bloch F. and Siegert A., *Phys. Rev.* **57** (1940), 522.
23. Landau L. D. and Lifshitz E. M., *Quantum Mechanics – Non Relativistic Theory*, Pergamon, Oxford, 1989.
24. Ainbinder N. E. and Furman G. B., *Sov. Phys. JETP* **58** (1983), 575.
25. Gromov I. and Schweinger A., *J. Magn. Reson.* **146** (2000), 110.

Hyperfine Interactions (2004) 159:173–179
DOI 10.1007/s10751-005-9102-7

Magnetism-Related Properties of CdSb Revealed by the Zeeman ^{121}Sb NQR Spectra

V. G. ORLOV[1,*], M. P. SHLIKOV[1], E. A. KRAVCHENKO[2], S. F. MARENKIN[2] and S. A. VARNAVSKII[2]
[1]*Russian Research Center "Kurchatov Institute," Kurchatov sq. 1, 123182 Moscow, Russia; e-mail: orlov@mbslab.kiae.ru*
[2]*Institute of General and Inorganic Chemistry, Russian Academy of Sciences, 117907 Moscow, Russia*

Abstract. The ^{121}Sb NQR spectra of CdSb single crystal were measured in the presence of weak (up to 500 Oe) external magnetic fields. As the analysis of the ^{121}Sb EFG symmetry showed, two magnetically nonequivalent Sb sites exist in the CdSb crystal lattice. An upper limit of the local magnetic field, which may exist in CdSb characterized with the observed ^{121}Sb NQR spectra, was estimated by modeling the Zeeman perturbed patterns. This amounted to H_{loc} ~10 G, hence being much less than earlier found local fields H_{loc} ~30–200 G in bismuth-based compounds.

Key Words: cadmium antimonide, computer modeling, nuclear quadrupole resonance.

PACS: 76.60.Gv, 76.60.Jx.

1. Introduction

Semiconductor CdSb has long been known for the valuable thermoelectric and optical properties. We examined the Zeeman ^{121}Sb NQR spectra of CdSb in connection with the NQR evidences that local magnetic fields (H_{loc}) exist in some Bi-based compounds comprising neither *d*- nor *f*-elements [1–3]. Moreover, the paramagnetism that depends on the conditions of cooling (FC or ZFC) a sample, and the longitudinal magnetoelectric effect were recently observed for α-Bi_2O_3 single crystal [4, 5]. The magnetic properties of CdSb are also of interest due, for instance, to a potential use of this II–V group semiconductor as a host for preparation of diluted magnetic semiconductors.

* Author for correspondence.

2. Experiment and modeling

The synthesis of stoichiometric CdSb was performed in an evacuated quartz ampoule at $T = 480°C$ from high purity reagents. Single crystals were grown in the Ar atmosphere by the Czochralski method (see the details in [6]). CdSb crystallizes in the orthorhombic lattice (space group Pbca (D_{2h}), $Z = 8$), both, Cd and Sb atoms occupying the eight-fold *c* sites of general type in the unit cell [7]. The CdSb single crystal, which was used as a sample, had a shape of a plate with a longer side nearly parallel to the [100] axis and a larger surface, parallel to the (010) plane.

A pulsed non-coherent NQR spectrometer, which enabled continuous scanning through the frequency range required, was used for the experiments. The crystal was successively turned over angle α around the radio-frequency field (H_{rf}) direction, which coincided with the [100] axis of the crystal. The starting position of the sample in the coil ($\alpha = 0°$) was determined by external magnetic field ($0 < H_e < 500$ Oe) being perpendicular to the surface of the plate. $\boldsymbol{H_e}$ and $\boldsymbol{H}_{rf}$ were perpendicular to each other during the measurements. The ^{121}Sb NQR spectral patterns were recorded for each of the transitions ν_1 ($\Delta m = 1/2 - 3/2$) (see Figure 1a) and ν_2 ($\Delta m = 3/2 - 5/2$) (see Figures 1b and 2) between the levels of the nuclear spin $I = 5/2$.

In agreement with the X-ray data [7], the zero-field ^{121}Sb NQR spectrum [8] evidenced for the crystallographic equivalence of the Sb sites in the CdSb structure. The multiplicity of the Zeeman perturbed ^{121}Sb NQR spectra of the CdSb single crystal revealed the magnetic nonequivalence of the Sb sites (see Figure 2), so that the analysis of the ^{121}Sb EFG symmetry properties was carried out. As it was found within the crystal-field potential model [9], four types of the spatially nonequivalent ^{121}Sb EFG tensors exist in CdSb (see [3]). Using the procedure described in [3], the symmetry relations were found for the Euler angles (see Figure 4 in [3]) ϕ_k, θ_k, ψ_k ($k = 1, \ldots, 4$) that determine the orientation of the principal axes (x', y', z') for each EFG tensor with respect to the crystallographic axes (x, y, z) of CdSb:

$$\begin{aligned}
&\phi_1, \theta_1, \psi_1;\\
&\phi_2 = \pi - \phi_1, \theta_2 = \theta_1, \qquad \psi_2 = \pi/2 - \psi_1;\\
&\phi_3 = \phi_1, \qquad \theta_3 = \pi - \theta_1, \psi_3 = 3/2\pi - \psi_1;\\
&\phi_4 = \pi - \phi_1, \theta_4 = \pi - \theta_1, \psi_4 = \pi + \psi_1.
\end{aligned} \tag{1}$$

If H_e and H_{loc} are zero, the energies and eigen functions of the ^{121}Sb nuclear spin can be found by diagonalization of the nuclear electric quadrupole Hamiltonian

$$\hat{H}_Q = \frac{eQq}{4I(2I-1)}\left[3I_{z'}^2 - I(I+1) + \frac{\eta}{2}\left(I_+^2 + I_-^2\right)\right], \tag{2}$$

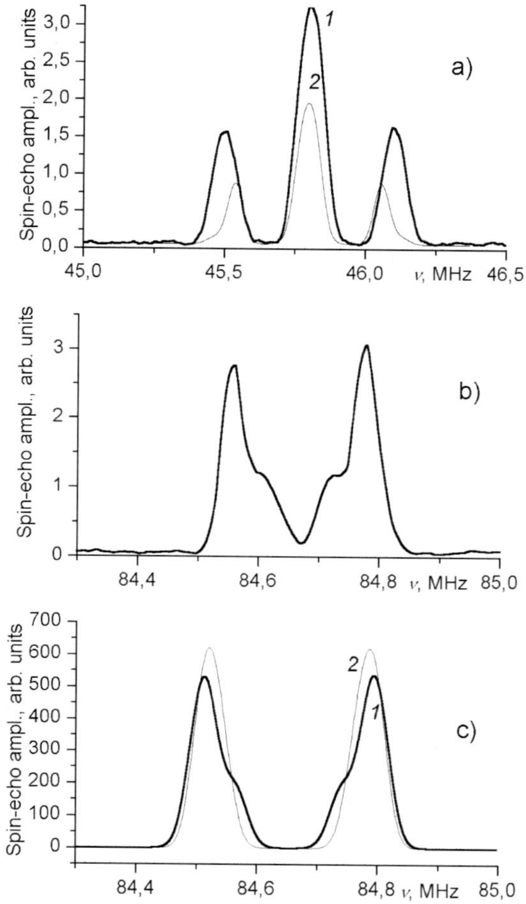

Figure 1. The observed and modeled Zeeman perturbed ν_1 and ν_2 line shapes in the ^{121}Sb NQR spectrum of the CdSb single crystal in $H_e = 225$ Oe, $\alpha = 45°$: a) ν_1, 1 – observed, 2 – modeled; b) ν_2, experimental data; c) ν_2, modeling, 1 – $H_{loc}\cos\theta_l = 10$ Oe, 2 – $H_{loc} = 0$.

where $I_\pm = I_{x'} \pm I_{y'}$. The Zeeman Hamiltonians $\hat{H}_{loc}$ and $\hat{H}_e$ describe, respectively, the interaction of local and external magnetic fields with the magnetic moment of the Sb nucleus

$$\begin{aligned}\hat{H}_{loc} &= -g_{Sb}\mu_B \boldsymbol{H}_{loc}\boldsymbol{I}\\ &= -g_{Sb}\mu_B H_{loc}\Big(\sin\theta_i\cos\phi_i I_{x'} + \sin\theta_i\sin\phi_i I_{y'} + \cos\theta_i I_{z'}\Big),\end{aligned} \qquad (3)$$

where θ_i and ϕ_i determine the direction of $\boldsymbol{H}_{loc}$ with respect to the EFG principal axes, $g_{Sb} = \mu_{Sb}/(I\mu_B)$, μ_B is the nuclear Bohr magneton, $\mu_{Sb} = 3.36\mu_B$ is the magnetic moment of ^{121}Sb nucleus, and

$$\hat{H}_e = -g_{Sb}\mu_B \boldsymbol{H}_e\boldsymbol{I} = -g_{Sb}\mu_B H_e\Big(R_{13}^{(k)} I_{x'} + R_{23}^{(k)} I_{y'} + R_{33}^{(k)} I_{z'}\Big), \qquad (4)$$

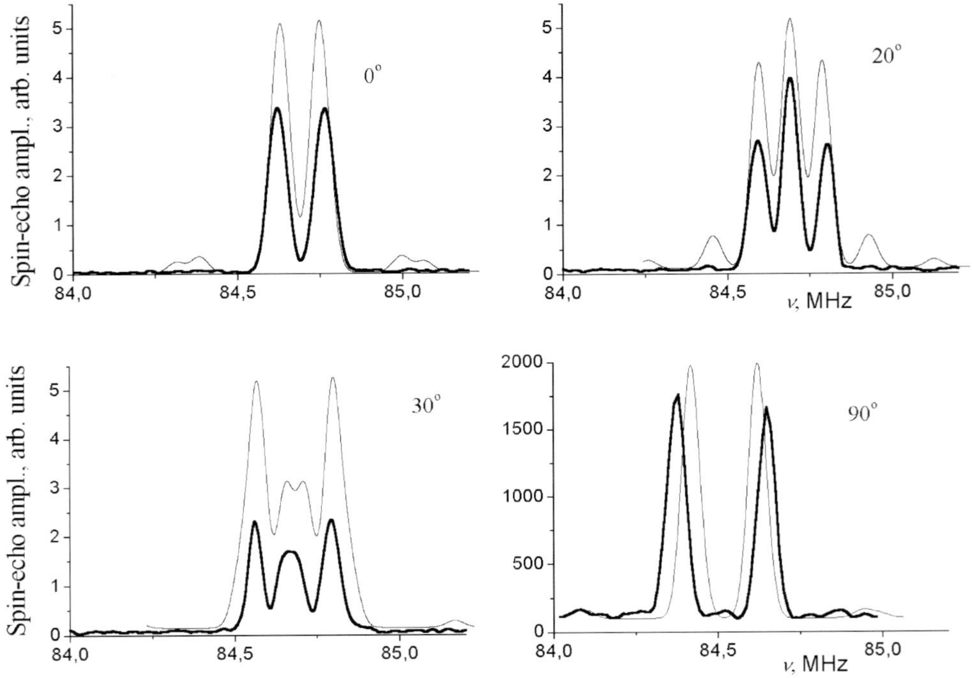

Figure 2. The observed (*thick*) and modeled (*thin*) Zeeman perturbed line shapes assigned to the transitions in the ^{121}Sb NQR spectrum of the CdSb single crystal in H_e = 225 Oe. The rotation angles (α) are given in degrees.

where $R_{i3}^{(k)}$, $i = 1, 2, 3$ are the components of the rotation matrix, which determines the projection of $\boldsymbol{H_e}$ on the ^{121}Sb EFG principal axes, k refers to the number of a set of the angular parameters in Equation (1).

At the starting moment, the crystallographic system of axes (x, y, z) was assumed to coincide with the laboratory frame $(\widetilde{x}, \widetilde{y}, \widetilde{z})$, whose axes are determined by the directions of the external fields as $\boldsymbol{H_e} \| \widetilde{z}$ and $\boldsymbol{H_{rf}} \| \widetilde{x}$. The turning of the crystal with respect to the laboratory frame is described by the matrix $R_{ij}^{(k)}$, which is a product of several matrices, those describing the rotation of the crystal in the coil and the others, corresponding to the rotation over the Euler angles ϕ_k, θ_k and ψ_k. The matrices in the Hamiltonian Equation (4), which relate to the geometry used in the experiments, are:

$$R_{13}^{(k)} = -\cos\psi_k \sin\phi_k \cos\alpha - \sin\psi_k (\cos\theta_k \cos\phi_k \cos\alpha + \sin\theta_k \sin\alpha); \tag{5}$$

$$R_{23}^{(k)} = \sin\psi_k \sin\phi_k \cos\alpha - \cos\psi_k (\cos\theta_k \cos\phi_k \cos\alpha + \sin\theta_k \sin\alpha); \tag{6}$$

$$R_{33}^{(k)} = \sin\theta_k \cos\phi_k \cos\alpha - \cos\theta_k \cos\alpha; \tag{7}$$

It is easy to show that the following symmetry relations are valid for the matrices $R_{ij}^{(k)}$ Equations (5–7):

$$R_{i3}^{(4)} = -R_{i3}^{(1)};\ R_{i3}^{(3)} = -R_{i3}^{(2)} \neq R_{i3}^{(1)};\ (i = 1, 2, 3), \tag{8}$$

where the upper index refers to a certain set of the angular parameters in Equation (1). It means that there are two magnetically nonequivalent Sb sites in the CdSb crystal structure.

If H_e and H_{loc} are nonzero, the eigen functions and energy levels of the ^{121}Sb nuclear spin can be found by diagonalization of the Hamiltonian, which is a sum of the Hamiltonians Equations (2–4):

$$\hat{H} = \hat{H}_Q + \hat{H}_{loc} + \hat{H}_e. \tag{9}$$

To model the complicated Zeeman ^{121}Sb NQR spectra of CdSb, it is necessary to know ten parameters, which enter the Hamiltonian Equation (9): eQq, η, H_{loc}, ϕ_i, θ_i, H_e, ϕ, θ, ψ and α. The first two parameters ($eQq/h = 285.4$ MHz and $\eta = 0.252$) were found in [8]. H_e and α are defined by the experimental conditions. Due to low (if nonzero) value of H_{loc} in CdSb, the angle ϕ_i has no influence on the Zeeman NQR spectra, and the local magnetic field enters as a projection of H_{loc} on the EFG z' principal axis $H_{loc}\cos\theta_i$. Three parameters, $H_{loc}\cos\theta_i$, ϕ and θ can be determined by the modeling of the Zeeman perturbed pattern assigned to the ν_2 line (see Figure 2). When H_e and H_{loc} are nonzero, the ν_2 line is a superposition of two quartets, as it follows from the symmetry properties of the rotation matrix Equation (8). Eigen functions $|Y_i>$ $(i = 1, \ldots, 6)$ of the Hamiltonian Equation (9) are linear combinations

$$|Y_i\rangle = a_i^1\left|\frac{5}{2}\right\rangle + a_i^2\left|\frac{3}{2}\right\rangle + a_i^3\left|\frac{1}{2}\right\rangle + a_i^4\left|-\frac{1}{2}\right\rangle + a_i^5\left|-\frac{3}{2}\right\rangle + a_i^6\left|-\frac{5}{2}\right\rangle. \tag{10}$$

In the presence of magnetic field, coefficients a_i^j are complex numbers. For the two higher levels ($i = 5, 6$), the imaginary parts of six $a_i^j (j = 1, \ldots, 6)$ and real parts of five out of six coefficients a_i^j are nearly zero, because the parameter $g_{Sb}\mu_B H/h \cong 1.03\cdot 10^{-3}$ H MHz is small compared to $eQq/h \cong 285.4$ MHz. Therefore, because of nearly zero probabilities of some transitions and due to the symmetry properties Equation (8), the ν_2 line is a superposition of two doublets. A separation of the doublet components can be calculated with an accuracy of several percent (neglecting the contribution of H_{loc}) using the formula

$$\Delta_k \cong 2g_{Sb}\mu_B H_e\left|R_{33}^{(k)}\right|/h = 2.05\cdot 10^{-3} H_e\left|\sin\theta_k\cos\phi_k\cos\alpha - \cos\theta_k\sin\alpha\right|\text{MHz}, \tag{11}$$

where H_e is measured in Oe and k refers to the number of a set of the angular parameters in Equation (1). Due to Equation (8), the splitting of the doublets satisfies the relationship $\Delta_1 = \Delta_4$, $\Delta_3 = \Delta_2 \neq \Delta_1$. But, as follows from Equations

(1) and (11), the splitting of all the doublets is the same for some positions of the sample in the rf-coil (for example, when $\alpha = 0°$ or $90°$). Such a spectrum was actually observed (see Figure 2).

The angle $\theta \cong 60° \pm 3°$ was found by comparing the values of Δ measured at $\alpha = 90°$ with those calculated using Equation (11). Taking into account as found value of θ, the angle $\phi \cong 60° \pm 5°$ was calculated using the data at $\alpha = 0°$. To verify these angular parameters and to determine the parameter $H_{\mathrm{loc}}\cos\theta_i$, a number of Zeeman perturbed ν_2 patterns were measured and modeled for various α and H_e. The procedure of the line shape modeling was described in [2, 3]. There was a good agreement between the measured and calculated ν_2 patterns (see Figure 1b,c) for $H_{\mathrm{loc}}\cos\theta_i \cong 10$ Oe. This is an upper limit of H_{loc}, which may exist in CdSb characterized with the observed NQR spectra. It appeared to be much less than H_{loc} earlier found in the bismuth-based compounds (30–200 G) [1–3]. The origin of these fields is still controversial [1, 3–5], but they definitely have the electronic nature: the magnetic fields induced by the nuclear magnetic moments are much weaker (of the order of 1–2 G).

The Euler angle ψ cannot be found by analyzing the shape of the Zeeman perturbed ν_2 line, because Δ in Equation (11) does not depend on ψ. No simple formula similar to Equation (11) can be written for the estimation of the ν_1 line splitting which essentially depends on ψ. Moreover, even very weak magnetic field makes nonzero most of the coefficients a_i^j in the eigen functions Equation 10) for the four ^{121}Sb lower levels ($i = 1, \ldots, 4$). Therefore, one can expect, taking into account Equation (8), that the ν_1 line is generally a superposition of two quartets. The multiplicity of the experimental ν_1 spectra is however much lower. For instance, only one triplet was observed at $\alpha = 45°$ (see Figure 1a). Having varied the ψ values in the interval $0 < \psi < 90°$ we found that the ν_1 spectra are satisfactorily reproduced for $\psi = 45° \pm 5°$.

3. Conclusions

The experimental Zeeman ^{121}Sb NQR spectra evidenced for magnetic nonequivalence of the Sb sites in the CdSb crystal structure, the analysis of the EFG symmetry revealing at least two Sb sites. The angular parameters that determine the orientation of the ^{121}Sb EFG principal axes with respect to the crystallographic axes of CdSb were found by modeling the Zeeman ^{121}Sb NQR spectra. An upper limit of the local magnetic field ($\sim$10 G), which is consistent with the observed Zeeman ^{121}Sb NQR spectra, was estimated in CdSb to appear much less than H_{loc} earlier found in the bismuth-based compounds [1–3].

Acknowledgements

The authors are grateful to the Russian Academy of Sciences (The Fundamental Research Program of the Department of Chemistry and Materials Sciences) and

Russian Foundation for Basic Research (project 02-03-33280) for the support. V.G.O. and M.P.S. are grateful to the Council for the Support of the Leading Scientific Schools of Russia under grant NS-1572.2003.2 for partial support.

References

1. Kravchenko E. A. and Orlov V. G., *Z. Naturforsch.* **49a** (1994), 418.
2. Kravchenko E. A., Orlov V. G., Suan H. F. and Kargin Y. F., *Z. Naturforsch.* **53a** (1998), 504.
3. Kravchenko E. A., Kargin Y. F., Orlov V. G., Okuda T. and Yamada K., *J. Magn. Magn. Mater.* **224** (2001), 249.
4. Kharkovskii A. I., Nizhankovskii V. I., Kravchenko E. A. and Orlov V. G., *Z. Naturforsch.* **51a** (1996), 665.
5. Nizhankovskii V. I., Kharkovskii A. I. and Orlov V. G., *Ferroelectrics* **279** (2002), 157.
6. Asheulov A. A., Voronka N. K., Marenkin S. F. and Rarenko I. M., *Neorgan. Mater. (Russ. Inorganic Materials)* **32** (1996), 1049.
7. Amin K. E., *Acta Chem. Scand.* **2** (1948), 400.
8. Buslaev Y. A., Kravchenko E. A., Lazarev V. B. and Marenkin S. F., *Phys. Stat. Solidi* **47b** (1971), P.K. 125.
9. Orlov V. G., *J. Magn. Magn. Mater.* **61** (1986), 337.

Hyperfine Interactions (2004) 159:181–185
DOI 10.1007/s10751-005-9094-3

NMR Study of the Dimerized State in $CuIr_2S_4$

KEN-ICHI KUMAGAI[1,*], MAYUMI SASAKI[1], KOSUKE KAKUYANAGI[1] and SHOICHI NAGATA[2]
[1]*Division of Physics, Graduate School of Science, Hokkaido University, Sapporo 060-0081, Japan; e-mail: kumagai@phys.sci.hokudai.ac.jp*
[2]*Department of Materials Science and Engineering, Muroran Institute of Technology, Muroran 050-8585, Japan*

Abstract. We have investigated the metal-insulator transition (MIT) of $CuIr_2S_4$ by a high resolution NMR measurement. The Cu-NMR spectrum below T_{MI} is broadened and split into four Cu signals with sizable electric quadrupole interactions. The NMR results are consistent with the charge ordering of Ir^{3+} and Ir^{4+} and the spin dimerization of Ir^{4+} spins, as revealed by a recent X-ray study.

Key Words: $CuIr_2S_4$, metal–insulator transition, NMR, spinel.

1. Introduction

In calcogenide spinel $CuIr_2S_4$, the metal–insulator transition (MIT) at T_{MI} ~230 K [1] has attracted much attention. Since NMR [2] and photoemission [3] measurements show that the Cu ions are monovalent in the insulating state and thus the nominal valence of the Ir atoms is 3.5, the nonmagnetic ground state below T_{MI} is puzzling for possible Ir^{3+} ($S = 0$) and Ir^{4+} ($S = 1/2$) configurations. There are discussions about the charge order of Ir^{3+} ($S = 0$) and Ir^{4+} ($S = 1 / 2$) ions, and consequently possible spin singlet dimers in the insulating state. This MIT is accompanied by a structural transition from cubic to tetragonal (triclinic) symmetry with volume contraction of 0.7%. [4] Recently, a precise X-ray measurement [5] has shown that a charge ordering of Ir^{3+} and Ir^{4+} and spin dimerization of Ir^{4+} ions occurs simultaneously below T_{MI}. The Ir sublattice consists of two type of Ir bi-capped hexagonal rings which described as Ir^{3+} and Ir^{4+} octamers. In order to elucidate the origin of the MIT in the Cu-spinels, it is important to accumulate the information about the valence state and magnetic properties in the insulating state from a microscopic point of view. Here, we investigate the evolution of the electronic state associated with MIT of $CuIr_2S_4$ by a Cu-NMR study using a high resolution NMR spectroscopy.

* Author for correspondence.

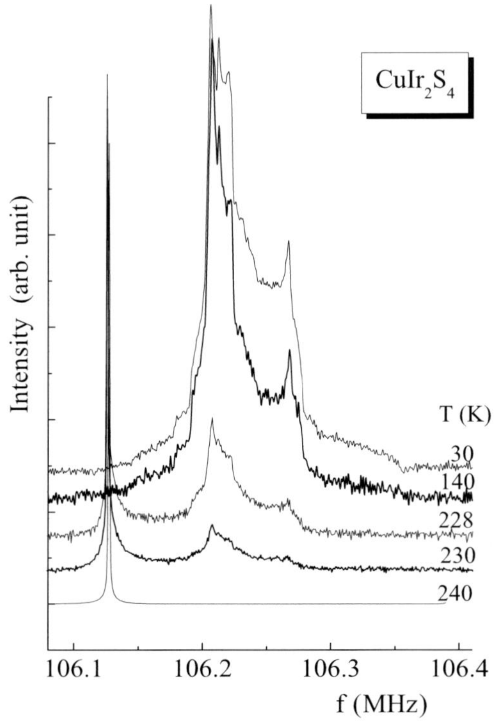

Figure 1. ^{63}Cu-NMR spectra of $CuIr_2S_4$ at various temperatures. The spectra at T = 230 K and T = 140 K are shown by *bold lines*.

2. Experimental

Polycrystalline samples were prepared by solid state reaction [1]. NMR was measured by the conventional phase coherent pulse method with a highly-homogeneous superconducting magnet. Spectra of ^{63}Cu were obtained by the Fourier Transformed-NMR with a constant magnetic field (H = 9.4T).

3. Results and discussion

As shown in Figure 1, we obtained very narrow ^{63}Cu spectra above T_{MI} in $CuIr_2S_4$. The line width is less than 5 kHz without any notable anisotropic Knight shift and nuclear quadrupole interactions. The ^{63}Cu NMR spectrum becomes broad below T_{MI}, and shows several peaks within the narrow range of 100 kHz. As the MIT is of a first order transition [1], NMR signals in the metallic and insulating phase coexist for between 230 and 220 K. The isotropic part of the Knight shift of −0.08% at the metallic state indicates that there exists a negative contribution of the Knight shift from core polarizations of the Cu 3*d* electrons. Below T_{MI}, the opening of a band gap reduces the Knight shift. The absence of apparent temperature dependence of the Knight shift below T_{MI} indicates that the

Table I. Calculated EFG parameters of insulating state in $CuIr_2S_4$

	Cu(1)	Cu(2)	Cu(3)	Cu(4)
V_{zz}^{cal}	4.23	2.27	2.31	3.43
η^{cal}	0.34	0.85	0.74	0.93

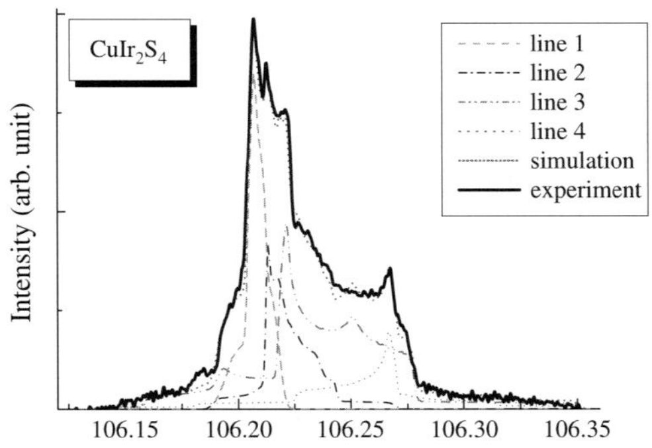

Figure 2. ^{63}Cu-NMR spectrum of $CuIr_2S_4$ at $T = 100$ K. Simulated spectrum which is summed up with NMR lines with reasonable values of quadrupole frequency and Knight shift for inequivalent Cu site are shown.

electronic state is basically nonmagnetic in the insulating state, suggesting that both Cu and the Ir atoms in $CuIr_2S_4$ are in nonmagnetic states.

Because of the lattice distortion from cubic to triclinic symmetry below T_{MI}, an appreciable electric field gradient (EFG) at the Cu site is expected. Thus, the nuclear quadrupole interaction for the Cu nuclei ($I = 3/2$) is not negligible below T_{MI}. We have calculated the EFG at the Cu site with a point charge model. For this purpose, we use atomic positions and ionic configurations of Ir^{3+} and Ir^{4+} obtained by the recent precise X-ray measurement [6]. The calculation reveals that only one equivalent Cu site in the cubic symmetry above T_{MI} changes to four nonequivalent Cu sites. The calculated parameters of V_{zz}, and η are given in Table I. The number of the Cu sites correspond to the ones assigned in [5].

We simulate the split spectrum by taking into consideration the perturbation of the nuclear quadrupole interaction for the Zeeman field. The best fitting is achieved with super-position of four components of the spectrum with the slightly-different Knight shift as shown in Figure 2. The NMR parameters for the best fittings are listed in Table II. From the differences of the quadrupole frequency, ν_Q, and anisotropy parameter, η, we assign each NMR component to the signal from the Cu atoms defined in [5]. As seen in Figure 2, the fitting to the experimental data is satisfactory. This indicates that the charge separation and the possible dimerization of Ir^{4+} proposed by *X*-ray study are reasonably confirmed by the NMR study at a microscopic point view.

Table II. Fitting parameters of NMR shift and nuclear quadrupole interactions

	Line (1)	Line (2)	Line (3)	Line (4)
K^{iso} (%)	0.12	0.104	0.09	0.13
ν_Q (kHz)	88	26	14	89
η	0.34	0.86	0.74	0.92

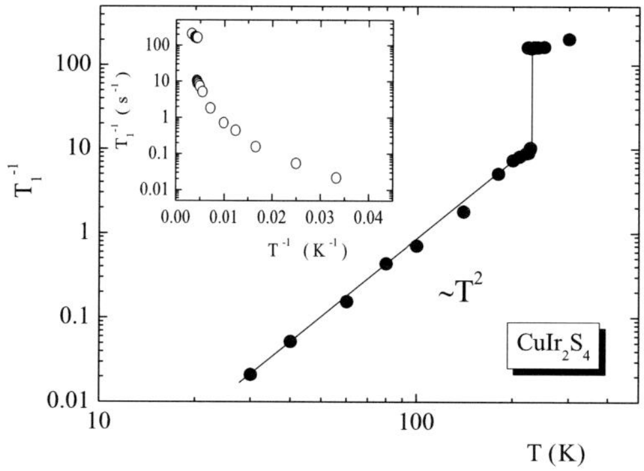

Figure 3. Temperature dependence of nuclear spin-lattice relaxation rate, $(T_1T)^{-1}$ in $CuIr_2S_4$. The inset shows the T_1^{-1} as a function of T^{-1}.

Finally, we show the nuclear spin lattice relaxation time, T_1, of ^{63}Cu in Figure 3. At the metallic state above T_{MI}, $(T_1T)^{-1}$ obeys the Korringa relation (temperature-independent), and decreases largely at T_{MI} due to the opening of the band gap in the insulating state. An interesting feature is that the temperature dependence of $(T_1T)^{-1}$ obeys a power law (the T^2 -relation) and not a thermally activated (exponential relation) behavior. This result indicates that the band gap in the insulating state of $CuIr_2S_4$ is quite anisotropic and that spin fluctuations are in very unusual nature.

4. Summary

We have investigated ^{63}Cu-NMR in $CuIr_2S_4$. ^{63}Cu-NMR spectrum consists of four components of nonequivalent Cu-NMR signals with different Knight shift and nuclear quadrupole interaction in the insulating state. The charge ordering of the Ir^{3+} and Ir^{4+} and the dimerization of Ir^{4+} spins in $CuIr_2S_4$ are verified by the NMR study.

Acknowledgement

The authors would like to thank Dr. Y. Horibe for fruitful discussions and kind informing the atomic coordination.

References

1. Nagata S., Hagino T., Seki Y. and Bitoh T., *Physica* **B194–196** (1994), 1077; Hagino T., Tojo T., Atake T. and Nagata S., *Philos. Mag.* **B71** (1995), 881.
2. Kumagai K., Tsuji S., Hagino T. and Nagata S., In: Fujimori A. and Tokura Y. (eds.), *Spectroscopy of Mott Insulator and Correlated Metals*, Springer, Berlin Heidelberg New York, 1995, p. 255.
3. Matsuno J., Mizokawa T., Fujimori A., Zatsepin D. A., Galakhov V. R., Kurmaev E. Z., Kato Y. and Nagata S., *Phys. Rev.* **B55** (1997), R15979.
4. Furubayashi T., Matsumoto T., Hagino T. and Nagata S., *J. Phys. Soc. Jpn.* **63** (1994), 3333.
5. Radaelli P. G., Horibe Y., Gutmann M. J., Ishibashi H., Chen C. H., Ibberson R. M., Koyama Y., Hor Y. S., Kiryukhin V. and Cheong S. W., *Nature* **416** (2002), 155.
6. Horibe Y., Private communication.

Hyperfine Interactions (2004) 159:187–191
DOI 10.1007/s10751-005-9095-2

Nuclear Spin Relaxation Studied by β-NMR of ^{12}N Implanted in TiO_2

M. MIHARA[1,*], Y. NAKASHIMA[1], S. KUMASHIRO[1], H. FUJIWARA[1], Y. N. ZHENG[2], M. OGURA[1], T. SUMIKAMA[1,†], T. NAGATOMO[1], K. MINAMISONO[3], M. FUKUDA[1], K. MATSUTA[1] and T. MINAMISONO[1,‡]
[1]*Graduate School of Science, Osaka University, Toyonaka, Osaka 560-0043, Japan; e-mail: mihara@vg.phys.sci.osaka-u.ac.jp*
[2]*China Institute of Atomic Energy, Beijing 102413, People's Republic of China*
[3]*TRIUMF, Vancouver Canada V6T 2A3*

Abstract. The β-NMR of short-lived β-emitter $^{12}N(I^{\pi} = 1^{+}, T_{1/2} = 11ms)$ in a rutile TiO_2 single crystal has been measured as functions of temperature and external magnetic field. Atomic motion induced spin lattice relaxation was observed for two known sites, O substitutional and interstitial sites. The data were analyzed in terms of the thermal atomic jump, which suggests that the motion of defects around the substitutional ^{12}N atom for O, and of the interstitial ^{12}N atom are attributed to the spin lattice relaxation. The electric field gradients have shown temperature dependence for both sites, which is probably due to the thermal expansion of rutile.

Key Words: β-NMR, impurities and defects, nuclear spin relaxation, thermal atomic jump.

1. Introduction

Spin polarized β-emitting nuclei implanted into crystals are quite useful probes to study the electronic structure or the dynamical behavior of impurities in crystals, through the hyperfine interactions. Various β-emitters have been implanted into a rutile TiO_2 single crystal as probes and their β-NMR has been detected. These measurements showed that the initial polarization generated through the nuclear reaction is almost totally maintained in the crystal at room temperature for a much longer time than the lifetime of the probe nuclei [1]. This result allows us to do the systematic studies on the hyperfine interactions of various impurities in TiO_2.

TiO_2 is known as an excellent photocatalytic material under ultra violet (UV) light irradiation. A lot of efforts on impurity doping into TiO_2 have been made in

* Author for correspondence.
† Present address: RIKEN, Wako, Saitama 351-0198, Japan.
‡ Present address: Fukui University of Technology, Fukui 910-8505, Japan.

order to shift the optical response from UV to visible light range, which can realize more extensive application of the TiO_2 photocatalyst. Especially, the substitutional doping of N for O into TiO_2 has been found to be effective for this purpose [2]. The β-NMR of ^{12}N implanted in TiO_2 has been measured previously [3], showing that about 100% of polarization is maintained at room temperature and the implanted ^{12}N nuclei occupy equally two locations. The electric field gradient (EFG) for each site was determined and the location was evaluated in comparison with the KKR band structure calculations, which proposed that first site (site I) is the O substitutional site and the other (site II) is the octahedral interstitial site of the unit cell [3, 4]. In the present study, the dynamical behavior such as diffusion of N in TiO_2, which is quite important information to synthesize the N doped TiO_2 photocatalyst, was studied by observing the nuclear spin relaxation of ^{12}N in TiO_2 by means of the β-NMR method.

2. Experimental

The experimental method used in this work was nearly the same as the previous ones [1, 3]. Polarized ^{12}N nuclei were produced through the nuclear reaction $^{10}B(^3He, n)^{12}N$. A ^3He beam of 3.0 MeV and an intensity of 10–15 μA from the 5-MV Van de Graaff accelerator at Osaka University was used to bombard the 90% enriched ^{10}B target. The ^{12}N nuclei ejected from the target at angles in between 12 and 28 degrees relative to the ^3He-beam direction were selected with a collimator to obtain spin polarization of ~20%. The polarized ^{12}N nuclei were implanted into a square plate of $0.3 \times 20 \times 20\text{-mm}^3$ rutile single crystal whose sides are along with the crystallographic c and $< 110 >$ axes, respectively. The sample was synthesized by the flame fusion method, and then annealed in air from 1900°C to room temperature [5]. To maintain the polarization of ^{12}N and to detect its NMR signals, a static magnetic field of H_0 = 0.35–0.73 T was applied parallel to the polarization direction. Beta rays emitted from the ^{12}N nuclei were detected by two sets of plastic-scintillation-counter telescopes located above and below the sample to detect the asymmetry in the β-ray angular distribution. A radiofrequency (rf) magnetic field H_1 was applied perpendicular to H_0 to induce transition between substates. In order to observe the NMR signals efficiently where the quadrupole interactions exist, multi rf was applied in the rf period (β-NQR method) [6]. Each site was distinguished from the frequency split by the quadrupole interaction and its fraction was obtained from the degree of β-ray asymmetry change.

3. Temperature dependence of electric field gradients

The EFGs of ^{12}N in TiO_2 are temperature dependent. As shown in Figure 1, observed splittings of the ^{12}N NMR lines, in case of the crystallographic c and $\langle 110 \rangle$ axes parallel to H_0 (= 0.73 T), vary with the temperature for both site I and

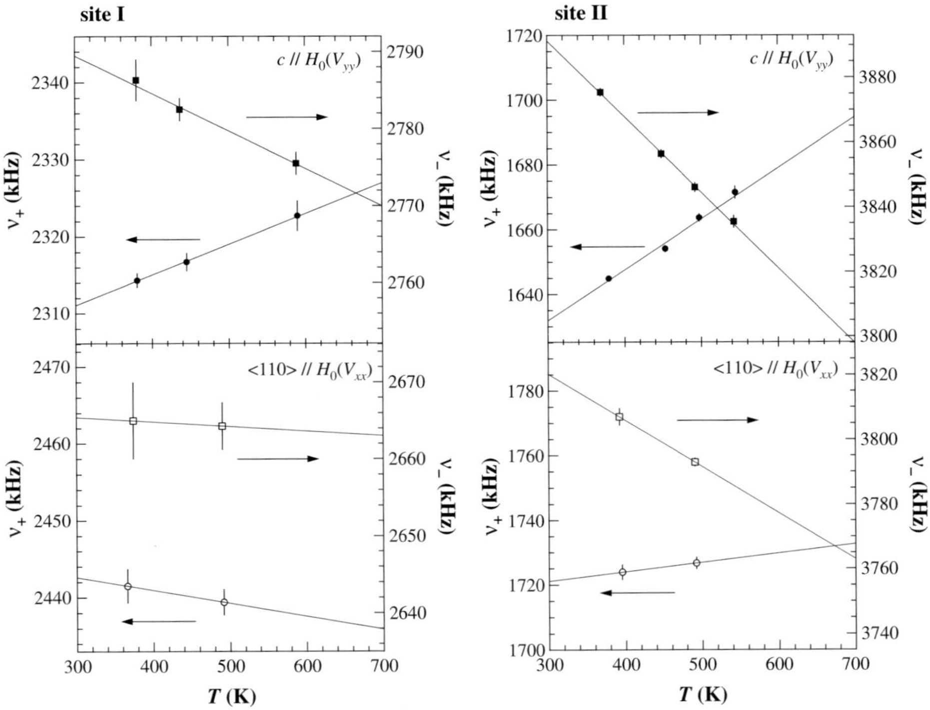

Figure 1. Temperature dependence of the splittings of *NMR lines* for ^{12}N in a rutile TiO_2 single crystal, measured at H_0 of 0.73 T. The *fitted lines* to the data are also shown.

site II. The separation frequency $\Delta\nu = \nu_+ - \nu_-$ gives the EFG tensor V_{ii} as $\Delta\nu = (3/2)eV_{ii}Q/h$ ($i = x, y, z$, $\Sigma V_{ii} = 0$), where ν_+ and ν_- denote the transition frequencies corresponding to the $m = +1 \leftrightarrow 0$ and $m = -1 \leftrightarrow 0$ transitions, respectively. The principle axes of the EFG tensor were previously determined, as the x- or z-components are parallel to the $\langle 110\rangle$ or $\langle 1\bar{1}0\rangle$ axes and the y-component is parallel to the c axis, for both sites [3]. In the present case, V_{yy} and V_{xx} were directly determined from the separation frequencies for c // H_0 and $\langle 110\rangle$ // H_0, respectively. As a result, the largest eigenvalue q of EFG (V_{zz}) and the asymmetry parameter $\eta = (V_{xx} - V_{yy})/V_{zz}$ were deduced as a linear function of temperature, as $|q| = 2.04 \times 10^{21}$ $(1 - 1.1 \times 10^{-4}$ T) V/m^2 and $\eta = 0.39 \times (1 - 2.4 \times 10^{-4}$ T) for site I, and $|q| = 1.27 \times 10^{22}$ $(1 - 1.2 \times 10^{-4}$ T) V/m^2 and $\eta = 0.05 \times (1 - 8.4 \times 10^{-4}$ T) for site II. Such a temperature dependence of q, is probably due to the thermal expansion of the lattice surrounding the ^{12}N atom. The temperature dependence of η can be attributed to the anisotropic lattice expansion [7].

4. Nuclear spin relaxation and dynamical behavior of ^{12}N in rutile

The polarization of ^{12}N in TiO_2, measured at $H_0 = 0.35$–0.73 T as a function of temperature in the range from 370 K to 900 K is shown in Figure 2. The c axis

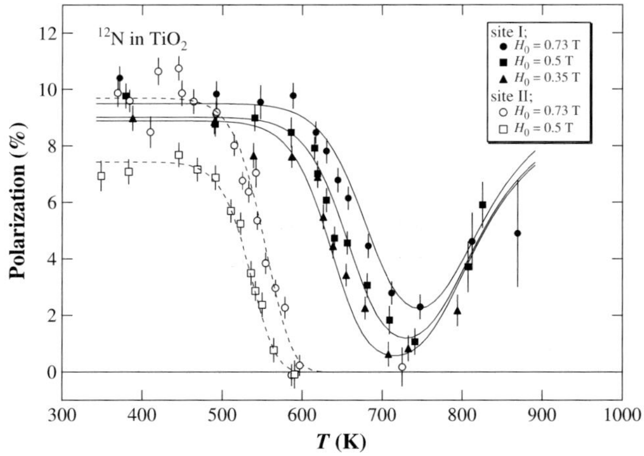

Figure 2. Temperature dependence of the observed polarization for ^{12}N in a rutile TiO_2 single crystal. The crystalline c axis was set parallel to H_0. The *solid* and *dashed lines* are the best fit curves to the data.

was set parallel to H_0. Considering the above temperature dependent splittings, frequency sets were properly adjusted to detect polarization. At above 500 K, the fraction of site II rapidly decreases and vanishes at 600 K. The fraction of site I also rapidly decreases with increasing temperature at above 600 K, however, it begins to restore at around 700–800 K. The reduction of the observed polarization was shown to be due to the fast spin–lattice relaxation. From the time dependence of polarization, the spin–lattice relaxation time T_1 was confirmed to vary by temperature as the observed polarization.

The present results were analyzed in terms of thermal atomic jumps. Motion of the ^{12}N probe itself or defects such as O vacancies, dislocations of Ti ions or impurities in rutile [8], are possible to cause the spin–lattice relaxation due to fluctuating electric field gradients. In this point of view, the spin–lattice relaxation rate $1/T_1$ is described as [8, 9]

$$1/T_1 = \left(\pi^2/2\right)\langle eqQ/h\rangle^2\{k_1(\omega_0) + 4k_2(2\omega_0)\}, \tag{1}$$

$$k_n(\omega_0) = (1/4)2\tau_c\Big/\left\{1 + (n\omega_0\tau_c)^2\right\}, \tag{2}$$

where $\langle eqQ/h\rangle$ is the magnitude of the mean fluctuating quadrupole field to the ^{12}N nucleus, and ω_0 is the Larmor frequency. τ_c is the correlation time, i.e. the mean time of atomic jumps between two identical sites, which is expressed as $1/\tau_c = \nu_0\exp(-E_a/kT)$, where ν_0 is the jump frequency at infinite temperature, and E_a is the activation energy. The observed polarization is connected with the relaxation rate considering the polarization reduction during the beam (16 ms), rf (12 ms) and β-ray counting period of 60 ms, due to the spin–lattice relaxation. Assuming the initial polarization is constant over the examined temperature range, the data for site I was analyzed by a χ^2-fitting procedure using Equations (1, 2) and the relation between the polarization and $1/T_1$ [9]. The best fit curves to the data are shown in Figure 2. From this fit, values of $\langle eqQ/h \rangle = (15 \pm 1)$ kHz, $\log_{10}(\nu_0$ in kHz$) = 12 \pm 1$, and $E_a = (1.1 \pm 0.1)$ eV were deduced for site I. The fluctuating quadrupole field for site I was found to be much smaller than the static quadrupole field of ~470 kHz [3], which suggests that the spin–lattice relaxation of ^{12}N for site I is attributed to the motion of defects around ^{12}N and consequently that substitutional N impurities for O in TiO_2 are stable up to at least ~900 K. In the case of site II, only the activation energy $E_a = \sim 1$ eV was deduced, because restoration of the polarization was not observed due to much larger fluctuation of the quadrupole field compared with site I. One possibility for this difference is the diffusion of interstitial N impurities which gives $\langle eqQ/h \rangle = \sim 2.9$ MHz for this site [3].

One of the authors was partly supported by the Tokui scholarship.

References

1. Ogura M., Minamisono K., Sumikama T., Nagatomo T., Iwakoshi T., Miyake T., Hashimoto K., Kudo S., Arimura K., Ota M., Akutsu K., Sato K., Mihara M., Fukuda M., Matsuta K., Akai H. and Minamisono T., *Hyperfine Interact.* **136–137** (2001), 195.
2. Asahi R., Morikawa T., Ohwaki T., Aoki K. and Taga Y., *Science* **293** (2001), 269.
3. Minamiosono T., Sato K., Akai H., Takeda S., Maruyama Y., Matsuta K., Fukuda M., Miyake T., Morishita A., Izumikawa T., and Nojiri Y., *Z. Naturforsch* **53a** (1998), 293.
4. Matsuta K., Miyake T., Minamisono K., Mihara M., Fukuda M., Sato K., Zhu S. Y. and Minamisono T., *Hyperfine Interact.* **136–137** (2001), 189.
5. Nakazumi Y., Suzuki K. and Yazima T., *J. Phys. Soc. Jpn* **17** (1962), 1806.
6. Mihara M., Hashimoto K., Arimura K., Kudo S., Akutsu K., Minamisono K., Miyake T., Fukuda M., Matsuta K. and Minamisono T., *Hyperfine Interact.* **136–137** (2001), 339.
7. Kanert O. and Kolem H., *J. Phys. C. Solid State Phys.* **21** (1988), 3909.
8. Kolem H. and Kanert O., *Z. Met.kd.* **80** (1989), 227.
9. Izumikawa T., Matsuta K., Tanigaki M., Miyake T., Sato K., Fukuda M., Zhu S. Y. and Minamisono T., *Hyperfine Interact.* **136–137** (2001), 599.

Hyperfine Interactions (2004) 159:193–197
DOI 10.1007/s10751-005-9096-1

NQR Study of Phase Transition and Cationic Motion in $4\text{-}NH_2C_5H_4NHBiBr_4 \cdot H_2O$

HARUO NIKI[1,*], HIROAKI KINJOU[1], TAICHI FUKUMURA[1], HIROMITSU TERAO[2], MAMI YOSHIDA[2] and MASAO HASHIMOTO[3]
[1]*Department of Physics, Faculty of Science, University of the Ryukyus, Nishihara, Okinawa 903-0213, Japan; e-mail: niki@sci.u-ryukyu.ac.jp*
[2]*Department of Chemistry, Faculty of Integrated Arts and Sciences, Tokushima University, Minamijosanjima-cho, Tokushima 770-8502, Japan*
[3]*Department of Chemistry, Faculty of Science, Kobe University, Nadaku, Kobe 657-8501, Japan*

Abstract. Four ^{81}Br NQR lines in $4\text{-}NH_2C_5H_4NHBiBr_4 \cdot H_2O$ were observed in the temperature range between 77 and ca. 380 K; with increasing temperatures the respective sets of higher and lower two resonance lines coalesced into single lines discontinuously at 274 K, showing the occurrence of a first-order type phase transition of this crystal. The transition was confirmed with heat anomaly on a DTA curve. Each higher and lower line of high-temperature phase is assignable to the terminal Br atoms and the bridging ones of one-dimensional poly anions $(BiBr_4^-)_n$ in the crystal structure (*C*2/c), which was investigated by a X-ray structure analysis at room temperature. The $1/T_1$ temperature dependence of ^{81}Br NQR follows the usual T^2 law in the temperature range between 77 and ca. 140 K, being explained by fluctuation of the EFG at Br nucleus due to lattice vibrations. The T_1 vs. $1/T$ curve in the temperature range between about 160 and 190 K was describable by the exponential curves, allowing us the estimation of activation energies. These exponential behaviors of T_1 of ^{81}Br NQR are attributable to the fluctuations caused by the thermal motion of $4\text{-}NH_2C_5H_4H^+$ ions. Echo signals of the ^{81}Br NQR could not be detected above 190 K owing to poor S/N with very short T_2.

Key Words: ^{81}Br NQR, $4\text{-}NH_2C_5H_4NHBiBr_4 \cdot H_2O$, phase transition, spin–lattice relaxation times, spin–spin relaxation times.

1. Introduction

Pyridinium ions in the crystals of RMX_4 (R = pyridinium; M = Sb, Bi; X = halogen) are often found piled up with their molecular planes in parallel within the tunnel made by the one-dimensional anion chains of $(MX_4^-)_n$. In the tunnel, the pyridinium ions are connected to the wall of $(MX_4^-)_n$ through N–H···X hydrogen bonds [1–3]. In such a circumstance, $C_5H_5NHSbX_4$ (X = Cl and Br),

* Author for correspondence.

abbreviated as $PyHSbX_4$, were found to exhibit a unique phase transition [1–3]. The phase transitions of these compounds are expected to correlate to the reorientations of the cations, on which the schema of pyridine N–H···X hydrogen bonds has been supposed to play an important role in their mechanisms.

It is interesting to replace the $C_5H_5NH^+$ (PyH^+) by $4\text{-}NH_2C_5H_4NH^+$ ($4\text{-}NH_2PyH^+$) because of the existence of two functional groups of $-NH_2$ and pyridine N–H groups, both of which can make the hydrogen bonds, as well as the increased size in $4\text{-}NH_2PyH^+$. Thus, a phase transition of $4\text{-}NH_2PyHSbBr_4$ was found to occur at a temperature of 224 ± 1 K, using X-ray diffraction, ^{81}Br and ^{121}Sb NQR, ^1H NMR, and thermal analysis [4, 5].

In the course of the study of phase transitions related to the motions of the pyridinium ions we tried to prepare the crystals of $4\text{-}NH_2PyHBiBr_4 \cdot H_2O$. A preliminary investigation of X-ray crystal analysis revealed that the present crystal is basically isostructural with $4\text{-}NH_2PyHSbBr_4$ [4], though a hydration water was included in this crystal. In the present study, the temperature dependence of spin–lattice relaxation times (T_1) and spin–spin relaxation times (T_2) as well as the resonance frequencies of ^{81}Br NQR was measured to discuss the possible phase transition and the crystal dynamics.

2. Experimental

Tin-yellowish needle crystals of $4\text{-}NH_2PyHBiBr_4 \cdot H_2O$ were obtained from a concentrated hydrobromic acid solution of 4-aminopyridine and bismuth trioxide in which the former was adjusted to a half of a stoichiometric amount, since the contamination with $(4\text{-}NH_2PyH)_2BiBr_5$ occurred in an equimolecular amount. Water contents were estimated by a weight loss in a vacuum desiccator. The results of elemental analyses; C, H, and N analyses; found/calc.; weight percent: C: 9.48/9.36; H: 1.24/1.41; N: 4.46/4.37 for $NH_2C_5H_4NHBiBr_4 \cdot H_2O$.

The NQR spectra were recorded in the frequency range between 31 and 140 MHz by the cw method using a super-regenerative spectrometer. The NQR frequency measurements were carried out in the temperature range of 77 to 380 K. The accuracy of frequency measurement is estimated to be within ± 0.02 MHz. T_1 and T_2 of ^{81}Br NQR were measured with a conventional pulse spectrometer. T_1 was determined by the echo sequences given by $90°–t_1–90°–t_2–180°$ and T_2 by a $90°–t–180°$ pulse sequences.

3. Results and discussions

The present $4\text{-}NH_2PyHBiBr_4 \cdot H_2O$ crystal has a monoclinic structure similar to that of $4\text{-}NH_2PyHSbBr_4$ [4], except for the existence of a hydration water; space group $C2/c$, $Z = 4$, $a = 13.612(4)$, $b = 13.984(4)$, $c = 7.440(2)$ Å, $\beta = 119.09(4)°$. Thus, the structure consists of infinite chains of $(BiBr_6)_n$ made up by two edges

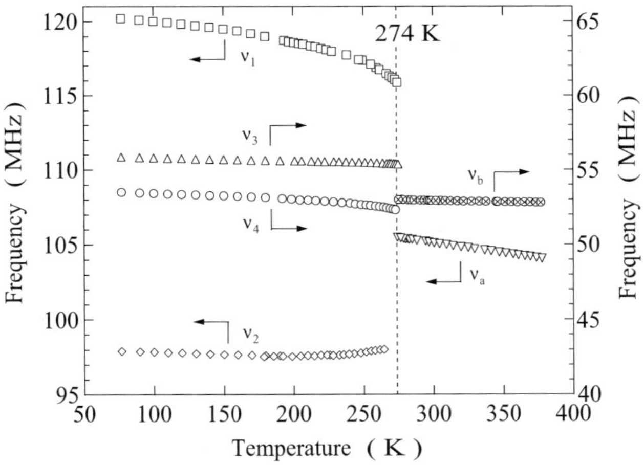

Figure 1. Temperature dependence of ^{81}Br NQR frequencies in 4-$NH_2PyHBiBr_4{\cdot}H_2O$.

Table I. ^{81}Br NQR frequencies at 77 K and activation energies for each Br site obtained by Equation (1)

Site	ν at 77 K (MHz)	E_a (J/mol)
1	120.19	18.4
2	97.91	18.0
3	55.86	24.5
4	53.51	25.3

sharing of respective $BiBr_6$ octahedrons, cations, and hydration water molecules. The cations are stacked into a tunnel-like space made with the anion chains, and water molecules may also exist in it, though they were not located in the present X-ray analysis.

Figure 1 shows the temperature dependence of ^{81}Br NQR resonance frequencies measured in the temperature range between 77 and ca. 380 K. The frequencies of four ^{81}Br NQR lines detected at 77 K are listed in Table I. These lines could be unambiguously assigned to ^{81}Br by the observation of the corresponding ^{79}Br lines. With increasing temperature these four lines abruptly disappeared at T_c = 274 K and instead new two lines appeared as shown in Figure 1. The NQR lines can be divided into two gropes. One group contains the line 1, 2, and a and the second group contains the lines 3, 4, and b. These groups can be assigned to the terminal and the bridging Br atoms, respectively. The frequency jump at T_c shows the occurrence of a first-order type phase transition in this crystal. The first-order transition nature was also recognized by the peak shapes and also a slight hysteresis followed on heating and cooling processes in differential thermal analysis (DTA) measurement. It is noticed that the value of ν_a is rather smaller than the mean value of ν_1 and ν_2 at T_c, and the same situation

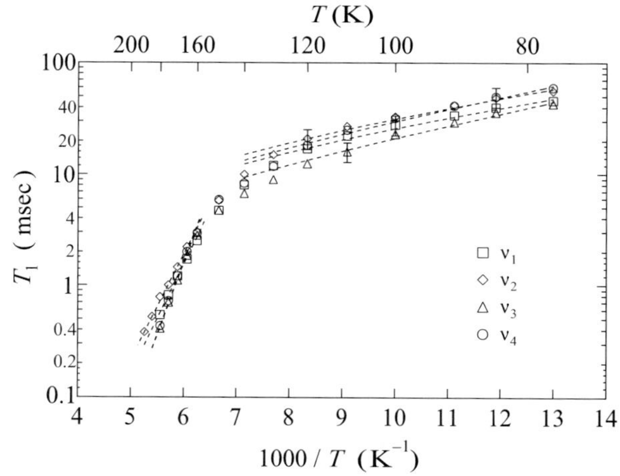

Figure 2. Temperature dependence of T_1 of ^{81}Br NQR in 4-NH_2PyHBi$Br_4 \cdot H_2O$. The *broken lines* in the figure are the calculated straight lines on Equation (1) and the *broken curves* are the obtained ones by using the T^2 law (see text).

is also kept for ν_b, and ν_3 and ν_4, being different from the phase transitions found before in the family of RMX_4.

Temperature dependence of T_1 of ^{81}Br NQR is shown in Figure 2. All of the recovery curves of ^{81}Br nuclear magnetizations can be described by an exponential function of time within the experimental errors. In the temperature range between 77 and 160 K, the temperature dependence of $1/T_1$ of ^{81}Br NQR follows the usual T^2 law [6, 7]. T_1 in this temperature region is dominated by fluctuation of the EFG at Br nucleus due to lattice vibrations [6, 7]. The T_1 vs. $1/T$ curves in the temperature range between 160 and 190 K can be described by the equation [3, 6],

$$T_1 = b\exp(E_a/RT). \tag{1}$$

The slope of the ln (T_1) vs. $1/T$ curve in this temperature range gives activation energies for each Br site as listed in Table I.

These behaviors are attributable to the modulation effect caused by the reorientation of the 4-NH_2PyH^+ ions as mentioned below. The 4-NH_2PyH^+ ions are vibrating in the lower temperature phase below T_c (abbreviated as LTP), while they are rotating around one symmetrical axis in the upper temperature phase above T_c (abbreviated as RTP). This behavior can be attributed the fact that the hydrogen bonds between 4-NH_2PyH^+ ions and the bromine atoms in the anions become to be broken by the thermal motions as expected in the case of 4-NH_2PyHSbBr_4 [4, 5]. It is noticed that the activation energies derived from T_1 of Br NQR are higher than those in 4-NH_2PyHSbBr_4, indicating the difference of the phase transition temperatures in both compounds (4-NH_2PyHBi$Br_4 \cdot H_2O$: $T_c = 274$ K; 4-NH_2PyHSbBr_4: $T_c = 224$ K). Therefore, the T_1 behaviors of ^{81}Br

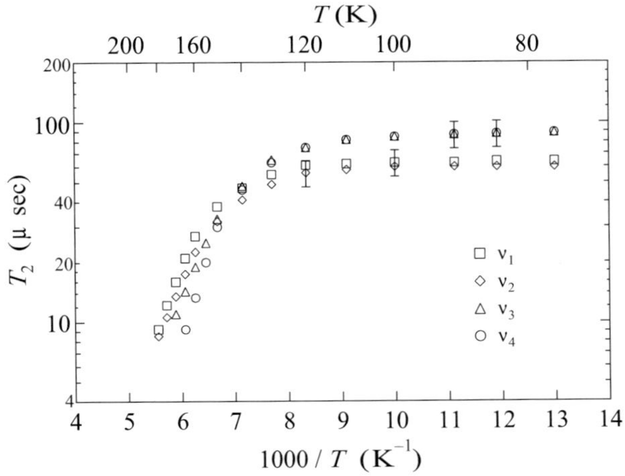

Figure 3. Temperature dependence of T_2 of ^{81}Br NQR in 4-$NH_2PyHBiBr_4{\cdot}H_2O$.

NQR in 4-$NH_2PyHBiBr_4{\cdot}H_2O$ above 160 K is attributable to the modulation effect caused by the fluctuations through thermal motions of 4-NH_2PyH^+ ions. Thermal motions of water molecules in 4-$NH_2PyHBiBr_4{\cdot}H_2O$ would make a contribution to T_1 mechanisms. Therefore, we must consider the effect of their motions. Further investigation including proton NMR is needed in order to fully understand the nature of the phase transition in 4-$NH_2PyHBiBr_4{\cdot}H_2O$.

Temperature dependence of T_2 of ^{81}Br NQR is shown in Figure 3. The values of T_2 of ^{81}Br NQR are considerately short in the low temperature region, which decrease further with increasing temperature. Therefore, the echo signals of Br NQR above 190 K could not be detected with poor S/N ratios caused by very short T_2. The reason why echo signals could not be detected for ν_a and ν_b in RTP would be also attributable to short T_2.

References

1. Yamada K., Ohtani T., Shirakawa S., Ohki H., Okuda T., Kamiyama T. and Oikawa K., *Z. Naturforsch.* **51a** (1996), 739.
2. Okuda T., Yamada K., Ishihara H., Hiura M., Gima S. and Negita H., *J. Chem. Soc. Commun.* (1981), 979.
3. Okuda T., Aihara Y., Tanaka N., Yamada K. and Ichiba S., *J. Chem. Soc. Dalton Trans.* (1989), 631.
4. Hashimoto M., Hashimoto S., Terao H., Kuma M., Niki H. and Ino H., *Z. Naturforsch.* **55a** (2000), 167.
5. Niki H., Yogi M., Tamanaha M., Seto U., Hashimoto M. and Terao H., *Z. Naturforsch.* **57a** (2002), 469.
6. Woessner D. E. and Gutowsky H. S., *J. Chem. Phys.* **39** (1963), 440.
7. Das T. P. and Hahn E. L., *Nuclear Quadrupole Resonance Spectroscopy, Solid State Physics Supplement 1*, Academic, New York, 1958.

Hyperfine Interactions (2004) 159:199–203
DOI 10.1007/s10751-005-9097-0

Comparative Nuclear Magnetic Resonance Study of As-Grown and Annealed $LiInSe_2$ Ternary Compounds

A. M. PANICH[1,*], A. P. YELISSEYEV[2], S. I. LOBANOV[2] and L. I. ISAENKO[2]
[1]*Department of Physics, Ben-Gurion University of the Negev, P.O. Box 653, 84105 Beer Sheva, Israel; e-mail: pan@bgumail.bgu.ac.il*
[2]*Design & Technological Institute for Monocrystals SB RAS, Novosibirsk 630058, Russia*

Abstract. We report on a nuclear magnetic resonance study of the nonlinear crystal $LiInSe_2$. This ternary compound shows a considerable variation of its color depending on the growth or annealing conditions. Room temperature ^{7}Li and ^{115}In NMR spectra of as-grown and annealed single crystalline and powdered samples have been measured. Analysis of the NMR spectra, along with the X-ray diffraction and optical spectroscopy data, leads to a conclusion that the change in color upon annealing in In_2Se_3 vapor is due to a change in the number of point defects.

Key Words: chalcogenide, NMR, nonlinear crystal, quadrupole interaction.

1. Introduction

$LiInX_2$ (X = S, Se) crystals, which were first studied by Boyd and co-workers in the 1970s [1], are now of renewed interest due to their attractive optical properties, such as their large transparency range from 0.35 to 13 μm and their high nonlinear susceptibility, as well as due to their potential usefulness in nonlinear optical devices. One of the features of these ternary compounds is a considerable variation of their color depending on the growth or annealing conditions. Such coloration may be due to some change in the crystal structure or due to phase transitions, and affects strongly the main output parameters including nonlinear efficiency. In the present paper, we studied the as-grown and annealed $LiInSe_2$ samples using ^{7}Li and ^{115}In nuclear magnetic resonance (NMR) and some other structurally sensitive techniques. We note that both ^{7}Li (I = 3/2) and ^{115}In (I = 9/2) are quadrupolar nuclei, and their

* Author for correspondence.

Table I. Lattice parameters for $LiInSe_2$ single crystals of different color determined in the present work

Color	Yellow	Greenish	Rose	Dark red
Space group	$Pna2_1$	$Pna2_1$	$Pna2_1$	$Pna2_1$
a [Å]	7.1917(8)	7.192	7.1939(8)	7.1934(10)
b [Å]	8.4116(10)	8.412	8.4163(10)	8.4159(11)
c [Å]	6.7926(8)	6.793	6.7926(8)	6.7971(9)
V [Å^3]	410.90(8)	410.97	411.27(8)	411.49(9)

line shapes are sensitive to the variations of the structure of compound and to appearance of defects. Our NMR spectra, along with the X-ray diffraction and optical spectroscopy data, lead to the conclusion that the color variation upon annealing in In_2Se_3 vapor is due to a change in the number of point defects.

2. Experimental

Sizable $LiInSe_2$ single crystals of high optical quality were grown by the Bridgman–Stockbarger technique. Details of the growth were reported elsewhere [2, 3]. The as-grown samples are yellow to greenish in color and contained a lot of small inclusions, which make them smoky. A high temperature annealing in In_2Se_3 vapor makes the $LiInSe_2$ samples highly transparent in the range from 0.7 to 13 μm but the short-wave transparency edge shifts to longer wavelengths (~700 nm) and crystals become dark red or even opaque.

The crystal structure determination of $LiInSe_2$ was performed using CAD4 diffractometer along with the SHELXL97 structure determination/refinement program. For all samples under study identical structure (space group $Pna2_1$, wurtzite-type lattice) was established. The lattice parameters are nearly unchanged after annealing (Table I).

For the NMR study, single crystals of $LiInSe_2$ (3.5 × 3.5 × 5 mm in size) were placed into an NMR coil of 5 mm inner diameter. The powder samples, prepared by grinding the crystals, were also studied. The ^7Li and ^{115}In spectra were measured with a Tecmag pulse NMR spectrometer in an applied magnetic field of 8.0196 T (at resonance frequencies of 132.68 and 74.84 MHz for ^7Li and ^{115}In nuclei, respectively), using Fourier transformation of the spin echo signals accumulated with the 16-phase cycled sequence. The length of $\pi/2$ pulse was 4 μs for ^7Li and 5 μs for ^{115}In. Angular dependence was measured when the crystals were rotated around their c-axis, which was perpendicular to the applied magnetic field. We note that the samples demonstrate a very long spin–lattice relaxation time (around 1 h), which is characteristic for high purity compounds.

3. Results and discussion

The ^{7}Li and ^{115}In NMR spectra of $LiInSe_2$ single crystals and powders are given in Figures 1–3, respectively. Due to the orthorhombic symmetry of the crystal structure, both ^{7}Li ($I = 3/2$) and ^{115}In ($I = 9/2$) should show quadrupolar perturbed NMR signals. In a single crystal the NMR spectrum of ^{7}Li nucleus should consist of three lines. Besides the central $1/2 \rightarrow -1/2$ transition at frequency ν_0, two satellites with the first-order shift of

$$\nu_1 = e^2qQ(3\cos^2\theta - 1)/4h, \quad (1)$$

should occur. Here q is the electric field gradient (EFG), Q is the quadrupole moment of the nucleus, θ is the angle between the applied magnetic field and the principal axis of the EFG tensor, and h is Planck's constant. The aforementioned satellites are clearly seen in the ^{7}Li spectrum of the single crystal (Figure 1). In the powder sample (Figure 2), the central line and the satellites are almost not resolved due to their overlap caused by the angular dependence of the resonance frequency given by Eq. (1). For the ^{115}In nucleus with $I = 9/2$, nine satellites should occur. However, these are barely excited with our $\pi/2$ pulse, and thus we show and analyze only the central line corresponding to the $1/2 \rightarrow -1/2$ transition (Figure 3).

The line width of the quadrupolar perturbed NMR signals is caused by dipole–dipole interaction of nuclei and the distribution of electric field gradients existing in an imperfect crystal. The nearest neighbors of In and Li atoms are Se atoms. Due to the low natural abundance of the ^{77}Se isotope (7.6%) having nuclear spin and the rather large Li–Li, In–In and Li–In distances (around 4.1 Å), the dipole–dipole interactions among the nuclei are small; their contribution to the second moment of the resonance line is calculated to be around 1 kHz^2 for both ^{7}Li and ^{115}In. Moreover, the dipole–dipole interactions should be the same for the as-grown and annealed samples. Thus the main contribution to the line width results from the distribution of EFGs, and the difference in the line width is caused by a variation of this distribution on annealing. From Figures 1–3, one can see that the powder ^{7}Li and ^{115}In NMR spectra of as-grown and annealed $LiInSe_2$ are very similar, though the resonance of the powder annealed sample is a little bit broader than that of the as-grown one. Two reasons that yield such an effect can be suggested. First, some difference in the structural parameters of two samples might yield the EFG values in the annealed sample to be larger than in the as-grown one. However, such effect would lead to a difference in the quadrupolar splitting of the NMR spectra of two crystals, which is not obtained in the single crystalline spectra (Figure 1). Moreover, XRD data show that the difference in lattice parameters for the as-grown and annealed samples is negligibly small. Second, one can suggest that the EFG distributions on Li and In nuclei are broader in the annealed sample since it is less perfect than the as-grown one. It may be, for example, that annealing in In_2Se_3 vapor produces

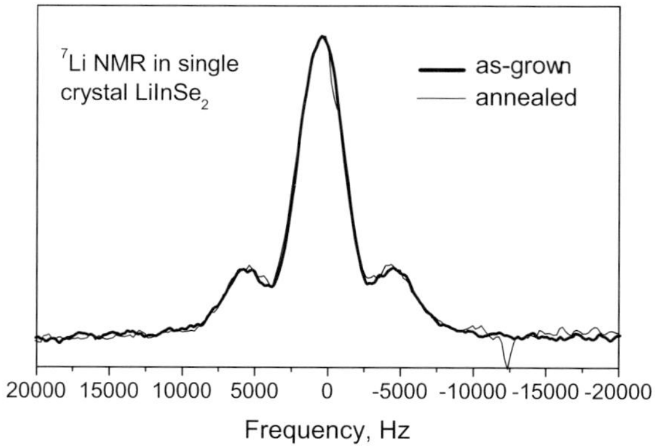

Figure 1. Room temperature ^{7}Li NMR spectra of as-grown (thick line) and annealed (thin line) $LiInSe_2$ single crystals. The orientation of external magnetic field B_0 ($B_0 \perp c$, angles (B_0, a) = 22.5°, (B_0, b) = 67.5°) corresponds to the maximal splitting of the lines.

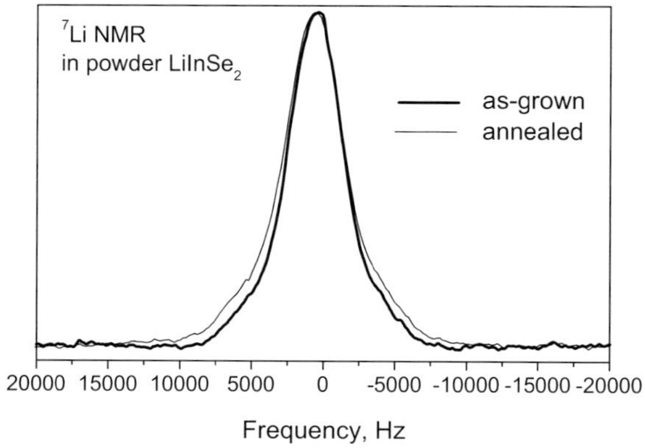

Figure 2. Room temperature ^{7}Li NMR spectra of as-grown (thick line) and annealed (thin line) powder samples of $LiInSe_2$.

additional defects (such as lithium vacancies). Such an imperfection of the crystal creates an additional electric field gradient at the nucleus; these gradients vary from site to site and yield additional line broadening. Thus one is led to conclusion that both as-grown and annealed $LiInSe_2$ samples have identical crystal structures, and that the difference in NMR spectra can be related to point defects.

This conclusion is supported by the optical spectroscopy data. Optical transmission spectra of as-grown $LiInSe_2$ yields a band gap E_g = 3.04 and 2.86 eV at 80 and 300 K, respectively [5], but annealed samples show some intense additional bands with maxima in the 500 to 550 nm range: these bands are responsible for the coloration of bulk $LiInSe_2$ samples. As a result the annealed thin plate is red in color. It is important to note that illumination by a visible light with λ ~400–500 nm from a 1 kW Xe lamp through a MDR2 diffraction monochromator was found to remove the red color of annealed $LiInSe_2$ crystals and to make them yellow, the effect being reversible: the red color is restored

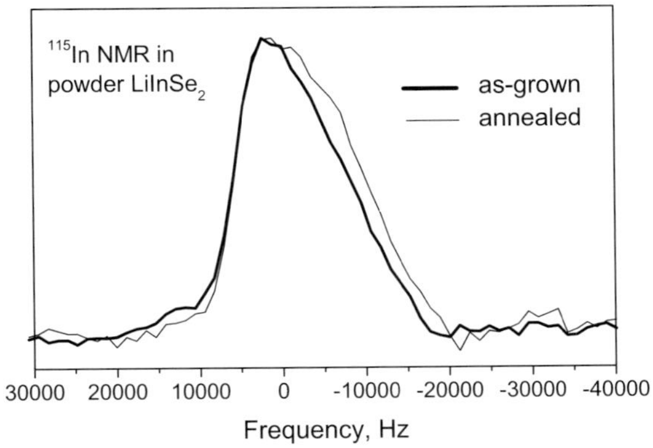

Figure 3. Room temperature ^{115}In NMR spectra of as-grown (*thick line*) and annealed (*thin line*) powder samples of $LiInSe_2$.

both at illumination with IR light ($\lambda > 700$ nm) and heating up to $\sim$700 K. Such effects are typical for the recharging of point defects in solids.

4. Conclusion

Comparative investigation of the as-grown and annealed $LiInSe_2$ crystals shows the identity of their structure and allows us to associate the coloration on annealing in In_2Se_3 vapor with point defects.

References

1. Boyd G. D., Kasper H. M. and McFee J. H., *J. Appl. Phys.* **44** (1973), 2809.
2. Isaenko L., Vasilyeva I., Yelisseyev A., Lobanov S., Malakhov V., Dovlitova L., Zondy J. J. and Kavun I., *J. Cryst. Growth* **218** (2000), 313.
3. Isaenko L., Yelisseyev A., Lobanov S., Petrov V., Rotermund R., Slekys G. and Zondy J. J., *J. Appl. Phys.* **91** (2002), 9475.
4. Kamijoh T. and Kuriyama K., *J. Cryst. Growth* **51** (1981), 6.
5. Isaenko L., Yelisseyev A., Zondy J.-J., Knippels G., Thenot I. and Lobanov S., *Opto-electron. Rev.* **9** (2001), 135.
6. Knippels G. M. H., van der Meer A. F. G., MacLeod A. M., Yelisseyev A., Isaenko L., Lobanov S., Thenot I., Zondy J. J., *Opt. Lett.* **26** (2001), 617.

Hyperfine Interactions (2004) 159:205–209
DOI 10.1007/s10751-005-9098-z

^{14}N Relaxation Study of 2-Nitrobenzoic Acid

D. STEPHENSON
Chemistry Department, University of the West Indies, St. Augustine, Trinidad and Tobago; e-mail: dstephenson@fsa.uwi.tt

Abstract. This paper describes a redesigned double resonance spectrometer allowing precise control of cross relaxation contacts. The apparatus has been used to determine the three relaxation times for the three level ^{14}N system in two crystalline modifications of 2-nitrobenzoic acid. The relaxation times are used to discuss the possible relaxation mechanisms in the two crystalline modifications. Also demonstrated are the much narrower peaks that can be achieved using double contact cross relaxation in the modified double resonance spectrometer described.

Key Words: ^{14}N, 2-nitrobenzoic acid, NQR, relaxation.

1. Introduction

Double contact cross relaxation is a relatively under utilised method for determining NQR resonance frequencies. Its sensitivity shows no frequency dependence making it particularly valuable at low frequencies. Line shapes are also considerably sharper than those seen in normal cross relaxation spectra. This is due to two characteristics of the technique. Firstly the magnetization contacts can be very short so that energy exchange will not occur if there is significant frequency mismatch between the protons and quadrupole levels. In a normal cross relaxation spectrum contacts are quite long as the initial energy exchange is followed by a long mutual relaxation phase. Secondly the two magnetization transfers between protons and quadrupole systems make the line shape the product of two normal cross relaxation line shapes. The technique can also be used to carry out a complete analysis of the relaxation behaviour of multi-level systems [1].

The fact that an n-level quadrupole system shows $(n-1)$ exponential behaviour has only rarely been acknowledged [2]. Difficulties in separating multiple exponents have led to this unpleasant fact being largely ignored. Indeed one respected text [3] argues that quadrupole relaxation is single exponential based on a statistical thermodynamics argument. The experimental evidence does not support this proposition [1, 4]. Furthermore if the populations of two levels in a multi-level system are disturbed, a single spin temperature cannot be applied to the system. It has to be conceded however that often the time constants of the exponents are relatively similar. Only when there is a large asymmetry in the relaxation, distinct multi-exponential behaviour can be observed in a simple relaxation experiment. Double contact cross relaxation can however be used to determine even small differences in the relaxation rates. This extra information can be used to give a better idea of the relaxation mechanisms at work. The simplest system to study is ^{14}N, because of the small number of quadrupole levels and the fact that the cross relaxation contacts can be carried out without the problem of Zeeman broadening in the quadrupole system. Such a system typically shows biexponential relaxation. Only in the special case where all three relaxation rates are equal single exponential behaviour would be seen. This paper attempts to show that there is much more information available by carrying out a complete relaxation analysis of two crystal forms of 2-nitrobenzoic acid. The improved resolution of the double contact technique is demonstrated by distinguishing subtle frequency shifts occurring for 2-nitrobenzoic acid as it changes between crystalline forms.

2. Experimental

2.1. THE SAMPLE

2-nitrobenzoic acid was obtained from BDH chemicals (laboratory reagent > 99%). Sample preparation involved taking the sample to just above its melting temperature and pouring it into the spectrometer sample tube. This was allowed to cool at room temperature. A spectrum of this sample is labelled ‘new sample’ in Figure 1. After a few days the spectrum and relaxation rates changed noticeably. After two weeks the sample’s ^{14}N resonance frequencies and relaxation times were indistinguishable from those of a sample that had not undergone the melting procedure. The spectrum of this sample is labelled ‘old sample’ in Figure 1.

2.2. THE SPECTROMETER

The double resonance spectrometer used to record all spectra was based on pneumatic transport of the sample from high to low field. In low field the cross relaxation contacts were provided by a solenoid linked to a digital power supply. The available current was from 0 to 10.23 A in 10 mA increments which covers cross relaxation frequencies up to 4.5 MHz. Power to the solenoid could also be pulsed to assist proton relaxation during the evolution phase. Spectra took typically 15 h to record. The long recording times are due to the long proton T_1 in these compounds. A long high field polarization time of about 3 min is required for each step in the spectrum. The two contacts of the double resonance cycle were achieved using field switching. The sample was first allowed to relax in a high field permanent magnet. It was then transferred to a solenoid that was energised to a proton resonance frequency of 4.5 MHz. This prevented relaxation of the sample prior to the first magnetisation contact; it also reduced any non-specific level crossing which may otherwise have occurred during sample

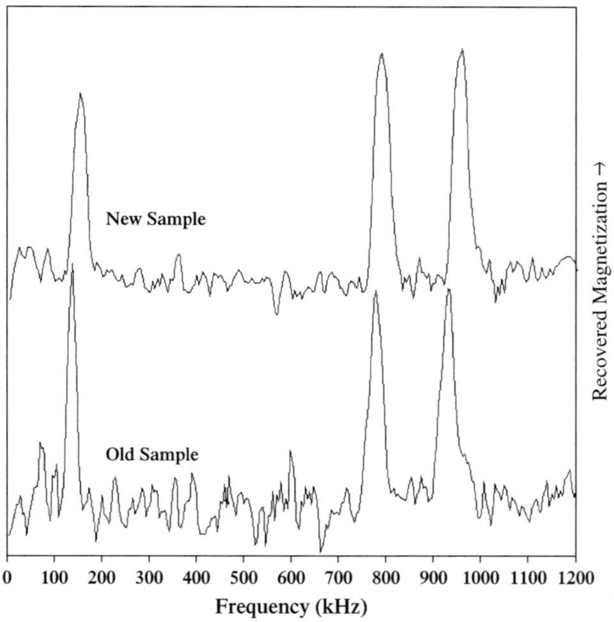

Figure 1. Double contact cross relaxation spectra of 2-nitrobenzoic acid. Upper spectrum of a sample freshly crystallised from the melt (spectrum scan time 15 h). Lower spectrum the same sample after two weeks (spectrum scan time 50 h).

transfer. The current supplied to the solenoid was then reduced to give the first magnetization transfer between the protons and a pair of the ^{14}N levels. Switching time was less than 1 ms and the level crossing condition was held for 10 ms. The field was then reduced to zero to cause the protons to relax fully. The process was accelerated by pulsing a small current through the solenoid at a rate of about 4 kHz. The pulsing speeded up the proton relaxation, but its effectiveness was limited by the small r.f. field that could be generated due to the large inductance of the solenoid. After 0.1 s the protons were fully relaxed then the cycle was carried out in reverse. The level crossing field was re-applied for 10 ms the current was switched up to 10 A (4.5 MHz ^{1}H resonance frequency) and the sample returned to the permanent magnet where the ^{1}H signal was measured at 20 MHz. The switching was all under computer control. The computer 'tick rate' was adjusted from its default value (51 ms) to 10 ms for controlling the cycle times and the short relaxing pulses were derived from a software delay loop.

2.3. RELAXATION TIMES

The three relaxation rates consistent with a three level system were determined using the method previously described [1]. The major obstacle to obtain precise values is the need to know the peak intensities at zero delay time. The best that could be achieved experimentally was 100 ms as the protons have to be relaxed completely to obtain meaningful data. Various methods were tested to extrapolate to time zero. The methods were evaluated using computer generated data. The most accurate method was found to involve using the shortest delay time as a first guess for the zero time value. This was then used to generate relaxation rates from which a new estimate of the zero time intensities could be calculated. The process could then be repeated. Generally satisfactory convergence could be achieved after three such iterations. In theory all the relaxation information could be calculated from the measurement of two double contact spectra; one spectrum using a very short delay time to obtain the zero delay intensities and one with a delay similar to the relaxation times under study. Practically there is often a wide range of relaxation rates even within a single sample so often measurements using several delay times are necessary to determine all the relaxation rates with acceptable accuracy. The method involves slowly scanning through the spectrum, and so a large amount of time is spent recording the baseline. A much better approach is to carry out measurements at the three peak maxima and some baseline points. Typically 10 measurements were made at each peak maximum and two points on the baseline. The values were averaged to obtain the peak intensity data (removing any obvious outliers).

3. Results and discussion

2-nitrobenzoic acid crystallises from the melt in an unstable crystalline form. This gradually changes to a more stable crystalline modification. Complete conversion takes about two weeks at 25°C. ^{14}N NQR spectra of the two crystalline forms are shown in Figure 1. The two crystalline forms show slightly different ^{14}N resonance frequencies. The freshly crystallised sample shows transitions at 155, 789 and 944 kHz (QCC = 1155 kHz, $\eta = 0.268$). After two weeks this changes to 143, 786 and 929 kHz (QCC = 1143 kHz, $\eta = 0.250$). The relaxation behaviour of the two crystalline forms is however quite different. Using the method of two-dimensional two contact double resonance spectroscopy [1] the three relaxation rates consistent with a three level can be determined. A sample freshly crystallised from the melt shows the relaxation between the levels -1 to $+1$ (corresponding to the ν_0 transition) to be fast. The other two relaxation rates 0 to $+1$ and 0 to -1 (corresponding to the ν_+ and ν_- transitions respectively) are much slower and of a similar magnitude. This is the opposite of what is found for the 4-nitrobenzoic acid (which also shows asymmetric relaxation) where the ν_0 relaxation is very long [1] compared to the other two relaxation rates. For an aged sample of 2-nitrobenzoic acid the ν_0 relaxation rate decreases to about a eighth of its previous value in a freshly prepared sample whereas the other two relaxation rates are little changed. The results are summarised in Table I.

The crystal structure [5] of 2-nitrobenzoic acid indicates that the plane of the nitro group is tilted about 55° from the plane of the benzene ring which is

Table I. Relaxation times for 2-nitrobenzoic acid

Sample	Relaxation times (s)		
	ν_0	ν_-	ν_+
2-nitrobenzoic acid (new)	2.5	14.3	13.8
2-nitrobenzoic acid (old)	20.0	17.9	19.9
4-nitrobenzoic acid [1]	303	11.5	13.2

The new sample is freshly crystallised from the melt, the old sample is the same sample after two weeks at 25°C. The approximate error in the values is ±20% or 1 s whichever is the greatest. The relaxation rates for 4-nitrobenzoic acid are shown for comparison.

considerably more than the 14° in 4-nitrobenzoic acid [6]. A possible explanation for the change in crystal structure is an adjustment of this tilt angle following solidification, the initial unstable form having more torsional freedom in the nitro group. As the principal component of the efg tensor lies along the C–N bond, torsional oscillations will modulate the q_{xx} and q_{yy} components of efg, this could lead to enhanced relaxation between the levels associated with the ν_0 transition. As the angle of the nitro group becomes fixed, the main cause of relaxation is now magnetic oscillations caused by dynamic proton disorder of the carboxyl dimer. Neither the inter or intra molecular carboxyls are at any special angle to the nitro group so the relaxation becomes symmetrical, that is at the same rate between all pairs of levels. This is consistent with the relaxation behaviour of 4-nitrobenzoic acid. In 4-nitrobenzoic acid (unlike 2-nitrobenzoic acid) the nitro groups are well away from any intramolecular carboxyl interaction [6] and the intermolecular carboxyls are nearly aligned along the q_{zz} component of the efg. This geometric arrangement should predominantly induce relaxation between levels associated with the ν_+ and ν_- transitions in agreement with the experimentally observed relaxation behaviour [1].

References

1. Stephenson D., *Z. Naturforsch.* **55a** (2000), 79.
2. Fukushima E. and Roeder S. B. W., *Pulsed NMR: A Nuts and Bolts Approach*, Addison-Westley, Reading, Massachusetts, USA, 1981.
3. Schlicter C. P., *Principles of Magnetic Resonance*, 3rd Edition, Springer, Berlin Heidelberg New York, 1990.
4. Stephenson D. and Smith J. A. S., *J. Chem. Soc. Faraday Trans. II* **83** (1987), 2123.
5. Sakore T. D., Tavale S. S. and Pant L. M., *Acta Crystallogr.* **22** (1967), 720.
6. Tavale S. S. and Pant L. M., *Acta Crystallogr.* **B27** (1971), 1479.

Hyperfine Interactions (2004) 159:211–216
DOI 10.1007/s10751-005-9099-y

NQR, NMR and Crystal Structure Studies of $[C(NH_2)_3]_3Sb_2Br_9$

HIROMITSU TERAO[1,*], YOSHIHIRO FURUKAWA[2], SATOMI MIKI[1], FUKUE TAJIMA[3] and MASAO HASHIMOTO[3]
[1]*Faculty of Integrated Arts and Sciences, Tokushima University, Tokushima 770-8502, Japan; e-mail: terao@ias.tokushima-u.ac.jp*
[2]*Graduate School of Education, Hiroshima University, Higashi-Hiroshima 739-8524, Japan*
[3]*Faculty of Science, Department of Chemistry, Kobe University, Nadaku, Kobe 657-8501, Japan*

Abstract. The crystal structure of $[C(NH_2)_3]_3Sb_2Br_9$ was determined at 143 K: monoclinic, space group $C2/c$, $Z = 4$, $a = 15.695$ (3), $b = 9.039(2)$, $c = 18.364(3)$ Å, $\beta = 96.94(1)°$. The structure consists of two crystallographically independent guanidinium ions and two-dimensional corrugated sheets of $(Sb_2Br_9{}^{3-})_n$, in which $SbBr_6$ octahedra are connected through three bridging Br atoms each other. One of the cations situates in a cavity of the $(Sb_2Br_9{}^{3-})_n$ layer with statistical disorder, while the other situates between the layers without disorder. Three ^{81}Br NQR resonance lines were assignable to terminal Br atoms, while only one line was found for two inequivalent bridging Br atoms. All the ^{81}Br NQR resonance lines were subjected to fade-out at low temperatures. The temperature dependence curve of ^{1}H NMR T_1 showed well defined two minima, which were explained by postulating the C_3 reorientations of two types of cations with very different activation energies. The DTA (DSC) measurement revealed a phase transition of a first-order type at 444 K.

Key Words: ^{81}Br NQR, $[C(NH_2)_3]_3Sb_2Br_9$, crystal structure, ^{1}H NMR T_1, molecular motion.

1. Introduction

A number of studies have been concerned on the crystals of alkylammonium halogenoantimonates(III) and bismuthates(III) owing to the interests about their structures and physical properties. Their structures change widely depending on the kinds of cations and halogens [1]. Furthermore, many of compounds exhibit interesting physical properties such as ferroelectricity, ferroelasticity, and pyroelectricity.

Recently, Zaleski and Pietraszko [2, 3] clarified the crystal structure of $[C(NH_2)_3]_3Sb_2Cl_9$ at room temperature and found a phase transition at 364 K by a DSC measurement. The anions in the crystal form a so-called two-dimensional corrugated sheet $(Sb_2Cl_9{}^{3-})_n$ in which $SbCl_6$ octahedra are connected through three bridging Cl atoms each other. In this structure there exist two inequivalent

* Author for correspondence.

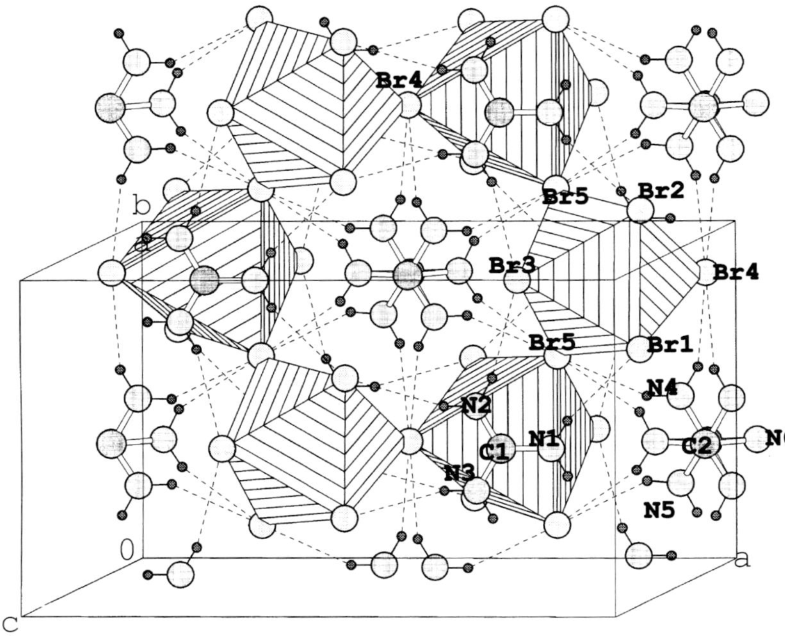

Figure 1. Crystal structure of $[C(NH_2)_3]_3Sb_2Br_9$ at 143 K. Crystal data: monoclinic, space group $C2/c$, $a = 15.695$ (3), $b = 9.039(2)$, $c = 18.364(3)$ Å, $\beta = 96.94(1)°$, and $Z = 4$. The $SbBr_6$ octahedra are connected through three bridging Br atoms each other. The dashed lines represent the possible N–H···Br hydrogen bonds.

cations. Two of three guanidinium cations of a $[C(NH_2)_3]_3Sb_2Cl_9$ unit situate equivalently between $(Sb_2Cl_9{}^{3-})_n$ layers without disordering, while the remaining one in the cavity within the layer with a statistical disorder.

According to the study of T_1 as well as the second moment by ^{1}H NMR Grottel *et al.* [3] clarified that the disorderd guanidinium ions experience much lower barrier for C_3 reorientation than that of the ordered one. This feature manifested itself in clear two step reduction of the second moment and well separated T_1 minima.

The formation of two-dimensional corrugated sheet structures may need matching in type as well as size between the relevant cation and cavity. We have interested in such condition of cation–anion combinations and in the cation motions as well. Then, in the present work we investigated the dynamics in the crystals of $[C(NH_2)_3]_3Sb_2Br_9$ by means of the ^{81}Br NQR and ^{1}H NMR T_1. The crystal structure of this compound was also determined to get the information related to the dynamics.

2. Experimental

$[C(NH_2)_3]_3Sb_2Br_9$ crystals were prepared by dissolving an equimolecular amount of $[C(NH_2)_3]$ Br and Sb_2O_3 in a hot concentrated hydrobromic acid followed by

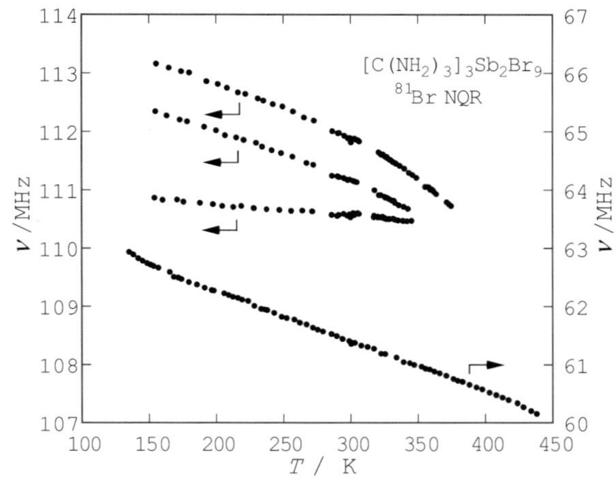

Figure 2. Temperature dependence of ^{81}Br NQR frequencies of $[C(NH_2)_3]_3Sb_2Br_9$.

cooling. The crystals were recrystallized from the same solvent in the form of yellow hexagonal plates. The starting material $[C(NH_2)_3]$ Br was prepared by an addition of hydrobromic acid to an aqueous solution of $[C(NH_2)_3]_2CO_3$. The results of C, H, and N elemental analyses; found/calc.; weight %: C: 3.09/3.15; H: 1.54/1.59; N: 10.94/11.03 for $[C(NH_2)_3]_3Sb_2Br_9$.

The NQR spectra were measured by using a super-regenerative spectrometer. The signals were recorded on a recorder through a lock-in amplifier with Zeeman modulation. The accuracy of frequency measurement is estimated to be within ± 0.02 MHz.

Spin-lattice relaxation times T_1 of ^1H NMR were measured by the inversion recovery method on a standard pulsed NMR spectrometer.

The crystal structure was determined at 143 K. CCDC 249999 contains the supplementary crystallographic data for this paper. These data can be obtained free of charge via http://www.ccdc.cam.ac.uk/data_request/cif, by emailing data_request@ccdc.cam.ac.uk, or by contacting The Cambridge Crystallographic Data Centre, 12, Union Road, Cambridge CB2 1EZ, UK; fax: +44 1223 336033.

3. Results and discussion

The DTA and DSC measurements showed a first-order type phase transition at 444 K before its melting at 508 K. The heat of transition was 21.8 kJ mol^{-1}.

The crystal structure is shown in Figure 1, which is isostructural with the corresponding chlorine compound [2, 3]. There are two crystallographically inequivalent guanidinium cations. One lies in the cavity formed by six $SbBr_6$ octahedra in the $(Sb_2Br_9{}^{3-})_n$ layer and the other lies between $(Sb_2Br_9{}^{3-})_n$ layers. As exactly same as seen in $[C(NH_2)_3]_3Sb_2Cl_9$ the former is statistically disordered at two sites with the same occupancy, while the latter is ordered. There are three inequivalent terminal Br atoms with short Sb–Br distances

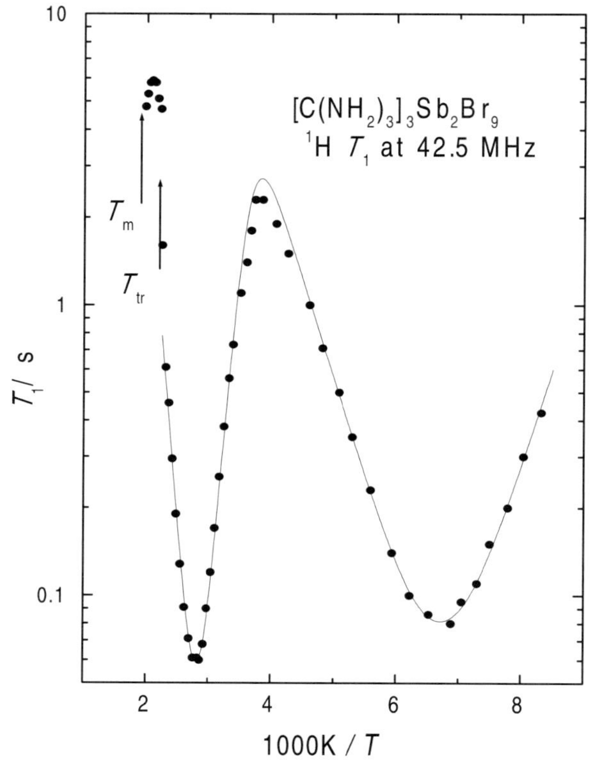

Figure 3. Temperature dependence of ^{1}H T_1 at 42.5 MHz in $[C(NH_2)_3]_3Sb_2Br_9$.

(2.621, 2.609 and 2.615 Å for Br1, Br2 and Br3, respectively), while two inequivalent bridging Br atoms with longer Sb–Br distances (3.068 Å (twice) for Br4, and 3.051 and 3.073 Å for Br5). Possible hydrogen bonds N–H···Br are depicted with dashed lines in Figure 1. Short N···Br distances are ca. 3.6 Å for the disordered cation and 3.4 Å for the ordered one. Thus, the hydrogen bonds are expected stronger for the ordered cation than for the disordered one. It is noted that the disordered cations may take part in the hydrogen bonding with only the bridging Br atoms and the ordered ones with only the terminal Br atoms. It is also understandable in Figure 1 how the disordered cations adjust themselves well into the cavities of the poly anion layers as to the size as well as the hydrogen bonding scheme.

As shown in Figure 2 four ^{81}Br NQR resonance lines (61.40, 110.57, 111.16, 111.90 MHz at 298 K) were detected in the investigated range (77 to ca. 380 K for three high frequency lines; 77 to ca. 440 K for the lowest one). Three resonance lines around 112 MHz are assignable to the inequivalent terminal Br atoms, then the remaining one at 61 MHz to the one of the two inequivalent bridging Br atoms. As to this observation that one line is seemingly absent, one should notice the resemblance in the NQR spectra between the present compound and $Cs_3Bi_2Br_9$ or $Cs_3Sb_2I_9$ which has a similar structure. $Cs_3Bi_2Br_9$

Table I. Motional parameters obtained from ^{1}H NMR T_1 for $[C(NH_2)_3]_3Sb_2Br_9$ and $[C(NH_2)_3]_3Sb_2Cl_9$ [3]

Compound (cation)	C/s^2	τ_0/s	E_a/kJ mol^{-1}	Reference
$[C(NH_2)_3]_3Sb_2Br_9$				
$[C(NH_2)_3]^+$ (disordered)	2.30×10^9	6.14×10^{-14}	13.0	This work
$[C(NH_2)_3]^+$ (ordered)	3.16×10^9	1.00×10^{-16}	50.2	This work
$[C(NH_2)_3]_3Sb_2Cl_9$				
$[C(NH_2)_3]^+$ (disordered)	–	1.6×10^{-14}	20.5	Ref. [3]
$[C(NH_2)_3]^+$ (ordered)	–	1.3×10^{-16}	54.5	Ref. [3]

undergo a phase transition of a second-order type at $T_c = 96$ K [4, 5] and $Cs_3Sb_2I_9$ a sequence phase transitions below at $T_c = 85$ K [5, 6]. Above T_c a single ^{81}Br or ^{127}I NQR resonance line was observed for the respective terminal and bridging halogen atoms in accordance with the room temperature structure (P3m1) of $Cs_3Bi_2Br_9$ [7] and $Cs_3Sb_2I_9$ [8], in which all terminal as well as bridging halogen atoms are equivalent. Below T_c the terminal line split into three, while the bridging line does not split just as in the case of $[C(NH_2)_3]_3Sb_2Br_9$. Furthermore, the crystal structures of the low temperature phases of $Cs_3Bi_2Br_9$ and $Cs_3Sb_2I_9$ are expected isostructural ($C2/c$) with $[C(NH_2)_3]_3Sb_2Br_9$ as a result of the X-ray crystal analyses [6]. A possible reason for the missing of one line is a very close proximity of two lines corresponding to the bridging Br atoms, but it seems difficult to keep the close proximity in the wide temperature region as observed (ca. 135–440 K). Thus, the reason is left open. At low temperatures all the resonance lines were subjected to disappearance. With increasing temperatures the signal of bridging Br atoms became detectable above ca. 135 K, while those of terminal Br atoms did above ca. 150 K. This observation may be connected to the difference of the strength of hydrogen bonds between the disordered cations and the ordered ones.

The temperature dependence of ^{1}H NMR T_1 is shown in Figure 3. Well defined two minima were observed with the minima of 60 ms at 355 K and 80 ms at 150 K. Because the planar $[C(NH_2)_3]^+$ ions usually undergo C_3 reorientation in crystalline solids [9, 10], the curve was analyzed by postulating the C_3 reorientations of the cations. From the numbers of inequivalent cations in the unit cell, the lower temperature T_1 minimum was assigned to the disordered cations, and the other was assigned to the ordered ones. The temperature dependence of T_1 was analyzed by using the BPP type equation and the Arrhenius relation for the correlation time τ of the reorientation.

$$T_1^{-1} = C\left[\tau/\left(1 + \omega^2\tau^2\right) + 4\tau/\left(1 + 4\omega^2\tau^2\right)\right], \tag{1}$$

and

$$\tau = \tau_0 \exp\left(E_a/RT\right). \tag{2}$$

By least-squares calculation, the motional constant C, the correlation time at infinite temperature τ_0, and the activation energy E_a were determined as listed together with those of $[C(NH_2)_3]_3Sb_2Cl_9$ [3] in Table I. It is reasonable that both E_a values in the Br compound are lower than the Cl compound owing to the expected weaker hydrogen bonds of the N–H···Br than N–H···Cl. The difference between the E_a values of the ordered cations and the disordered ones is also consistent with the strengths of their hydrogen bonds.

References

1. Sobczyk L., Jakubas R. and Zaleski J., *Pol. J. Chem.* **71** (1997), 265.
2. Zaleski J. and Pietraszko A., *Z. Naturforsch.* **49a** (1994), 895. Though the chemical formulae of the crystal was described as $[C(NH_2)_3]_3Sb_2Cl_9 \cdot 0.9H_2O$ in this paper, it has been referred in Ref. 3 that new refinement excluded the existence of water.
3. Grottel M., Pajak Z. and Zaleski J., *Solid State Commun.* **120** (2001), 119.
4. Terao H., Ishihara H., Okuda T., Yamada K. and Weiss Al., *Z. Naturforsch.* **47a** (1992), 1259.
5. Ivanov Y. N., Sukhovskii A. A., Lisin V. V. and Aleksandrova I. P., *Inorg. Mater.* **37**(6) (2001), 623.
6. Aleksandrova I. P., Burriel R., Bartolome J., Bagautdinov B. S., Blasco J., Sukhovsky A. A., Torres J. M., Vasiljev A. D. and Solovjev L. A., *Phase Transit.* **75**(6) (2002), 607.
7. Lazarini F., *Acta Crystallogr.* **B33** (1977), 2961.
8. Bagautdinov B. Sh., Novikova M. S., Aleksandrova I. P., Blomberg M. K. and Chapuis G., *Solid State Commun.* **111** (1999), 361.
9. Pajak Z., Grottel M. and Koziol A., *J. Chem. Soc., Faraday Trans.* **78**(2) (1972), 1529.
10. Gima S., Furukawa Y. and Nakamura D., *Ber. Bunsenges. Phys. Chem.* **88** (1984), 939.

Hyperfine Interactions (2004) 159:217–226
DOI 10.1007/s10751-005-9103-6

MSR Studies in the Progress Towards Diamond Electronics

S. H. CONNELL[1,*], I. Z. MACHI[2] and K. BHARUTH-RAM[3]
[1]*Schonland Research Institute for Nuclear Sciences, University of the Witwatersrand, Johannesburg, South Africa e-mail: connell@src.wits.ac.za*
[2]*Physics Department, University of South Africa, Cape town, South Africa*
[3]*School of Pure and Applied Physics, University of KwaZulu Natal, Durban, South Africa*

Abstract. The recent development of device quality synthetic diamond dramatically increases the potential of diamond as a wide band gap semiconductor. A remaining obstacle is the lack of shallow n-type dopants. Molecular dopant systems have been shown theoretically to lead to the shallowing of levels in the band gap. Some of these systems involve defect-hydrogen complexes. This, and other phenomena, motivate the study of the chemistry and dynamics of hydrogen in diamond. Much information on this topic has been obtained from Muon Spin Rotation (MSR) experiments. These experiments view the muonium (Mu $\equiv \mu^+e^-$) atom as a light chemical analogue of hydrogen. Data on isolated muonium in diamond is reviewed, and evidence on formation of N-Mu-N (a shallow dopant candidate), the trapping of Mu at B-dopants, and fast quantum diffusion of muonium are discussed.

Key Words: diamond, diffusion, dopants, muonium.

1. Introduction

Diamond has for a long time been known to have a theoretical performance as an electronic material with figures of merit (FoM) for various active electronic applications that can exceed those of silicon by one to four orders of magnitude, depending on the application [1–4]. In fact, it is advantageous for extreme electronic applications, such as high power, high temperature, and high frequency devices. Of particular significance are the high breakdown field and combined charge carrier mobility together with the highest room temperature thermal conductivity and extremely low intrinsic resistivity. Diamond is also radiation hard. Besides these properties that are directly relevant to device performance, diamond also exhibits a range of other excellent optical, physical, thermal, mechanical and chemical properties. For example, the spectral transparency is the widest of any material. The low thermal expansion coefficient together with the previously mentioned high thermal conductivity renders it very resistant to thermal stresses.

* Author for correspondence.

The main reason for the slow development of applications has been the small size, cost and defected nature of natural and synthetic material. True grain boundaries, dislocations and significant concentrations of inhomogenously incorporated single substitutional nitrogen are recognised as important defects. Recently, however, the situation regarding sample quality and size has improved, both with respect to the high pressure high temperature (HPHT) synthesis and the chemical vapour deposition (CVD) synthesis of diamond. For example, away from the seed, HPHT diamonds appropriately processed attain X-ray rocking curves close to the Darwin line width [5]. Single crystal (SC) CVD diamonds exhibit charge carrier lifetimes in the microsecond range, more than two orders of magnitude better than selected nitrogen free natural diamonds [6]. Charge collection distances are therefore larger than typical device dimensions.

The realisation of device quality diamond dramatically increases the potential to exploit this material as a wide band gap semi-conductor. A reasonably shallow p-type dopant exists, this being the boron acceptor with an activation energy of 0.37 eV. There is not yet a shallow n-type dopant. Nitrogen forms a very deep donor with an activation energy of 1.7 eV. It is possible that phosphorous yields an acceptor level with an activation energy of 0.52 eV, however, it readily complexes with the vacancy, and activation levels are very low. Other possibilities for elemental dopants have been considered theoretically and experimentally, but they are not yet reliably established [7, 8]. There are effectively only intrinsic or unipolar electronic devices. The possibility of shallow molecular dopant systems in diamond is intriguing. Several of these have been predicted, see for example [9–11]. Complexation of certain defects appears to be favourable in diamond, and it is expected that this can be exploited to produce carefully selected molecular species.

There is therefore an active interest for research which increases the understanding of factors leading to improved diamond quality and developing shallow n-type dopants. In this regard, hydrogen plays an important role. For example, diamonds are always contaminated with hydrogen. Hydrogen is known to passivate acceptors and donors, and possibly also other compensating centres themselves. There may be a mechanism by which it can facilitate the formation of extended defects. Hydrogen surface termination leads to surface conductivity. There is at least one electrically active hydrogen defect (H1). Finally, hydrogen complexation with certain impurities is expected to lead to the shallowing of donor states (see [7, 11] and references therein).

Much of the information on hydrogen in diamond has come from Muon Spin Rotation/Relaxation/Resonance (MSR) studies, and this method also provides the only unambiguous data [11, 12]. These experiments view the muonium (Mu $\equiv \mu^+e^-$) atom as a light chemical analogue of hydrogen. Chemical information is related to the spectroscopy of the muonium states formed. Caveats apply due to the lighter mass of muonium in comparing dynamical information.

This work will briefly review the situation for isolated muonium in diamond. Following this several recent experiments are of interest. A possible shallow molecular dopant system is N-H-N [9] and a candidate for this system has been observed in MSR experiments. The passivation of the boron dopant in diamond by hydrogen has been observed in electrical measurements. MSR measurements in p-type diamond evidence trapping behaviour which is consistent with passivation. These processes involve diffusion. It is therefore of interest to discuss studies of the fast quantum diffusion of muonium in diamond.

2. Isolated muonium in diamond

There is not yet any information on isolated hydrogen in diamond, and therefore MSR studies are especially important here. The population of different isolated muonium states formed in diamond are affected by the sample quality. In the purest of diamonds (HPHT samples with the concentration of nitrogen $C_N <$ ppm), the following species are observed, where f is the prompt formation fraction.

Mu$_T$: Muonium at the tetrahedral interstitial site ($f_{Mu_T} \approx 70\%$).
Mu$_{BC}$: Bond centered muonium ($f_{Mu_{BC}} \approx 20\%$).
μ_D^+ : The muon in a diamagnetic environment ($f_{u_D^+} \approx 10\%$).

In this case there is no missing fraction ($f_{MF} \approx 0$). The hyperfine (hf) interaction of Mu_T is isotropic with a hf constant $A = 3711$ MHz. This is 83% of the vacuum muonium value which is amongst the largest measured in any material. This species is known to diffuse rapidly, as in defected diamond, it has a rapidly relaxing amplitude.

The hf interaction of Mu_{BC} is anisotropic with 111 axial symmetry. The fermi contact and dipolar hf coupling constants are small and of opposite sign, $A_s = -205$ MHz and $A_p = 186$ MHz respectively [13]. The nuclear hf coupling constants for the nearest and next nearest neighbours have also been measured [14]. Taken together this indicates that the unpaired electron spin density is dominantly localised on the two carbon neighbours (92%) somewhat analogous to a muonated radical. The correspondence to theoretical models which include relaxation and electron correlations is good (see [11, 14] and references therein). The neighbouring carbon atoms relax away from each other by about 40%. This may be seen rather as a rehybridisation of the valence orbitals of these carbon atoms from sp^3 to sp^2 so that less energy is required than an elastic distortion. The temperature stability of this species (measured up to 1080 K) as well as the preservation of anisotropy and the very narrow line widths are indicative of an immobile state.

There is a thermally activated first order transition from $\mathrm{Mu}_T \rightarrow \mathrm{Mu}_{BC}$ at 450 K. The Arrenhius prefactor for this is compatible with the optical phonon frequency so that only one hop is required. Mu_{BC} is therefore the most stable form of isolated muonium in diamond.

In comparing the situation to hydrogen, one expects to be able to identify both H_T and H_{BC}. However, zero point energies would be lower so that the barrier for the transition $H_T \rightarrow H_{BC}$ would be higher. In addition, the time-window for MSR is about 12 μs wide immediately following implantation of muons at a concentration so low that they would not be expected to interact with each other. In contrast, later and longer time windows for H measurements, together with the mobility of H_T suggest that it traps competitively with other defects and impurities present in the diamond. In addition, auto-trapping configurations are possible (H_2^* e.t.c.). Nonetheless, despite searches, H_T and H_{BC} have not yet been seen in diamond.

3. N-H complexes – possible shallow molecular dopants

Transverse Field (TF)-μSR measurements on nitrogen-rich type Ia diamond containing aggregated A- and B-centres of up to 1000 ppm exhibit a substantial missing fraction (MF), a reduced and rapidly spin relaxing Mu_T fraction and no Mu_{BC} [15–17]. An A-centre is formed by two adjacent nitrogen atoms at substitutional sites [18], and a B-centre is believed to be formed by four nearby nitrogen atoms at substitutional sites surrounding a vacancy [19–21]. The spin relaxation rate of the Mu_T state has been interpreted as due to diffusion and trapping at these nitrogen-aggregated centres [14, 15]. The MF was conjectured to be a muonium state associated with nitrogen aggregates, and may therefore be revealed in Longitudinal Field (LF)-μSR measurements. This turned out to be the case. In fact almost 100% recovery of the spin polarization was possible already at a field of 0.4 T for the extrapolated initial muon fraction [22–24]. The corresponding TF-μSR data for this sample were used to assist in extracting the components due to Mu_T, Mu_{BC} and μ^+ from the net LF-μSR repolarisation curve. The field quenching behaviour of the remaining 60% (previously missing) fraction was then revealed as shown in Figure 1. The new configuration was termed Mu_X and would represent muonium trapped in some way at an A or B-centre. The hf interaction symmetry is just less than axial, and its parameters are given as $A_s = 4158(1500)$ MHz, $A_p = 248(130)$ MHz and $\delta A = -28(50)$ MHz.

Various muonium complexes with A and B centres have been explored theoretically. A likely candidate is the muonium trapped in the 'bond centre' position of the A-centre. Neutral hydrogen was found to be more stable within an A-centre than within the bond-centred site by about 2.4 eV [25]. In addition, the hydrogen, though centrally located between the nitrogen atoms in the A-centre, has a small off axis displacement which may account for the observed less than axial symmetry of the hf parameters at this site.

The N-H-N system is a possible shallow molecular dopant [9]. Future experiments will be optimised to study the structure and possible ionisation of this system.

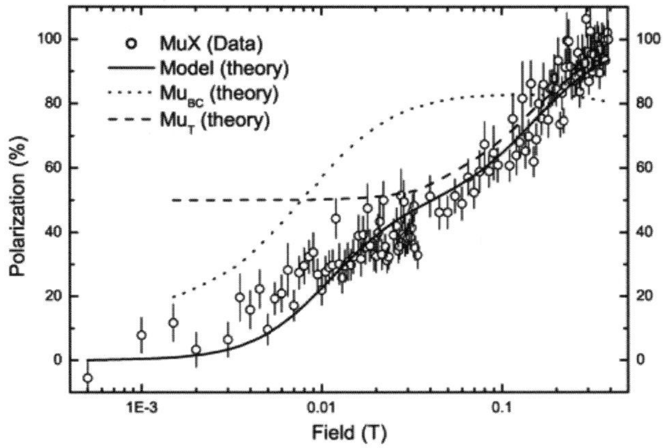

Figure 1. Repolarisation curve for muonium configuration Mu_X, in nitrogen-rich diamond. The known behavior (from TF-μSR measurements) for Mu_T and μ_D^+ have been removed. Curves for Mu_T and Mu_{BC} are included for comparison. (Asymmetries referenced to 100% for each species).

4. Muonium interactions in p-type diamond

As has been mentioned before, low nitrogen content diamonds are expected to exhibit a large non-relaxing Mu_T fraction. Natural semi-conducting diamond has sufficiently low nitrogen and other defects that a significant proportion of boron acceptors are uncompensated. The boron acceptors would therefore be expected to be the dominant trap in such a diamond. TF-μSR measurements were conducted in a natural semi-conducting diamond which contained 0.28 ppm uncompensated boron acceptors. The external magnetic field was 7.5 mT and the temperature was varied in the range 3-300 K. The spin relaxation rate λ_{Mu_T}, for the Mu_T component, was determined in the usual way [26], and is displayed in Figure 2.

The relaxation rate is high at low temperatures, and has a power law fall off from 100-300 K. Its final value is consistent with the relaxation expected for a pure diamond. The relaxation is explained by rapid diffusion of the Mu_T followed by deep trapping at a defect where the muon(ium) experiences a substantially different hyperfine environment. This is a T_2 mechanism where the muon ensemble dephases due to different arrival times at the traps. Effectively, the relaxation rate is proportional to the diffusion constant $D_c(T)$. The faster diffusion at lower temperatures has been discussed elsewhere in the context of quantum diffusion. What is relevant here is clear evidence of the interaction of Mu_T with the boron dopants, following diffusion [27, 28].

5. Quantum diffusion of muonium in diamond

The rapid diffusion at low temperature, seen in the previous section, led to a continuing study of the motional behavior of the tetrahedral interstitial muonium (Mu_T) in diamond. The presence of traps in the previous study complicate

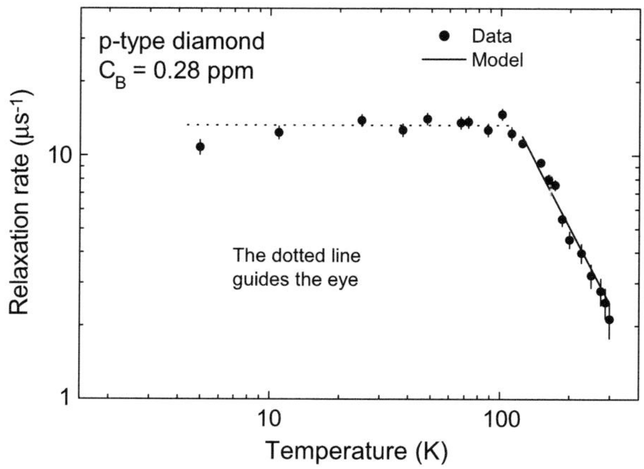

Figure 2. Spin relaxation rate of Mu_T, as a function of temperature, in p-type semi-conducting type IIb diamond.

extraction of the diffusion constant as the trap radius is itself also temperature dependent. Clearly, one needs to observe the motion of the Mu_T in a pure diamond. This requires a pure synthetic diamond consisting dominantly of ^{13}C. Mu_T diffusion experiments were carried out using both the TF-μSR and LF-μSR methods. In the previous case, and in the following two cases, the **relaxation mechanism** by which the diffusion constant is related to the relaxation rate λ_{Mu_T} is summarised.

TF-μ SR: diffusion, deep trapping; This has been explained above. The sample must have defects. Faster relaxation corresponds to faster diffusion. This is a T_2 mechanism.

TF-μ SR : motional narrowing; The sample is pure, the diffusion is against a background of randomly oriented ^{13}C nuclear moments. A static or trapped ensemble experiences different local fields and each member of the ensemble precesses at a slightly different rate, leading to the dephasing of the spin polarisation of the ensemble. If the Mu_T is sufficiently mobile, such that it averages the local field to a single well-defined value on a time scale which is short compared to the period of its precession, then the ensemble does not dephase. In this mode, faster diffusion leads to smaller relaxation of the spin polarisation. This is a T_2 mechanism.

LF-μ SR: induced transitions; The sample is pure, the diffusion is against a background of randomly oriented ^{13}C nuclear moments. This is equivalent to illumination of the Mu_T by radio-frequency radiation with a Lorentzian power spectrum, centered at zero with a width dependent on the diffusion constant. This effective radiation field induces transitions in the hyperfine levels of the Mu_T,

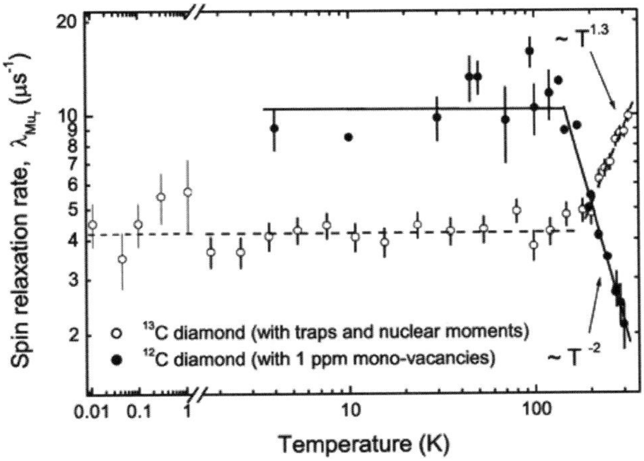

Figure 3. Spin relaxation rate of Mu_T, as a function of temperature, in diamond as described in the text.

which depolarize the Mu_T ensemble. There is not a monotonic relation between the relaxation rate and the diffusion constant. This is a T_1 mechanism.

Light particle diffusion in diamond would be expected to proceed through different mechanisms in different temperature regimes. This would be fast quantum diffusion of Mu_T in a relatively delocalised Bloch state at low temperatures and fast classical over-barrier hopping at high temperatures. The intermediate regime is associated with slower quantum diffusion due to phononic mechanisms that cause a loss of coherence (see [29, 30] and many references therein). Ideally, one would observe a minimum in the diffusion constant between these two regimes. The lower temperature side of this minimum is related to a two-phonon mediated tunneling between adjacent interstitial sites of the small polaron represented by the weakly self-trapped Mu_T due to its own local lattice distortion. This section is expected to follow a negative power law temperature behaviour. The higher temperature side of this minimum is related to effects like the fluctuational preparation of the barrier (FPB) and the polaron effect (PE), which are single phonon mediated and have an Arrenhius-like signature even though they are still related to quantum tunneling. In diamond, the period of quantum diffusion is expected to be maintained to rather high temperatures [12].

The relaxation rate data for the first two mechanisms mentioned above is presented in Figure 3. It is clear that the diffusion constant would be high in both cases, and then follow a power law decrease in the region 100–300 K. It is expected that the ^{13}C sample is not perfectly pure as the relaxation rate is not sufficiently low in the rapid diffusion regime. There would be therefore a mixture of the two mechanisms for this sample. In addition, the temperature dependence of the trap radius would weaken the power law behaviour in the 100–300 K regime for both measurements. These two measurements of the relaxation rate,

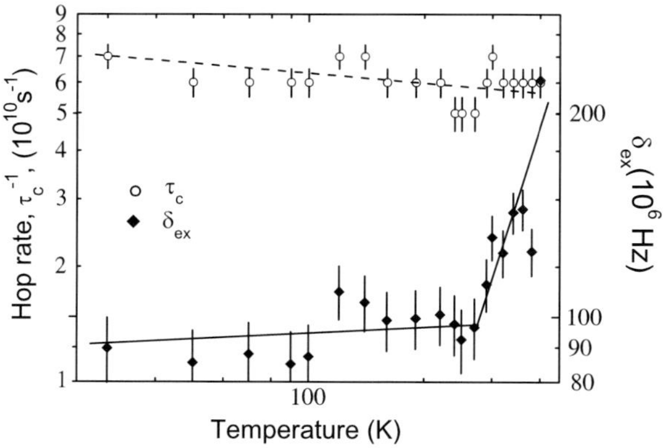

Figure 4. Hop rates, τ_c^{-1}, and nuclear hyperfine interaction, δ_{ex}, of Mu_T in ^{13}C diamond, as a function of temperature. Dotted and solid lines guide the eye.

with very different mechanisms, therefore seem to corroborate each other. Further details of the experiment can be found in [27, 29].

The relaxation rate data for the third mechanism mentioned above is presented in Figure 4. To produce this data, conventional LF-μSR time spectra were recorded at various temperatures (ranging from 10 K to 400 K) and magnetic fields (ranging from 20 mT to 0.4 T). The extraction of the diffusion constant from the relaxation rate is not unique unless the fitting of the temperature dependence of the time differential muon asymmetry is done simultaneously for different magnetic fields. This fitting is performed using the Redfield analysis method [32]. Moreover the choice of the magnetic fields must be carefully made. The procedures used to select the fields and extract the inverse correlation time τ_c^{-1} (which is proportional to the diffusion constant) are discussed in detail in [31]. This mechanism has the more reliable model for the extraction of the diffusion constant, and it is also quantitative in the inverse correlation time, τ_c^{-1}. An additional parameter extracted is the nuclear hf parameter, δ_{ex}, which is a measure of the combined effect of all the nuclear spins in the vicinty of the muonium interacting with the muon via the correlated electron.

One notes from Figure 4 that the inverse correlation time τ_c^{-1} is essentially constant at the rather high value of 6×10^{10} s^{-1}. This is inconsistent in its temperature behaviour with classical diffusion, and its magnitude confirms the very fast diffusion appropriate for a quantum mechanism. However, the power law behaviour from 100–300 K seen in the previous two types of measurement is not evident. Instead, the nuclear hf parameter, δ_{ex}, appears to display a power law behaviour. The phenomena resembles that observed in NaCl [32], which was attributed to dynamical effects. The measurements could not be pursued to higher temperatures in order to continue to map out the diffusion behaviour due to the onset of the thermal conversion of $Mu_T \rightarrow Mu_{BC}$.

6. Conclusions

Hydrogen chemistry and dynamics in diamond are important in understanding the material in the context of ultra-high quality diamond as well as electronic diamond. Here, MSR studies have been essential. The isolated forms of muonium in diamond have been reviewed. Recent measurements elaborate the frontiers of these studies where the data is consistent with the formation of possible shallow molecular dopants, as well as the direct observation of passivation. Finally, it appears that diamond does indeed support the quantum diffusion of muonium up to at least room temperature, the highest temperature for this diffusion mode of any material.

Acknowledgements

The authors would like to thank the SA National Research Foundation (grant numbers: 2053846, 2053305, 2053464 and 2053323) and Element six for supporting this work. Authors are also grateful to Prof Thomas Anthony Ramdas for loaning us the ^{13}C diamond samples used in some of the experiments. The expertise of Mr Mik Rebak in sample preparation was essential.

References

1. Coe S. E., Hall C. E., McClymont M. R., Pickles C. S. J., Twitchen D. J., Whitehead A. J. and Wort C. J. H., CVD Diamond: Advances and Applications, *Proc. of the Intertech 2003*, Vancouver, Canada, July 28–August 1 2003, Industrial Diamond Association of America Inc., http://www.superabrasives.org.
2. Yoder M., Semiconducting diamond technology, *Naval Res. Rev.* **39** (1987), 27.
3. Kohn E., Adamschik M., Schmid P., Denisenko A., Aleksov A. and Ebert W., *J. Phys. D: Appl. Phys.* **34** (2001), R77–R85.
4. Isberg J., Hammersberg J., Twitchen D. J. and Whitehead A. J., *Diamond Relat. Mater.* **13** (2004), 320–324.
5. Sellschop J. P. F., Freund A., Hoszowska J., Connell S. H., Rebak M. and Burns R. C., *Phys Status Solidi A* **193/3** (2002), 415–422.
6. Isberg J., Hammersberg J., Johansson E., Wikstrom T., Twitchen D. J., Whitehead A. J., Coe S. E. and Scarsbrook G. A., *Science* **297**(5587) (2002), 1670–1672.
7. Goss J. P., Briddon P. R., Jones R. and Sque S., *Diamond Relat. Mater.* **13** (2004), 684–690.
8. Collins A. T., *Diamond Relat. Mater.* **8** (1999), 1455–1462.
9. Miyazaki T., Okushi H. and Uda T., *Phys. Rev. Lett.* **88/6** (2002), 066402(4).
10. Lowther J. E., *Phys. Rev. B* **67/11** (2004), 115206.
11. Goss J. P., *J. Phys., Condens. Matter* **15** (2003), R551–R580.
12. Stoneham M., In: Field J. E. (eds.), *The Properties of Natural and Synthetic Diamond*, Chapter 1, Academic Press Ltd, 1992.
13. Odermatt W., Baumeier H. P., Keller H., Kundig W., Patterson D. B., Schneider J. W., Sellschop J. P. F., Stemmet M. C., Connell S. H. and Spencer D. P., *Phys. Rev., B* **38** (1988), 4388–4393.

14. Schneider J. W., Kiefl R. F., Chow K. H., Johnston S., Sonier J., Estle T. L., Hitti B., Lichti R. L., Connell S. H., Sellschop J. P. F., Smallman C. G., Anthony T. R. and Banholzer W. F., *Phys. Rev. Lett.* **71** (1993), 557–560.
15. Holzschuh E., Kundig W., Meier P. F., Patterson B. D., Sellschop J. P. F., Stemmet M. C. and Appel H., *Phys. Rev., A* **25** (1982), 1272–1286.
16. Bharuth-Ram K., Scheuermann R., Machi I. Z., Connell S. H., Major J., Sellschop J. P. F. and Seeger A., *Hyperfine Interact.* **105** (1997), 339.
17. Spencer D. P., Fleming D. G. and Brewer J. H., *Hyperfine Interact.* **17–19** (1984), 567.
18. Evans T., In: Field J. E. (ed.), *The Properties of Natural and Synthetic Diamond*, Academic Press, 1992, p. 168.
19. van Wyk J. A. and Loubser J. H. N., *J. Phys. Cond. Matter.* **7** (1993), 3019.
20. van Wyk J. A. and Woods G. S., *J. Phys. Cond. Matter.* **5** (1995), 5901.
21. Holliday K., He X.-F., Fisk P. T. H. and Manson N. B., *Optics Letters* **15** (1990), 983.
22. Baker J. M., Machi I. Z., Connell S. H., Bharuth-Ram K., Butler J. E., Cox S. F. J., Fischer C. G., Jestadt T., Murphy P., Nilen R. W. N. and Sellschop J. P. F., *Hyperfine Interact.* **120/121** (1999), 579–583.
23. Machi I. Z., Connell S. H., Baker M., Sellschop J. P. F., Bharuth-Ram K., Butler J. E., Cox S. F. J., Fischer C. G., Jestadt T., Murphy P. and Nilen R. W. N., *Physica B* **289–290** (2000), 507–510.
24. Machi I. Z., Connell S. H., Sellschop J. P. F. and Bharuth-Ram K., *Hyperfine Interact.* **136/3** (2001), 727–730.
25. Goss J., Jones R., Briddon P. R., Oberg S., Proceedings of the Diamond Conference, Bristol, England (1997).
26. Patterson B. D., *Rev. Mod. Phys.* **60** (1988), 70.
27. Machi I. Z., Connell S. H., Sellschop J. P. F., Bharuth-Ram K., Major J., Doyle B. P. and Maclear R. D., *Physica B* **289/290** (2000), 468–472.
28. Machi I. Z., Connell S. H., Major J., Smallman C. G., Sellschop J. P. F., Bharuth-Ram K., Maclear R. D., Doyle B. P., Butler J. E., Scheuermann R. and Seeger A., *Hyperfine Interact.* **120/121** (1999), 585–589.
29. Machi I. Z., Connell S. H., Dalton M., Sithole M. J., Bharuth-Ram K. and Cox S. F. J. *et al.*, *Diam. and Relat. Mater* **13** (2004), 909.
30. Storchak V. G. and Prokof'ev N. V., *Rev. Mod. Phys.* **70** (1998), 929.
31. Gxawu D., Machi I. Z., Connell S. H., Bharuth-Ram K., Sithole M. J., Cox S. F. J., Presentation to the 14th European Conference on Diamond, Diamond-Like Materials, Carbon Nanotubes, Nitrides and Silicon Carbide, DIAMOND 2003, 7–12 September 2003, Salzburg, Austria.
32. Kadono R., *Hyperfine Interact.* **64** (1990), 615.

Hyperfine Interactions (2004) 159:227–234
DOI 10.1007/s10751-005-9104-5

Thin Film, Near-Surface and Multi-Layer Investigations by Low-Energy μ^+SR

T. PROKSCHA[1,*], E. MORENZONI[1], A. SUTER[1], R. KHASANOV[1,2], H. LUETKENS[1,3], D. ESHCHENKO[1,2], N. GARIFIANOV[1,4], E. M. FORGAN[5], H. KELLER[2], J. LITTERST[3], C. NIEDERMAYER[1,6] and G. NIEUWENHUYS[7]

[1]*Paul Scherrer Institute, PSI, CH-5232 Villigen, Switzerland; e-mail: thomas.prokscha@psi.ch*
[2]*Universität Zürich, CH-8057 Zürich, Switzerland*
[3]*Technische Universität Braunschweig, D-38106 Braunschweig, Germany*
[4]*Kazan Physical-Technical Institute, Kazan 420029, Russian Federation*
[5]*University of Birmingham, Birmingham B15 2TT, UK*
[6]*Universität Konstanz, D-78434 Konstanz, Germany*
[7]*Leiden University, 2300 RA Leiden, The Netherlands*

Abstract. At the Paul Scherrer Institute (PSI, Villigen, Switzerland) the beam of low-energy positive polarised muons (LE-μ^+) with tunable energy between 0.5 and 30 keV allows the extension of the muon-spin-rotation technique (μSR) to studies on thin films and multi-layers (LE-μ^+SR). The range of these muons in solids covers the near-surface region up to implantation depths of about 300 nm. As a sensitive local magnetic probe with a complementary observational time window to other techniques LE-μ^+SR offers the unique possibility to gain new insights in these nano-scale objects. After outlining the current status of the LE-μ^+ beam line we demonstrate the potential of this new technique by presenting the results of recent experiments: i) the direct observation of non-local effects in a superconducting Pb film, ii) the oxygen isotope effect on the in-plane penetration depth in optimally doped $YBa_2Cu_3O_{7-\delta}$, and iii) the first observation of the conduction electron spin polarisation in the Ag spacer of a Fe/Ag/Fe tri-layer.

Key Words: local probe, low-energy muons, multi-layers, muon-spin-rotation, thin films.

1. Introduction

The availability of a low-energy muon beam (LE-μ^+) opens the fascinating possibility for depth dependent investigations of magnetic properties on a nanometer scale by means of a local magnetic probe. Unlike the "standard" muon-spin-rotation technique (μSR) [1] which makes use of muons with energies larger than 3 MeV – resulting in penetration depths and widths in solids of the order of fraction of mm and more, and thus limiting that technique to the study of bulk matter properties – LE-μ^+ with tunable energies between 0.5 and 30 keV

---

* Author for correspondence.

allow investigations of near-surface regions, thin films, interfaces and multi-layers up to depths of 300 nm [2].

The last decades have witnessed a steadily growing interest in thin film and multi-layer studies, because the controlled reduction of dimensionality can lead to new physical phenomena and technological applications. Techniques making use of polarised electrons, neutrons, photon and positrons are not able to provide *direct* magnetic depth sensitivity with nm resolution. Polarised LE-μ^+ as a local magnetic probe offer a novel experimental tool to gain new insights in fundamental physical mechanisms.

About 10 years ago the development of a continuous LE-μ^+ beam at the Paul Scherrer Institute (PSI, Villigen, Switzerland) was started to extend the well developed and established μSR technique to nanometer scale studies. The application of LE-μ^+ as an extension of μSR (LE-$^+$SR) has reached now a maturity that allows its use as a standard tool in condensed matter research. Before we present recent exemplary experiments in Section 3 which demonstrate the capability of this young technique we briefly review the generation and properties of the LE-μ^+ beam at PSI in Section 2.

2. Generation of low-energy polarised positive muons

The LE-μ^+ beam at PSI is generated by moderation of a continuous 4-MeV, nearly 100%-polarised μ^+ beam in a thin layer of a few hundred nanometers of a van der Waals bound cryosolid (s-Ne, s-Ar, s-N_2) deposited on the downstream side of a cold metal substrate [3]. Due to the lack of efficient inelastic energy loss mechanisms in these large band gap (>14 eV) insulators at epithermal energies below 50 eV, epithermal muons below that energy have a large escape depth (several tens of nanometers). Therefore, a fraction of 10^{-5}–10^{-4} of the incoming beam may escape into vacuum with a mean energy of about 20 eV and an RMS spread of same order. Since the moderation process is very fast (~*ps*) the initial polarisation of the μ^+ is conserved [4] which is a mandatory requirement for their usage in μSR investigations. These epithermal μ^+ form the source of the LE-μ^+ beam, which is generated by electrostatic acceleration (by applying up to +20 kV to the electrically insulated moderator) of the epithermal μ^+. The LE-μ^+ are transported by electrostatic einzel lenses to the sample. The sample is located at a distance of about 2.3 m from the moderator position, after a 90° deflection by an electrostatic mirror. It is mounted electrically insulated on the cold head of a LHe flow cryostat. This allows the application of an accelerating or decelerating potential of up to ±12.5 kV to the sample. During transport the LE-μ^+ pass through a special detector with an ultra-thin 2-μg/cm^2 carbon foil, which cause a RMS spread in energy of about 400 eV. This so-called trigger detector generates the start signal of the LE-μSR measurement by deflecting secondary electrons – released by the muons when passing through the carbon foil – onto a microchannel plate detector. The high-intensity continuous μ^+ beams at PSI with a currently useable rate of 3×10^7/s for LE-μ^+ production translates into a LE-μ^+ rate of about 3000/s at the moderator. Of these, up to 1000/s reach the sample position. Accepted good event rates range between 100/s and 500/s, depending on the experimental conditions such as LE-μ^+ energy, sample and setup geometry. The mean implantation energy can be varied between 0.5 and 32 keV by choosing the appropriate high voltages at the moderator and the sample. Due to the energy spread of 400 eV the time resolution is limited to 5 ns. For details, we refer to [2, 5, 6].

3. Applications

The knowledge of the shape of the stopping distribution of LE-μ^+ is a prerequisite for applications in *depth-resolved* studies on thin films and multi-layers. Note, that not all LE-μ^+SR experiments require necessarily detailed information about the stopping distribution: for example, in studies of the energy dependence of muon charge differentiation in insulators [7], or in muon diffusion experiments in metal layers with thicknesses of order 30 nm [8] the details of the stopping distribution are of minor importance. In cases, where detailed knowledge about the implantation profile is mandatory the Monte-Carlo simulation TRIM.SP [9] can be used to calculate the stopping profile with sufficient accuracy for LE-μSR investigations [10]. This procedure has been utilised in experiments to determine the magnetic profile beneath the surface of a superconductor in the Meissner state. We describe briefly two recent studies that addressed different physical questions: the direct observations of non-local effects in Pb and the oxygen isotope effect in optimally doped $YBa_2Cu_3O_{7-\delta}$. In these experiments the muon implantation depth is controlled by the variation of the muon energy, and the magnitude of field is determined by the muon-spin precession frequency. As a last example the sensitivity of μ^+ is used to probe for the first time the spatially oscillating conduction electron spin polarisation in the non-magnetic Ag spacer of a Fe/Ag/Fe trilayer.

3.1. DIRECT OBSERVATION OF NON-LOCAL EFFECTS IN Pb

The electromagnetic response of a superconductor results in the expulsion of the magnetic field beneath the surface on a scale of several 10 to several 100 nanometers (Meissner–Ochsenfeld effect). The spatial field dependence $B(z)$ – with z the distance to the surface – yields valuable information about its nature. If the Cooper pairs can be treated as point-like the field penetrates exponentially with a typical length λ_L (London penetration depth): this is the result of the well known *local* relationship

$$j(\boldsymbol{r}) = -\frac{1}{\mu_0 \lambda_L^2} A(\boldsymbol{r}) \tag{1}$$

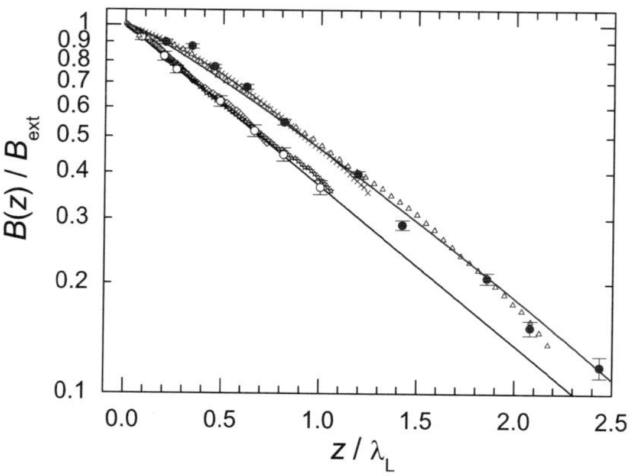

Figure 1. Upper data set: measured magnetic penetration profile $B(z)$ on a logarithmic scale of Pb in the Meissner state at $T = 3.05$ K ($T_c = 7.21$ K), external field $B_{ext} = 8.8$ mT. The shape clearly deviates from exponential. *Lower data set*: $B(z)$ in $YBa_2Cu_3O_{7-\delta}$ at 20 K ($T_c = 87.5$ K): for an extreme type-II superconductor ($\xi \ll \lambda_L$) such as YBCO the local description manifested in the purely exponential shape of $B(z)$ applies. The *solid lines* represent fits to the data.

between the supercurrent density $j(\boldsymbol{r})$ and the vector potential $A(\boldsymbol{r})$, where μ_0 is the permeability of vacuum. In general, the relationship is more complex allowing for a variation of the magnetic field over the size of a Cooper pair. This non-local effects are important, for example, in conventional type-I superconductors like Pb, where the coherence length $\xi > \lambda_L$ [11, 12]. In this case, the field penetrates non-exponentially into the superconductor. A direct experimental verification of this BCS prediction has been lacking up to now. Using LE-μ^+ we were able to observe directly for the first time the non-exponential shape of $B(z)$ in Pb [13], see Figure 1, which represents a direct, model-independent proof of the non-local response of Pb. The Pb samples with thicknesses 1055(50) nm and 430(20) nm were sputtered onto a sapphire disk with 50 mm diameter where the thickness was controlled offline by RBS. Thin oxide layers of 6 and 16 nm were found, respectively. The critical temperature $T_c = 7.21(1)$ K was determined by means of resistivity and susceptibility measurements. After zero-field cooling the samples an external field $B_{ext} = 8.8$ mT was applied parallel to the sample surface. For the Pb data the analysis yields $\xi^{Pb} = 90(5)$ nm and $\lambda_L^{Pb} = 57(2)$ nm for $T = 0$ K. For comparison, data of the type-II superconductor $YBa_2Cu_3O_{7-\delta}$ are shown in Figure 1 as well. The profile is purely exponential, as expected, and the analysis yields $\lambda_L^{YBCO} = 146(3)$ nm for T = 0 K.

3.2. OXYGEN ISOTOPE EFFECT IN $YBa_2Cu_3O_{7-\delta}$

By means of LE-μ^+SR the penetration depth can be determined to an accuracy of better than 1%. This allows investigations of small effects on the penetration depth in the high-temperature superconductor (HTSc) $YBa_2Cu_3O_{7-\delta}$ due to complete oxygen isotope exchange. Up to now a theory describing the pairing mechanism in HTSc's is still missing, and one of the open questions is, if the supercarriers couple to the lattice. The measurement of λ_L is well suited for that study in optimally doped $YBa_2Cu_3O_{7-\delta}$, because it is directly related to the effective mass m^* of the supercarriers:

$$1/\lambda^2 = \mu_0 e^2 n_s / m^*, \tag{2}$$

where e is the elemental charge and n_s the superconducting charge carrier density. Another important parameter of a superconductor, the superconducting transition temperature T_c, shows only a tiny oxygen-isotope shift in optimally doped $YBa_2Cu_3O_{7-\delta}$. In a recent experiment the oxygen-isotope effect on the in-plane penetration depth λ_L^{ab} was directly measured by means of LE-μ^+SR [14]. In this measurement the ^{16}O of the sample was completely substituted by ^{18}O, and λ_L^{ab} was measured before and after oxygen-exchange. It is found that λ_L^{ab} is 2.8(1.0)% larger in the ^{18}O substituted sample. There is strong evidence that this is mainly an effect due to a change of m^*_{ab} since an independent NQR measurement reveals that n_s does not change due to oxygen substitution. The result indicates that the mass of the superconducting carriers is not decoupled from the lattice in optimally doped cuprate superconductors. This is not expected in a BCS model within the adiabatic Migdal approximation, which is applicable for most of the classical superconductors.

3.3. CONDUCTION ELECTRON SPIN POLARISATION IN THE Ag SPACER OF A Fe/Ag/Fe TRILAYER

The capability of controlled growth of heterostructures on a nanometer scale opened the possibility of tailoring materials with new physical properties. In particular, multi-layers of magnetic and non-magnetic materials led to the discovery of new physical phenomena, such as giant magnetoresistance (GMR) [15] or the interlayer exchange coupling (IEC) [16]. The IEC between two ferromagnetic films separated by a non-magnetic intermediate layer depends on the spacer thickness: it oscillates and changes sign when varying the thickness. Theoretical approaches to describe IEC are RKKY and quantum-well models. The magnetic coupling between the two magnetic layers is mediated by the induced spin-polarisation of conduction electrons in the non-magnetic layer. This polarisation is hardly accessible experimentally because it rapidly decays with increasing distance x to the magnetic interface, and because of its smallness compared to the large magnetic moments of the magnetic layers. This requires a local magnetic probe with large sensitivity. The LE-μ^+SR technique offers this sensitivity and is capable to investigate the non-magnetic layer sandwiched between the ferromagnetic films by selective implantation of the muons in the spacer. This has been demonstrated in a recent experiment [17] where for the first time the conduction electron spin polarisation has been measured in the Ag layer of an epitaxial 4 nm Fe/20 nm Ag/4 nm Fe(001) trilayer on a MgO(001)

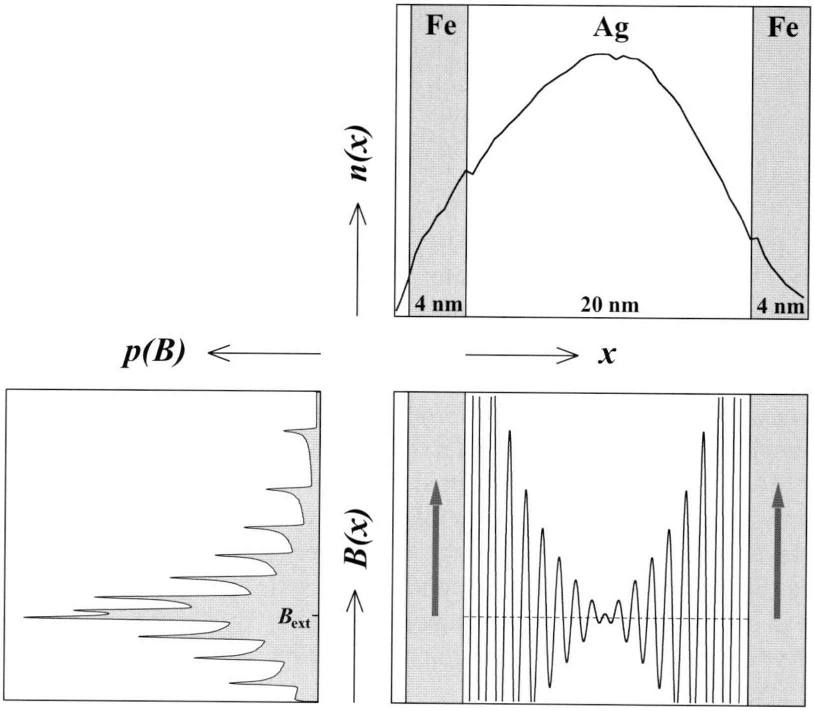

Figure 2. Principle of the LE-μ^+SR measurement of the oscillating electron spin polarisation in the Ag intermediate layer of a 4 nm Fe/20 nm Ag/4 nm Fe(001) film on a MgO(001) substrate. The maximum of the muon stopping profile $n(x)$ is placed close to the centre of the Ag layer by proper adjustment of the LE-μ^+ energy. The field distribution $p(B)$ as measured by the muon ensemble directly reflects the oscillating polarisation in the Ag spacer which is proportional to the local field $B(x)$ sensed by the μ^+. The distinct peaks in $p(B)$ correspond to maxima or minima of the oscillating $B(x)$.

substrate. The principle of the measurement is shown in Figure 2. The sample was first magnetised at room temperature in a field of 0.1 T. In the remnant state it was then cooled to 20 K before applying a field of 8.8 mT parallel to the easy axis and parallel to the sample surface to carry out the LE-μ^+SR measurements. The implantation energy was 3 keV to stop most of the muons in the Ag layer. The experimental data are in good agreement with the theoretical shape of $p(B)$ shown in Figure 2. The fit of the theoretical $p(B)$ to the data yields an attenuation of $x^{-0.8(1)}$ of the oscillating polarisation in the Ag layer perpendicular to the interface. Here, the local magnetic field $B_{\mathrm{loc}}(x)$ as sensed by the muon is modelled by the superposition of the magnetic field profiles induced by each of the two Fe/Ag interfaces [$B_{\mathrm{loc}}(x) = B(x) + B(D - x)$, $D = 20$ nm the Ag layer thickness]. The observed attenuation indicates a different spatial dependence than the IEC which follows x^{-2}. This unexpected result cannot be explained within a RKKY description of IEC. It may be explained in a theory that considers the full confinement of the electrons within the intermediate layer, as it was found for the Co/Cu/Co trilayer [18].

4. Outlook

Over the past years the continuous developments on the surface muon beam line, the epithermal muon source and the LE-μ^+ apparatus have led to a steady increase of the LE-μ^+ flux at the sample up to 1000/s. This affords LE-μ^+SR as a serious tool for condensed matter investigations. However, this novel technique is still limited in statistics, especially if compared with the intensities of several tens of kilohertz available at existing μSR facilities. It is therefore evident that an intensity increase of the LE-μ^+ rate is of primary importance. A significant straightforward advancement can not be expected by improvements of the existing setup, whereas a corresponding increase of the surface muon flux can lead to an enhancement by one order of magnitude. A new, large-acceptance surface muon beam line is currently under construction at PSI and will be ready for user operation in 2005. It was specially designed for the generation of a low-energy muon beam [19]. The new beam line will yield a raise of LE-μ^+ rate by a factor of seven, thus allowing the full exploitation of the potential of polarised muons as nano-scale probes.

Acknowledgements

This work was performed at the Swiss Muon Source, Paul Scherrer Institute, Villigen, Switzerland. The long-term technical support by H.P. Weber is gratefully acknowledged. We thank the German BMBF, the UK EPSRC, and the Swiss National Science Foundation for financial support.

References

1. Blundell S. J., *Cont. Phys.* **40** (1999), 175.
2. Bakule P. and Morenzoni E., *Cont. Phys.* **45** (2004), 203.
3. Harshman D. R., Mills Jr. A. P., Beveridge J. L., Kendall K. R., Morris G. D., Senba M., Warren J. B., Rupaal A. S. and Turner J. H., *Phys. Rev. B* **36** (1987), 8850.
4. Morenzoni E., Kottmann F., Maden D., Matthias B., Meyberg M., Prokscha T., Wutzke T. and Zimmermann U., *Phys. Rev. Lett.* **72** (1994), 2793.
5. Morenzoni E., Glückler H., Prokscha T., Weber H. P., Forgan E. M., Jackson T. J., Luetkens H., Niedermayer C., Pleines M., Birke M., Hofer A., Litterst J., Riseman T. and Schatz G., *Physica B* **289–290** (2000), 653.
6. Prokscha T., Morenzoni E., David C., Hofer A., Glückler H., Scandella L., *Appl. Surf. Sci.* **172** (2001), 235.
7. Prokscha T., Morenzoni E., Garifianov N., Glückler H., Khasanov R., Luetkens H., Suter A., *Physica B* **326** (2003), 51.
8. Luetkens H., Korecki J., Morenzoni E., Prokscha T., Garifianov N., Glückler H., Khasanov R., Litterst J., Slezak T., Suter A., *Physica B* **326** (2003), 545.
9. Eckstein W., *Computer Simulation of Ion-Solid Interactions*, Springer, Berlin, Heidelberg New York, 1991.
10. Morenzoni E., Glückler H., Prokscha T., Khasanov R., Luetkens H., Birke M., Forgan E. M., Niedermayer C. and Pleines M., *NIM B* **192** (2002), 254.

11. Pippard A. B., *Proc. R. Soc. London, A* **216** (1953), 547.
12. Bardeen J., Cooper L. N. and Schrieffer J. R., *Phys. Rev.* **108** (1957), 1175.
13. Suter A., Morenzoni E., Khasanov R., Luetkens H., Prokscha T. and Garifianov N., *Phys. Rev. Lett.* **92** (2004), 087001.
14. Khasanov R., Eshchenko D. G., Luetkens H., Morenzoni E., Prokscha T., Suter A., Garifianov N., Mali M., Roos J., Conder K. and Keller H., *Phys. Rev. Lett.* **92** (2004), 057602.
15. Baibich M. N., Broto J. M., Fert A., Nguyen Van Dau F., Petroff F., Eitenne P., Creuzet G., Friederich A. and Chazelas J., *Phys. Rev. Lett.* **61** (1988), 2472.
16. Grünberg P., Schreiber R., Pang Y., Brodsky M. B. and Sowers H., *Phys. Rev. Lett.* **57** (1986), 2442.
17. Luetkens H., Korecki J., Morenzoni E., Prokscha T., Birke M., Glückler H., Khasanov R., Klauss H. H., Slezak T., Suter A., Forgan E. M., Niedermayer C. and Litterst F. J., *Phys. Rev. Lett.* **91** (2003), 017204.
18. Mathon J., Umerski A., Villeret M. and Muniz R. B., *Phys. Rev.* **B59** (1999), 6344.
19. Prokscha T., Morenzoni E., Deiters K., Foroughi F., George D., Kobler R. and Vrankovic V., these proceedings.

Hyperfine Interactions (2004) 159:235–238
DOI 10.1007/s10751-005-9105-4

Quadrupole Moments of Na Isotopes

M. OGURA[1,*], T. NAGATOMO[1], K. MINAMISOMO[2,a], K. MATSUTA[1], T. MINAMISONO[1], Y. NAKASHIMA[1], C. D. P. LEVY[2], T. SUMIKAMA[1,†], M. MIHARA[1], H. FUJIWARA[1], S. KUMASHIRO[1], M. FUKUDA[1], J. A. BEHR[2], K. P. JACKSON[2], S. MOMOTA[3], Y. NOJIRI[3], T. OHTSUBO[4], M. OHTA[4], A. KITAGAWA[5], M. KANAZAWA[5], M. TORIKOSHI[5], S. SATO[5], M. SUDA[5], J. R. ALONSO[6], G. F. KREBS[6] and T. M. SYMONS[6]

[1]*Department of Physics, Osaka University, Toyonaka, Osaka 560-0043, Japan; e-mail: ogura@phys.sci.osaka-u.ac.jp*
[2]*TRIUMF, 4004 Wesbrook Mall, Vancouver B.C., Canada, V6T 2A3*
[3]*Kochi University of Technology, Tosoyamada, Kochi 782-8502, Japan*
[4]*Department of Physics, Niigata University, Niigata 950-2181, Japan*
[5]*National Institute of Radiological Sciences, Inage, Chiba 263-0024, Japan*
[6]*Lawrence Berkely Laboratory, Berkely CA 94720, USA*

Abstract. The electric quadrupole coupling constants eqQ/h of ^{20}Na ($I^{\pi} = 2^+$, $T_{1/2} = 449.7$ ms), ^{21}Na ($I^{\pi} = 3/2$, $T_{1/2} = 22.49$ s) and ^{25}Na ($I^{\pi} = 5/2$, $T_{1/2} = 59.6$ s) in single crystal ZnO and TiO_2 were precisely measured by applying the β-NMR technique. The ratio of the quadrupole moments between ^{20}Na (^{25}Na) and ^{21}Na was determined to be 0.728 ± 0.023 (0.011 ± 0.002).

Key Words: β-NMR, Na isotopes, Quadruple moment.

1. Introduction

The quadrupole moments of nuclei in the long chain of Na isotopes have been measured for the study of the systematic change in the nuclear structure [1, 2]. For some Na isotopes, however, the measurements of the quadrupole moments show rather large errors and hence it is important to remeasure them for detailed discussions. In the present study we measured the electric quadrupole coupling constants eqQ/h of ^{20}Na ($I^{\pi} = 2^+$, $T_{1/2} = 449.7$ ms), ^{21}Na ($I^{\pi} = 3/2^+$, $T_{1/2} = 22.49$ s) and ^{25}Na ($I^{\pi} = 5/2$, $T_{1/2} = 59.6$ s), implanted in ZnO and TiO_2 single crystals, by use of the β-NMR technique. The precise ratios of the quadrupole moments of these nuclei, determined from these quadrupole coupling constants, are discussed.

---

* Author for correspondence.
† Present address: RIKEN, 2–1 Hirosawa, Wako Saitama 351-0198, Japan.
a Present address: NSCL/MSU, 1 Cyclotron, East Lansing, MI 48824, USA.

2. Experiment

The experiments for the measurement of $eqQ(^{20,21}\text{Na})/h$ were performed at the radioactive beam facility ISAC/TRIUMF. The proton beam from TRIUMF cyclotron was used to bombard a SiC production target coupled to the surface ionization source. Then ^{20,21}Na ions were extracted at an energy of 40.8 keV, mass separated and polarized by the laser-pumping method [3]. The polarized Na beam was implanted in the catchers (NaF, ZnO and TiO_2 single crystals) placed under an external magnetic field H_0 of 0.5 T. The direction of H_0 was parallel/anti-parallel to the polarization direction with positive/negative laser-helicity light. The β rays from the stopped Na were detected by two sets of plastic-scintillation counter telescopes placed 0° and 180° relative to the polarization direction. In order to check the polarization created through the laser pumping method, we compared the β-ray counting ratio between 0° and 180° counters with positive and negative laser-helicity light. About 50% and 34% nuclear polarization were achieved in a NaF for ^{20,21}Na, respectively. For the measurement of the β-NMR spectra, ^{20,21}Na were implanted into ZnO and TiO_2. The rf oscillating magnetic fields H_1 was applied perpendicular to H_0.

The experiment for the measurement of $eqQ(^{25}\text{Na})/h$ was performed at HIMAC/NIRS. The ^{25}Na nucleus was produced through the projectile fragmentation process in the 100 A-MeV $^{26}\text{Mg}_{+}{}^{9}\text{Be}$ (2.0 mm thick) collision. The ^{25}Na beam was separated by a fragment separator installed in the secondary beam course of HIMAC and polarized by selecting the ejection angle (1.0 ± 0.6)% and momentum (+0.8 ± 0.5)%. After suitable energy degradation, the ^{25}Na beam was implanted into a TiO_2, placed under the H_0 magnetic field of 0.5 T. The detector system and the rf system for the β-NMR technique are similar to the experiments for ^{20,21}Na.

3. Results and discussions

The measured β-NQR spectra of ^{20,21}Na in ZnO are shown in Figure 1(a,b). In the figure, the effective asymmetry change defined by the difference between the asymmetry measured with positive and negative laser-helicity lights is shown as a function of eqQ/h. We found two resonances, which might be explained by two final locations of Na isotopes in ZnO. Because the measurement of the smaller resonance (Sub) of ^{20}Na is incomplete, the larger resonances (Main) will be discussed. The results are summarized in Table I. The first errors in eqQ/h in Table I are the statistical errors and the second errors are the systematic errors caused by the unknown origin of Sub resonance. One-tenth of the width (±HWHM) of Main resonances is considered as the systematic errors [4].

The β-NQR spectra of ^{21,25}Na in TiO_2 are shown in Figure 2(a,b), where the asymmetry change is shown as a function of eqQ/h. The results are summarized

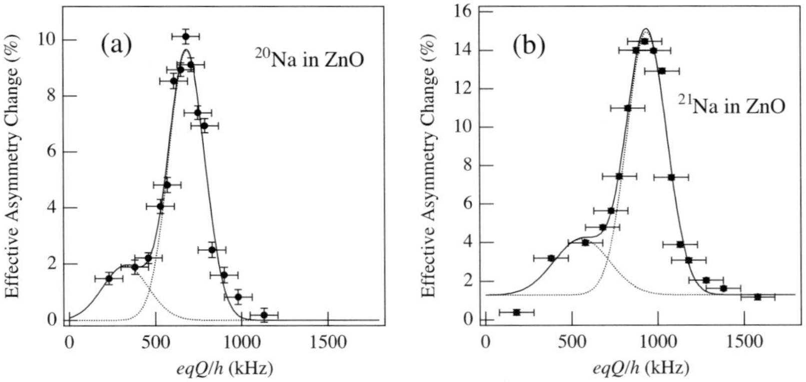

Figure 1. β-NQR spectrum of ^{20}Na (a) and ^{21}Na (b) in ZnO ($c \perp H_0$). In the spectrum (a), the resonance line from the subcomponent was fixed considering fitting results to the spectrum (b).

Table I. Electric quadrupole coupling constants of ^{20,21,25}Na. The quadrupole moments in the last line were extracted taking $Q(^{20}\text{Na})$ [6] as a standard

		^{20}Na	^{21}Na	^{25}Na
$\vert eqQ/h \vert$ (kHz)	In ZnO	683.7 ± 2.7 ± 12.4	939.3 ± 1.9 ± 19.3	–
	In TiO_2	–	5,200 ± 24 ± 80	58 ± 3 ± 7
$Q/Q(^{21}\text{Na})$		0.728 ± 0.023	1	0.011 ± 0.002
Q (literature) (mb)		90 ± 10 [6]	58 ± 33 [1, 2]	64 ± 44 [2]
$Q/Q(^{21}\text{Na})$		1.5 ± 0.9	1	1.1 ± 1.0
Q_{theory} (mb) (OXBASH)		+84	+110	−2.7
$Q/Q(^{21}\text{Na})$		0.76	1	0.025
Q (mb)		90 ± 10	124 ± 14	1.36 ± 0.29

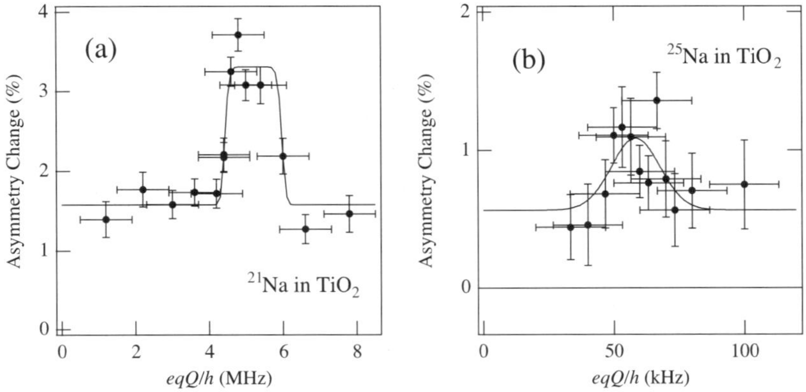

Figure 2. β-NQR spectrum of ^{20}Na (a) and ^{21}Na (b) in TiO_2 ($c // H_0$).

in Table I. In the table, the systematic errors are based on the variation of the center frequency depending on the fitting functions. Compared with the β-NQR spectra of ^{20,21}Na in ZnO, the spectra of ^{21,25}Na in TiO_2 are incomplete and hence additional measurements are planed.

The ratio of the quadrupole moments $|Q(^{20}\mathrm{Na})/Q(^{21}\mathrm{Na})|$ and $|Q(^{25}\mathrm{Na})/Q(^{21}\mathrm{Na})|$ were determined to be 0.728(9) and 0.011(2), respectively. In Table I, the previous values and the shell model predictions calculated by the OXBASH code [5] are also listed. The accuracy of the ratios are improved greatly. The ratio $|Q(^{20}\mathrm{Na})/Q(^{21}\mathrm{Na})|$ agrees with the shell model prediction, but $|Q(^{25}\mathrm{Na})/Q(^{21}\mathrm{Na})|$ does not. For the detailed discussion, the quadrupole moments have to be extracted from the ratios. For the purpose, we are planning measurements of a reference value of eqQ/h in $NaNO_3$, in which q at the Na site is well known.

Acknowledgements

The present work was performed under the Research Project at TRIUMF (TRIUMF Experiment 871 and 971) and at NIRS/HIMAC (Program # P-026). We are grateful to the staffs of TRIUMF and HIMAC for their technical support. The present study was partly supported by the 21st century COE program named 'Towards a new basic science'.

References

1. Toucherd T. *et al.*, *Phys. Rev. C* **25** (1982), 2756.
2. Keim M. *et al.*, *Eur. Phys. J. A* **8** (2000), 31.
3. Levy C. D. P. *et al.*, Proc. 9th Int. Workshop on Polarized Source and Targets, World Scientific, 2002, 334.
4. Minamisono T. *et al.*, *Phys. Lett. B* **420** (1998), 31.
5. Brown B. A., Etchegoyen A. and Rae W. D. M., MSUCL Report 524 (1986).
6. Keim M. *et al.*, ENAM98, AIP Conference Proceedings 445 (2000), 40.

Hyperfine Interactions (2004) 159:239–243
DOI 10.1007/s10751-005-9106-3

Hyperfine Fields of Sr and Y in Ferromagnetic Hosts, and Magnetic Moment of ^{93}Y

K. NISHIMURA[1,*], S. OHYA[2], T. OHTSUBO[2], T. IZUMIKAWA[3], M. SASAKI[4], I. SATO[4], M. SUZUKI[4], J. GOTO[5], M. TANIGAKI[6], A. TANIGUCHI[6], Y. OHKUBO[6], Y. KAWASE[6] and S. MUTO[7]
[1]*Faculty of Engineering, Toyama University, Toyama 930-8555, Japan; e-mail: nishi@eng.toyama-u.ac.jp*
[2]*Faculty of Science, Niigata University, Niigata 950-2181, Japan*
[3]*Radioisotope Centre, Niigata University, Niigata 951-8510, Japan*
[4]*Graduate School of Science, Niigata University, Niigata 951-8510, Japan*
[5]*Japan Atomic Energy Research Institute, Tokai 319-1195, Japan*
[6]*Research Reactor Institute, Kyoto University, Kumatori, Osaka 590-0494, Japan*
[7]*Neutron Science Laboratory, KEK, Tsukuba 305-0801, Japan*

Abstract. Recent studies of hyperfine fields B_{HF} of Sr and Y in Fe are briefly reviewed. The B_{HF} value of Y in Ni is given by NMRON of $^{91\mathrm{m}}$Y*Ni*. The nuclear magnetic moment of the ground state of ^{93}Y is extracted from the β-ray detected NMRON of ^{93}Y*Fe*.

Key Words: β-ray detection, hyperfine field, NMRON, nuclear g-factor, Sr, Y.

If atoms are embedded as dilute impurities in a ferromagnetic host lattice such as Fe and Ni, the impurity nuclei are subject to a large magnetic hyperfine interaction. The magnetic hyperfine fields B_{HF} experienced by probe nuclei have been intensively studied experimentally and theoretically over the past three decades [1]. The radioactive detection techniques of low-temperature nuclear orientation (LTNO) and NMR on oriented nuclei (NMRON) have been well established, being applied for a wide range of alloy systems and at extremely dilute impurities in a given host. Use of these techniques has led to a more complete series of experimental data, especially for 3d and 4d impurities in Fe. Cottenier and Haas have performed *ab-initio* band calculations for $Fe_{15}X$ supercells, taking the lattice relaxation effect into consideration, for 4d and 5sp elements in Fe [2]. The calculations reproduced the experimental B_{HF} values quantitatively very well; notable exceptions were Rb and Sr. Recently, we have observed the NMRON resonance of ^{91}Sr*Fe* [3] with β-ray detection and ion

* Author for correspondence.

implantation of the radioactive precursor isotopes. The resultant precise B_{HF} value has removed the discrepancy between the experimental and theoretical studies.

The first report of the B_{HF} value of YFe originates from a NMR measurement of the stable isotope, ^{89}Y; B_{HF} = (−)28.6(6) T [4]. Ion-implantation of the radioactive precursor Rb by the ISOLD facilities led to the first observation of NMRON of ^{91m}Y*Fe*, which resulted in B_{HF} = −30.67(36) T estimated from the resonance shift vs. the external field, neglecting the possible existence of nonzero Knight-shift [5]. The magnitudes of these two values are inconsistent given the quoted errors. We commenced NMRON of the ground state of ^{91}Y (I = 1/2) in Fe to clear up this inconsistency. A combination of the derived zero-field frequency with the known nuclear g-factor of ^{91}Y [6] has brought a precise B_{HF} value, −29.4(2) T, and the Knight-shift value, 0.0(8)% [7]. It is a consequence of these studies that the disagreement in the B_{HF} values can be attributed to a large hyperfine anomaly, $^{91}\Delta^{91m}$ = −4.2(8)%, between the I = 1/2 and 9/2 nuclear states of Y in Fe [7].

As demonstrated in those studies, β-/γ-ray detected NMRON using samples prepared by ion implantation has enabled us to investigate hyperfine interactions of Sr and Y in ferromagnetic hosts. In this paper we report experimental results of NMRON of ^{91m}Y in Ni and ^{93}Y in Fe, including a preliminary result of ^{89}Sr in Ni.

The samples were prepared by the ion-implantation of Rb (A = 89, 91, 93) isotopes into pure Fe or Ni foils using the online isotope separator at Research Reactor Institute, Kyoto University. The precursor Rb nuclei were produced as fission products, and implanted into the foils with an accelerating voltage of 100 kV [8]. The activated part of the foil was soft-soldered to the copper cold finger of a ^{3}He/^{4}He dilution refrigerator, and cooled down to about 10 mK at Niigata University. The sample temperature was monitored using a ^{54}MnNi nuclear thermometer. A geometrical arrangement of Si detectors for β-ray detections and HP Ge detectors for γ-ray has been given elsewhere [9]. A radio-frequency oscillating field with frequency modulation (FM) at 100 Hz was applied perpendicular to the external field. The frequency and modulation width were carefully monitored with a spectrum analyser during the measurements.

The metastable ^{91m}Y (I^{π} = 9/2$^+$) de-excites to the ground state via M4 γ-transition of energy 556 keV. NMRON of ^{91m}Y in Ni was carried out monitoring the γ-ray at external fields B_0 up to 1.0 T. The observed NMRON spectra are shown in Figure 1(a). The measuring time at each data point was 900 s, separated by intervals of 180 s to allow the nuclear spin–lattice relaxation. The effective nuclear spin–lattice relaxation time [10] was estimated to be about 140(30) s with B_0 = 0.2 T and at about 10 mK, by switching on and off the FM at the resonant centre frequency ν which was obtained by Gaussian fits of the data. The FM width was kept constant (±0.2 MHz) in the measurements, and the observed resonance line widths were found to be almost the same: about 1.07(6) MHz. From a

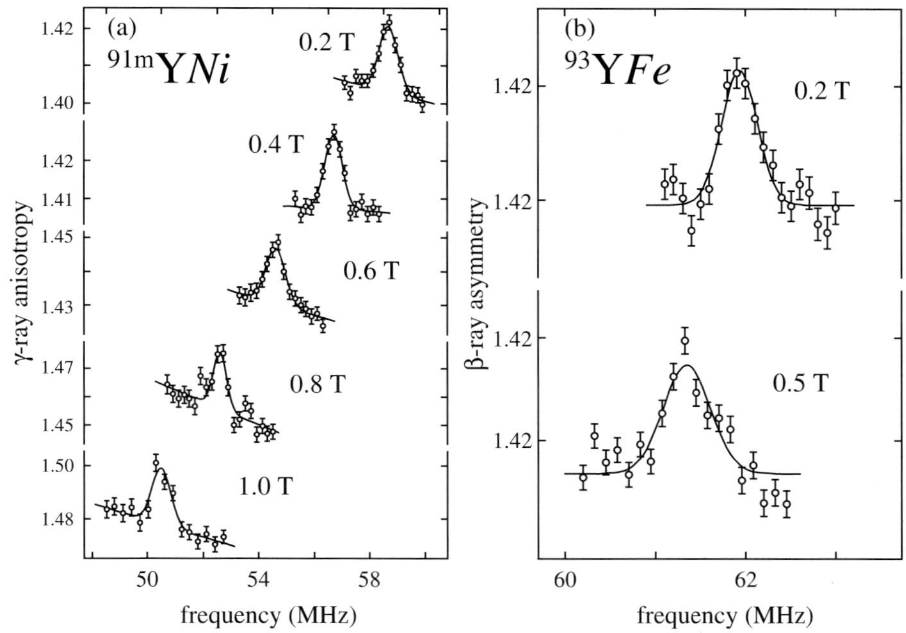

Figure 1. (a) NMRON spectra of ^{91m}Y*Ni* via γ-ray transition of 556 keV at fields up to 1.0 T. (b) NMRON spectra of ^{93}YFe via β-ray detections at fields of 0.2 and 0.5 T.

linear fit of ν vs. B_0 data, the extracted values of the zero-field frequency ν_0 and the field shift dν/dB_0 are $\nu_0 = 60.21(4)$ MHz and dν/d$B_0 = -10.18(21)$ MHz/T.

NMRON spectra of ^{93}Y ($I^{\pi} = 1/2^-$) in Fe were taken by β-ray detections with $B_0 = 0.2$ and 0.5 T, as shown in Figure 1(b). Resonant centre frequencies were obtained as $\nu = 61.91(4)$ MHz at 0.2 T, and $\nu = 61.36(4)$ MHz at 0.5 T. The effective nuclear spin–lattice relaxation time was found to be about 150(90) s with $B_0 = 0.2$ T. In the previous NMRON of ^{91}Y*Fe* experiments, the resonant frequency at 0.2 T was reported as $\nu = 73.09(5)$ MHz [7]. The nuclear g-factor of the ^{93}Y ground state can be estimated from the ratio; $g(^{93}\mathrm{Y}) = g(^{91}\mathrm{Y})\nu(^{93}\mathrm{Y})/\nu(^{91}\mathrm{Y}) = -0.2780(17)$, using the known value of $g(^{91}\mathrm{Y})$ [6].

The ground state of ^{89}Sr ($I^{\pi} = 5/2^+$) decays via the first-forbidden beta decay with probability 99.99% to the ground state of ^{89}Y. NMRON of ^{89}Sr*Ni* by β-ray detection has been undertaken. A reliable resonance spectrum was acquired at an external field of 0.2 T, in which the resonance frequency was derived to be 20.86(2) MHz, which resulted in the B_{HF} value of ($-$)6.16(2) T, assuming the negative sign and $K = 0$. (Uncertainty of K was considered to be 2%, which was taken into account in the resultant error given in Table I) This value is different from that of TDPAD experiments of ^{86}Sr*Ni*: $-5.42(11)$ T [11].

The present results, including recently reported ones, from NMRON experiments of Sr and Y isotopes in Fe and Ni are listed in Table I. Using the extracted g-factor value of ^{91m}Y for the zero-field frequency and the field shift of

Table I. Results of zero-field frequency ν_0; field shift $d\nu/dB_0$; nuclear g-factor g; magnetic hyperfine field B_{HF}

Sample	ν_0 [MHz]	$d\nu/dB_0$ [MHz/T]	g-factor	B_{HF} [T]
^{91}Sr*Fe*	64.30(12)	−2.85(34)	−0.3547(3)[14)]	−23.78(7)
^{89}Sr*Ni*	21.56(2)[a]	–	−0.45952(28)[14)]	−6.16(2)
^{91m}Y*Fe*	310.005(23)	−10.111(38)	−1.327(16)[b]	−30.7(4)
^{91}Y*Fe*	73.566(22)	−2.501(17)	−0.3282(16)[14)]	−29.4(2)
^{91m}Y*Ni*	60.21(4)	−10.18(21)	−1.327(16)	−5.95(8)
^{93}Y*Fe*	62.28(7)	−1.84(17)	−0.2780(17)[c]	−29.4(2)

[a] Corrected value for an external field $B_0 = 0.2$ T, assuming the Knight shift $K = 0$.
[b] Recalculated value using $d\nu/dB_0$ and $K = 0.0(8)\%$.
[c] Estimated value from the ratio of resonant frequencies of ^{91}Y*Fe* and ^{93}Y*Fe* at $B_0 = 0.2$ T.

^{91m}Y*Ni*, the B_{HF} value and the Knight shift are estimated to be −5.95(8) T and 0.6(33)%, respectively. The accurate B_{HF} value found in this work confirms the result from the DPAC experiment of ^{86}Y*Ni* (5.7(4) T) [1] and fits in with a systematic variation reported experimentally [1] and theoretically [12]. It is surprising that the g-factor of ^{93}Y differes from that of ^{91}Y by about 20%, even though both nuclei possibly have the same nuclear configuration: π(p1/2). The magnitudes of the g-factors are about half of the Schmidt value: −0.527. The configuration mixing, generally explaining deviations between experimental g-factors and the Schmidt values, is little effective for the π(p1/2) configuration [13]. There is a relatively small number of studies for the g-factors in π(p1/2) state. It is hoped that the present result will stimulate a systematic study in experiment and theory for the g-factors in the π(p1/2) state.

Acknowledgement

The authors would like to acknowledge the support by the Grant-in-Aid for Science Research (No. 09440102).

References

1. Rao G. N., *Hyperfine Interact.* **26** (1985), 1119.
2. Cottenier S. and Haas H., *Phys. Rev.*, **B62** (2000), 461; Cottenier S. and Haas H., *Hyperfine Interact.* **133** (2001), 239.
3. Nishimura K., Ohya S. and Ohtsubo T. *et al.*, *Phys. Rev., B* **68** (2003), 012403-1.
4. Kontani M. and Itho J., *J. Phys. Soc. Jpn.* **22** (1967), 345.
5. Hinfurtner B., Hagn E., Zech E., Eder R. and ISOLDE, *Phys. Rev. Lett.* **66** (1991), 96.
6. Petersen F. R. and Shugart H. A., *Phys. Rev.* **128** (1962), 1740.
7. Nishimura K., Ohya S. and Ohtsubo T. *et al.*, *Phys. Rev.*, **B70** (2004, in press).
8. Kawase Y., Okano K. and Funakoshi Y., *Nucl. Instrum. Methods* **A241** (1985), 305.
9. Ohtsubo T., Cho D. J., Yanagihashi Y. and Ohya S., *Phys. Rev.*, **C54** (1996), 554.

10. Bacon F., Barclay J. A., Brewer W. D., Shirley D. A. and Templeton J. E., *Phys. Rev.,* **B5** (1972), 2397.
11. Raghavan P., Ding Z. Z., and Raghavan R. S., *Hyperfine Interact.* **15/16** (1983), 317.
12. Akai H. and Akai M. *et al.*, *Prog. Theor. Phys. Suppl.* **101** (1990), 11.
13. Noya H., Arima A. and Horie H., *Prog. Theor. Phys. Suppl.* **8** (1958), 33.
14. Firestone R. B. and V. S. Shirley (eds.), *Table of Isotopes*, 8th Edition., Wiley, New York, 1996, Appendix E.

Hyperfine Interactions (2004) 159:245–249
DOI 10.1007/s10751-005-9113-4

Measurement of the g-Factor of the 27^- High-Spin Isomer State of ^{152}Dy

M. FUJITA[1,*], T. ENDO[1], A. YAMAZAKI[1], T. SONADA[1], T. MIYAKE[1], E. TANAKA[1], T. SHINOZUKA[1], T. SUZUKI[2], A. GOTO[2], Y. MIYASHITA[2], N. SATO[2], Y. WAKABAYASHI[3], N. HOKOIWA[3], M. KIBE[3], Y. GONO[3], T. FUKUCHI[4] and A. ODAHARA[5]

[1]*Cyclotron and radioisotope center, Tohoku University, Aramaki-Aoba, Sendai, Miyagi 980-8578, Japan; e-mail: mafujita@cyric.tohoku.ac.jp*

[2]*Department of Physics, Tohoku University, Aramaki-Aoba, Sendai, Miyagi 980-8578, Japan*

[3]*Department of Physics, Kyushy University, Hakozaki, Fukuoka 812-8581, Japan*

[4]*Center for Nuclear Study, University of Tokyo, RIKEN campus, Wako, Saitama 351-0198, Japan*

[5]*Nishinippon Institute of Technology, Kanda, Fukuoka 800-0394, Japan*

Abstract. The g-factor of the 27^- isomer state of ^{152}Dy has been measured using the Time-Integral Perturbed Angular Distribution (TIPAD) method. The high-spin states of ^{152}Dy have been populated by ^{141}Pr(^{16}O,p4n)^{152}Dy reaction at E = 115 MeV from the AVF cyclotron at CYRIC. The paramagnetic correction factor of Dy ions in Pr has been determined to be 4.2(5) by the Time-Differential Perturbed Angular Distribution (TDPAD) measurement of the 21^- state of ^{152}Dy. As a result, the g-factor of the 27^- isomer state of ^{152}Dy has been obtained to be +0.09(5). This shows the smaller value than the expected one of +0.39 deduced from a fully aligned configuration of $\pi(h^2_{11/2}) \otimes \nu(f^2_{7/2}h_{9/2}i_{13/2})$.

Key Words: high-spin isomer, the nuclear g-factor, the paramagnetic correction factor, TDPAD, TIPAD.

1. Introduction

As a result of recent study, high-spin isomers have been observed systematically around the neutron magic number $N = 82$ [1]. These states are called 'yrast trap' and are considered to be caused by an irregularity in the yrast line formed by states of single particle structure.

The nuclear g-factor can provide useful information about these isomers because it is directry affected by the microscopic nuclear structures. However it is generally difficult to measure the nuclear g-factor in rare-earth region because of the large ambiguity of the effective magnetic field caused by the paramagnetic effect [3].

---

* Author for correspondence.

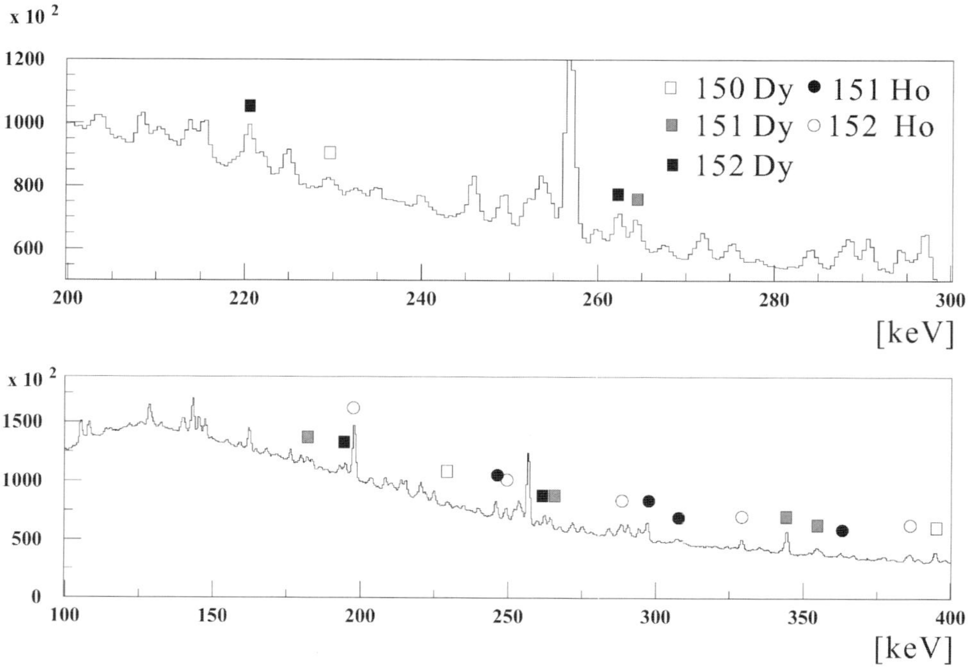

Figure 1. γ-ray spectra measured by the 230 cm^2 HPGe detector.

We report the first measurement of the g-factor of the 27^- high-spin isomer of ^{152}Dy, which belongs to an island of high-spin isomers around $N = 82$ and has relatively short half life ($T_{1/2} = 1.6 \pm 0.2$ ns) [2].

2. Experiment

The nuclear g-factor of the 27^- high-spin state of ^{152}Dy was measured by the TIPAD method. The high-spin isomer states of ^{152}Dy were populated by the reaction ^{141}Pr(^{16}O,p4n)^{152}Dy at $E = 115$ MeV from the AVF cyclotron at CYRIC in Tohoku University. An enriched ^{141}Pr target of 6 mg/cm^2 thickness was placed in an external magnetic field (B_{ext}) of 2.03 T applied perpendicularly to the beam-detector plane. Figure 1 shows the energy spectra of γ-rays.

The time-integral perturbed angular distributions (TIPAD) of the 220.6 keV in-beam γ-rays emitted from the 27^- isomer were measured by a 230 cm^3 HPGe detector at five angles between 70° and 115° with respect to the beam direction (see Figure 2). The deflection of the beam is usually an inherent difficulty in the in-beam TIPAD measurement using an external magnetic field. In our experimental system, the change of position and direction of the beam due to the PAD magnet is compensated by a pair of electromagnetic dipoles attached to the entrance and exit of the PAD magnet [4]. The residual beam deflection on target achieved is estimated to be within ±0.1°.

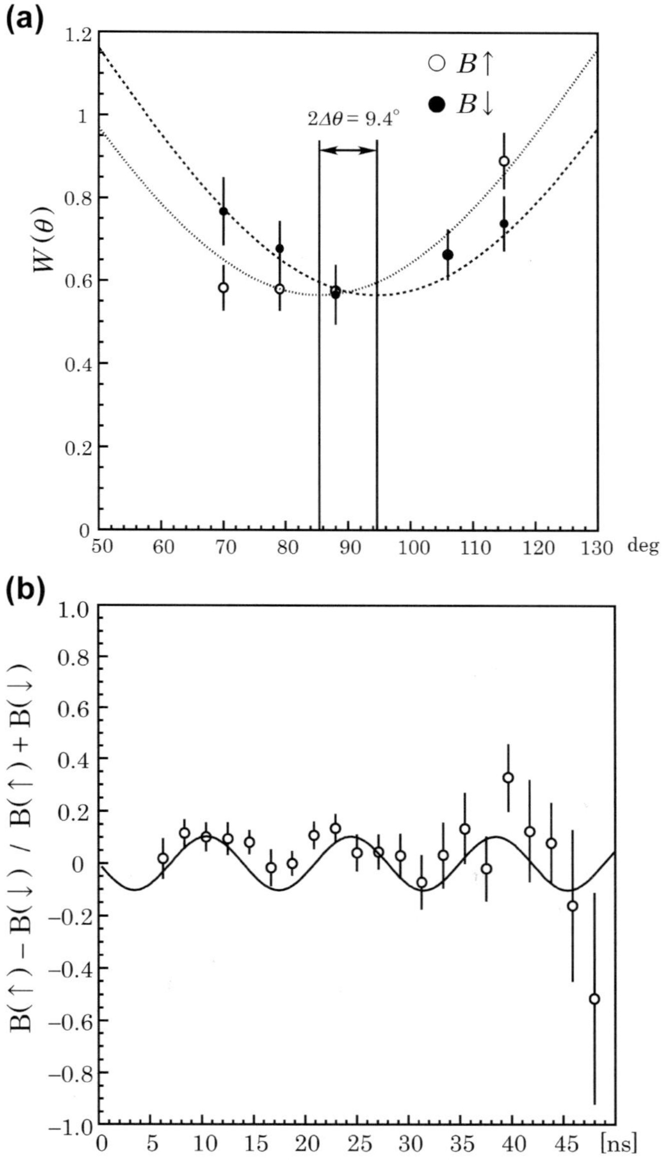

Figure 2. (a) TIPAD spectrum of 220.6 keV γ-rays emitted from the 27^- isomer state of ^{152}Dy, and (b) TDPAD spectrum of 262.4 keV γ-rays emitted from the 21^- isomer state of ^{152}Dy.

The TIPAD data were fitted to an expression:

$$W(\theta \pm B_{\text{eff}}) = A_0 + A_2 P_2(\cos(2(\theta \mp \Delta\theta)))$$

where $\Delta\theta$ is the angular shift due to the Larmor precession, and given by the Larmor anglar velocity (ω_L) and mean life τ of the isomeric state as $\tan 2\Delta\theta = 2\omega_L\tau$.

Table I. Expected g-factors for possible configurations, excitation energies and deformation parameters calculated by DIPM

Configuration	g (expected)[a]	E_{cal} [MeV]	β_{cal}
(experimental value $g = +0.09$ (5) $E_{\mathrm{ex}} = 7.88$)			
$\pi(h^2_{11/2}) \otimes \nu(f^2_{7/2}h_{9/2}i_{13/2})$	+0.36 (5)	6.79	−0.097
$\pi(h_{11/2}d_{3/2}) \otimes \nu(f_{7/2}h_{9/2}i^2_{13/2})$	+0.15 (5)	9.02	−0.091

[a] The expected g-factors are evaluated using the experimental g-factors for single-particle (or hole) components as below [7];
$g(\pi h_{11/2}) = +1.288(3)$ (^{145,147,149}Eu), $g(\pi d_{3/2}) = +0.109(1)$,
$g(\nu f_{7/2}) = -0.304(2)$ (^{143}Nd), $g(\nu i_{13/2}) = -0.15(2)$ (^{205}Pb),
$g(\nu h_{9/2}) = -0.08(3)$ (Derived from the 21^- state in ^{152}Dy [2]).

The nuclear g-factor is deduced from the Larmor angular velocity as $g = \hbar\omega_L / \mu_N B_{\mathrm{eff}}$, where μ_N is the nuclear magneton and B_{eff} is the effective magnet field at the site of the Dy nuclei in the target.

The effective magnetic fields for rare-earth ions are known to be quite different from the external field by the large paramagnetic effect [3]. The B_{eff} is written as $B_{\mathrm{eff}} = \beta B_{\mathrm{ext}}$, where β is the paramagnetic correction factor which must be measured independently to obtain the nuclear g-factor.

The g-factor of the 21^- isomer state of ^{152}Dy has been known to be $+0.55 \pm 0.06$ [2] and fortunately this state has been populated simultaneously in the present experiment. The mean life of this isomer state is relatively long ($\tau = 13.7 \pm 1.0$ ns [2]), therefore we have measured the Larmor angular velocity of this state by the TDPAD method and have deduced the effective magnetic field in our experimental condition. Figure 2 shows the result of the TDPAD measurement of the 21^- isomer state of ^{152}Dy. The effective magnetic field has been obtained to be 8.50 ± 0.95 T, which corresponds to $\beta = 4.2 \pm 0.5$. As a result, the g-factor of the 27^- high-spin isomer state is obtained to be $+0.087 \pm 0.042$.

3. Discussion

Possible configurations for the 27^- state in ^{152}Dy are listed in Table I together with expected g-factors which are calculated by means of experimental g-factors for single particle (or hole) components. In addition, the excitation energies and the deformation parameters from the deformed independent particle model (DIPM) calculation [5, 6] are listed. This model dealt with independent particle configurations in an axially symmetric deformed potential. Energies of single-particle orbits were calculated as a function of deformation with a Woods–Saxon potential based on empirically deduced spherical single-particle energies.

The yrast configuration of the 27^- state is expected to be $\pi(h^2_{11/2}) \otimes \nu(f^2_{7/2}h_{9/2}i_{13/2})$ in the framework of the DIPM calculation. The present result is in rather good agreement with the expected g-factor for the $\pi(h_{11/2}d_{3/2}) \otimes \nu(f_{7/2}h_{9/2}i^2_{13/2})$ configuration, however the excitation energy of this configuration is more than 2 MeV higher than that of the yrast one. In addition to perform a more precise calculation including configuration mixing effects, systematic measurements in the high-spin isomer states of Dy isotope are needed to understand this contradiction, and g-factor measurements of the high-spin state in other Dy isotopes are planned.

References

1. Pederson J. *et al.*, *Phys. Rev. Lett.* **39** (1977), 990.
2. Meerdinger J. C. *et al.*, *Phys. Rev. Lett.* **42** (1979), 23.
3. Günther C. and Lindgen I., *Perturbed Angular Correlations*, North-Holland, Amsterdam, 1964, p. 357.
4. Kawamura N. *et al.*, *Hyperfine Interact.* **15/16** (1983), 1057.
5. Døssing T., Neergård K. and Sagawa H., *Phys. Scr.* **24** (1981), 258.
6. Neergård K., Døssing T. and Sagawa H., *Phys. Lett., B* **99** (1981), 191.
7. Raghaban P., *At. Data Nucl. Data Tables* **42** (1989), 189.

Hyperfine Interactions (2004) 159:251–255
DOI 10.1007/s10751-005-9114-3

Measurement of the Magnetic Moment of the First Excited State in ^{93}Sr Using On-Line TDPAC Technique

T. SASANUMA†, A. TANIGUCHI*, M. TANIGAKI, Y. OHKUBO and Y. KAWASE
Research Reactor Institute, Kyoto University, Kumatori-cho, Sennan-gun, Osaka 590-0494, Japan; e-mail: taniguti@rri.kyoto-u.ac.jp

Abstract. The g-factor of the first excited state of ^{93}Sr (E = 213 keV, $T_{1/2}$ = 4.6 ns) was measured by an on-line TDPAC technique with use of the strong hyperfine field in Fe metal. The Larmor frequency $\omega_L = (2.60 \pm 0.15) \times 10^8$ rad/s was obtained. The g-factor is derived as $g = -0.227 \pm 0.013$ from $g = -\hbar\omega_L/B_{hf}\mu_N$. If the spin of the first excited state of ^{93}Sr is assumed to be 3/2, the g-factor is predicted by a simple core-excitation model as $g = -0.22$, which is in good agreement with the present experimental result.

Key Words: ^{93}Sr, g-factor, ion implantation, on-line isotope separator, TDPAC.

1. Introduction

The odd-mass nuclei with a few nucleons outside a spherical even–even core have been investigated with the interest that quite different modes of excitation can lead to the low-lying levels: single particle transitions to high-lying orbits, coupling of an odd particle with an even–even core, interactions among several particles outside a closed shell, etc. It was pointed out [1] that the appearance of the low-lying anomalous coupling (AC) states with spin $I = j - 1$ of odd-mass nuclei in a spherical region can be regarded as a typical phenomenon in which the dressed three-quasi-particle mode manifests itself as a relatively pure eigenmode. The experimental studies on these nuclei can help to understand the mechanism of excitations and properties of excited levels more clearly.

The magnetic moments of the first excited states of ^{91}Sr, ^{95}Mo, ^{99}Ru and ^{101}Ru have been measured and evaluated by several authors [1–5]. These nuclei have three or five nucleons outside an even-even core and are expected to have similar excitation modes. In the previous TDPAC experiments on ^{91}Sr [2], the observed angular correlation coefficient turned out to be very small and

---

* Author for correspondence.

† Graduate School of Science, Kyoto University, Kyoto 606-8502, Japan.

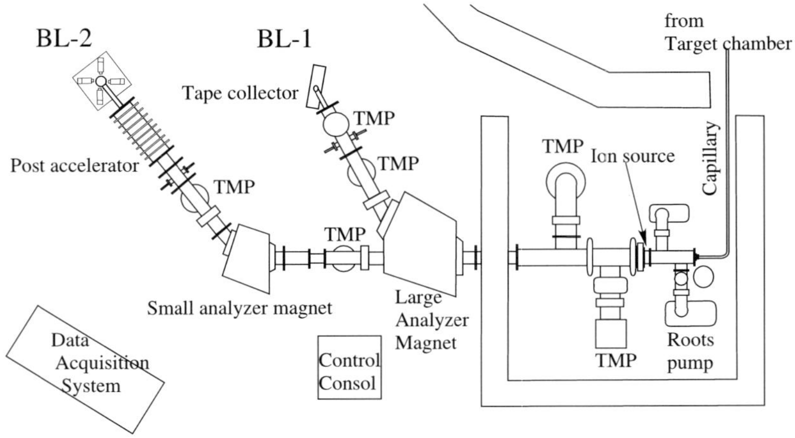

Figure 1. Layout of the KUR-ISOL.

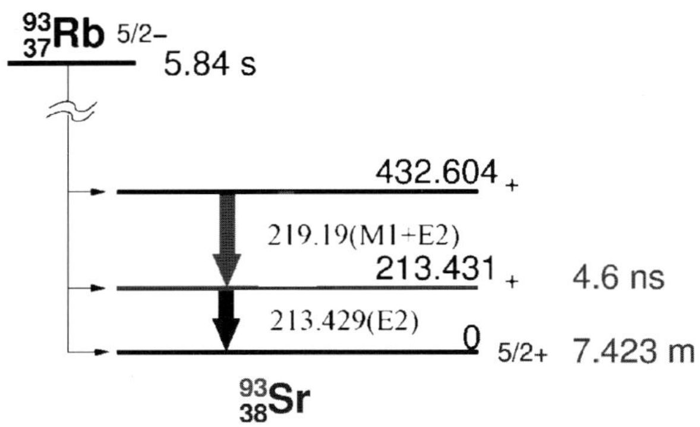

Figure 2. Simplified decay scheme of ^{93}Rb.

the reported value of the *g*-factor of the first excited state has remained unsatisfactory.

In the present report, we employed an on-line TDPAC method to determine the *g*-factor of the first excited state in ^{93}Sr using a beam line of the on-line isotope separator (KUR-ISOL) installed at the research reactor of Kyoto University.

2. Source preparation

Since the mother nucleus ^{93}Rb, which decays to ^{93}Sr with a half-life of 5.8 s, can be produced by nuclear fission of ^{235}U, the on-line mass separator was used to prepare the source activity. The 93% enriched ^{235}U of 50 mg in total was irradiated by a thermal neutron flux of 3×10^{12} n/cm^2/s in the target chamber. The fission products were stopped in 1–2 atm carrier gas at the center of the target chamber and transported through a stainless steel capillary of 1.5 mm in diameter by N_2–He mixed jet with PbI_2 aerosols to the skimmer chamber located 11 m apart from the target chamber.

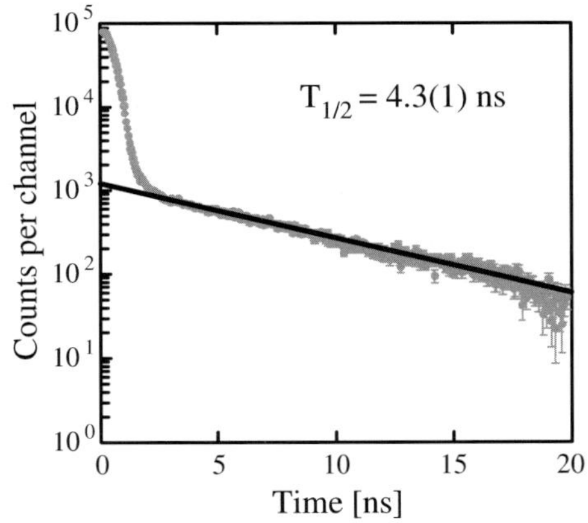

Figure 3. Half-life of the first excited level in ^{93}Sr.

A surface ionization ion source was employed to ionize elements with low ionization potentials such as Rb and Cs at very high efficiency. After being extracted by 14 kV high voltage potential, radioactive ions were accelerated to 30 keV and focused by the electrostatic lens. The mass-analyzed ^{93}Rb ions were implanted on an Fe foil in a sample chamber where an on-line TDPAC measurement was carried out.

The recent layout of the on-line isotope separator (KUR-ISOL) [6] is shown in Figure 1: The BL-1 beam-line is mainly used for nuclear spectroscopic works on short-lived nuclei and another beam-line (BL-2) is specially prepared for solid state physics.

3. Half-life measurement

The half-life of the first excited level in ^{93}Sr is reported [7] to be 4.6(3) ns as shown in Figure 2. We measured the half-life with BaF_2 detectors to confirm a γ-cascade relation of the 219−213 keV transition for the TDPAC experiment. The time resolution was about 0.9 ns. Because the γ-ray energies of this cascade are very close, a doublet peak of about 216 keV was chosen by the BaF_2 detectors. We obtained 4.3(1) ns, in good agreement with the reported value [Figure 3].

4. TDPAC experiment

The TDPAC experiment was carried out at the BL-2 of KUR-ISOL [Figure 1] in an on-line mode. An Fe foil was put at the center of the sample chamber to catch radioactive ^{93}Rb ions. Three BaF_2 detectors were arranged to make angles of +90°, +135° and −135° surrounding the Fe catcher as shown in Figure 4. A magnetic field of 0.2 T was applied by a small permanent magnet to produce a saturated field at the source position. Time spectra for required detector

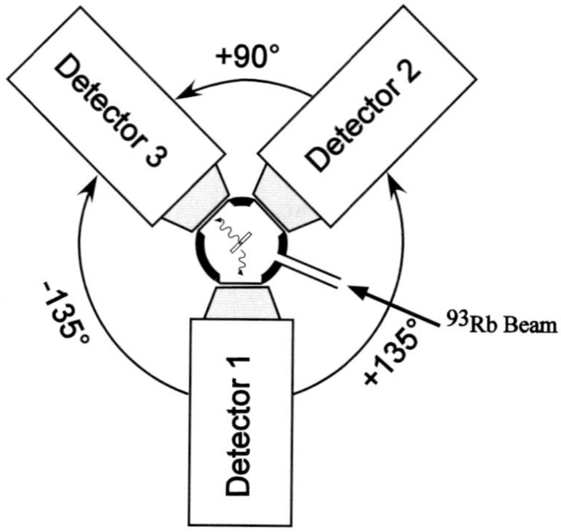

Figure 4. Detector arrangement of on-line TDPAC.

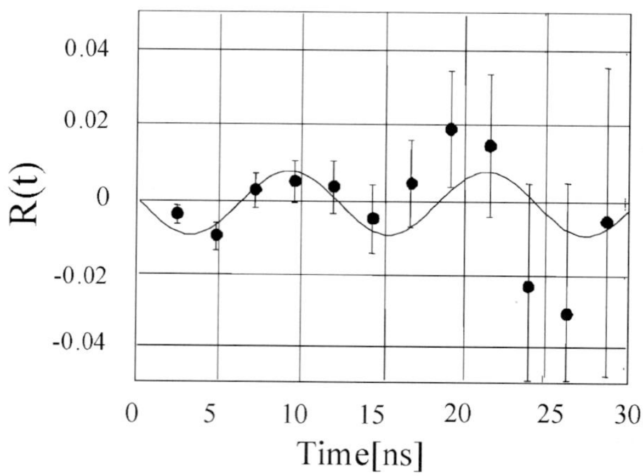

Figure 5. The result of the TDPAC for ^{93}Sr.

combinations were recorded with a conventional electronic circuit. The ^{93}Rb source strength was monitored to keep the count rate at optimum. Because the angular correlation coefficient of the 219–213 keV cascade is very small, one week measurement was necessary to obtain sufficient coincidence counts.

5. Result and discussion

The time-differential perturbed angular correlation of γ-rays, $W(\theta, t)$, is described as a function of the emission interval time t and the angle θ between the γ-rays. Calculating the ratio $R(t) = [W(-135°, t) - W(+135°, t)]/[W(+135°, t) + W(-135°, t)]$, as given in the literature, the form becomes a simple function; $R(t) = 3/4 \cdot A_{22}\sin(2\omega_L t)$ in which A_{22} is the angular correlation coefficient. The obtained $R(t)$ values are plotted as a function of the delay time in Figure 5. Although the observed Larmor precession is unclear due to the very small A_{22}

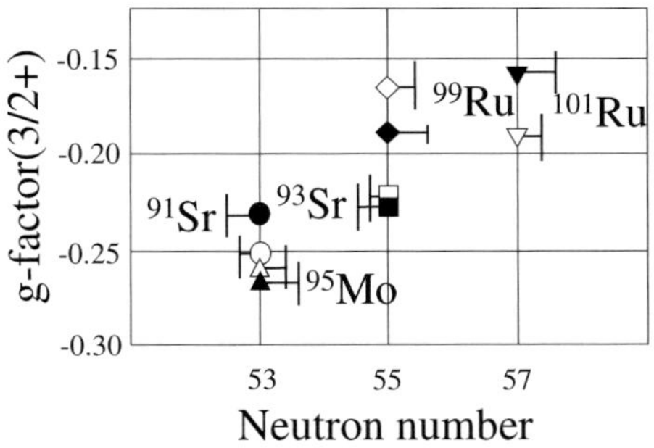

Figure 6. Comparison of experimental g-factors (*closed*) with calculated values (*open*).

value, we obtained the Larmor frequency $\omega_L = (2.60 \pm 0.15) \times 10^8$ [rad/s]. The g-factor is determined as -0.227 ± 0.013, derived from $g = -\hbar\omega_L/B_{hf}\mu_N$. We used a reported value of the hyperfine magnetic field B_{hf} at Sr in Fe, $B_{hf} = -23.83 \pm 0.07$ T, at low temperature [8].

In the $N = 53$–57 region, a core excitation model [9] is thought to explain the magnetic moments of excited states. If the spin of the first excited state of ^{93}Sr is assumed to be 3/2, we can calculate the g-factor predicted by this model, $g = -0.22$, which is in good agreement with the present experimental result. The comparison of experimental g-factors with values calculated by this model in this mass region is shown in Figure 6.

However, the energy of this state, 213 keV, is too low to be explained by a coupling of a particle with an even–even core, because the 2+ state energy of the neighboring ^{92}Sr is as high as 815 keV. Kuriyama *et al.* [1] made the g-factor calculations for the 3/2+ AC-like states in odd-mass Mo and Ru isotopes by a systematic microscopic theory. For example, they gave $g = -0.29$ for ^{99}Ru with the excitation energy of $E_{3/2}{\sim}200$ keV, low enough to explain the experimental value of 89 keV. Calculations for Sr isotopes are required to discuss the present result more deeply.

References

1. Kuriyama A., Marumori T. and Matsuyanagi K., *Suppl. Prog. Theor. Phys.* **58** (1975).
2. Kawase Y., Uehara S. and Nasu S., *Hyperfine Interact.* **84** (1994), 329.
3. Wolf A., Gill R. L., Brenner D. S., Berant Z., Schuhmann R. B. and Zamfir N. V., *Phys. Rev. C* **48** (1993), 562.
4. Matthias E., *Phys. Rev.* **139** (1965), B532.
5. Alzner A., Bodenstedt E., Gemünden G., Herrmann C., Münning H., Reif H., Rudolf H. J., Vianden R. and Wrede U., *Z. Phys.* **A317** (1984), 107.
6. Okano K., Kawase Y., Kawade K., Yamamoto H., Hanada M., Katoh T. and Fujiwara I., *Nucl. Instrum. Methods* **186** (1981), 115.
7. Firestone R. B. and Shirley V. S. (eds.), *Table of Isotopes*, 8th Edition, Wiley, 1996.
8. Nishimura K., Ohya S., Ohtubo T., Sasaki M., Goto J., Izumikawa T., Tanigaki M., Taniguchi A., Ohkubo Y., Kawase Y. and Muto S., *Phys. Rev. B* **68** (2003), 012403.
9. deShalit A. *et al., Phys. Rev.* **122** (1961), 1530.

Hyperfine Interactions (2004) 159:257–260
DOI 10.1007/s10751-005-9115-2

Magnetic Moment and Spin of the Extremely Proton-Rich Nucleus ^{23}Al

K. MATSUTA[1,*], Y. NAKASHIMA[1], T. NAGATOMO[1], A. OZAWA[2], K. YAMADA[3], M. MIHARA[1], S. KUMASHIRO[1], H. FUJIWARA[1], S. MOMOTA[4], M. OTA[5], T. OHTSUBO[5], K. YOSHIDA[6], T. SUMIKAMA[1,†], M. OGURA[1], K. MINAMISONO[7], M. FUKUDA[1], T. MINAMISONO[1], Y. NOJIRI[4], T. SUZUKI[8], T. IZUMIKAWA[5], I. TANIHATA[9], J. R. ALONSO[10], G. F. KREBS[10] and T. J. M. SYMONS[10]

[1]*Department of Physics, Osaka University, Osaka 560-0043, Japan;*
e-mail: matsuta@vg.phys.sci.osaka-u.ac.jp
[2]*Institute of Physics, University of Tsukuba, Tsukuba, Ibaraki 305-8571, Japan*
[3]*College of Science, Rikkyo University, Toshima-ku, Tokyo 171-8501, Japan*
[4]*Kochi University of Technology, Tosayamada, Kochi 782-8502, Japan*
[5]*Department of Physics, Niigata University, Niigata 950-2181, Japan*
[6]*RIKEN, Wako, Saitama 351-0198, Japan*
[7]*TRIUMF, 4004 Wesbrook Mall, Vancouver, B.C. V6T 2A3, Canada*
[8]*Department of Physics, Saitama University, Saitama 338-8570, Japan*
[9]*Argonne Natsional Laboratory, Argonne, IL 60439, USA*
[10]*Lawrence Berkeley Laboratory, Berkeley, CA 94720, USA*

Abstract. The g-factor of the exteremely proton-rich nucleus ^{23}Al($T_{1/2}$ = 0.47 s) has been measured for the first time, applying β-NMR technique on this nucleus implanted in Si. The obtained $|g| = (1.58 \pm 0.2)$ suggests that the spin of the ground state of ^{23}Al is 5 / 2. The magnetic moment is determined as $|\mu| = (3.95 \pm 0.55)\ \mu_N$.

Key Words: ^{23}Al, β-NMR, ground state spin parity, magnetic moment, proton halo.

1. Introduction

The experimental study of electromagnetic moments had been limited to the nuclei close to the stability line. In the light mass region, mostly isospin $T = 1/2$, and some $T = 1$ nuclei had been studied. Although, there has been a few $T_z = 3/2$ nuclei with known magnetic moment (μ), only the two $T_z = -3/2$ proton-rich nuclei ^{9}C and ^{13}O were studied [1]. In the present study, the magnetic moment of the $T_z = -3/2$ extremely proton-rich nucleus ^{23}Al($T_{1/2}$ = 0.47 s) has been measured for the first time.

---

* Author for correspondence.
† Present Address: RIKEN, Wako, Saitama 351-0198, Japan.

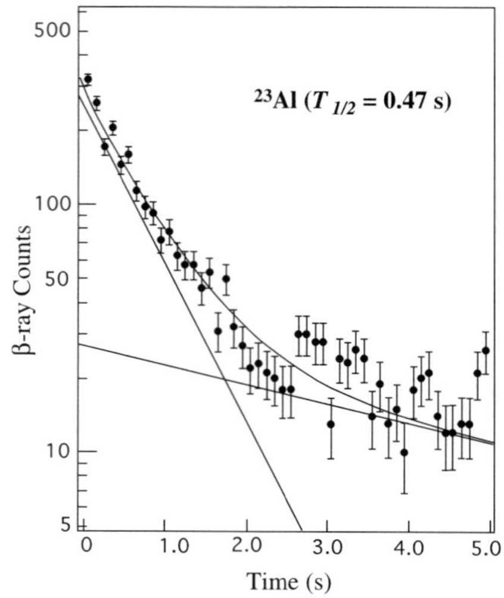

Figure 1. Typical β-ray decay curve spectrum.

Even the spin and parity of the ground state of ^{23}Al has not been determined yet. The nuclear structure of ^{23}Al has drawn attention recently because this nucleus is unique proton-halo candidate with a largely extended single proton in the valence orbit. It is noted that the single halo proton might be in an abnormal orbit, which is reflected in various properties of this nucleus, especially in its spin. Such a case has been found only in the neutron halo in ^{11}Be, which has the abnormal spin parity $1/2^+$. Also in the present ^{23}Al case, it may have abnormal spin 1/2 instead of normal 5/2, which is expected from its very small separation energy (125 keV), and from the recently found enhanced experimental reaction cross section [2, 3], although the error is still large. The nuclear spin of ^{23}Al can be clearly identified from the present measurement. The *g*-factor, which is related with the magnetic moment as $\mu = g\mu_N \boldsymbol{I}$, is totally different whether the nuclear spin is 5 / 2 (shell model prediction value $g_{SM}(I = 5/2) = 1.58$) or 1 / 2 ($g_{SM}(I = 1/2) = 3.40$). The clear spin assignment should lead to the better understanding of the structure of this unstable nucleus.

2. Experimental procedure

The ^{23}Al nuclei were produced through the projectile fragmentation process in the collision of ^{28}Si on ^{9}Be target at energies of about 135 *A* MeV at RIKEN ring cyclotron. The target thickness was 4.0 mm. Only the fragments ejected at the angles $\theta = 1.0° \pm 0.6°$ were separated by the fragment separator RIPS. In the separator, the momentum of the fragments was analyzed, and only the fragments with momentum $(2.0 \pm 0.5)\%$ larger than the central momentum were selected. The projectile fragments were polarized with this selection of the reaction angle

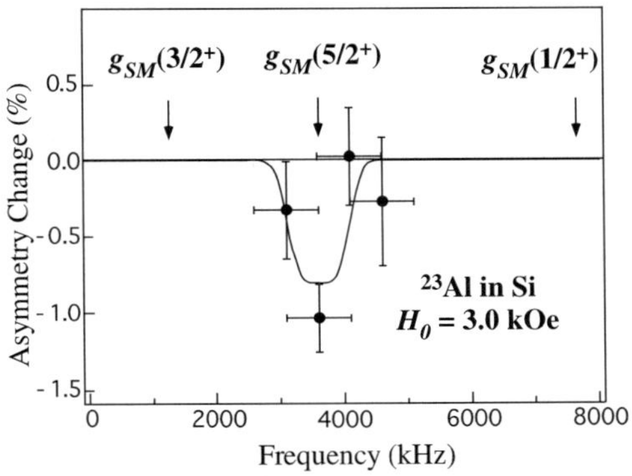

Figure 2. Typical β-NMR spectrum.

and momentum. At the achromatic focus F2 of the RIPS separator, an RF deflector was installed to improve the $S\,/\,N$ ratio of the secondary radioactive beam.

The selected fragments were slowed down by an energy degrader and were implanted in a Si catcher of 0.6 mm in thickness placed in a strong magnetic field H_0 of 0.3 T, in order to maintain polarization. The space from the final vacuum window to the entrance of the catcher chamber, was filled with He gas to minimize the polarization destruction by the gas collisions. Beta rays emitted from the stopped ^{23}Al were detected during the beam-off time after beam bombardment, by two sets of plastic-scintillation-counter telescopes placed above and below the catcher relative to the reaction plane. 1.0-mm thick Cu absorbers were placed in between the first and the second scintillators of the telescopes to improve the $S\,/\,N$ ratio in the β-ray counts. The S/N ratio in the ^{23}Al beam was improved seven times from 0.024 to 0.17 by the RF deflector. The S/N ratio in the β-ray counts was improved to 1.9 by the Cu absorber, to get a clean decay curve as shown in Figure 1. The polarization of the ^{23}Al is detected by the β-ray asymmetry in the counter system. An rf magnetic field H_1 of about 1.3 mT was applied to induce transitions between magnetic sub levels, resulting in the inversion of the initial polarization.

3. Results and discussion

A typical NMR spectrum is shown in Figure 2. The frequency modulation (FM) was ±500 kHz. Data points are limited and the error bars of some data points are large. So the full FM range of the second data point which significantly deviates from zero, was selected for the resonance frequency. Thus the resonance frequency is determined to be (3600 ± 500) kHz. From the resonance frequency, the g-factor was determined to be $|g| = 1.58 \pm 0.22$.

The present experimental g-factor is well reproduced by the shellmodel value $g_{SM}(5/2^+) = 1.58$ for the ground state spin and parity $I^\pi = 5/2^+$ as shown in

Figure 2. The shell-model value $g_{SM}(1/2^+) = 3.40$ for $I^\pi = 1/2^+$, strongly deviates from the present experimental value. Here, the nucleon g-factors in free space were used for the prediction. This suggests that the spin I of the ground state of ^{23}Al is 5/2. Thus the magnetic moment μ is determined to be $|\mu(^{23}\mathrm{Al})| = |g\mu_N I| = (3.95 \pm 0.55)\ \mu_N$. K. Peräjärvi *et al.* [4] measured the branching ratios of the transitions from ^{23}Al. The log *ft*-values to the first $(5/2^+)$ and the second $(7/2^+)$ excited levels of ^{23}Mg are 5.18 and 5.4, respectively, which suggests that these are the allowed transitions and the present spin and parity assignment $5/2^+$ for the ground state of ^{23}Al is favored, compared with $I^\pi = 1/2^+$. The spin expectation value $\langle\sigma_z\rangle$ can be determined from the present result combined with the μ of the mirror partner ^{23}Ne; $\mu(^{23}\mathrm{Ne}) = -1.08(1)\ \mu_N$, as,

$$\mu(^{23}\mathrm{Al}) + \mu(^{23}\mathrm{Ne}) = J + \left(\mu_p + \mu_n - \frac{1}{2}\right)\langle\sigma_z\rangle. \tag{1}$$

The sign of $\mu(^{23}\mathrm{Al})$ is assumed to be positive, which is reasonable from the theoretical consideration. Thus, the spin expectation value is determined as $\langle\sigma_z\rangle = 1.0 \pm 1.4$. The present error is still large for further discussion. In the next experiment, the magnetic moment will be measured with better precision. The Q-moment will also be measured in the future. This may allow us to discuss the proton halo structure in the nucleus.

Acknowledgements

The present work was performed at RIKEN Ring Cyclotron. The authors are grateful to the staff of RIKEN for their technical support. Support was also given by a Grant in Aid for Scientific Research from the Ministry of Education, Culture and Science, Japan.

References

1. Matsuta K. *et al.*, *Nucl. Phys.*, *A* **588** (1995), 153c.
2. Cai Y. Z. *et al.*, *Phys. Rev.*, *C* **65** (2002), 02610.
3. Zhang H. Y. *et al.*, *Nucl. Phys.*, *A* **707** (2002), 303.
4. Peräjärvi K. *et al.*, *Phys. Lett.*, *B* **492** (2000), 1.

Hyperfine Interactions (2004) 159:261–264
DOI 10.1007/s10751-005-9107-2

Production of Nuclear Polarization of Na Isotopes at ISAC/TRIUMF and its Hyperfine Interaction

K. MINAMISONO[1,2,*,†], K. MATSUTA[3], T. MINAMISONO[3], C. D. P. LEVY[1], T. NAGATOMO[3], M. OGURA[3], T. SUMIKAMA[3,‡], J. A. BEHR[1], K. P. JACKSON[1], H. FUJIWARA[3], M. MIHARA[3] and M. FUKUDA[3]
[1]*TRIUMF, 4004 Wesbrook Mall, Vancouver, B.C., Canada V6T 2A3; e-mail: minamiso@nscl.msu.edu*
[2]*Research Abroad, Japan Society for the Promotion of Science, Tokyo, Japan*
[3]*Department of Physics, Osaka University, 1-1 Toyonaka, Osaka, 560-0043, Japan*

Abstract. Hyperfine interactions of Na isotopes in single crystals have been studied using highly nuclear polarized ^{20,21,26,27,28}Na beams provided by ISAC/TRIUMF. The degree of polarization kept in the crystals, the spin-lattice relaxation times, the electric quadrupole coupling constants and the initial distribution of the populations were measured. Such knowledge is indispensable for the application of the hyperfine interactions in the study of precision measurements such as the nuclear structure through nuclear moments and the fundamental symmetries.

Key Words: ^{20,21,26,27,28}Na, hyperfine interaction, laser pumping, polarized beam.

1. Introduction

Hyperfine interactions of Na isotopes in single crystals have been studied using highly nuclear polarized ^{20,21,26,27,28}Na beams provided by ISAC/TRIUMF. The final objectives of the study are to test the *G*-parity symmetry, one of the fundamental symmetries in weak nucleon current, and to measure quadrupole moments of Na isotopes, which have poor statistics and/or have not been measured yet. For such studies, the knowledge of the hyperfine interaction, such as implantation media which can keep the polarization, the degree of the effective polarization, the relaxation time and so on, are very much important. In this paper, such fundamental data are reported. A letter on the production of polarization of Na isotopes at ISOLDE/CERN can be found in [1].

---

* Author for correspondence.
† Present address: NSCL/MSU, 1 Cyclotron, East Lansing, MI 48824, USA.
‡ Present address: RIKEN, 2-1 Hirosawa, Wakao, Saitama, 351-0918, Japan.

Table I. Polarizations of Na isotopes measured at ISAC (H_0 = 5 kOe at RT). The μ are from ref. [3] and Q from ref. [1] except for ^{20}Na

	^{20}Na	^{21}Na	^{26}Na	^{27}Na	^{28}Na
$T_{1/2}$	447.9 (ms)	22.49 (s)	1.072 (s)	301 (ms)	30.5 (ms)
I^{π}	2^+	$3/2^+$	3^+	$5/2^+$	1^+
A	0.33	0.81	−0.94	−0.88	−0.76
$\mu(\mu_N)$	+0.3694(2)	+2.3861(1)	+2.851(2)	+3.895(5)	+2.426(3)
Q (mb)	+90 ± 10	+124 ± 14	−5.3 ± 0.2	−7.2 ± 0.3	+39.5 ± 1.2
	NaF (cubic)				
AP (%)	16.0 ± 0.4	20.8 ± 0.2	−46.6 ± 1.1	−37.1 ± 1.9	−35.9 + 2.5
T_1 (s)	9.9 ± 3.1	9.0 ± 0.2	24.6 ± 4.2	8.1 ± 1.8	–
P_0 (%)	51.0 ± 1.3	33.5 ± 0.5	52.8 ± 1.3	45.3 ± 2.4	47.3 ± 3.3
	TiO_2 (rutile)				
AP (%)	5.3 ± 0.3	13.7 ± 0.3	−44.8 ± 0.5	−42.9 ± 1.0	−34.1 ± 2.5
T_1 (s)	3.4 ± 1.3	13.0 ± 0.5	32 ± 11	13.5 ± 2.6	–
P_0 (%)	18.3 ± 1.5	24.0 ± 0.5	46.8 ± 0.9	50.9 ± 1.2	44.9 ± 3.3
$\|eqQ/h\|$ (MHz)	–	5.20 ± 0.03	–	–	–
η		0.33 ± 0.03			
	$LiNbO_3$ (ilmenite)				
AP (%)	4.3 ± 0.3	5.3 ± 1.6	−40.4 ± 0.6	–	−26.3 ± 3.0
T_1 (s)	1.8 ± 0.5	1.3 ± 0.3	5.3 ± 0.5	–	–
P_0 (%)	17.0 ± 1.7	56 ± 28	51.8 ± 1.5	–	34.7 ± 4
	MgF_2 (rutile)				
AP (%)	3.0 ± 0.6	–	−18.6 ± 0.6	–	–
T_1 (s)	5.9 ± 0.2	–	7.4 ± 2.0	–	–
P_0 (%)	10.0 ± 2.1	–	22.3 ± 1.3	–	–
	ZnO (hcp)				
AP (%)	11.2 ± 0.3	19.7 ± 0.3	–	−33.0 ± 2.0	–
T_1 (s)	9.0 ± 0.5	9.63 ± 0.09	–	9.5 ± 2.5	–
P_0 (%)	37 ± 1	29.7 ± 0.4	–	39.9 ± 2.5	–
$\|eqQ/h\|$ (kHz)	684 ± 13	939 ± 19	–	–	–
	Mg (hcp)				
AP (%)	11.0 ± 0.2	–	–	–	–
T_1 (s)	4.5 ± 0.1	–	–	–	–
P_0 (%)	39.3 ± 0.8	–	–	–	–
$\|eqQ/h\|$ (kHz)	36.7 ± 0.7	–	–	–	–
	Pt (ccp)				
AP (%)	18.3 ± 0.3	–	−14.5 ± 0.4	–	–
T_1 (s)	22.0 ± 1.9	–	0.78 ± 0.08	–	–
P_0 (%)	57.0 ± 1.1	–	55.0 ± 5.9	–	–

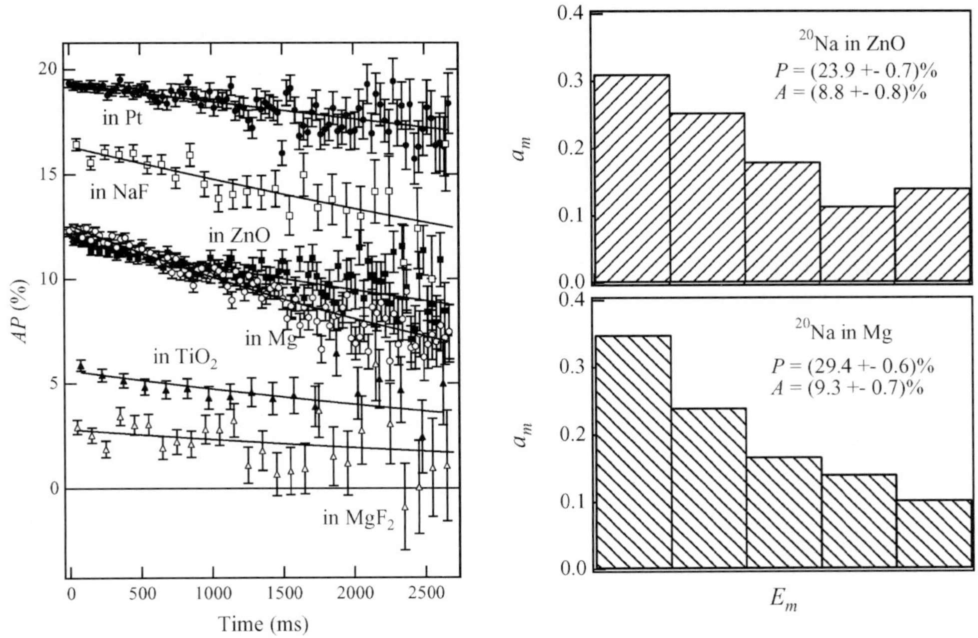

a. Spin-lattice relaxation time. b. a_m vs E_m in ZnO and Mg.

Figure 1. Part of the results of ^{20}Na.

2. Experiment

The experiment was performed at the radioactive beam facility ISAC/TRIUMF. For the production of ^{20,21,27}Na(^{26,28}Na) ions, an intense proton beam was used to bombard a thick SiC(Ta) target coupled to a surface ionization source. Then Na ions were extracted at an energy of 40.8 keV, mass separated and transported to the polarizer beam line, where Na atoms were polarized by pumping on the D_1 transition with circularly polarized laser light. Both ground state hyperfine levels were pumped to achieve high polarization using electro-optic modulators [2]. Finally, the polarized Na beam was stopped on the surface of a single crystal placed in an external magnetic field $H_0 \sim 5$ kOe (parallel to the polarization). The nuclear polarization was extracted by comparing the β-ray asymmetric angular distribution with positive-helicity laser light and that with negative one. The β-rays were detected by a set of thin plastic-scintillation-counter telescopes placed at 0° and 180° relative to the polarization axis. For the measurement of electric quadrupole coupling constants, the β-NMR technique was employed.

3. Results

All the measurements done refer to the ground state of Na isotopes were performed at room temperature (RT). The results are summarized in Table I.

Here A is the β-decay asymmetry parameter integrated over measured decay branches. P_0 is a typical uncorrected asymmetry change, which depends on the spin-lattice relaxation time T_1 and the timing program. P_0 is a corrected polarization for A and T_1. For ^{28}Na, a value corrected for A is shown. Electric quadrupole coupling constants $|eqQ/h|$ were also listed for ^{20,21}Na in TiO_2, ZnO and Mg. ZnO would be suitable for the measurement of eqQ/h because of the appropriate size of the electric field gradient q. Mg has a relatively small and TiO_2 a very large q. Taking the measured quadrupole moment of ^{20}Na [1] as a standard, $Q(^{21}$Na) was extracted. The precision is greatly improved. An asymmetry parameter η of q is extracted from ^{21}Na in TiO_2.

A typical result of T_1 of ^{20}Na is shown in Figure 1(a). The annealed Pt can keep the polarization well with long T_1 even better than NaF, because of the small magnetic moment of ^{20}Na. Generally, in Pt low temperature is required otherwise T_1 is short as in ^{20}Na. ZnO and Mg can keep the polarization with long T_1 (shorter in Mg) whereas the polarization in TiO_2, MgF_2 and $LiNbO_3$ is small and has shorter T_1.

The initial distribution of the population, a_m, of ^{20}Na magnetic sub-levels, m, in ZnO and Mg were determined, although the atomic and nuclear polarization created by laser pumping method were not known. For the extraction of a_m, four different partially depolarized polarizations by rfs in the NMR technique were measured. All 5 a_m ($I = 2$) can be deduced from these four conditions with the normalization $\sum_{m=-2}^{+2} a_m = 1$. Typical results of the deduced populations as a function of m (energy) are shown in Figure 1(b), where P and alignment $\mathcal{A}$ of the distribution are also listed. Different distributions were obtained for ^{20}Na in ZnO and Mg. Both have a relatively large positive alignment. The distribution in ZnO is smaller and slightly parabolic, which might be explained by the energy-level crossing during the implantation process because of the larger eqQ/h than that in Mg. Note that relaxations of the higher orientations than P were not considered in the analysis.

Acknowledgements

The present work was the Research Project at TRIUMF (E871, 971). We are grateful to the staffs of TRIUMF. The study was partly supported by the 21st century COE program ‘Towards a new basic science: depth and synthesis’.

References

1. Keim M. *et al.*, *Eur. Phys. J., A* **8** (2000), 31; ENAM98, *AIP Conference Proceedings* **445** (2000), 50.
2. Levy C. D. P. *et al.*, In: *Proc. 9th Int.* Workshop on Polarized Source and Targets, 2002, p. 334.
3. Huber G. *et al.*, *Phys. Rev., C* **18** (1978), 2342.

Hyperfine Interactions (2004) 159:265–268
DOI 10.1007/s10751-005-9108-1

β-Ray Angular Distribution from Purely Nuclear Spin Aligned ^{20}Na

K. MINAMISONO[1,*,†,‡], K. MATSUTA[2], T. MINAMISONO[2], C. D. P. LEVY[1], T. NAGATOMO[2], M. OGURA[2], T. SUMIKAMA[2,§], J. A. BEHR[1], K. P. JACKSON[1], H. FUJIWARA[2], M. MIHARA[2] and M. FUKUDA[2]
[1]*TRIUMF, 4004 Wesbrook Mall, Vancouver, B.C., V6T 2A3, Canada;*
e-mail: minamiso@nscl.msu.edu
[2]*Department of Physics, Osaka University, 1-1 Toyonaka, Osaka 560-0043, Japan*

Abstract. The alignment correlation term in the β-decay angular distribution from purely nuclear spin aligned ^{20}Na has been measured for the first time. The final objective is to test the G parity symmetry, one of the fundamental symmetry in the weak nucleon current. For artificial creation of the alignment, the knowledge of the hyperfine interaction of ^{20}Na implanted in a single-crystal ZnO was utilized.

Key Words: ^{20}Na, β NMR, alignment correlation term, G parity.

1. Introduction

Hyperfine interactions of dilute impurities in crystal lattice is an important tool for nuclear spin related precision measurements. In the present study, the alignment correlation term of ^{20}Na(2^+, 449.7 ms) in the β-ray angular distribution has been measured in order to test the G-parity symmetry in the weak nucleon current. The knowledge of the hyperfine interaction of ^{20}Na implanted in a single-crystal ZnO (hcp) has been applied to artificially create a nuclear-spin alignment. Toward the detailed study of the G-parity symmetry, the present study is a natural extension of the previous study in the $A = 12$ system [1].

G operation is a product of the charge conjugation and the charge symmetry. The proton β-decay current is transformed to that of the neutron. Many experimental and theoretical works have been performed and it is shown that there is no large G-parity violation in the weak nucleon current. However,

* Author for correspondence.
† Research Abroad, Japan Society for the Promotion of Science.
‡ Present address: NSCL/MSU, 1 Cyclotron, East Lansing, MI 48824, USA.
§ Present address: Riken, 2-1 Hirosawa, Wako, Saitama 351-0198, Japan.

there still remains a possible small violation caused by the mass difference between the proton and neutron, or more fundamentally, between the up and down quarks inside the nucleon. A recent review can be found in Ref. [2].

The β-ray angular distribution from an aligned nucleus is given by $W(E,\theta) \sim 1 + \mathcal{A}\hat{B}_2(E)P_2(\cos\theta)$, where E is the β-ray energy, θ the angle between the direction of the β-ray and the orientation axis, $\mathcal{A}$ the alignment and $\hat{B}_2(E)$ the alignment correlation term given by $\hat{B}_2(E) = -2\hat{H}_2(E,0)/3$ [3], where

$$\hat{H}_2(E,s) = \frac{E}{2M}\left\{1 - \frac{d \pm d_{\mathrm{II}}}{c} \pm \frac{b}{c} + (-1)^s\delta_1(E,\pm) + \delta_2(E)\right\}.$$

Here $\pm$ is for the electron and positron decays, respectively, M the nuclear mass, b the weak magnetism, c the Gamow–Teller matrix element, d the time component in the main axial vector current, d_{II} the G-parity irregular induced tensor term and $\delta_1(E, \pm)$, $\delta_2(E)$ the higher order terms. Taking advantage of the mirror symmetry in the β decays of ^{20}F and ^{20}Na, d_{II} can be extracted from the difference of the alignment correlation term as $\Delta\hat{B}_2(E) = -2E(b - d_{\mathrm{II}} - \Delta\delta_1(E,\pm))/(3Mc)$, where $2\Delta\delta_1(E, \pm) = \delta_1(\mathrm{E}, -) - \delta_1(E, +)$. $\delta_1(E, \pm)$ are not well known experimentally. On the other hand, the angular correlation between the β and subsequent γ ray from the first excited state in ^{20}Ne is also sensitive to the G-parity symmetry. The counterpart of the alignment term is $p(E) = \hat{H}_2(E,1)/2$, where $W(\theta_{\beta\gamma}) = 1 + a(E)\cos\theta_{\beta\gamma} + p(E)\cos^2\theta_{\beta\gamma}$, which provides us a chance to study and cancel the higher order terms.

2. Experiment

The experiment was performed at the radioactive beam facility ISAC/TRIUMF. For the production of ^{20}Na ions, an intense proton beam bombarded a thick SiC target coupled to a surface ionization source. Then ^{20}Na ions were extracted at an energy of 40.8 keV, mass separated and transported to the polarizer beam line, where Na atoms were polarized by pumping on the D_1 transition with circularly polarized laser light. Both ground state hyperfine levels were pumped to achieve high polarization [4]. Finally, the polarized ^{20}Na was stopped at the surface of a single crystal exposed to an external magnetic field $H_0 \sim 5$ kOe (parallel to the polarization). The nuclear polarization was checked by comparing the β-ray asymmetric angular distribution with positive-helicity laser light and that with negative one. Typically (50 ± 1.3)% nuclear polarization with long spin-lattice relaxation time $T_1 = 9.9 \pm 3.1$ s was achieved, where single-crystal NaF was used.

For the measurement of the alignment correlation term, a single crystal ZnO(hcp) was used as an implantation medium. In ZnO, (35.5 ± 0.2)% polarization and $T_1 = 9.0 \pm 0.5$ s were observed and the polarization was converted

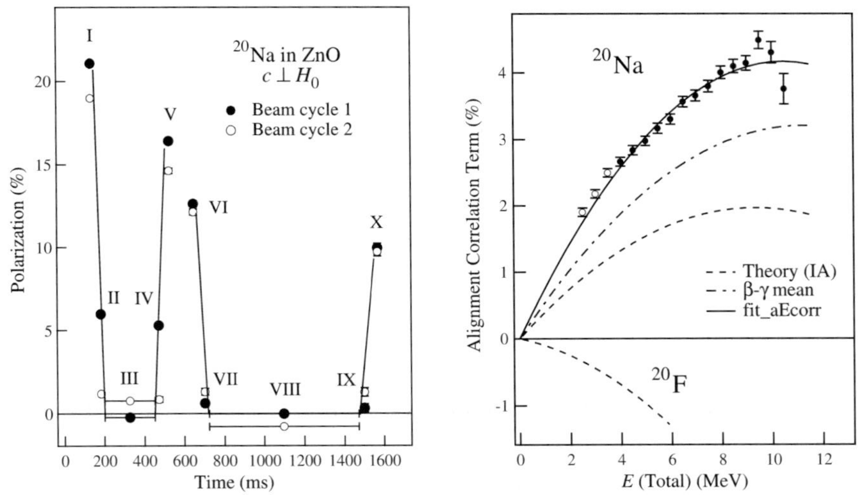

a. Spin manipulation. b. Alignment correlation terms of ^{20}Na.

Figure 1. Results of the experiment.

to an alignment by the spin manipulation technique. Since ZnO has a well defined electric field gradient q, the interaction between the nuclear moments of ^{20}Na and the electromagnetic field changes the energy of each magnetic sublevel as $E_m = -h\nu_L m + h\nu_Q\,(3\cos^2\beta - 1)\,\{3m^2 - I(I+1)\}/12$ so that 4 NMR lines appear. Here, m is the magnetic sub-levels, ν_L the Larmor frequency, $\nu_Q = 3eqQ/\{2I\,(2I-1)h\}$, β the Euler angle of q relative to H_0 and q is assumed to be symmetric. In this condition, a transition between adjacent two magnetic sublevels can be selectively induced by applying an rf field in the NMR technique. A typical result of the spin manipulation is shown in Figure 1a. A pulsed beam method was employed. The polarization in counting section named I was converted into alignment in section III and converted back again to polarization in section V. The alignment was calculated from the polarizations in sections I, II, IV, V and the efficiency of the spin manipulation measured in a separate run. In section VIII, an alignment of the opposite sign was created. The full circles are the beam cycle named 1, in which the alignments were created in the sequence from negative to positive ones. To compensate the relaxation of the polarization and alignment, the alignments were created in a different order in beam cycle 2, which is shown by the open circles.

Finally, β-ray energy spectra from aligned ^{20}Na were measured by a set of plastic-scintillation counter telescopes placed at 0° and 180° relative to the polarization axis and the alignment correlation terms were extracted as $\hat{B}_2(E) \sim (N(\mathcal{A}_1^+, E)/N(\mathcal{A}_2^-, E) \cdot N(\mathcal{A}_2^+, E)/N(\mathcal{A}_1^-, E) - 1)/\hat{\mathcal{A}}$ with $\hat{\mathcal{A}} = \mathcal{A}_1^+ - \mathcal{A}_2^- + \mathcal{A}_2^+ - \mathcal{A}_1^-$. Here, $A_i^{\pm}$ is the degree of positive or negative alignment in beam cycle i and $N(\mathcal{A}_i^{\pm}, E)$ the β-ray counts from the aligned nucleus.

3. Result and discussion

The extracted alignment correlation terms is shown in Figure 1b., where systematic corrections were applied. In the figure, full circles were used for the fit and open circles were not because of a large scattering effect on the surface of the catcher at the lower energy region. The solid line is the result of the fit and the broken lines are the theoretical predictions [5] based on the impulse approximation (IA). Together with the present data, the average of three sets of data of β–γ angular correlations [6] are shown by the dot-dashed line. The original data of β–γ angular correlations were multiplied by $-4/3$ to be compared with the present data. There is a huge enhancement compared with the IA value, which might be explained by the meson exchange effect inside the nucleus as in the case of the $A = 12$ system [7]. We also found a discrepancy between the present result and the result of β–γ angular correlation experiments. Though a detailed analysis of the systematic corrections and errors is now in progress, it shows that the contribution from higher order terms, $\delta_1(E, \pm)$ and $\delta_2(E)$, are important in the extraction of the matrices in the $A = 20$ system. In order to reduce the scattering effect on the surface of the catcher, which is one of the largest source of the systematic corrections and errors, a new experiment with thin single crystal ZnO and Mg has been done, recently. For the extraction of very small d_{II}, the alignment correlation term of the mirror partner, ^{20}F, has to be measured, which will be done soon.

Acknowledgements

The present work was the Research Project at TRIUMF (E871). We are grateful to the staffs of TRIUMF. The study was partly supported by the 21st century COE program "Towards a new basic science: depth and synthesis."

References

1. Minamisono K. *et al.*, *Phys. Rev., C* **65**, 015501.
2. Wilkinson D. H., *Eur. Phys. J., A* **7** (2000), 307.
3. Holstein B. R., *Rev. Mod. Phys.* **46** (1974), 789.
4. Levy C. D. P. *et al.*, In: *Proc. 9th Int. Workshop on Polarized Source and Targets*, 2002, p. 334.
5. Calaprice F. P., Chung W. and Wildenthal B. H., *Phys. Rev., C* **15** (1977), 2178.
6. Dupuis-Rolin N. *et al.*, *Phys. Lett.*, **79B** (1978), 359; Tribble R. E. *et al.*, *Phys. Rev., C* **23** (1981), 2245; Rosa R. D. *et al.*, *Phys. Rev., C* **37** (1988), 2722.
7. Minamisono K. *et al.*, *Phys. Rev., C* **65**, 015209.

Hyperfine Interactions (2004) 159:269–272
DOI 10.1007/s10751-005-9109-0

Precise Nuclear Quadrupole Moments of ^{8}B and ^{13}B

T. NAGATOMO[1,*], T. SUMIKAMA[1,†], M. OGURA[1], K. MATSUTA[1], Y. NAKASHIMA[1], K. AKUTSU[1], T. IWAKOSHI[1], H. FUJIWARA[1], T. MINAMISONO[1], M. FUKUDA[1], M. MIHARA[1], K. MINAMISONO[2], T. MIYAKE[1], S. MOMOTA[3], Y. NOJIRI[3], A. KITAGAWA[4], M. SASAKI[4,‡], M. TORIKOSHI[4], M. KANAZAWA[4], M. SUDA[4], M. HIRAI[4], S. SATO[4], S. Y. ZHU[5], J. Z. ZHU[5], Y. J. XU[5], Y. N. ZHENG[5], J. R. ALONSO[6], G. F. KREBS[6] and T. M. SYMONS[6]

[1]*Department of Physics, Osaka University, Osaka 560-0043, Japan; e-mail: nagatomo@vg.phys.sci.osaka-u.ac.jp*
[2]*TRIUMF, Vancouver Canada V6T 2A3*
[3]*Kochi University of Technology, Kochi 782-8502, Japan*
[4]*National Institute of Radiological Sciences, Chiba 263-0024, Japan*
[5]*China Institute of Atomic Energy, Beijing 102413, P.R. China*
[6]*Lawrence Berkeley Laboratory, Berkeley, CA 94720, USA*

Abstract. The electric quadrupole coupling constants eqQ/h of ^{8}B ($I^\pi = 2^+$, $T_{1/2}$ = 769 ms) and ^{13}B ($I^\pi = 3/2^-$, $T_{1/2}$ = 17.4 ms) in single crystal TiO_2 have been precisely measured by the β-NQR technique. The ratios of these Q moments to $Q(^{12}B)$ were determined as $|Q(^8B)/Q(^{12}B)|$ = 4.882(32) and $|Q(^{13}B)/Q(^{12}B)|$ = 2.768(24).

Key Words: ^{8}B, ^{13}B, β-NQR, eqQ, Q moment, TiO_2.

1. Introduction

The nuclear quadrupole moment Q is one of the suitable probes for the study of nuclear structures, especially for the halo structures of unstable nuclei, because the Q moments directly reflect the nuclear matter-density distributions. Concerning the Q moment of ^{8}B ($I^\pi = 2^+$, $T_{1/2}$ = 769 ms), the abnormally large $Q(^8B)$ is a strong evidence of the proton-halo structure [1]. However, the quadrupole coupling constant eqQ in the Mg crystal was not large enough to eliminate the non-negligible contributions from the double quantum transitions in the previous work. In the ^{13}B case, the eqQ in Mg was determined [2] from the β-NMR spectra with the c-axis parallel and perpendicular to the external magnetic field.

---

* Author for correspondence.
† Present address: RIKEN, Saitana 351-0198, Japan.

However, only one transition was observed among the three in both measurements. In the present work, the ^{8,13}B isotopes were implanted in a TiO_2 single crystal, in which the electric field gradient q for B isotopes was well determined [3], and their quadrupole coupling constants eqQ were measured by the β-NQR method precisely.

2. Experiments

The $eqQ(^8B)$ measurement was performed at the Van de Graaff accelerator facility of Osaka University. The polarized ^{8}B nuclei were produced through the nuclear reaction ^{6}Li(^{3}He, n)^{8}B. The energy of ^{3}He was 4.7 MeV. Recoiling ^{8}B nuclei in the ejection angle 20° ± 10° were selected so as to be polarized, and were implanted into a single crystal of TiO_2. In the ^{13}B case, the experiment was performed at HIMAC/NIRS. The ^{13}B nuclei were produced through the projectile fragmentation process in the 100A-MeV ^{15}N on Be collisions and were separated from other fragments by the secondary beam line installed in HIMAC. For large yield and polarization, the momentum deviated from the center of the momentum distribution and the ejection angle relative to the projectile direction of the fragments were set at +167(28) MeV/c and 1.5(6)°, respectively. Then the ^{13}B were implanted into a TiO_2 crystal. In both cases, the orientation of the c-axis of TiO_2 was set parallel to the external magnetic field H_0. The H_0 was set at 3000 Oe for ^{8}B and 5000 Oe for ^{13}B and was parallel to the direction of polarization. For the main site where the 90% of B isotopes were implanted, the electric field gradient $q = V_{zz}$ is parallel to the c-axis, and is well determined to be $+37.06(75) \times 10^{19}$ V/m^2 [3]. The β-rays emitted from the implanted B isotopes were detected by two sets of plastic-scintillation-counter telescopes placed 0° and 180° relative to the polarization direction. The NQR spectra were obtained by the detection of the β-ray-asymmetry change with the certain set of rf magnetic fields H_1, whose frequencies were defined as a function of eqQ/h (the β-NQR method).

3. Results and discussion

The obtained β-NQR spectra are shown in Figure 1(a) and (b) for ^{8}B and ^{13}B in TiO_2, respectively, where ν_Q is defined as $\nu_Q = \frac{3}{2I(2I-1)} eqQ/h$. From the Gaussian fitting to these spectra, the quadrupole coupling constants were precisely determined as

$$\left| \frac{eqQ}{h} (^8\mathrm{B}) \right| = 578.0 \pm 1.6 \pm 2.2(\mathrm{kHz}), \tag{1}$$

$$\left| \frac{eqQ}{h} (^{13}\mathrm{B}) \right| = 327.7 \pm 0.9 \pm 2.2(\mathrm{kHz}). \tag{2}$$

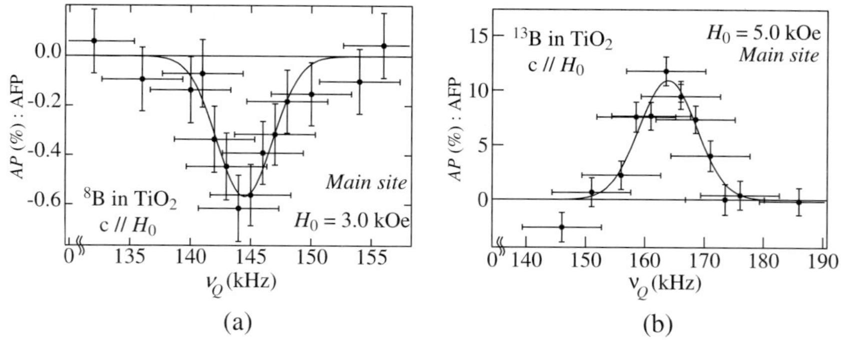

Figure 1. The obtained NQR spectra of ^{8}B (a) and ^{13}B (b).

Table I. Electric quadrupole coupling constants of ^{8,13}B and ^{12}B. The present Q moments were derived using the known $Q(^{12}\text{B})$ from the references [1, 4]

	^{8}B	^{13}B	^{12}B
$\vert eqQ/h\vert$ (kHz) in TiO$_2$	578.0 ± 1.6 ± 2.2	327.7 ± 0.9 ± 2.2	118.4 ± 0.5[3]
$\vert Q/Q(^{12}\text{B})\vert$	4.882 ± 0.032	2.768 ± 0.024	1
Q (mb) Old values	+68.3 ± 2.1[1, 4]	(+)37.4 ± 40[2]	+13.21 ± 0.26[1, 4]
Q (mb) Present	+64.5 ± 1.3	+36.6 ± 0.8	–
Q_{OXBASH} (mb)	+39.2	+34.7	+14.0

The first error is caused by the statistical error and the second error is systematical error, which is 1/10 of the FWHM, caused by the unknown distributions of defects etc. From the ratios to the coupling constant of ^{12}B [3], the ratios of Q moments to that of ^{12}B were determined as $\vert Q(^{8}\text{B})/Q(^{12}\text{B})\vert = 4.882(32)$ and $\vert Q(^{13}\text{B})/Q(^{12}\text{B})\vert = 2.768(24)$. From the known $\vert Q(^{12}\text{B})\vert = 13.21(26)$ mb [1], the Q moments were precisely determined as $\vert Q(^{8}\text{B})\vert = 64.5(13)$ mb and $\vert Q(^{13}\text{B})\vert = 36.6(8)$ mb. Compared with the previous value $Q(^{8}\text{B}) = +68.3(21)$ mb determined by the NQR on B isotopes in Mg [1] and the α-ray anisotropy measurement [4], the present value is more precise and is slightly smaller. The difference may be caused by the contribution from the double quantum transition effects in the previous NQR spectrum. In the case of ^{13}B, the present Q moment is consistent with the known value of $\vert Q(^{13}\text{B})\vert = 37.4(40)$ mb [2] but is about five times more accurate. In addition, the sign of the $Q(^{13}\text{B})$ was determined to be positive taking the polarization mechanism of the fragmentation process into account [5]. These results are summarized in Table I.

The theoretical prediction from the shell model calculation by the OXBASH code [6] are shown in Table I. The effective charges of a proton and a neutron were taken to be $1.3e$ and $0.5e$, respectively. The Q moments of ^{12,13}B nuclei are well reproduced by the shell model calculation, while in the case of ^{8}B, the abnormal enhancement of the Q moment was confirmed, i.e., the experimental Q moment is about 1.5 times as large as the shell model prediction. This result indicates that the ^{8}B has an exotic halo structure and the extended distribution of the last proton can reproduce the enhancement [1]. One of other theoretical calculations is the microscopic cluster model. From the three cluster model [7], the predicted nucleon density distributions of ^{8}Li and ^{8}B are quite similar, not supporting the halo structure of ^{8}B. Although this calculation well reproduce the large $Q(^{8}\mathrm{B})$ ($Q_{\mathrm{theory}}(^{8}\mathrm{B})$ = 73 mb), it cannot reproduce the magnetic moment $\mu_{\mathrm{experiment}}(^{8}\mathrm{B}) = 1.0355(3)\mu_N$ ($\mu_{\mathrm{theory}}(^{8}\mathrm{B}) = 1.45\mu_N$). In conclusion, we obtained the much reliable and precise Q moments of ^{8}B and ^{13}B experimentally and we have to investigate the reliability of the theoretical nuclear models.

Acknowledgements

The present work was performed under the Research Project with Heavy Ions at NIRS-HIMAC (Program # P-026). The authors are grateful to the staff of HIMAC for their technical support. Support was also given by the 21st Century COE Program named "Towards a new basic science: depth and synthesis," by the Grant in Aid for Scientific Research from the Ministry of Education, Culture and Science, Japan, and by the Japan Society for Promotion of Science and the National Natural Science Foundation of China under Grant No. 10175088.

References

1. Minamisono T. *et al.*, *Phys. Rev. Lett.* **69** (1992), 2058.
2. Haskell R. C. *et al.*, *J. Phys. Soc. Jpn., Suppl.* **34** (1973), 167.
3. Ogura M. *et al.*, *Hyperfine Interact.* **136/137** (2001), 195.
4. Yamaguchi T. *et al.*, *Hyperfine Interact.* **120/121** (1999), 689.
5. Nagatomo T. *et al.*, Nuclear Spin Orientation Created in Heavy Ion Collisions and the Sign of the Q Moment of ^{13}B, in Proceedings for the HFI/NQI 2004, Bonn, Germany, R. Vianden, ed., unpublished.
6. Brown B. A. *et al.*, MSUSL Report Number 524, (1988).
7. Baye D. *et al.*, *Nucl. Phys.* **A588** (1995), 147c.

Hyperfine Interactions (2004) 159:273–276
DOI 10.1007/s10751-005-9110-7

Nuclear Spin Orientation Created in Heavy Ion Collisions and the Sign of the Q Moment of ^{13}B

T. NAGATOMO[1,*], K. MATSUTA[1], Y. NAKASHIMA[1], T. SUMIKAMA[1,†], M. OGURA[1], K. AKUTSU[1], T. IWAKOSHI[1], H. FUJIWARA[1], T. MINAMISONO[1], M. FUKUDA[1], M. MIHARA[1], T. MIYAKE[1], K. MINAMISONO[2], S. MOMOTA[3], Y. NOJIRI[3], A. KITAGAWA[4], M. SASAKI[4,‡], M. TORIKOSHI[4], M. KANAZAWA[4], M. SUDA[4], M. HIRAI[4], S. SATO[4], S. Y. ZHU[5], J. Z. ZHU[5], Y. J. XU[5], Y. N. ZHENG[5], J. R. ALONSO[6], G. F. KREBS[6] and T. M. SYMONS[6]

[1]*Department of Physics, Osaka University, Osaka 560-0043, Japan; e-mail: nagatomo@vg.phys.sci.osaka-u.ac.jp*
[2]*TRIUMF, Vancouver, V6T 2A3, Canada*
[3]*Kochi University of Technology, Kochi 782-8502, Japan*
[4]*National Institute of Radiological Sciences, Chiba 263-0024, Japan*
[5]*China Institute of Atomic Energy, Beijing 102413, People's Republic of China*
[6]*Lawrence Berkeley Laboratory, Berkeley, CA 94720, USA*

Abstract. The momentum dependences of the nuclear spin polarization P and alignment A of ^{13}B($I^{\pi} = 3/2^{+}$, $T_{1/2}$ = 17.36 ms) produced in the 100A MeV ^{15}N + Be collisions have been measured by detecting β-ray asymmetry. Because both the P and A were significantly smaller than the prediction from a simple kinematical model, some relaxation mechanisms must be take into account. Comparing the signs of the observed alignment of ^{12}B, the sign of the quadrupole coupling constant eqQ of ^{13}B in TiO_2 was determined to be positive.

Key Words: ^{13}B, β-NMR, alignment, eqQ, heavy ion collision, polarization, TiO_2.

1. Introduction

Studies of the polarization phenomena in heavy ion collisions are important to elucidate details of the reaction mechanism, which leads to useful knowledge to establish the technique for producing polarized unstable nuclear beams for the studies of nuclear moments and hyperfine interactions of dilute impurities in materials. The motion of the fragment produced through the projectile

* Author for correspondence.
† Present address: RIKEN, Saitama 351-0198, Japan.
‡ Present address: Department of Photonics, Ritsumeikan University, Shiga 525-8577, Japan.

fragmentation process in heavy ion collisions in the energy region of more than 100 MeV/nucleon, reflects the motion of nucleons in the projectile just before they are removed by the target. Because the momentum and the angular momentum of the nucleon cluster to be removed should be strongly related with each other kinematically, the angular momentum of the fragment hence the polarization can be determined by selecting the momentum (the kinematical model). The polarization mechanism in the heavy ion collisions in the intermediate energy has been studied well [1–3]. But the experimental study of the alignment [4] is very scarce, partly because the alignment is hard to be determined (e.g., have to detect the γ-ray anisotropy). If we could determine nuclear polarization and alignment at the same time, it would be very much helpful for the study of the reaction mechanism. The β-NMR on ^{13}B($I^{\pi} = 3/2^{-}$, $T_{1/2} = 17.36$ ms) nuclei in a crystal TiO_2 with a well determined electric field gradient makes it possible to determine the nuclear spin orientation including the alignment. By combining this results for ^{13}B with the previous results for ^{12}B($I^{\pi} = 1^{+}$, $T_{1/2} = 20.20$ ms) fragments [5], the sign of the quadrupole coupling constant eqQ, hence the sign of the Q moment of ^{13}B can be determined.

2. Experimental procedure

The present experimental setup and procedure were similar to the previous work [5]. The ^{13}B fragments were produced in the ^{15}N + Be collision at the secondary beam line of HIMAC (Heavy Ion Medical Accelerator in Chiba) at NIRS (National Institute of Radiological Sciences). The incident energy of the projectile was 100 MeV/nucleon and the thickness of the Be target was 2.08 mm. Fragments with a certain momentum were separated by the fragment separator installed in the secondary beam line of HIMAC. The ejection angle θ was set at 1.5(6)° for large yield and polarization. Then the fragments were implanted into a TiO_2 single crystal. The hyperfine interactions of B isotopes in TiO_2 were well determined [6]. From [6], the electric field gradient for the main site, where about 90% of B were implanted, was determined as $q = V_{zz} = +37.1(8) \times 10^{19}$ V/m^2 with the asymmetry $\eta \sim 0$, and $q//c$-axis. Applying the external magnetic field H_0 of 5 kOe parallel to the reaction normal and the c-axis, the NMR frequency is split into three frequencies, e.g., $f_1 = 7907.5$ kHz, $f_2 = 8071.5$ kHz and $f_3 = 8235.5$ kHz. By detecting the β-ray asymmetry changes with applying rf magnetic field H_1, we obtained a set of three partial polarization changes which are defined as $\Delta P_f \equiv P_0 - P_f$, where f is the frequency of H_1, P_0 is the initial polarization and P_f is the polarization after applied H_1. The ΔP_f is proportional to the difference between the populations of the magnetic sub-levels involved in the transition induced by the H_1. From the set of the ΔP_f, we obtained the sub-state population, hence the spin orientation produced in the fragmentation process.

Table I. The partial polarization changes ΔP_{f_i} and the P_0 (90% of the initial polarization) were obtained by β-NMR and β-NQR, respectively. The alignments A were obtained with an assumption of $eqQ > 0$

$p - p_0$ (MeV/c)	ΔP_{f_1} (%)	ΔP_{f_2} (%)	ΔP_{f_3} (%)	P_0 (%)	A (%)
−111 ± 28	1.82 (97)	−1.52 (97)	−1.26 (98)	−0.43 (41)	4.60 (92)
0 ± 28	−1.10 (47)	−1.70 (97)	−4.52 (96)	−5.42 (50)	5.17 (92)
167 ± 28	−1.23 (88)	−4.25 (88)	−4.76 (87)	−11.69 (41)	5.41 (85)

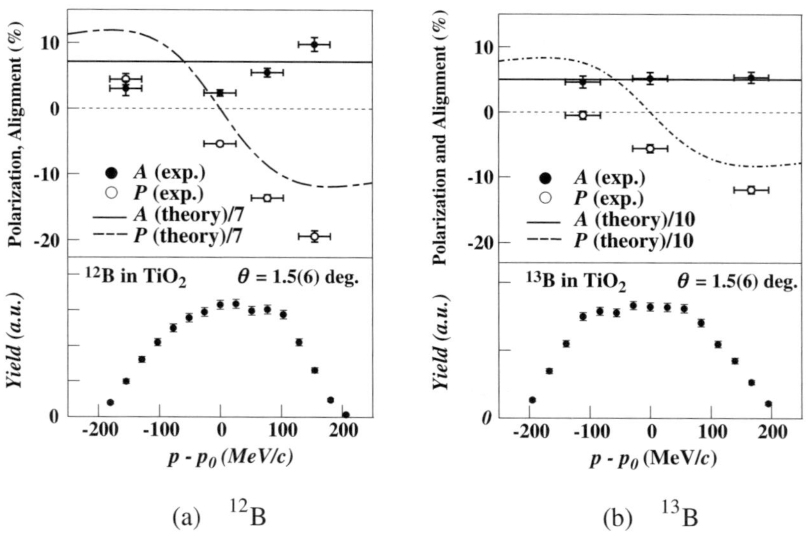

Figure 1. The momentum dependences of polarization and alignment with the momentum distributions.

3. Results and discussion

The partial polarization changes ΔP_{f_i} are summarized in Table I. In Table I, p is the actual momentum of the fragment and p_0 is the momentum corresponding to the primary beam velocity. The polarization P_0, which was obtained by the β-NQR on the ^{13}B nuclei in the main site, is 90% of the initial polarization produced by the collision. The ΔP_{f_i} and P_0 are consistent with each other in terms of the trivial equation described as $P_0 = \Delta P_{f_1} + \frac{4}{3}\Delta P_{f_2} + \Delta P_{f_3}$, which is independent of the eqQ sign. When we assume a positive (negative) sign for $eqQ(^{13}\text{B})$, the polarization changes are described as $\Delta P_{f_1(f_3)} = a_{+3/2} - a_{+1/2}$, $\Delta P_{f_2(f_2)} = a_{+1/2} - a_{-1/2}$ and $\Delta P_{f_3(f_1)} = a_{-1/2} - a_{-3/2}$, where a_m is the population of the magnetic sub-level m. From the definition of the alignment as $A \equiv$

$a_{+3/2} - a_{+1/2} - a_{-1/2} + a_{-3/2}$, the sign of the alignment has one-to-one correspondence with the sign of eqQ. From the previous result of the momentum dependence of the alignment of ^{12}B as shown in Figure 1(a) [5], the alignment had a positive sign in the full range of the momentum distribution. Because the polarization mechanism strongly depends on the orbital angular momentum of the removed nucleon and both ^{12}B and ^{13}B were produced through a similar process, i.e., nucleons in the p shell are removed, the alignment of ^{13}B most possibly has the same trend as that of ^{12}B. Hence, the sign of alignments of ^{13}B should also be positive. From the espondence between the signs of the A and the eqQ, the $eqQ(^{13}\mathrm{B})$ also has a positive sign. From the $q > 0$ for B isotopes in TiO_2 for this site [6], we determined the sign of $Q(^{13}\mathrm{B})$ as positive. The degrees of alignment A are obtained as shown in Table I from the populations which were derived from the fitting to the above equations with the constraints $P_0 = a_{3/2} + a_{1/2}/3 - a_{-1/2}/3 - a_{3/2}$ and $\sum a_m = 1$. The obtained momentum dependences of the initial polarization and alignment are shown in Figure 1(b). Compared with the theoretical prediction by the kinematical model [2, 7] in Figure 1, the degrees of both polarization and alignment are much quenched by factors of about 1/7 for ^{12}B and 1/10 for ^{13}B. The spin–lattice relaxation time in TiO_2 was obtained as $T_1 = 810^{+1900}_{-340}$ ms and it was so long that it cannot cause the reduction of P. The similar quenching factors for both A and P may indicate that the reaction model itself needs to be modified, or the co-existence of the γ de-excitation after the collisions and the admixture of the trajectory of the fragments cancels out the difference in the quenching factor for A and P.

Acknowledgements

The present work was performed under the Research Project with Heavy Ions at NIRS-HIMAC (Program # P-026). The authors are grateful to the staff of HIMAC for their technical support. Support was also given by the 21st Century COE Program named "Towards a new basic science: depth and synthesis," by the Grant in Aid for Scientific Research from the Ministry of Education, Culture and Science, Japan, and by the Japan Society for Promotion of Science and the National Natural Science Foundation of China under Grant No. 10175088.

References

1. Asahi K. *et al.*, *Phys. Lett B* **251** (1990), 488.
2. Okuno H. *et al.*, *Phys. Lett., B* **355** (1994), 29.
3. Matsuta K. *et al.*, *Nucl. Instrum. Methods Res. A* **402** (1998), 229.
4. Asahi K. *et al.*, *Nucl. Phys. A* **488** (1988), 83c.
5. Nagatomo T. *et al.*, *Hyperfine Interact.* **136/137** (2001), 233.
6. Ogura M. *et al.*, *Hyperfine Interact.* **136/137** (2001), 195.
7. Hüfner J. *et al.*, *Phys. Rev. C* **23** (1981), 2538.

Hyperfine Interactions (2004) 159:277–280
DOI 10.1007/s10751-005-9111-6

Magnetic Moment of the $3/2^-$ Ground State of ^{185}W

T. OHTSUBO[1,*], S. OHYA[1] and S. MUTO[2]
[1] *Department of Physics, Niigata University, Niigata 950-2181, Japan; e-mail: tohtsubo@np.gs.niigata-u.ac.jp*
[2] *Neutron Science Laboratory, KEK, Tsukuba 305-0801, Japan*

Abstract. Nuclear magnetic resonance measurement have been performed for ^{185}W oriented at 8 mK in an Fe host. The magnetic hyperfine splitting frequency at an external magnetic field of 0.1 T was determined to be 196.6(2) MHz. With the known hyperfine field of $B_{hf} = -71.4(18)$ T, the nuclear magnetic moment of ^{185}W is deduced as $\mu(^{185}W) = +0.543(14)\ \mu_N$.

Key Words: deformed nuclei, NMR-ON, nuclear magnetic moment.

1. Introduction

Nuclear magnetic resonance on oriented nuclei (NMR-ON) is a powerful method for precise measurements of the nuclear magnetic moments. Magnetic moments are good probes for studying the electromagnetic nuclear structure. Previously we have measured the magnetic moment of ^{187}W by the NMR-ON method. The observed value is $|\mu\ (^{187}W)| = 0.621(15)\ \mu_N$ [1]. It is almost consistent with the theoretical value [2] for the $3/2^-$ [512] state of the Nilsson model. The ground state of ^{185}W is also $3/2^-$ [512]. Systematic studies of electromagnetic moments are useful to test the nuclear models. Figure 1 shows the decay scheme of ^{185}W. Without γ ray emission, 99.9% of ^{185}W decays to the ground state of ^{185}Re. Therefore we observed the NMR-ON with β-ray asymmetry change.

2. Experiment

The sample of ^{185}WFe was prepared by the thermal neutron irradiation method. Thin alloy foils of WFe (0.15 at.% of natural tungsten) were irradiated in a reactor at the Japan Atomic Energy Research Institute. After irradiation, the sample was annealed in vacuum for 1 h at 800°C. It was soldered to the copper cold finger in a $^3He/^4He$ dilution refrigerator and was cooled down to about 8 mK. An external magnetic filed (B_0) was applied to polarize Fe foil. The β rays were detected by two Si detectors of 50 mm^2 area and 0.5 mm thickness. The Si

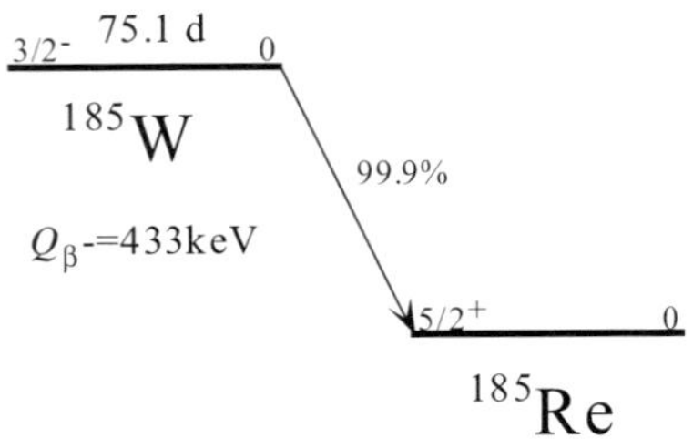

Figure 1. Decay scheme of ^{185}W.

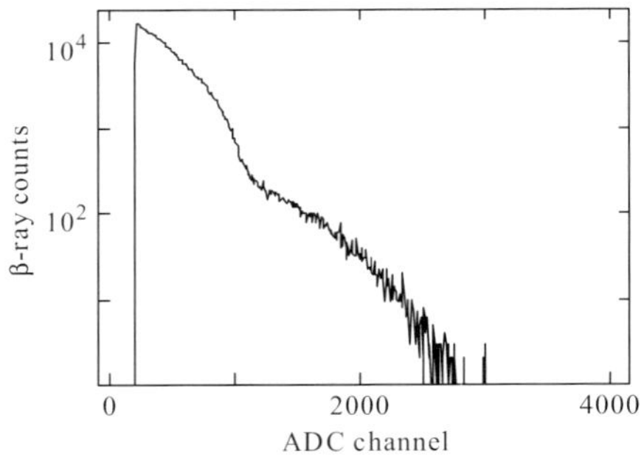

Figure 2. β-ray energy spectrum.

detectors were mounted on a heat shield of 0.7 K inside the refrigerator at 0° and 180° with respect to B_0. The γ rays were detected by pure Ge detectors placed outside at 0° and 180°. The sample temperature was monitored by the anisotropy of γ rays from a ^{60}CoCo thermometer.

The NMR-ON spectrum was observed by detecting the β-ray asymmetry change with the radio frequency (rf) oscillating field. The rf field was applied to the foil plane and perpendicular to B_0. The rf was modulated at a rate of 100 Hz. The width of the frequency modulation was ±1.5 MHz.

The maximum energy of β rays of interests is 433 keV. A typical β-ray energy spectrum is shown in Figure 2. The β-ray spectrum contains contributions from other activities produced simultaneously by the neutron irradiation, mainly ^{59}Fe, for which the maximum energy of β rays is 466 keV. Since the resonance frequency of ^{59}FeFe is known [2], we can distinguish the resonance signal of ^{185}W from that of ^{59}Fe.

3. Result and discussion

Figure 3 shows typical NMR-ON spectra of ^{185}WFe obtained at external fields of 0.1, 0.2 and 0.4 T. A vertical axis is the β-ray asymmetry in arbitrary unit. The solid curves in the figure are the results of least-squares fits assuming a resonance with a Gaussian shape and linear background. Table I shows a summary of experimental results.

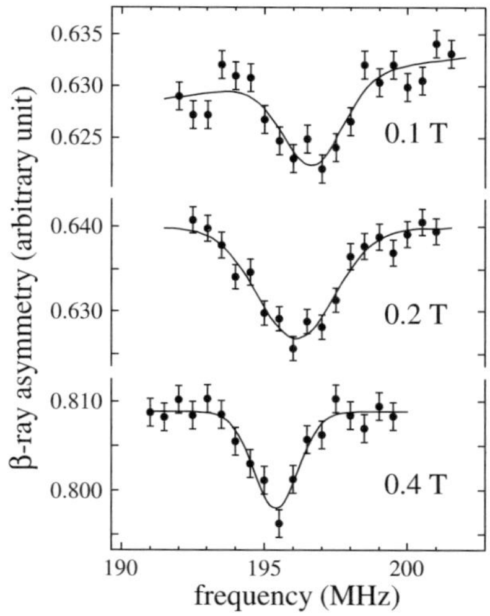

Figure 3. NMR-ON spectra for ^{185}WFe.

Table I. Experimental results of ^{185}WFe

B_0 (T)	Resonance frequency (MHz)
0.1	196.6(2)
0.2	196.1(2)
0.4	195.4(2)

Table II. Comparison of experimental and theoretical magnetic moments

Isotope	I^π	μ_{exp} (μ_N)	μ_{th} (μ_N)	
			g_s^{free}	$0.6g_s^{free}$
^{185}W	3/2$^-$	+0.543(14)	+0.86	+0.57
^{187}W	3/2$^-$	(+)0.621(15)	+0.83	+0.55

For a pure magnetic interaction, the resonance condition is given by

$$h\nu = |g\{B_{hf} + (1+K)B_0\}|\mu_N.$$

B_{hf} of WFe system has been measured by NMR-ON method for ^{187}W previously as $B_{hf}(^{187}\text{WFe}) = -71.4(18)$ T [1]. We neglected possible hyperfine anomalies between ^{185}W and ^{187}W, because they have the same $I^\pi = 3/2^-$. Neglecting the Knight shift factor, the magnetic moment of ^{185}W was deduced as:

$$\mu(^{185}W) = +0.543(14)\mu_N.$$

The sign of the magnetic moment can be determined by the β-ray asymmetry change at the resonance frequency.

Ekstroem *et al.* calculated the magnetic moments of isotopes in the region Hf–Ir using the modified oscillator model [3]. They calculated the magnetic moments of the $3/2^-[512]$ states of ^{185}W and ^{187}W with the effective g_s values of g_s^{free} and $0.6g_s^{free}$, respectively. Table II shows the experimental and theoretical magnetic moments. These values for ^{185}W are consistent with the shell-model calculation on deformed nuclei by Lamm [4].

The present result is consistent with the theoretical value with an effective g_s value of 0.6 g_s^{free}. The experimental value of the magnetic moment increases from ^{185}W to ^{187}W, however the theoretical prediction is opposite. Such a tendency can be seen in ^{189}Os and ^{193}Os [5]. These results show that the effective *g*-factor increases with increasing the neutron number to the magic number of 126 in this region. Recently Stuchbery discussed the magnetic properties of rotational states in the pseudo-Nilsson model [6]. The full pseudo-Nilsson model gives a better description of the *g*-factor of ^{189}Os than the pseudo-SU(3) model.

Acknowledgement

This work was supported by the Inter-University Program for the Joint Use of JAERI Facilities.

References

1. Ohya S., Nishimura K. and Mutsuro N., *Hyperfine Interact.* **36** (1987), 219.
2. Ohtsubo T. *et al.*, *Phys. Rev.* **C54** (1996), 554.
3. Ekstroem C. and Rubinsztein H., *Phys. Scr.* **14** (1976), 199.
4. Lamm I.-L., *Nucl. Phys.* **A125** (1969), 504.
5. Berkes I. *et al.*, *J. Phys.* **G11** (1985), 287.
6. Stuchbery A. E., *Nucl. Phys.* **A700** (2002), 83.

Hyperfine Interactions (2004) 159:281–284
DOI 10.1007/s10751-005-9112-5

Nuclear Spin Alignments and Alignment Correlation Terms in Mass $A = 8$ System

T. SUMIKAMA[1,*,†], T. IWAKOSHI[1], T. NAGATOMO[1], M. OGURA[1], Y. NAKASHIMA[1], H. FUJIWARA[1], K. MATSUTA[1], T. MINAMISONO[1], M. MIHARA[1], M. FUKUDA[1], K. MINAMISONO[2] and T. YAMAGUCHI[3]
[1]*Department of Physics, Osaka University, Toyonaka, Osaka 560-0043, Japan; e-mail: sumikama@riken.jp*
[2]*TRIUMF, 4004 Wesbrook Mall, Vancouver, British Columbia, V6T 2A3, Canada*
[3]*Department of Physics, Saitama University, Saitama 338-8570, Japan*

Abstract. The pure nuclear spin alignments of ^{8}Li and ^{8}B were produced from the nuclear spin polarization applying the β-NMR method. The alignment correlation terms in the β-ray angular distribution were observed to test the G parity conservation in the nuclear β decay.

Key Words: β-NMR, G parity, induced tensor term, nuclear spin orientation, second class current.

1. Introduction

We can test the fundamental symmetry in nuclear β decay through the small induced term by use of the knowledge of the hyperfine interactions. The induced tensor term $g_{\rm II}$ in the weak nucleon current, which induces nuclear β decay, is a strange term, since it violates the G symmetry, which is, in a simple word, the symmetry between proton and neutron. So, the induced tensor term might be finite through the mass difference between up and down quarks. Deep knowledge on hyperfine interactions of ^{8}Li and ^{8}B in crystals provides us with the powerful technique to approach this important but difficult problem. In the present study, the induced tensor term has been studied through the alignment correlation term in the β-ray angular distribution from spin aligned ^{8}Li and ^{8}B. The polarization could be converted to the pure alignment by applying the β-NMR technique based on the well-determined hyperfine interactions.

* Author for correspondence.
† Present address: RIKEN, 2-1 Hirosawa, Wako, Saitama 351-0198, Japan.

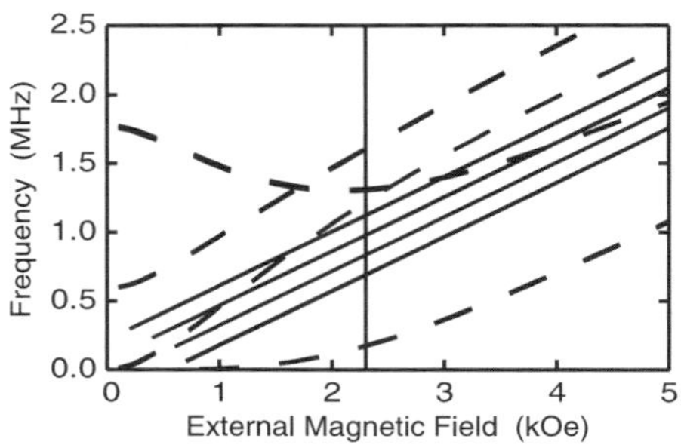

Figure 1. External magnetic field dependence of the resonance frequencies of ^{8}B in TiO_2. The *solid and dashed lines* are the frequencies of site 1 and site 2, respectively.

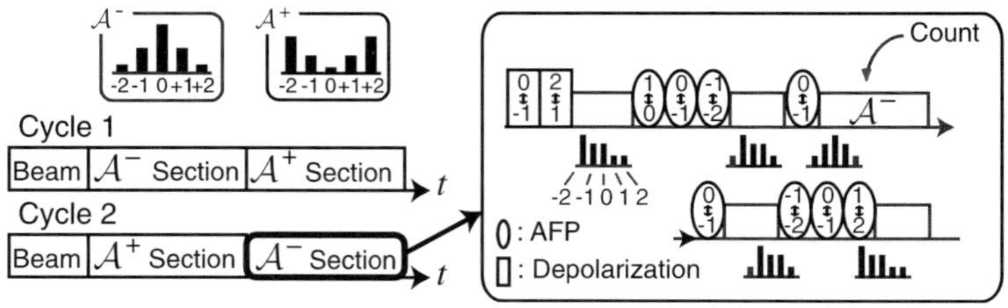

Figure 2. Timing program of alignment production. The procedure of the β-NMR techniques is shown for a negative alignment production.

2. β-ray angular distribution

The β-ray angular distribution from purely aligned nuclei is given by $dW \propto pE(E-E_0)B_0(E)\,[1+\frac{B_2(E)}{B_0(E)}\mathcal{A}(3\cos^2\theta-1)/2]\,dEd\Omega$ where $\mathcal{A}$ is the alignment, E and E_0 are the β-ray energy and end-point energy, respectively, p is the β-ray momentum and θ is the β-ray ejection angle. The difference δ between the alignment correlation term B_2/B_0 of ^{8}Li and ^{8}B is formulated by Holstein [1] as

$$\delta = -\frac{2E}{3M_n}\left[\frac{b}{Ac} - \frac{d_{\mathrm{II}}}{Ac} + \frac{3}{\sqrt{14}}\frac{f}{Ac} + \sqrt{\frac{3}{28}}\frac{g}{A^2c}\frac{E_0-E}{M_n}\right], \tag{1}$$

where b is the weak magnetism, c is the Gamow–Teller term, $d_{\mathrm{II}}/Ac = g_{\mathrm{II}}/g_{\mathrm{A}}$ is the ratio of the induced tensor term to the axial–vector coupling constant, f and g is the second forbidden matrix elements of the vector current, M_{n} is the nucleon mass and A is the mass number. The g_{II} term is given by combining the present alignment correlation term with the weak magnetism [2] and the β–α correlation term [3] where the f and g terms contribute in opposite directions to Equation (1).

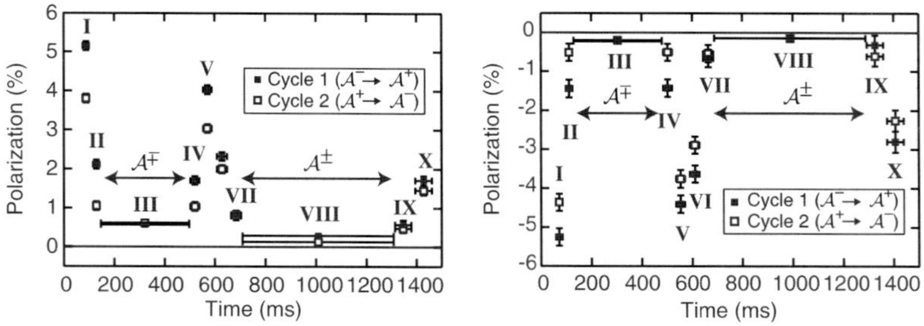

(a) ^{8}Li run. (b) ^{8}B run.

Figure 3. Polarization change of the alignment production cycle.

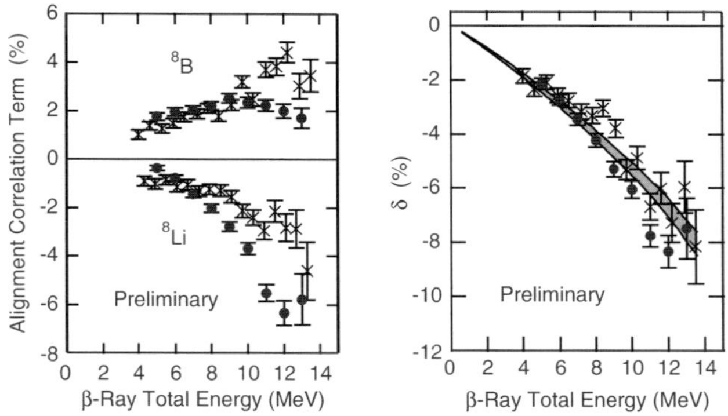

(a) Alignment correlation terms. (b) The difference of the alignment correlation terms.

Figure 4. The *circles* and *crosses* are the present data and the β–α angular correlation term [3]. The shadow shows the weak magnetism with $\pm 1\sigma$ bands [2].

3. Experiment

The experimental procedure and setup are essentially the same as those used in the previous works [4]. The Van de Graaff accelerator at Osaka University provided a deuteron (^{3}He) beam at 3.5 MeV (4.7 MeV) energy. The ^{8}Li (^{8}B) nuclei were produced through the reaction ^{7}Li(d,p)^{8}Li, (^{6}Li(^{3}He,n)^{8}B), and the nuclear polarization was produced by selecting the recoil angle of the reaction product. The ^{8}Li or ^{8}B nuclei were implanted into single crystals of Zn and TiO_2, respectively, placed in an external magnetic field H_0 to maintain the polarization. Each crystal c axis was set parallel to H_0. The Larmor frequency splits into four resonance frequencies due to the quadrupole coupling eqQ between the electric quadrupole moment Q of the implanted nuclei and the electric field gradient q in the crystal. The hyperfine interaction must be well known for all implantation sites, so that the obtained initial polarization can be converted into positive and negative alignments applying the β-NMR technique for each resonance frequency. In the case of ^{8}B, two implantation sites were known from the β-NQR spectrum of ^{12}B implanted into TiO_2 [5]. The relative populations are 90% for site 1 and 10% for site 2. The measurement of the alignment correlation term requires that only the spins of ^{8}B in site 1 have to be controlled and the spins of ^{8}B in site 2 should not been affected. The four resonance frequencies of each site must be well separated not to affect each other at a certain value of H_0. The electric field gradient tensor in the site 2 was determined very precisely from ^{12}B in TiO_2 [6]. Based on this knowledge, the frequencies of ^{8}B were plotted as a function of H_0 in Figure 1 to look for the best condition. In the figure, errors are shown by the thickness of the lines. The relatively large error of $Q(^{8}\mathrm{B})$ is not included, because Q is common for site 1 and site 2. For the best line separation, $H_0 = 2.3$ kOe was chosen not to overlap the applying frequency ranges including FM (Frequency Modulation) as shown in Figure 1. We could control the nuclear spin of ^{8}B only in site 1 in TiO_2 on this condition for the first time. Figure 2 shows the procedure of the alignment production from the initial polarization using two different NMR techniques. The one is the Adiabatic First Passage (AFP) method which interchanges two populations between neighboring magnetic substates. The other is the depolarization method which equalizes two populations. Alignments of both signs are produced in turn, in the same beam cycle to reject a systematic error caused by the beam fluctuation. The pure alignments were produced in count section III and VIII as shown in Figure 3.

The alignment correlation terms and the difference δ were extracted from the β-ray energy spectra of the purely aligned ^{8}Li and ^{8}B as shown in Figure 4. The $\beta{-}\alpha$ correlation term [3] are also plotted in the same figure, after multiplied by $-2/3$ to compare with the alignment correlation terms. The average of two kinds of the correlation terms was well reproduced only by the weak magnetism [2]. The upper limit of the G-parity violating induced tensor term was preliminarily extracted to be $|d_{\mathrm{II}}/\overline{b}| < 0.06$, where $\overline{b}$ is the energy average of the weak magnetism.

Acknowledgement

Thanks for the financial assistance from the Special Postdoctoral Researcher Program of RIKEN.

References

1. Holstein B. R., *Rev. Mod. Phys.* **46** (1974), 789.
2. De Braeckeleer L., Adelberger E. G. *et al.*, *Phys. Rev. C* **51** (1995), 2778.
3. McKeown R. D., Garvey G. T. *et al.*, *Phys. Rev. C* **22** (1980), 738.
4. Minamisono K., Matsuta K. *et al.*, *Phys. Rev. C* **65** (2001), 015501.
5. Ogura M., Minamisono K., *et al.*, *Hyperfine Interact.* **136/137** (2001), 195.
6. Sumikama T., Ogura M., Nakashima Y., Iwakoshi T., Mihara M., Fukuda M., Matsuta K., Minamisono T., Akai H., Electric Field Gradients of B in TiO_2, in Proceedings for the HFI/NQI 2004, Bonn, Germany, R. Vianden, ed., unpublished.

Hyperfine Interactions (2004) 159:285–291
DOI 10.1007/s10751-005-9116-1

The Binding of Iron to Perineuronal Nets: A Combined Nuclear Microscopy and Mössbauer Study

M. MORAWSKI[1], T. REINERT[2], G. BRÜCKNER[1], F. E. WAGNER[3], Th. ARENDT[1] and W. TRÖGER[2,*]

[1]*Paul Flechsig Institute für Hirnforschung, Universität Leipzig, Leipzig, Deutschland*
[2]*Fakultät für Physik und Geowissenschaften, Universität Leipzig, Linnestraße 5, 04103, Leipzig, Deutschland; e-mail: troeger@physik.uni-leipzig.de*
[3]*Physik-Department E15, Technische Universität München, 85747 Garching, Deutschland*

Abstract. A specialized form of extracellular matrix (ECM) surrounds subpopulations of neurons termed 'perineuronal nets' (PNs). These PNs form highly anionic charged structures in the direct microenvironment of neurons, assumed to be involved in local ion homeostasis since they are able to scavenge and bind redox-active iron ions. The quantity and distribution of iron-charged PNs of the extracellular matrix in the rat brain areas of the cortex and the red nucleus was investigated using the powerful combination of Particle-Induced X-ray Emission (PIXE) and Mössbauer spectroscopy. These studies reveal that the iron is bound to the PNs as Fe(III). PNs in both brain regions accumulate up to three to five times more Fe^{3+} than any other tissue structure in dependency on the applied Fe concentration with local amount maximums of 480 mmol/l Fe at PNs.

Key Words: brain research, iron, life sciences, Mössbauer effect, particle induced X-ray emission, perineuronal nets.

1. Introduction

Perineuronal nets (PNs) are a lattice-like amassment of extracellular matrix (ECM) components, originally described by Golgi [20] as a reticular structure covering the cell bodies and proximal dendrites of certain neurons. They are formed on different types of neurons in certain regions of the brain of many vertebrate species including man [3, 8]. PNs are molecularly heterogeneous, consisting primarily of chondroitin-sulphate proteoglycans (CSPG) of the aggrecan family complexed with hyaluronan [1, 8, 12].

The glycosaminoglycan chains (GAG) of PNs provide a highly negatively charged structure in the direct microenvironment of neurons that might be involved in local ion homeostasis since it can act as 'spatial buffer' for physiologically relevant ions of calcium, potassium and sodium around highly active types of neurons [8]. The GAG might also interact with ions involved in the generation of oxidative stress such as Fe^{3+}. PNs might be able to reduce the local oxidative potential in the neuronal microenvironment through scavenging and binding of redox-active Fe^{3+} and therefore providing some neuroprotection to those neurons which are ensheathed by PNs.

Iron is the most abundant transition metal in the brain, and in biology in general. It is found in active centers of many enzymes. It is a principal component of cytochromes and the iron-sulphur complexes of the oxidative chain and therefore important for producing adenosine triphosphate (ATP). Also the synthesis of neuronal transmitters is iron-dependent. Iron is a cofactor for the generation of dopamine, which decreases in Parkinson's disease (PD). The capability of iron to catalyse the generation of free radicals by the Fenton reaction in biological systems is well investigated (for a review see [4]). Generally, iron is almost always complexed with proteins, but usually about 2% of iron ions can also be present in a labile iron pool (LIP) of chelatable and redox-active iron [6]. The LIP comprises both ionic forms of iron (Fe(II) and Fe(III)) associated with different ligands such as organic anions, polypeptides and membrane components. Furthermore, it reacts with hydrogen peroxide and superoxide to form highly reactive hydroxyl radicals causing lipid peroxidation, DNA strand breaks and degradation of biomolecules [5] and is supposed to play a major role in Alzheimer's disease (AD) [11]. Post mortem studies in AD brain tissue display a disturbance in the iron distribution and a pathological increase in neuritic plaques [7].

In a previous study, we could show that in AD brains cortical areas highly enriched in PNs are less frequently affected by neurofibrillary degeneration while PNs are much less abundant in vulnerable areas [2]. Moreover, we could extend the findings on the presumptive neuroprotective capacity of ECM components by demonstrating that neurons associated with PNs are less frequently affected by lipofuscin accumulation than neurons devoid of PNs [8]. In the present study, we investigate these potential neuroprotective effects of PNs for iron ions by studying the iron binding capacity using PIXE and the chemical state of the bound iron using Mössbauer spectroscopy.

2. Experimental

2.1. ION BEAM ANALYSIS

The Ion beam analysis was carried out at the Leipzig microprobe laboratory LIPSION [14] using a 2.25 MeV proton beam focused to approximately 1 μm spot size at beam currents of about 100 pA. Particle Induced X-ray Emission (PIXE) was used for simultaneous multi-elemental analysis of the elements above sodium in the periodic table. The elemental distributions, especially of phosphorus (cell indicator) and iron were analysed. Simultaneously, Rutherford Backscattering Spectroscopy (RBS) was used to obtain information on the matrix composition, on the sample thickness and on the accumulated beam charge by fitting the RBS-spectrum (yielded by the total scanned area) using the RUMP code [13].

Since hydrogen is always present in organic material, a ratio between carbon and hydrogen of 0.5 (C_5H_{10}) was assumed for the brain tissue. The information on matrix composition and accumulated charge was used as input for the quantitative analysis by the PIXE data analysis program GeoPIXE II. This software package allows quantitative elemental imaging and various image processing tools for μPIXE measurements [10].

2.2. MÖSSBAUER SPECTROSCOPY

^{57}Fe Mössbauer spectra were measured at room temperature (RT), 160 K and 4.2 K for a cortex and a red nucleus specimen. The absorbers had a thickness of 123 mg/cm^2 for the cortex and of 142 mg/cm^2 for the red nucleus sample. The source of ^{57}Co in rhodium was always kept at the temperature of the absorber. All isomer shifts will be given as measured, i.e., with respect to the source having the same temperature as the absorber. The spectra were fitted with appropriate superpositions of Lorentzian lines allowing for a Gaussian distribution of the magnetic hyperfine field or the electric quadrupole splitting, respectively.

3. Results

In the present study, we analysed the iron binding capacity of perineuronal nets as well as the chemical state of the bound iron in rat brain.

3.1. ION BEAM ANALYSIS

The PN ensheathed neurons are explicitly different from the neurons devoid of PNs, which are clearly identifiable by the superposition of the cellular P– and Fe– distribution (Figure 1). In general the phosphorus map demonstrates the localisation of the cell somata of all cells by highlighting the cytosol.

Elemental profiles and average concentrations in selected brain regions of interest were extracted. In order to determine the region of the PN, a threshold iron concentration of 40% of the maximum value was chosen. The region defining the neural tissue devoid of cell bodies and PNs, the so-called neuropil, was chosen outside the PNs and outside the somata of PN-devoid neurons

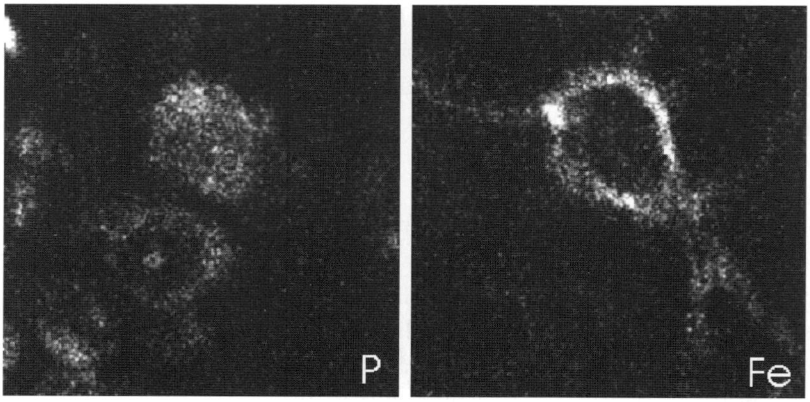

Figure 1. PIXE phosphorus (*left*) and iron (*right*) distribution map. The neurons can be localised by their P-rich somata. One neuron possesses a perineuronal net that clearly shows iron accumulation. $50 \mu n \times 50~\mu m$.

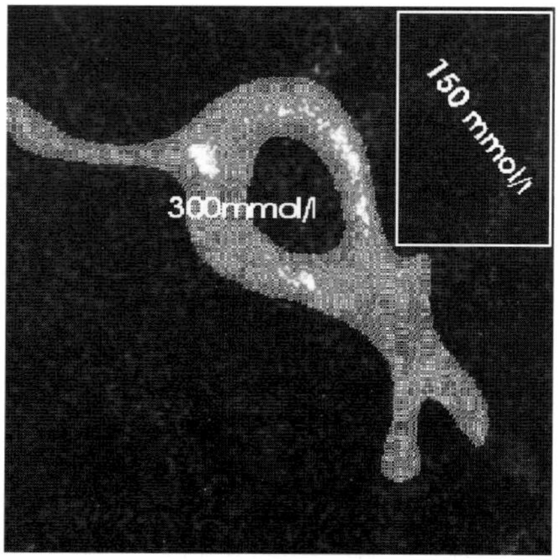

Figure 2. Distribution map of iron. The regions are marked from which the quantitative concentrations were extracted.

represented by higher phosphorus concentrations (Figure 2). PNs accumulate three to five times more iron than any other tissue structure in the investigated brain areas depending on the applied Fe concentration (Figure 3). For a detailed description see [9].

3.2. MÖSSBAUER SPECTROSCOPY

The Mössbauer spectra of both the cortex and the red nucleus specimen exhibit six-line magnetic hyperfine patterns at 4.2 K and quadrupole doublets at 160 K and RT. Spectra of the cortex material taken at 4.2 K and 160 K are shown in Figure 4. The magnetic hyperfine patterns were fitted allowing for a Gaussian distribution of the magnetic hyperfine fields. The median field values were found

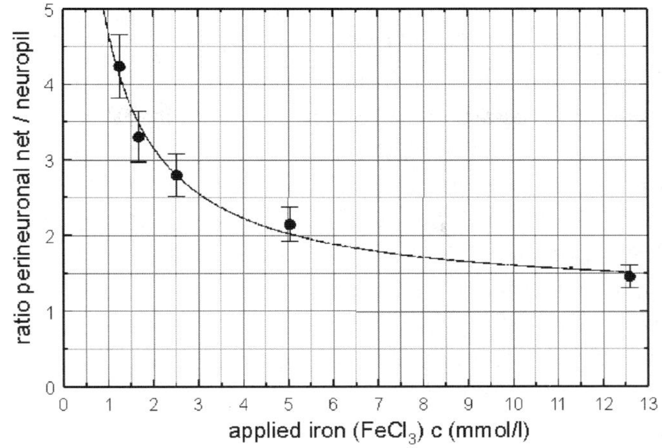

Figure 3. The relation between the initial $FeCl_3$ loading and the ratio of concentrations perineuronal net/neuropil. It can be described by a hyperbolic function with an offset of 1.

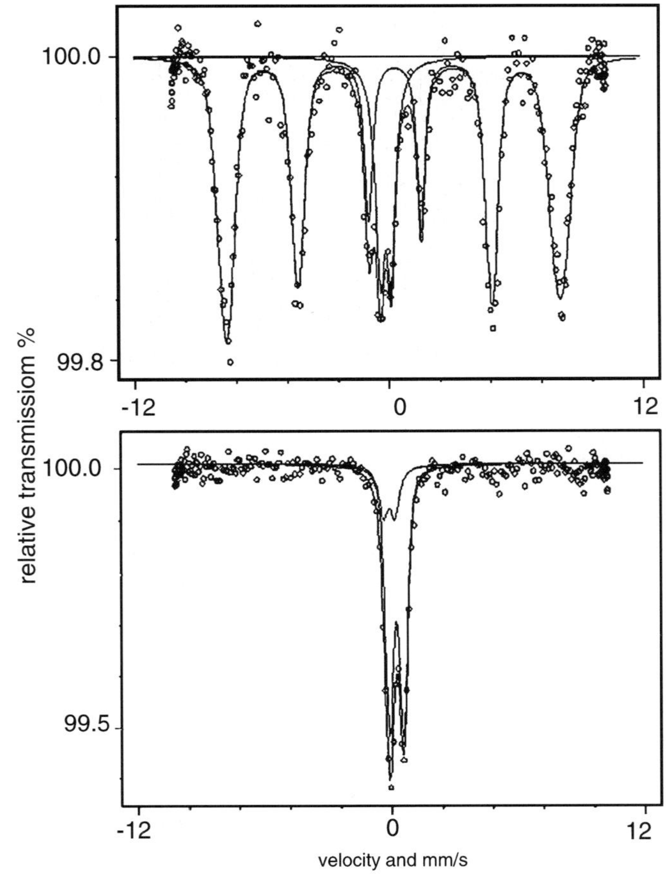

Figure 4. Mössbauer spectra of the cortex specimen taken at 4.2 K (*top*) and at 160 K (*bottom*).

to be 48.9(1) and 49.0(1) T for the cortex and the red nucleus, respectively, with variances of 2.2(1) and 1.9(1) T. The isomer shifts of 0.24(1) mm/s of the magnetically split patterns are typical for trivalent iron. The quadrupole splittings in the magnetic patterns are close to zero.

The quadrupole doublet visible in the 4.2 K spectra (Figure 4) has a shift of −0.14(2) mm/s and a splitting of 0.55(2) mm/s. It is attributable to iron impurities in the beryllium window of the proportional counter and thus has no relation to the properties of the iron in the brain samples. When this component is properly taken into account in the fitting of the spectra taken at 160 K and RT, the quadrupole doublets of the iron in the brain specimens all have median isomer shifts of 0.23(2) mm/s, while the quadrupole splittings have median values of 0.68(2) mm/s and variances of near 0.3 mm/s. The Mössbauer spectra show no trace of divalent iron, which would be expected to give rise to a quadrupole doublet with a shift of about 1 mm/s and a splitting of more than 2 mm/s.

4. Discussion

The locally resolved, quantitative nuclear microscopy was successfully used to investigate the accumulation of iron in PNs in rat brain. It could be shown that PNs have a higher ability to bind a large amount of iron concentrated close to the neuron than any other tissue components in the rat brain.

The Mössbauer spectra of both the cortex and the red nucleus specimen show that the iron is bound to the extracellular matrix as an Fe(III) oxihydroxide. The magnetic hyperfine splitting observed at 4.2 K reveals magnetic ordering at low temperatures. At 160 K the iron exhibits only a quadrupole doublet and hence is already paramagnetic or superparamagnetic. This behaviour is similar to that observed, for instance, for iron in ferritin [15–17] or ferrihydrite [18, 19]. In ferritin the iron forms oxidic Fe(III) clusters of about 8 nm size which begin to exhibit superparamagnetic relaxation effects at about 10 K and exhibit completely collapsed spectra above about 50 K [18, 19]. More details on the state of the iron in the brain specimens might therefore be obtained by a more detailed study of the temperature dependence of the Mössbauer spectra between 4.2 K and about 100 K.

Acknowledgements

The authors wish to thank Mrs Ute Bauer and Mrs Hildegard Gruschka for their excellent technical assistance. This study was supported by the Deutsche Forschungsgemeinschaft (grant AR 200/6-1), by the Deutsches Bundesministerium für Bildung, Forschung und Technologie (BMBF NBL3/01ZZ 0106), the

Interdisziplinäres Zentrum für Klinische Forschung (IZKF) Leipzig at the Faculty of Medicine of the University of Leipzig (C1).

References

1. Brückner G., Bringmann A., Härtig W., Köppe G., Delpech B. and Brauer K., *Exp. Brain Res.* **121** (1998), 300–310.
2. Brückner G., Hausen D., Härtig W., Drlicek M., Arendt T. and Brauer K., *Neuroscience* **92** (1999), 791–805.
3. Celio M. R. and Blümcke I., *Brain Res. Brain Res. Rev.* **19** (1994), 128–145.
4. Connor J. R., Menzies S. L., Burdo J. R. and Boyer P. J., *Pediatr. Neurol.* **25** (2001), 118–129.
5. Halliwell B., *Drugs Aging* **18** (2001), 685–716.
6. Kakhlon O. and Cabantchik Z. I., *Free Radic. Biol. Med.* **33** (2002), 1037–1046.
7. Lovell M. A., Robertson J. D., Teesdale W. J., Campbell J. L. and Markesbery W. R., *J. Neurol. Sci.* **158** (1998), 47–52.
8. Morawski M., Brückner M. K., Riederer P., Brückner G. and Arendt T., *Exp. Neurol.* **188** (2004), 309–315.
9. Reinert T., Morawski M., Arendt T. and Butz T., *Nucl. Instr. Meth. B* **210** (2003), 395–400.
10. Ryan C. G., *Nucl. Instr. Meth. B* **181** (2001), 170–179.
11. Smith M. A., Nunomura A., Zhu X., Takeda A. and Perry G., *Antioxid. Redox. Signal.* **2** (2000), 413–420.
12. Yamaguchi Y., *Cell. Mol. Life Sci.* **57** (2000), 276–289.
13. Doolittle L. R., *Nucl. Instr. Meth. B* **9** (1985), 344.
14. Vogt J., Flagmeyer R. H., Heitmann J., Lehmann D., Reinert T., Jankuhn S., Spemann D., Tröger W. and Butz T., *MiKrochim. Acta* **133** (2000), 105.
15. St. Pierre T. G., Dickson D. P. E., Webb J., Kim K. S., Macey D. J. and Mann S., *Hyperfine Interact.* **29** (1986), 1427.
16. Bell S. H., Weir M. P., Dickson D. P. E., Gibson J. F., Sharp G. A. and Peters T. J., *Biochimica et Biophysica Acta* **787** (1984), 227.
17. Dickson D. P. E., *Hyperfine Interact.* **111** (1998), 171.
18. Jambor J. L. and Dutrizac J. E., *Chem. Rev.* **98** (1998), 2549.
19. Murad E., Bowen L. H., Long G. J. and Quin T. G., *Clay Minerals* **23** (1988), 161.
20. Golgi C., *Rendiconti della Reale Accademia dei Lincei (21 maggio)* **2** (1893), 443–450.

Hyperfine Interactions (2004) 159:293–304
DOI 10.1007/s10751-005-9137-9

DFT Study of HFI in Halogen-Containing Gold, Silver and Copper Complexes

O. K. POLESHCHUK[1,*], E. L. SHEVCHENKO[1], V. BRANCHADELL[2], A. SCHULZ[3], B. NOGAJ[4] and B. BRYCKI[4]
[1] *Tomsk State Pedagogical University, 634041, Komsomolskii 75, Tomsk, Russia; e-mail: poleshch@tspu.edu.ru*
[2] *Universitat Autonama de Barcelona, Barcelona, Spain*
[3] *Ludwig-Maximilians-Universität München, München, Germany*
[4] *Adam Mickiewicz University, Poznań, Poland*

Abstract. We have analyzed the nuclear quadrupole coupling constants in copper, silver and gold halides and related compounds an the basis of the calculations with use of pseudo-potentials. The geometrical parameters and NQR halogen quadrupole constants obtained by these calculations substantially corresponded to the data of microwave spectroscopy in the gas phase. The analysis of the quality of the calculations with use of pseudo-potentials and the expanded basis set for the copper compounds was carried out. The ZORA model is shown to be a viable alternative to the computationally demanding B3LYP/SDD model for the calculation of halogen coupling constants in molecules. Besides the ZORA model as against BXYP/SDD model have been caused to realistic values of gold nuclear quadrupole coupling constants. In this case of the gold compounds the main contribution of the chemical Mössbauer shift comes from the 6*s*-orbiral population of the gold atoms.

1. Introduction

The properties of the coordination compounds of d^{10} monovalent ions continue to receive much attention. Antes *et al.* [1] investigated the stability, structural properties and electron distribution on the formation of carbonyl complexes of the group 11 chlorides with *ab initio* calculations using relativistic and non relativistic energy adjusted pseudo-potentials for the metal atoms. Structural data and vibrational frequencies were obtained with very good agreement with experimental data. A molecular orbital analysis shows that metal *d*- and metal *p*-contribution are important in metal ligand bonding in contrast to the interpretation of Mössbauer data [2]. Microwave spectra of the MXL (M = Au, Ag, Cu; X = Cl, Br; L = ligands) complexes and nuclear quadrupole coupling constants (NQCC) for Cu, Au, Br and CI nuclei of studied compounds were presented by Gerry and coworkers [3–7].

* Author for correspondence.

In this work we discuss the calculated quadrupole coupling constants and Mössbauer isomeric shifts of the MClL (M = Cu, Ag, Au) complexes depending on the strength of different Lewis bases L. In our recently published studies [8, 9] compounds of non-transition and transition elements containing tin, antimony, titanium and niobium atom with several organic ligands have been intensively investigated. Parameters such as the quadrupole splitting (QS) or the quadrupole coupling constant (NQCC) of the nuclei ^{127}I, ^{35}C1, ^{81}Br, ^{93}Nb, ^{121}Sb and ^{119}Sn were determined. Moreover, recently some AuClL complexes were investigated by nuclear quadrupole resonance (NQR) and Mössbauer effect analyses and their electronic structure has been discussed depending on the ligands. The influence of the donor ability of the base on the chemical shift and QS [10, 11] has been modeled.

Many chemical applications of Mössbauer spectroscopy use the sensitivity of the Mössbauer parameters to investigate changes in the electron density at the nucleus [12]. The isomer shift is a function of the nuclear and the electronic properties of the molecular systems, which are combined in such a way that independent quantitative information on both these properties cannot be obtained by Mössbauer spectroscopy alone. Since the electronic properties are usually of interest and because the nuclear parameters are constant, the hyperfine parameters are most frequently used to compare the electronic properties of different molecules. The covalence effects and the shielding of one set of electrons by another also influences the electronic environment of the nucleus and may be reflected in changes in the isomer shift [13].

The QS involves a nuclear quantity, the nuclear quadrupole moment, and an electronic quantity, the electric field gradient (EFG). The EFG contains a number of different contributions. The principal of these arises from the valence electrons of the Mössbauer atom itself and is associated with asymmetry in the electronic structure. This asymmetry results from partly filled electronic shells occupied by the valence electrons. Another contribution stems from the lattice and arises from the asymmetric arrangement of the ligand atoms. Molecular orbitals can also contribute to the EFG. The effects of these contributions at the Mössbauer nucleus are modified by the polarization of the core electrons of the Mössbauer atom which may reduce or enhance the EFG. It is very important to realize that the usefulness of the Mössbauer parameters for yielding significant information is strongly dependent on whether the nuclear parameters are sufficiently favourable to allow differences in the electronic environment to be reflected in significant and interpretable changes in the spectra.

Two-fold linear co-ordination is one of the simplest geometries known for metal complexes, and gold(I), copper(I) and silver(I) are unusual in forming a large number of such complexes. Two different bonding schemes have been proposed for such complexes [2]. The M^+ is formally nd^{10} ion, and the simpler explanation supposes donation from the ligands into the empty 6s and 6p_z orbitals of the metal atom (the z axis being the chlorine–metal-ligand axis). An alternative scheme, involving the formation of two sd_{z2} hybrid orbitals (one empty, one occupied), supposes the occupied hybrid to lie in the x–y plane, and donation to take place into the empty hybrid orbital directed along the z axis. These two schemes predict different trends for the quadrupole splitting of gold: the first predicts charge donation into the p_z orbital of a spherical d^{10} ion, giving an increasing QS with increasing σ-donor power of the ligands. The second model assumes a donation from ligand into the sd_{z2} hybrid orbital and as a result it predicts that the QS will decrease with increasing σ-donor power of the ligands. In both models π acceptance can take place from d_{xy} and d_{yz} orbitals.

For the benefit of the first model, an increasing QS with increasing σ-donor power of the ligand in a number of gold (I) complexes accounts. However, the comparison between QS and proton ability of complexes with ligands such as Cl^-, CO and PPh_3 published by Jones *et al.* [2] displayed an opposite trend. Besides, many results of the calculations of MXCO complexes show that the *np* orbital occupations are close to zero.

It is well known that for Mössbauer atoms the magnitude of the isomer shift depends simultaneously on the *s* and *p* orbital populations of these atoms [13]. The multiparameter dependence between the Mössbauer isomer shifts and the *s* and *p* populations of antimony, tin and iodine atoms are characteristic for this class of compounds and both one-parameter as well as multiparameter dependencies have been investigated in previous works [8, 9]. Earlier we have used the multiparameter dependencies for all nuclei considered, which include the direct effect of the valence-shell *s* electrons and their shielding of the core electrons, and also the other shielding effects. For various Mössbauer atoms very good correlations between isomeric shifts and orbital populations have been found. Thus, the main contribution to isomer shift stems from the *s* orbital population for iodine compounds, but for tin and antimony compounds a considerable contribution arises from the shielding by *p* orbitals.

Unfortunately, the interpretation of the NQCC's in the complexes is difficult for transition metal containing species, especially for the heavier metal such as Au, Ag and Cu. The NQCC values for ^{35}Cl, ^{79}Br, ^{63}Cu and ^{197}Au in MXL, complexes have been reported and discussed previously [3–7]. Above all, we are interested in the changes of copper, chlorine and bromine NQCC values upon complex formation. These changes can be expected to be quite large since the EFG can be quite sensitive to small changes in the electron density around the metal-ligand centre.

2. Computational details

Most of the calculations were carried out with the Gaussian 98 program package [14]. For metals we have used the small-core (6*s*5*p*3*d*) Stuttgard–Dresden basis set – relativistic effective core potential combination supplemented by (2f1g) functions for transition metals [15]. The use of all electron basis set (e.g., $6-31+G(d)$) for light atoms is better with respect to accuracy and efficiency. The B3LYP functional together with described above basis sets were optimal. On the other hand it is know that almost all of the density functionals perform poorly with the possible exceptions of the BHandHLYP hybrid functional, where the correct Hartree–Fock exchange is incorporated on a 50/50 basis in the exchange functional [16]. Therefore, we have been used BHandHLYP/$6-31l$+G($3df$,$3pd$) for the calculation of NQCC values of copper compounds. The NQCC values were obtained from the principal components of the electric field gradient tensor along the principal axes. The experimental values for the electric quadrupole moments were taken from [17]. Weinhold's natural bond orbital approach (NBO) [18] was used to determine the natural charges of the complexes and for the calculation of the gold orbital populations.

Gold complexes have also been studied using the ADF'2004 program [19, 20]. We have used the OPTX exchange functional [21] combined with PBE correlation functional [22] with an uncontracted STO triple-ζ + polarization basis set using the frozen core approximation to treat the inner electrons. We have done both non-relativistic and scalar relativistic ZORA calculations [23–25].

3. Results and discussion

3.1. BOND DISTANCES

The calculated metal–ligand and metal–chlorine distances for all ClML species are reported in Table I together with available experimental data. It is well known that the DFT approach tends to underestimate the strength of atom–atom interaction in such kind of system. However, in this case good agreement between theory (B3LYP) and experiment (accurate gas phase structural data for CuCl, CuBr, CuClCO, CuBrCO, AuCl, AuClCO, AgCl, and AgClCO are available [3–7]) was found. The same holds were for the MP2 results. For example, MP2 values reported by Antes [1] and Fortunelli [26] for the Au–Cl bond distances in AuClCO is 2.266 and 2.265 Å, respectively, and the Au–C bond length are 1.872 and 1.864 Å, respectively. MP2 values of the analogous silver complex AgClCO [1] are 2.254 Å for the Ag–Cl and 1.947 Å for the Ag–C bond. The B3LYP data set used here slightly overestimates (about 0.03–0.07 Å) the bond lengths in comparison with the experimental values. Moreover, the data of Table I show an anomalous trend not only for the metal–ligand bonds, that was noticed by Antes [1], but also for the metal–chlorine bonds. It is known that the longest bond distances are usually measured for the silver compound in group 11 [27]. It can be assumed that this anomalous trend can be partly attributed to the different magnitude of the relativistic effects along the series Cu, Ag and Au complexes [28]. Since relativistic effective core potentials have been used, the most important relativistic effect has been introduced.

Table I. The bond distances (Å) in MClL compounds obtained from rotational spectroscopy and calculated at the B3LYP/SDD level for Au and Ag compounds, and the BHandHLYP/6−311+*G*(3*df*, 3*pd*) level for Cu compounds

Compound	$R^{exp.}_{M-X}$	$R^{cal.}_{M-X}$	$R^{exp.}_{M-L}$	$R^{cal.}_{M-L}$
AuF	1.918	1.961		
AuFCO	1.909	1.928	1.847	1.866
AuCl	2.199	2.268		
AuClCO	2.217	2.259	1.884	1.901
AuClAr	2.198	2.261	2.469	2.655
AuClKr	2.210	2.264	2.522	2.657
AuBr	2.318	2.390		
AuBrCO	2.337	2.318	1.892	1.910
AgCl	2.281	2.324		
AgClCO	2.253	2.284	2.013	2.009
AgClAr	2.280	2.313	2.610	2.756
AgBr	2.393	2.436		
AgBrCO	2.373	2.400	2.027	2.022
AgBrAr	2.393	2.427	2.640	2.787
CuF	1.745	1.781		
CuFCO	1.736	1.758	1.764	1.833
CuFAr		1.773	2.220	2.325
CuCl	2.051	2.105		
CuClCO	2.056	2.096	1.796	1.871
CuClAr		2.101	2.270	2.388
CuBr	2.173	2.232		
CuBrCO	2.182	2.225	1.802	1.822
CuBrAr		2.228	2.300	2.410

From Table I it can also be seen that in the most complexes the Au–C1 bond distance increases upon complex formation (e.g., about 0.02 Å in AuXCO), whereas Ag–C1 and Cu–F bond lengths decrease in the corresponding carbonyl complexes (AgXCO: −0.02 to −0.03 Å; CuFCO: −0.009 Å). For the Cu–Cl and Cu–Br carbonyl complexes only a very small increase of the Cu–X bond lengths was found (C1: 0.005, Br: 0.009 Å). Actually, in weakly bound donor acceptor systems such as the MXCO species, a M–X bond length increase would be expected in all cases upon complex formation. The above discussed results can only be explained by an underestimation of the donor–acceptor interactions by DFT methods (changing of metal atom hybridization and redistribution of the atomic charges upon complex formation).

At the same time a comparison of the geometrical parameters calculated by us with the experimental data of the free molecules and complexes displays that the bond lengths have been overestimated. An analysis leads to the following correlation between the calculated and experimental bond lengths for the compounds studied:

$$R^{cal.} = -0.3 + 1.15\, R^{exp.} (r = 0.990; s = 0.04; n = 23) \quad (1)$$

Table II. The available experimental data (NQR and rotational spectroscopy) and the results of the calculations at B3LYP/SDD level NQCC (MHz) for Au and Ag compounds

Compound	NQCC ^{35}Cl, ^{79}Br, ^{127}I exp.	NQCC ^{35}Cl, ^{79}Br, ^{127}I calc.
AuCl	−61.99	−62.36
AuClCO	−36.39	−36.23
$AuCl_2^-$	−35.0	−34.61
AuClAr	54.05	57.19
AuClKr	52.01	54.09
AuBr	492.3	438.0
AuBrCO	285.1	284.4
AuBrAr	428.5	443.0
$AuBr_2^-$	202.3	263.0
$AuClPMe_3$	28.7[a]	33.5
$AuClPPh_3$	30.74[a]	37.64
$AuClPOMe_3$	29.4[a]	33.3
$AuClSMe_2$	34.0[a]	40.0
AuClPy	35.4[a]	41.3
AuI	−1708	−1535[b]
AgClCO	−28.15	−28.9
AgCl	−36.45	−40
$AgCl_2^-$	−16.7	−24.3
AgClAr	34.48	38.84
AgBr	297.1	324.8
AgBrCO	223.9	230.7
AgBrAr	278.9	308.8

[a] NQR data in the assumption of zero asymmetry parameter.
[b] 3–21*G*(*d*) basis set for iodine nucleus.

for Au and Ag compounds and

$$R^{cal.} = -0.1 + 1.1\ R^{exp.}(r = 0.994; s = 0.03; n = 12) \quad (2)$$

for Cu compounds. In these and subsequent correlation equations r is a correlation coefficient, s is the standard curve fit error, and n is the number of compounds.

It is necessary to note, that these correlations are valid for all compounds studied, that demonstrates quality carried out of the calculations.

3.2. HYPERFINE INTERACTIONS

It is well known that rotational spectroscopy is a powerful and precise method of determining molecular properties in the gas phase. Traditionally, it has been the source of geometries, force field, electric dipole moments and electric field gradients. The spectroscopic, and as a consequence molecular, properties obtained from rotational spectra when observed in this way refer to the molecule in isolation

Table III. The available experimental NQCC data (MHz) (NQR and rotational spectroscopy) and the results of the calculations BHandHLYP/6−311+*G*(3*df*, 3*pd*) for Cu compounds

Compounds	NQCC ^{35}Cl, ^{79}Br exp.	NQCC ^{35}Cl, ^{79}Br calc.	NQCC ^{63}Cu exp.	NQCC ^{63}Cu calc.
CuClCO	−21.5	−19.1	70.8	41.4
CuCl	−32.1	−27.8	16.2	10.1
$CuCl_2^-$	−19.3	−16.6	61.4	34.9
CuClAr	−28.0	−24.9	33.2	17.0
CuBr	261.2	−258	12.8	7.8
$CuBr_2^-$	152.8	147	57.7	33.3
CuBrCO	171.6	179	67.5	39.6
CuBrAr	225.6	232	29.9	15.2
CuF			22.0	17.4
CuFCO			75.4	45.9
CuFAr			38.1	21.9

and are therefore more appropriate for a comparison with the results of *ab initio* calculations.

The rotational spectra of chemical compounds allow to obtain the hyperfine structure arising from interactions between the electric quadrupole moment of the nucleus and EFG [29]. It is also possible to obtain the NQCC from NQR and Mössbauer spectroscopy data. If any significant variation in the electronic structure, occurs as for example the formation of a new chemical bond, then their changes will be reflected by the EFG's at the quadrupole nuclei, and also in their NQCC's.

Antes *et al.* [1] pointed out that the chlorine EFG in metal–carbonyl complexes are most sensitive to changes in the molecular environment and even predict correctly the trend in the metal–ligand bond stability. Though the agreement with the experiment was excellent for the AuClCO and CuClCO complexes, it was rather poor for the monomers, so the agreement for the complexes may be somewhat fortuitous. Tables II and III contain the NQCC's of ^{35}Cl and ^{79}Br for a number of Au(I), Ag(I) and Cu(I) species. The NQCC's of some halogen nuclei of related systems are also included in Table II, which were obtained using NQR. Theoretically obtained NQCC's are in better agreement with experimental ones in the cases of carbonyl complexes than in the case of free acceptor as well as with those presented in Antes' study [1]. Probably, this is connected to the more significant deviation in the metal–halogen bond lengths in the acceptors, than in carbonyl complexes.

For the complexes of all three metals the magnitudes of the halogen coupling constants decrease very significantly upon complex formation. This is consistent with electron donation from the donor ligand, e.g., CO to the MX acceptor fragment. For carbonyl complexes the change is again considerably larger than for the noble gas–MX complexes, although somewhat less than those found for the MX_2^-. These trends nicely agree with the experimental observations. A comparison of theoretically obtained and experimentally observed NQR-^{35}Cl frequencies for $AuClPMe_3$, $AuC1PPh_3$, $AuClPOMe_3$, AuClPy and $AuClSMe_2$ complexes can be found in Table II. In this case the agreement is rather poor, because the NQR spectra are sensitive to the intermolecular interactions in solid. However the overall trend in the NQR frequency values is the same.

On the basis of these results the following correlations between the experimental (from rotational spectroscopy [3–7] and NQR [10]) and calculated NQCC values for all halogen atoms was derived:

$$\mathrm{NQCC(cal.)} = 0.1 + 1.03\ \mathrm{NQCC(exp.)} \tag{3}$$

$$(\mathrm{r} = 0.999; \mathrm{sd} = 10; \mathrm{n} = 20)$$

for Au and Ag compounds

$$\mathrm{NQCC^{Cu}(cal.)} = 0.3 + 0.6\ \mathrm{NQCC^{Cu}(exp.)} \tag{4}$$

$$(\mathrm{r} = 0.990;\ \mathrm{sd} = 2; \mathrm{n} = 11)$$

$$\mathrm{NQCC^{X}(cal.)} = 3.0 + 0.99\ \mathrm{NQCC^{X}(exp.)} \tag{5}$$

$$(\mathrm{r} = 0.999;\ \mathrm{sd} = 5; \mathrm{n} = 8)$$

for Cu compounds.

It is necessary to note, that such correlations are valid for all metal compounds studied by the same method of the calculations.

The calculated NQCC values for Au atoms obtained from our B3LYP/SDD calculations were close to zero and are therefore not shown. Practically zero calculated NQCC values of the central atom is apparently due to the polarization of internal, filled orbitals (the Sternheimer effect) [30]. In this case the revolving electrostatic potential near quadrupole nucleus breaks the spherical symmetry of the closed environment and contributes to the total quadrupole moment. The interaction between valence electrons and this induced quadrupole moment results in a change of the NQCC. Moreover, the effect of the valence electrons contributes to the electric field gradient on the nucleus. The influence of the Sternheimer factor is especially important for ionic crystals, i.e., for compounds with a large degree of bond ionicity. For the investigated complexes we observed a significant ionicity in all donor–acceptor bonds of up to 80% as estimated by NBO calculations. The relativistic pseudo potential probably does not reproduce the effect of the internal occupied orbitals very well. From our calculations it is possible to conclude that the NQCC is highly dependent on the quality of the used basis set and the level of the core treatment. At the same time, the use of

Table IV. The available experimental NQCC data (MHz) (rotational spectroscopy) and the results of the calculations OPTX/TZP for Au compounds

Compound	NQCC ^{197}Au exp.	NQCC ^{197}AU calc. non-relativistic	NQCC ^{197}Au calc. ZORA	NQCC ^{35}Cl, ^{79}Br, ^{127}I exp.	NQCC ^{35}Cl, ^{79}Br, ^{127}I calc. ZORA	NQCC ^{35}Cl, ^{79}Br, ^{127}I calc. non-relativistic
AuF	−53	−151	248	–	–	–
AuFCo	−1006	−625	−552	–	–	–
AuFAr	−333	−229	109	–	–	–
AuCl	9.6	−86	209	−61.99	−59.5	−31.9
AuClCO	−1026	−602	−574	−36.39	−34.5	−19.4
$AuCl_2^-$	−765	−437	−300	−35.0	−35.6	−21.2
AuClAr	−260	−167	95	54.05	−54.8	−30.1
AuClKr	−250	−213	16	52.01	−52.6	−29.1
AuBr	37	−63	200	492.3	535	298.5
AuBrCO	−999	−578	−554	285.1	304	178.0
$AuBr_2^-$	−790	−406	−282	202.3	301	185.3
AuBrAr	−217	−140	106	482.5	490	278.2
AuI	78	−32	188	−1708	−1710	−919

Table V. Population analysis of the Au atom from NBO approach

Compound	N_s	N_p	N_d	δ, mm/s, relative to ^{197}Au in Pt
AuCl	0.62	0.02	9.85	−1.42
$AuCl_2^-$	0.81	0.00	9.77	0.54
AuClCO	0.90	0.03	9.55	1.88
AuClCNPh	0.93	0.03	9.58	2.92
$AuClP(OMe)_3$	0.97	0.01	9.72	2.63
$AuClPPh_3$	0.98	0.02	9.76	2.87
$AuClSMe_2$	0.88	0.00	9.74	1.26
AuClPy	0.86	0.01	9.67	1.70
$AuClPCl_3$	0.86	0.01	9.69	1.80
$AuClPMe_3$	0.92	0.02	9.73	2.61
AuBr	0.65	0.01	9.88	−1.07
$Au(CN)_2^-$	1.02	0.01	9.63	3.25

extended all electron basis sets for copper species gives acceptable NQCC values compared to experimental data.

The values of the NQCC of the halogens and gold nuclei from ADF calculations are presented in Table IV. The formation of complexes causes major changes in the Au NQCC. These changes are strong indicators of formation of new chemical bonds even when the ligands are noble gases [6]. When CO is the ligand, the changes are much greater and also are comparable to changes of the

AuX_2^- ions. This is perhaps to be expected because the Au–C bond order is much higher, than Au-noble gas once. A similar trend of the central atom was observed and for copper complexes (Table III). From our data we can see that the agreement between experimental and calculated halogen NQCC ZORA values a little bit worse, than for those obtained at the B3LYP/SDD level of calculations (Equation (3)):

$$\mathrm{NQCC^X(cal.)} = 24 + 1.02\ \mathrm{NQCC^X(exp.)} \qquad (6)$$

$$(\mathrm{r} = 0.999; \mathrm{s} = 33; \mathrm{n} = 10).$$

The comparison between non-relativistic and ZORA results (Table IV) shows that relativistic effects are important for halogen electric field gradient due to the large relativistic change in the molecular electron density [1]. Moreover, it is especially important that the calculation of NQCC of gold nuclei with use ADF program as against B3LYP/SDD calculations are close experimental data (Table IV):

$$\mathrm{NQCC^{Au}(cal.)} = 239 + 0.7\ \mathrm{NQCC^{Au}(exp.)}\ (\mathrm{r} = 0.979;\ \mathrm{s} = 67; \mathrm{n} = 13) \qquad (7)$$

for ZORA results and

$$\mathrm{NQCC^{Au}(cal.)} = 68 + 0.5\ \mathrm{NQCC^{Au}(exp.)}\ (\mathrm{r} = 0.987; \mathrm{s} = 36; \mathrm{n} = 13) \qquad (8)$$

for a non-relativistic basis set Probably not so high quality of (7) and (8) correlations (the large standard curve fit errors) is due to the great sensitivity of the EPG's to the size of the basis set especially in the core region [31] and to deficiencies in the exchange–correlation potential [32].

Table V presents the experimental values of the Mössbauer isomer shifts [2, 11] and the calculated populations of the 6*s*-, 6*p*- and 5*d*-orbitals of gold atom from NBO approach. It is necessary to note, that the population of the 6*p* orbital of a gold atom is close to zero. Earlier for various so-called Mössbauer atoms very good correlations between isomeric shifts and orbital populations have been found [8]. For iodine compounds the main contribution to isomeric shift comes from the 5*s*-orbital population, but for tin and antimony compounds a considerable contribution comes from the shielding by 5*p*-orbitals. On the basis of our theoretical results the following correlation can be deduced:

$$\delta = 11\mathrm{N_s} - 1.6\mathrm{N_d} + 7.5 \quad (\mathrm{r} = 0.965; \mathrm{s} = 0.3; \mathrm{n} = 12) \qquad (9)$$

that includes the direct effect of the valence-shell *s* electrons and their shielding of the *p* electrons. According to Equation (9) it is possible to confirm the conclusion about the greater contribution of the 6*s*-orbital than 5*d*-orbital of a gold atom to the isomeric shift. Thus chemical bonding in gold compounds is determined basically by *s*- and to a lower extent by *d*-orbitals of the central atom.

References

1. Antes J., Dapprich S., Frenking G. and Schwerdtfeger P., *Inorg. Chem.* **35** (1996), 2089.
2. Jones P. G., Maddock A. G., Mays M. J., Muir M. M., Williams A. F., *J. Chem. Soc., Dalton Trans.* (1977), 1434.
3. Walker N. R. and Gerry M. C. L., *Inorg. Chem.* **40** (2001), 6158.
4. Walker N. R. and Gerry M. C. L., *Inorg. Chem.* **41** (2002), 1236.
5. Evans C. J., Lesarri A. and Gerry M. C. L., *J. Am. Chem. Soc.* **122** (2000), 6100.
6. Evans C. J., Reynard L. M. and Gerry M. C. L., *Inorg. Chem.* **40** (2001), 6123.
7. Evans C. J. and Gerry M. C. L., *J. Mol. Spectrosc.* **203** (2000), 105.
8. Poleshchuk O. K., Latosinska J. N. and Yakimov V. G., *Phys. Chem. Chem. Phys.* **2** (2000), 1877.
9. Poleshchuk O. K, Latosinska J. N. and Nogaj B., *Z. Naturforsch.* **55a** (2000), 271.
10. Jones P. G. and Williams A., *J. Chem. Soc., Dalton Trans.* (1977), 1430.
11. Dickson D. P. E and Berry, F. J. (eds.), *Mössbauer Spectroscopy*, Cambridge University Press, 1986.
12. Parish R. V., *Prog. Inorg. Chem.* **15** (1972), 101.
13. Parish R. V., *Coord. Chem. Rev.* **42** (1982), 1.
14. Gaussian 98 (Revision A.6 and Revision A.7), Frisch M. J., Trucks G. W., Schlegel H. B., Scuseria G. E., Robb M. A., Cheeseman J. R., Zakrzewski V. G., Montgomery J. A., Stratmann R. E., Burant J. C., Dapprich S., Millam J. M., Daniels A. D., Kudin K. N., Strain M. C., Farkas O., Tomasi J., Barone V., Cossi M., Cammi R., Mennucci B., Pomelli C., Adamo C., Clifford S., Ochterski J., Petersson G. A., Ayala P. Y., Cui Q., Morokuma K., Malick D. K., Rabuck A. D., Raghavachari K., Foresman J. B., Cioslowski J., Ortiz J. V., Stefanov B. B., Liu G., Liashenko A., Piskorz P., Komaromi I., Baboul A. G., Gomperts R., Martin R. L., Fox D. J., Keith T., Al-Laham M. A., Peng C. Y., Nanayakkara A., Gonzalez C., Challacombe M., Gill P. M. W., Johnson B. G., Chen W., Wong M. W., Andres J. L., Head-Gordon M., Replogle E. S. and Pople J. A., Gaussian, Inc., Pittsburgh PA, 1998.
15. Parr R. G. and Yang W., *Density-Functional Theory of Atoms and Molecules*, Oxford University Press, Oxford, 1989.
16. Schwerdtfeger P., Pernpointer M., Laerdahl J. K., *J. Chem. Phys.* **111** (1999), 3357.
17. Seliger J., In: Holmes J. L. (ed.), *Encyclopedia of Spectroscopy and Spectrometry*, Academic Press, London, 2000, p. 1672.
18. Glendening E. D., Reed A. E., Carpenter J. E. and Weinhold F., NBO Version 3.1.
19. te Velde G., Bickelhaupt F. M., van Gisbergen S. J. A., Guerra C. F., Baerends E. J., Snijders J. G. and Ziegler T., *J, Comput. Chem.* **22** (2001), 931.
20. ADF2004.01, SCM, Theoretical Chemistry, Vrije Universiteit, Amsterdam, The Netherlands, http://www.scm.com.
21. Handy N. C. and Cohen A. J., *Mol. Phys.* **99** (2001), 403.
22. Perdew J. P., Burke K. and Ernzerhof M., *Phys. Rev.* **77** (1996), 3963.
23. van Lenthe E., Baerends E. J. and Snijders J. G., *J. Chem. Phys.* **99** (1993), 4597.
24. van Lenthe E., Baerends E. J. and Snijders J. G., *J. Chem. Phys.* **101** (1994), 9783.
25. van Lenthe E., Baerends E. J. and Snijders J. G., *J. Chem. Phys.* **110** (1999), 8943.
26. Fortunelli A. and Germano G., *J. Phys. Chem., A* **104** (2000), 10834.
27. Huber K. P. and Herzberg G., *Molecular Spectra and Molecular Structure Constants of Diatomic Molecules*, Van Nostrand, New York, 1979.

28. Klapötke T. M. and Schulz A., with an invited Chapter of R. D. Harcourt about VB Theory in: *Ab initio Methods in Main Group Chemistry*, Wiley, New York, 1998.
29. Legon A. C., *Angew. Chem., Int. Ed.* **38** (1999), 2686.
30. Sternheimer R. M., *Phys. Rev.* **95** (1954), 736.
31. Schwerdtfeger P., Aldridge L. P., Boyd P. D. W. and Bowmaker G. A., *Struct. Chem.* **1** (1990), 405.
32. van Lenthe E. and Baerends E. J., *J. Chem. Phys.* **112** (2000), 8279.

Hyperfine Interactions (2004) 159:305–311
DOI 10.1007/s10751-005-9120-5

Atomic Arrangement in B2 FeAl Prepared by Self-Propagated High-Temperature Synthesis at Varying Al Content and Annealing

J. FEDOTOVA[1,*], A. ILYUSCHENKO[2], T. TALAKO[2], A. BELYAEV[2], A. LETSKO[2], A. ZALESKI[1], J. STANEK[3] and A. PUSHKARCHUK[4]
[1]*NC PHEP BSU, 153, M.Bogdanovich str., 220040 Minsk, Belarus; e-mail: julia@hep.by*
[2]*Research Institute of Powder Metallurgy, 41, Platonova str., 220071 Minsk, Belarus*
[3]*Institute of Physics, Jagiellonian University, 4, Reymonta str., 30-059 Cracow, Poland*
[4]*Institute of Physical-Organic Chemistry, 13 Surganova str., 220072 Minsk, Belarus*

Abstract. Atomic arrangement in B2 FeAl prepared by self-propagated high temperature synthesis (SHS) as a function of Al concentration and annealing temperature has been studied by ^{57}Fe Mössbauer spectroscopy (MS) and X-ray diffraction (XRD). The increase of B2 FeAl isomer shift (IS) and lattice parameter (a) with Al concentration in the whole concentration range has been detected. This may originate from the formation of Al antisite atoms. Calculation of s-electron populations against number of Al antisites using a cluster approach and MO LCAO method supported this assumption. Annealing resulted in atomic rearrangements both in near-stoichiometric and Al-rich B2 FeAl.

Key Words: B2 FeAl, Mössbauer spectroscopy.

1. Introduction

Considerable interest in studying the B2 FeAl defect structure is conditioned with its strong relationship with mechanical and physical properties [1]. Experimental study of thermal and constitutional defects formed in B2 FeAl has been carried out by XRD, perturbed angular correlation (PAC), ^{57}Fe MS [1–3]. XRD investigations of a in B2 FeAl within stoichiometric composition elucidated its triple-defect structure [1]. It means that thermal defects in B2 FeAl consist of Fe sublattice vacancies and antisite atoms, while constitutional defects consist of Fe antisites in the Fe-rich side and Fe vacancies in the Al-rich side. The triple-defect structure of B2 FeAl has been confirmed also by MS [3] and PAC analysis [2]. It is worth noticing here that no traces of Al-antisites (or vacancies) were detected. A theoretical study of point defects in B2 FeAl showed

* Author for correspondence.

that the formation of Fe vacancies is favored [4]. Also a triple defect in B2 FeAl has been predicted as a Fe antisite atom and two vacancies in the Fe sublattice that strongly supported referred experimental interpretations.

It should be mentioned that values of *a* as well as of IS attributed to B2 FeAl varied noticeably for different methods of sample preparation. This is evidence for the interplay between synthesizing conditions and final atomic arrangement in B2 FeAl.

The present paper is aimed at a systematic study by ^{57}Fe MS and XRD of the atomic arrangement in B2 FeAl prepared in far from equilibrium conditions by SHS after mechanical activation of Fe and Al powders.

2. Experimental

$Fe_{100-x}Al_x$ alloys (x = 36–64 at.%) containing B2 FeAl has been prepared by SHS with prior mechanical activation (MA) of Al and Fe powders during 3 h. MA has been performed using A2 attritor at impeller rotation speed of 190 RPM. Relation between mass of spheres and powders comprised 15:1. Annealing has been performed in vacuum (10^{-4} Pa) in the temperature range of 300–700°C.

The local Fe states in $Fe_{100-x}Al_x$ alloys have been studied by ^{57}Fe transmission MS at room temperature using a ^{57}Co/Rh source (25 mCi). Spectra have been evaluated using the MOSMOD program [5]. All IS are given with respect to α-Fe with experimental error not exceeded 0.006 mm/s. The variation of the Fe s-electron population with the number of incorporated Al antisite atoms has been calculated using a cluster approach and a MO LCAO method in PM3 approximation [6].

Values of *a* for B2 FeAl have been determined in (310) reflection by a high resolution D8 ADVANCE diffractometer using complex software DIFRAC^{plus}. The experimental error did not exceed 0.0003 Å. The Al concentration in B2 FeAl has been verified by a scanning microscope CAMSCAN.

3. Results and discussion

Mössbauer spectra of as-sintered $Fe_{100-x}Al_x$ powders are presented in Figure 1. The spectrum of $Fe_{64}Al_{36}$ sample was fitted assuming a single line and a doublet. The single line with IS = 0.264 mm/s has been assigned to B2 FeAl [7] while the doublet – to non-stoichiometric Fe_2Al_5 [8]. The spectrum of the $Fe_{48}Al_{52}$ powder displayed a superposition of the B2 FeAl singlet (IS = 0.262 mm/s) and the Fe_2Al_5 doublet [8]. Spectra of $Fe_{42}Al_{58}$ and $Fe_{36}Al_{64}$ powders have been fitted in the assumption of a doublet and a single line. The observed doublets have parameters close to those of $FeAl_2$ [7, 8]. The single line could be attributed again to B2 FeAl although the fitted IS values (0.279 and 0.294 mm/s for $Fe_{42}Al_{58}$ and $Fe_{36}Al_{64}$, respectively) exceeded that of the stoichiometric compound.

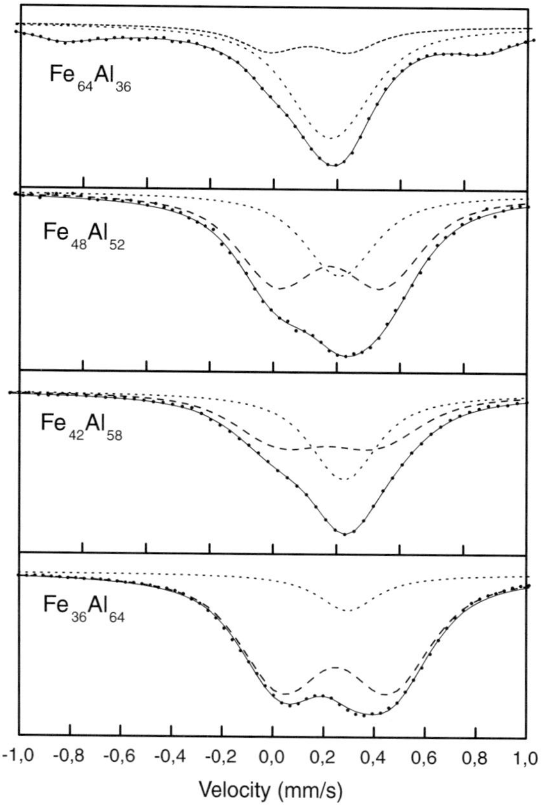

Figure 1. Mössbauer spectra of $Fe_{100-x}Al_x$ powders.

As evidenced from their spectral evaluation samples represented predominantly two-phase alloys. Variations with Al concentration of the IS extracted from B2 FeAl subspectra and corresponding values of a shown in Figure 2 offer evidence for their sequential increase.

In order to verify the variation of the s-electron density within the Fe nuclei that influences the IS value, a computer simulation of a B2 FeAl cluster with incorporated of Al antisite atoms has been performed. A cluster model of B2 FeAl has been developed by computer generation of a $FeAl_8$ cluster representing nine atoms in typical BCC structure (see Figure 3a). Boundary conditions have been imposed by the method of charge clusters introduced by Messmer and Whatkins in [9]. Dangling chemical bonds were saturated with additional electrons incorporated into the cluster.

In order to simulate the B2 FeAl lattice containing different Al concentrations a set of $FeAl_n$ (n = 8–15) clusters has been generated where additional Al atoms substituted Fe atoms. A typical cluster with three Al antisites is shown in Figure 3b. For all generated clusters the s-atomic orbital electron population within the Fe nuclei has been calculated (see Figure 3c). As can be seen, it decreases with

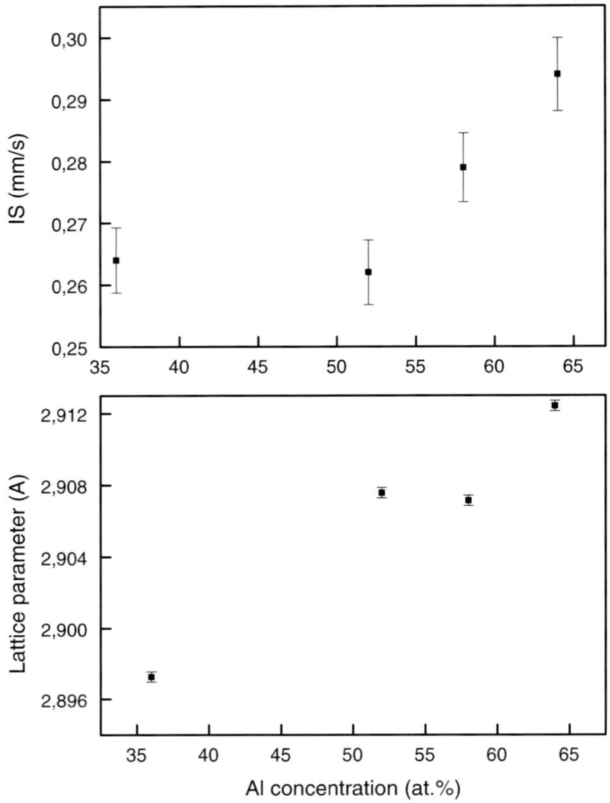

Figure 2. Variation of IS and lattice parameter of B2 FeAl with Al content.

the number of added Al antisites and correlates with IS increase obtained from the MS study.

Mössbauer spectra recorded on $Fe_{100-x}Al_x$ samples after annealing (not shown here) demonstrated that no evidence of phases other than in the as-sintered powders was detected. This fact was confirmed by a XRD study [10]. However, in some cases annealing caused noticeable changes in the IS of B2 FeAl. IS variations with annealing temperature for alloys with different Al concentrations are presented in Figure 4a. As can be seen, the most pronounced changes of IS are observed for $Fe_{64}Al_{36}$ and $Fe_{36}Al_{64}$ alloys. Fluctuations of B2 FeAl a with annealing temperature (see Figure 4b) show that at lower Al concentration it varies within experimental error while at higher Al concentration a evidently decreases with annealing temperature.

The variation of the IS and lattice parameter characterizing B2 FeAl reflects atomic rearrangements in the lattice with growing Al content. According to an estimation of the atomic Al concentration in B2 FeAl it comprised 50 ± 1 and 54 ± 1 at.% for $Fe_{64}Al_{36}$ and $Fe_{36}Al_{64}$ alloys, respectively. Regarding the model of triple-defect structure the decrease of a and possibly IS should be expected for

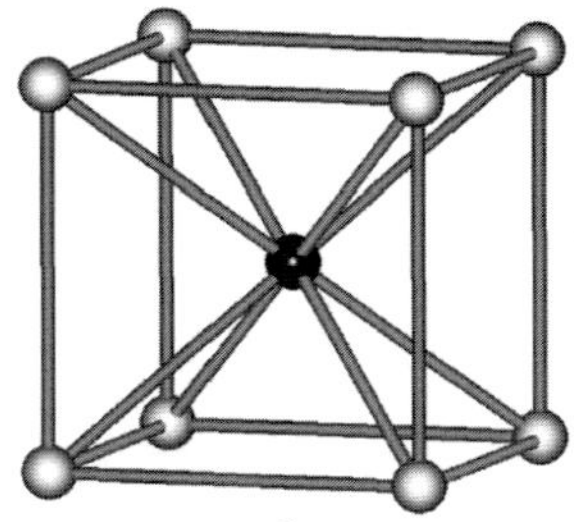

a

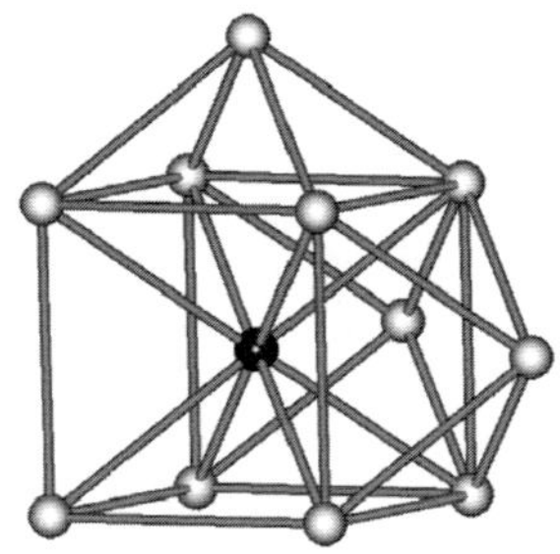

b

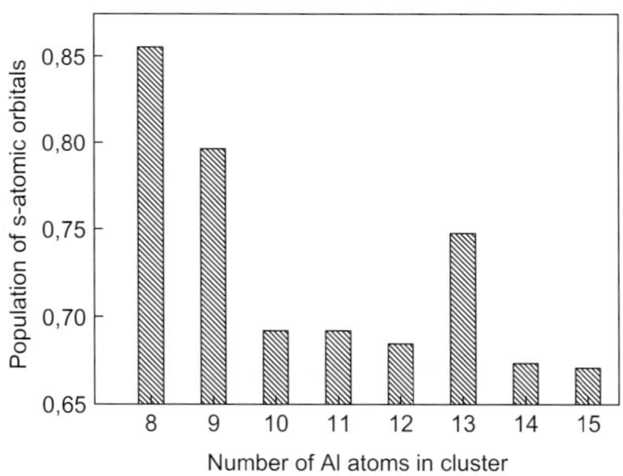

c

Figure 3. $FeAl_8$ (a) and $FeAl_{11}$ (b) clusters (*black* – Fe atoms, *white* – Al atoms). Population of Fe s-atomic orbital electrons vs. number of Al atoms in $FeAl_n$ ($n = 8 – 15$) cluster (c).

the $Fe_{36}Al_{64}$ composition. However, the growth of both *a* and IS has been observed in the whole concentration range studied in the present paper. According to [1], an *a* increase should be attributed to the formation of Al antisites. Moreover, the observed increase of the IS and consequently the decrease of the electronic density within Fe nuclei also supports the Al enrichment of B2 FeAl. In fact, the decrease of electronic density may be the result of enhanced Fe 4s-electrons screening due to Al sp-electrons charge transfer to the 3d-electron Fe atomic shell [11]. The latter is consistent with results of MO LCAO modeling of an $FeAl_n$ cluster revealing the growth of the s-electrons population within Fe nuclei when Al antisites were incorporated.

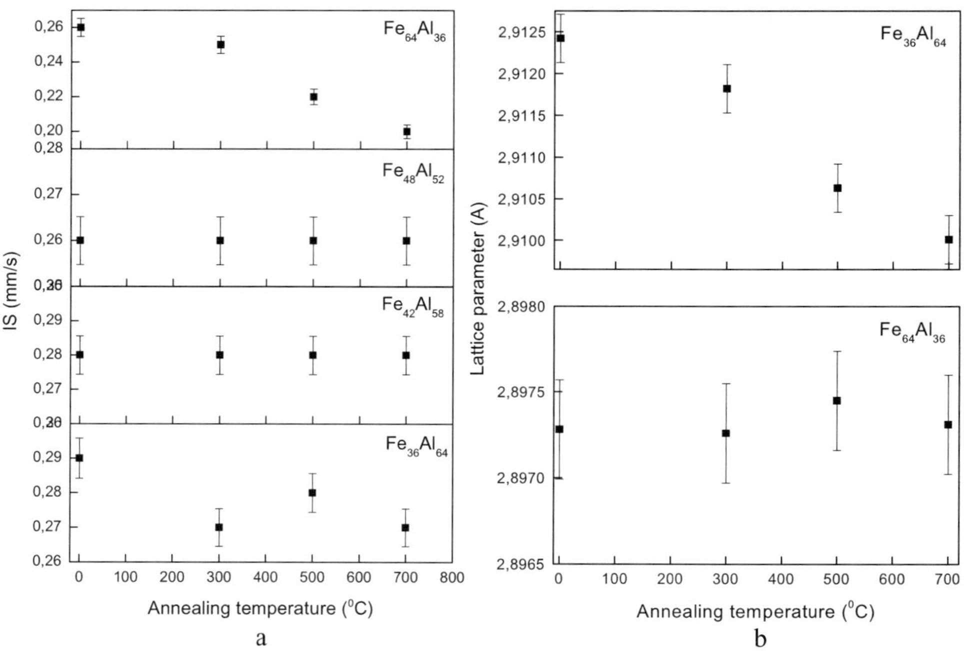

Figure 4. Transformation of B2 FeAl IS (a) and lattice parameter (b) with annealing temperature.

Annealing caused noticeable changes of the IS only for alloys with $Fe_{64}Al_{36}$ and $Fe_{36}Al_{64}$ nominal compositions. In both cases the IS decreased revealing the growth of the electronic density within Fe nuclei. As for B2 FeAl *a* value, it evidently dropped for $Fe_{64}Al_{36}$ alloys but for $Fe_{36}Al_{64}$ it remained unchanged. A decrease of B2 FeAl IS in Al-rich alloys seems quite natural as it may result from thermally stimulated recovery of crystalline lattice and migration of Al excess atoms. At the same time, the reason of B2 FeAl IS decrease in near-stoichiometric Al content ($Fe_{64}Al_{36}$) is unclear.

In doing so, the present results are not totally consistent with previously reported data where Fe vacancies in Al-rich B2 FeAl were proven to be the dominating defect [1]. Several reasons could be responsible for this discrepancy. It should be remembered that the previously referred results have been obtained for singe-phase B2 FeAl samples subjected to long annealing bringing crystalline lattice in relatively equilibrium state. The preparation process of $Fe_{100-x}Al_x$ powders studied in the present paper is supposed to be non-equilibrium. It has been proven in [12] that MA prior to the SHS process increases residual lattice distortion and possible ion substitution occurred between interacting Fe and Al particles. So, it could be suggested that such pre-existing disorder may have affected the final defect structure of B2 FeAl. On the other hand, the atomic arrangement of B2 FeAl may also be influenced by the multiphase composition of analyzed powders as compared to monophase B2 FeAl studied in [1–4]. However, this assumption needs additional study.

The main result of present study is that atomic arrangement of B2 FeAl sintered by nonequilibrium SHS with prior MA involves Al antisite atoms. Atomic rearrangements during annealing are governed by the Al contents in as-sintered B2 FeAl.

References

1. Pike L. M., Chang Y. A. and Liu C. T., In: Schneibel J. H. and Crimp M. A. (eds.), *Processing, Properties and Applications of Iron Aluminides, The Minerals, Metals and Materials Society*, 1994, p. 217.
2. Collins G. S., Peng L. S.-L. and Zacate M. O., *Def. Diff. Forum* **107** (2003), 213–215.
3. Collins G. S. and Peng L. S.-L., *Il Nuovo Cimento* **18D** (1996), 329.
4. Kellou A., Feraoun H. I., Grosdidier T., Coddet C. and Aourag H., *Acta Mater.* **52** (2004), 3263.
5. Rancourt D. G. and Ping J. Y., *Nucl. Instrum. Methods B* **58** (1991), 85.
6. James J. P. and Stewart J., *Comput. Chemistry* **10** (1989), 209.
7. Van der Kraan A. M. and Buschow K. H. J., *Phys. Rev., B* (1986), 55.
8. Li Y., Vagizov F., Gorugandi V. and Ross J. H., Cond-mat/0308471.
9. Messmer R. P. and Whatkins G. D., *Phys. Rev., B* **7** (1973), 2568.
10. Fedotova J., Ilyuschenko A., Talako T., Belyaev A. and Letsko A., In: Gorlich E. A., Krolas K. and Pedziwiatr A. (eds.), *Proc. Int. Conf. Condensed Matter Studies with Nuclear Methods*, Institute of Physics, Jagellonian University, Cracow, 2003, p. 233.
11. Azez K. A. and Al-Omari I. A. *et al.*, *Physica B* **321** (2002), 178.
12. Charlot F., Gaffet E., Zeghmati B., Bernard F. and Niepce J. C., *Mat. Sci. Eng. A* **262** (1999), 279.

Hyperfine Interactions (2004) 159:313–322
DOI 10.1007/s10751-005-9121-4

The Nuclear Quadrupole Interaction of ^{204m}Pb in Lead Oxides

S. FRIEDEMANN[1], F. HEINRICH[1], H. HAAS[2] and W. TRÖGER[1,*]
[1]*Institut für Experimentelle Physik II, Universität Leipzig, Linnestraße 5, 04103, Leipzig, Germany; e-mail: troeger@physik.uni-leipzig.de*
[2]*CERN, 1211 Geneve 23, Switzerland*

Abstract. The nuclear quadrupole interaction of ^{204m}Pb in lead oxides has been measured by γ–γ time differential perturbed angular correlation. *Ab-initio* calculations of the electric field gradients and X-ray diffraction allowed the assignment of the detected nuclear quadrupole interactions to the different Pb sites in the PbO phases litharge and massicote as well as in Pb_3O_4. The TDPAC probe ^{204m}Pb was produced with a ^{204}Bi/^{204m}Pb-generator at the home laboratory at the University of Leipzig. The use of a high performance liquid chromatography system increased significantly the yield, the specific activity of ^{204m}Pb, and reduced the acidic concentration of the eluate.

Key Words: ^{204}Bi/^{204m}Pb-generator, NQI, nuclear quadrupole interaction, ^{204m}Pb, PbO, Pb_3O_4, TDPAC, time differential perturbed angular correlation.

1. Introduction

For the study of the nuclear quadrupole interaction (NQI) via time differential perturbed angular correlation (TDPAC) in macromolecules, e.g. proteins, mainly the isomeric TDPAC probes ^{111m}Cd ($t_{1/2}$ = 49 min) and ^{199m}Hg ($t_{1/2}$ = 43 min) are used because after-effects are negligible [1, 2]. There is one more isomeric probe for TDPAC spectroscopy: ^{204m}Pb ($t_{1/2}$ = 67 min). Here, the 912 keV − 375 keV cascade (see Figure 1) has an effective anisotropy of 18%, the intermediate state (nuclear spin $I = 4$, $t_{1/2}$ = 256(10) ns) has a nuclear quadrupole moment of $Q = 0.44(2)$ barn and a magnetic dipole moment of $\mu = +0.225(4)\ \mu_N$. The intermediate state of this cascade has an integer nuclear spin I which leads to linear dependency of the eigenvalues of the nuclear quadrupole Hamiltonian operator on the asymmetry parameter η for small values of η, whereas in the case of half integer spins this dependence is quadratic, e.g., for ^{111m}Cd and ^{199m}Hg both with $I = 5/2$. Therefore, ^{204m}Pb is endowed with a much greater sensitivity to deviations from axial symmetry than the half-integer spin TDPAC probes. Furthermore, the quite long half-life provides a much better frequency resolution.

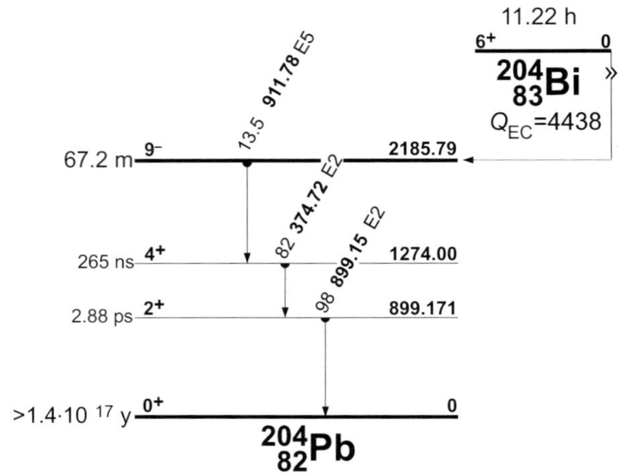

Figure 1. The simplified decay schemes of the TDPAC-probe ^{204m}Pb and the ^{204}Bi/^{204m}Pb-generator-mother-isotope ^{204}Bi. For a more detailed decay-scheme see [3].

Hence, this isotope should have a broad range of applications in the field of nuclear solid state physics and molecular biophysics.

In an early effort the NQIs of the isotope ^{204m}Pb in various metals as well as in insulating solids were determined by γ–γ-TDPAC measurements using ^{204m}Pb as produced by a ^{204}Bi/^{204m}Pb-generator [4]. However, TDPAC experiments with ^{204m}Pb were lacking for the following 30 years. In 2002, ^{204m}Pb was produced for the first time at the on-line isotope separator ISOLDE/CERN with the resonance laser ionization source RILIS [5, 6] and used for the determination of the NQI of ^{204m}Pb in cadmium metal [7]. Later on, a ^{204m}Pb-NQI-study of lead titanate ($PbTiO_3$) followed [8].

Here, we present the results of our ^{204m}Pb-TDPAC experiments in several lead oxides. In order to interpret the results of the TDPAC experiments *ab initio* density functional theory calculations with the WIEN code [9] were performed.

2. Experimental methods and results

The theory of TDPAC is well described in standard textbooks [10]. For the description of a state-of-the-art TDPAC spectrometer ("PAC-Camera"), combining extremely high efficiency and excellent time resolution, see [11]. A description of the data analysis is found in [12].

In the present study the TDPAC-probe ^{204m}Pb was produced by an improved version of the ^{204}Bi/^{204m}Pb-generator described in [13, 14]. The parent isotope ^{204}Bi ($t_{1/2}$ = 11.22 h) was produced at the Ionenstrahllabor ISL of the Hahn-Meitner-Institut (Berlin) via the reaction ^{206}Pb(p, 3n)^{204}Bi by irradiating a Pb-target with 33 MeV protons. For the proton irradiation of the Pb target a special target holder was developed [15]. This target holder consisted of a graphite

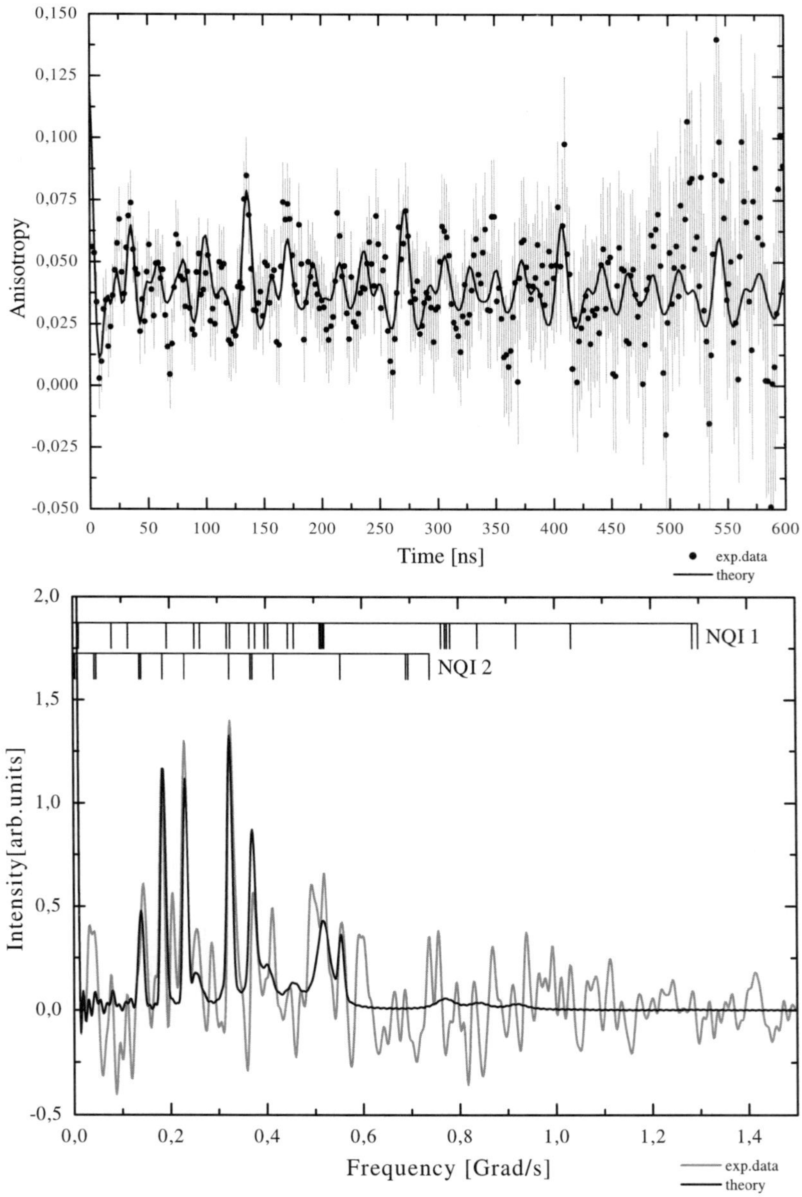

Figure 2. The TDPAC time spectrum (*upper*) and its cosine transform (*lower*) of ^{204m}Pb in PbO at room temperature. The *black lines* show the least-squares-fitting for the NQI parameters listed in Table I. The frequency *bar* in the cosine spectrum indicates the nuclear precession frequencies.

moderator in order to use proton beams with energies up to 68 MeV which are routinely used at the ISL and a specially designed copper support for the lead target in order to avoid unwanted activations of the target holder and beam line. After a cooling period of at least 4 h the irradiated lead target was transported to our home laboratory at the University of Leipzig. There, the irradiated target was

Table I. The parameters of NQIs 1 and 2 obtained by least-square-fitting of the experimental data of ^{204m}Pb in PbO at room temperature as displayed in Figure 2

	Amplitude	Line width δ	NQI frequency ω_Q [Mrad/s]	Asymmetry η	Fraction [%]
NQI 1	0.05(2)	0.03(2)	25.3(4)	0.51(3)	48(16)
NQI 2	0.06(1)	0.003(4)	15.41(4)	0.01(1)	52(16)

dissolved in hot concentrated nitric acid. In order to separate the ^{204}Bi activity from the target material the lead was precipitated as $PbCl_2$ with concentrated HCl. After a short centrifugation the supernatant was removed and evaporated to a small volume ($\approx$ 0.5 ml) of azeotropic mixture (20.2% HCl). This mixture, containing the ^{204}Bi activity, was diluted with deionized water to 0.3 M HCl and injected into a high performance liquid chromatography (HPLC) column filled with an ion exchanger. The HPLC column used was made from metal free PEEK (polyether ether ketone) and filled with the ion exchanger resin Dowex 1×8 in the Cl^--form [16, 17]. This resin binds Bi ions whereas Pb ions can be eluted with 0.03 M HCl. The decay of the ^{204}Bi produces ^{204m}Pb continously (see Figure 1) and it is possible to 'milk' the ion exchange column every 2 h. A ^{204m}Pb activity of about 1.5 MBq is obtained in $\approx$ 1 ml of 0.03 M HCl.

The preparation of the lead oxide PbO was carried out according to the procedures in [18]. First, lead carbonate was precipitated from the solution containing ^{204m}Pb and dissolved ammonium carbonate ($(NH_4)_2CO_3$) by adding a concentrated lead nitrate ($Pb(NO_3)_2$) solution. Prior to the precipitation the pH value of the eluated ^{204m}Pb/HCl solution was raised to pH 9 to 11 by adding sodium hydroxide solution (NaOH). After the removal of the supernatant the white precipitate, $PbCO_3$, was carefully heated in air until the color changed to a pale yellow. The subsequent ^{204m}Pb-TDPAC experiment was performed at room temperature and the obtained spectrum is shown in Figure 2. For data presentation the spectrum was compressed from 800 to 400 channels. An analysis of the time spectrum by a least-squares-fitting routine yields NQI 1 and NQI 2 whose parameters are given in Table I. After the decay of the ^{204m}Pb activity of the sample a X-ray diffraction analysis of this sample was carried out, revealing that the two PbO phases massicote and litharge were present besides traces of lead carbonates and hydroxides. Later, we will assign the two NQIs to the PbO phases massicote and litharge.

In order to obtain the lead oxide Pb_3O_4 we followed the procedure reported in [19] and used a PbO sample prepared as mentioned above. This PbO sample was inserted into a furnace mounted between the detectors of a PAC-Camera as described in [20]. The TDPAC experiment was executed at a temperature of 500°C in a continuous flow of oxygen. The first TDPAC data were stored 15 min after the insertion of the sample and later in irregular time intervals of 10 to 40 min which allowed to follow the conversion from PbO to Pb_3O_4. In order to

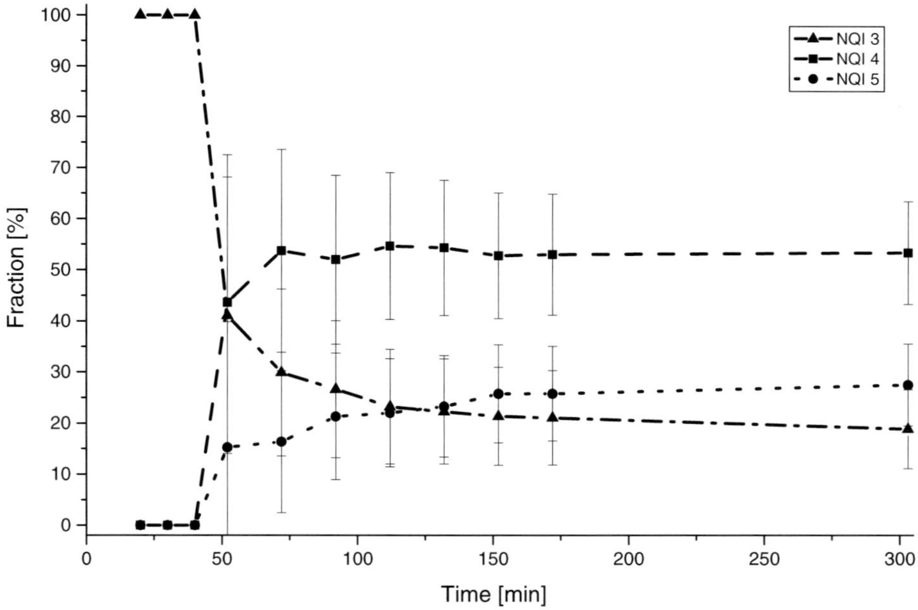

Figure 3. The development of the fractions of the NQIs identified in the TDPAC spectra stored at different times for the second Pb_3O_4 sample.

Table II. The parameter of the NQIs 3, 4, and 5 obtained by least-square-fitting of the experimental data of ^{204m}Pb in PbO and Pb_3O_4 at 500°C

	Amplitude	Line width δ	NQI frequency ω_Q [Mrad/s]	Asymmetry η	Fraction [%]
NQI 3	0.03(1)	0.000(4)	14.47(8)	0.00(0)	20(5)
NQI 4	0.12(1)	0.000(1)	16.94(2)	0.49(1)	67(5)
NQI 5	0.06(1)	0.000(4)	9.63(3)	0.40(1)	33(5)

improve the statistical quality this experiment was repeated. Since the TDPAC spectra displayed no difference within the error margins corresponding spectra of the two experiments were added. In total, three different NQIs were detected which are listed in Table II. In Figure 3 the time development of the three NQIs of this series of spectra is depicted. Obviously the conversion from PbO to Pb_3O_4 occurs in the first 80 min of this measurement. Whereas NQI 3 is dominant during the first 30 min of this measurement, NQIs 4 and 5 are becoming prevalent later. The TDPAC time spectrum and its cosine transform of Figure 4 correspond to the final state of the sample. This TDPAC time spectrum was generated from the data collected between 80 min and 300 min from the

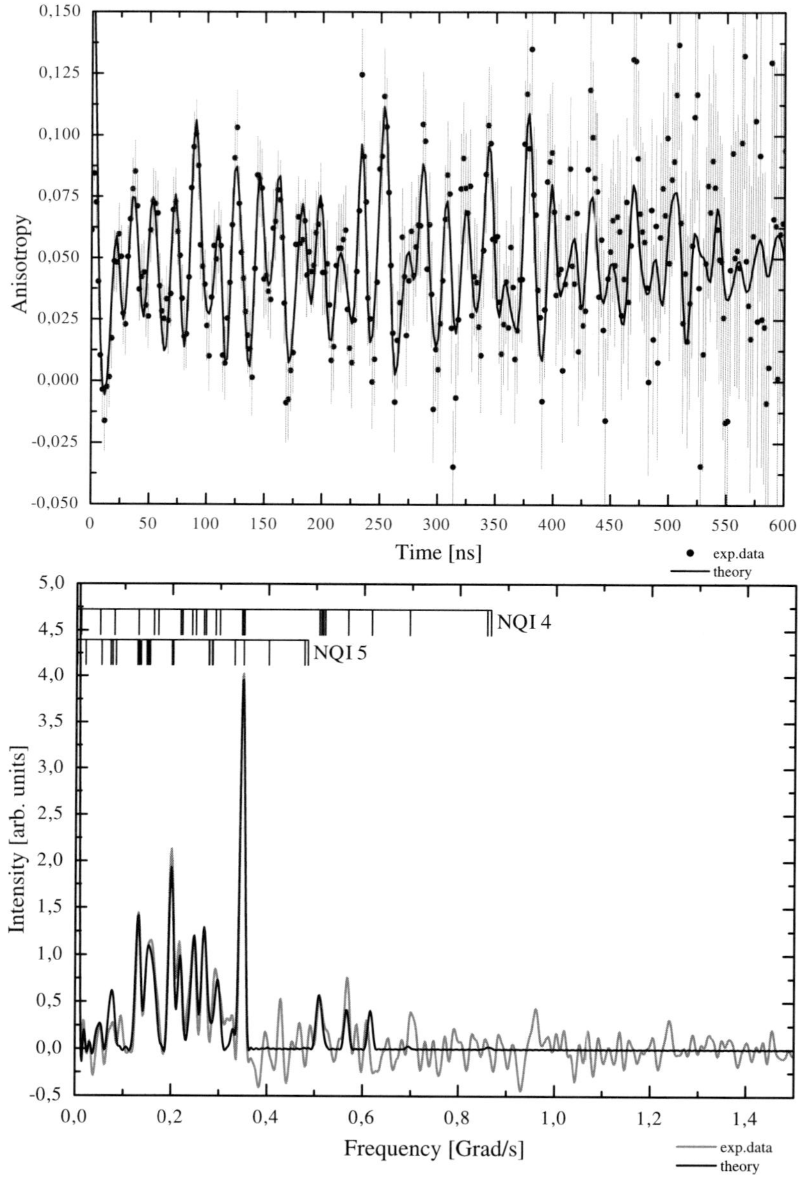

Figure 4. The TDPAC time spectrum (*upper*) and its cosine transform (*lower*) of ^{204m}Pb in Pb_3O_4 measured at 500°C. The *black lines* show the least-squares-fitting for the NQI parameters listed in Table II. The frequency *bar* in the cosine spectrum indicates the nuclear precession frequencies.

beginning of the measurement. It contains the NQIs 4 and 5 with a ratio of 1:2 which are listed in Table II and have to be assigned to Pb_3O_4. The NQI 3 was detected at the starting point of the experiment and, therefore, has to be assigned to PbO. The X-ray diffractogram taken after the TDPAC experiment shows only the presence of Pb_3O_4 in this sample.

Table III. The parameters of the NQIs calculated with the DFT code WIEN [9] in comparison with the experimental data and their site assignments

Compound		Site	Experiment			WIEN code		
			ω_Q [Mrad/s]	η	Frac. [%]	ω_Q [Mrad/s]	η	Frac. [%]
PbO	NQI 1	Massicote	25.3(4)	0.51(3)	48(16)	24.12	0.49	–
	NQI 2	Litharge	15.41(4)	0.01(1)	52(16)	14.89	0	–
Pb_3O_4	NQI 4	Pb(II)	16.94(2)	0.49(1)	67(5)	17.94	0.57	67
	NQI 5	Pb(IV)	9.63(3)	0.40(1)	33(5)	9.46	0.62	33

The Pb_3O_4 experimental data were obtained at 500°C, whereas PbO was measured at room temperature.

3. *Ab-initio* electric field gradient calculations

Ab-initio calculations of the electric field gradient (EFG) for the Pb sites in PbO and Pb_3O_4 were performed with the full potential linearized augmented plane waves code WIEN97 [9]. All calculations were carried out at the experimentally known lattice parameters and performed with the most modern density functional using generalized gradient corrections. For the final calculations, 30 k-points in the irreducible wedge of the Brillouin zone were employed. In Table III the calculated NQIs for the tetragonal PbO phase litharge and the orthorhombic PbO phase massicote are displayed. In Pb_3O_4 there are two inequivalent Pb sites: one third of all Pb sites are occupied by Pb(IV) in PbO_6 octahedra and two thirds are occupied by Pb(II) in a pyramidal arrangement of three oxygen atoms [19].

4. Discussion and conclusion

4.1. PbO

The X-ray diffractogram of the first PbO sample is dominated by the PbO phases massicote and litharge. The results of the *ab-initio* EFG calculations for litharge and massicote are in extremely good agreement with the two NQIs detected in the TDPAC experiment (see Table III). Therefore, we assign the NQI 1 to massicote and the NQI 2 to litharge. For the TDPAC experiment with Pb_3O_4 additional PbO samples were prepared which show only one NQI at 500°C before conversion to Pb_3O_4, i.e., NQI 3, with a slightly lower frequency than NQI 2 and axial symmetry. Taking into account that typical temperature dependencies of NQIs are in the range of $\frac{d \ln \omega_Q}{dT} = -1 \ldots 1.4 \cdot 10^{-4} K^{-1}$ the difference between the frequencies of NQI 2 and NQI 3 can easily be explained by the different temperatures at which the experiments were carried out. Assuming that this is correct, the temperature dependence of the NQI in PbO phase litharge is $\frac{\Delta \ln \omega_Q}{\Delta T} = -1.2 \cdot 10^{-4} K^{-1}$. It is reported in the literature that litharge starts to convert very slowly to massicote at temperatures higher than

488°C [18]. According to our experiments both phases of PbO, litharge and massicote, can be obtained by heating $PbCO_3$ and are stable at room temperature. It may very well be that in the first preparation of PbO higher temperatures were reached than in the second preparation of PbO resulting in a higher fraction of the high temperature phase massicote in the PbO phase. Since the conversion of litharge to massicote is described as quite slow it is assumed that in the TDPAC experiment at 500°C the time was too short that a significant fraction of the litharge could convert to massicote.

4.2. Pb_3O_4

The NQIs 4 and 5 together with their fractions in the TDPAC spectrum detected in Pb_3O_4 are listed in Table II. Since these experimentally determined fractions are in perfect agreement with the fractions of the two different Pb sites in Pb_3O_4 reported in the literature [19] we assign NQI 4 to the Pb(II) site and NQI 5 to the Pb(IV) site. The *ab-initio* EFG calculations support this assignment, albeit the agreement with the experimental data is less impressive than in the case of PbO. The reason for this larger deviation compared to the *ab-initio* calculation for PbO remains up to now unclear. A possible explanation are deficiencies of the structural data.

A summary of the assignments is given in Table III. Our result for the PbO phase litharge agrees well with a previous TDPAC experiment which yielded the following NQI: ω_Q = 16.7(9) Mrad/s and η = 0 [4].

4.3. ^{204}Bi/^{204}Pb-GENERATOR

We would like to emphasize the fact that the present study clearly demonstrates that the use of the ^{204}Bi/^{204m}Pb-generator with a HPLC system has several advantages: (1) the experimenter is independent of the rare beam times of on-line isotope separators like ISOLDE/CERN, (2) the ^{204m}Pb activity is obtained in an aqueous HCl solution with high specific activity every 2 h, (3) the total amount of the ^{204m}Pb activity of one eluation is sufficient for two or three simultaneous TDPAC experiments, (4) the acidity of the aqueous HCl solution is an order of magnitude lower than described in [4, 14]: 0.03 N instead of 0.3 N, (5) a single irradiation of the lead target allows to perform TDPAC experiments up to four days, (6) the HPLC ion exchanger column is remotely controlled and – due to its small volume ($\approx$ 4 ml) – is easily covered by an adequate lead shield which reduces the radiation exposure of the experimenter significantly, (7) there is no need to use an isotopically enriched lead target for the irradiation.

Due to these advantages against the other isomeric TDPAC probes ^{111m}Cd and ^{199m}Hg, we expect that the ^{204m}Pb TDPAC probe will enjoy a multitude of fascinating applications in material and life sciences.

Acknowledgements

The authors thank the ISL team at the Hahn-Meitner-Institut (Berlin) for their technical help, and especially Dr. W.-D. Zeitz for his unlimited patience and support during the beam times. Financial support for this work was provided by grants of the Deutsche Forschungsgemeinschaft, Bundesministerium für Bildung und Forschung and the Fonds der Chemischen Industrie, Germany.

References

1. Tröger W., Nuclear probes in life sciences, *Hyperfine Interact.* **120/121** (1999), 117–128.
2. Tröger W. and Butz T., Inorganic biochemistry with short-lived radioisotopes as nuclear probes, *Hyperfine Interact.* **129** (2001), 511.
3. Firestone R. B., *Table of Isotopes CD*, 1996.
4. Haas H. and Shirley D. A., Nuclear quadrupole interaction studies by perturbed angular correlations, *J. Chem. Phys.* **58**(8) (1973), 3339–3355.
5. Fedoseyev V. N., Huber G., Köster U., Lettry J., Mishin V. I., Ravn H., Sebastian V. and the ISOLDE Collaboration, The ISOLDE laser ion source for exotic nuclei, *Hyperfine Interact.* **127** (2000), 409–416.
6. Kugler E., The ISOLDE facility, *Hyperfine Interact.* **129** (2000), 23–42.
7. Tröger W., Dietrich M., Araujo J. P., Correia J. G. and Haas H., The nuclear quadrupole interaction of ^{204}Pb in cadmium monitored by γ–γ-perturbed angular correlations, *Z. Naturforsch.* **57a** (2002), 586–590.
8. Dietrich M., Deicher M., Haas H., Dorda U. and the ISOLDE-Kollaboration, ^{204m}Pb in $PbTiO_3$: Der elektrische Feldgradient auf dem A-Platz in Perovskiten. In: *Abstract-Band zum Arbeitstreffen "Forschung mit nuklearen Sonden und Ionenstrahlen"*. ISSN 0936-0891, 2003.
9. Blaha P., Schwarz K. and Luitz J., WIEN97, a FLAPW package for calculating crystal properties, ISBN 3-9501031-0-4, 1997.
10. Frauenfelder H. and Steffen R. M., Angular distribution of nuclear radiation, In: Siegbahn K. (ed.), *Alpha-, Beta- and Gamma-Ray Spectroscopy*, Vol. 2, Chapter 29, North-Holland Publishing Company, Amsterdam, 1965, pp. 997–1198.
11. Butz T., Saibene S., Fraenzke T. and Weber M., A "TDPAC-Camera," *Nucl. Instrum. Methods* **A284** (1989), 417.
12. Butz T., Nuclear quadrupole interactions studied by time differential perturbed angular correlations of gamma-rays, *Z. Naturf.* **51a** (1996), 396.
13. Gibson W. M., *The Radiochemistry of Lead*, National Academy of Sciences, Washington, USA, 1961.
14. Stockendal R., McDonell J. A., Schmorak M. and Berström I., Nuclear isomerism in odd lead isotopes, *Arkiv för Fysik* **11** (1956), 165.

15. Friedemann S., ^{204m}*Pb:Die Wiederentdeckung einer isomeren TDPAC–Sonde in der Festkörperphysik*, Diplomarbeit, Fakultät für Physik und Geowissenschaften, Universität Leipzig, Leipzig, 2004.
16. Dow Chemical Company, Data sheet Dowex 1 × 8 Anion Resin, http://www.dow.com/liquidseps/prod/dx 1x8.htm, 2003.
17. Kraus K. A., Anion-exchange studies. XI. Lead(II) and Bismuth(III) in Chloride and Nitrate Solutions, *J. Am. Chem. Soc.* **76** (1954), 5916.
18. Holleman A. F. and Wiberg E., *Lehrbuch der Anoganischen Chemie*, Walter de Gruyter, Berlin, 1976.
19. Wells A., *Structural Inorganic Chemistry*, Oxford Science Publications, Oxford, 1984.
20. Tröger W. and Butz T., Oxygen diffusion in $YBa_2Cu_3O_{7-\delta}$ studied via ^{133}Ba(EC)^{133}Cs nuclear quadrupole interaction, *Zeitung für Naturforschung* **47a** (1992), 12–20.

Hyperfine Interactions (2004) 159:323–329
DOI 10.1007/s10751-005-9122-3

PAC Studies of BSA Conformational Changes

J. GRYBOŚ[1,*], M. MARSZAŁEK[1], M. LEKKA[1], F. HEINRICH[2] and W. TRÖGER[2]
[1]*The Henryk Niewodniczański Institute of Nuclear Physics Polish Academy of Sciences, Radzikowskiego 152, 31-342 Kraków, Poland; e-mail: Joanna.Grybos@ifj.edu.pl*
[2]*Institute of Experimental Physics II, University of Leipzig, Linnéstraße 5, 04103 Leipzig Germany*

Abstract. The structure of biological molecules plays an important role in many biological processes. The variation of ambient parameters such as pH, temperature or pressure can influence important properties of protein molecules like conformation or stability. By means of the perturbed angular correlation (PAC) technique and atomic force microscopy (AFM) a conformation change of bovine serum albumin (BSA) molecules has been studied as a function of the pH value of the BSA solution. The observed decrease of the rotational correlation time and the increase of molecule diameter as a function of pH were attributed to conformational changes of bovine serum albumin induced by different pH values of the BSA solution.

Key Words: albumin conformation, atomic force microscopy, AFM, dynamic hyperfine interaction, nuclear quadrupole interaction, NQI, rotational correlation time, perturbed angular correlation of γ-rays, PAC.

1. Introduction

Bovine serum albumin is one of the most abundant proteins in blood circulatory systems. They provide many functions like, e.g., maintaining osmotic pressure or pH. Albumins participate in the transport of a variety of ligands whose affinity depend on the state of the protein which is affected by the pH and the calcium concentration in the blood. Since these conditions vary in different tissues and organs, the characterization of albumin isoforms may help for a better understanding of its functionality [1].

The primary structure of BSA is constituted from a single polypeptide chain of 582 amino acid residues and its amino acid sequence is well known [2]. Its secondary structure, in the native form, is formed by 67% of α-helixes with six turns, 33% of β-sheets, and 17 disulfide bridges. The protein was modeled as an ellipsoid with axes of 14 nm × 4 nm, with three domains in line.

It is known that albumin undergoes pH dependent conformational changes in alkaline and acidic solutions [3]. These conditions influence the forms of the BSA molecules and they are classified as expanded (E, below pH 2.7), fast (F, below pH 4.3), native (N, pH 7.0), basic (B, above pH 8.0) and aged (A, above pH 10.0). These five forms are characterized by a different content of α-helixes, β-sheets and random coils [4]. At a pH value between 2.7 and 4.3, albumin is in the fast form which is characterized by an increase of the intrinsic viscosity, and a significant loss of the helical content is observed [5]. The native form with a globular shape is predominant in the pH range 4.5–7.0 [3].

In this paper, two different techniques, perturbed angular correlation of γ-rays and atomic force microscopy, were applied to study the size and dynamics of BSA molecules in aqueous solutions at different pH. By AFM the diameters, by PAC the rotational correlation times of the BSA molecules were determined. Both parameters depend on pH.

2. Samples preparation and experiment

The BSA molecules for the AFM imaging were immobilized on glass coverslips activated with a 2.5% glutaraldehyde (supplied by Sigma) aqueous solution for 30 min. Before activation, they were silanized in 4% solution of 3-amino-propyltriethoxysilane (APTES, Sigma) in toluene. The protein immobilization was done by substrate immersion in 0.5 mg/ml aqueous protein solution for one hour.

For the PAC experiments a radioisotope, the PAC probe, has to be attached to the BSA molecule. Here, we used the PAC probe ^{111}In(EC)^{111}Cd, which decays from ^{111}In to ^{111}Cd via an electron capture (EC) before the γ-γ-cascade occurs which is used for the PAC experiments. The samples for the PAC experiments were prepared in the following way: first, a 200 μM aqueous solution of the bovine serum albumin (Sigma, M_w = 69 kDa) was prepared which was incubated with a solution of $^{111}InCl_3$ (no carrier added) at 30°C for 1 h in order to assure better binding of the In^{3+} ions. The sample activity was in the range of 100 to 400 kBq. The total sample volume was about 1–2 ml. The pH value of the BSA solution varied from 2 to 10 and it was checked before and after each measurement. PAC experiments were performed at room temperature and at liquid nitrogen temperature.

The AFM images were recorded using silicon nitride cantilevers with a spring constant of 0.03 N/m and the radius of curvature was about 50 nm. The measurements were carried out at room temperature in deionized water using a "liquid cell" setup. The pH value of the solution was adjusted to 2.0, 3.5, 5.0, 7.0 and 10.0. The size of the albumin molecules can be calculated from the dependence [6] between the measured diameter S of the investigated structures (assumed to be circular) and their real diameter D, taking into account the radius R of curvature of the AFM tip:

$$S = 2\sqrt{R\cdot D + \frac{D^2}{4}} \tag{1}$$

The angular correlations of the 172–245 keV γ-ray cascade in the PAC probe $^{111}In(EC)^{111}Cd$ were measured using a 6 detector PAC spectrometer equipped with 6 BaF_2 detectors [7]. Thirty coincidence spectra $N_i(\Theta,t)$ (i=1,...,30) were collected simultaneously at angles $\Theta = 90°$ and 180° between the detectors. For a detailed description of the data analysis see [8].

In the case of a static nuclear quadrupole interaction the data were fitted by the perturbation function $G^{22}_{static}(t)$:

$$G_{22}^{static}(t) = \sum_{n=0}^{3} s_{2n}(\eta)\cos(\omega_n(\eta)t)\exp(-\delta\omega_n(\eta)t) \tag{2}$$

where s_{2n} are the amplitude coefficients and ω_n are the frequencies, which are functions of the asymmetry parameter $\eta=(V_{xx}-V_{yy})/V_{zz}$ of the electric field gradient (tensor components V_{ii} (i=x,y,z)). $\delta\omega_n$ is the half width at half-maximum of the Lorentzian frequency distribution around the mean value ω_n. Such distributions arise due to variations of the charge environment of the probe nuclei. This is typical for frozen solutions, where the EFG is determined not only by the regular charge distribution in the molecule, but also by random arrangements of the solvent molecules.

In the case of molecules dissolved in a liquid a relaxation effect is produced by the rotational diffusion of the molecules. This can be described by the following perturbation function:

$$G_{22}(t) = e^{-\lambda_s t}\cdot G_{22}^{static}(t) \tag{3}$$

with the slow relaxation constant $\lambda_s = \frac{1}{\tau_s}$ and with the assumption that $\omega_n\tau_s > 1$. In the case of a very rapid molecular motion which corresponds to the condition $\omega_n\tau_s < 1$, the perturbation function $G_{22}(t)$ becomes in accordance with the Abragam and Pound theory [9] a simple exponential form:

$$G_{22}(t) = e^{-\lambda_f t} \tag{4}$$

where λ_f is the fast relaxation constant. For the nuclear spin $I = \frac{5}{2}$ of the used PAC probe $^{111}In(EC)^{111}Cd$, λ_f becomes $\lambda_f \approx 100.8\omega_Q^2\tau_c$, where τ_c is the rotational correlation time [10] describing the mobility of molecules in a solution, ω_Q being the nuclear quadrupole frequency.

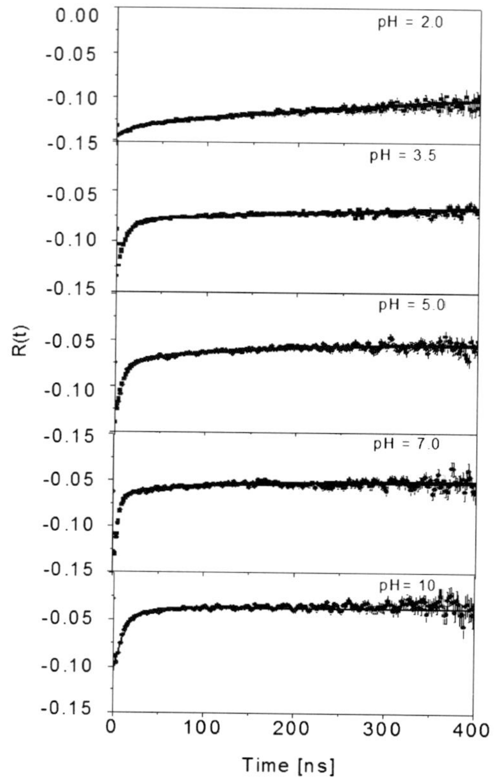

Figure 1. The PAC spectra of the ^{111}In-BSA complex at different pH measured at room temperature.

Table I. The hyperfine parameters for ^{111}In-BSA samples at different pH value

pH	f_0 [%]	f_1 [%]	λ_s [ns^{-1}]	τ_s [ns]	f_2 [%]	λ_f [ns^{-1}]	τ_f [ns]
2.0	34(7)	58(1)	0.001(0)	925(35)	8(1)	0.044(10)	0.9(5)
3.5	33(11)	26(11)	0.001(1)	909(495)	41(1)	0.102(3)	2.1(5)
5	38(1)	16(1)	0.008(1)	122(13)	46(1)	0.124(5)	3(1)
7	32(1)	9(1)	0.013(1)	77(10)	59(1)	0.202(10)	4(2)
10	35(0)	0(0)	0(0)	0(0)	65(1)	0.077(1)	1.6(2)

3. Results and discussion

The PAC spectra recorded at room temperature and at different pH values are shown in Figure 1. The spectra were fitted with a linear combination of formulae (3) and (4) and an unperturbed, time independent PAC signal (a_0):

$$G_{22,total}(t) = a_0 + a_1 e^{-\lambda_s t} \cdot G_{22}^{static}(t) + a_2 e^{-\lambda_f t} \tag{5}$$

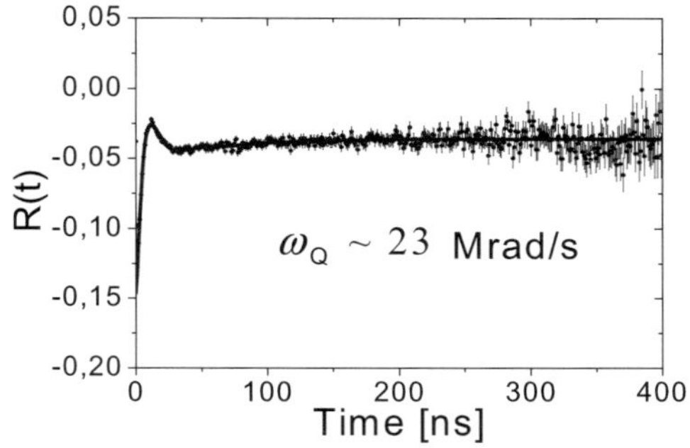

Figure 2. The PAC spectrum of the ^{111}In-BSA complex at pH 7 measured at 77 K.

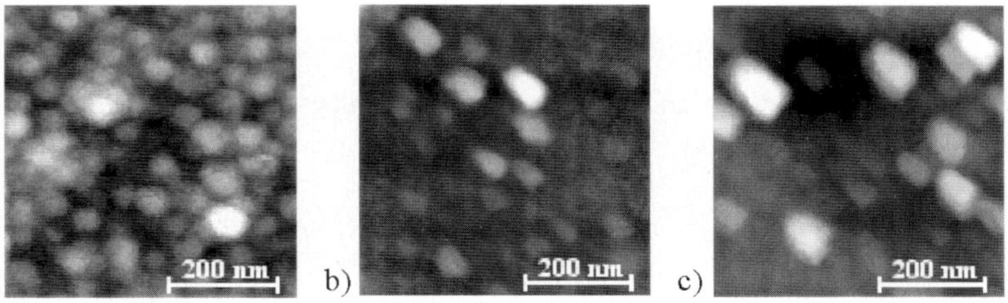

Figure 3. The AFM images of BSA molecules measured for three pH: (a) 3.5, (b) 5.0, (c) 7.0. (Image size 427 nm × 427 nm).

The amplitudes a_0, a_1 and a_2 were treated as free parameters. These amplitudes contain the anisotropy of the cascade, the solid angle correction factors, and the population of the different states by the PAC probe. In order to compare the populations directly the fractions $f_i = \frac{a_i}{\sum_{i=0}^{2} a_i}$ were calculated and shown together with the hyperfine parameters obtained from the fitting procedure in Table I. The fraction f_1 of the slowly rotating species was assigned to In^{3+} ions bound to BSA molecules whereas the fraction f_2 with the fast rotating species was attributed to In^{3+} ions bound to hydroxyl groups or other small molecules. The unperturbed fraction f_0, which changed only little with pH, was assigned to unbound probe ions in the solution. In order to determine the fast rotational correlation time τ_c, it is necessary to know the parameters of the static nuclear quadrupole interaction. These parameters were determined from PAC measurements at the liquid nitrogen and were very similar for all pH values: quadrupole frequency $\omega_Q \approx$ 22–23 Mrad/s, asymmetry parameter $\eta \approx 0.5$ and a large Lorentzian frequency distribution. The static PAC spectrum of the ^{111}In-BSA complex at pH 7 measured at 77 K is shown in Figure 2. The rotational correlation time of slow molecular motion is simply given as a reciprocal of the λ_s value. The values of both rotational correlation times are also shown in Table I.

The decrease of the fraction of slow relaxation (f_1, λ_s) accompanied by the increase of the number of indium ions bound to hydroxyl groups (f_2, λ_f) with

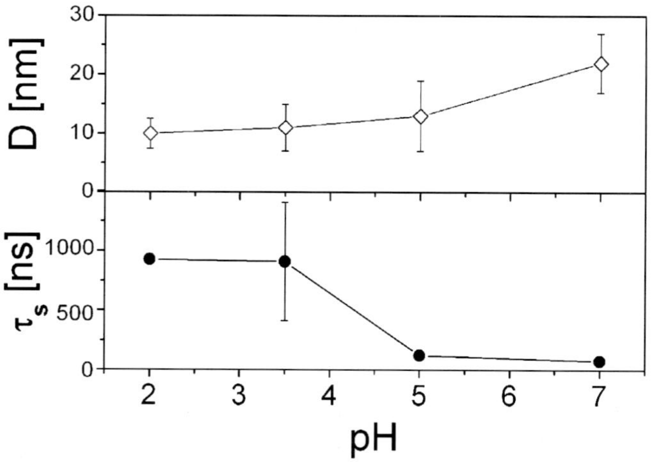

Figure 4. The BSA diameter and the rotational correlation time as a function of pH of the BSA solution.

increasing pH can be explained by conformational changes of the BSA molecules. The only amino acid of BSA which can bind In, cysteine 34, is located in the center of the elongated BSA structure, and for the F-conformation of albumin this binding site is easily accessible for the indium probe. However, an increase of the pH leads to a conformational change of BSA into the more oval structure (N), which results in a reduced accessibility of the cysteine 34 for In^{3+} ions. Such a structural change is also reflected by the rotational time τ_f which decreases with the increase of pH. For pH 10, the parameters f_1, λ_s and τ_s, indicate that the indium ions did not bind to the BSA molecule at all. Previously reported rotational correlation times for ^{111}In-BSA system were different from those obtained here [11]. The discrepancy can be explained by the different conditions of the PAC experiments (shorter time scale and less statistical quality of the PAC spectra of [11]) and by different procedures of the sample preparation.

The second fraction f_2 represents $^{111}In^{3+}$ ions bound to hydroxyl groups or other smaller molecules. These smaller molecules may result from the electron capture after-effects accompanying the radioactive decay of ^{111}In to ^{111}Cd leading to a partial or entire disintegration of the BSA molecules. This fraction f_2 increases with pH at the expense of the fraction f_1 and becomes dominant for pH values larger than 5 since the amount of hydroxyl groups in the solution increases with increasing pH.

The AFM images of albumin molecules immobilized on a glass surface for three chosen pH values are presented in Figure 3. The average value of the diameter of a single albumin molecule was determined by fitting Gaussian distributions to the histograms of the molecule diameters. Taking into account the effect of a topography convolution with AFM tip (Equation 1), the real diameter of the BSA molecule is 10 ± 2.5 nm, 11 ± 4 nm, 13 ± 6 nm and 22 ± 5

nm for pH 2.0, 3.5, 5.0 and 7.0, respectively. It was observed that the molecule diameter increases as a function of pH. Figure 4 shows the comparison of the BSA molecule diameter and rotational correlation times.

These results obtained by two different methods confirmed that albumin molecules undergo conformational changes at different pH value. These structural properties were reflected by the decrease of the rotational correlation time (PAC) and the increase of the molecule diameter seen in AFM images.

Acknowledgments

The authors acknowledge the support of a Polish–German Scientific and Technical Collaboration (WTZ No 4440/2002) from the Bundesministerium für Bildung und Forschung and State Committee for Scientific Research.

References

1. Carter D. C. and Ho J. X., *Adv. Protein Chem.* **45** (1994), 153.
2. Burova T. V., Grinberg N. V., Golubeva I. A., Mashkevich A. Y., Grinberg V. Y. and Tolstoguzov V. B., *Food Hydrocoll.* **13** (1999), 7.
3. Harmsen B. J. M. and Braam W. J. M., *Int. J. Protein Res.* **1** (1969), 225.
4. Reed R. G., Feldhoff R. C., Clute O. L. and Peters T. Jr., *Biochemistry* **14** (1975), 4578.
5. Harrington W. F., Johnson P. and Ottewill R. H., *Biochem. J.* **62** (1956), 569.
6. Engel A., Schoenenberger C. A. and Müller D. J., *Curr. Opin. Struct. Biol.* **7** (1997), 279.
7. Butz T., Saibene S., Fraenzke T. and Weber M., *Nucl. Instrum. Methods A* **284** (1989), 417.
8. Butz T. *Z., Naturforsch.* **51a** (1996), 396.
9. Abragam A. and Pound R. V., *Phys. Rev.* **92** (1953), 943.
10. Marshall A. G. and Meares C. F., *J. Chem. Phys.* **56** (1972), 1226.
11. Martin P. W. and Kalfas C. A., *Nucl. Instrum. Methods* **171** (1980), 603.

Hyperfine Interactions (2004) 159:331–335
DOI 10.1007/s10751-005-9123-2

In situ ^{54}Mn NMRON Studies of the Mixed Halide $Mn(Br_XCl_{1-X})_2 \cdot 4H_2O$ in Applied Magnetic Fields

W. D. HUTCHISON[1,*], S. J. HARKER[1,a], D. H. CHAPLIN[1] and G. J. BOWDEN[2]
[1]*School of Physical, Environmental and Mathematical Sciences, The University of New South Wales at the Australian Defence Force Academy, Canberra, Australia; e-mail: w.hutchison@adfa.edu.au*
[2]*School of Physics and Astronomy, University of Southampton, SO17 1BJ, Southampton, UK*

Abstract. ^{54}Mn NMRON in the mixed halide antiferromagnet, $Mn(Br_XCl_{1-X})_2 \cdot 4H_2O$, for varying external field, is reported. Significant qualitative differences are found between these NMRON transitions and those of the two terminal compounds, especially in respect to line widths. A tentative assignment is made to most of the observed NMRON transitions out to 0.8 T. An unidentified lower frequency inhomogeneous signal, present only at the lowest temperatures, with no counterpart in the terminal compounds, is also reported.

Key Words: antiferromagnets, magnetic field dependence, mixed halides, NMRON.

1. Introduction

For many years, *in situ* ^{54}Mn doped in $MnCl_2 \cdot 4H_2O$ ($T_N = 1.62$ K) has been used, as an archetypal test bed for low temperature nuclear orientation (LTNO) and NMR on oriented nuclei (NMRON), in anti-ferromagnetic insulators [1–3]. More recently $MnCl_2 \cdot 4H_2O$ has been used as a LTNO host for on-line implantation [4]. Rather surprisingly, despite being monoclinic, with four Mn ions per unit cell and measurable biaxial anisotropy, $MnCl_2 \cdot 4H_2O$ behaves like a simple uniaxial two sub-lattice antiferromagnet (AF), with a well-defined single stage spin–flop B_{SF} and paramagnetic transition B_P [3, 5, 6]. This is in marked contrast with the nominally isomorphic compound $MnBr_2 \cdot 4H_2O$ ($T_N = 2.12$ K), which is characterised by a two-stage spin–flop [7]. But, in a companion paper to this one [8], it is shown that the spin–flop mechanism in the mixed halides of the form $Mn(Cl_xBr_{1-x})_2 \cdot 4H_2O$ is essentially interpolative between the two

---

* Author for correspondence.

[a] Present address: School of Physics and Materials Science, Monash University, Melbourne, Australia.

terminal compounds, with the single stage spin–flop mechanism persisting well into the bromide rich configuration.

From a DC magnetic point of view therefore it would appear that the mixed halides are relatively straightforward and well understood. However, in [9] we showed that in a mixed halide, the ^{54}Mn spins cool much faster, to a lower base temperature, than in either terminal compound. Thus partial halide substitution considerably enhances the nuclear spin lattice relaxation rate (NSLR). This paper focuses on the dynamics of the mixed halide through NMRON line-width studies. In addition, we highlight the startling influence of very small non-resonant radio frequency (RF) fields on the reduction of the ^{54}Mn γ-ray anisotropy, again not observed in the terminal compounds. Finally, the field dependence of the centre frequencies is compared and contrasted with those of the terminal compounds.

2. Experimental details

^{54}Mn radioactively doped $Mn(Br_{0.46}Cl_{0.54})_2 \cdot 4H_2O$ single crystals were grown per [8]. The relative halide concentrations are by mass. The results shown in this paper are predominantly from one single crystal (A), which was the most radioactive (19.4 μCi) of all the crystals made. This was necessary to ensure good counting statistics required for the NMRON line shape details. The sample was limited in base temperature when cooling in zero fields due to the non-negligible radioactive self-heating but this was of little consequence in small applied DC fields as by far the greater influence in decreasing the working γ-anisotropy was the deleterious effect of very small non-resonant RF fields. As in [8, 9], a ^{60}CoCo(hcp) thermometer was placed in close proximity to the mixed halide sample, to monitor the non-resonant RF heating of the Cu cold finger.

3. Results and discussion

Marked differences between the mixed halides and their terminal counterparts were found particularly at low magnetic fields ($\lesssim$0.2 T). Line widths are very much broader with a large homogeneous component. However, at higher fields, substantial line narrowing and suppression of the non-resonant RF degradation of the γ-anisotropy [9] occurs which allowed tentative assignment to be made of several of the anticipated transitions, including the lesser populated $|-2\rangle \leftrightarrow |-1\rangle$ branch. In zero applied field it was noticeable that RF fields as low as −14 dBm (nominal) could seriously degrade the working ^{54}Mn γ-ray anisotropy from −23% to the order of −10% in ~2 h. But, despite the reduced γ-ray anisotropy, NMRON signals could still be clearly discerned, over typical frequency sweeps from 485 to 515 MHz. Optimal NMRON experiments were performed using RF fields in the range −10 to −14 dBm. For these very low RF

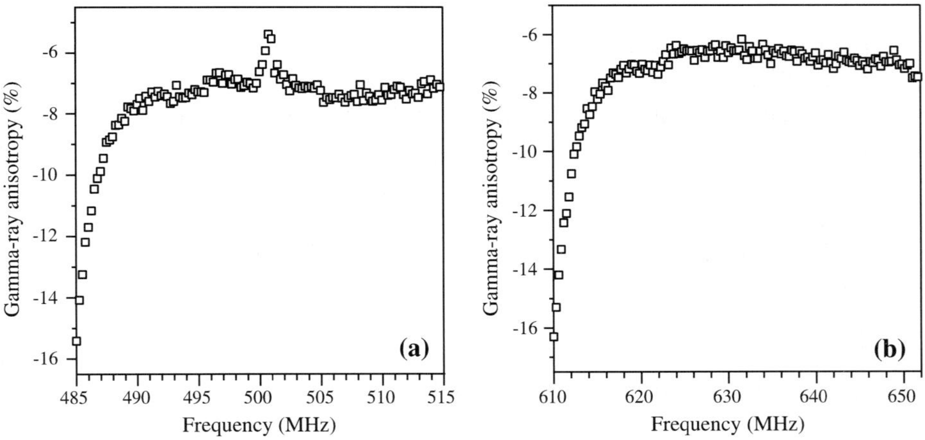

Figure 1. (a) Stepping up over a frequency range corresponding to ^{54}Mn NMRON, 0 T. (b) Stepping up over a frequency range corresponding to ^{55}Mn NMR-TDNO, 0 T.

fields the simultaneously recorded ^{60}CoCo γ -anisotropy varied by only ~0.5%. In Figure 1, the generic reduction of the working γ-anisotropy is seen to dominate in zero field. The experiments shown generously span the anticipated ^{54}Mn NMRON and ^{55}Mn NMR- TDNO frequency ranges and reveal only small amplitude (narrow) resonances near 501 and 624 MHz, respectively.

Figure 2(a,b) illustrates the difficulty of detecting all the broad low field resonances against the sloping background. Figure 2(a) has FM of ±200 kHz, which can only enhance the amplitude of the genuine inhomogeneous NMRON signals, while Figure 2(b) has no FM. The major difference between these two runs is actually the much lower starting frequency of 470 MHz in Figure 2(b). The existence of such a low frequency homogeneous and broad resonance at ~487 MHz in Figure 2(b) was not anticipated. We tentatively suggest that this is due to a lifetime limited magnon-nuclear mode. Further investigation at 0 T with no FM (not shown) revealed a second broad homogeneous resonance with FWHM of ~10 MHz centred at ~500 MHz overlaying precisely the expected degenerate NMRON sub resonances of (^{54}Mn) $Mn(Br_{0.46}Cl_{0.54})_2 \cdot 4H_2O$. Thus the NMRON signal in Figure 2(a) is therefore not a normal NMRON signal but is dressed possibly by magnons and/or localised excitations that are heavily lifetime broadened and therefore homogeneous (see for example [10]).

In Figure 3 we show the zero field results for two other crystals B and C of comparable Cl concentration but significantly less radioactive doping at 8.1 μCi and 7.9 μCi, respectively. The lower frequency resonance, ~497 MHz, was inhomogeneous, disappearing with no FM (not shown). It was not observable in the higher temperature runs with the 19.4 μCi (see Figure 1(a)). It is sufficiently low in frequency that it is not readily identified with calculations based on molecular field theory [3, 7]. Finally, in Figure 4 we compare the applied field

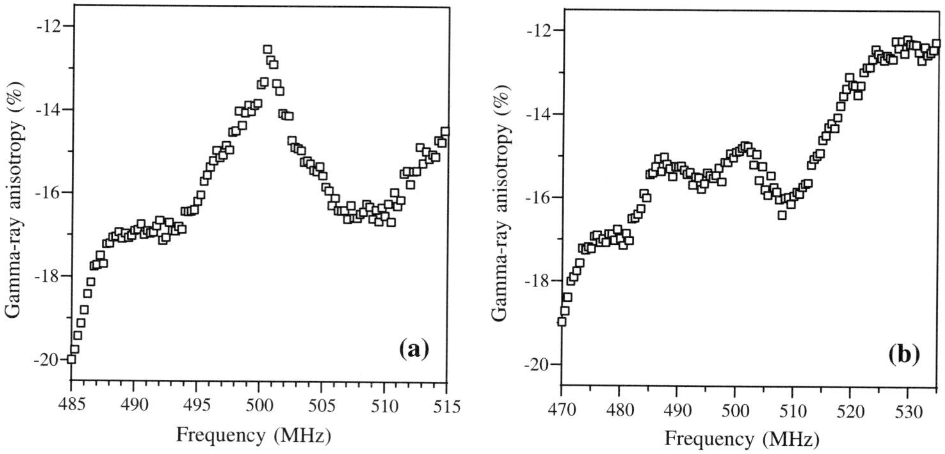

Figure 2. (a) Stepping up over a frequency range corresponding to ^{54}Mn NMRON, (a) 485 to 515 MHz (FM ± 200 kHz), (b) 470 to 540 MHz (FM zero), in 0.2 T applied field.

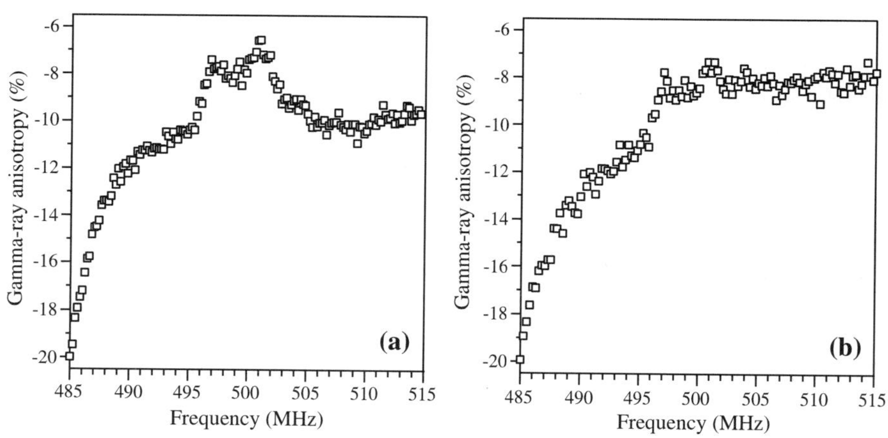

Figure 3. Stepping up over a frequency range corresponding to ^{54}Mn NMRON in zero field, (a) crystal B, (b) crystal C, respectively.

dependence of the genuine NMRON signals (three lines in all) with the $|-3\rangle$ to $|-2\rangle$ NMRON of the terminal compounds. Theoretical predictions, based on an interpolative molecular field model, for this lowest sub resonance are also shown. A fuller description of the results, analysis and discussion will be given elsewhere. But for the moment we note that NMRON studies offer a new way to study the properties of substitutionally disordered antiferromagnets, which offer the prospect of (1) probing the interplay between localised and de-localised excitations [10], and (2) faster cooling of nuclei in insulators.

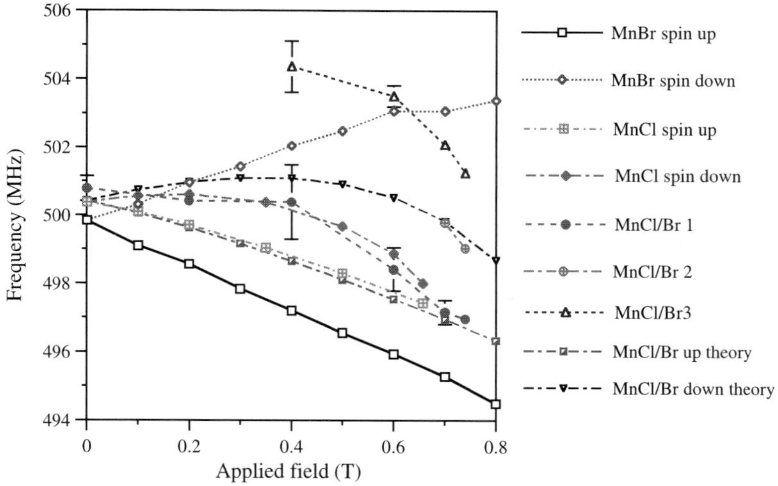

Figure 4. Magnetic field dependence of the mixed halide crystal A centre frequencies.

Acknowledgements

This work was supported by an ARC Large grant. One of us, SJH, acknowledges financial support from this grant for the position of ARC Research Associate. GJB also wishes to acknowledge support from the UNSW @ ADFA Visiting Fellows Fund.

References

1. Kotlicki A. and Turrell B. G., *Hyperfine Interact.* **11** (1981), 197.
2. Allsop A. L., deAraujo M., Bowden G. J., Clark R. G. and Stone N. J., *J. Phys. C* **17** (1984), 915.
3. Kotlicki A., McLeod B. A., Shott M. and Turrell B. G., *Phys. Rev. B* **29** (1984), 26.
4. Pond J. *et al., Phys. Rev. B* **62** (2000), 12241.
5. Spence R. D. and Nagarajan V., *Phys. Rev.* **149** (1966), 191.
6. Rives J. E. and Benedict V., *Phys. Rev. B* **12** (1975), 1908.
7. Prandolini M. J., Hutchison W. D., Lieb J., Chaplin D. H. and Bowden G. J., *Hyperfine Interact.* **C1** (1996), 84.
8. Harker S. J., Hutchison W. D., Chaplin D. H. and Bowden G. J., this conference.
9. Chaplin D. H. and Hutchison W. D., *Hyperfine Interact.* **136** (2001), 239.
10. Tahir-Kheli A. R., Fujiwara T. and Elliott R., *J. Phys. C. Solid State Phys.* **11** (1978), 497.

Hyperfine Interactions (2004) 159:337–343
DOI 10.1007/s10751-005-9124-1

The Relationship of Mössbauer Hyperfine Parameters and Structural Variations of Iron Containing Proteins and Model Compounds in Biomedical Research

M. I. OSHTRAKH*
Division of Applied Biophysics, Faculty of Physical Techniques and Devices for Quality Control, Ural State Technical University – UPI, Ekaterinburg, 620002 Russian Federation; e-mail: oshtrakh@mail.utnet.ru

Abstract. The relationship of Mössbauer hyperfine parameters and structural variations of iron containing proteins and model compounds was considered as applied to biomedical research. It was shown that Mössbauer hyperfine parameters give new information about structural changes of proteins during molecular diseases or effect of environment factors as well as about structural peculiarities of pharmaceutical compounds.

Key Words: biomedical research, hyperfine parameters, iron containing proteins, Mössbauer spectra, pharmaceutical compounds, structural variations.

1. Introduction

Iron containing proteins play a very important role in living systems. These proteins realize enzyme functions, electron, oxygen and iron transport, oxygen and iron storage. Each type of proteins has a wide range of functional and structural variations in nature related to the evolution process. Variations of the protein primary structure in some cases may lead to the changes of iron environment and the iron electronic structure. Moreover, some pathological changes in the body may be caused or accompanied by the synthesis of proteins with structural changes or by protein destruction. Therefore, Mössbauer hyperfine parameters may reflect some variations of iron electronic structure and environment in iron containing proteins in normal and pathological cases. We consider so-called 'drastic' and 'small' changes of the iron electronic structure. 'Drastic' changes imply change of the iron valence and/or spin state, for instance, Fe(II, $S = 2$) → Fe(III, $S = 3/2$) or Fe(II, $S = 2$) → Fe(II, $S = 0$). 'Small' changes imply changes of the iron electronic structure without changes of the iron valence and spin state but with changes of the energies of the ground and low-lying excited iron electronic states [1–3]. In this

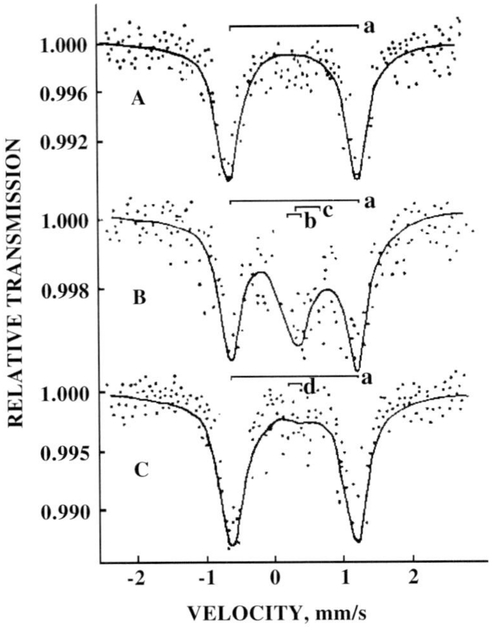

Figure 1. Mössbauer spectra of RBC from a normal person (A) and patients with hemoglobin E/β-thalassemia (B) and hemoglobin G Coushatta/β-thalassemia (C); **a** HbO_2, **b**, **c** and **d** new components. $T = 77$ K [4].

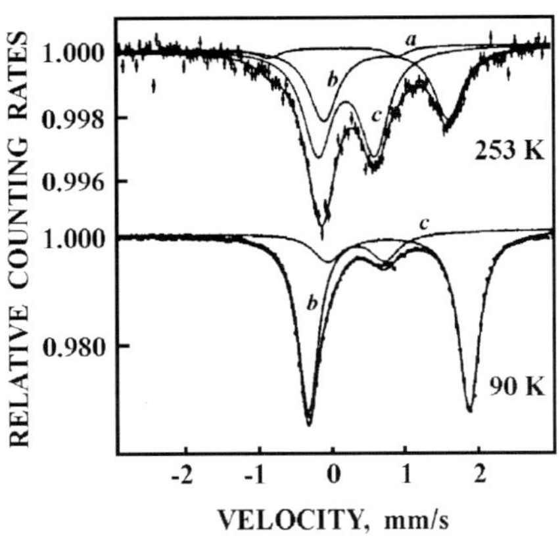

Figure 2. Mössbauer spectra of RBC infected with malarial parasite; *a* HbO_2, *b* Hb, *c* malarial pigment iron [5].

review the results of our and other authors' studies are briefly presented and discussed.

2. 'Drastic' changes

'Drastic' changes of the iron electronic structure were observed using Mössbauer spectroscopy in a comparative study of hemoglobin in red blood cells (RBC)

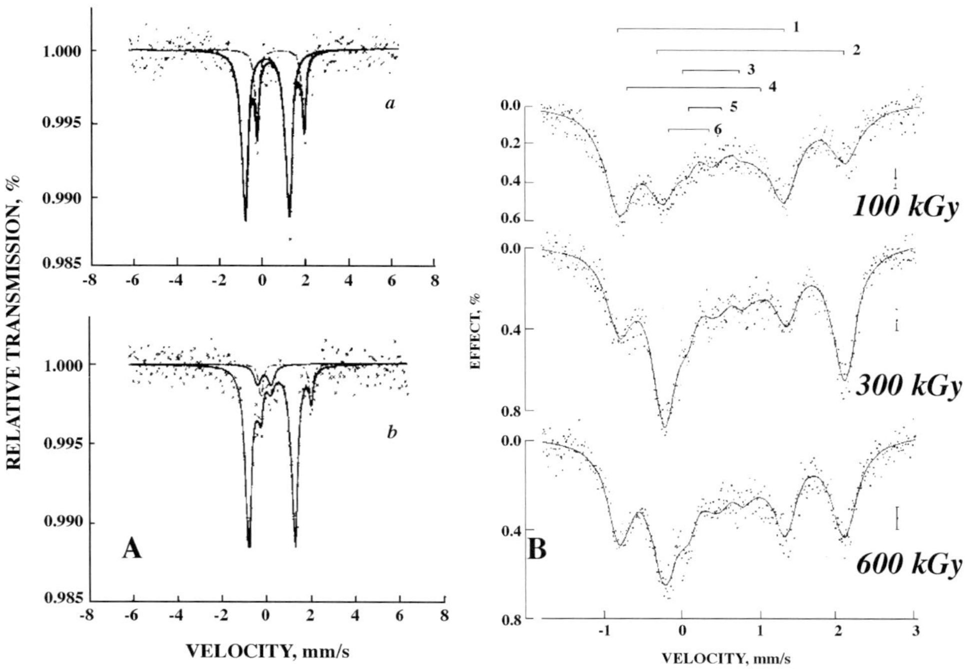

Figure 3. Mössbauer spectra of: (A) rat RBC: *a* initial (HbO_2 and Hb), *b* after feeding of 20 mg $CeCl_3$/kg body weight/day for 70 days (HbO_2, Hb and new component), $T = 77$ K [6]; (B) human oxygenated RBC irradiated with γ-rays (**1** HbO_2, **2** Hb, **3, 4, 5** products of hemoglobin radiolysis). $T = 87$ K [7].

from a normal person and patients with various hemoglobinopathies [4]. In Figure 1 an appearance of new components in addition to oxyhemoglobin (HbO_2) resulting from anomalous protein destruction is clearly seen. Studies of RBC infected with malaria parasite also demonstrated deoxyhemoglobin (Hb) destructions and the appearance of new component related to a malarial pigment iron (Figure 2) [5]. 'Drastic' changes were well observed in Mössbauer studies of hemoglobin effected by chemicals [6] and radiation [7] (Figure 3).

3. 'Small' changes

'Small' changes of Mössbauer hyperfine parameters were well defined in Figure 4 for the temperature dependences of the quadrupole splitting (ΔE_Q) for normal and anomalous Hb [8] and in Figure 5 in the plot of ΔE_Q and isomer shift (δ) for HbO_2 from normal people and patients with blood system malignant diseases [9]. Small variations of Mössbauer hyperfine parameters were also observed for ferritin-like iron in human normal and patient's tissues [10] (Figure 6) and in tissues from normal chicken (see also [11]) and chicken with leukemia (Figure 7).

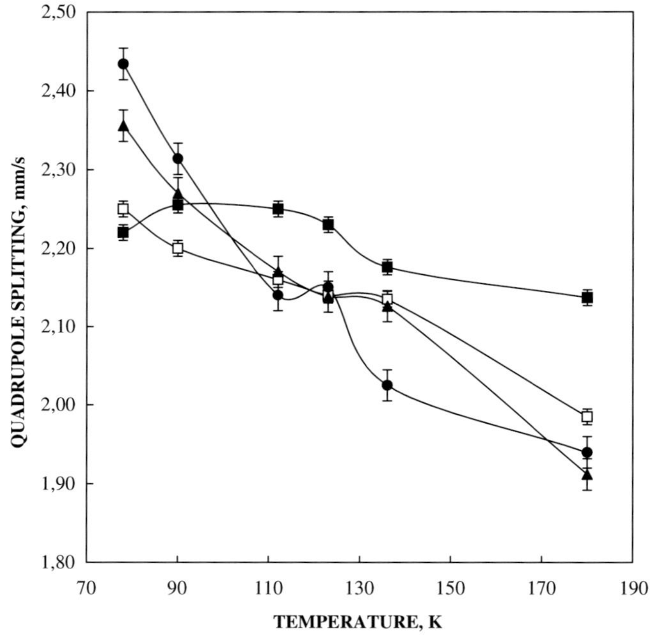

Figure 4. Differences of temperature dependences of quadrupole splitting for deoxy-forms of normal adult Hb (□), Hb E (■), Hb Guangzhou (▲) and Hb Queens (●). Data were taken from [8].

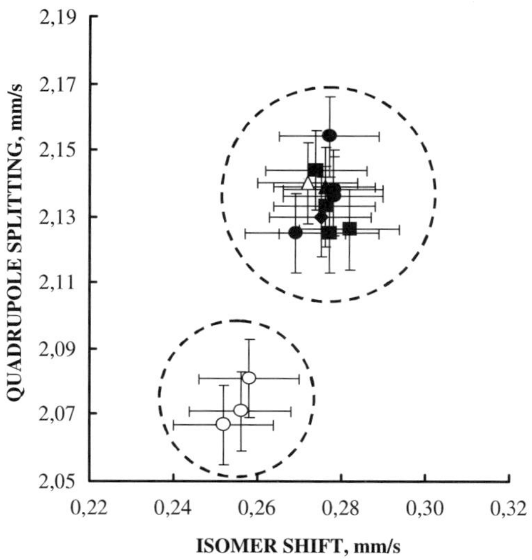

Figure 5. Differences of Mössbauer hyperfine parameters for HbO_2 of normal adult (○), normal fetal (△), HbO_2 from patients with chronic myeloleukemia (▲), acute myeloblastic leukemia (●), acute myelomonoblastic leukemia (♦) and erythremia (■); $T = 87$ K [9].

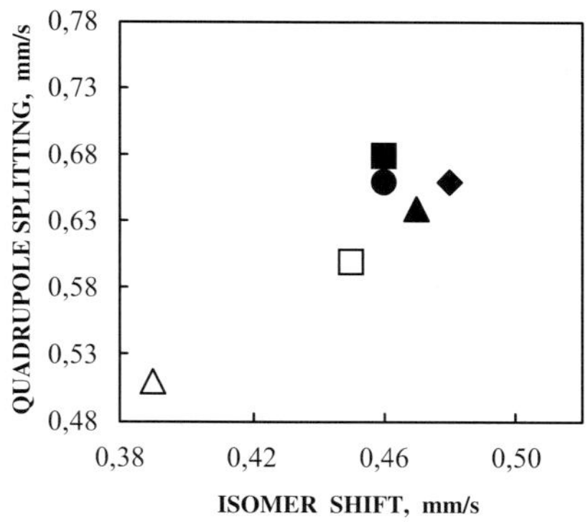

Figure 6. Differences of Mössbauer hyperfine parameters plotted as mean values of quadrupole splitting and isomer shift for human normal spleen (△), normal liver (□), pancreas from Thai patients with β-thalassemia/Hb E (▲), liver from Thai patients with β-thalassemia/Hb E (◆), spleen from Thai patients with β-thalassemia/Hb E (●), spleen from Australian patients with β-thalassemia (■) at 78 K. Data were taken from [10].

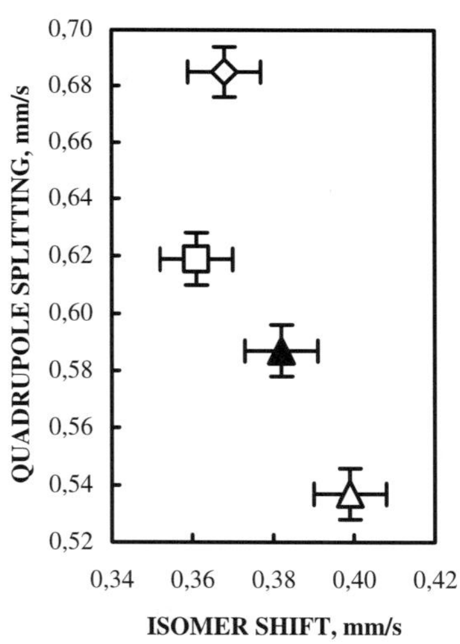

Figure 7. Differences of Mössbauer hyperfine parameters plotted as quadrupole splitting and isomer shift for human liver ferritin (◇), chicken liver (□), normal (△) and leukemic (▲) spleen. $T = 295$ K.

Small variations of ΔE_Q were found in the study of normal human HbO_2 in solution and in lyophilized form in comparison with those of oxygenated blood substitute based on human hemoglobin modified by pyridoxal-5′-phosphate (PLP) and glutaraldehyde (GA) in solution and in lyophilized form [12] (Figure 8). Some small differences of ΔE_Q were observed for several iron dextran complexes (IDC) which were pharmaceutically important ferritin models [13] (Figure 9).

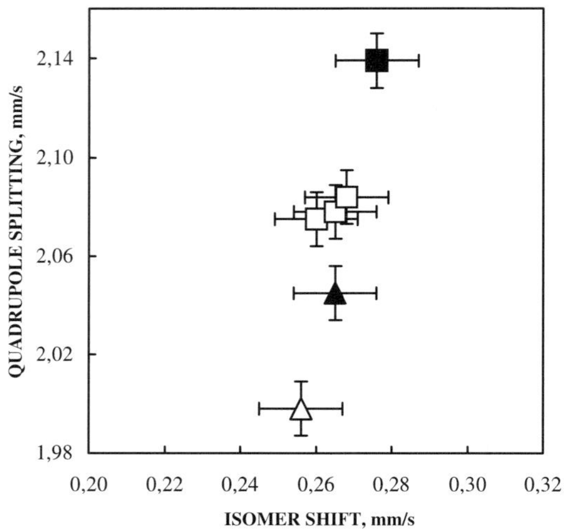

Figure 8. Differences of quadrupole splitting for: □ – HbO_2 in frozen solution, ■ – HbO_2 in lyophilized form, △ – HbO_2 modified by PLP + GA in frozen solution, ▲ – HbO_2 modified by PLP + GA in lyophilized form. $T = 87$ K. Data were taken from [12].

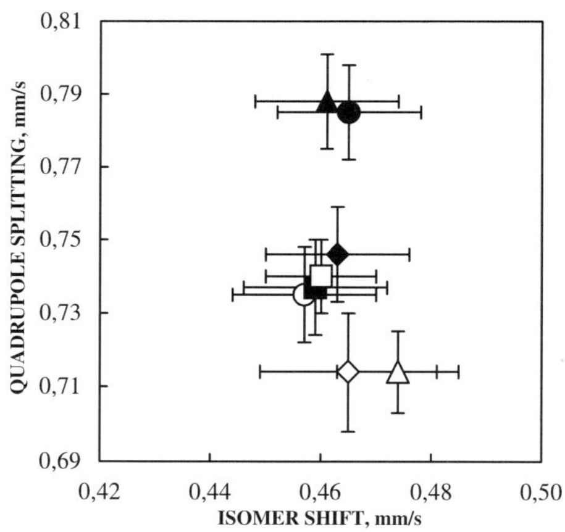

Figure 9. Differences of quadrupole splitting for IDC and human liver ferritin in lyophilized form at 87 K: ♦ – IDC3P; ■ – IDC4P; ▲ – IDC6P; • – IDC8P; ◊ – IDC87; ○ – Ferridextran (Lečiva, ČSSR); □ – Imferon (Fisons, UK); △ – Ferritin [13].

4. Conclusion

The results considered demonstrate possibilities of Mössbauer spectroscopy to analyze structural changes of hemoglobin resulting from protein destruction due to molecular diseases such as hemoglobinopathies or an effect of environmental factors. Mössbauer hyperfine parameters were very sensitive to the 'drastic' changes of the iron electronic structure accompanied by hemoglobin deoxygenation, oxidation and destruction as well as appearance of new iron containing compounds. Therefore, it is possible to determine new iron containing compounds basing on Mössbauer hyperfine parameters and evaluate the changes of hemoglobin molecules during pathological processes, effect of chemicals and protein radiolysis.

Mössbauer hyperfine parameters were also sensitive to the 'small' changes of the iron electronic structure; however, detection of these changes is more complicated and requires the high sensitive, precision and stable Mössbauer spectrometer. Variations of the primary structure in several abnormal hemoglobins resulting from the protein biosynthesis disturbance influenced the heme iron as it was shown by Mössbauer spectroscopy. Small variations of Mössbauer hyperfine parameters were found for oxyhemoglobin from patients with leukemia and erythremia. The structural changes in hemoglobin from these patients are unknown yet, however, small variations of the heme iron stereochemistry were supposed on the base of Mössbauer hyperfine parameters. Small variations of Mössbauer hyperfine parameters found for iron storage proteins in different human and animal tissues in normal and pathological cases indicated small changes in the iron cores of these proteins as a result of the level of core crystallinity and packing variations.

Thus, Mössbauer hyperfine parameters give new information about the changes of the iron electronic structure and stereochemistry in proteins that may be a result of pathological processes or an effect of environmental factors and in pharmaceutical compounds resulting from preparation.

References

1. Oshtrakh M. I., *Z. Naturforsch.* **51a** (1996), 381.
2. Oshtrakh M. I., *Spectrochim. Acta, Part A* **60** (2004), 217.
3. Oshtrakh M. I., *Faraday Discuss.* **126** (2004), 119.
4. Ding X. Q. and Hsia Y. F. *et al.*, *Hyperfine Interact.* **42** (1988), 893.
5. Bauminger E. R. *et al.*, *Hyperfine Interact.* **15/16** (1983), 885.
6. Cheng Y. *et al.*, *Chem–Biol. Interact.* **125** (2000), 191.
7. Oshtrakh M. I., *Nucl. Instrum. Meth. Phys. Res.* **B185** (2001), 129.
8. Zeng X. S., Lian Y. and Ren B. Z., *Jinan Li-yi Xuebao* **3** (1988), 27.
9. Oshtrakh M. I. and Semionkin V. A., *Stud. Biophys.* **139** (1990), 157.
10. St. Pierre T. G. and Chua–anusorn W. *et al.*, *Biochim. Biophys. Acta* **1407** (1998), 51.
11. Oshtrakh M. I., Milder O. B. and Semionkin V. A. *et al.*, *Hyperfine Interact.* **156–157** (2004), 279.
12. Oshtrakh M. I., Milder O. B. and Semionkin V. A. *et al.*, *Int. J. Biol. Macromol.* **28** (2000), 51.
13. Oshtrakh M. I. and Semionkin V. A. *et al.*, *Int. J. Biol. Macromol.* **29** (2001), 303.

Hyperfine Interactions (2004) 159:345–350
DOI 10.1007/s10751-005-9117-0

The Features of Mössbauer Spectra of Hemoglobins: Approximation by Superposition of Quadrupole Doublets or by Quadrupole Splitting Distribution?

M. I. OSHTRAKH[1,*] and V. A. SEMIONKIN[2]
[1]*Division of Applied Biophysics, Faculty of Physical Techniques and Devices for Quality Control, Ural State Technical University – UPI, Ekaterinburg, 620002 Russian;*
e-mail: oshtrakh@mail.utnet.ru
[2]*Faculty of Experimental Physics, Ural State Technical University – UPI, Ekaterinburg, 620002 Russian*

Abstract. Mössbauer spectra of hemoglobins have some features in the range of liquid nitrogen temperature: a non-Lorentzian asymmetric line shape for oxyhemoglobins and symmetric Lorentzian line shape for deoxyhemoglobins. A comparison of the approximation of the hemoglobin Mössbauer spectra by a superposition of two quadrupole doublets and by a distribution of the quadrupole splitting demonstrates that a superposition of two quadrupole doublets is more reliable and may reflect the non-equivalent iron electronic structure and the stereochemistry in the α- and β-subunits of hemoglobin tetramers.

Key Words: hemoglobins, Mössbauer spectra, quadrupole splitting.

1. Introduction

Mössbauer spectra of the oxy-form of hemoglobins (HbO_2) demonstrate non-Lorentzian asymmetric line shape in the range of liquid nitrogen temperature while Mössbauer spectra of the deoxy-form of hemoglobins (Hb) have Lorentzian line shape in this temperature range (see refs. [1, 2]). The approximation of HbO_2 Mössbauer spectra using one quadrupole doublet was not satisfactory in contrast to Hb Mössbauer spectra. Therefore, several approaches considered in [1, 2] were supposed to explain the features of Mössbauer spectra and to fit a non-Lorentzian line shape. The approximation of a non-Lorentzian line shape using a superposition of quadrupole doublets gives a better fit of HbO_2 Mössbauer spectra. This approximation implies the presence of different iron sites in hemoglobin. On the other hand, an approximation of HbO_2 Mössbauer

* Author for correspondence.

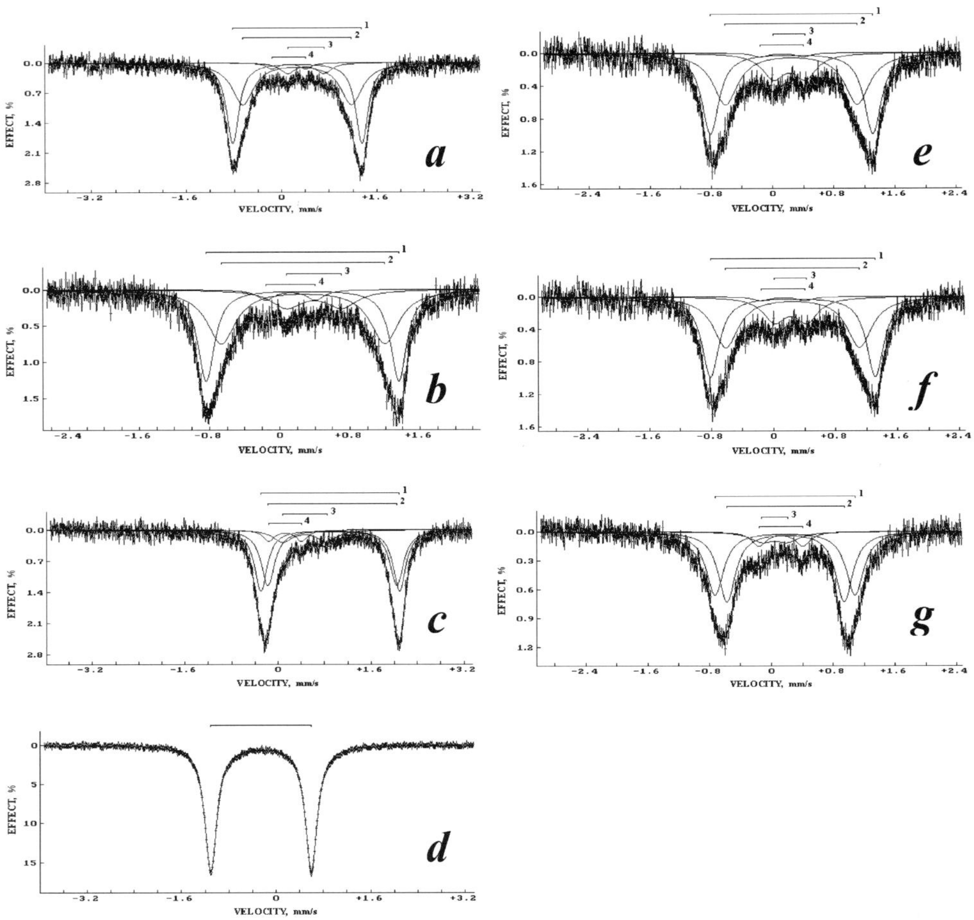

Figure 1. Mössbauer spectra of: ***a*** – HbO_2 (*FS*, $T = 87$ K), ***b*** – HbO_2 (*FS*, $T = 87$ K), ***c*** – Hb (*FS*, $T = 87$ K), ***d*** – SNP (295 K), ***e*** – Hb(PLP + GA)O_2 (*FS*, $T = 87$ K), ***f*** – Hb(PLP + GA)O_2 (*LF*, $T = 87$ K), ***g*** – Hb(PLP + GA)O_2 (*LF*, $T = 295$ K). 1 – α-subunits in tetramer, 2 – β-subunits in tetramer, 3 – unknown Fe^{3+} compound, 4 – Be(^{57}Fe).

spectra using a quadrupole splitting distribution was also used [3]. This approach may imply, for instance, a number of conformational substates of the hemoglobin molecule. In this work we compare both approximations for fitting of hemoglobin Mössbauer spectra measured with a high precision and sensitive spectrometer.

2. Materials and methods

The preparation of human HbO_2 and Hb in frozen solutions (FS) and oxygenated human hemoglobin modified by pyridoxal-5′-phosphate and glutaraldehyde (Hb(PLP + GA)O_2) in lyophilized form (LF) and frozen solution was described

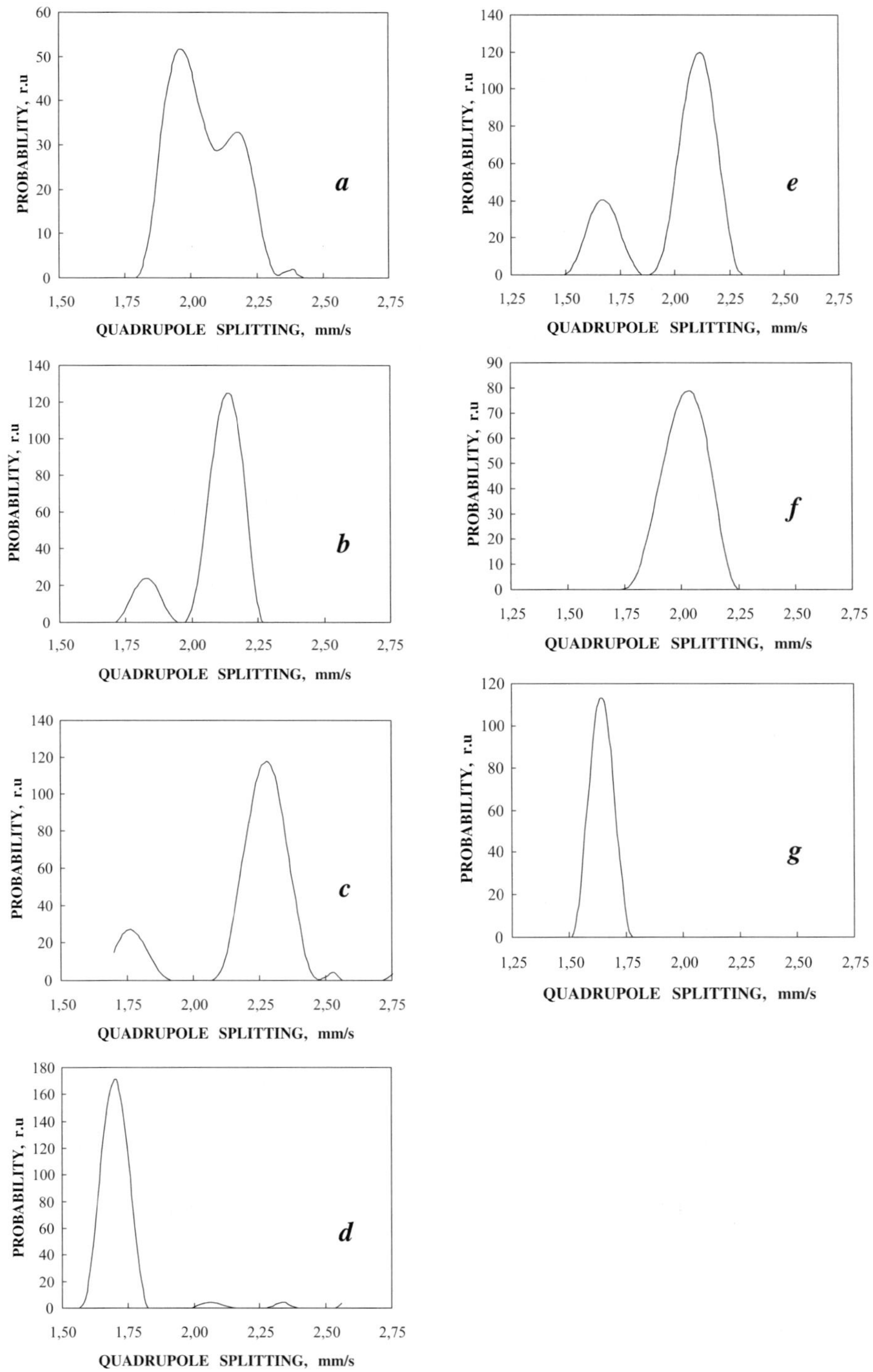

Figure 2. Distributions of quadrupole splitting for Mössbauer spectra of: ***a*** – HbO_2 (*FS*, $T = 87$ K), ***b*** – HbO_2 (*FS*, $T = 87$ K), ***c*** – Hb (*FS*, $T = 87$ K), ***d*** – SNP (295 K), ***e*** – Hb(PLP + GA)O_2 (*FS*, $T = 87$ K), ***f*** – Hb(PLP + GA)O_2 (*LF*, $T = 87$ K), ***g*** – Hb(PLP + GA)O_2 (*LF*, $T = 295$ K).

Table I. Results of the Mössbauer spectra fitting using 1 quadrupole doublet and a superposition of 2 quadrupole doublets[a]

Sample (spectrum in Figure 1)	Γ, (mm/s)	Γ_1, (mm/s)	Γ_2, (mm/s)	S_1, (%)	S_2, (%)	δ_1, (mm/s)	δ_2, (mm/s)	ΔE_{Q1}, (mm/s)	ΔE_{Q2}, (mm/s)
HbO_2, (***a***)	0.377	0.256	0.450	53	47	0.267	0.264	2.174	1.830
HbO_2, (***b***)	0.388	0.230	0.402	50	50	0.261	0.268	2.191	1.859
Hb (***c***)	0.335	0.270	0.291	51	49	0.899	0.936	2.385	2.215
SNP (***d***)	0.246	–	–	–	–	–	–	–	–
Hb(PLP+GA)O_2 (***e***)	0.425	0.253	0.422	44	56	0.256	0.252	2.137	1.776
Hb(PLP+GA)O_2 (***f***)	0.391	0.246	0.437	52	48	0.264	0.264	2.142	1.757
Hb(PLP+GA)O_2 (***g***)	0.383	0.280	0.270	48	52	0.182	0.190	1.814	1.519

[a] Experimental errors for Γ were from ± 0.019 to ± 0.028 mm/s, for δ and ΔE_Q were from ± 0.009 to ± 0.014 mm/s.

in [4]. Samples of these hemoglobins and the standard absorber of sodium nitroprusside (SNP) were measured using a high precision, sensitive and stable spectrometer SM-2201 with characteristics given in [4]. Hemoglobin samples were measured at 87 K, the lyophilized Hb(PLP + GA)O_2 sample was measured at room temperature as well. The standard absorber SNP was measured at room temperature. Mössbauer spectra were computer fitted with the least squares procedure using Lorentzian line shape and with the distribution of quadrupole splitting. Mössbauer parameters (quadrupole splitting ΔE_Q, isomer shift δ, line width Γ, subspectrum area S) were determined from the least squares fit. The values of Mössbauer parameters of the ^{57}Fe in the beryllium window of the scintillator detector Be(^{57}Fe) were determined from an independent measurement and fixed during the hemoglobin spectra fitting. The values of the isomer shift are given relative to α-Fe at 295 K.

3. Results and discussion

Mössbauer spectra of hemoglobin samples and SNP as well as distributions of quadrupole splitting are shown in Figures 1 and 2. Mössbauer parameters obtained from 1 and a superposition of two quadrupole doublets (except SNP) are given in Table I. The approximation by two quadrupole doublets with similar areas related to the small structural differences of the heme iron stereochemistry in α- and β-subunits of hemoglobin tetramers is illustrated in Figure 3.

Theoretical quantum chemical calculations [5] for the heme models in α- and β-subunits of Hb showed different ΔE_Q temperature dependencies which were in agreement with ΔE_{Q1} and ΔE_{Q2} differences at 87 K. We also pointed out that Γ values for HbO_2 and Hb Mössbauer spectra were higher than Γ for SNP (1 doublet fit). This broadening of HbO_2 and Hb Mössbauer spectra lines may also reflect a superimposed nature of these spectra. The larger values of Γ_2 may reflect an O_2 rotation in β-subunits and a distribution of Fe–O–O angles frozen at 87 K.

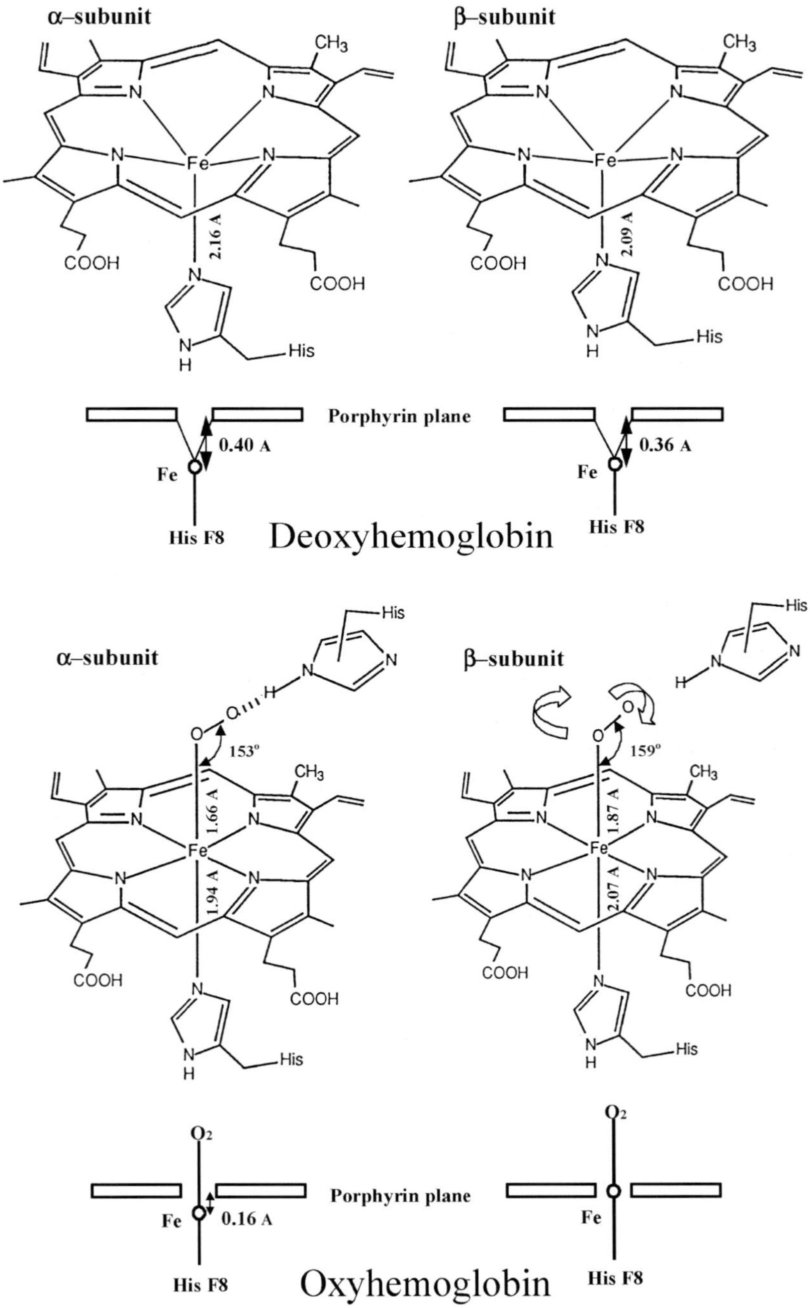

Figure 3. Differences of the heme iron stereochemistry in α- and β-subunits in deoxy- and oxyhemoglobins.

The distributions of ΔE_Q for HbO_2 Mössbauer spectra (*a*, *b*, *e* and *f*) with the same asymmetry revealed two-peak distributions with different peak intensities (*a*, *b* and *e*) and broad one-peak distribution (*f*). Mössbauer spectra of two human HbO_2 samples in frozen solutions were identical while its distributions of ΔE_Q were different, although we cannot expect any differences of the conformational substates in these hemoglobins. On the other hand, similar distributions of ΔE_Q for Mössbauer spectra of one non–modified human HbO_2 (sample *b*) and sample of modified Hb(PLP + GA)O_2 in frozen solutions should not be considered as evidence of the similar conformational substates in native and cross-linked hemoglobins. Distributions of ΔE_Q for Mössbauer spectra of Hb, SNP and Hb(PLP + GA)O_2 (*c*, *d*, and *g*) with symmetrical lines were narrow (*d* and *g*) and broad (*c*) one-peak distributions. The distribution of a ΔE_Q for the Mössbauer spectrum of the lyophilized modified hemoglobin measured at room temperature appeared to be narrower than that of hemoglobin in frozen solution, although we can expect an increase of conformational substates of hemoglobin at room temperature. Thus, this fitting does not permit us to explain different distributions for similar spectra in terms of conformational substates of hemoglobins.

4. Conclusion

The comparison of various fittings of hemoglobin Mössbauer spectra shows that the approximation using a superposition of two quadrupole doublets with similar areas seems to be more reliable and correlates with differences of the heme iron electronic structure and stereochemistry in α- and β-subunits of hemoglobin tetramers.

References

1. Oshtrakh M. I., *J. Inorg. Biochem.* **56** (1994), 221.
2. Oshtrakh M. I., *Z. Naturforsch.* **53a** (1998), 608.
3. Cai S. Z., Ortalli I. and Pedrazzi G. In: Ortalli I. (ed.), *ICAME 95 Conf Proc*, vol. 50, SIF, Bologna, 1996, p. 827.
4. Oshtrakh M. I., Milder O. B., Semionkin V. A., Berkovsky A. L., Azhigirova M. A. and Vyazova E. P., *Int. J. Biol. Macromol.* **28** (2000), 51.
5. Khleskov V. A., Burykin B. N., Smirnov A. B. and Oshtrakh M. I., *Biochem. Biophys. Res. Communs.* **155** (1988), 1255.

Hyperfine Interactions (2004) 159:351–356
DOI 10.1007/s10751-005-9118-z

TDPAC Study of the Hydrogen Uptake Process in HfNi

I. YAAR*, D. COHEN, I. HALEVY, S. KAHANE, H. ETTEDGUI, R. ASLANOV and Z. BERANT
Nuclear Research Center Negev (NRCN), P.O. Box 9001, Beer-Sheva 84190, Israel; e-mail: iyaar@bgu.ac.il

Abstract. The electronic properties of the HfNi-hydrogen system has been investigated, as a function of the hydrogen composition ratio (x), using combined Time Differential Perturbed Angular Correlation (TDPAC) technique and standard full-potential Linearized-Augmented-Plane-Wave method. The experimental TDPAC data confirm the presence of a two-step hydrogenation process in this system, with the octahedral holes filled first. The major part of the electric field gradient at the hafnium site is from p–p contribution, shifted down in energy by hydrogen s-states contribution.

Key Words: hydrogen, hyperfine interaction, LAPW, TDPAC.

1. Introduction

The two-step hydrogen absorption process in the CrB-type intermetallic compound HfNi has been studied in the past using volumetric analysis, X-ray diffraction technique and thermo-analytical measurements [1–3]. In the first step, a mono hydride is formed $\left(\mathrm{HfNi} + \frac{1}{2}\mathrm{H}_2 \rightarrow \mathrm{HfNiH}_1\right)$ and a ternary hydride in the second $(\mathrm{HfNiH}_1 + \mathrm{H}_2 \rightarrow \mathrm{HfNiH}_3)$. Two different types of holes, large enough to contain hydrogen atoms, exist in the CrB-type crystalline lattice of HfNi, with the number of tetrahedral (Oy1/4) ones being twice as large as that of the octahedral (0yz) ones [4]. Since the CrB-type structure of the system is retained on hydrogen uptake, and since the second plateau of the pressure–composition isotherm is about twice as long as the first one, it is reasonable to assume that the octahedral holes are occupied in the first step and the tetrahedral holes in the second [1].

The hyperfine interaction of ^{181}Ta in $\mathrm{Hf}_x\mathrm{Ni}_{1-x}$ was studied in the past using Time Differential Perturbed Angular Correlation (TDPAC) technique [5]. The spectrum of the compound containing 50.21 at.% hafnium (close to the HfNi phase) was fitted using a combined electric and magnetic interaction. This fit

* Author for correspondence.

gave a hyperfine magnetic field value of 2.44 ± 0.04 T and an electric field gradient (efg) main component (V_{zz}) value of 1.16 ± 0.02 × 10^{21} V/m^2, with asymmetry parameter $\eta = 0$.

In this work, the two-step hydrogenation process in the HfNi-hydrogen system is evaluated using combined TDPAC technique and full potential Linearized Augmented Plane Wave (LAPW) calculations [6, 7].

2. Experimental

The HfNi sample was made by arc melting of metallic Hf (97% Hf and 3% Zr), with metallic Ni (99.9% pure). The sample was irradiated to obtain radioactive probes with ≈50 μCi activity of ^{181}Hf ($T_{1/2} = 42$ d), and annealed at 1200 K in vacuum for a week. The sample was sealed in an evacuated stainless steel reactor and out-gassed for several hours at 550 K in high vacuum. After cooling to 300 K, purified hydrogen gas of a measured volume and known pressure was admitted into the reactor. The sample went through 10 hydrating-dehydrating cycles, and was then hydrogenated to different hydrogen composition ratios at 300 K, by a standard pressure–composition-isotherms (PCT) procedure.

X-ray diffraction measurements for HfNi and $HfNiH_3$, taken with Cu – K_α monochromator, indicated a single orthorhombic CrB-type (63 Cmcm) structure. The lattice constants (in angstroms (Å)), obtained at room temperature after exposing to normal air pressure are as follows: $a = 3.2189$, $b = 9.7913$, $c = 4.1243$ for HfNi, and $a = 3.4945$, $b = 10.4136$, $c = 4.3019$ for $HfNiH_3$, in good agreement with previously published results [1].

TDPAC measurements were carried out at room temperature using the ^{181}Ta 133–482 keV $\gamma_1 - \gamma_2$ cascade ($1/2^+ \rightarrow 5/2^+ \rightarrow 7/2^+$) with an intermediate state half-life of 10.8 ns. This cascade is obtained via the decay of the 18 μs 615 keV state in the ^{181}Ta probe, populated in the β decay of ^{181}Hf → ^{181}Ta [8]. The 18 μs delay before the excited level under observation is populated, gives sufficient time for hydrogen to readjust its distribution toward equilibrium around the probe, and for electronic reorganization according to the valence change from ^{181}Hf to ^{181}Ta [9]. The time spectra at π and $\pi/2$ were recorded simultaneously using a 4-BaF_2 detector system with time resolution of 600 ps and a conventional multi-parameter acquisition system MPA-2 (Fast-ComTec).

3. Results

The experimental TDPAC spectra for different hydrogen content with x values of 0, 1, 3 are depicted in Figure 1. As expected from a two-step process, each of the spectra is fitted using a single efg, the parameters of which are compared in Table I with LAPW calculated results. The calculated V_{zz} value for $x = 0$ and $x = 1$ are in reasonable agreement with the experimental results, while

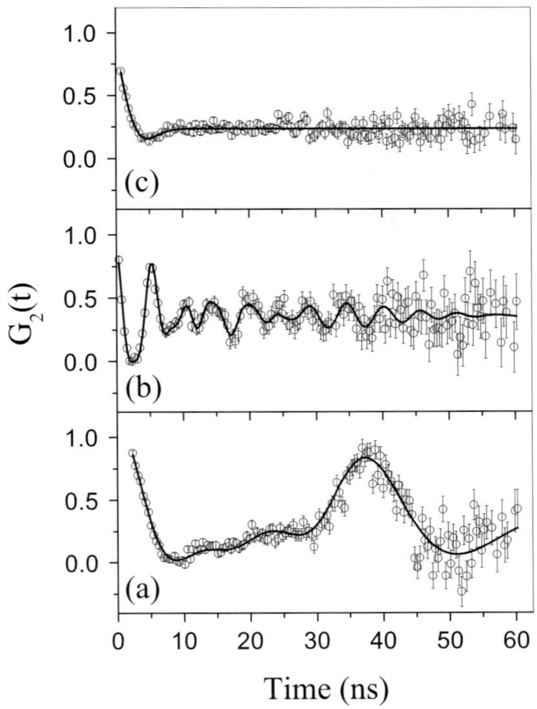

Figure 1. TDPAC spectra of $HfNiH_x$ for x values of (a) $x = 0$, (b) $x = 1$ and (c) $x = 3$. The *solid lines* are the best fit obtained for the experimental data.

Table I. Measured and calculated efg parameters, V_{ZZ} (10^{21} V/m^2) and η, for the $HfNiH_x$ ($x = 0, 1, 3$) system

x	Measured		Calculated			
	Hf site		Hf Site		Ni Site	
	V_{ZZ}	η	V_{ZZ}	η	V_{ZZ}	η
0	3.0 ± 0.05	0.25 ± 0.05	−4.6	0.28	3.4	0.22
1	12.8 ± 0.1	0.73 ± 0.11	12.0	0.14	1.5	0.80
3	7.9 ± 0.3	0.61 ± 0.18	17.2	0.78	2.4	0.74

the origin for the factor two deviation in the calculated V_{ZZ} value for $x = 3$ has to be further investigated. The discrepancy between experimental and calculated value of η can only be attributed to vacancies and impurities existing in the experimental sample.

A combination of two efgs was needed to fit the TDPAC spectra in the intermediate range of $0.2 \leq x \leq 0.9$ and $1.2 \leq x \leq 2.5$, with V_{ZZ} values in the

Table II. The s–d, p–p, d–d and the lattice contribution to the V_{zz} value at the Hafnium site, for x values of 0, 1 and 3

x	Contribution to the V_{zz} value at the hafnium 4(c) site (10^{21} V/m^2)				
	s–d	p–p	d–d	Lattice	V_{zz}
0	0.036	−5.036	0.396	0.040	−4.6
1	0.115	10.695	1.188	0.074	12.0
3	0.264	13.864	3.017	0.205	17.2

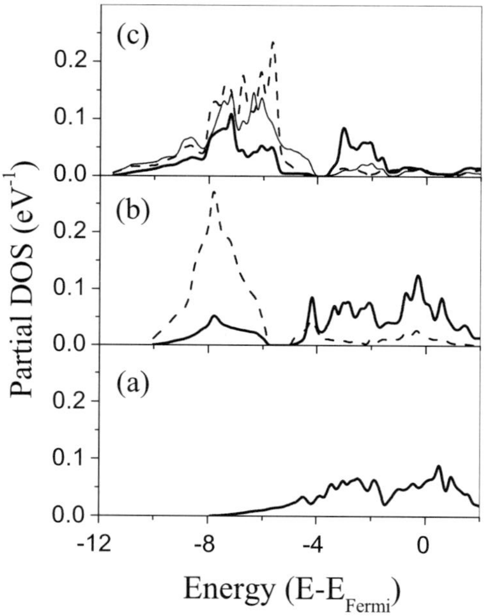

Figure 2. Partial density of states (DOS) for: (a) p-states at the hafnium site (–) in HfNi, (b) p-states at the hafnium site (–) and s-states at the hydrogen octahedral site (···) in $HfNiH_1$ and (c) p-states at the hafnium site (–) and s-states at the hydrogen octahedral site (···) and tetrahedral site (–) in $HfNiH_3$.

range from 4.07 to 25.49 × 10^{21} V/m^2, and η values of 0.20–0.74. The change in the efg values as a function of x in the intermediate range is a result of the uniform distribution of hydrogen atoms in the octahedral holes and then in the tetrahedral holes, a process in which different probe atom are surrounded with different number of close hydrogen neighbors. This process is different from the one observed in the single-step hydrogenation process in ZrCo [7], were only a metal ZrCo site and a ternary $ZrCoH_3$ site were present, indicating a hydrogenation process in which new hydrogen atoms are absorbed in the crystal sites close to the sites that are already occupied with hydrogen atoms.

4. Calculations

The calculations were done using the full potential linearized plane wave (LAPW) method, as embodied in the WIEN97 code [6, 10], using the generalized gradient approximation (GGA) [11]. All of the calculations were performed on pure HfNi, $HfNiH_1$ and $HfNiH_3$ crystals, and not for a tantalum atom located at one of the hafnium sites in the crystal. The value of V_{zz} in these calculations is given by an integral of the charge density over the unit cell. In most cases, the charge asymmetry inside the atomic sphere, where the V_{zz} value is being calculated, determines more than 90% of the V_{zz} value. Inside this sphere, the contribution to V_{zz} can be further divided into s–d, p–p and d–d components, so that the physical origin of the efg can be analyzed [12].

The calculations for the HfNi and $HfNiH_3$ compounds were made for the experimental X-ray diffraction values measured by us, and the crystal constants of $HfNiH_1$ were evaluated from the above measured values assuming a linear volume increase as a function of hydrogen content. The final results of the calculated V_{zz} and η values, after refinement of the atomic positions using minimum energy considerations, are compared in Table I with the experimental TDPAC values.

The partial valence s–d, p–p and d–d and the interstitial or lattice contribution to the efg at the hafnium site are listed in Table II, for x values of 0, 1 and 3. From these results, it is evident that the increase in the V_{zz} value as a function of hydrogen content x is attributed mainly to the increase in the p–p contribution next to the probe nucleus [13].

The partial p-electron density of state (DOS), at the hafnium site, was therefore calculated and plotted in Figure 2, as a function of energy for x values of 0, 1 and 3, together with the partial s-electrons DOS at the hydrogen octahedral and tetrahedral sites. As the hydrogen octahedral site is filled, the hafnium p-states are shifted down in energy by about 2 eV relative to their origin location in HfNi, to the same energy were the hydrogen s-states are formed. With the addition of the hydrogen atoms at the tetrahedral site, the hafnium p-states and first hydrogen s-states are shifted farther down in energy by about 1.5 eV, due to the formation of new s-states around this energy.

5. Summary

The TDPAC spectra of the HfNi hydrogen system were measured and compared with LAPW calculated results. The experimental TDPAC data confirm the presence of a two-step hydrogenation process in this system, as demonstrated by previously volumetric and X-ray measurements [1, 2]. From the comparison between the experimental and the calculated efg parameters at the Hf site in $HfNiH_1$, a confirmation of the assumption that the octahedral holes in this structure are occupied before the tetrahedral ones is given. The dominant p–p contribution to the efg at the hafnium site in this system is demonstrated. The increase in the V_{zz} value as a function of hydrogen content is attributed to the hybridization between hafnium p-states and hydrogen s-states, formed at an energy of 8 ± 2 and 10 ± 1 eV below E_{Fermi} for $HfNiH_1$ and $HfNiH_3$, respectively.

References

1. Van Essen R. M. and Buschow K. H. J., *J. Less-Comm. Met.* **64** (1979), 277.
2. Nemirovskaya I. E., Alekseev A. M. and Lunin V. V., *J. Alloys Compd.* **177** (1991), 1–15.
3. Nemirovskaya I. E., *J. Alloys Compd.* **209** (1994), 93–97.
4. Griessen R. and Driessen A., *J. Alloys Compd.* **103** (1984), 235.
5. Gerdau E., Winkler H., Gebert W., Giese B. and Braunsfurth J., *Hyperfine Interact.* **1** (1976), 459–467.
6. Blaha P., Schwarz K. and Luitz J., Wien97-Code (K. Schwarz, TU Wien, ISBN 3-9501031-0-4), 1999.
7. Yaar I., Gavra Z., Cohen D., Levitin Y., Kimmel G., Kahane S., Hemy A. and Berant Z., *Hyperfine Interact.* **120/121** (1999), 563.
8. Leaderer C. M. and Shirley V. (eds.), *Table of Isotopes*, Wiley, NY, 1978, p. 1135.
9. Yaar I., Gavra Z., Cohen D., Levitin Y., Feuerlicht J., Mints M. H. and Berant Z., *J. Alloys Compd.* **260** (1997), 1.
10. Blaha P., Schwarz K., Faber W. and Luitz J., *Hyperfine Interact.* **126** (2000), 389–395.
11. Perdew J. P., Burke K. and Ernzerhof M., *Phys. Rev. Lett.* **77** (1996), 3865–3868.
12. Blaha P., Dufek P., Schwarz K. and Haas H., *Hyperfine Interact.* **97/98** (1996), 3–10.
13. Andreev A. V., Zadvorkin S. M., Bartashevich M. I., Goto T., Karmarad J., Arnold Z. and Drulis H., *J. Alloys Compd.* **267** (1998), 32–36.

Hyperfine Interactions (2004) 159:357–362
DOI 10.1007/s10751-005-9119-y

High Pressure Study of HfNi Crystallographic and Electronic Structure

I. HALEVY[1,2,*], S. SALHOV[1], A. F. YUE[2], J. HU[3] and I. YAAR[1]
[1]*Nuclear Research Center – Negev, P.O. Box 9001, Beer-Sheva, Israel; e-mail: ihalevy@bgu.ac.il*
[2]*Department of Materials Science California Institute of Technology, Pasadena, CA 91125, USA*
[3]*Geophysical Laboratory of Carnegie Institution of Washington, Washington, DC 20015, USA*

Abstract. The crystallographic structure and electronic properties of HfNi were studied as a function of pressure by combining X-ray diffraction results with the full potential linearized augmented plane wave (LAPW) calculations. No phase transition was observed up to a pressure of 35.3 GPa, with a total volume contraction of $V/V_0 = 0.85$, a bulk modulus value of $B_0 = 52 \pm 3$ GPa and $B'_0 = 1.29 \pm 0.26$. The calculated linear increase in the V_{zz} value as a function of the pressure induced volume reduction at the hafnium site was attributed mainly to the p–p contribution, while in the nickel site, a non negligible d–d contribution to V_{zz} is also observed, and attributed to the high 3d-partial DOS near the nickel nucleus. Based on the total electronic DOS at E_{Fermi} calculated for 0 K ($N(E^0_{\mathrm{Fermi}})$), a value of 6.85 and 5.03 (mJ/mol/k^2) was calculated for the band contribution (γ_{band}) to the electronic specific heat coefficient (γ) at a pressure of 0 and 35.3 GPa, respectively.

Key Words: bulk modulus, DOS, high pressure, LAPW, specific heat.

1. Introduction

The study of the electronic properties of the BCr-type intermetallic compound HfNi has been motivated by its ability to absorb large quantities of hydrogen gas [1–4]. In the hydrogen uptake process, the system retains its BCr-type structure, with a volume increase of $V/V_0 \approx 1.176$ [1]. The Hyperfine Interaction properties of the Hf_xNi_{1-x} system were studied in the past at ambient pressure using the Time Differential Perturbed Angular Correlation (TDPAC) technique [5]. In the above study, the spectrum of the compound containing 50.21 at percent hafnium (close to the HfNi phase) was fitted using a combination of electric and magnetic interactions.

In this work, the electronic properties of HfNi were evaluated as a function of applied pressure using combined X-ray powder diffraction technique and full potential Linearized Augmented Plane Wave (LAPW) calculations [6]. The

---

* Author for correspondence.

results obtained in this study will give us a better understanding and a full picture of the electronic properties of this compound as a function of V/V_0.

2. Experimental

Samples with HfNi nominal composition were prepared by arc-melting stoichiometric amounts of high purity hafnium 99.5% and nickel 99.95% lumps under dry argon atmosphere. The samples were annealed in an evacuated quartz tube and treated at 850°C for four days. The sample microstructure was then examined by a scanning electron microscope (SEM) and the phases were analyzed by Energy-Dispersive-System (EDS), to determine the chemical compositions.

The X-ray powder diffraction measurement at ambient pressure was taken using Cu-$K_{\alpha 1}$ radiation. The high-pressure angle dispersive X-ray diffraction studies, up to a pressure of 10 GPa, were performed at the Advanced Photon source (APS) beam-line located at the Argon National Laboratory (ANL) (http://www.aps.anl.gov/aps/frame_home.html), and the high-pressure energy dispersive X-ray diffraction studies up to a pressure of 35.3 GPa were performed at the X17 beamline of the National Synchrotron Light Source (NSLS) (http:www.bnl.gov/x17c/).

The pressure was applied using a Merrill-Bassett, 'Tel-Aviv'-type diamond–anvil-cell (DAC) [7], and measured by the fluorescence Ruby technique [8]. The pressure distribution inside the sample volume was checked at different locations and was determined to vary by less than 5%. The energy dispersive data at NSLS was collected with a Ge detector at a fixed Bragg angle ($2\theta = 13°$). The angle dispersive data at APS was collected with the image-plate technique, using a wave length of $\lambda = 0.4228$ Å. The image plate diffraction patterns were first analyzed by the FIT2D software [9], for correction of the image-plate tilt in the different axis, and then Rietveld analysis was performed [10].

Based on the refined unit cell parameter, the value of the electric field gradient (efg) parameters and the partial density of states (DOS) at the hafnium and nickel sites, were calculated, utilizing the full potential linearized augmented plane wave (LAPW) code WIEN97 [6, 11].

3. Results

The morphology of the as-cast HfNi sample was identified as a single phase using SEM. This result was confirmed by X-ray diffraction pattern, taken at ambient pressure. In this measurement, a single HfNi phase, BCr-type (Cmcn) orthorhombic structure was observed, with the unit cell parameters of: a = 3.2182(5), b = 9.7885(5) and c = 4.1179(5)Å, in good agreement with Dwight [12].

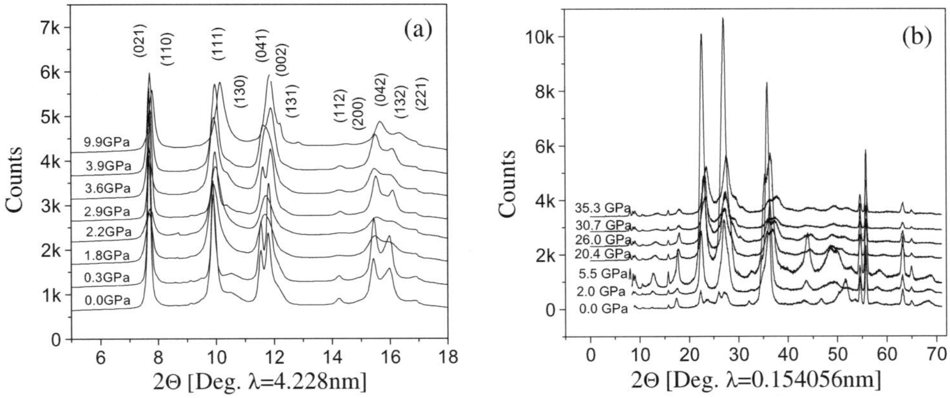

Figure 1. X-ray powder diffraction of HfNi as a function of pressure taken at (a) APS and (b) NSLS. The peaks are indexed for the orthorhombic BCr structure.

The high-pressure X-ray powder diffraction measurements taken at APS and NSLS are depicted in Figure 1(a) and in Figure 1(b), respectively. The diffraction data was analyzed using the Rietveld Powder Cell refinement software package [13]. All of the refined results indicated a single orthorhombic BCr-type (Cmcm) HfNi phase.

The change in the unit cell parameters as a function of pressure is shown in Figure 2(a). As the pressure is increased, no phase change is observed, and the unit cell parameters *a*, *b* and *c* are monotonically decreasing with a slope of 1.31, 3.82 and 1.04 $Å^{-1}$, respectively. From this data, it appears that the unit cell parameter *b* is more sensitive to the change in pressure relative to the unit cell parameters *a* and *c*. The volume-pressure curve, calculated from the X-Ray data, is depicted in Figure 2(b). The relationship between pressure and volume change was determined by the modified Vinet (MV) equation of state [14]. This constitutive correlation was shown to be universally valid for all solids under a wide range of pressure values [14].

Any attempts to fit the data to other equation-of-state formalism, like Holzapfel [15] or Birch [16] gave the same results. No hysteresis was observed while the pressure was increased or decreased. From the results of the MV fit, depicted in Figure 2(b), a volume contraction of $V/V_0 = 0.85$ at 35.3 GPa, and a bulk modulus value of $B_0 = 52 \pm 3$ GPa and $B'_0 = 1.29 \pm 0.26$ were calculated.

4. Calculations and discussion

The calculations were performed using the full potential linearized plane wave (LAPW) method, as embodied in the WIEN97 code [6, 11], using the generalized gradient approximation (GGA) [17]. All of the calculations were performed on a pure HfNi crystal, and not for a tantalum atom located at one of

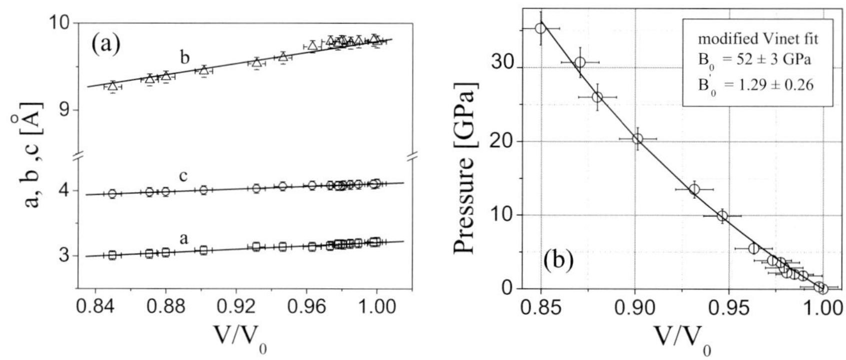

Figure 2. HfNi (a) unit cell parameters a, b and c and (b) relative volume–pressure curve, derived from the X-ray high-pressure diffraction Rietveld refinement.

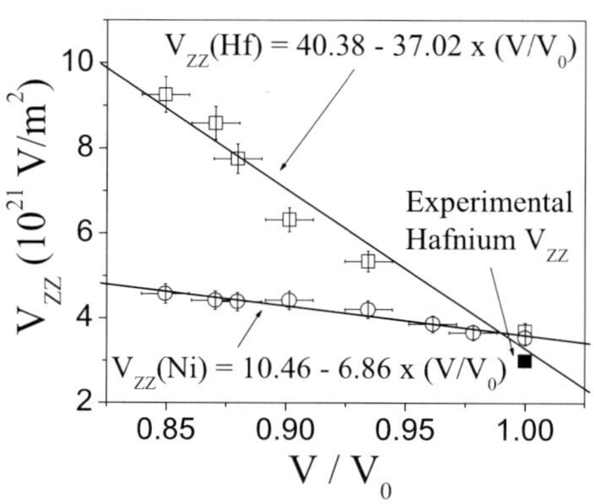

Figure 3. The calculated $|V_{zz}|$ value at the hafnium [□] and nickel [○] sites, as a function of V/V_0, and the ambient pressure experimental [■] V_{zz} value at the Hafnium site.

the hafnium sites in the crystal. Total and partial s–d, p–p and d–d contribution to V_{zz} were calculated, so that the physical origin of the efg can be analyzed [18]. The atomic muffin-tin sphere radius in all of the calculations was taken as 1.40 and 0.98 Å for hafnium and nickel atoms, respectively. The calculated V_{zz} value at the hafnium and nickel sites is depicted in Figure 3 as a function of V/V_0. The straight line connecting the points is the result of a linear fit made to the experimental data, the parameters of which are listed inside the figure.

The s, p and d-partial electron DOS at the nickel and hafnium sites are plotted in Figure 4, as a function of energy for pressure values of 0 and 35.3 GPa. As the pressure increases from 0 to 35.3 GPa, the states below E_{Fermi} in both atoms are shifted down in energy by about 0.6 eV, the states above E_{Fermi} are shifted up in energy by about 0.4 eV, and the charge becomes more localized in narrow bands.

Based on the total electronic DOS at the Fermi level calculated for 0 K ($N(E^{0}_{\mathrm{Fermi}})$), the band contribution (γ_{band}) to the value of the electronic specific

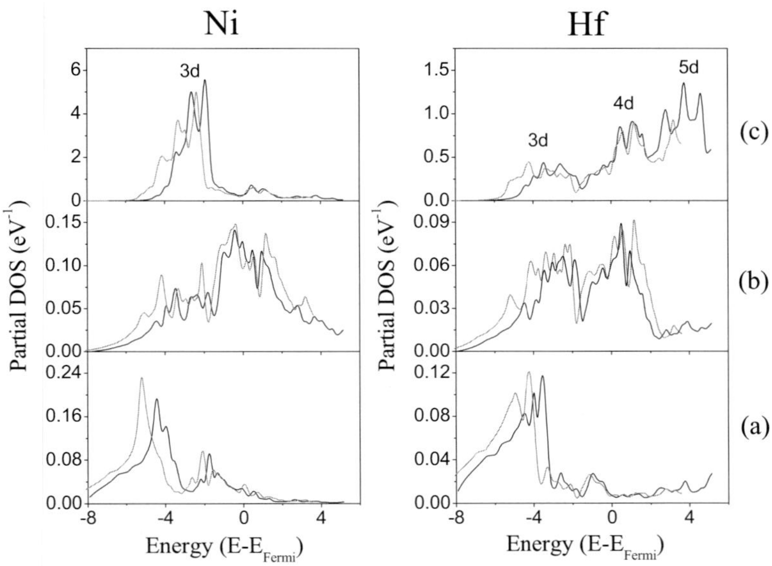

Figure 4. Partial (a) s, (b) p and (c) d density of states (DOS) in the Ni (*left*) and Hf (*right*) sites for pressure values of 0 (—) and 35.3 (—) Gpa.

Table I. The s–d, p–p, d–d and the lattice partial contributions to the V_{zz} value at the hafnium and nickel sites, for pressure values of 0, 13.5 and 35.3 GPa

Pressure (Gpa)	Partial contribution to the V_{zz} value (10^{21} V/m^2)									
	Hafnium site					Nickel site				
	s–d	p–p	d–d	Lattice	V_{zz}	s–d	p–p	d–d	Lattice	V_{zz}
0	0.036	−5.036	0.396	0.040	−4.6	−0.003	5.015	−1.592	−0.081	3.4
13.5	0.226	−7.787	0.147	0.115	−7.4	−0.001	5.791	−1.632	−0.049	4.2
35.3	0.103	−10.003	0.880	0.089	−9.0	0.008	6.422	−1.772	0.079	4.7

heat coefficient (γ) can be evaluated [19]. Based on the calculated results, a value of 39.5 and 29.0 (states/Ry/f.u.) is calculated for ($N(E^0_{\text{Fermi}})$) at a pressure of 0 and 35.3 GPa, respectively. This gives, according to equation 2, a γ_{band} coefficient value of 6.85 and 5.03 (mJ/mol/k^2) for pressure values of 0 and 35.3 GPa, respectively.

The partial valence s–d, p–p and d–d and the interstitial or lattice contribution to the efg at the hafnium and nickel sites are listed in Table I, for pressure values of 0, 13.5 and 35.3 GPa. From these results, it is evident that the increase in the V_{zz} value as a function of pressure in the hafnium site can be attributed mainly to the negative p–p contribution, while in the nickel site the V_{zz} value is determined by the joined positive p–p and negative d–d contributions. The relatively large d-d contribution to the V_{zz} value at the nickel site can be attributed to the high 3d-partial DOS, observed in Figure 4(a) in the energy window of −5 to −3 eV below E_{Fermi}.

5. Summary

The crystallographic structure of the intermetallic compound HfNi was measured and refined as a function of external pressure, using a 'Tel-Aviv'-type diamond-Anvil Cell [7]. No phase transition was observed up to a pressure of 35.3 GPa, with a total volume contraction of $V/V_0 = 0.85$, and a bulk modulus value of $B_0 = 52 \pm 3$ GPa. The refined cell parameters were then used to calculate the efg parameters of this compound as a function of the pressure induced volume reduction. The linear increase in the V_{zz} value as a function of the pressure induced volume reduction at the hafnium site is attributed mainly to the p–p contribution, while at the nickel site, a non negligible d–d contribution to V_{zz} is also observed, and attributed to the high 3d-partial DOS near the nickel nucleus. Based on the total electronic DOS at E_{Fermi} calculated for 0 K ($N(E^0_{\mathrm{Fermi}})$), a value of 6.85 and 5.03 (mJ/mol/k2) was calculated for the band contribution (γ_{band}) to the electronic specific heat coefficient (γ) at a pressure of 0 and 35.3 GPa, respectively.

References

1. Van Essen R. M. and Buschow K. H. J., *J. Less-Comm. Met.* **64** (1979), 277.
2. Nemirovskaya I. E., Alekseev A. M. and Lunin V. V., *J. Alloys Comp.* **177** (1991), 1–15.
3. Nemirovskaya I. E., *J. Alloys Comp.* **209** (1994), 93–97.
4. Griessen R. and Driessen A., *J. Alloys Comp.* **103** (1984), 235.
5. Gerdau E., Winkler H., Gebert W., Giese B. and Braunsfurth J., *Hyp. Interact.* **1** (1976), 459–467.
6. Blaha P., Schwarz K. and Luitz J., Wien97-Code (K. Schwarz, TU Wien, ISBN 3-9501031-0-4) 1999.
7. Sterer E., Pasternak M. P. and Taylor R. D., *Rev. Sci. Instr.* **61** (1990), 1117–1119.
8. Forman R. A., Piermarini G. J., Barnett J. D. and Block S., *Science* **176** (1972), 284–285.
9. Hammersley A., FIT2D software (ESRF) 1997.
10. Young R. A., Sakthivel A., Moss T. S. and Paiva-Santos C. O., *Rietveld Analysis of X-ray and Neutron Diffraction Patterns* (1994), Georgia Institute of Technology Atlanta, GA 30332, and Rietveld H. M., *J. Appl. Crystallogr.* **2** (1969), 65–71.
11. Blaha P., Schwarz K., Faber W. and Luitz J., *Hyp. Interact.* **126** (2000), 389–395.
12. Dwight, *Trans. Am. Soc. Met.* **53** (1961), 479.
13. Kraus W. and Nolze G., PowderCell software, *J. Appl. Cryst.* **29** (1996), 301–303.
14. Vinet P., Ferrante J., Rose J. H. and Smith J. R., *J. Geophys. Res.* **92** (1987), 9319.
15. Holzapfel W. B., Reports on Progress in Physics **59** (1996), 29–90.
16. Birch F. J., *Geophysical. Res.* **95** (1978), 1257–1268.
17. Perdew J. P., Burke K. and Ernzerhof M., *Phys. Rev. Lett.* **77** (1996), 3865–3868.
18. Blaha P., Dufek P., Schwarz K. and Haas H., *Hyp. Interact.* **97/98** (1996), 3–10.
19. White R. M., *Quantum Theory of Magnetism*, Berlin, Heidelberg, New York, Springer, 1983.

Hyperfine Interactions (2004) 159:363–372
DOI 10.1007/s10751-005-9125-0

Recent Emission Channeling Studies in Wide Band Gap Semiconductors

U. WAHL[1,2,*], J. G. CORREIA[1,2,3], E. RITA[1,2], E. ALVES[1,2], J. C. SOARES[2], B. DE VRIES[4], V. MATIAS[4], A. VANTOMME[4] and THE ISOLDE COLLABORATION[3]
[1]*Instituto Tecnológico e Nuclear, Department Física, Estrada Nacional 10, 2686-953, Sacavém, Portugal; e-mail: uwahl@itn.pt*
[2]*Centro de Física Nuclear da Universidade de Lisboa, 1649-003, Lisbon, Portugal*
[3]*CERN-PH, 1211, Geneva 23, Switzerland*
[4]*Instituut voor Kern- en Stralingsfysica, 3001, Leuven, Belgium*

Abstract. We present results of recent emission channeling experiments on the lattice location of implanted Fe and rare earths in wurtzite GaN and ZnO. In both cases the majority of implanted atoms are found on substitutional cation sites. The root mean square displacements from the ideal substitutional Ga and Zn sites are given and the stability of the Fe and rare earth lattice location against thermal annealing is discussed.

Key Words: Fe doping, implantation, lattice location, rare earths, semiconductors.

1. Introduction

Wide band gap semiconductors (WBGS) can be roughly defined as semiconductors with an energy gap E_G corresponding to the blue or ultraviolet (UV) range of the optical spectrum, i.e., above 2.5 eV. Typical examples are SiC (E_G = 2.4–3.2 eV, depending on polytype), ZnSe (2.8 eV), GaN (3.4 eV), ZnO (3.4 eV), diamond (5.5 eV), or AlN (6.2 eV). They are all tetrahedrally coordinated and crystallize in the cubic diamond and zinc blende (3C–SiC, ZnSe), or hexagonal wurtzite (2H–SiC, GaN, AlN, ZnO) and related hexagonal (e.g., 4H–SiC) structures.

There exist already many technological applications for WBGS, such as light emitting diodes or lasers in the blue and UV [1, 2], visible–blind detectors for UV light [1–3], high-temperature high-power transistors [1, 2, 4], or gas sensors [5, 6]. Recently, WBGS have also emerged as candidates for multiple-color electroluminescent devices [7] suitable for displays, and in the field of spintronics [8]. Both cases are based on the incorporation of impurities, either

* Author for correspondence.

rare earth (RE) elements as optical dopants, or *3d* transition metals (TM) such as Cr, Mn, Fe, Co or Ni as ferromagnetic dopants. In the case of RE doping, an empirical rule was established [9] that the quenching of the RE luminescence at room temperature is less pronounced in semiconductors with a wide band gap. The major quest in order to realize spintronic devices is to find semiconductors which exhibit ferromagnetism at or above room temperature, and theoretical considerations predict the diluted magnetic semiconductors $Ga_{1-x}TM_xN$ and $Zn_{1-x}TM_xO$ and TM-doped diamond to be most promising for that purpose [10, 11].

A crucial point with regards to any kind of doping in semiconductors, electrical, optical or magnetic, is the activation of the dopants. Not only is it required to incorporate the dopants into the appropriate lattice sites but also to control possible interactions with additional crystal defects. The successful implementation of ion implantation as a means of doping would considerably enhance the possibilities of WBGS technology and is currently an active field of research [2, 4, 12] but faces some inherent problems. While high resistance against the formation of radiation damage is a common feature of WBGS, due to the fact that a large band gap is correlated with strong atomic bonds, it is on the other hand extremely difficult to anneal defects once they have been created. Crystal defects and their interaction with dopants are therefore of particular relevance in these materials.

While the emission channeling method provides direct information on the lattice sites of implanted impurities, hyperfine interaction methods such as Mössbauer spectroscopy (MS) or the perturbed angular correlation (PAC) techniques are in many cases suitable in order to investigate the local environment around probe atoms and can hence be applied in a complementary way. Although our group is engaged in lattice location studies of a broad range of implanted impurities in a number of WBGS, mainly GaN, AlN, ZnO, diamond, and recently also SiC, we will present here only the cases of RE and Fe impurities in GaN and ZnO, since they are of particular relevance with respect to the applications mentioned above and also accessible to investigation by hyperfine interaction techniques. Apart from GaN and ZnO there have also been emission channeling experiments reported on Fe in diamond [13], rare earths in InGaN [14], AlN [15], SiC [16] and cubic BN [17].

There exist a number of RE probe nuclei whose applicability for Mössbauer studies has been discussed, e.g., ^{141}Pr, ^{149}Sm, ^{151}Eu, ^{155}Gd, ^{161}Dy, ^{166}Er, ^{169}Tm and ^{170}Yb [18–20]. However, except for the work of Masterov et al on ^{169}Tm in amorphous Si [21], almost nothing has been published on MS from rare earth probes in semiconductors. Likewise, RE are not among the most common PAC probes, but some applications on the study of electric field gradients of ^{160}Dy, ^{169}Tm, and ^{172}Yb in metals have been reported [22–25], and first studies on ^{172}Yb PAC in WBGS [26] are being presented at this conference. Quite some experimental work has been done on Mössbauer effect of ^{57}Fe in semi-

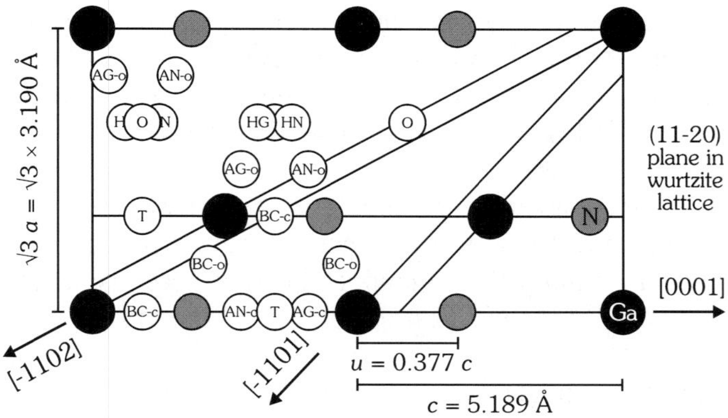

Figure 1. The (11$\bar{2}$0) plane in the GaN wurtzite lattice, showing the Ga and N lattice positions and some of the possible interstitial sites; "-c" denotes sites within and "-o" off the c-axis.

conductors, including absorber experiments and source experiments from the decay of ^{57}Mn or ^{57}Co or following Coulomb excitation of ^{57}Fe [27–29]. Investigations on wide band gap semiconductors, however, are found scarcer, with diamond being the best studied host [30] and only a small number of experiments reported on ZnO [31, 32] and GaN [33].

2. Method

The electron emission channeling method makes use of the fact that β^- particles and conversion electrons emitted from radioactive isotopes in a single crystal experience channeling or blocking effects that depend characteristically on the lattice site(s) of the emitter atom. Several reviews on emission channeling can be found in the literature [34–37].

The production of radioisotopes and implantation into single-crystalline ZnO and thin film GaN samples at 60 keV and fluences around 1–3 $\times$ 10^{13} cm^{-2} was done at CERN's ISOLDE facility [38, 39]. A position-sensitive electron detector was used in order to measure the angle-dependent electron emission yield around various crystallographic directions [40]. Quantitative results were achieved by comparing the experimental data with theoretical patterns calculated by means of the "manybeam" formalism [34–37] for a wide range of possible lattice sites. The fitting procedures used for that purpose have been described in detail previously [37, 40]. The GaN and ZnO crystallographic parameters and room temperature atomic displacements used in the manybeam simulations can be found in Refs. [41] and [50].

Examples for possible lattice sites of higher symmetry in the GaN wurtzite structure are shown in Figure 1, including substitutional Ga (S_{Ga}) and N (S_N) sites, bond-centered (BC) and anti-bonding (AB) sites both within and off the c-axis, the "hexagonal" sites HG and HN, and the so-called T- and O-sites [41].

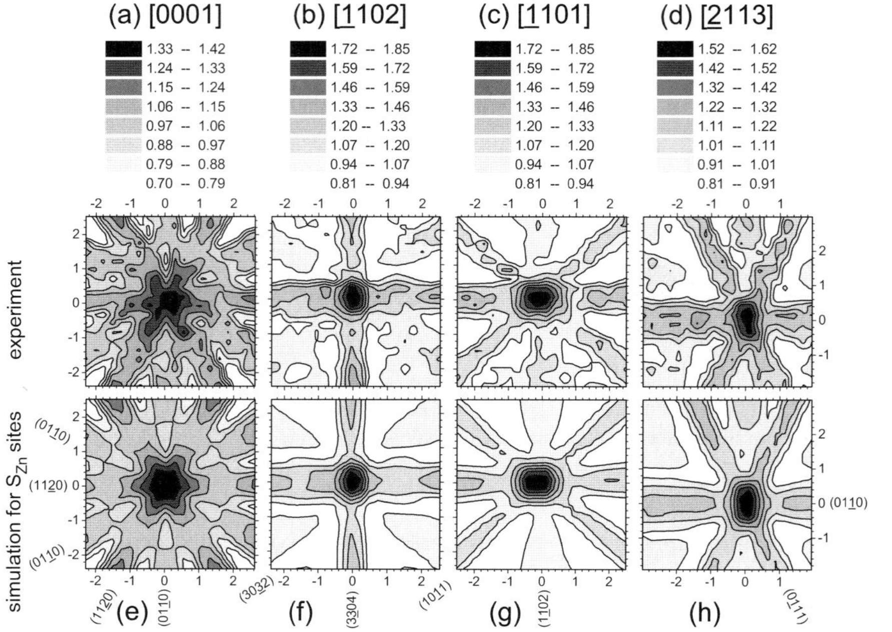

Figure 2. Angular distribution of β^- emission yields from ^{59}Fe in ZnO in the as-implanted state. The best fits of the channeling patterns for each direction are also shown, yielding an average of 89(11)% of Fe atoms at S_{Zn} sites.

With respect to hyperfine interaction studies it is worth while to point out that in an *ideal* wurtzite structure (i.e., with c/a = 1.633 and c-axis bond length parameter $u = 0.375\ c$) the symmetry of the *intact first atomic neighbour shells* around perfect S_{Ga} and S_N positions is tetrahedral cubic (T_d). Hence impurities in substitutional positions experience comparably small electrical field gradients which are mainly due to the non-cubic 2nd nearest neighbour shell or non-ideal u parameters. Lattice sites along the c-axis (e.g., T, BC-c) and along the main interstitial axis parallel to the c-axis (e.g., O, HG, HN) possess strict trigonal (C_{3v}) symmetry, while all sites off these axes are of lower symmetry.

3. Results

3.1. Fe IN GaN AND ZnO

As an example, Figures 2(a–d) show the β^- emission channeling patterns from ^{59}Fe in ZnO, directly following room temperature implantation. The theoretical patterns resulting from the best two-fraction fits are shown in Figure 2(e–h). They were obtained by considering only Fe on substitutional S_{Zn} sites and varying its root mean square (rms) displacement u_1(Fe), and Fe on random sites. Note that the random fraction accounts for Fe atoms on sites contributing with an

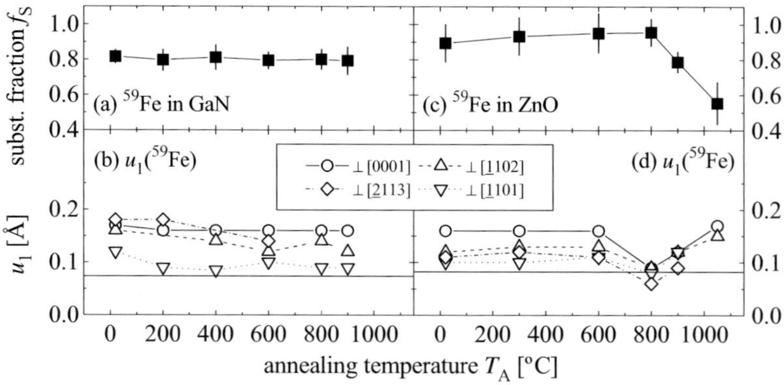

Figure 3. Fraction of Fe atoms on substitutional cation sites in GaN (a) and ZnO (c) and their rms displacements (b, d) following thermal annealing (up to 900 °C for 10 min in vacuum, the 1050 °C step for ZnO was 30 min in air). The straight lines indicate the room temperature rms displacements of Ga and Zn atoms, respectively.

isotropic emission yield, which are sites of very low crystal symmetry or in heavily damaged surroundings. In the as-implanted state, the ^{59}Fe rms displacements from S_{Zn} sites which gave the best fit were 0.16 Å, 0.12 Å, 0.10 Å and 0.11 Å, perpendicular to the [0001], [$\bar{1}$102], [$\bar{1}$101] and [$\bar{2}$113] directions, respectively, and the corresponding fractions on S_{Zn} sites were 108%, 84%, 84% and 81%. The fact that the fraction derived from the [0001] data is larger than 100% can be explained by variations in the background from scattered electrons for the different sample-detector orientations. The background is estimated by means of Monte Carlo simulations, but the corresponding errors are around 10–15%.

Figure 3 compares the Fe fractions on substitutional cation sites and their rms displacements as a function of annealing temperature for GaN [42] and ZnO. In both cases 80–90% of Fe atoms were found on S_{Ga} or S_{Zn} sites already in the as-implanted state, but with rms displacements exceeding the thermal vibration amplitude of the Ga or Zn atoms by up to a factor of two. In GaN, annealing up to 900°C practically did not change this situation. In ZnO, on the other hand, the 800°C anneal induced a significant decrease in the Fe rms displacements, reducing them to more or less the same values as expected for the Zn host atoms. Apart from the random fractions we found no evidence for ^{59}Fe located at other lattice sites than S_{Zn} or S_{Ga}, respectively.

The high substitutional fractions illustrate that the implantation damage in both materials was low, while rms displacements exceeding those of the host atoms can be explained by slight displacements of Fe atoms due to the presence of nearby defects. The fact that annealing of the ZnO sample at 800 °C resulted in perfect substitutional incorporation of Fe shows that it is possible to remove most of the damage in this material following low fluence implantation, in contrast to GaN, where the only visible effect of annealing was a slight decrease in the Fe

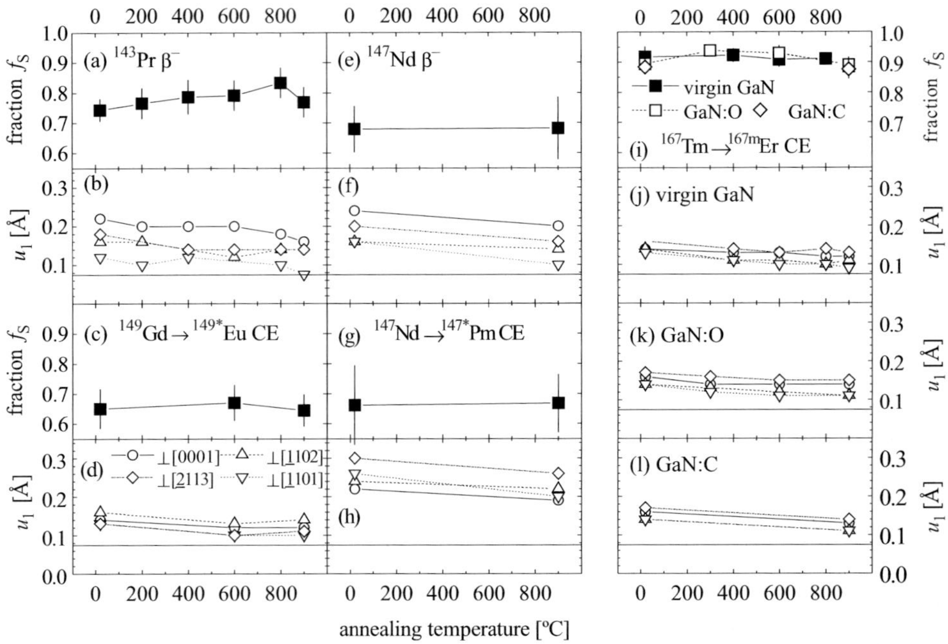

Figure 4. Fractions on substitutional Ga sites and rms displacements as function of annealing temperature for various rare earth probe isotopes in GaN.

rms displacements. In ZnO, annealing at 900 °C and 1050 °C caused the Fe rms displacements to increase again, accompanied by a significant decrease in the fraction on S_{Zn} sites. This indicates that the high temperature annealing has introduced crystal defects in the near-surface layers, which can interact with the Fe atoms, causing them to occupy displaced substitutional positions or random sites.

3.2. RARE EARTHS IN NITRIDES AND ZnO

The first emission channeling experiments on RE-implanted GaN were done by Dalmer *et al.* [43] who examined the conversion electrons emitted by ^{167m}Er resulting from the decay of ^{167}Tm and by ^{169*}Tm resulting from ^{169}Yb. However, only *c*-axis angular scans were compared to simulations, with the best fits obtained for 90(10)% of RE atoms on sites showing displacements of 0.25 Å from the *c*-axis, both in the as-implanted state and following annealing at 800°C.

More recently we have done emission channeling experiments using the β^- particles from ^{143}Pr [41, 44] and the conversion electrons (CE) emitted by ^{149*}Eu following the decay of ^{149}Gd [44, 45]. In addition we used the isotope ^{147}Nd, which allows to do lattice location studies of both the β^- emitting mother ^{147}Nd and CE emitting daughter ^{147*}Pm [44, 46]. An additional experiment with CE from ^{167m}Er was done on virgin GaN and GaN co-implanted with O or C [47].

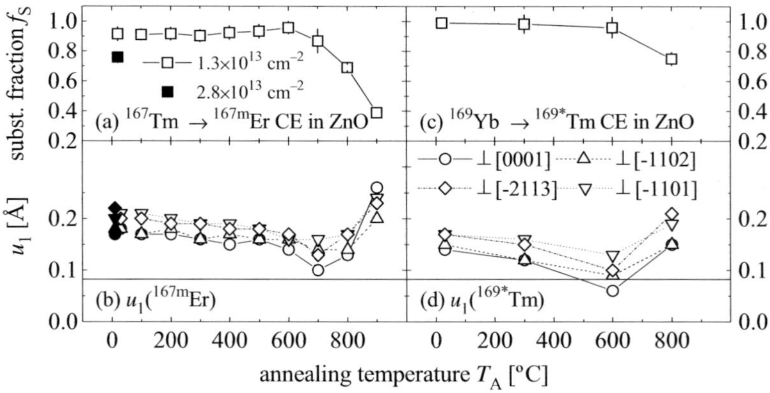

Figure 5. Fractions on Zn sites as a function of annealing temperature for ^{167m}Er (a) and ^{169*}Tm (c) in ZnO and corresponding rms displacements (b), (d). All annealings were for 10 min in vacuum, except for the 800°C step in (c) and (d), which was under flowing O_2.

In all cases the majority of RE atoms (around 65–95%) were found on substitutional Ga sites (Figure 4), the substitutional fractions being highest in the case of ^{143}Pr and ^{167}Tm → ^{167m}Er, and lower for ^{147}Nd, ^{147}Nd → ^{147*}Pm and ^{149}Gd → ^{149*}Eu. Whereas the rms displacements from the S_{Ga} sites decreased with annealing temperature, the substitutional fractions hardly changed following 10-min annealing sequences up to 900 °C under vacuum. Variations in the rms displacements perpendicular to the four different channeling directions for each experiment and from isotope to isotope could indicate that different rare earth atoms experience small static displacements along well-defined crystallographic directions. However, more experiments with additional isotopes will be needed to confirm whether there are systematic trends in that respect. Note that in the conversion electron experiments we have so far not found any indication that there are major changes in the lattice sites due to nuclear recoil or after-effects from radioactive decay.

It has been reported in the literature that co-doping with O or C enhances the RE luminescence in GaN [48] and it is known that the formation of complexes between O and Er changes the lattice location of Er in Si [49]. We therefore also investigated the lattice sites of ^{167m}Er in two GaN samples that were pre-implanted with O and C. Our results [Figure 4 (i–k)] show that the presence of O and C, even if it exceeds the RE concentration by a factor of 14, has no effect on the lattice site of ^{167m}Er in GaN [47].

In ZnO the rare earths showed a somewhat different behaviour. As in GaN, already immediately following implantation 75–100% of RE atoms occupied cation Zn sites. However, 600–700°C annealing resulted in a significant decrease of the rms displacements of both ^{167m}Er [50] and ^{169*}Tm probes [51], while for higher annealing temperatures the substitutional RE fractions decreased and their rms displacements increased again, even more pronounced than in the case of Fe (Figure 5). It is known from Rutherford backscattering experiments on higher-dose implanted samples that Er atoms in ZnO diffuse around 1050°C [52]. Hence it seems possible that in our case short-range RE diffusion may have contributed to some extent in its pairing with other defects at 800–900°C.

4. Conclusions

Following low dose implantations ($1–3 \times 10^{13}$ cm^{-2}) of Fe and rare earth elements into GaN and ZnO, the majority of the implants occupy substitutional cation sites. However, rms displacements which are substantially larger than those of the host atoms indicate that there are defects present in the immediate neighbourhood of the implanted probes. In GaN, only small reductions in the impurity rms displacements were observed up to 900°C, showing that a substantial part of them remains in distorted Ga sites. In ZnO best incorporation on substitutional Zn sites was obtained after 800°C annealing for Fe and 600–700 °C annealing for RE elements. At higher annealing temperatures, the probe atoms experience again increased levels of disorder. The mechanism of this is not yet entirely clear but probably involves the formation of defect complexes as a result of defect creation in near surface layers during annealing and/or short-range diffusion of the implanted species.

We would like to point out that some of the open questions with respect to RE and TM impurities in wide band gap semiconductors are accessible to hyperfine interaction techniques. As it is clear from our studies, following ion implantation the trapping of defects by RE or Fe and the annealing of defect complexes should be observable. It has been proposed that defects play an important role in the luminescence excitation of RE in GaN [53], which begs for an assessment of crystal defects in the RE neighbourhood. In a similar way, it has been pointed out [8] that more information is needed on the microstructure of TM-related defects in diluted magnetic semiconductors in order to explain the nature of the observed magnetism and the possible role of second phases.

Acknowledgements

This work was funded by the FCT, Portugal (project POCTI-FNU-49503-2002) and by the European Union (Large Scale Facility contract HPRI-CT-1999-00018). U. Wahl and E. Rita acknowledge their fellowships supported by the FCT, Portugal.

References

1. Mohammad S. N. and Morkoç H., *Prog. Quantum Electron.* **20** (1996), 361.
2. Pearton S. J., Zolper J. C., Shul R. J. and Ren F., *J. Appl. Phys.* **86** (1999), 1.
3. Monroy E., Omnès F. and Calle F., *Semicond. Sci. Technol.* **18** (2003), R33.
4. Pearton S. J., Ren F., Zhang A. P. and Lee K. P., *Mater. Sci. Eng., R* **30** (2000), 55.
5. Kohl D., *J. Phys., D, Appl. Phys.* **34** (2001), R125.

6. Pearton S. J., Kang B. S., Kim S., Ren F., Gila B. P., Abernathy C. R., Lin J. and Chu S. N. G., *J. Phys., Condens. Matter* **16** (2004), R961.
7. Steckl A. J. and Zavada J. M., *Mater. Res. Soc. Bull.* **24** (1999), 23.
8. Pearton S. J., Abernathy C. R., Overberg M. E., Thaler G. T., Norton D. P., Theodoropoulou N., Hebard A. F., Park Y. D., Ren F., Kim J. and Boatner L. A., *J. Appl. Phys.* **93** (2003), 1.
9. Favennec P. N., L'Haridon H., Moutonnet D., Salvi M. and Gauneau M., *Mater. Res. Soc. Symp. Proc.* **301** (1993), 181.
10. Dietl T., *Semicond. Sci. Technol.* **17** (2002), 377.
11. Sato K. and Katayama-Yoshida H., *Hyperfine Interact.* **136** (2001), 737.
12. Kucheyev S. O., Williams J. S., Jagadish C., Zou J., Evans C., Nelson A. J. and Hamza A. V., *Phys. Rev., B* **67** (2003), 094115.
13. Bharuth-Ram K., Wahl U. and Correia J. G., *Nucl. Instrum. Methods B* **206** (2003), 941.
14. Alves E., Correia R., Pereira S., Wahl U., De Vries B. and Vantomme A., *Nucl. Instrum. Methods B* **206** (2003), 1042.
15. Vetter U., Reid M. F., Hofsäss H., Ronning C., Zenneck J., Dietrich M. and the ISOLDE Collaboration, *Mater. Res. Soc. Symp. Proc.* **743** (2003), L 6.16.
16. Vetter U., Hofsäss H., Wahl U., Dietrich M. and the ISOLDE Collaboration, *Diamond and Related Materials* **12** (2003), 1883.
17. Vetter U., Taniguchi T., Wahl U., Correia J., Müller A., Ronning C., Hofsäss H., Dietrich M. and the ISOLDE Collaboration, *Mater. Res. Soc. Symp. Proc.* **744** (2003), M 8.38.
18. Stewart G. A., *Materials Forum* **18** (1994), 177.
19. Dalmer M., Vetter U., Restle M., Stötzler A., Hofsäss H., Ronning C., Moodley M. K., Bharuth-Ram K. and the ISOLDE Collaboration, *Hyperfine Interact.* **120** (1999), 347.
20. Cadogan J. M. and Ryan D. H., *Hyperfine Interact.* **153** (2004), 25.
21. Masterov V. F., Nasredinov F. S., Seregin P. P., Kudoyarova V. K., Kuznetsov A. N. and Terukov E. I., *Appl. Phys. Lett.* **72** (1998), 728.
22. Rasera R. L. and Li-Scholz A., *Phys. Rev., B* **1** (1970), 1995.
23. Thomé L., Bernas H., Abel F., Bruneaux M., Cohen C. and Chaumont J., *Phys. Rev., B* **14** (1976), 2787.
24. Dogra R., Bhati A. K. and Bedi S. C., *Hyperfine Interact.* **108** (1997), 515.
25. Vianden R., *Hyperfine Interact.* **35** (1987), 1079.
26. Nédélec R., Vianden R. and the ISOLDE Collaboration, this conference, contr. C.7
27. Schwalbach P., Laubach S., Hartick M., Kankeleit E., Keck B., Menningen M. and Sielemann R., *Phys. Rev. Lett.* **64** (1990), 1274.
28. Langouche G., *Hyperfine Interact.* **72** (1992), 217.
29. Weyer G. and the ISOLDE Collaboration, *Hyperfine Interact.* **129** (2000), 371.
30. Bharuth-Ram K., *Hyperfine Interact.* **151** (2003), 21.
31. Goya G. F., Stewart S. J. and Mercader R. C., *Sol. State. Comm.* **96** (1995), 485.
32. Goya G. F. and Leite E. R., *J. Phys., Condens. Matter* **15** (2003), 641.
33. Alves E., Liu C., Waerenborgh J. C., Da Silva M. F. and Soares J. C., *Nucl. Instrum. Methods B* **175** (2001), 241.
34. Hofsäss H. and Lindner G., *Phys. Rep.* **210** (1991), 121.
35. Hofsäss H., Wahl U. and Jahn S. G., *Hyperfine Interact.* **84** (1994), 27.
36. Hofsäss H., *Hyperfine Interact.* **97** (1996), 247.
37. Wahl U., *Hyperfine Interact.* **129** (2000), 349.
38. Kugler E., *Hyperfine Interact.* **129** (2000), 23.
39. Deicher M., Weyer G., Wichert T. and the ISOLDE Collaboration, *Hyperfine Interact.* **151** (2003), 105.
40. Wahl U., Correia J. G., Czermak A., Jahn S. G., Jalocha P., Marques J. G., Rudge A., Schopper F., Vantomme A., Weilhammer P. and the ISOLDE Collaboration, *Nucl. Instrum. Methods A* **524** (2004), 245.

41. Wahl U., Vantomme A., Langouche G., Araujo J. P., Peralta L., Correia G. and the ISOLDE Collaboration, *J. Appl. Phys.* **88** (2000), 1319.
42. Wahl U., Vantomme A., Langouche G., Correia J. G., Peralta L. and the ISOLDE Collaboration, *Appl. Phys. Lett.* **78** (2001), 3217.
43. Dalmer M., Restle M., Stötzler A., Vetter U., Hofsäss H., Bremser M. D., Ronning C. and Davies R. F., *Mater. Res. Soc. Symp. Proc.* **482** (1998), 1021.
44. Wahl U., Alves E., Lorenz K., Correia J. G., Monteiro T., De Vries B., Vantomme A. and Vianden R., *Mater. Sci. Eng., B* **105** (2003), 132.
45. De Vries B., Wahl U., Vantomme A., Correia J. G. and the ISOLDE Collaboration, *Mater. Sci. Eng., B* **105** (2003), 106.
46. De Vries B., Wahl U., Vantomme A., Correia J. G. and the ISOLDE Collaboration, *Phys. Status Solidi* **(c) 0** (2002), 453.
47. De Vries B., Matias V., Vantomme A., Wahl U., Rita E. M. C., Alves E., Lopes A. M. L., Correia J. G. and the ISOLDE Collaboration, *Appl. Phys. Lett.* **84** (2004), 4304.
48. Torvik J. T., Qiu C. H., Feuerstein R. J., Pankove J. I. and Namavar F., *J. Appl. Phys.* **81** (1997), 6343.
49. Wahl U., Vantomme A., De Wachter J., Moons R., Langouche G., Marques J. G., Correia J. G. and the ISOLDE Collaboration, *Phys. Rev. Lett.* **79** (1997), 2069.
50. Wahl U., Rita E., Alves E., Correia J. G., Araújo J. P. and the ISOLDE collaboration, *Appl. Phys. Lett.* **82** (2003), 1173.
51. Rita E. M. C., Wahl U., Lopes A. L., Araújo J. P., Correia J. G., Alves E., Soares J. C. and the ISOLDE Collaboration, *Mater. Res. Soc. Symp. Proc.* **774** (2003), M 3.7.
52. Alves E., Rita E., Wahl U., Correia J. G., Monteiro T. and Boemare C., *Nucl. Instrum. Methods B* **206** (2003), 1047.
53. Braud A., Doualan J. L., Moncorge R., Pipeleers B. and Vantomme A., *Mater. Sci. Eng., B* **105** (2003), 101.

Hyperfine Interactions (2004) 159:373–377
DOI 10.1007/s10751-005-9126-z

NMR with Hyperpolarised Protons in Metals

A. ENGELBERTZ*, P. ANBALAGAN, C. BOMMAS, P.-D. EVERSHEIM, D. T. HARTMAN and K. MAIER
Helmholtz- Institut fuer Strahlen und Kernphysik, University of Bonn, Bonn, Germany; e-mail: engelbert@iskp.uni-bonn.de

Abstract. Proton pulse NMR, established as a versatile method in Solid State Physics, Chemistry, Biology and Medical Science, requires on the order of 10^{18} nuclei to detect an electromagnetic signal in a free induction decay (FID). The main cause for this small sensitivity is the low polarisation in the order of a few ppm due to the Boltzmann distribution in the magnetic field. Thus, NMR experiments on hydrogen are limited to metals with extremely high hydrogen solubility like Pd near room temperature. Using a polarised proton beam, a NMR signal is possible with as few as 10^{13} implanted nuclei. For the first time spin–spin and spin–lattice relaxation times were measured in Au and W with this technique at the Bonn cyclotron.

Key Words: gold, hydrogen, hyperpolarisation, ion implantation, nuclear magnetic resonance, tungsten.

1. Introduction

NMR is an established method for analysing the properties of hydrogen in materials. The spin–spin (T_2) and spin–lattice (T_1) relaxation times are useful in the analysis of the environmental and spatiotemporal behaviour (for example diffusion) of hydrogen in different materials. In metals there are a number of problems encountered in NMR measurements. The conductance is high, therefore the skindepth is very shallow. Consequently the hydrogen solubility must be high, but apart from a few elements like Pd the hydrogen solubility in metals is very low [3]. For this reason we chose hyperpolarised protons for our NMR measurements. With this choice we were able to reduce the number of protons needed for a signal from 10^{18} in conventional NMR to 10^{13} in polarised beam (PB)-NMR [2].

2. Experiment and discussion

The transversely polarised protons are introduced into the Bonn cyclotron, where they are accelerated to an energy of 7.5 MeV. The asymmetry in elastic scattering of polarised protons by a thin carbon foil is then used to measure the level

* Author for correspondence.

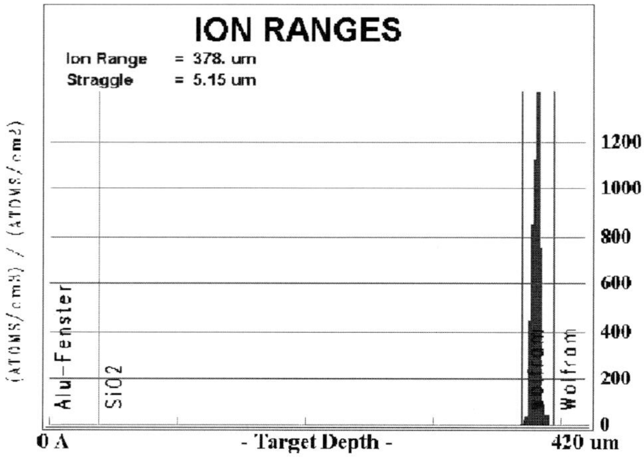

Figure 1. Implantation profile of polarised protons in a tungsten foil. The 7.5 MeV proton beam penetrates a 30 μm thick Al-foil and a quartz-degrader with a thickness of 350 μm. The protons are stopped in the first tungsten foil. The second foil is needed only to guarantee that if the density of SiO_2 varies all protons will still be stopped inside tungsten [1].

of polarisation. The ratio of polarised to unpolarised p^+ is 4:1 at a current of 0.3 μA. The target was mounted (Figure 1) in a superconducting magnet which produces a homogenous magnetic field B_o of 1.2 T parallel to the beam axis. To implant the protons, it is necessary that the beam is longitudinally polarised. This is realised with a solenoid which turns the direction of polarisation from vertical to horizontal but still perpendicular to the beam. Then a bending magnet is used which aligns the spin with the beam and sends the beam to the NMR target. The bending angle was chosen to be $\alpha = 90° \left(\frac{\gamma_p m_p}{e-1}\right)^{-1}$ so that the vertical magnetic field could reorientate the polarisation parallel to the beam during the time of flight [4]. By switching a weak or intermediate field transition in the atomic beam in the polarised ion source, the beam polarisation is set parallel or antiparallel to the magnetic field. Our samples were positioned in a vertical NMR-RF coil and conventional NMR- pulse techniques were used to measure the free induction decay (FID). The samples were composed of a 350 μm thick SiO_2 degrader and two 25 μm thick tungsten foils in a Teflon carrier. We calculated the thickness of the degrader such that the Bragg peak for 7.5 MeV protons would be in the first tungsten foil. To be certain that all protons stopped within the tungsten foil in spite of fluctuations in the SiO_2-density, two foils were used. Before the tungsten sample was irradiated, the sensitivity of the apparatus was tested with a liquid D_2O sample. These measurements affirmed the beam positioning, the polarisation, and provided data for the magnetic field drift correction. Then the tungsten was cooled to $T \approx 50°$K and irradiated for 1 s with (Figure 2) a repetition period of 6s.

After every irradiation period the polarity of the beam was changed and after every second irradiation period the delay time (the time between the beam end

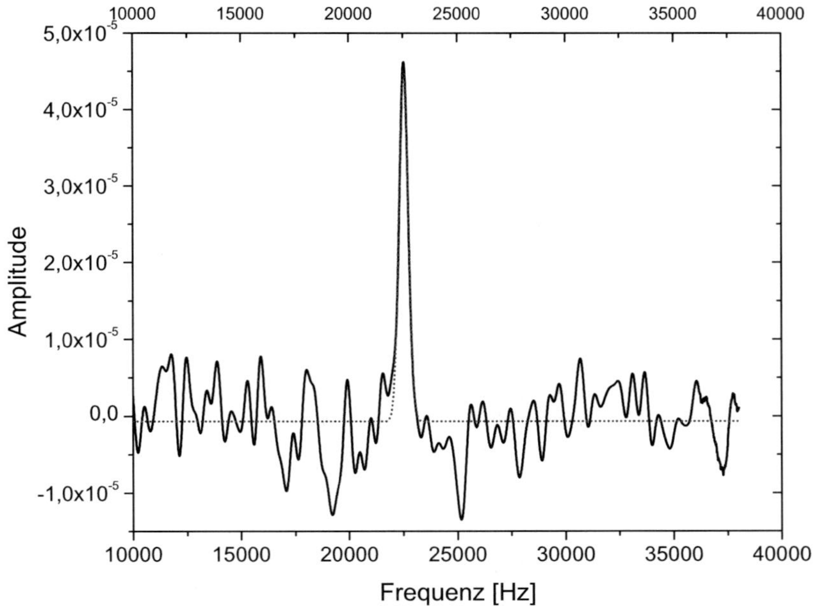

Figure 2. Wolfram spectrum (*black line*) of difference between parallel and antiparallel polarisation in the mixing frequency with a Gaussian fit (*blue line*) to determine the line width. For the tungsten measurement, the line width was 290 ± 23 Hz [1].

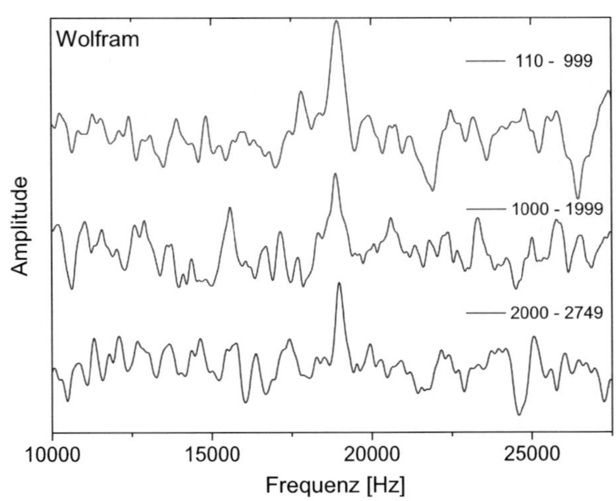

Figure 3. Development of the resonance-signal in a tungsten foil during irradiation with protons. The *first line* shows the signal summed over the periods 110–999, the *second* from 1,000–1,999 and the *third line* from 2,000–2,749. The amplitude of resonance-signal level is of the same value over all periods. This shows that the trapping in the defects did not grow with the proton irradiation [1].

and the 90°-pulse). To improve the signal to noise ratio the drift corrected FFT of the FID for each polarity and delay time was summed over a large number of irradiation cycles. To be sure that we analysed only implanted proton signals, we took the difference between the signals of inplanted protons with spin orientation

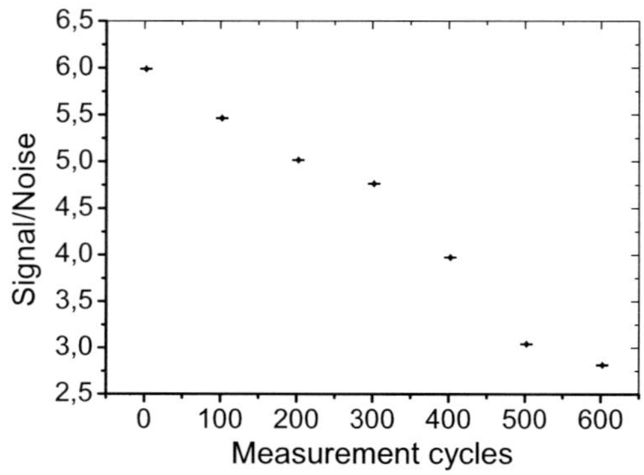

Figure 4. The FID amplitude reduces in gold with each measurement cycle, do to the accumulating defects. The *diagram* shows the signal to noise, after summation over 1,500 cycles as a function of the skipped cycles previous to summation.

parallel and antiparallel to the magnetic field. The line width in the Fourier transformation of the FID is determined by the spin- spin relaxation time T_2. Analysis of the (T_2) spectra shows that the implanted protons at 50°K in the motional averaging region are not fixed in the metal. The narrow line width of 250 Hz is an indication of high proton mobility. If they were fixed the dipole interaction with the magnetic moment of the ^{183}W would broaden the line width. Another interesting point to be considered is that, with the implantation of the polarised protons, radiation defects are produced in the tungsten sample but the NMR line width and amplitude remain constant for a large number of measurement cycles, which means that with an implanted proton dose of about 10^{14} protons/cm^2 there is still no trapping observed in the defects in tungsten. Whereas in other metals like gold for example the number of defects grows with each measurement cycle and eliminates (Figure 3) the polarised proton signal over time. The line width of trapped protons is not measurable with our present set-up. Trapping leads to a reduction of signal amplitude of free protons with increased implantation dose. With measurements at two different delay times we are able to calculate the spin-lattice-relaxation time T_1 in tungsten. An exponential fit through the amplitude at the two different delay times delivers the T_1 time [1].

3. Conclusion

Proton NMR in metals with low hydrogen solubility is possible with a polarised beam. In order to accumulate enough polarisation in the sample the spin lattice relaxation time at the implantation temperature should be at least 500 ms. Our first results in tungsten indicate that we observe free diffusing protons at 50°K. However in gold radiation damage creates trapping centres for the implanted protons. This is clearly visible in the dose dependent amplitude of the NMR signal. To get a general limitation for the dose in gold we need a better statistic on different gold foils (Figure 4).

Acknowledgements

The authors thank the cyclotron personal S. Birkenbach, A. Henny, B. Kann, S. Lehmann O. Rast, for the cooperation. Günther Majer from the Max-Plank-Institue für Metallforschung in Stuttgart for a lot of help and for support we thank the Deutsche Forschungsgesellschaft (contract nr MA 1689/8).

References

1. Bommas C., *Puls NMR mit hyperpolarisierten Protonen in Metallen*, PhD thesis, Rheinischen Friedrich-Willhelms Universitt Bonn, 2004.
2. Mathews H. G., Krüger A. and S. P. A., The new high intensity polarized proton and deuteron source for the Bonn isochronous cyclotron, *Nucl. Instr. Meth. Phys. Res.* **B22** (1980), 2226.
3. Wipf H., *Hydrogen in Metalls III*, 1st edn. Springer, Berlin Heidelberg New York, 1997.
4. Schüth J., Eversheim P.-D., Herzog P., Maier K., Majer G., Meyer P. and Roduner E., NMR on protons from a polarized cyclotron beam. *Chem. Phys. Lett.* **303** (1999), 453.

Hyperfine Interactions (2004) 159:379–383
DOI 10.1007/s10751-005-9128-x

A New Generation TDPAC Spectrometer

CHRISTIAN H. HERDEN[1,*], MAURO A. ALVES[1], KLAUS D. BECKER[1] and JOHN A. GARDNER[2]
[1]*Technische Universität Braunschweig, Braunschweig, Germany; e-mail: Cherden@viewplus.de*
[2]*Oregon State University, Corvallis, USA; e-mail: John.Gardner@viewplus.com*

Abstract. Time Differential gamma–gamma Perturbed Angular Correlation spectroscopy has traditionally been done using scintillation detectors along with constant–fraction discriminators, spectroscopy amplifiers, single channel analyzers, and time to amplitude detectors. We describe a new generation spectrometer where these electronics are replaced by high speed digital transient recorders that record the output from each scintillation detector. The energy and time-of-arrival of gamma rays in any detector can be determined accurately. Many experimental difficulties related to electronics are eliminated; the number of detectors can be increased with no increase in complexity of the apparatus; coincidences among any two detectors are measurable; and coincidences separated by as little as a ns are detectable in principle within one detector. All energies are collected, and energy windows are imposed by software filtering, permitting both high energy resolution and high data-gathering power.

Key Words: digital spectrometer, time differential perturbed angular correlation.

1. Introduction and Description of the Spectrometer

Gamma–gamma time Differential Perturbed Angular Correlation (TDPAC) spectroscopy is a well-established experimental technique [1, 2] providing information about condensed matter [3] through measurement of gamma ray emission during decay of ^{111}In, ^{181}Hf or other radioactive nuclei [4–6] commonly introduced at trace levels into the material of interest.

Most conventional TDPAC gamma-ray spectrometers [7–15] utilize spectroscopy amplifiers and single channel analyzers to determine the energy of a gamma ray in parallel with some type of time to digital conversion to determine the difference δt between the time-of-arrival of the first and second gamma ray. In Figure 1, we show a block diagram of a fundamentally different design TDPAC spectrometer. This first model has four BaF_2 [13] scintillators mounted in a plane in the conventional 4-detector TDPAC geometry [8]. A dynode or anode signal of each detector is recorded by a 1 GHz transient recorder, model DP110 from Acqiris, Inc. of Geneva, Switzerland. In our configuration these recorder boards

* Author for correspondence.

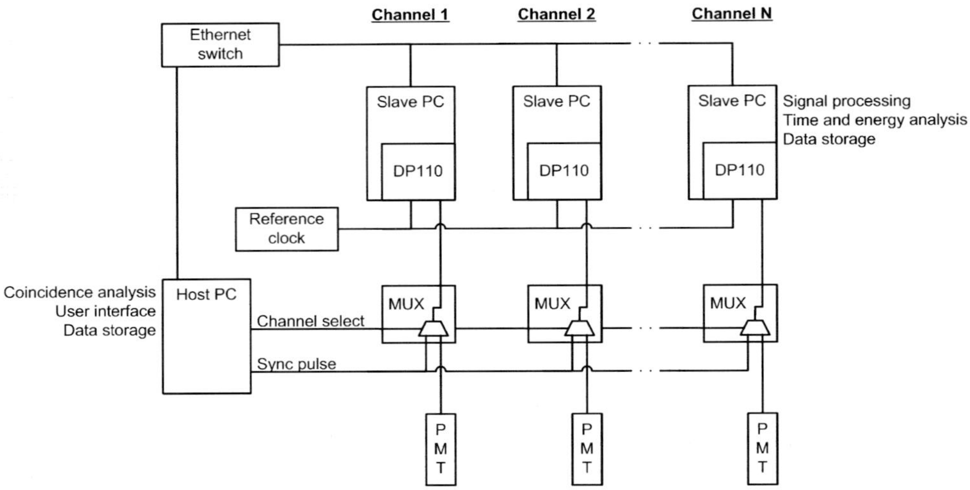

Figure 1. Block diagram of the new generation TDPAC spectrometer.

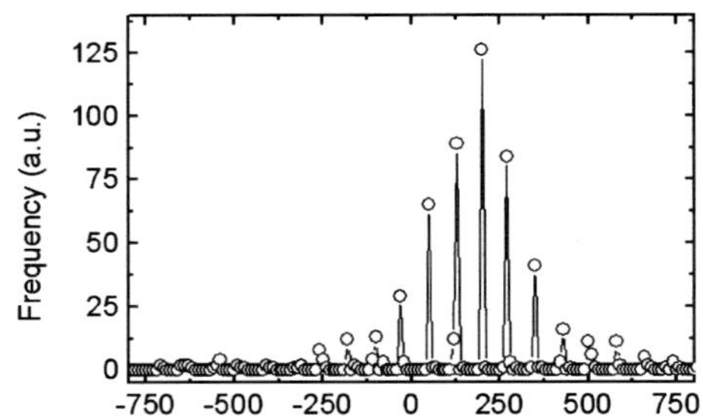

Figure 2. 180° prompt peak for ^{22}Na with a FWMH of less than 400 ps. The presence of peaks at 80 ps intervals is due to the quantization of the digitizer clock.

have a capacity of 8 MB. They can be programmed to record an arbitrary number of increment windows of arbitrary length in "sequence" mode with very low dead time. Typically we record 1000 ns length increments when using BaF_2 scintillator crystals. No additional pulse shaping equipment or amplifier is used.

Each recorder board is mounted in its individual "slave" PC computer. The slave computers perform low level analysis of the pulses and then pass the parameters of that analysis to a separate host computer. A home-built electronic board in the host computer drives the clock synchronization inputs on each recorder board from a common source. The host board also provides a gate signal and a trigger pulse signal for each of the slave units. Each slave computer has a home-built electronic board with an Analog Devices Model AD8180 analog multiplexer (MUX) chip. The signal from each detector is routed through the MUX and then to its recorder board. A recording session is initialized by gating

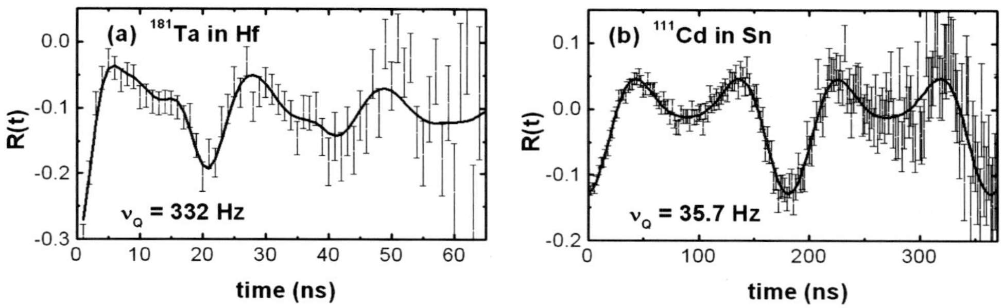

Figure 3. TDPAC spectra measured with the new spectrometer. (a) ^{181}Hf/Ta spectrum in high-purity Hf thin-foil metal. (b) ^{111}In/Cd PAC spectrum in Sn metal.

the slave MUX chips off so that no signals can pass from the detectors, then initializing each recorder board, and sending a common trigger signal that is recorded as the first data increment by each board. The MUX gates are then turned on so that pulses from the detectors are recorded until the memory of the recorder boards fills. The initial trigger pulse gives a common time base.

When the recorder board has filled its memory, it flags its slave computer and then downloads its data to that slave computer. When all four recorders have downloaded their data, the host computer institutes a new recording session. In addition to the analog recording of the pulses, the trigger time initializing each data increment is recorded. This time is interpolated between clock pulses and has a precision of order 80 ps. The slave computer can, in principle, fit the pulse shapes and determine a goodness-of-fit value, then perform further analysis to find multiple events. Currently it only finds the integral of the pulse, which is proportional to the gamma ray energy. It passes the integral and the time-of-arrival of each event to the host computer. For modest count rates, of order 10 k/s, the dead time is estimated to be less than 20%.

The host computer compares the times of arrival for events in each detector pair and records coincidence events when the times of arrival differ by less than some set amount, e.g. 500 ns for ^{111}In or 100 ns for ^{181}Hf. It then creates a table of events with columns for \delta t and the energies of the two gamma rays. Histograms of counts vs. time for any set of energy windows can be constructed from this table. In many cases, no energy windowing is needed, and one obtains an order of magnitude more data than when conventionally windowed. One can still use narrow windows for some purposes, e.g. finding the proper behavior near $t = 0$ for ^{181}Hf data.

A "prompt peak" signal is shown in Figure 2. This is a histogram of events vs. the time difference between gamma ray trigger events in two detectors with a ^{22}Na sample between them. ^{22}Na decays by positron emission, and the two gamma rays are the 511 keV gamma rays emitted simultaneously at approximately 180° with respect to each other when the positron annihilates with an electron in the material containing the ^{22}Na. This method is conventionally used for calibrating time response of TDPAC spectrometers.

The full width at half maximum of approximately 400 ps is close to the theoretical value for BaF_2 detectors and is considerably smaller than what is normally achieved in conventional spectrometers.

Figure 3(a) shows conventional TDPAC R(*t*) ^{181}Hf/Ta spectrum in high-purity Hf metal foil, and Figure 3b ^{111}In/Cd PAC spectrum in Sn metal. Each are created from histograms and fitted using conventional 4-detector techniques [8]. The fitted A_2 parameter is approximately what has been obtained previously [15, 16] for these detectors in a conventional spectrometer. The frequencies are in acceptable agreement with those previously published [4, 18] for these materials. The geometrical factors for the ^{111}In/Cd spectrum in Sn metal show evidence that the grains are not randomly oriented, but otherwise is normal.

Acknowledgements

The authors wish to thank Dr. Viktor Hungerbuhler and his colleagues at Acqiris for assistance in learning how to get the proper performance from the recorder boards. They also thank Dr. James Sommers of ATI Wah Chang for providing the Hf metal foil used for the sample of Figure 3(a). They thank Dr. Matthew Zacate and Prof. Gary Collins for the Sn metal sample used for Figure 3(b). They thank the Oregon State University Reactor Center for neutron irradiation of the Hf foil and Prof. Kenneth Krane for arranging that irradiation. This project was funded in part by the German Research Foundation (Deutsche Forschungsgemeinschaft DFG) grants BE 972/21-2 and BE 927/21-1.

References

1. Frauenfelder H. and Steffen R. M. In: Siegbahn K. (eds), *Alpha-, Beta-, and Gamma-ray Spectroscopy*, Vol. 2, North Holland, Amsterdan, 1965, p. 995.
2. Schatz G. and Weidinger A., *Nuclear Condensed Matter Physics*, John Wiley and Sons, New York, 1996.
3. Catchen L., *J. Mater. Educ.* **12** (1990), 253.
4. Vianden R., *Hyperfine Interact.* **16** (1983), 1081.
5. Rots M., van Cauteren J., de Doncker G. and Rao G. N., *Hyperfine Interact.* **35** (1987), 967.
6. Lerf A. and Butz T., *Hyperfine Interact.* **36** (1987), 275.
7. Rinneberg H. H., *At. Energy Rev.* **17** (1979), 47.
8. Arends A. R., Hohenemser C., Pleiter F., de Waard H., Chow I. and Sutter R. M., *Hyperfine Interact.* **8** (1980), 191.
9. Baudry A., Boyer P., Choulet, S., Gamrat C., Peretto P., Perrin D. and Van Zurk R., *Zurk, Nucl. Instr. Meth.* **A260** (1987), 160.
10. Jaeger H., Gardner J. A., Su H. T. and Rasera R. L., *Rev. Sci. Instrum.* **58** (1987), 1694.
11. Schatz G., Klas T., Platzer R., Voigt J. and Wesche R., *Hyperfine Interact.* **34** (1987), 555.
12. van Cauteren J., Rao G. N. and Rots M., *Nucl. Instrum. Methods* **A243** (1986), 445.
13. Laval M., Moszynski M., Allemand R., Cormoreche E., Guinet P., Odru R. and Vacher J., *Nucl. Instrum and Methods* **206** (1983), 169.
14. Marszałek M., Grębosz J., Jaworski J., Stachura Z., Ziębliński M. and Sulkio-Cleff B., *Z. Naturforsch.* **55a** (2000), 74.

15. Warnes W. H. and Gardner J. A., *Phys. Rev. B* **40** (1989), 4276.
16. Mommer N., Lee T., Gardner J. A. and Evenson W. E., *Phys. Rev. B* **61** (2000), 162.
17. Rasera R. L., Butz T., Vasquez A., Ernst H., Shenoy G. K., Dunlap B. D., Reno R. C. and Schmidt G., *J. Phys. F*, **8** (1987), 1581.
18. Christiansen J., Heubes P., Keitel R., Klinger W., Loeffer W., Sandner W. and Whitthuhn W., *Z. Phys., B* **24** (1976), 177.

Hyperfine Interactions (2004) 159:385–388
DOI 10.1007/s10751-005-9129-9

A New High-Intensity, Low-Momentum Muon Beam for the Generation of Low-Energy Muons at PSI

T. PROKSCHA*, E. MORENZONI, K. DEITERS, F. FOROUGHI, D. GEORGE, R. KOBLER and V. VRANKOVIC
Paul Scherrer Institute, PSI, CH-5232, Villigen, Switzerland; e-mail: thomas.prokscha@psi.ch

Abstract. At the Paul Scherrer Institute (PSI, Villigen, Switzerland) a new high-intensity muon beam line with momentum $p < 40$ MeV/c is currently being commissioned. The beam line is especially designed to serve the needs of the low-energy, polarized positive muon source (LE-μ^+) and LE-μ SR spectrometer at PSI. The beam line replaces the existing μE4 muon decay channel. A large acceptance is accomplished by installing two solenoidal magnetic lenses close to the muon production target E that is hit by the 590-MeV PSI proton beam. The muons are then transported by standard large aperture quadrupoles and bending magnets to the experiment. Several slit systems and an electrostatic separator allow the control of beam shape, momentum spread, and to reduce the background due to beam positrons or electrons. Particle intensities of up to 3.5×10^8 μ^+/s and 10^7 μ^-/s are expected at 28 MeV/c beam momentum and 1.8 mA proton beam current. This will translate into a LE-μ^+ rate of 7,000/s being available at the LE-μSR spectrometer, thus achieving μ^+ fluxes, that are comparable to standard μSR facilities.

Keywords: low-energy muons, muon beam.

At present the Paul Scherrer Institute (PSI, Villigen, Switzerland) is operating the most powerful proton accelerator at medium energies (590 MeV) which is used for the generation of high-intensity, low-momentum muon beams at 100% duty cycle. These particles are used in stopped experiments which cover a large spectrum of physical problems, from fundamental particle physics experiments to condensed and soft matter investigations [1]. For the condensed and soft matter applications a large research program for muon-spin-rotation (μSR) [2] has been established over the last 20 years. A new important extension of the μ SR program has been achieved by the recent development of a low-energy μ^+ beam (LE-μ^+, 'LEM') with μSR spectrometer (LE-μSR) [3–6]. Whereas standard muon beams are produced with energies of order MeV the LE-μ^+ beam has a tunable energy between 0.5 and 30 keV, thus allowing the controlled implantation at depths on the nanometer scale below the surface of a sample. The beam is generated by the moderation of a surface muon beam (momentum $p < 29.8$ MeV) in an appropriate solid rare gas film. Up to now LE-μSR has been intensity limited. In order to use the full capacity of this new technique an increase of

*Author for correspondence.

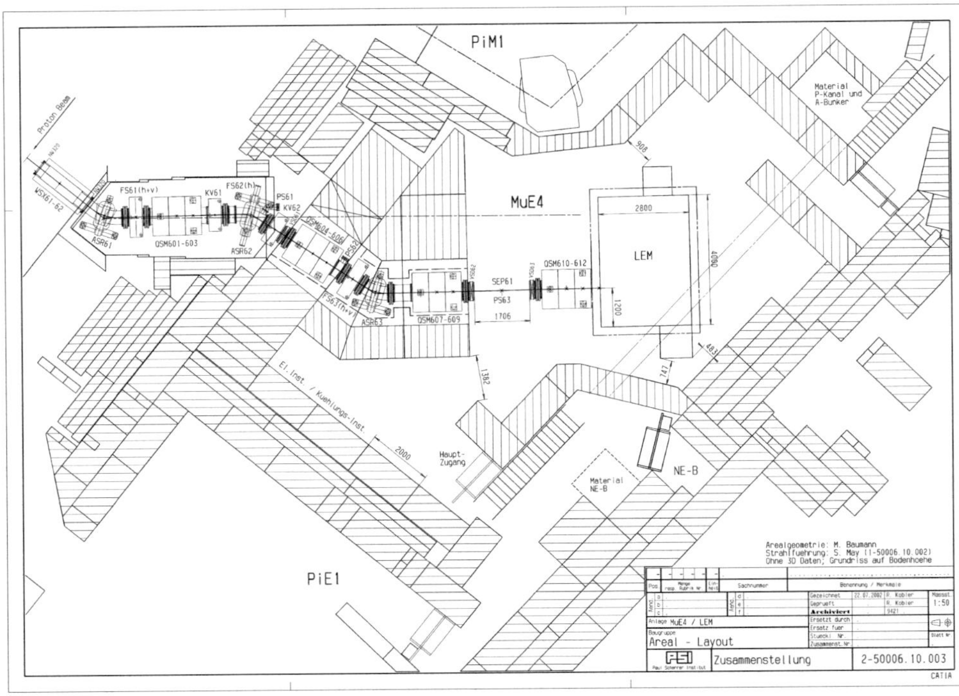

Figure 1. Layout of the new μE4 beam line and experimental area. LEM refers to the low-energy muon beam and LE-μSR spectrometer.

muon beam intensity by about one order of magnitude at a beam momentum of ∼28 MeV/c is required. To achieve this purpose the existing μE4 channel at PSI has been completely redesigned to provide maximum μ^+ intensity at 28 MeV/c. The beam line can deliver μ^- as well, with the maximum momentum restricted to 40 MeV/c due to limitations of the first focusing elements.

In order to increase significantly the acceptance of the beam line the existing standard quadrupoles had to be replaced by a solenoid as the first focusing element. The rotational symmetry of the solenoid allows to focus the beam in all directions. Disadvantages are the higher electric power consumption compared to a quadrupole lens with same focal length, and a mixing of the horizontal and vertical phase spaces. The new μE4 beam line uses two solenoids (`WSX61`, `WSX62`) where the first one can be operated at fields up to 0.37 T, which limits the beam momentum to 40 MeV/c. Both solenoids consist of normal conducting, radiation hard coils. By a proper combination of both elements a phase space rotation close to 90° can be achieved, thus exchanging the initial horizontal and vertical phase space and keeping the mixing small. The solenoids are followed by large aperture (240 mm) bending magnets (`ASR61–63`) and quadrupole tripletts (`QSM601–612`, aperture 400 mm). A static $E \times B$ filter (`SEP61`) with 180 mm electrode distance and up to 400 kV

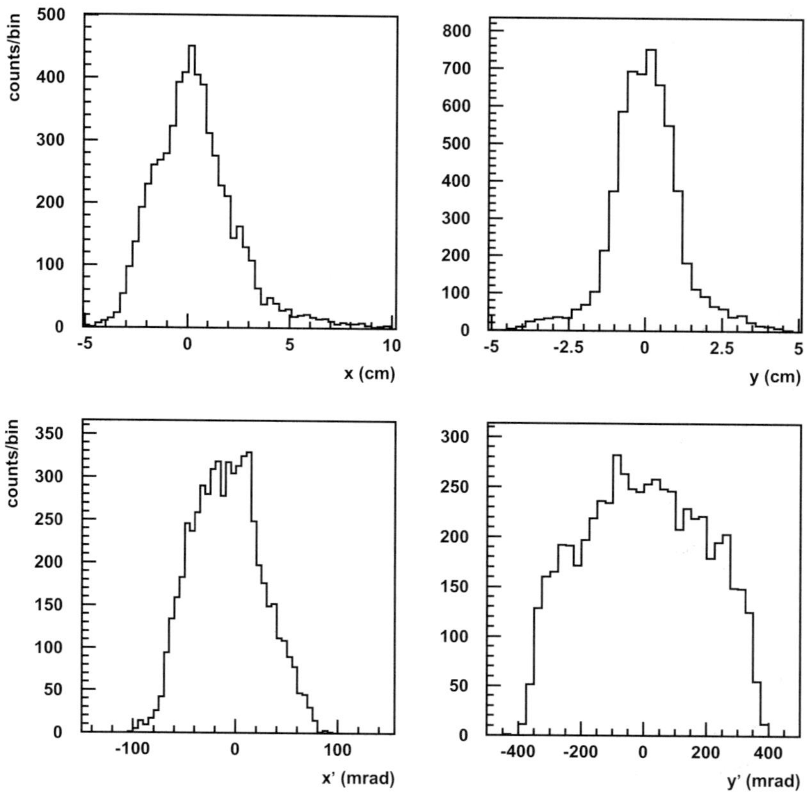

Figure 2. Phase space parameter of the new μ E4 beam line at the LEM moderator position as calculated by `TURTLE` [7]. x and y are the horizontal and vertical coordinates, respectively.

voltage difference will efficiently separate beam-contaminating positrons from the muons. Three-slit systems (`FS61–63`) allow the control of the beam intensity. The first and third device dispose of horizontal and vertical slits; the second one with horizontal slits is installed in the second bending magnet to adjust the beam momentum spread between 4.5% and 10% (FWHM). In a special operating mode, the beam line can achieve a momentum resolution of better than 2% at the expense of intensity. An overview of the new beam line with the area layout is shown in Figure 1.

Calculations of the beam line optics have been performed by the programs `TRANSPORT` and `TURTLE` [7] and by the ray-tracing program `TRACK` using calculated and measured field maps [8]. The calculated phase space parameters at the position of the LEM moderator are shown in Figure 2. A comparison with calculations for the existing π E3 beam – where the LEM facility was running up to now – shows an anticipated increase of the LE-μ^+ intensity by a factor of 7, see Table I. This will allow the exploitation of the full capacity of LE-μSR. Routine operation is planned to start in 2005.

Table I. Comparison of beam line parameters for the existing πE3 – that was used for LE-μ^+ generation up to now – and the new μE4 channel

	πE3 achromatic mode	New μE4
Horizontal emittance	60 π cm mr	350 π cm mr
Vertical emittance	24 π cm mr	50 π cm mr
Accepted solid angle	17 msr	135 msr
Δp / p (FWHM)	7.5%	4.5%–10%
On a 3 × 3 cm^2 area (LEM moderator)	7.2%	4.5%–7.5%
μ^+ intensity	50 × 10^6/s	350 × 10^6/s
μ^+ intensity on 3 × 3 cm^2	25 × 10^6/s	180 × 10^6/s
Low-energy μ^+ rate (at moderator)	3,000/s	21,000/s
Low-energy μ^+ rate (at sample)	1,000/s	7,000/s
x–y beam spot (FWHM) at end	2.0 × 3.0 cm^2	4.0 × 2.5 cm^2
x–y beam divergence (FWHM) at end	120 × 50 mr	150 × 450 mr
Channel length	15.5 m	18.7 m
Positron contamination after separator	2%	2%

μ^+ rates are quoted for a muon momentum of 28 MeV/c and a primary proton beam current of 1.8 mA.

Acknowledgements

This work was performed at the Swiss Muon Source, Paul Scherrer Institute, Villigen, Switzerland. The construction of the beam line received important financial contributions from the German BMBF (TU Braunschweig and U Konstanz), the UK EPSRC (U Birmingham), and from the Universities of Zürich and Leiden.

References

1. Petitjean C., In: Jacob M. and Schopper H. (eds.), *Large Facilities in Physics*, World Scientific, Singapore, 1995, p. 316.
2. Blundell S. J., *Cont. Phys.* **40** (1999), 175.
3. Morenzoni E. *et al.*, *Phys. Rev. Lett.* **72** (1994), 2793.
4. Morenzoni E. *et al.*, *Physica B* **289–290** (2000), 653.
5. Bakule P. and Morenzoni E., *Cont. Phys.* **45** (2004), 203.
6. Prokscha T., *et al.*, these proceedings.
7. PSI Graphic `TURTLE` Framework by U. Rohrer, based on a CERN-SLAC-FERMILAB version by K.L. Brown et al.; PSI Graphic `TRANSPORT` Framework by U. Rohrer based on a CERN-SLAC-FERMILAB version by K.L. Brown et al.
8. `TRACK`: Three-dimensional Raytracing Analysis Computational Kit, developed by PSI Magnet section.

Hyperfine Interactions (2004) 159:389–393
DOI 10.1007/s10751-005-9130-3

A New Method to Obtain Frequency Offsets in NQR Multi-Pulse Sequences★

LUCAS M. C. CERIONI[1] and DANIEL J. PUSIOL[1,2,*]
[1]*Fa.M.A.F. – U.N.C., Medina Allende s/n, X5016LAE Córdoba, Argentina*
[2]*Spinlock S.R.L., C. Arenal 1020, X5000GZU Córdoba, Argentina;*
e-mail: dpusiol@spinlock.com.ar

Abstract. A new method to improve the signal to noise ratios in multi-pulse nuclear quadrupole resonance (NQR) sequences was developed. In this method, called TONROF, the transmitter frequency is set on-resonance, while the receiver frequency is offset. Experimental results of using TONROF in conjunction with the SSFP sequence with RDX sample are presented. Elevated signal levels were obtained without losses in transmission efficiency. The results are compared with those obtained with the typically used off-resonance version of SSFP, called strong off-resonance comb (SORC) sequence.

Key Words: frequency offset, multi-pulse, NQR, TONROF.

1. Introduction

Multi-pulse sequences are widely used for signal enhancement in nuclear quadrupole resonance (NQR) [1–3]. The simplest and most widely used multi-pulse technique is the Carr sequence and its modifications, especially the steady-state free precession (SSFP) sequence, consisting of a long train of phase-coherent RF pulses of equal duration applied with constant pulse spacing τ [4]. In many cases, the transmitter frequency is offset from the exact resonance to improve the intensity of the NQR signals [5], as in the case of the off-resonance version of SSFP, called strong off-resonance comb (SORC) sequence. The off-resonance condition, however, implies suboptimal excitation. In this work, we propose a new method to improve the NQR signals levels in SSFP sequence without the losses in excitation efficiency. In this method, the transmission is performed on-resonance, while the receiver frequency is offset. We called this method TONROF (transmission on – reception off). Experiments were conducted to optimize the method's parameters, such as frequency offset and inter-pulse time spacing, and the data are presented for hexahydro-1,3,5-trinitro-*s*-triazine (RDX).

★ Argentinian and International patents pending.
* Author for correspondence.

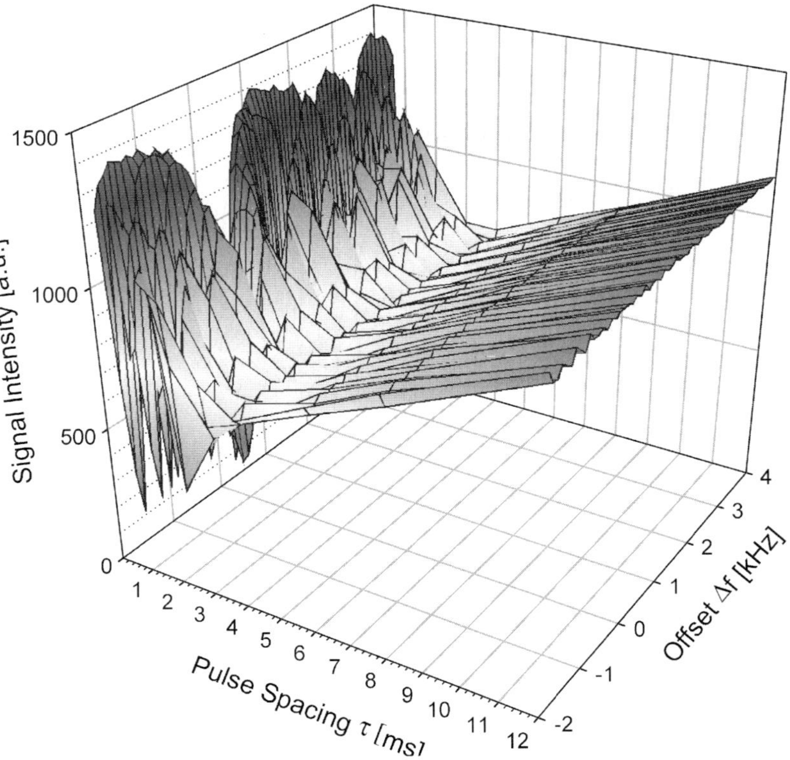

Figure 1. FT NQR signal intensity of RDX as a function of the pulse spacing τ and frequency offset Δf obtained with the SSFP pulse sequence using the TONROF method.

The results obtained using the TONROF method and thetypically used SORC sequence are compared.

2. Experimental

The experiments were carried out using an NQR spectrometer designed in our laboratory to operate in the frequency bandwitdh of 3–6 MHz. The probe has a standard design and contains a solenoidal coil with the diameter of 5 cm and the length of 8 cm, and matching elements shielded inside a cubic cupper box with the edge length of 50 cm. The sample used in the experiments was 150 g of policrystalline RDX. The transition frequency of ν_+ = 5.193 MHz (at room temperature) was utilized in all experiments. The line–shape parameter T_2^* in this sample is about 2 ms, as estimated from the FID signal. The duration of a single pulse producing maximum nuclear induction amplitude was about 40 μs. The delay between the end of an RF pulse and the beginning of the acquisition was 0.25 ms. Experiments with the SSFP pulse sequence using the TONROF method and the SORC sequence were carried out. In the case of TONROF, the

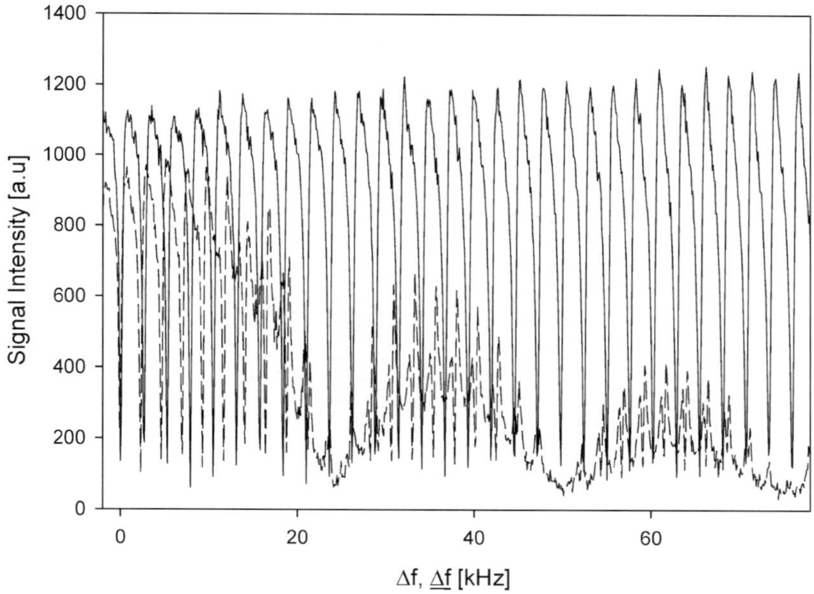

Figure 2. Solid line: FT NQR signal intensity obtained with the SSFP pulse sequence using the TONROF method as a function of the frequency offset between the resonance frequency and the reception frequency. *Dashed line*: FT NQR Signal intensity obtained with the normal SORC pulse sequence as a function of the frequency offset between the resonance frequency and the transmission/reception frequency. The interval between pulses τ was 0.42 ms in both cases.

transmitter frequency was always set to the ν_+ transition frequency. The intensities of fourier transformed (FT) NQR signals in the intervals between pulses of SSFP sequence were studied as a function of $\Delta f = \nu_{ref} - \nu_+$, where ν_{ref} is the reference frequency of the receiver. In the case of the SORC pulse sequence, the intensities of the NQR signals were studied as a function of $\underline{\Delta f} = \nu_{t/r} - \nu_+$, where $\nu_{t/r}$ is the transmitter frequency, which is always equal to frequency of the receiver. In both cases, pulse spacing τ was incremented from 0.42 ms to 12 ms and the pulse width was set to produce maximum excitation.

3. Results and discussion

The dependence of the FT NQR signal intensities on Δf and τ was experimentally investigated by applying the TONROF method with the SSFP sequence. This dependence is presented in Figure 1. As in the SORC sequence [4], intensity anomalies (due to interference between the FID component and the Spin Echo component of the SSFP signal) were encountered for short τ values. These anomalies were found to depend, among other parameters, on the value of the reference frequency of the receiver. In the case of TONROF, however, there

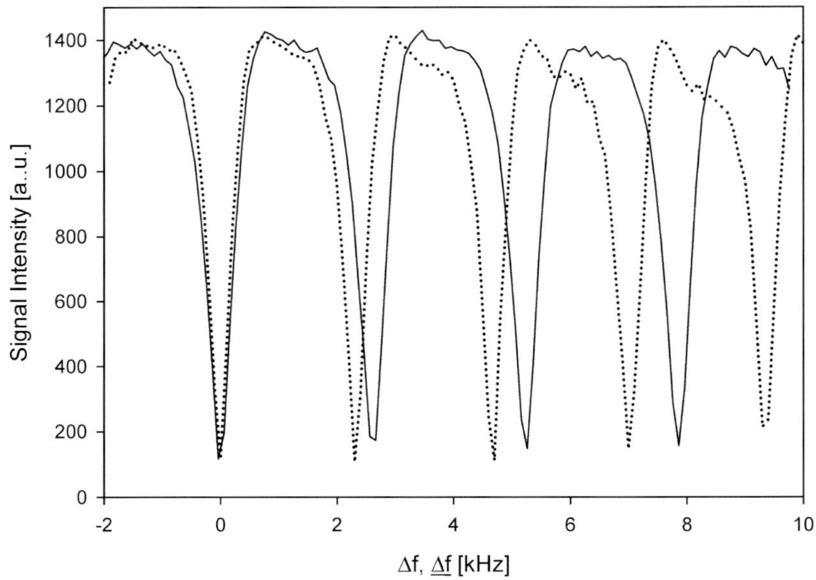

Figure 3. NQR signal intensity obtained with the SSFP pulse sequence using the TONROF method (*solid line*) and with the SORC sequence (*dotted line*) as a function of the frequency offset. The interval between pulses τ was 0.42 ms.

were no losses in transmission efficiency since the transmitter frequency is always set on-resonance. This effect can be observed in Figure 2, where it can be seen that the SORC signal amplitude is modulated by a sinc function, representing the excitation efficiency profile, while the SSFP signal amplitude with the TONROF method is not.

An additional difference between the TONROF method and the SORC sequence was found in the positioning of signal minima for short τ, as illustrated in Figure 3. It was experimentally determined that in the SORC sequence the minima are spaced by $1/\tau$, whereas in the TONROF method the minima are spaced by $1/\tau_{\text{eff}}$, where $\tau_{\text{eff}} = \tau - t_w$, and t_w is the pulse width.

4. Conclusions

A new method, called TONROF, used to obtain high signal intensities in NQR SSFP pulse sequences is presented. In this method transmission efficiency losses encountered in the typically used SORC sequence are not encountered since the samples are always irradiated on-resonance. As well as in SORC sequences, intensity anomalies were encountered for short τ values. These anomalies depend on the value of the reference frequency of the receiver. It was experimentally demonstrated that the anomalies where the signal is minimum are spaced by $1/\tau_{\text{eff}}$, where $\tau_{\text{eff}} = \tau - t_w$, and t_w is the pulse width.

References

1. Marino R. A. and Klainer S. M., *J. Chem. Phys.* **67** (1977), 3388.
2. Schiano J. L., Routhier T., Blauch A. J. and Ginsberg M. D., *J. Magn. Reson.* **139** (1999), 139.
3. Osokin D. Y. and Shagalov V. A., *Solid State Nucl. Magn. Reson.* **10** (1997), 63.
4. Rudakov T. N. and Mikhaltsevich V. T., *Chem. Phys. Lett.* **324** (2000), 69.
5. Rudakov T. N., Mikhaltsevich V. T. and Selchikhin O. P., *J. Phys., D, Appl. Phys.* **30** (1997), 1377.

Hyperfine Interactions (2004) 159:395–400
DOI 10.1007/s10751-005-9131-2

A New Method of Mössbauer Spectra Treatment Based on the Method of Self-Organisation of Mathematical Models

A. TIMOSHEVSKII* and V. YEREMIN
Institute of Magnetism, Kiev, Ukraine; e-mail: tim@imag.kiev.ua, tim@ukron.kiev.ua

Abstract. A new technique for the treatment of high noisy measurement data is proposed based on the methods of self-organisation. This technique has been used to develop two methods for data treatment: the extraction of deterministic component from high noisy data with a controlled accuracy, and the Mössbauer spectra component expansion (Lorentz functions). These two methods were applied correspondingly to the treatment of X-ray Photoelectron Spectroscopy (XPS) measurements and to the model Mössbauer spectra with a different noise level. It is shown that these methods are highly effective.

Key Words: doublet, Mössbauer spectra treatment, principles of self-organisation, singlet.

1. Introduction

In experimental studies of the crystal structure, short range order, electronic structure and other important characteristics of solids it is necessary to perform additional treatment of the experimental data, because changes in the data due to changes in characteristics to be measured (for example, alloying) may be comparable with the noise level. Thus, it is highly necessary to develop an approximation method that is 'self-regulating' with respect to the initial data and real noise level. Furthermore, such method should not depend on the 'human factor,' i.e., it has to provide correct results irrespectively of the researcher's personal abilities. We have developed exactly such methodology applicable for the treatment of experimental data with high noise level and based on the method of self-organisation of mathematical models [1]. In this work, we have used this methodology for the development of algorithms for processing XPS and Mössbauer spectra. For Mössbauer spectra, processing implies the component expansion (by the Lorentz functions).

* Author for correspondence.

2. The method and its application

The method of self-organisation of mathematical models is described more detailed in [1]. The heart of the method is epy generation of a mathematical model for the object under study; in our case data arrays that form the experimental spectrum. From a large number of mathematical models, describing the spectrum, we have to choose an optimal one that provides the best description. As the main selection criterion, we have used the criterion of a minimum mean-square error on a separate data checking sequence (regularity criterion). The regularity criterion requires division of the set of initial data into the 'training sequence' and 'checking sequence.' The choice of this criterion is based on the consideration that the new points are similar to data that are already available. Available data could be divided into the training and checking sequences in different ways depending on the specific task, so there could be a lot of regularisation techniques. Let us consider the best one. This method implies that the interpolation points are divided into the training and checking sequences in such a way that the training sequence includes $(N-1)$ points and checking sequence includes only one point. We build the model of a given complexity on the training sequence and find a mean-square error in a test point; then we repeat this process for all of the N points, i.e., we eliminate the points of a given sequence one by one and thereby verify prognosis properties of the model on these points. Let us introduce a dimensionless measure of the model prognosis quality based on the determination coefficient $D(m)$:

$$D(m) = \left\{ 1 - \frac{\sum_{i=1}^{n} [y_i - y_i(m)]^2}{\sum_{i=1}^{n} [y_i - \bar{y}]^2} \right\} 100\,\%$$

here y_i – experimental values of the function, $y_i(m)$ – values obtained by m-th model, $\bar{y}$ – mean value.

For processing of XPS spectra, we have used Chebyshev's polynomial as the basic function. The determination coefficient characterizes the percentage of the deterministic component in the present data set and given method. A purposeful search of the model within the set of the model-pretenders that are m-powered polynomials ($m \sim 4$–22) allows for finding the best unique solution for a given regularity criterion. Let us explain how it is to be done. For any specific m-power of polynomial, we find an average error that is a summarized error of all of the scans divided by the number of the scans. From the set of the models, we find a model with a less average error and select it as the optimal one.

2.1. X-RAY PHOTOELECTRON SPECTROSCOPY (XPS)

The XPS spectrum contains important information regarding the energy of the inner atomic shells and energy structure of the valence band (valence electron

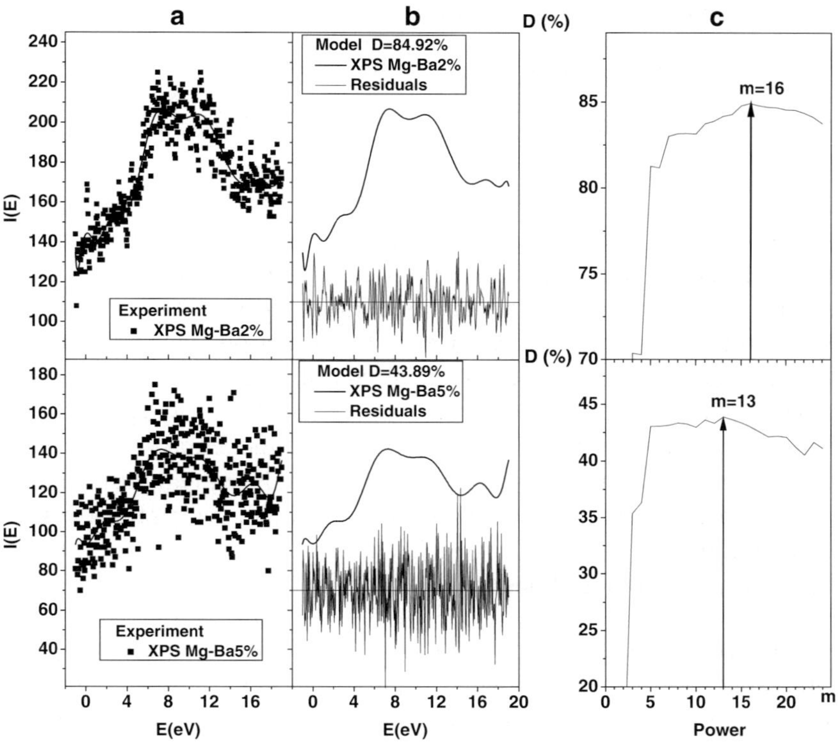

Figure 1. (a) experimental XPS spectra of the valence band of MgBa alloys having different Ba atoms' content; (b) deterministic constituents obtained by our method; (c) results of calculations of optimal power of Chebyshev's polynomial.

energy distribution) in compounds and alloys. In many cases, XPS valence band spectra are highly noisy, and it is difficult to derive a real curve of the valence electrons energy distribution for the experimental spectrum. The method of these spectra treatment is described in more details in [2]. In Figure 1, XPS spectra of the valence band of Mg–Ba alloys having 2% and 5% Ba content are shown together with the results of the calculations of deterministic constituents. Figure 1c shows the results of calculations on finding an optimal spectrum mathematical model that provides largest determination coefficient. Parameter characterising this model is Chebyshev's polynomial power *m*. The curve of dependence of determination coefficient on this value allows us to define optimal *m* value that corresponds maximal determination coefficient.

When the Ba content equals 5%, the determination coefficient is 47.3% and the optimal Chebyshev's polynomial power is 13. For Mg-2%-Ba alloy, the determination coefficient is 86.1% (much less than noise level) and the polynomial power is 16. The optimal values of polynomial power have been determined from the regularisation plots shown in Figure 1 for XPS spectra of Mg-2% Ba and Mg-5%-Ba alloys. This result is very important for studying the electronic structure peculiarities of this alloy. Our method is highly effective for the treatment of various X-ray and neutron diffraction measurements, where quite often reliable information should be derived from highly noisy spectrum intervals.

2.2. MÖSSBAUER SPECTROSCOPY

Mössbauer spectroscopy is a unique experimental method to derive information about local atomic short range order. When studying atomic structure of an interface in multilayer systems, widely used in spintronics, it is extremely difficult to obtain information about the atomic structure from the Mössbauer spectra, because the spectrum, as a rule, has a number of singlet and doublet lines within a narrow speed range. So, we face with the problem of extracting from the spectra maximum information within the controlled accuracy. For processing of measured spectra it is important also to avoid or reduce greatly the human factor: the results of processing must not depend on initial assumptions made by different researchers.

The solution algorithm consists of a number of stages, as follows:

The first stage – the search for additional correlations between the parameters. The main point of this stage is generation of a system of equations assuming in each observation point of the spectrum Lorentz lines. They are defined within x_1, x_2 window and within $\Gamma_1 - \Gamma_2$ width range. Into the equation set, we add one more condition, as follows: the sum of k-order momentums from generated lines equals some constant. Determination of the constant is carried out by the regularization method. In our case, this is just the sample division into two parts, while using correlations between the first six momentums (from 0th to 5th ones). In that way, we have determined additional relations between the system's parameters.

The second stage – optimisation of the initial grid of the lines arrangement. We generate the grid of the lines arrangement, as follows: from maximal lines number (in each observation point) to minimal lines number with corresponding step. As the optimal grid we consider that one which will get minimum regularisation criterion; in present case, this is regularisation with one by one points deleting.

The third stage – defining the most informative lines set. On this stage, using additional relations and optimal grid of the lines arrangements, we have used an iteration procedure for finding the most stable lines. The main point of this stage is selection of optimal lines with positive intensities and estimation of their stability. Stability is characterised by the parameters' ability to change within some limits (at any regularisation). For the less variability coefficient, the considered parameter is more stable (more informative). Using this fact, we build an iteration procedure that at each step removes the most variable line. An informative parameters set is determined by minimum of regularisation criterion.

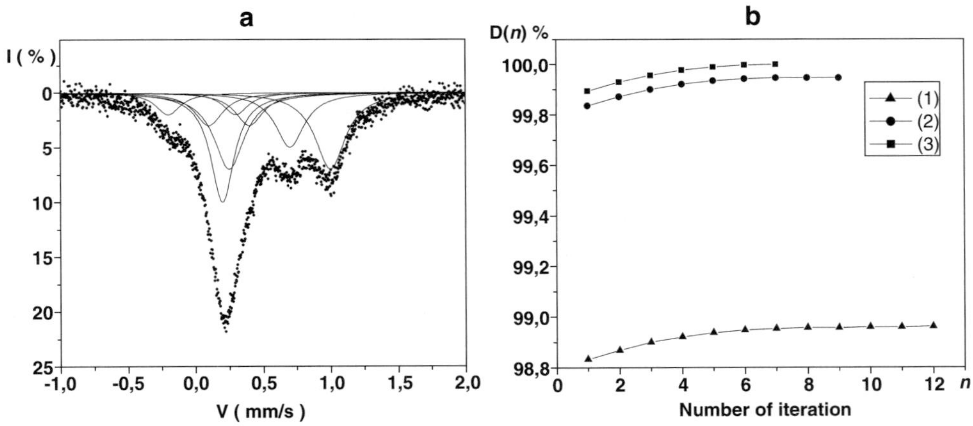

Figure 2. (a) Model spectrum with a great statistical dispersion and background value of $5 \cdot 10^4$. (b) Curves of dependence of determination coefficient $D(n)$ on a number of iterations n for three model spectra with a different noisy level: (1) background $A_0 = 5 \cdot 10^7$; (2) background $A_0 = 10^6$; (3) background $A_0 = 5 \cdot 10^4$.

The fourth stage – iterative determination of parameters. We have used a standard procedure based on gradient methods.
The fifth stage – determination of singlet and doublet lines which is made by iteration procedure on the base of the matrix of the integrals of correlations between the lines.

We have used our method for decomposition of a model spectrum with a high noise level consisting of three doublet and two singlet lines lying within the interval from $-0{,}1$ to 1,0 mm/s. Model spectra with a different noise level (statistical dispersion) were synthesized over different algorithms which simulate obtaining a real Mössbauer spectra.

$$I_i = y_i + \sqrt{y_i} \cdot \mathrm{rand}\big|_0^1 (-1)^{\mathrm{rand}(1,2)}, \text{ here } y_i = A_0 \left(1 - \sum_{m=1}^{n} \frac{A_m \Gamma_m^2}{4(x_m - x_i)^2 + \Gamma_m^2} \right),$$

Here $A_m = \frac{A_0 - y_m}{A_0}$, y_m – value of function in x_m point, A_0 – coefficient that characterizes background, Γ_m – line half-width, n – number of the Lorenz lines.

In this case, statistical dispersion (spectra noisiness) unambiguously relates to the background value and the spectral intensity. When exposure time increases, the background rises and statistical dispersion decreases.

Figure 2a shows a synthesized model spectrum with a great statistical dispersion and background value of $A_0 = 5 \cdot 10^4$ (high noisy spectrum). In the Figure 2b, we have shown regularization process, namely, the process of finding optimal set of the Lorenz lines and parameters that characterize them. At each iteration, we define the lines which removing will result in the growth of determination coefficient $D(n)$. Thus, starting from some iteration, $D(n)$ value

Table I. Calculation results for the model Mössbauer spectra with a different noise level

N	Actual			Calculated								
				Background $A_0 = 5 \cdot 10^4$			Background $A_0 = 10^6$			Background $A_0 = 5 \cdot 10^7$		
	X	Γ	I	X	Γ	I	X	Γ	I	X	Γ	I
1	−0.20	0.22	2	−0.20	0.22	2	−0.20	0.22	2	−0.2	0.22	2
2	0.10	0.25	3	0.10	0.21	3	0.10	0.21	3	0.10	0.24	3
3	0.20	0.21	10	0.20	0.21	9	0.20	0.21	10	0.20	0.21	10
4	0.25	0.27	7	0.25	0.26	9	0.25	0.25	7	0.25	0.27	7
5	0.30	0.22	2	0.30	0.21	0.3	0.30	0.25	2	0.30	0.21	2
6	0.40	0.25	3	0.40	0.23	3	0.40	0.25	3	0.40	0.25	3
7	0.70	0.26	5	0.70	0.26	5	0.70	0.25	5	0.70	0.26	5
8	1.00	0.27	7	1.00	0.27	7	1.00	0.27	7	1.00	0.27	7

X – line position (mm/s); Γ – line half-width (mm/s); intensity (% relative to background).

becomes maximal. As it is seen from Figure 2, a number of iterations needed depend greatly on a quality of experimental spectrum (i.e., its noise level). If spectrum is of a more quality, i.e., has a less noise level, then $D(n)$ coefficient increases accordingly.

We have synthesized two more spectra with a reduced noise level (background: $A_0 = 10^6$ and $A_0 = 5 \cdot 10^7$) and expanded them into components. The treatment of this spectrum using our method provides excellent results shown in the Table I.

3. Summary

We have developed a new method for the treatment of high noisy experimental data, which is based upon the method of self-organisation. We have used this method for the treatment of the model Mössbauer spectra having a different noise levels. From our calculations we can conclude that our method is well effective for the treatment of even very noisy spectra. It should be noted that this method supplies the researcher with a powerful tool for spectra processing but, at the same time, it excludes the 'human factor' and its accuracy does not depend on personal decisions. Spectra processing does not require also availability of preliminary information about the number of the components and their positions in the spectrum.

References

1. Timoshevskii A., Yeremin V. and Kalkuta S., *Ecol. Model.* **162**(1–2) (2003) 1.
2. Timoshevski A. N., Yeremin V. I., Kalkuta S. A. and Tymoshevska L. V., In: *Proc. Int. Conf. on Computational and Mathematical Methods on Science and Engineering*, Vol. 2, Alicante, Spain, 20–25 September 2002, p. 388.

Hyperfine Interactions (2004) 159:401–405
DOI 10.1007/s10751-005-9132-1

Nuclear Spin Maser Oscillation of ^{129}Xe by Means of Optical-Detection Feedback

A. YOSHIMI[1,*], K. ASAHI[1,2], S. EMORI[2], M. TSUKUI[2] and S. OSHIMA[2]
[1]*RIKEN, Hirosawa 2-1, Wako, Saitama, 351-0198, Japan; e-mail: yoshimi@rarfaxp.riken.jp*
[2]*Department of Physics, Tokyo Institute of Technology, Oh-okayama 2-12-1, Meguro, Tokyo, 152-8851, Japan*

Abstract. We have developed the nuclear spin maser oscillating at a low frequency of 34 Hz with highly polarized nuclear spins of the noble gas element ^{129}Xe. The system is advantageous for detecting a small frequency shift of the nuclear spin precession. We are thus planning to apply this system to the search for an atomic electric dipole moment of ^{129}Xe. We here report the development of the system and its performance.

Key Words: nuclear spin polarization, optical pumping, spin maser.

1. Introduction

The precise measurement of nuclear spin precession allows the detection of extraordinary small energy acting on the nuclei. The experimental sensitivity of a frequency shift of 1 mHz leads to a detection of the energy shift of 4×10^{-18} eV. Such a measurement of the small energy shift is required for the search for the permanent electric dipole moment (EDM) [1]. For the precise measurement of precession frequency, it is crucial how long the spin precession is observed and how high a sensitivity of nuclear spin detection can be realized. A noble gas nuclear spin can be largely polarized using optical pumping method and its spin relaxation time is quite long, thus it can be an important element for such experiments. For the precessing nuclear spin, a limitation of observation time due to a finite transverse relaxation time T_2 may be avoided by introducing a spin maser technique [2–5]. We thus developed the nuclear spin maser with spin polarized ^{129}Xe nuclei ($I = 1/2$), which is operated through the optical detection of the nuclear spin direction leading to the high sensitivity and low-frequency operation.

2. Nuclear spin maser with optical spin detection

The nuclear spin maser, where the nuclear spins continuously precess without losing their transverse polarization, can oscillate when transverse magnetic field,

* Author for correspondence.

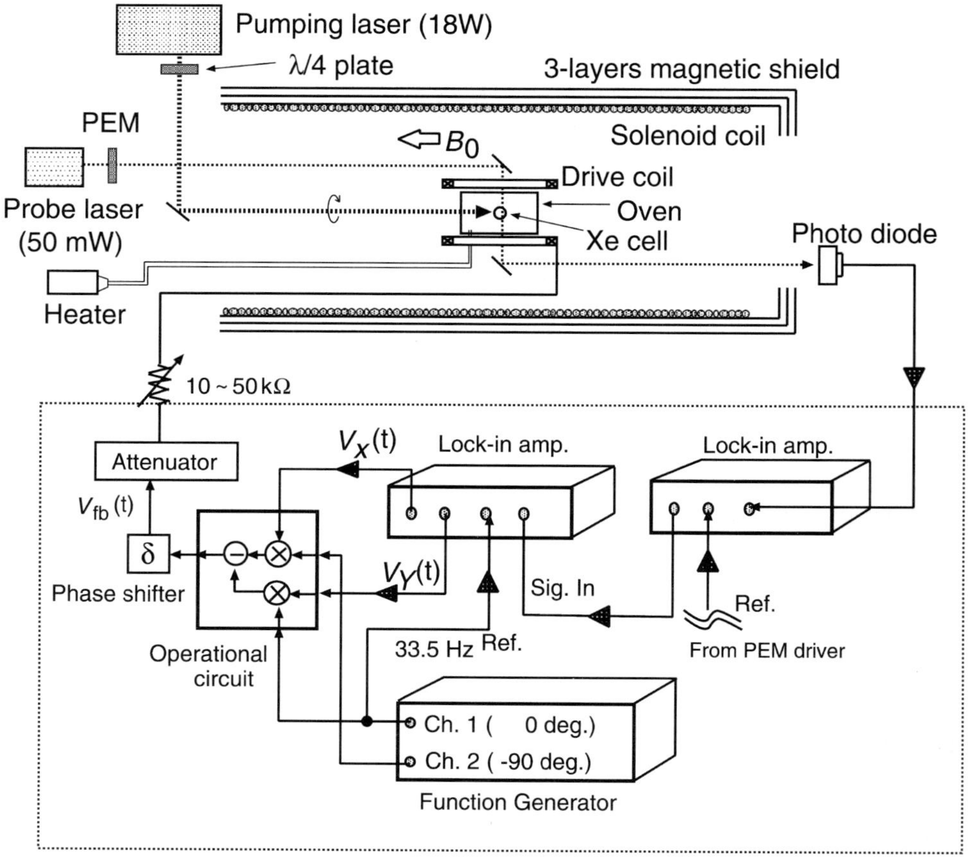

Figure 1. Conceptual setup of a spin maser with the optical spin detection feedback.

whose phase is shifted by 90° to the transverse polarization component and whose amplitude is proportional to that of the transverse polarization, is applied to the precessing spin system [4, 5]. In order to produce the 'feedback' transverse field which rotates in synchronism with the nuclear spins, the instantaneous direction of the transverse spin polarization must be known. In a conventional spin maser scheme, a radio frequency coil whose resonance frequency is tuned to the spin precession frequency automatically produces the feedback field. In the present system, we adopted the optical spin detection system in order to obtain the instantaneous transverse polarization direction.

The experimental apparatus is shown in Figure 1. A Xe gas (79% enriched ^{129}Xe) was confined in a spherical glass cell of 20 mm diameter. The cell was irradiated by a circular polarized laser beam from a diode laser (18 W output power) which was tuned to the D1 absorption line of a Rb atom. The Rb atoms coexisting with the Xe atoms are thus polarized, and the spin polarization is transfered to the Xe nuclei through the hyperfine interactions [6]. We typically

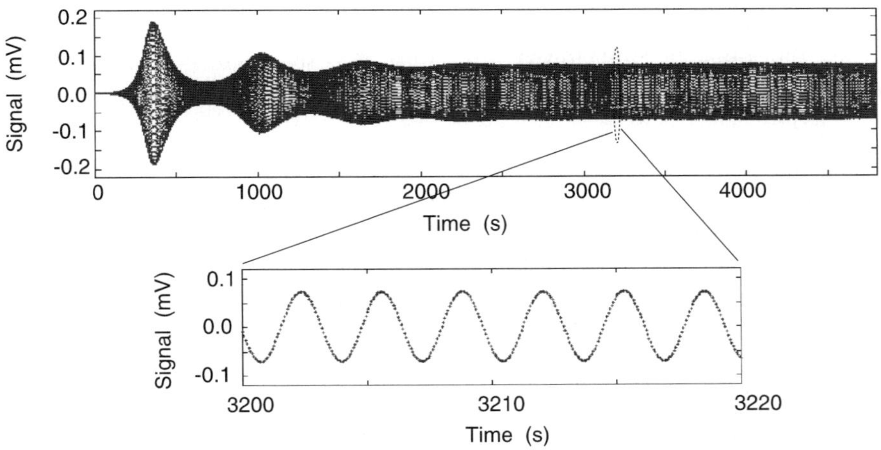

Figure 2. The observed spin maser oscillation signal. (Observed by lock-in detection with a reference frequency of 33.7 Hz.)

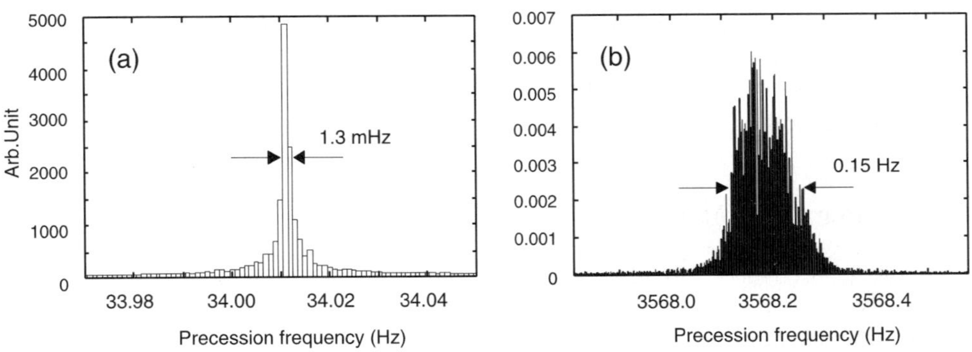

Figure 3. Fourier spectrum of measured continuous oscillation. (a) The spectrum obtained from the operation of the optical spin detection-feedback spin maser described in the letter. (b) The spectrum obtained from the conventional spin maser.

obtained a polarization of $P_0 \approx 10\%$ for the ^{129}Xe density of 6×10^{18} atoms/cm^3.

The cell was mounted in a cylindrical three-layer magnetic shield made of Permalloy to suppress the stray magnetic field and its gradient. A static field $B_0 =$ 28.7 mG applied to the Xe cell was produced by a solenoid coil installed inside the shield. A probe laser beam from a single-frequency diode laser passes through the cell in a direction perpendicular to the B_0 axis. With the probe beam and a photoelastic modulator (PEM), the transverse polarization of Rb atoms is detected through the circular dichroism. Since the transverse polarization of the ^{129}Xe nuclei induces a slight transverse polarization of Rb atomic spins (repolarization) through spin exchange, the ^{129}Xe precession signal appeared in the lock-in detection of the transmitted probe beam. In order to operate the nuclear spin maser, the optically detected precession signal is converted to the feedback signal with the lock-in amplifier and the operational circuit [5].

The typically measured oscillation signal is shown in Figure 2. The precession signal appeared after the feedback system became active, and on a time scale of some thousand seconds the signal approaches the steady-state oscillation which eventually set in. The transient pattern of the maser amplitude toward the steady state oscillation was well reproduced by numerical simulations based on the modified Bloch equations.

The present nuclear spin maser is capable of operation at frequencies as low as $\nu = 34.0$ Hz. Such low-frequency operation provides the narrow absolute frequency width. The Fourier spectrum, obtained from one-hour continuous oscillation with the apparatus described above and from the conventional spin maser setup with the spin-coil coupling operated earlier, are respectively shown in Figure 3(a,b). The static fields in these experiments are respectively 28.7 mG and 3.01 G. The frequency width in the spin maser with the optical spin detection approaches 1.3 mHz, which is smaller by a factor of 100 than the conventional spin maser. This is due to a suppression of the static field fluctuations coming from the current instability in the solenoid coil owing to the lower static field.

3. Summary and future

We have developed the nuclear spin maser with polarized ^{129}Xe nuclei by using the optical detection of instantaneous nuclear spin direction, leading to maser operation at the low frequency of 34 Hz. The frequency width of the spin maser became two orders narrower than the conventional spin maser. We are now planning to apply this system to the search for the EDM in ^{129}Xe atom. We have already constructed four-layer magnetic shield with a length of 1.60 m whose internal residual magnetic field is almost 50 μG. In order to stabilize the B_0 field, a high-sensitivity magnetometer utilizing nonlinear magneto-optical rotation in Rb atoms [7] is under development. The static field is expected to be stabilized to a level of 10^{-13} G, which indicates that frequency fluctuations of ^{129}Xe precession can be reduced to below 0.1 nHz. Such a sensitive experiment will provide us interesting information about the mechanism of CP-violation.

References

1. Kriplovich I. B. and Lamoreaux S. K., *CP Violation Without Strangeness*, Springer-Verlag, Heidelberg, 1997.
2. Richards M. G., Cowan B. P., Secca M. F. and Machin K., *J. Phys., B* **21** (1988), 665.
3. Chupp T. E., Hoare R. J., Walsworth R. L. and Wu B., *Phys. Rev. Lett.* **72** (1994), 2363.

4. Asahi K., Yoshimi A., Sakai K., Ogawa H., Suzuki T. and Nagakura M., *Czechoslovak J. Phys.* **50** (2000), 179.
5. Yoshimi A., Asahi K., Sakai K., Tsuda M., Yogo K., Ogawa H., Suzuki T. and Nagakura M., *Phys. Lett. A* **304** (2002), 13.
6. Happer W., Miron E., Schaefer S., Schreiber D., van Wijngaarden W. A. and Zeng X., *Phys. Rev., A* **29** (1984), 3092.
7. Budker D., Kimball D. F., Rochester S. M., Yashchuk V. V. and Zolotorev M., *Phys. Rev., A* **62** (2000), 043403.

Hyperfine Interactions (2004) 159:407–409

Keyword Index to Volume 159 (2004)

Hyperfine Interactions (2004) 158:407–411
DOI 10.1007/s10751-005-9067-6

Implantation of the ^{111}In/Cd Probe as InO^- Ion for Radioisotope Tracer Studies

SANTOSH K. SHRESTHA[1,*], HEIKO TIMMERS[1], AIDAN P. BYRNE[2,3], WAYNE D. HUTCHISON[1], DON H. CHAPLIN[1] and RAKESH DOGRA[2,4]
[1]*School of Physical, Environmental and Mathematical Sciences, University of New South Wales at the Australian Defense Force Academy, Canberra, ACT 2600, Australia;*
e-mail: Santosh@uow.edu.au
[2]*Department of Nuclear Physics, Research School of Physical Sciences and Engineering, Australian National University, Canberra, ACT 0200, Australia*
[3]*Department of Physics, Faculty of Science, Australian National University, Canberra, ACT 0200, Australia*
[4]*Department of Electronic Materials Engineering, Research School of Physics Sciences and Engineering, Australian National University, Canberra, ACT 0200, Australia*

Abstract. A radioisotope ion implanter has been developed using a cesium-sputtering, negative ion source, which offers versatility and sustained operation. Employing the molecular $^{111}In^{16}O^-$ ion, μCi activities of the radioisotope probe ^{111}In/Cd have been implanted into different material hosts. The implanted tracer activity has been shown to be sufficient for LTNO, NMRON and PAC. A new NMRON resonance for ^{111}InAg was observed at 75.08 MHz. In_2O_3 powder performed well as the radioisotope carrier in the ion source, with the ratio of radioisotope and parasitic ion current being typically 4×10^{-4}.

Key Words: ion source, LTNO, NMRON, PAC, radioisotope implantation.

1. Introduction

Nuclear condensed matter techniques such as Perturbed Angular Correlation (PAC) spectroscopy, Low-Temperature Nuclear Orientation (LTNO) and Nuclear Magnetic Resonance on Oriented Nuclei (NMRON) require radioisotope tracers to be introduced into a material host as probing nuclei. Though radioisotopes can be diffused into a material, or recoil-implanted following fusion synthesis, the controlled ion implantation of a radioisotope probe has several advantages. They include generally minimal sample damage due to the low implantation energy (50–200 keV), accurate dosimetry of the number of probing nuclei, and a high degree of control over the implantation depth.

For the low energy ion implantation of radioisotope probes, positive ion sources are routinely employed. Since a cesium-sputtering, negative ion source

* Author for correspondence.

[1, 2] offers greater versatility and extended operation, a 50–155 keV ion implanter based on such a source [3] has been equipped and commissioned for radioisotope implantation. While cesium-sputtering ion sources can generally provide large ion outputs, it is not clear, how significant outputs are best achieved for a standard radioisotope probe such as ^{111}In/Cd. Middleton e.g., reports an output of 10 μA of molecular stable $^{115}In^{16}O^-$ ions using indium oxide (In_2O_3) as material for the sputter cathode, whereas the best output of elemental $^{115}In^-$ ions was only 700 nA and obtained using indium metal [2]. Due to the often short half-life of the radioisotope probe, e.g., $t_{1/2} = 2.81$ d for ^{111}In/Cd, and the minute quantities involved, different sputter cathodes are required than those typically used for the production of stable ion beams [2]. Suitable cathode designs and materials have not been reported previously.

In this work the use of a cesium-sputtering, negative ion source for the implantation of the radioisotope probe ^{111}In/Cd has been studied. The efficacy of this approach for nuclear condensed matter physics has been tested using LTNO, NMRON and PAC.

2. Production and implantation of $^{111}In^-$ and $^{111}In^{16}O^-$ ions

Cesium-sputtering ion sources have been described and studied by Middleton [1, 2]. In such a source Cs^+ ions strike a cooled cathode and sputter material which partially forms negative ions by accepting electrons from cesium atoms. In our experiments both copper and aluminium cylindrical cathode holders (length = 20.8 mm, ∅ = 8 mm) have been tried inside a SNICS-II ion source [3]. The holders have a 7 mm deep recess with an inner diameter of 4.7 mm, which from a depth of 3 mm onwards narrows conically. This matches the sputter crater of the Cs^+ ions.

As cathode material and carrier of the ^{111}In radioisotope, powders of Al, Al_2O_3 and In_2O_3 have been investigated. Compressed Al powder, even with admixtures of Al_2O_3, is not sufficiently absorbent to function as carrier. While it avoids the stable indium isotopes ^{113,115}In and their molecular ions, the slow sputter-rate of Al_2O_3 and the extremely high $^{16}O^-$ output has proven this also a bad choice. The use of In_2O_3 as cathode material has produced the best results. Figure 1(a) shows a mass spectrum for this cathode. The relative yields of the two molecular ions $^{113}In^{16}O^-$ and $^{115}In^{16}O^-$ reflect the natural abundance of the two stable indium isotopes. Notably, the yield of the elemental $^{115}In^-$ ion is three orders of magnitude smaller than that of the molecular ion $^{115}In^{16}O^-$. The $^{113}In^-$ current was not observed.

The In_2O_3 powder was pressed into the cathode holder recess. Approximately 1.5 mm were left above the powder surface to facilitate the loading of the radioisotope solution. Activities of 5 mCi of ^{111}In/Cd were purchased as $^{111}InCl_3$ in 0.05 molar HCl[1]. The volume of the solution was reduced to typically 50 mm^3

[1] PerkinElmer Life and Analytical Sciences, Boston, USA.

using a 150 W heat lamp at a distance of 10 cm. In several steps the solution was then micro-pipetted onto the surface of the pressed In_2O_3 carrier, which absorbs the solution quickly. Experiments with diluted ink and radiographs of used cathodes suggest that the loaded activity is essentially homogeneously distributed throughout the carrier. Following 45 min drying under a 150 W heat lamp, the activated cathode was placed inside the ion source. Measurements have shown that 90% of the activity are loaded into the cathode. The missing fraction remains in the delivery bottle and the pipette.

The ion source was operated with typical parameters [3]. Negative ions were extracted from the source, accelerated to an energy of 125 keV, and focused through a quadratic aperture (4 mm $\times$ 4 mm) into a Faraday cup, with which the source output current I_s was measured. Mass analysis was achieved with a 90° dipole magnet. Individual mass numbers A were selected (Figure 1(a)) using an aperture (3 mm $\times$ 3 mm) after the magnet. The analysed current I_a was measured with a second Faraday cup. Typical beam currents were $I_s = 100\ \mu$A and $I_a = 6\ \mu$A for ^{115}In^{16}O$^-$ ions. During radioisotope implantation the two apertures were opened further (5 mm $\times$ 5 mm) for increased transmission.

The field of the magnet was calibrated using ^{16}O$^-$, ^{113}In^{16}O$^-$, ^{115}In^{16}O$^-$ and ^{133}Cs$^-$ beams. Both ^{111}In$^-$ and ^{111}In^{16}O$^-$ ions have been implanted into a variety of materials. As suggested by Figure 1(a) the source output of elemental ^{111}In$^-$ ions was too small to achieve practically useful implanted activities. Hence the ^{111}In/Cd probe has been implanted as molecular ^{111}In^{16}O$^-$ ion. Implanted activities were of the order of 1 μCi, with activity increases in the range of 0.02–0.15 μCi/h, and a maximum implanted activity of 2.3 μCi. Figure 1(b) shows a γ-ray spectrum for this sample. The analysed current I_a for the mass number of ^{111}In^{16}O$^-$ ($A = 127$) was dominated by parasitic ions, as it is demonstrated in Figure 1(c). The ratio of radioisotope to parasitic current was typically 4×10^{-4}. Implanted fluences, including the parasitic ions, were several 10^{14} ions/cm^2. In the literature e.g., implanted fluences of the order of 10^{13} ions/cm^2 are quoted for the low energy implantation of the ^{111}In/Cd probe for PAC using a positive ion source and ^{111}In$^+$ ions [4]. While ^{111}Cd^{16}O$^-$ ions are likely to constitute a fraction of the parasitic ion current, a much larger fraction must have an origin unrelated to the ^{111}In/Cd probe. The identification and suppression of this latter fraction would improve the technique.

3. Radioisotope tracer studies with ^{111}In/Cd

The efficacy of the implantation of ^{111}In^{16}O$^-$ ions to introduce the ^{111}In/Cd probe has been demonstrated with LTNO, NMRON and PAC measurements on a variety of samples. Importantly, the co-implantation of a significant number of parasitic ions does generally not interfere with the application of these techniques. Samples of the radiation-sensitive semiconductor indium nitride were, however, severely depleted of nitrogen in the implanted volume [5].

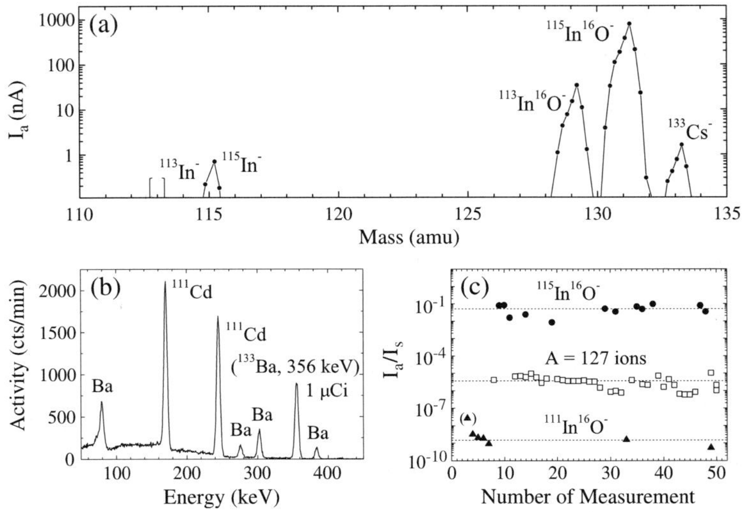

Figure 1. (a) Partial mass spectrum for a Cs-sputtering ion source using an In_2O_3 cathode. (b) γ-ray spectrum from a sample implanted with $^{111}In^{16}O^-$ ions demonstrating an implanted activity of 2.3 μCi. Additional lines are from a calibrated ^{133}Ba source. (c) The $^{111}In^{16}O^-$ current compared to that of the parasitic ions (A = 127) and that of $^{115}In^{16}O^-$ from many measurements. The $^{111}In^{16}O^-$ current has been derived from the implanted activity. Currents have been divided by the source output I_s.

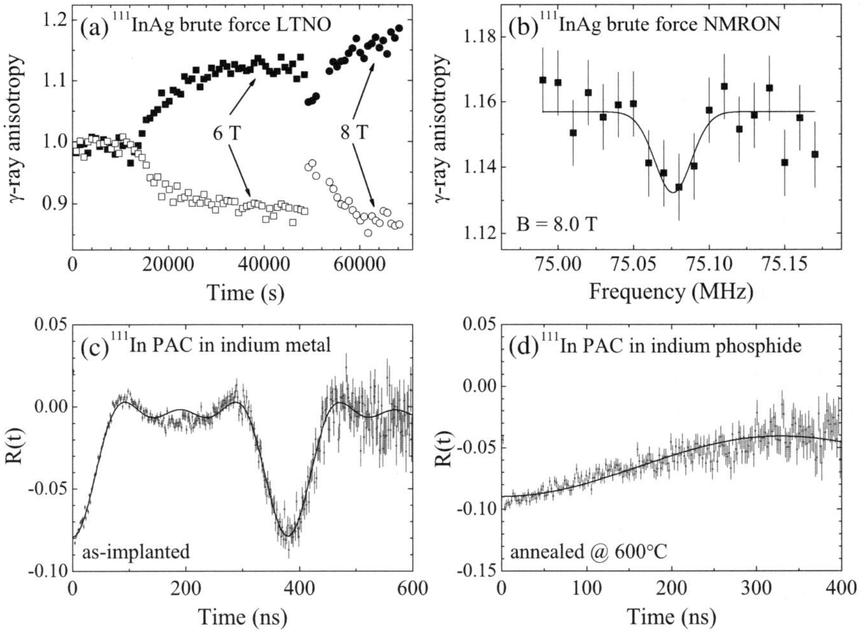

Figure 2. (a) γ-ray anisotropies from ^{111}InAg brute force LNTO as a function of cool-down time. (b) γ-ray anisotropy from ^{111}InAg brute force NMRON. A gaussian fit indicates a resonance at 75.08 MHz. (c) PAC spectrum and fit for ^{111}In/Cd in indium as measured at room temperature. (d) PAC spectrum and fit for ^{111}In/Cd in indium phosphide.

Figure 2(a) shows the results of a brute force LTNO measurement on silver (^{111}InAg) following the implantation of ^{111}InO$^-$ ions to an activity of 0.8 μCi. Comparison with ^{111}InCu data [6] indicates that the ^{111}In nuclei are well thermalised and therefore sited in the host in metallic form. Following a second implantation, resulting in a probe activity of 2.3 μCi, the first brute force NMRON measurement with ^{111}In/Cd was performed on silver in a magnetic field of 8.0 T. The anisotropy for the 171 keV γ-ray line of ^{111}Cd is shown in Figure 2(b). Though statistics are poor, the resonance may be identified at 75.08 MHz. This frequency is in agreement with that predicted by the ^{111}InCu slope to three significant figures.

For PAC spectroscopy, among other hosts, indium and indium phosphide were implanted with ^{111}In^{16}O$^-$ ions. The anisotropy ratios $R(t)$, determined as described in Ref. [7], are shown in Figure 2(c) and (d), respectively. Least-squares-fits with the appropriate theoretical perturbation function are also shown. The observed quadrupole interaction frequency of ν_Q = 16.6 MHz for indium agrees with earlier measurements [8]. As expected, for indium phosphide a cubic environment of the probe nuclei is seen [7, 9].

4. Conclusions

Low energy ion implantation of the radioisotope tracer ^{111}In/Cd has been demonstrated using molecular, negative ^{111}In^{16}O$^-$ ions delivered by a cesium-sputtering source. A suitable design for a sputter cathode has been developed with In_2O_3 powder as the radioisotope carrier. The ratio of the ^{111}In^{16}O$^-$ ion current to parasitic ion current of the same mass was found to be typically 4×10^{-4} for the set-up used. Efforts to reduce the parasitic fraction of the ion flux by employing alternative radioisotope carriers are underway. Measurements for several material hosts have shown that μCi ^{111}In/Cd activities implanted using this technique are sufficient for reliable LTNO, NMRON and PAC. A new NMRON resonance for ^{111}InAg was observed at 75.08 MHz.

References

1. Middleton R., *Nucl. Instrum. Methods* **122** (1974), 329.
2. Middleton R., *A Negative Ion Cookbook*, University of Pennsylvania, USA, 1989 [http://tvdg10.phy.bnl.gov/COOKBOOK/]; *Nucl. Instrum. Methods* **214** (1983), 139.
3. Instruction Manual SNICS-II, Model 2JA001110, NEC, Middleton, Wisconsin, USA.
4. Lorenz K. and Vianden R., *Phys. Status Solidi (C)* **1** (2002), 413.
5. Timmers H., Shrestha S. K. and Byrne A. P., *J. Cryst. Growth* **269** (2004), 50.
6. Nuytten C. *et al.*, *Phys. Rev. Lett.* **49** (1982), 347.
7. Bezakova E. *et al.*, *Appl. Phys. Lett.* **75** (1999), 1923.
8. Budtz-Jørgensen C. *et al.*, *Phys. Rev., B* **8** (1973), 5411.
9. Pfeiffer W. *et al.*, *Appl. Surf. Sci.* **50** (1991), 154.

Hyperfine Interactions (2004) 159:411–414

Author Index to Volume 159 (2004)

Hyperfine Interactions (2004) 158:429–436
DOI 10.1007/s10751-005-9071-x

Hf_3Al_2 and Zr_3Al_2 Isostructural Aluminides Studied by PAC with ^{181}Ta and ^{111}Cd Probes

P. WODNIECKI[1,2,*], A. KULIŃSKA[1,2], B. WODNIECKA[1], M. UHRMACHER[2] and K. P. LIEB[2]
[1]*IFJ PAN, Kraków, Poland; e-mail: pawel.wodniecki@ifj.edu.pl*
[2]*II. Physikalisches Institut, Universität Göttingen, D-37077 Göttingen, Germany*

Abstract. The electric quadrupole hyperfine interactions for ^{181}Hf/^{181}Ta and for ^{111}In/^{111}Cd probes in polycrystalline Zr_3Al_2 and Hf_3Al_2 compounds were measured in the temperature range 24–1100 K. The hyperfine quadrupole interaction parameters were determined after different doping techniques and heat treatments of the samples. ^{181}Hf/^{181}Ta was found to substitute the Hf/Zr sites and the ^{111}In/^{111}Cd impurities appear to substitute both the 8(*j*) Al sites and the three nonequivalent Hf/Zr sites.

Key Words: aluminides, electric field gradients, lattice location.

1. Introduction

Aluminum forms a large number of high-melting intermetallic compounds with various transition metals, which have gained special attention, due to their desired mechanical properties and oxidation resistance at high temperatures. Our previous systematic PAC experiments on hafnium and zirconium aluminides with different structures and Al concentrations were performed with ^{181}Ta and ^{111}Cd probes in a wide temperature range [1–6]. In this paper we focus on Hf_3Al_2 and Zr_3Al_2 compounds occurring in tetragonal *tP20* structure ($P4_2/mnm - D^{14}{}_{4h}$) [7]. The Al atoms are in 8(*j*) position of low symmetry while Hf/Zr atoms occupy one axially symmetric 4(*d*) site and two (4(*f*) and 4(*g*)) sites of low *m.2m* symmetry (see Figure 1). The position of ^{181}Hf/^{181}Ta probes can be predicted due to the chemical similarities of Hf and Zr but the ^{111}In/^{111}Cd probes have to be treated as impurities in the $(Hf/Zr)_3Al_2$ lattices. Our previous experiments had shown that ^{111}In/^{111}Cd probes, matching well the electronegativity of Al, but oversized, can occupy both Al and/or Hf/Zr sites and that site preferences can change with temperature.

---

* Author for correspondence.

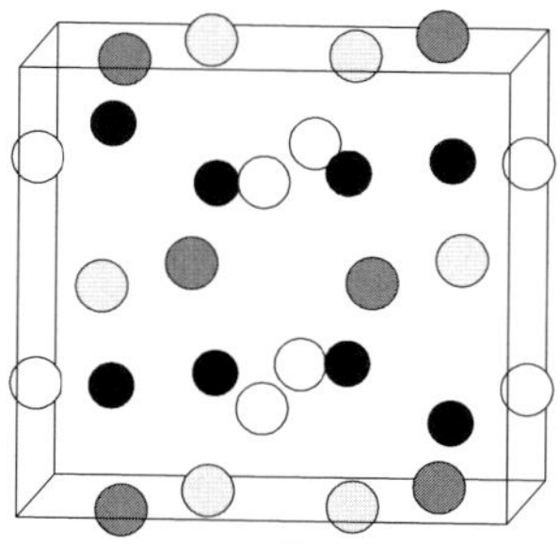

Figure 1. Unit cell of Zr_3Al_2 compound.

2. Experiment

Samples were produced by multiple arc melting, under argon atmosphere, followed (in most cases) by two weeks of annealing at 1000°C in evacuated and sealed quartz tubes. Phase analysis carried out by X-ray powder diffraction stated the structure of the investigated samples corresponding to the starting mixture composition with small traces of $(Hf/Zr)_4Al_3$ phases. Doping of the samples with the ^{181}Hf activity was done by neutron irradiation in the pile of the Świerk reactor. Doping with ^{111}In/^{111}Cd probes was executed either via implantation (at 400 or 250 keV) into ca. 0.5 mm thick sample slices using the Göttingen heavy ion implanter IONAS [8], or by adding the carrier-free ^{111}In activity during the sample melting. Annealing of the samples was than applied at different temperatures. Most PAC experiments covered the temperature range between 30 and 1100 K.

For a static EFG present at the probe nuclei, the perturbation factor $G_2(t)$ [9]

$$G_2(t) = \sum_{i=1}^{k} f_i \sum_{n=0}^{3} s_{2n}(\eta_i)\cos\left(g_n(\eta_i)\nu_{Qi}t\right)\exp\left(-g_n(\eta_i)\delta_i\nu_{Qi}t\right) \tag{1}$$

was applied to analyse the experimental R(t) spectra. The least squares fits of the perturbation factor to the experimental data allow to extract the fractions f_i of probes exposed to different EFGs, determined by the quadrupole frequency ν_{Qi}, asymmetry parameter η_I and the V_{zz} spread described by a Lorentzian shape with a relative width δ_i.

The EFG values were calculated from the quadrupole frequencies ν_Q measured with the ^{111}Cd or ^{181}Ta probe adopting quadrupole moment values, equal to 0.83(13) b for ^{111}Cd [10] and 2.36(5) b for ^{181}Ta [11].

The measured temperature variations of the quadrupole frequencies was fitted using a linear $V_{zz}(T) = V_{zz}(0)\,[1 - a\,T]$ dependence.

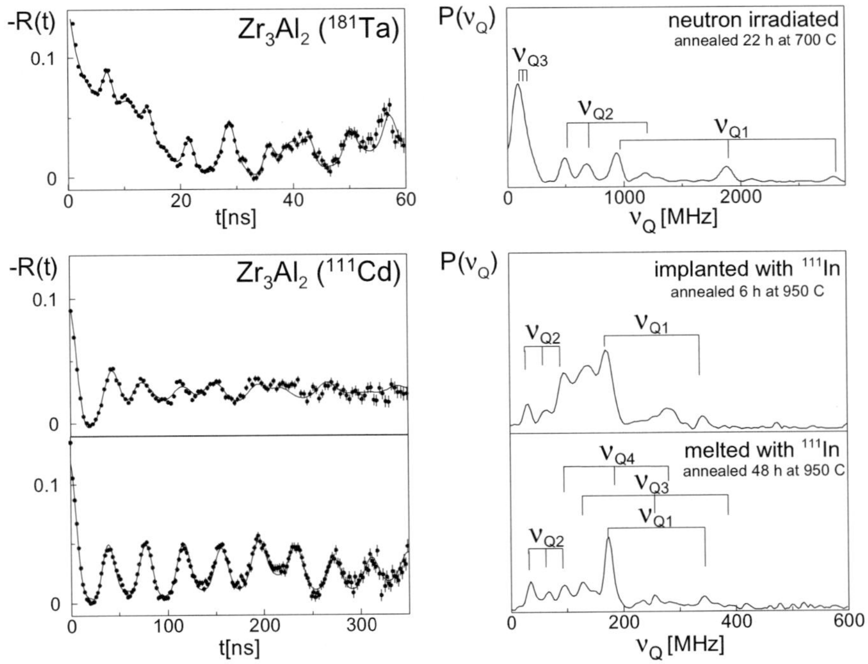

Figure 2. The room temperature PAC spectra with corresponding Fourier transforms taken at ^{181}Ta and ^{111}Cd probes for Zr_3Al_2 sample.

3. Results and discussion

The upper panels of Figures 2 and 3 show room temperature PAC spectra of ^{181}Ta in Zr_3Al_2 and Hf_3Al_2 samples. In both spectra three quadrupole frequencies (whose parameters are listed in Table I), connected with three substitutional 4(d), 4(f) and 4(g) Hf/Zr lattice sites of the ^{181}Hf/^{181}Ta probes, are observed. For Hf_3Al_2 samples a relative fraction ratio of 1:1:1 was found for probe atoms randomly distributed on the equally populated Hf sites, while for Zr_3Al_2 the ratio is 1:1:2, reflecting probe site preference in this phase. All EFGs are characterized by linear temperature dependencies (see Figure 4 and Table I). Particularly interesting is the rather unusual rising character of the EFGs corresponding to $\nu_Q = 65(1)$ MHz for Zr_3Al_2 and $\nu_Q = 117(1)$ MHz for Hf_3Al_2. In the other Zr and Hf aluminides studied previously [1–6], both linear and $T^{3/2}$-dependences have been encountered at different sites, in all but three cases with negative slopes as expected. Only in the case of ^{111}Cd probes in Zr_3Al [6] and ^{181}Ta in $ZrAl_2$ and $HfAl_2$ [3] the quadrupole frequencies increasing with the temperature were found. Also for ^{111}Cd in the HfV_2 Laves phase [12] the increase of V_{zz} with temperature was observed and related to the high density of the V d-states at the Fermi energy. The positive slopes of $\nu_Q(T)$ were in all these cases correlated with small EFG values.

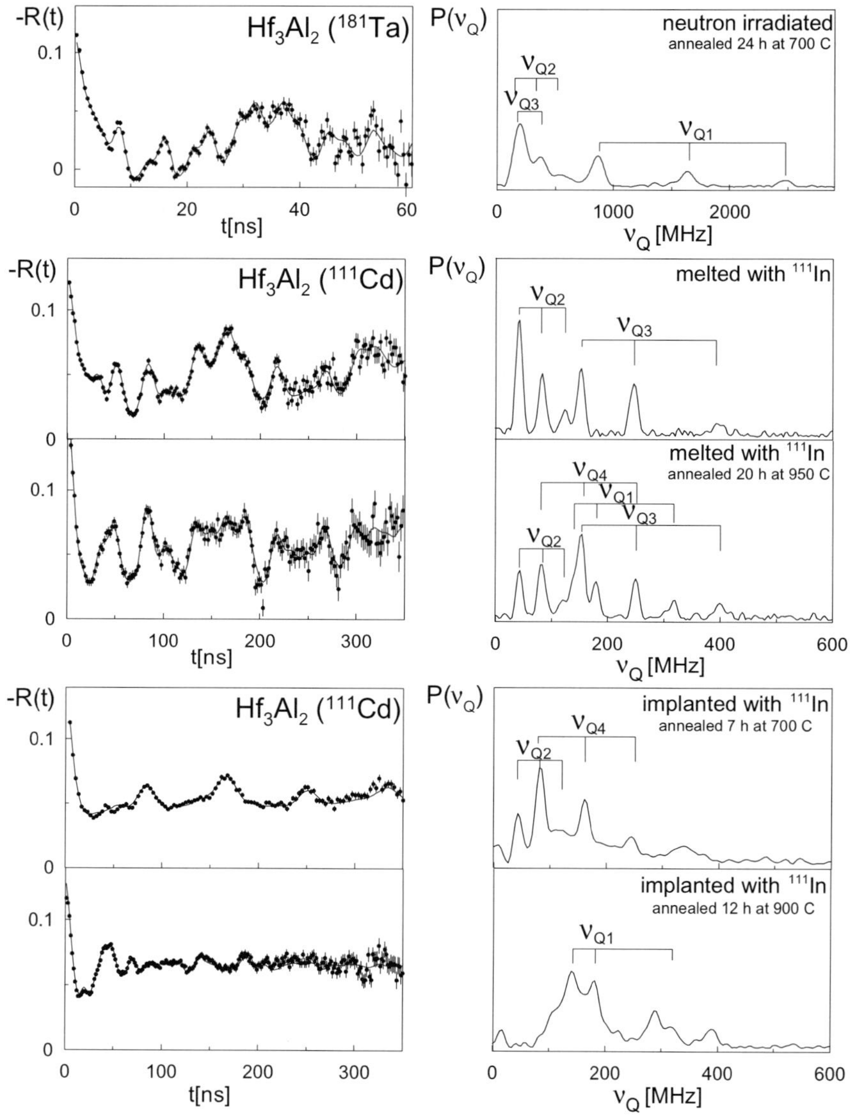

Figure 3. The room temperature PAC spectra with corresponding Fourier transforms taken at ^{181}Ta and ^{111}Cd probes for Hf_3Al_2 sample.

Taking into account site symmetries and available space, connected with nearest neighbor distances, combined with Ta probe preference in Zr_3Al_2, we believe that the EFG equal to 1.14×10^{17} Vcm^{-2} can be most probably ascribed to ^{181}Ta in 4(d) site of Zr_3Al_2 lattice. Although the fitted asymmetry parameters for lower measurement temperatures differ from zero – the value expected for this crystallographic site, they become equal to zero above ca. 500 K. Similarly small η values were observed $HfAl_2$ and $ZrAl_2$ compounds for Ta probes in 4(f)

Table I. Quadrupole interaction parameters of ^{111}Cd and ^{181}Ta probes in Zr_3Al_2 and Hf_3Al_2: quadrupole frequency ν_Q, asymmetry parameter η, room temperature EFG value at probe site V_{zz} and the slope parameters a of the linear EFG temperature dependences

Phase	Probe	Proposed lattice site	ν_Q [MHz]	η	$\lvert V_{zz}\rvert$ [10^{17} Vcm^{-2}]	a [10^{-4} K^{-1}]
Zr_3Al_2	^{181}Ta	$4(d)\bar{4}$..	928(3)	0.01(3)	16.26	2.02(6)
			362(2)	0.63(1)	6.34	3.5(2)
			65(1)	0.31(1)	1.14	−37(3)
	^{111}Cd	$8(j)$ ..m	97.3(5)	0.99(1)	4.48	1.95(1)
			33(1)	0.23(3)	1.64	1.6(1)
			127(1)	0.22(2)	6.33	1.71(1)
		$4(d)\bar{4}$..	93(5)	0.00(2)	4.66	2.54(8)
Hf_3Al_2	^{181}Ta	$4(d)\bar{4}$..	822(1)	0.22(1)	14.40	1.38(7)
			185(2)	0.00(1)	3.24	8.2(4)
			117(1)	1.00(1)	2.05	−2.7(5)
	^{111}Cd	$8(j)$ ..m	96.3(5)	0.72(1)	4.80	1.18(3)
			42(1)	0.08(2)	2.09	2.8(1)
			128(1)	0.44(1)	6.38	1.01(3)
		$4(d)\bar{4}$..	78(2)	0.00(5)	3.89	1.7(1)

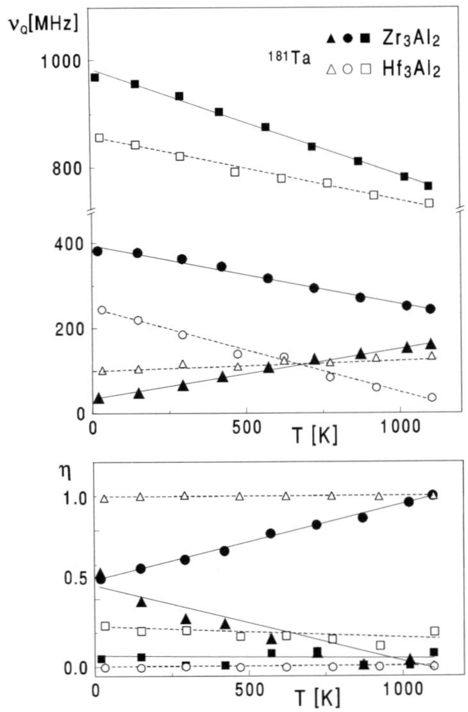

Figure 4. The temperature dependences of quadrupole interaction parameters for ^{181}Ta in Zr_3Al_2 and Hf_3Al_2 samples.

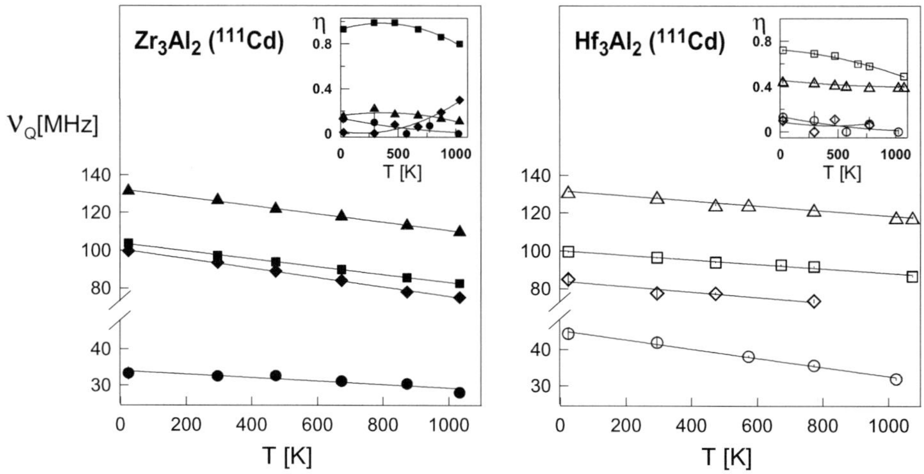

Figure 5. The temperature dependences of quadrupole interaction parameters for ^{111}Cd probes in Hf_3Al_2 and Zr_3Al_2 samples.

$3m$ sites [3]. The $V_{zz} = 3.24 \times 10^{17}$ Vcm^{-2} with $\eta = 0$ is most likely the EFG for 4(d) lattice site in the Hf_3Al_2 compound. Since Zr and Hf are isoelectronic, their compounds (alloys, oxides) show pronounced similarities in their lattice structures, hyperfine interactions and other properties. Surprisingly, the deduced hyperfine parameters (especially asymmetry parameters η) for ^{181}Ta in the Zr_3Al_2 and Hf_3Al_2 isostructural compounds are not very close to each other.

The experiments executed with ^{111}In/^{111}Cd probes after different doping techniques and heat treatments evidenced four well-defined EFGs corresponding to four non-equivalent probe sites in both Hf/Zr and Al sublattices. The fitted QI parameters are compiled in Table I. It should be noticed that the large number of probe fractions, reflecting the number of non-equivalent lattice sites, makes the interpretation of PAC spectra difficult, especially in the case of Cd probes, where the situation is strongly influenced by the sample treatment. The lower panels of Figures 2 and 3 show the representative ^{111}Cd PAC spectra recorded at room temperature. The number and type of the observed EFGs strongly depend on the sample preparation technique and annealing conditions. Since the sample slices used for In implantation were rather thick, the PAC spectra are perturbed by the absorption. Moreover, the annealing and measurements at higher temperatures caused probably the oxidation of the sample surface and a great fraction of probes distributed near the surface (especially after In implantation at 250 keV) were, therefore, exposed to the broad EFG distributions. In contrary, after melting of the ^{111}In activity with compound ingredients the smaller samples were used in PAC experiments and the probe atoms were randomly distributed in the whole sample volume, leading to the significantly clearer PAC patterns.

The definite assignment of the observed EFGs to the particular lattice sites was not possible. It was stated that the fraction of EFG with the greatest asymmetry parameter is not evidenced in as-prepared samples and rises with annealing temperature in both compounds. This EFGs equal at room temperature to ca. 4.8 $\times$ 10^{17} Vcm^{-2}, with $\eta = 0.99$ for Zr_3Al_2 and $\eta = 0.72$ for Hf_3Al_2 phase, were tentatively assigned to In probes substituting at the unique Al $8(j)..m$ sites since the point charge model calculations indicate the largest asymmetry parameter value for this site. Also the available space for the oversized In impurity is at this site smaller than at Hf/Zr sites. The more spacious Hf/Zr sites are therefore favoured by the oversized ^{111}In probe at low temperatures. At higher temperatures the thermally expanded lattice enables In to replace the Al sites – more favourable considering the electrochemical factor.

The EFGs of 4.66 $\times$ 10^{17} Vcm^{-2} for Zr_3Al_2 and 3.89 $\times$ 10^{17} Vcm^{-2} for Hf_3Al_2, with η values close to zero, could be ascribed to ^{111}In probes substituted at *4(d)* $\bar{4}..\mathrm{Hf}/\mathrm{Zr}$ sites.

The temperature dependences of quadrupole interaction parameters for ^{111}Cd probes in Hf_3Al_2 and Zr_3Al_2 samples are presented in Figure 5. In both compounds, a linear decrease of all quadrupole frequencies with the temperature was evidenced.

It should be noticed that, as can be seen in Table I, the experimental EFG values at *4*(*d*) lattice site on ^{181}Ta are smaller than those measured on ^{111}Cd nuclei, while the ratio of the corresponding antishielding factors for the free ions equals to ca. 2 [13]. Also the unusual rising character of EFGs at ^{181}Ta probes situated at these lattice sites manifests that electronic effects can play a significant role in this case.

Though the definite assignment of the lattice sites occupied by ^{111}In/^{111}Cd probes was not possible, nevertheless the substitution of both Al and Hf/Zr sublattices (similarly as in $HfAl_2$ [2] and Zr_3Al [6] compounds) in Hf_3Al_2 and Zr_3Al_2 compounds by In impurities seems to be rather unquestionable (although the assignment of the EFG of ca. 4.8 $\times$ 10^{17} Vcm^{-2} to probe location in the admixture phases cannot be absolutely excluded without an additional experiment).

Acknowledgements

The authors gratefully acknowledge the help of D. Purschke and Dr. L. Ziegeler during the ^{111}In implantations at IONAS. This work was funded partially by the Deutsche Forschungsgemeinschaft.

References

1. Wodniecki P., Wodniecka B., Kulińska A., Uhrmacher M. and Lieb K. P., *J. Alloys Compd.* **312** (2000), 17.
2. Wodniecki P., Wodniecka B., Kulińska A., Uhrmacher M., Lieb K. P., *Phys. Lett., A* **288** (2001), 227.
3. Wodniecki P., Wodniecka B., Kulińska A., Uhrmacher M., Lieb K. P., *J. Alloys Compd.* **335** (2002), 20.

4. Wodniecki P., Wodniecka B., Kulińska A., Uhrmacher M. and Lieb K. P., *Hyperfine Interact.* **136/137** (2001), 535.
5. Wodniecki P., Wodniecka B., Kulińska A., Uhrmacher M., Lieb K. P., *J. Alloys Compd.* **351** (2003), 1.
6. Wodniecki P., Wodniecka B., Kulińska A., Uhrmacher M. and Lieb K. P., *J. Alloys Compd.* **365** (2004), 52.
7. Villars P. and Calvert L. D., *Pearson's Handbook of Crystallographic Data for Intermetallic Phases*, ASM, Materials Park, Ohio, 1991; Okamoto H., *Phase Diagrams for Binary Alloys*, ASM, Material Park, Ohio, 2000.
8. Uhrmacher M., Neubauer M., Bolse W., Ziegeler L. and Lieb K. P., *Nucl. Instrum. Methods, B* **139** (1998), 306.
9. Frauenfelder H. and Steffen R. M., In: Karlsson K., Matthias E. and Siegbahn K. (eds.), *Perturbed Angular Correlations*, North Holland, Amsterdam, 1963.
10. Herzog P., Freitag K., Reuschenbach M. and Walitzki H., *Z. Phys. A* **294** (1980), 13.
11. Butz T. and Lerf A., *Phys. Lett.* **97 A** (1983), 217.
12. Jain H. C., Freise L. and Forker M., *J. Phys., Condens. Matter* **1** (1989), 2137.
13. Feiock F. D. and Johnson W. R., *Phys. Rev.* **187** (1969), 39.

Hyperfine Interactions (2004) 158:437–441
DOI 10.1007/s10751-005-9072-9

The Magnetic Response of Europium Implanted in Cerium and in Platinum as Investigated by the PAC-Method

W.-D. ZEITZ[1,*], S. UNTERRICKER[2], F. SCHNEIDER[2], V. SAMOKHVALOV[2], K. POTZGER[1], A. WEBER[1] and M. DIETRICH[3]
[1]*Hahn-Meitner-Institut Berlin GmbH, Bereich Strukturforschung, Glienicker Strasse 100, D-14109 Berlin, Germany; e-mail: zeitz@hmi.de*
[2]*TU Bergakademie Freiberg, Institut für Angewandte Physik, D-09596 Freiberg, Germany*
[3]*Technische Physik, Universität des Saarlandes, D-66041 Saarbrücken, Germany*

Abstract. The magnetic response of europium in γ-cerium and in platinum was studied by applying the perturbed angular correlation spectroscopy. The probe nuclei were ^{147}Eu(11/2$^-$) and ^{149}Eu(11/2$^-$). The response in γ-Ce was determined by the electronic $S = J = 7/2$ ground state of divalent Eu. In Pt, on the other hand, Eu is trivalent ($J = 0$ ground state). Here the magnetic contributions originate from Van Vleck terms of the whole multiplet system.

Key Words: magnetic moments, PAC method, rare earth impurities, valence.

PACS: 71.20.Gj, 75.20.Hr, 76.80.+y, 23.20.En, 07.05.Fb.

The magnetic behaviour of isolated rare earth atoms is predominantly determined by the local moments in the 4f-shell. Embedded in crystals, these moments may be subjected to various kinds of interactions which can lead to instabilities or to a change in valence. Especially europium is very sensitive to the environment [1–5]. When divalent, Eu will resemble Gd^{3+} and show the magnetic behaviour of a local 7/2 spin, whereas trivalent Eu has a $J = 0$ ground state and may have strong Van Vleck contributions [6]. In the current measurement with the PAC-method, we have used ^{147}Eu and ^{149}Eu nuclei to study isolated Eu impurities in γ-cerium and in platinum.

Europium may take two different valences in metal hosts. In Mössbauer experiments, divalent Eu was found in some alloys such as EuPd and $EuPt_2$, but trivalent Eu in $EuPt_3$ [7]. The trivalent configuration was present in Pt and in Pd [2] bulk materials, but in Ce and La isolated Eu atoms bear indications for divalent configurations [3]. By changing the composition of $EuNi_xZn_{5-x}$ and $EuNi_xCu_{5-x}$ alloys, a mixture of both valences was found except for $x = 0$ and

* Author for correspondence.

$x = 5$ [4]. Mainly from experiments with synchrotron radiation, Eu was deduced to be divalent at the surface of a Ni single crystal [5].

From these investigations we deduce that methods which determine averaged parameters are not appropriate when different valences are likely to occur. As the PAC-method has already demonstrated its capability to discriminate between different sites of Cd on the Ni surface [8], we started these measurements as a first step to study the magnetism of rare earth atoms in various local configurations.

The PAC technique calls for special nuclear properties: A radioactive precursor and a nucleus with a γ–γ-cascade are necessary. The lifetime of the implanted precursor should be long enough to allow annealing of the samples and the γ–γ-cascade in the probe nucleus should have an intermediate isomeric level with an appropriate lifetime (some nanoseconds (ns) to microseconds (μs)). There are more than 70 isotopes among the lanthanides which fulfil the demanded conditions in the level scheme [9].

As probes, we have chosen ^{149}Gd(^{149}Eu) and ^{147}Gd(^{147}Eu) nuclei. Although the β-decays from the precursors feed into 27 or 24 levels in ^{147}Eu or ^{149}Eu and the level schemes are constructed from 129 or 79 γ-transitions, respectively, both probe nuclei compare to most favourite cases: By 38% or 33.9%, the Gd precursors predominantly feed into one level in ^{147}Eu or ^{149}Eu, which are depopulated by the most prominent transitions to the $11/2^-$ isomeric states. The intermediate isomeric levels have half-lives of $T_{1/2}$ = 765 ns and $T_{1/2}$ = 2.45 μs. The precursors of the PAC-probes were provided by the ISOLDE on-line mass separator in Geneva/Switzerland and the ISL accelerator facility in Berlin/Germany.

In order to determine the local moment in the 4f-shell, the temperature dependence of the magnetic hyperfine fields were measured in response to an externally applied magnetic field. The external field of $B_{ext}(293\ \mathrm{K}) = 0.52(7)$ Tesla was produced by a small FeNdB permanent magnet. In the PAC-method, the measured magnetic field B_{eff} is determined through the Larmor precession frequency $\omega_L(T) = -g\mu_N B_{eff}(T)/\eta$. Here $g = 1.28$ and $g = 1.27$ are the known g-factors for ^{147}Eu($11/2^-$) and for ^{149}Eu($11/2^-$) respectively [9] and μ_N is the nuclear magneton. The quantity $\omega_L(T)$ is extracted from the spinrotation spectra like the one which is shown in Figure 1. In the analysis of the data we used the paramagnetic enhancement factor $\beta(T)$ which is defined by the ratio of the measured magnetic field B_{eff} and the externally applied field B_{ext}.

The moment of the divalent $^8S_{7/2}$–ground state was present for Eu in γ-Ce. As the first excited state lies about 4 eV above the ground state [10] and does not contribute to the magnetism, the paramagnetic enhancement factor $\beta(T)$ is expected to have a linear relation to the inverse temperature: $\beta(T) = 1 + C_{Cu}/T$. For divalent Eu the Curie constant C_{Cu} may be calculated from the total angular momentum J and the hyperfine field $B_J(0)$ at zero temperature: $C_{Cu} = g_J\mu_B(J + 1)B_J(0)/3k$. There are the Bohr magneton μ_B, the Boltzmann constant k and the Landé factor g_J in the formula.

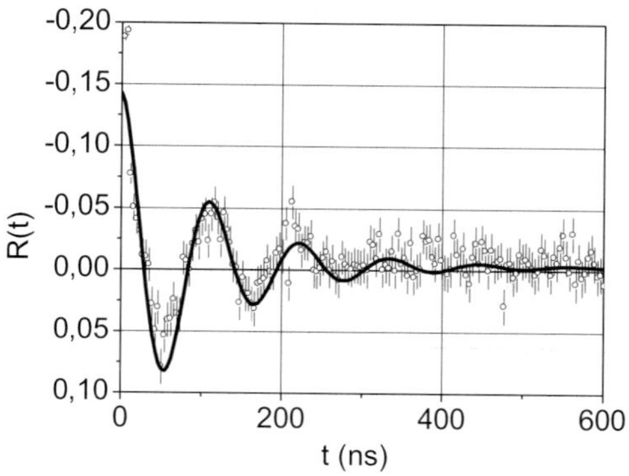

Figure 1. Perturbed angular correlation spectrum for ^{149}Eu in γ-Ce at 200 K.

Spinrotation spectra are obtained for Eu in γ-Ce for a whole set of temperatures, but in the α-phase of Ce, at low temperatures, no oscillations were seen. The damping of the amplitudes, which is visible in the spectra, might originate from the quadrupolar interaction in a slightly disturbed non-cubic environment or from a dynamical interaction of fluctuating local moments.

In the upper part of Figure 2 the paramagnetic enhancement factor for Eu in Ce is plotted versus the inverse temperature. From the slope of the fitted line, the measured hyperfine field is estimated to be $B_J(0) = -13(3)$ T. Disregarding the sign, this value is much smaller than the value $B_J(0) = -32.4$ T which was calculated for the free ion [11], and even lower than the value of $B_J(0) = -17(2)$ T for Eu in La [3]. The reason for this behaviour might be sought in the instability of the moment. Instabilities were already found for Sm in both γ- and α-Ce and even in La [12].

For Eu in Pt, the behaviour of the trivalent status was seen. Trivalent Eu has a non-magnetic ground state ($J = 0$) and the intervals between the multiplets of the fine structure levels are small enough to demand for the Van Vleck contributions to be included besides the Curie terms of all fine structure levels: $\beta_{\text{total}}(T) = \beta_{\text{Cu}}(T) + \beta_{\text{VV}}(T)$. In a case like this, the Curie term $\beta_{\text{Cu}}(T)$ is calculated from a weighted sum over the contributions of different J-levels (see refs. [6, 13, 14]):

$$\beta_{\text{Cu}}(T) = \frac{\sum_J [(2J+1) \cdot C_{\text{Cu}}(J)/T] \cdot \exp(-E_J/kT)}{\sum_J (2J+1) \cdot \exp(-E_J/kT)} \qquad (1)$$

Estimates for the hyperfine fields which are needed to calculate the Curie constants can be found from the procedure of Elliott and Stevens [15]. The positions E_J of the fine structure levels serve as free parameters and are estimated by the procedures determining the Van Vleck coefficients.

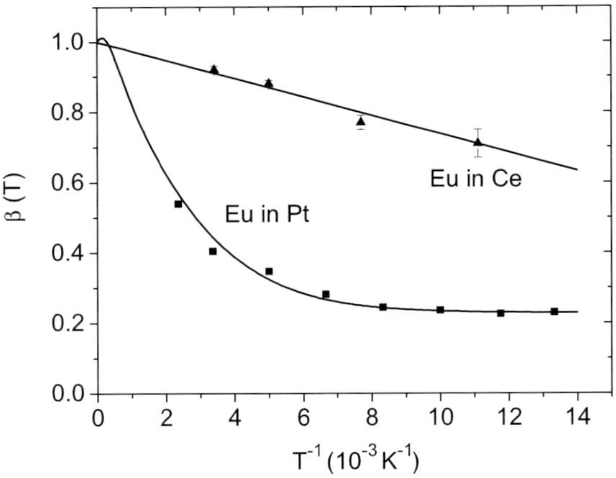

Figure 2. Paramagnetic enhancement factors *versus* inverse temperature for isolated Eu impurities in γ-Ce and in Pt. The solid are calculated for divalent Eu (in γ-Ce) and for trivalent Eu (in Pt) according to the procedures which are described in the text.

In the Van Vleck terms [2, 13, 14]:

$$\beta_{VV}(T) = \frac{\sum_J C_{VV}(J)\cdot[\exp(-E_{J-1}/kT) - \exp(-E_J/kT)]}{\sum_J (2J+1)\cdot \exp(-E_J/kT)} \tag{2}$$

the Van Vleck constants $C_{VV}(J)$ are defined in analogy to the Curie constants. They are estimated from $C_{VV}(J) = (2\mu_2^B \langle r_{4f}^{-3}\rangle F(J) N_{J,J-1})/(E_J - E_{J-1})$ according to the parametrisation of Gross [17]. In the formula the quantum mechanical average has to be taken for the 4f-wave functions, $\langle r_{4f}^{-3}\rangle = 7.555\ (1/a_0)^3$ [16]. The free parameters are the level positions E_J. They have the dominant influence on the shape of the curve. The first Van Vleck coefficient $C_{VV}(J = 1)$ is determined from the value of the measured curve at low temperatures, demanding the level position to be at $E_J(J = 1) = 53$ meV. In order to get the theoretical curve which is shown in Figure 2 for Eu^{3+} we included the Van Vleck contributions of all levels in the ^7F_J-multiplet and took level positions for $^7F_2$ up to $^7F_6$ by scaling the values which are given by Boyn [10]. Details of the analysis and of the measurement can be found in a preceding publication [18].

The measured data are in good agreement with the predictions from the model and this way provide evidence for the existence of divalent Eu in γ-Ce and trivalent Eu in Pt. Even for a configuration with a $J = 0$ ground state the Van Vleck contributions make sure to get a magnetic response. With these measurements the PAC probe nuclei ^{147}Eu and ^{149}Eu were successfully applied for the first time to determine the valences of isolated Eu atoms in Ce and Pt. They shall further be introduced to measurements of isolated impurities in bulk materials and on surfaces in order to find the parameters which determine the magnetic behaviour.

References

1. Gschneidner K. A. Jr., Eyring L., Lander G. H. and Choppin G. R., *Handbook of Physics and Chemistry of Rare Earths*, Vol. 17, Elsevier Science Publ., Amsterdam, 1993.
2. Biedermann K., PhD thesis, Freie Universität Berlin, Berlin, 1987.
3. Bertschat H. H., Haas H., Mahnke H.-E., Netz G., Barth J., Luszik-Bhadra M. and Riegel D., *J. Magn. Magn. Mater.* **47+48** (1985), 592.
4. Perscheid B., Sampathkumaran E. V. and Kaindl G., *Hyperfine Interact.* **28** (1986), 1059.
5. Wieling S., Molodtsov S. L., Laubschat C. and Behr G., *Phys. Rev., B* **65** (2002), 075415.
6. Van Vleck J. H., *The Theory of Electric and Magnetic Susceptibilities*. Oxford University Press, Oxford, 1932.
7. Potzel W., Kalvius G. M. and Gal J., Mössbauer Studies on the Electronic Structure of Intermetallic Compounds, In: Gschneidner K. A. Jr., Eyring L., Lander G. H. and Choppin G. R. (eds.), *Handbook of Physics and Chemistry of Rare Earths*, Vol 17, Elsevier Science Publ., Amsterdam, 1993.
8. Potzger K., Weber A., Bertschat H. H., Zeitz W.-D. and Dietrich M., *Phys. Rev. Lett.* **88** (2002), 247201.
9. Firestone R. B. and Shirley V. S., In: Baglin C. M., Frank Chu S. Y. and Zipkin J. (eds.), *Table of Isotopes*, 8th Edition. Wiley, New York, 1966.
10. Boyn R., *Phys. Status Solidi, B* **148** (1988), 11.
11. Stewart G. A., *Mater. Forum* **18** (1994), 177.
12. Müller W., Bertschat H. H., Haas H., Mahnke H.-E. and Zeitz W.-D., *Phys. Rev., B* **40** (1989), 9346.
13. Bodenstedt E., *Fortschr. Phys.* **10** (1962), 321.
14. Günther L. and Lindgren L., In: Karlsson E., Matthias E. and Siegbahn K. (eds.), *Perturbed Angular Correlations*. North Holland, Amsterdam, 1964, p. 357.
15. Elliott R. J. and Stevens K. W. H., *Proc.-Royal Soc.* **A218** (1953), 533.
16. Freeman A. J. and Desclaux J. P., *J. Magn. Mater.* **12** (1979), 11.
17. Gross K. D., PhD thesis, Freie Universität Berlin, Berlin, 1988.
18. Dietrich M., Zeitz W.-D., Weber A., Potzger K., Unterricker S., Schneider F., Samokhvalov V. and the ISOLDE-Collaboration, *J. Magn. Magn. Mater.* **294** (2005), 330.

Hyperfine Interactions (2004) 158:443–445

Keyword Index to Volume 158 (2004)

Hyperfine Interactions (2004) 158:447–449

Author Index to Volume 158 (2004)